Water Resources Monitoring, Management, and Sustainability

Application of Geostatistics and Geospatial Modeling

Developments in Environmental Science

Volume 15	**Urban Health** Edited by Lakshmi Sivaramakrishnan, Bharat Dahiya, Madhuri Sharma, Saswati Mookherjee, Ranita Karmakar
Volume 14	**Modeling and Mitigation Measures for Managing Extreme Hydrometeorological Events Under a Warming Climate** Edited by K.S. Kasiviswanathan, B. Soundharajan, Sandhya Patidar, Jianxun He, Chandra Shekhar Prasad Ojha
Volume 13	**Climate Change, Air Pollution and Global Challenges: Understanding and Perspectives from Forest Research** Edited by R. Matyssek, N. Clarke, P. Cudlin, T.N. Mikkelsen, J.-P. Tuovinen, G. Wieser, E. Paoletti
Volume 12	**Forest Monitoring** Edited by Marco Ferretti, Richard Fischer
Volume 11	**Alberta Oil Sands: Energy, Industry and the Environment** Edited by Kevin E. Percy
Volume 10	**Cross-Border Resource Management, 2nd edition** Edited by Rongxing Guo
Volume 9	**Air Quality and Ecological Impacts** Edited by Allan Legge
Volume 8	**Wildland Fires and Air Pollution** Edited by Andrzej Bytnerowicz, Michael Arbaugh, Allen Riebau, Christian Andersen
Volume 7	**Persistent Organic Pollutants in Asia** Edited by An Li, Shinsuke Tanabe, Guibin Jiang, John Giesy, Paul Lam
Volume 6	**Air Pollution Modeling and its Application XVIII** Edited by Carlos Borrego, Eberhard Renner
Volume 5	**Concepts and Applications in Environmental Geochemistry** Edited by Dibyendu Sarkar, Rupali Datta, Robyn Hannigan
Volume 4	**Cross-Border Resource Management, 1st edition** Edited by Rongxing Guo
Volume 3	**Air Pollution, Global Change and Forests in the New Millennium** Edited by D.F. Karnosky, K.E. Percy, A.H. Chappelka, C. Simpson, J. Pikkarainen
Volume 2	**Ozone Air Pollution in the Sierra Nevada – Distribution and Effects on Forests** Edited By A. Bytnerowicz, R. Alonso, M. Arbaugh
Volume 1	**Air Pollution Science for the 21st Century** Edited By J. Austin, P. Brimblecombe, W. Sturge

Developments in Environmental Science
Volume 16

Water Resources Monitoring, Management, and Sustainability

Application of Geostatistics and Geospatial Modeling

Edited by

Subodh Chandra Pal
Department of Geography, The University of Burdwan, India

Shouraseni Sen Roy
Department of Geography & Sustainable Development, University of Miami, United States

Asish Saha
Department of Geography, The University of Burdwan, India

Mohamed Abioui
Department of Earth Sciences, Ibnou Zohr University, Morocco

Series Editor

Uday Chatterjee

ELSEVIER

Elsevier
Radarweg 29, PO Box 211, 1000 AE Amsterdam, Netherlands
125 London Wall, London EC2Y 5AS, United Kingdom
50 Hampshire Street, 5th Floor, Cambridge, MA 02139, United States

ISBN: 978-0-443-23665-5

ISSN: 1474-8177

For Information on all Elsevier publications
visit our website at https://www.elsevier.com/books-and-journals

Publisher: Candice Janco
Acquisitions Editor: Jessica Mack
Editorial Project Manager: Soumya Yadav
Production Project Manager: Sruthi Satheesh
Cover Designer: Vicky Pearson Esser

Typeset by MPS Limited, Chennai, India

Contents

Section II
Water resources management

List of contributors

Nirupam Acharyya Department of Remote Sensing and GIS, Vidyasagar University, Midnapore, West Bengal, India

Pratibha Sunil Agrawal Department of Applied Chemistry, Laxminarayan Innovation Technological University, Nagpur, India

Rangel Arthur School of Technology (FT), University of Campinas (UNICAMP), Campinas, Brazil

Jatisankar Bandyopadhyay Department of Remote Sensing and GIS, Vidyasagar University, Midnapore, West Bengal, India

Girija Shankar Behera Department of Applied Geology, IIT (ISM), Dhanbad, Jharkhand, India

Pramod Belkhode Department of Mechanical Engineering, Laxminarayan Innovation Technological University, Nagpur, India

Pradyut Bera Department of Geography, Sidho-Kanho-Birsha University, Purulia, West Bengal, India

Gupinath Bhandari Department of Civil Engineering, Jadavpur University, Kolkata, West Bengal, India

Madhukar Govindrao Bhotmange Department of Food Technology, Laxminarayan Institute of Technology, RTM Nagpur University, Nagpur, India

Gouri Sankar Bhunia Independent Researcher, Paschim Medinipur, West Bengal, India

Idowu Racheal Bodunrin Department of Civil Engineering, College of Science, Engineering and Technology, University of South Africa, Florida, Johannesburg, South Africa

Rodrigo Bonacin Renato Archer Information Technology Center (CTI), Computing Methodologies Division-DIMEC, Campinas, Brazil

M.C. Chandan Department of Civil Engineering, The National Institute of Engineering, Mysore, India

S. Chandana Department of Civil Engineering, The National Institute of Engineering, Mysore, India

Priya Chatterjee Department of Zoology, The University of Burdwan, Purba Bardhaman, India

Soumendranath Chatterjee Department of Zoology, The University of Burdwan, Purba Bardhaman, India

Sweta Chatterjee School of Water Resources Engineering, Faculty of Interdisciplinary Studies, Law & Management (FISLM), Jadavpur University, Kolkata, West Bengal, India

Sarbajit Chowdhury Department of Remote Sensing and GIS, Vidyasagar University, Midnapore, West Bengal, India

Niladri Das Department of Geography, Hiralal Bhakat College, Nalhati, Birbhum, West Bengal, India

Nilanjan Das Ecology and Environmental Modelling Laboratory, Department of Environmental Science, The University of Burdwan, Purba Bardhaman, West Bengal, India

Suman Das Department of Remote Sensing and GIS, Vidyasagar University, Midnapore, West Bengal, India

Nirasindhu Desinayak Department of Geology, Ravenshaw University, Cuttack, Odisha, India

Praduyt Dey Department of Geography, Rajiv Gandhi University, Doimukh, Arunachal Pradesh, India

Sukhendu Dey Department of Environmental Science, The University of Burdwan, Burdwan, West Bengal, India

Suman Dutta Department of Chemical Engineering, Indian Institute of Technology (Indian School of Mines), Dhanbad, India

Fatima El Hammichi Natural Resources and Environment Laboratory, Polydisciplinary Faculty of Taza, Sidi Mohamed Ben Abdellah University - Fez, Morocco

Reinaldo Padilha França Renato Archer Information Technology Center (CTI), Computing Methodologies Division-DIMEC, Campinas, Brazil

Apurba Ratan Ghosh Department of Environmental Science, The University of Burdwan, Burdwan, West Bengal, India

Ranajit Ghosh Department of Geography, Suri Vidyasagar College, Suri, Birbhum, West Bengal, India

Sajal Ghosh Department of Geography, Hiralal Bhakat College, Nalhati, Birbhum, West Bengal, India

Jayanta Gour Department of Geography, Sambhu Nath College, Labpur, West Bengal, India

Sunil Kumar Gupta Department of Environmental Science and Engineering, Indian Institute of Technology (Indian School of Mines), Dhanbad, Jharkhand, India

Suyog Gupta Department of Environmental Science and Engineering, Indian Institute of Technology (Indian School of Mines), Dhanbad, Jharkhand, India; Department of Civil Engineering, Harcourt Butler Technical University, Kanpur, Uttar Pradesh, India

Bijay Halder Department of Remote Sensing and GIS, Vidyasagar University, Midnapore, West Bengal, India

Arpana Handique Center for Studies in Geography, Dibrugarh University, Dibrugarh, Assam, India

Aishwarya Hegde Department of Water Resource and Ocean Engineering, National Institute of Technology, Surathkal, Karnataka, India

Mallika Jana Department of Remote Sensing and GIS, Vidyasagar University, Midnapore, West Bengal, India

Gh Jeelani Department of Earth Sciences, University of Kashmir, Srinagar, India

Cathy Koloa School of Surveying and Land Studies, The Papua New Guinea University of Technology, Lae, Papua New Guinea

Rajesh Kumar Department of Geography, Sikkim University, Gangtok, India

Suhail A. Lone Department of Earth Sciences, University of Kashmir, Srinagar, India

Rajib Maity Department of Civil Engineering, Indian Institute of Technology Kharagpur, Kharagpur, West Bengal, India

Saroj Maity Indian Space Research Organization, Space Applications Center, Ahmedabad, Gujarat, India

Mrinal Mandal Department of Geography, Sidho-Kanho-Birsha University, Purulia, West Bengal, India

Paramita Mandal Department of Zoology, The University of Burdwan, Purba Bardhaman, India

Sudipto Mandal Ecology and Environmental Modelling Laboratory, Department of Environmental Science, The University of Burdwan, Purba Bardhaman, West Bengal, India

Susanta Mandi Department of Geography, Sidho-Kanho-Birsha University, Purulia, West Bengal, India

Dickson Okoree Mireku Department of Academic Planning, Directorate of Academic Planning and Quality Assurance, University of Cape Coast, Cape Coast, Ghana

Partha Modak Department of Geography, Sidho-Kanho-Birsha University, Purulia, West Bengal, India

Kareem Mohamed Soliman Department of Zoology, Faculty of Science, Ain Shams University, Cairo, Egypt

Ayan Mondal Ecology and Environmental Modelling Laboratory, Department of Environmental Science, The University of Burdwan, Purba Bardhaman, West Bengal, India

Prolay Mondal Department of Geography, Raiganj University, Raiganj, India

Ana Carolina Borges Monteiro Renato Archer Information Technology Center (CTI), Computing Methodologies Division-DIMEC, Campinas, Brazil

Mohammed Mourjane Natural Resources and Environment Laboratory, Polydisciplinary Faculty of Taza, Sidi Mohamed Ben Abdellah University - Fez, Morocco

Suren Nayak Department of Geology, Utkal University, Bhubaneswar, Odisha, India

Ednah Kwamboka Onyari Department of Civil Engineering, College of Science, Engineering and Technology, University of South Africa, Florida, Johannesburg, South Africa

Soumyadip Pal Ecology and Environmental Modelling Laboratory, Department of Environmental Science, The University of Burdwan, Purba Bardhaman, West Bengal, India

Manish Pandey University Center for Research and Development (UCRD), Chandigarh University, Mohali, Punjab, India; Department of Civil Engineering, Chandigarh University, Mohali, Punjab, India

Manju Pandey Department of Geography, Government College Barpali, Korba, Chhattisgarh, India

Binay Prakash Panigrahy Department of Environmental Science & Engineering, IIT (ISM), Dhanbad, Jharkhand, India

U. Pruthviraj Department of Water Resource and Ocean Engineering, National Institute of Technology, Surathkal, Karnataka, India

Seema Rani Department of Geography, Institute of Science, Banaras Hindu University, Varanasi, India

Harisankar Ray Ecology and Environmental Modelling Laboratory, Department of Environmental Science, The University of Burdwan, Purba Bardhaman, West Bengal, India

Padma Ruhidas Department of Geography, Sidho-Kanho-Birsha University, Purulia, West Bengal, India

Palas Samanta Department of Environmental Science, Sukanta Mahavidyalaya, University of North Bengal, Dhupguri, West Bengal, India

Sailesh Samanta School of Surveying and Land Studies, The Papua New Guinea University of Technology, Lae, Papua New Guinea

Megha Santra Department of Environmental Science, The University of Burdwan, Burdwan, West Bengal, India

Bappa Sarkar Department of Geography, Dinhata College, Koch Bihar, India

Saroj Senapati Department of Geography, Atal Bihari Vajpayee Viswavidyalaya, Chhattisgarh, India

Rahul Kumar Singh Geological Survey of India, Ranchi, Jharkhand, India

Rohini Singh Department of Chemical Engineering, R.N.G. Patel Institute of Technology, Bardoli, Gujarat, India

Surendra Singh Department of Geography, North-Eastern Hill University, Shillong, India

Mayank Suman Department of Civil and Infrastructure Engineering, Indian Institute of Technology Jodhpur, Jodhpur, Rajasthan, India

Subhasish Sutradhar NATMO, Rashtriya Atlas Bhavan, Kolkata, West Bengal, India

H.J. Syiemlieh Department of Geography, North-Eastern Hill University, Shillong, India

Hassan Tabyaoui Natural Resources and Environment Laboratory, Polydisciplinary Faculty of Taza, Sidi Mohamed Ben Abdellah University - Fez, Morocco

Wiame Waheed Emam Department of Zoology, Faculty of Science, Ain Shams University, Cairo, Egypt

Divyata Yadav School of Environmental Sciences, Jawaharlal Nehru University, New Delhi, India

Mayur Yenkie Department of Applied Chemistry, Laxminarayan Innovation Technological University, Nagpur, India

About the editors

Subodh Chandra Pal is an Assistant Professor in the Postgraduate Department of Geography at The University of Burdwan, Purba Bardhaman, West Bengal, India. He received his PhD in Geography from Visva-Bharati (Central University in India) in 2015. His main fields of expertise are fluvial geomorphology, pedo-geomorphology, land degradation, soil erosion, gully erosion, landslide vulnerability, climate change, and water resources monitoring and management and published nearly 200 international and national research articles in various renowned journals. He has authored a book on *Climate Change Impact on Soil Erosion in Sub-tropical Environment: Application of Empirical and Semi-empirical Models*. He is the Editor of four running edited books and is also an editorial board member of *MDPI Water* and *Frontiers of Earth Science* journals. His research work has been funded by the University Grants Commission (UGC) and the Department of Science and Technology (DST), Govt. of India. Dr. Pal has served as a reviewer for many international reputed journals. Since 2022, Stanford University has recognized Dr. Pal as one of the world's top 2% scientists.

Shouraseni Sen Roy is a Professor in the Department of Geography and Sustainable Development at the University of Miami, United States. Her core area of research over the last two decades has concentrated on expanding our understanding of long-term trends in climatic processes, in view of impending climate change. Her research methodology incorporates extensive spatial analysis using advanced geo-statistical techniques to analyze trends in climatic variables. In addition to researching climate processes over the Indian subcontinent, she has addressed targeted climate-related research questions in other regions of the world, including the United States, Myanmar, Burundi, China, and South Africa. Furthermore, she has recently started working on crime patterns in South Florida.

Asish Saha is a PhD research scholar (UGC-SRF) in the Department of Geography at The University of Burdwan, Purba Bardhaman, West Bengal, India. He received his master's degree in Geography from The University of Burdwan and postgraduate diploma in Geoinformatics from the Department of Science and Technology, Govt. of West Bengal, in collaboration with Maulana Abul Kalam Azad University of Technology, Kolkata. His fields of expertise are climate variability, climate change, regional climate modeling, hydro-climatic extremes, water resources management, water quality, hazard and risk assessment, statistical modeling, and machine learning applications. He has published more than 90 international research articles along with 5 book chapters in various renowned journals. His research work has been funded by the University Grants Commission (UGC), India. Alongside, he has served as a reviewer for many well-regarded international journals.

Mohamed Abioui is an Assistant Professor (habilitation) of sedimentology–stratigraphy and petroleum geology in the Department of Earth Sciences at Ibnou Zohr University, Morocco; and a collaborator member with MARE-Marine and Environmental Sciences Centre—Sedimentary Geology Group, Department of Earth Sciences, Faculty of Sciences and Technology, University of Coimbra, Portugal. In 2024 he became an Adjunct Professor at Universiapolis—International University of Agadir, Morocco. His research interests are related to the study of sedimentology, stratigraphy, petroleum geology (geomechanics and pore pressure), geoheritage, geoeducation, rock art, flood forecasting, uncertainty analysis, water resources monitoring, management, and sustainability, and flood risk management. Currently, he collaborates with international leading academics to develop novel approaches for tackling water management challenges that our human society is facing. He has contributed to many research papers published in various reputed national and international journals.

Foreword

It is with great pleasure that I introduce this comprehensive volume entitled *Water Resources Monitoring, Management, and Sustainability*, which delves into the multifaceted realm of water resources. As the world continues to evolve and face complex challenges, the need for in-depth understanding and innovative solutions becomes increasingly paramount. In this regard, the contributions assembled within this book represent a commendable effort to advance knowledge and foster dialogue within the field.

Water is the essence of life, a precious resource vital for the sustenance of ecosystems, economies, and societies. As the global population continues to grow and urbanize, the demand for water escalates, exerting unprecedented pressure on freshwater resources. In the face of climate change, pollution, and unsustainable water management practices, the need for effective monitoring, management, and sustainable utilization of water resources has never been more critical.

This book encompasses a diverse array of topics, ranging from hydrochemical analysis, groundwater potentiality, mangrove ecosystem, marine plastic pollution, drainage basin management, etc. This book emerges as a beacon of knowledge, offering comprehensive insights into the multifaceted challenges and innovative solutions shaping the field of water resources. Authored by leading experts and researchers from around the world, this book serves as a definitive guide for policymakers, practitioners, scientists, and students alike.

Spanning a broad spectrum of topics, the book delves into the latest advancements in water resources monitoring technologies, providing state-of-the-art methodologies for data collection, analysis, and interpretation. From hydrological analysis to geospatial modeling, the chapters offer a wealth of tools and techniques for enhancing our understanding of water dynamics at various scales. Moreover, this book explores radical approaches to water resources management including wastewater purification and assessment of groundwater recharge to climate change vulnerability of water resource management in showcasing integrated and adaptive strategies aimed at promoting water security, resilience, and equitable distribution. Central to the book's ethos is the theme of sustainability, underscoring the imperative of balancing human needs with the preservation of ecological integrity encompassing case studies from Indian Sundarbans to Jharia Coal field

Region, India. By examining the nexus between water and environmental systems, this book elucidates pathways toward achieving sustainable water management practices that foster social equity, economic prosperity, and environmental stewardship.

In an era marked by rapid technological advancements and globalization, interdisciplinary collaboration and cross-pollination of ideas are more essential than ever. This book serves as a testament to the power of collaboration, bringing together researchers, practitioners, and policymakers from various disciplines to address pressing challenges and explore new frontiers of knowledge.

I extend my heartfelt gratitude to the authors for their scholarly contributions, the editorial team for their dedication and expertise in bringing this volume to culmination, and the readers for their curiosity and engagement with the ideas presented herein. May this book serve as a catalyst for continued exploration, discovery, and innovation in the field of water resources. The authors' commitment to advancing knowledge, coupled with their dedication to practical solutions, makes this book an authoritative guide for anyone seeking to navigate the complex terrain of water resource challenges. May this volume inspire and equip a new generation of stewards committed to ensuring the resilience and vitality of our planet's most precious resource, that is, water.

Prof. (Dr.) Pankaj Kumar Roy, Director and Professor, School of Water Resources Engineering, Former Dean, Faculty of Interdisciplinary Studies, Law & Management Jadavpur University, India

Preface

Globally, freshwater supplies are becoming more and more challenging due to increased demand brought on by population growth, expanding industrialization as a result of rising living standards, pollution in surface and groundwater, and the effects of emerging climate change. Unsustainable way of water extraction for diverse consumptive uses and poor management practices further accelerate this scenario. In this regard, pioneering knowledge, novel strategies, and a comprehensive grasp of the underlying scientific, economic, social, and environmental issues are essential to overcome the substantial obstacles to the management of water resources. The quantity and quality of water resources are severely threatened by contemporary global changes. The management of water resources can benefit from improvements in environmental monitoring and modeling. Recently, data availability of in situ and remote sensing have been widely used in several fields, including water resources studies. A current and promising area of research involves the adaptation of remote sensing, GIS, hydrological modeling, and machine learning algorithms for sustainable management of water resources. Adopting a sustainable planning and management viewpoint in these domains is becoming indeed popular. In this quickly developing research area, the use of sophisticated mathematical and statistical models is crucial. In practice, models are key instruments for forecasting water availability and quality, in connection to many potential uses such as drinking, irrigation, and domestic purposes, as well as under both local and global perspectives.

Sustainable development is crucial for both the present and the future in all areas relating to the management of water resources. There are many issues, including environmental pollution occurrences and water shortages or flooding. Therefore the definition of rational water resource management concepts is beginning to include the environmental component. A comprehensive, systemic, and interdisciplinary perspective is necessary for this. To plan and manage water resources sustainably, the edited book introduces a variety of advanced modeling techniques like changing climate and land use/land cover, hydrological models, and machine learning approaches. These techniques are applied to address water resource and environmental monitoring, as well as planning and sustainable management of water resources. This book illustrates a geospatial technology approach to data mining techniques, data analysis, modeling, risk assessment and visualization, and management strategies for many elements of natural and societal hazards. This edited book delves into cutting-edge techniques based on open-source software and R statistical programming, Google Earth Engine, and modeling in modern artificial intelligence techniques, with a particular emphasis on recent trends in data mining techniques and robust modeling in water resource crisis-related hazards. Furthermore, this edited book also discusses how to manage water resources at various local, regional, and international levels while incorporating environmental considerations. This book contains works that address the connection between water, climate change, and the environment. These works include those that identify and evaluate policies, quantify the effects of public interventions, land use changes, or climate change on environmental water management, and describe the decision-making process and its role in modeling.

We hope that the breadth of this book is broad enough to introduce both novices and academics to many key themes at the forefront of current research on environmental and social hazards, risk assessment, and management. This book is specifically designed to serve the community of graduate and postgraduate students, early career researchers, MPhil and PhD scholars, research assistants, postdoctoral fellows, faculty members, scientists, and senior fellows in the fields of environmental sciences, geography, hydrology, natural hazards, geospatial sciences, remote sensing & GIS, agriculture, crop science, forestry, soil science, geology, humanistic & social sciences, etc.

Subodh Chandra Pal
Shouraseni Sen Roy
Asish Saha
Mohamed Abioui

Section I

Water resources monitoring

Chapter 1

Integrating modeling, monitoring, and management to minimize pollutants in water resources

Pratibha Sunil Agrawal[1], Mayur Yenkie[1], Madhukar Govindrao Bhotmange[2] and Pramod Belkhode[3]

[1]Department of Applied Chemistry, Laxminarayan Innovation Technological University, Nagpur, India, [2]Department of Food Technology, Laxminarayan Institute of Technology, RTM Nagpur University, Nagpur, India, [3]Department of Mechanical Engineering, Laxminarayan Innovation Technological University, Nagpur, India

1.1 Introduction

Phenols are used in the synthesis of industrial products such as paints, plastics, herbicides, petrochemicals, explosives, pharmaceuticals, and synthetic dyes (Chairam, Konkamdee, & Parakhun, 2017; Serrà, Artal, Pozo, Garcia-Amorós, & Gómez, 2020; Tranchant et al., 2020). The presence of phenols in industrial wastewater due to ever-increasing industrial discharge containing unacceptable concentrations produces toxicity of phenols to the receiving bodies. Phenols cause intensive toxic effects on human physiology and are considered priority pollutants by environmental protection agencies (Kalaimurugan et al., 2021; Khoshnamvand, Hao, Huo, & Liu, 2020; Neelam et al., 2020). Hence, it is imperative to apply effective treatment techniques to phenol-containing wastewater to eliminate or decrease its concentration, at least to the permissible limits. Technologies such as adsorption (Fu, Li, Wang, Luan, & Dai, 2021; Saheed, Oh, & Suah, 2021), chemical oxidation (Copete-Pertuz, Serna-Galvis, Plácido, Torres-Palma, & Mora-Martínez, 2021; Ma et al., 2021; Yang et al., 2021), and photo-degradation (Bisaria, Sinha, Singh, & Iqbal, 2021; Shi et al., 2021; Zhao, Li, Li, Huo, & Shi, 2021) have been developed and used. Among these methods, adsorption is promising, but it will only transfer the pollutant from wastewater to the adsorbent surface, causing secondary pollution. The excessive presence of

Water Resources Monitoring, Management, and Sustainability. DOI: https://doi.org/10.1016/B978-0-443-23665-5.00001-6

phenols (Singh & Balomajumder, 2021) and their derivatives, which are organic priority pollutants, in wastewater necessitates the implementation of effective treatment methods (Belkhode, Ganvir, Shelare, Shende, & Maheshwary, 2022). These methods aim to either eliminate or minimize the pollutants within the permissible limits set by regulatory agencies while avoiding their transfer to another phase.

Advanced oxidative processes (AOPs) have the potential to serve as a supplementary method for the oxidation of organic residues that have undergone conventional treatment processes (Loures et al., 2013). AOPs encompass a collection of methodologies that, in specific circumstances, have the potential to convert a significant portion of pollutants into carbon dioxide, water, and inorganic ions through oxidation reactions (Deshpande, Agrawal, & Yenkie, 2019a). These reactions involve the transient oxidant species $OH\cdot$ free radicals, which are generated in situ. The field of literature contains a wealth of information pertaining to AOPs; however, there is a scarcity of available reports that specifically address the kinetics, mechanism, and thermodynamics of this process (Yan, Li, Liu, Liu, & Brusseau, 2017).

There is a significant level of interest in developing a treatment technology that is both efficient and cost-effective in removing hazardous and refractory pollutants. AOPs present themselves as a highly effective treatment option for the comprehensive oxidation and mineralization of pollutants without the need for phase transfer (Deshpande et al., 2019a). During AOPs, hydroxyl ($OH^{\bullet}$) free radicals are generated in the reaction environment. These radicals can indiscriminately degrade organic contaminants (Charamba et al., 2018; He et al., 2020). During AOPs, hydroxyl ($OH\cdot$) free radicals with a significantly high oxidation potential of 2.8 V (Kulaksız et al., 2022; Sokol, Ratkiewicz, Tomaszewska, & Karpinska, 2021) are generated in the reaction environment. These radicals can indiscriminately degrade organic contaminants.

The main aspect of the present chapter was to analyze the oxidative degradation of some nitro derivatives of phenol, namely 2-Nitrophenol, 4-Nitrophenol, and 2,4,6-Trinitrophenol in water by classical Fenton ($H_2O_2 + Fe^{++}$) and ultraviolet (UV)-assisted Photo-Fenton ($UV + H_2O_2 + Fe^{++}$) oxidation (Table 1.1). The efficiency of Fenton-like processes ($H_2O_2 + Fe^{+++}$ and $UV + H_2O_2 + Fe^{+++}$) was also studied. A UV lamp of 8 Watt, 220–240 V, and 50 Hz Philips-made tubular shape was used as a UV source, which emitted approximately 87% UV rays at 254 nm. The optimization of Fenton processes was initially conducted through experimental investigation, as these processes exhibit a strong dependence on various variables such as pH, temperature, substrate, and oxidant concentration, as well as the concentrations of Fe^{++} and Fe^{+++} (Deshpande, Agrawal, & Yenkie, 2019b, 2019c). The present work investigates the impact of temperature on the oxidative degradation of nitrophenols. Specifically, the research was conducted at five distinct temperatures ranging from 293 to 313K. The results obtained in this study provide evidence that

TABLE 1.1 Physical properties of nitrophenols.

Compound	2-Nitrophenol	4-Nitrophenol	2,4,6-Trinitrophenol
Abbreviation	2-NP	4-NP	TNP
Molecular weight	139.11	139.11	229.11
Aq. Solubility at 25°C, mg/L	1697	16,000	14,000
M. pt. (°C)	44.8	114.0	122.5
B. pt (°C)	216.0	279.0	>300
Density (g/cm^3)	1.485	1.479	1.763
Acid dissociation constant, pKa	7.22	7.15	0.78
Vapor pressure (25°C) (Pa)	29.10	0.0044	4.68×10^{-5}
Enthalpy of fusion (kJ/mol)	17.44	18.25	19.50

TABLE 1.2 Values of λmax (maximum wavelength of absorption) and molar extinction coefficient for the pollutants under investigation.

No.	Pollutant	$l_{max,\ nm}$	Î, M/cm	Beer's law validity range
1	2-Nitrophenol	276	5343	0–0.00015 ML^{-1}
2	4-Nitrophenol	314	6298	0–0.00015 ML^{-1}
3	2,4,6-Trinitrophenol	357	14,090	0–0.0001 ML^{-1}
4	H_2O_2	420	868	0–0.002 ML^{-1}

the Fenton process demonstrates considerable potential for effectively treating effluents that contain nitrophenols (Qiao et al., 2021). The abbreviations employed in the text for the sake of convenience are as follows: 2NP, which stands for 2-Nitrophenol; 4NP, which stands for 4-Nitrophenol; and TNP, which stands for 2,4,6-Trinitrophenol (TNP) (Table 1.2). The different Fenton processes are used in wastewater treatment.

Formulation and development of mathematical models in the form of relationships among responses and causes involved, including optimization of deduced models, in addition to the determination of weightage for influencing variables, has also been focused.

1.2 Experimental methods

The UV-visible spectrophotometric technique was utilized to quantify the amount of pollutants in water. The LABMAN UV-VIS double-beam spectrophotometer was used to investigate the amount of pollutant remaining after degradation. Nitrophenols exhibit significant absorption bands in the UV region; hence, their absorbance measurements were conducted using quartz cuvettes with a path length of 1 cm, ensuring optimal matching (Deshpande, Yenkie, & Agrawal, 2018).

Based on Beer's law:

$$A = \varepsilon\, c\, l = \varepsilon\, c \tag{1.1}$$

where A = optical density or the absorbance, l = length of optical path in cm, ε = coefficient of molar extinction in $M^{-1}\ cm^{-1}$, and c = investigation sample concentration in ML^{-1}.

A plot was generated to depict the relationship between the concentration of nitrophenol in the aqueous phase and its absorbance at the wavelength corresponding to its maximum absorbance (λmax). The obtained experimental data were then analyzed using linear regression analysis, revealing a high degree of linearity within a specific concentration range. The reaction between hydrogen peroxide (H_2O_2) and potassium titanium oxalate sulfate solution results in the formation of a solution with a yellow color. This solution exhibits maximum absorption at a wavelength of 420 nm. The concentration of the stock solution of H_2O_2 was determined through a titration process using a standard $KMnO_4$ solution. Following the standardization of the prepared H_2O_2 solution, it was utilized to create multiple solutions within the concentration range of 6–60 mg H_2O_2/L through suitable dilutions of the stock solution. This was done to establish Beer's law plot. Table 1.2 presents a summary of the maximum absorbance wavelength (λmax), molar absorptivity (ε), and the range of validity for Beer's law for nitrophenols and hydrogen peroxide (H_2O_2). Regular testing was conducted to assess the validity of Beer's absorption law from a straight-line nature.

The percentage degradation of the substrate at any given instant of time was then calculated by the formula:

$$\eta = \frac{C0 - C}{C0} X100 \tag{1.2}$$

Where C and C0 are concentrations at time t = t and t = 0, respectively.

1.3 AOP for degradation of pollutants

All of the experiments were conducted within a low-pressure mercury lamp (8 W, UV-C), which was centered within a 1.0-L jacketed thermostatic batch photoreactor. This 220–240 V, 50 Hz Phillips-made tubular UV lamp emits approximately 87% UV rays at 254 nm, and about 98% of the rays at this wavelength can pass through quartz. The reactor also provides continuous and uniform stirring of the reaction mixture (Deshpande, Agrawal, Yenkie, & Dhoble, 2020) along with a thermostat/cryostat bath manufactured by Fourtech Systems, located in Mumbai, India, to maintain the constant temperature. Since high-energy UV radiation is harmful and hazardous to the eyes and skin, external exposure to harmful radiation was cut off by wrapping the reactor with aluminum foil and black century paper and placing the whole assembly in a wooden box designed and fabricated for these studies Fig. 1.1. During classic Fenton processes, the UV lamp was kept in switched-off mode, while during photo-Fenton studies, the lamp was switched on. The experiments were conducted in an 800 mL batch photoreactor. All experimental procedures were conducted under standard atmospheric conditions. All experiments were conducted using a batch-mode approach. The synthetic water-polluted solution with a concentration of 3×10^{-4} M was filled into a

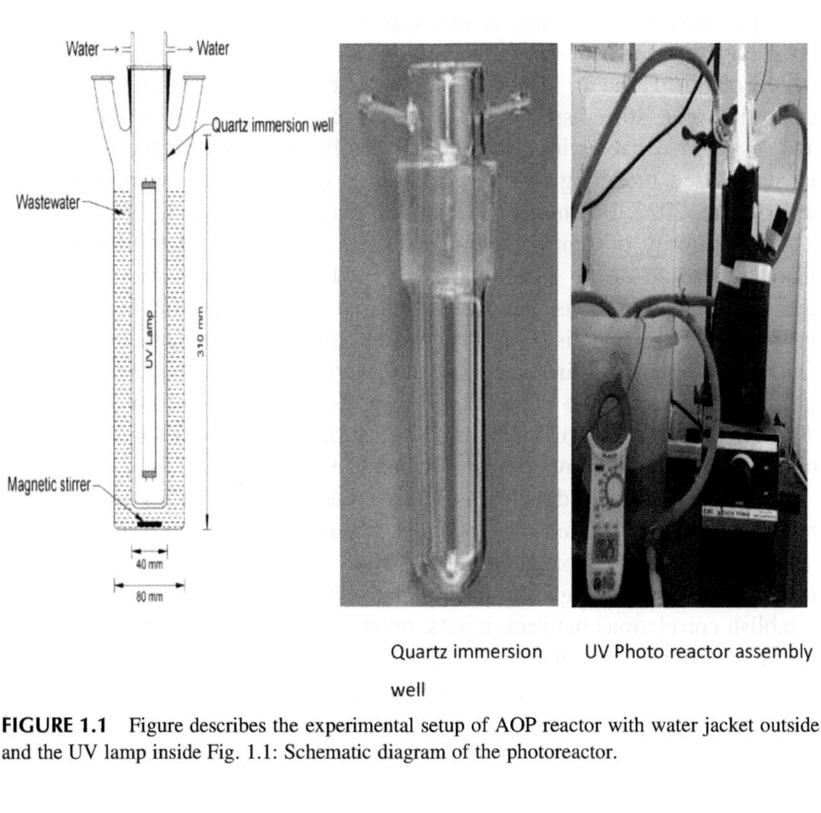

Quartz immersion well   UV Photo reactor assembly

FIGURE 1.1 Figure describes the experimental setup of AOP reactor with water jacket outside and the UV lamp inside Fig. 1.1: Schematic diagram of the photoreactor.

photoreactor. The pH of the solution was initially adjusted by employing dilute 0.01 N hydrochloric acid (HCl) or sodium hydroxide (NaOH). The necessary concentrations of iron (Fe^{2+}) in the form of either $FeSO_40.7H_2O$ or $Fe_2(SO_4)_30.7H_2O$ are required in the event that Fe^{3+} is added. In the course of the experiments, samples were extracted from the reactor using a syringe at consistent intervals to determine their concentration (Table 1.3). The researchers conducted experiments to optimize process parameters, including pH, H_2O_2 concentration, Fe^{2+} concentration, and Fe^{3+} concentration (Figs. 1.2–1.4).

To assess the thermodynamic energy parameters associated with the degradation of chlorophenol, a series of experiments were conducted at five distinct temperatures ranging from 293 to 313K. The chosen temperature range is based on the actual temperatures involved in the treatment process. The Systronics light-emitting diode (LED) display is a technological device that utilizes LEDs to provide visual information or content. pH determination was conducted using the digital pH meter Model No. MK-VI. The Remi 2 ML magnetic stirrer, specifically Model No. Q-19 A, was employed to agitate the system of the reactor throughout the experimental procedures.

1.4 Formulation of mathematical model

This part discusses the degradation rate and concentration dependence for removing the impurities, leading to low efficiency of AOP systems, poor data if obtained, and an increased number of efforts. The significance of concentration, pH, stirrer speed, time, and temperature as various independent and response variables involved in this specific operation for the experimental data-based modeling of this operation is discussed (Agrawal, Belkhode, & Rokhum, 2022; Belkhode, Modak, Vidyasagar, & Agrawal, 2021).

The majority of manufacturing processes are performed through manual labor as a result of the constraints associated with mechanization, including technological limitations and cost considerations. The purification of water in industrial settings is typically carried out through manual processes. Operators are engaged in the operation of various types of machine tools and process machines while being exposed to diverse environmental conditions. The experimental design for water purification varies depending on the specific constraints faced by operators (Gbolahan & Williams, 2013).

It is not feasible to develop a logical model that establishes a correlation between causes and effects for complex phenomena of this nature. The sole suitable methodology for this particular study is the utilization of experimental data-based modeling. The proposed model utilizes experimental data to establish correlations between inputs, outputs, and underlying causes through the formulation of a quantitative mathematical framework. The indices of the equation responses, specifically the mathematical model, serve to indicate the inputs that have the greatest influence. This correlation demonstrates the presence of both weaknesses and strengths in the process, which

TABLE 1.3 Data related to dimensions.

pH	H_2O_2 (Mmol/L)	Fe2 + (Mmol/L)	Temperature	Time (min)	Reactor (mL)	UV (Watts)	Stirrer (rpm)	C Mmol/L	% Degrade	IC mM/L
3	1.7	0.0023	303	15	0.8	8	11.666	1.669	44.37	3.00
3	1.7	0.0023	303	30	0.8	8	11.666	0.949	68.37	3.00
3	1.7	0.0023	303	45	0.8	8	11.666	0.416	86.13	3.00
3	1.7	0.0023	303	60	0.8	8	11.666	0.186	93.80	3.00
3	1.7	0.0023	303	70	0.8	8	11.666	0.084	97.20	3.00
3	1.7	0.0063	303	15	0.8	8	11.666	1.329	55.70	3.00
3	1.7	0.0063	303	30	0.8	8	11.666	0.679	77.37	3.00
3	1.7	0.0063	303	45	0.8	8	11.666	0.223	92.57	3.00
3	1.7	0.0063	303	60	0.8	8	11.666	0.09	97.00	3.00
3	1.7	0.0063	303	70	0.8	8	11.666	0.053	98.23	3.00
3	1.7	0.0093	303	15	0.8	8	11.666	1.014	66.20	3.00
3	1.7	0.0093	303	30	0.8	8	11.666	0.278	90.73	3.00
3	1.7	0.0093	303	45	0.8	8	11.666	0.053	98.23	3.00
3	1.7	0.0093	303	60	0.8	8	11.666	0.017	99.43	3.00
3	1.7	0.0093	303	70	0.8	8	11.666	0.01	99.67	3.00
3	1.7	0.012	303	15	0.8	8	11.666	0.96	68.00	3.00

(*Continued*)

TABLE 1.3 (Continued)

pH	H_2O_2 (Mmol/L)	Fe2 + (Mmol/L)	Temperature	Time (min)	Reactor (mL)	UV (Watts)	Stirrer (rpm)	C Mmol/L	% Degrade	IC mM/L
3	1.7	0.012	303	30	0.8	8	11.666	0.399	86.70	3.00
3	1.7	0.012	303	45	0.8	8	11.666	0.163	94.57	3.00
3	1.7	0.012	303	60	0.8	8	11.666	0.058	98.07	3.00
3	1.7	0.012	303	70	0.8	8	11.666	0.034	98.87	3.00
3	1	0.0093	303	15	0.8	8	11.666	1.438	52.07	3.00
3	1	0.0093	303	30	0.8	8	11.666	0.561	81.30	3.00
3	1	0.0093	303	45	0.8	8	11.666	0.236	92.13	3.00
3	1	0.0093	303	60	0.8	8	11.666	0.116	96.13	3.00
3	1	0.0093	303	70	0.8	8	11.666	0.072	97.60	3.00
3	1.4	0.0093	303	15	0.8	8	11.666	1.266	57.80	3.00
3	1.4	0.0093	303	30	0.8	8	11.666	0.399	86.70	3.00
3	1.4	0.0093	303	45	0.8	8	11.666	0.171	94.30	3.00
3	1.4	0.0093	303	60	0.8	8	11.666	0.079	97.37	3.00
3	1.4	0.0093	303	70	0.8	8	11.666	0.048	98.40	3.00
3	1.7	0.0093	303	15	0.8	8	11.666	1.014	66.20	3.00
3	1.7	0.0093	303	30	0.8	8	11.666	0.278	90.73	3.00
3	1.7	0.0093	303	45	0.8	8	11.666	0.053	98.23	3.00

3	1.7	0.0093	303	60	0.8	8	11.666	0.017	99.43	3.00
3	1.7	0.0093	303	70	0.8	8	11.666	0.01	99.67	3.00
3	2	0.0093	303	15	0.8	8	11.666	1.016	66.13	3.00
3	2	0.0093	303	30	0.8	8	11.666	0.354	88.20	3.00
3	2	0.0093	303	45	0.8	8	11.666	0.143	95.23	3.00
3	2	0.0093	303	60	0.8	8	11.666	0.047	98.43	3.00
3	2	0.0093	303	70	0.8	8	11.666	0.024	99.20	3.00
2	1.7	0.0093	303	15	0.8	8	11.666	1.14	62.00	3.00
2	1.7	0.0093	303	30	0.8	8	11.666	0.454	84.87	3.00
2	1.7	0.0093	303	45	0.8	8	11.666	0.161	94.63	3.00
2	1.7	0.0093	303	60	0.8	8	11.666	0.071	97.63	3.00
2	1.7	0.0093	303	70	0.8	8	11.666	0.056	98.13	3.00
2.5	1.7	0.0093	303	15	0.8	8	11.666	1.014	66.20	3.00
2.5	1.7	0.0093	303	30	0.8	8	11.666	0.278	90.73	3.00
2.5	1.7	0.0093	303	45	0.8	8	11.666	0.053	98.23	3.00
2.5	1.7	0.0093	303	60	0.8	8	11.666	0.017	99.43	3.00
2.5	1.7	0.0093	303	70	0.8	8	11.666	0.01	99.67	3.00
3	1.7	0.0093	303	15	0.8	8	11.666	1.121	62.63	3.00
3	1.7	0.0093	303	30	0.8	8	11.666	0.377	87.43	3.00

(*Continued*)

TABLE 1.3 (Continued)

pH	H_2O_2 (Mmol/L)	Fe2 + (Mmol/L)	Temperature	Time (min)	Reactor (mL)	UV (Watts)	Stirrer (rpm)	C Mmol/L	% Degrade	IC mM/L
3	1.7	0.0093	303	45	0.8	8	11.666	0.096	96.80	3.00
3	1.7	0.0093	303	60	0.8	8	11.666	0.043	98.57	3.00
3	1.7	0.0093	303	70	0.8	8	11.666	0.018	99.40	3.00
4	1.7	0.0093	303	15	0.8	8	11.666	1.358	54.73	3.00
4	1.7	0.0093	303	30	0.8	8	11.666	0.743	75.23	3.00
4	1.7	0.0093	303	45	0.8	8	11.666	0.321	89.30	3.00
4	1.7	0.0093	303	60	0.8	8	11.666	0.128	95.73	3.00
4	1.7	0.0093	303	70	0.8	8	11.666	0.063	97.90	3.00

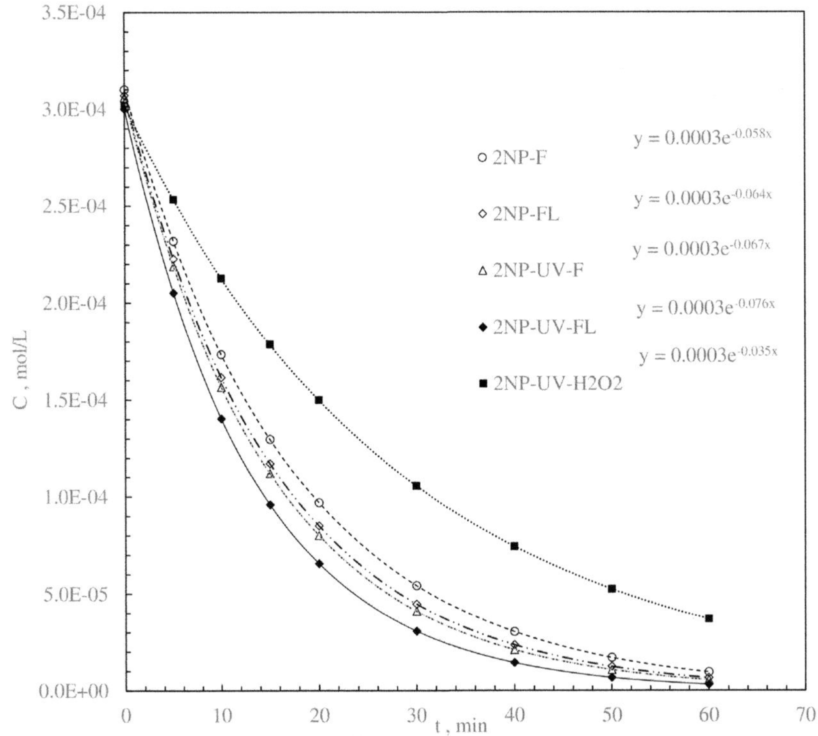

FIGURE 1.2 Exponential Concentration decay profile for 2NP-AOP systems. Concentration decay profile for 2NP-AOP systems.

contribute to enhancing the efficiency of the system. Therefore, it is imperative to establish data-driven causal connections that are conceptualized as models based on experimental data.

Previous studies (Ballard, Palada, Griffin, & Neal, 2021; Belkhode, Ganvir, Shende, & Shelare, 2021) proposed the experimentation theory as a means to understand the outcome of a given activity in relation to the different inputs of a particular phenomenon. Indeed, there is a prevailing sentiment that the current approach lacks the necessary depth to fully comprehend the intricacies of the operations at hand (Dhutekar, Mehta, Modak, Shelare, & Belkhode, 2021; Gondane, Belkhode, Joshi, Maheshwary, & Modak, 2022; Liu, Karniadakis, Tang, & Yvonnet, 2019). This approach ultimately establishes a model for the phenomenon that is based on experimental data.

The variables used in the current setup of AOP are as follows: (1) time; (2) pH; (3) initial concentration; (4) Fe; (5) temperature; (6) H_2O_2; (7) reactor; (8) stirrer; and (9) UV involved in experimentation, and the variables like concentration of pollutants after degradation and their percent degradation were measured as response variables.

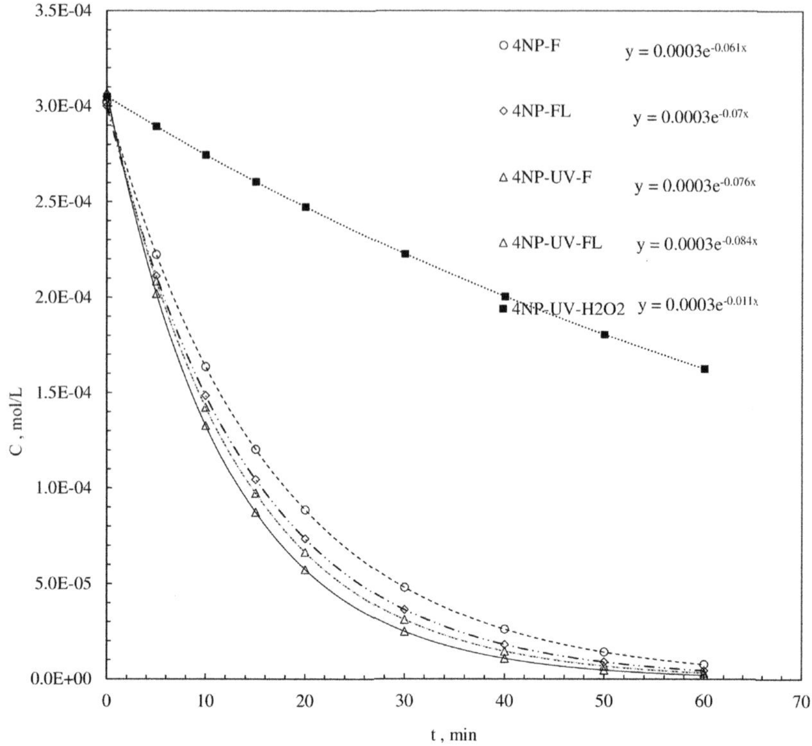

FIGURE 1.3 Exponential Concentration decay profile for 4NP-AOP systems. Concentration decay profile for 4NP-AOP systems.

1.4.1 Approach to formulate

The primary aim of this section is to elucidate the methodology employed in developing the mathematical model for the AOP system under investigation. The identification and study of the most effective systems are crucial steps in the development of a mathematical model. In this section, we have developed approximate generalized mathematical models for the AOP operations in 2NP systems, utilizing theories of experimentation (Belkhode, Sakhale, & Bejalwar, 2020; Neelam et al., 2020; Sonde, Belkhode, & Sakhale, 2014; Waghmare, Sakhale, Tembhurkar, & Shelare, 2020). The procedure that is commonly employed is as follows:

1. Study of existing experimentation, including the decrease in concentration and degradation rate.
2. The potential for developing a model to enhance water quality.

   Planning industrial activities can be a complex task that entails significant financial investments. Hence, it is imperative to develop a model that aids in the reduction of time. Data is gathered through the meticulous

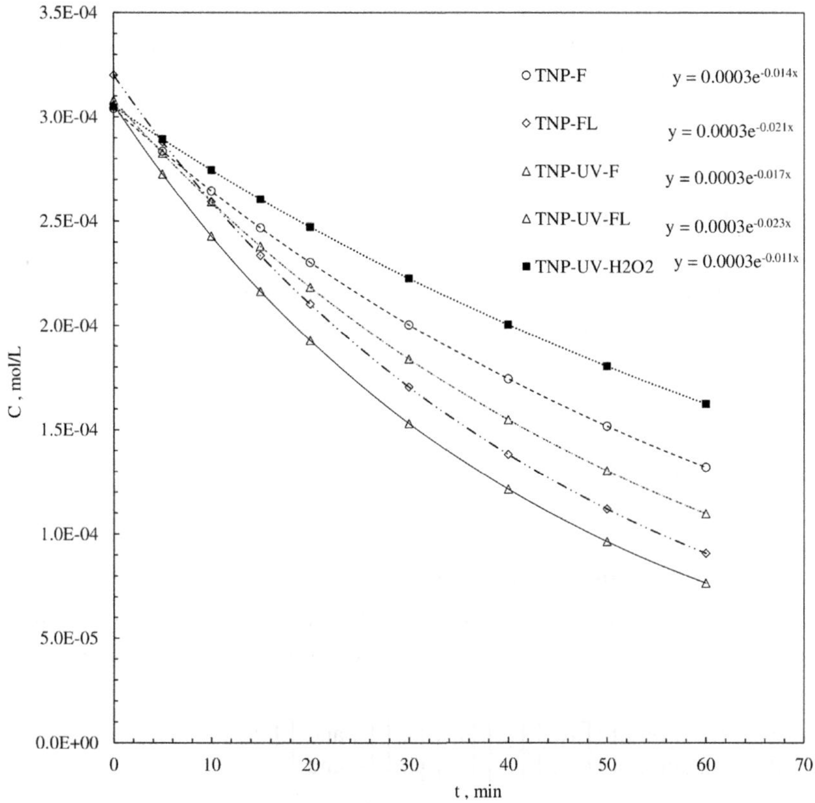

FIGURE 1.4 Exponential Concentration decay profile for TNP-AOP systems. Concentration decay profile for TNP-AOP systems.

application of the Adverse Outcome Pathway methodology *via* direct measurement. The determination of input and output variables is based on the provided data, and a model is constructed by developing a dimensionless equation through the application of regression analysis (Belkhode, 2019).

3. The study aims to study the usefulness and efficiency of the model.

Proper planning is necessary for the execution of the experiments in the research task. Typically, experimenting involves a substantial number of parameters. To ensure accuracy, it is anticipated that the economic impact of all variables should be examined through the application of a structured model that can be used toward the prediction of dependency on the input variables while studying the reaction system. The theory and the practical applications are a scholarly investigation into scientific phenomena, encompassing the examination and integration of said phenomena (Belkhode, 2019). The research may encompass both a theoretical framework and an experimental methodology.

Theoretical approaches involve the application of laws of mechanics and physics, which encompass the energy balance equation. The investigation of multiple independent variables and their impact on a complex process is conducted to develop a comprehensive understanding of the process (Belkhode et al., 2020).

1.4.2 Improvement of water quality

The main factor that is majorly contributing to the oxidative degradation process is the production of radicals, which in turn depends on various parameters, affecting the efficiency of the process. Those are pollutant concentration, catalyst concentration, time and details of the interacting physical system (details of effluents), and environmental (conditions) variables. The majorly considered dependent variables are:

1. Concentration (C).
2. Degradation (D).

1.4.3 Formulation of mathematical model

Six independent p terms ($\prod_1$, $\prod_2$, $\prod_3$, $\prod_4$, $\prod 4_5$, and $\prod_6$) and two dependent $\prod$ terms (Z_1 and Z_2) have been considered for model formulation.

In this, Z1 and Z2 are considered as:

Z_1 = function of ($\prod_1$, $\prod_2$, $\prod_3$, $\prod_4$, $\prod_5$, and $\prod_6$)
Z_2 = function of ($\prod_1$, $\prod_2$, $\prod_3$, $\prod_4$, $\prod_5$, and $\prod_6$)

The generalized equations for the resultant Z1 and Z2 parameters can be formulated commonly as

$$(Z) = K_*[pH]a, \left[Fe/H_2O\right]b, [\theta]c, [(H_2O)(Re)]d, \left[IC/H_2O\right]e, [(St)\ (Ti)]f \quad (1.3)$$

The analysis and build-up of the model are shown in Table 1.4. Exponential equations for both parameters will be of the form:

$$(Z_1) = K_{1*}\left[\left(\prod_1\right)^{a1} * \left(\prod_2\right)^{b1} * \left(\prod_3\right)^{c1} * \left(\prod_4\right)^{d1} * \left(\prod_5\right)^{e1} * \left(\prod_6\right)^{f1}\right] \quad (1.4)$$

$$(Z_2) = K_{2*}\left[\left(\prod_1\right)^{a2} * \left(\prod_2\right)^{b2} * \left(\prod_3\right)^{c2} * \left(\prod_4\right)^{d2} * \left(\prod_5\right)^{e2} * \left(\prod_6\right)^{f2}\right] \quad (1.5)$$

To find the unknown exponential values, the matrix method was applied. The exact form of the model obtained is as under:

$$(Z_1) = 0.1670_*\left(\prod_1\right)^{-0.1017} * \left(\prod_2\right)^{0.1052} * \left(\prod_3\right)^{0.0307} * \left(\prod_4\right)^{0.2745} * \left(\prod_5\right)^{0.0598} * \left(\prod_6\right)^{0.3295} \quad (1.6)$$

TABLE 1.4 Data related to pie terms and dependent terms.

P1	P2	P3	P4	P5	P6	Z1 = Degrad
3	0.0013529	303	1.36	1.76471	175	0.44
3	0.0013529	303	1.36	1.76471	350	0.68
3	0.0013529	303	1.36	1.76471	525	0.86
3	0.0013529	303	1.36	1.76471	700	0.94
3	0.0013529	303	1.36	1.76471	816.667	0.97
3	0.0037059	303	1.36	1.76471	175	0.56
3	0.0037059	303	1.36	1.76471	350	0.77
3	0.0037059	303	1.36	1.76471	525	0.93
3	0.0037059	303	1.36	1.76471	700	0.97
3	0.0037059	303	1.36	1.76471	816.667	0.98
3	0.0054706	303	1.36	1.76471	175	0.66
3	0.0054706	303	1.36	1.76471	350	0.91
3	0.0054706	303	1.36	1.76471	525	0.98
3	0.0054706	303	1.36	1.76471	700	0.99
3	0.0054706	303	1.36	1.76471	816.667	1.00
3	0.0070588	303	1.36	1.76471	175	0.68
3	0.0070588	303	1.36	1.76471	350	0.87
3	0.0070588	303	1.36	1.76471	525	0.95
3	0.0070588	303	1.36	1.76471	700	0.98
3	0.0070588	303	1.36	1.76471	816.667	0.99
3	0.0093	303	0.8	3	175	0.52
3	0.0093	303	0.8	3	350	0.81
3	0.0093	303	0.8	3	525	0.92
3	0.0093	303	0.8	3	700	0.96
3	0.0093	303	0.8	3	816.667	0.98
3	0.0066429	303	1.12	2.14286	175	0.58
3	0.0066429	303	1.12	2.14286	350	0.87
3	0.0066429	303	1.12	2.14286	525	0.94
3	0.0066429	303	1.12	2.14286	700	0.97
3	0.0066429	303	1.12	2.14286	816.667	0.98

(Continued)

TABLE 1.4 (Continued)

P1	P2	P3	P4	P5	P6	Z1 = Degrad
3	0.0054706	303	1.36	1.76471	175	0.66
3	0.0054706	303	1.36	1.76471	350	0.91
3	0.0054706	303	1.36	1.76471	525	0.98
3	0.0054706	303	1.36	1.76471	700	0.99
3	0.0054706	303	1.36	1.76471	816.667	1.00
3	0.00465	303	1.6	1.5	175	0.66
3	0.00465	303	1.6	1.5	350	0.88
3	0.00465	303	1.6	1.5	525	0.95
3	0.00465	303	1.6	1.5	700	0.98
3	0.00465	303	1.6	1.5	816.667	0.99
2	0.0054706	303	1.36	1.76471	175	0.62
2	0.0054706	303	1.36	1.76471	350	0.85
2	0.0054706	303	1.36	1.76471	525	0.95
2	0.0054706	303	1.36	1.76471	700	0.98
2	0.0054706	303	1.36	1.76471	816.667	0.98
2.5	0.0054706	303	1.36	1.76471	175	0.66
2.5	0.0054706	303	1.36	1.76471	350	0.91
2.5	0.0054706	303	1.36	1.76471	525	0.98
2.5	0.0054706	303	1.36	1.76471	700	0.99
2.5	0.0054706	303	1.36	1.76471	816.667	1.00
3	0.0054706	303	1.36	1.76471	175	0.63
3	0.0054706	303	1.36	1.76471	350	0.87
3	0.0054706	303	1.36	1.76471	525	0.97
3	0.0054706	303	1.36	1.76471	700	0.99
3	0.0054706	303	1.36	1.76471	816.667	0.99
4	0.0054706	303	1.36	1.76471	175	0.55
4	0.0054706	303	1.36	1.76471	350	0.75
4	0.0054706	303	1.36	1.76471	525	0.89
4	0.0054706	303	1.36	1.76471	700	0.96
4	0.0054706	303	1.36	1.76471	816.667	0.98

Time for Degradation rate (Z_1) =

(De) = $0.1670_*[pH]^{-0.1017}\left[Fe/H_2O\right]^{0.1052}[\theta]^{0.0307}[(H_2O)\,(Re)]^{0.2745}\left[IC/H_2O\right]^{0.0598}$
[(St)(Ti)]0.3295

1.5 Formulation of Kinetic model

Approximately 85%–90% of the substrate experienced oxidative degradation within a time frame of one hour. In the present investigation, the kinetic plots were carried out using Windows MS Excel, which is an excellent tool for such plots since it has the facilities for linear and non-linear regression of the data. The reliability of the kinetic data is understood by the closeness of the linear regression coefficient "R^2" value to 1. In the present studies, the R^2 values obtained were above 0.98, showing that the data adheres to the first-order kinetics with respect to the substrate concentration decay and 0th order with respect to the concentration decay of the oxidant H_2O_2. The possibility of adherence of the experimental rate data to second-order kinetics was also explored, but none of the systems studied obeyed the second-order rate equation. The overall rate expression for the degradation of pollutants may therefore be written as:

$$-d[C]/dt = k[\text{substrate}]^1[\text{Oxidant}]^\circ \quad (1.8)$$

The obtained rate data was therefore analyzed by first-order kinetics, which excellently adhered to linearity, particularly during the initial 50–60 minutes of contact time during which 80%–90% of oxidative degradation was observed. Thus, a first-order rate equation can be written as:

$$\ln\left[C_o/C_t\right] = k\,t \quad (1.9)$$

and plotting ln [C_o/C_t] values against the pseudo-first-order reaction rate constant "k" was evaluated. The obtained experimental data are summarized in Table 1.5. The ln [C_o/C_t] versus plots for 2-NP-Fenton, 4-NP-Fenton, and TNP-Fenton systems are shown in Figs. 1.2–1.4.

1.5.1 Activation energy

The activation energy refers to the minimum quantity of energy necessary for the creation of an activated complex, which subsequently facilitates the reaction in either direction while releasing energy. An increase in temperature results in an elevation of the kinetic energy possessed by the molecules of the reactants. Consequently, the likelihood of activated collisions occurring is heightened, ultimately leading to the formation of an activated complex. This phenomenon, in turn, leads to an escalation in the rate at which the reaction proceeds (Galvan et al., 2013). To assess the energy of activation, the pseudo-first-order rate constants 'k' were experimentally determined

TABLE 1.5 Effect of temperature on rate constant for AOPs studied.

No.	Temperature	$1/T \times 10^3$, K^{-1}					Eyring plot	
3.41	3.36	3.33	3.25	3.19				
	System	Rate constant k, min^{-1}					Slope	Intercept
1	$2NP + H_2O_2 + Fe^{2+}$	0.0412	0.0522	0.0583	0.0693	0.0819	2739.1	0.5003
	$2NP + H_2O_2 + Fe^{3+}$	0.0486	0.0574	0.0646	0.0781	0.0909	2557.1	0.0165
	$2NP + UV + H_2O_2 + Fe^{2+}$	0.0480	0.0581	0.0673	0.0791	0.0912	2612.2	0.2078
	$2NP + UV + H_2O_2 + Fe^{3+}$	0.0541	0.0654	0.0765	0.0880	0.1011	2552.4	0.1244
2	$4NP + H_2O_2 + Fe^{2+}$	0.0464	0.0569	0.0637	0.0751	0.0889	2591.5	0.1040
	$4NP + H_2O_2 + Fe^{3+}$	0.0512	0.0621	0.0712	0.0842	0.0987	2663.9	0.4463
	$4NP + UV + H_2O_2 + Fe^{2+}$	0.0538	0.0651	0.0762	0.0878	0.1023	2612.1	0.3034
	$4NP + UV + H_2O_2 + Fe^{3+}$	0.0595	0.0704	0.0838	0.0978	0.1105	2573.7	0.2872
3	$TNP + H_2O_2 + Fe^{2+}$	0.0082	0.0118	0.0141	0.0189	0.0246	4593.0	5.2175
	$TNP + H_2O_2 + Fe^{3+}$	0.0109	0.0148	0.0213	0.0276	0.0354	5166.8	7.4456
	$TNP + UV + H_2O_2 + Fe^{2+}$	0.0091	0.0131	0.0172	0.0224	0.0311	5231.9	7.4874
	$TNP + UV + H_2O_2 + Fe^{3+}$	0.0120	0.0178	0.0231	0.0312	0.0401	5158.7	7.5407

at distinct temperatures and then plotted in accordance with the Arrhenius equation.

$$\ln k = \ln A - (E_a/R).(1/T) \tag{1.10}$$

The data exhibited a high degree of linearity, as evidenced by the regression coefficient (R2 ≥ 0.97), depicted in Figs. 1.5–1.7. The energy of activation, denoted as Ea, was subsequently determined by analyzing the slope. The empirical findings pertaining to the influence of temperature on the reaction rate constant have been documented in Table 1.3. The Arrhenius equation, which was derived through linear regression analysis of the natural logarithm of the rate constant (ln k) plotted against the reciprocal of temperature (1/T), is depicted in Fig. 1.5–1.7 for the investigated AOPs.

1.5.2 Evaluation of thermodynamic parameters

The degradation processes involve the attack of substrate molecules by hydroxyl (OH.) free radicals that are generated in situ. This results in the

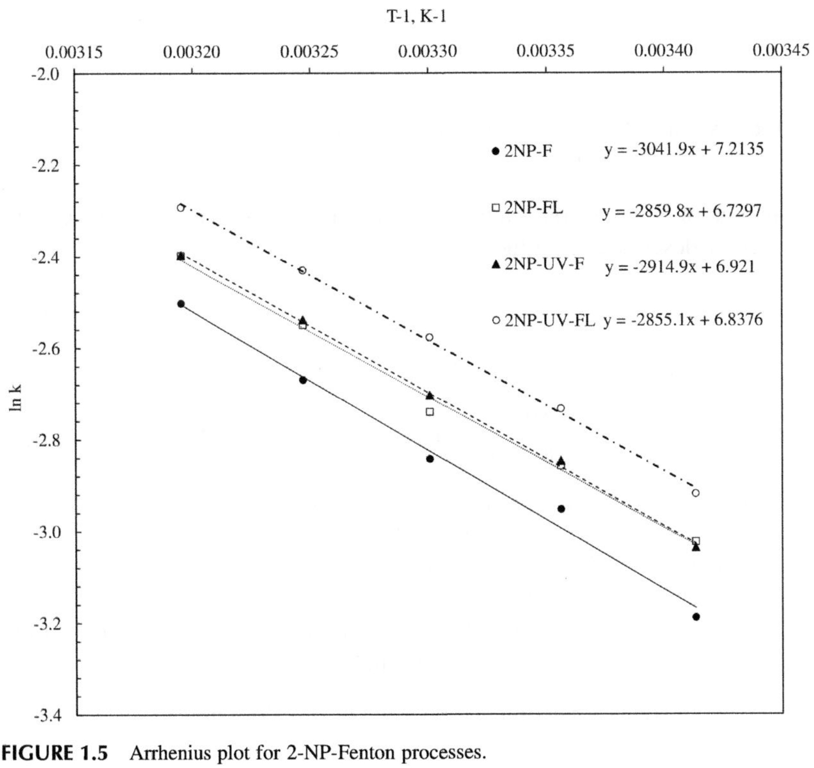

FIGURE 1.5 Arrhenius plot for 2-NP-Fenton processes.

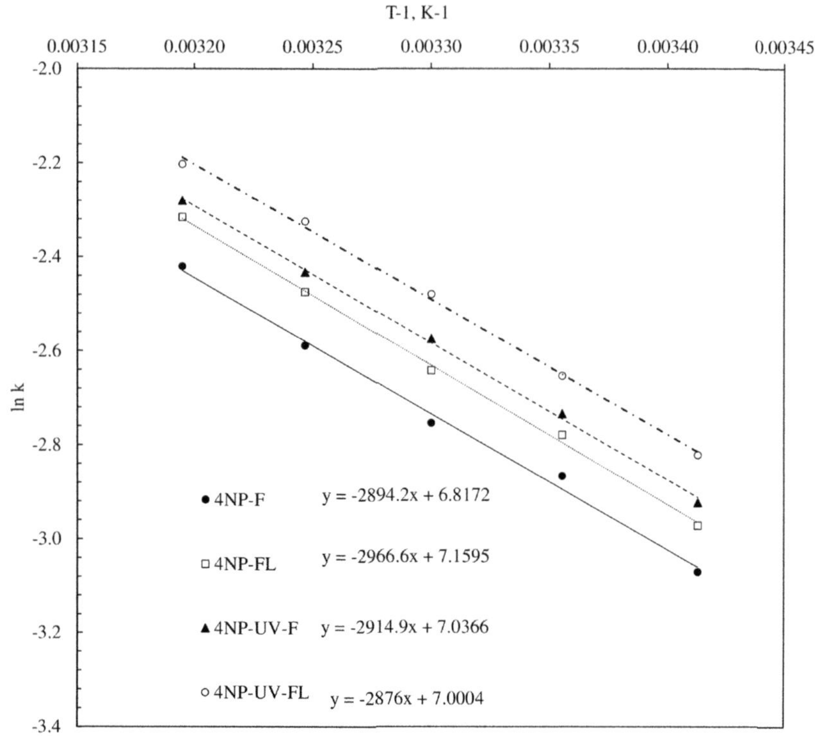

FIGURE 1.6 Arrhenius plot for 4-NP-Fenton processes.

formation of a transition state complex denoted as $SOH^{\#}$, which can be effectively described using the principles of transition state theory.

$$S + OH\cdot \underset{k'}{\overset{k}{\rightleftharpoons}} OH^{\#} \rightarrow P \tag{1.11}$$

where the equilibrium constant is calculated as:

$$K^{\#} = \frac{[SOH\#]}{[S][OH\cdot]} or [SOH\#] = \frac{k}{k'} = [k'][S][OH\cdot] \tag{1.12}$$

In Eq. (1.11), the equilibrium can lead to the formation of products or a reversion back to the initial reactants. The intermediate transition state complex, denoted as $SOH^{\#}$, exhibits a significantly brief existence. Consequently, it is reasonable to hypothesize that the sum of constants k′ and k2 is much greater than the rate constant k. Therefore, at any given moment, the concentration of $SOH^{\#}$ is expected to be exceedingly low.

$$\text{rate of reaction} = r = v.\left[SOH^{\#}\right] = v.[S].[OH^{\cdot}]k^{\#} \tag{1.13}$$

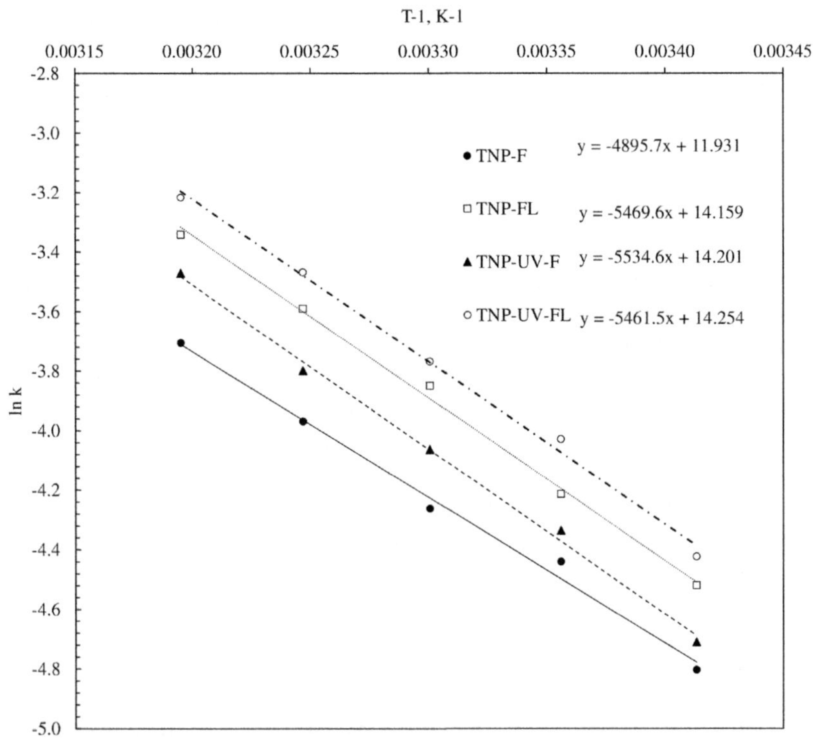

FIGURE 1.7 Arrhenius plot for TNP-Fenton processes.

From the above equations, the rate of the reaction:

$$r = \frac{kbT}{h} k^{\#}[S][OH^{\cdot}] \tag{1.14}$$

The Eyring equation, which is widely recognized in the field, is derived from the principles of transition state theory. It establishes a mathematical relationship between the rate of a chemical reaction and the Gibbs free energy of activation. It has proven to be a useful tool in the determination of various thermodynamic parameters, such as the standard enthalpy of activation ($\Delta H^{\#}$), the standard entropy of activation ($\Delta S^{\#}$), and the standard Gibbs energy of activation ($\Delta G^{\#}$). These parameters can be calculated for a given reaction, provided that its rate constant has been determined through experimental means. Gibbs free energy comprises both an entropic component and an enthalpic component. The entropic term in the Arrhenius equation is incorporated into the pre-exponential factor A. The linearized form of the Eyring equation is denoted as (Eq. 1.15):

$$\ln\frac{k}{T} = -\frac{\Delta H^{\#}}{R}\frac{1}{T} + \ln\frac{kb}{h} + \frac{\Delta S^{\#}}{R} \tag{1.15}$$

The thermodynamic parameters were then evaluated, as shown in Fig. 1.8–1.10. The slope of the plot $-\frac{\Delta H^{\#}}{R}$ and intercept $\ln\frac{kb}{h} + \frac{\Delta S^{\#}}{R}$ values given in Table 1.3 were then used to calculate the standard thermodynamic parameters for the AOPs studied, which are reported in Table 1.6.

1.6 Artificial neural network for the AOP

Mathematical models refer to quantitative relationships that are recognized in the field of study. Two distinct types of models have been established, namely the mathematical model, which utilizes the least-squares multiple regression method. The primary focus of the operator is to strategically organize inputs to achieve desired outcomes (Milling, 1995). Ultimately, significant conclusions are derived from the examination of the models.

The present study involves the analysis of a model designed to investigate the hierarchical impact of different inputs (causes) on outputs (effects). The primary objective is to determine the relative influence of these causes on the resulting effects. Additionally, the study aims to provide an interpretation of

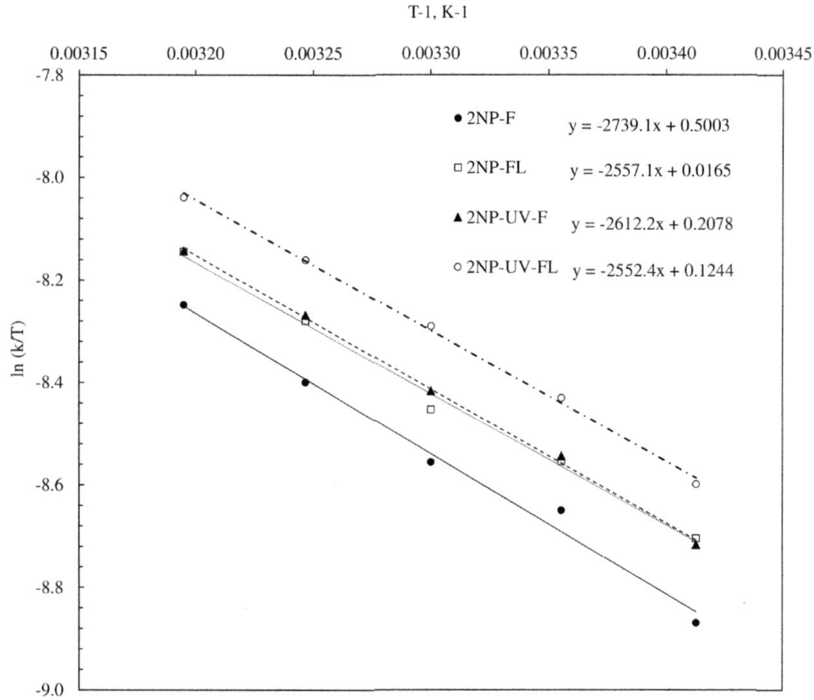

FIGURE 1.8 Arrhenius plot for 2-NP-Fenton processes.

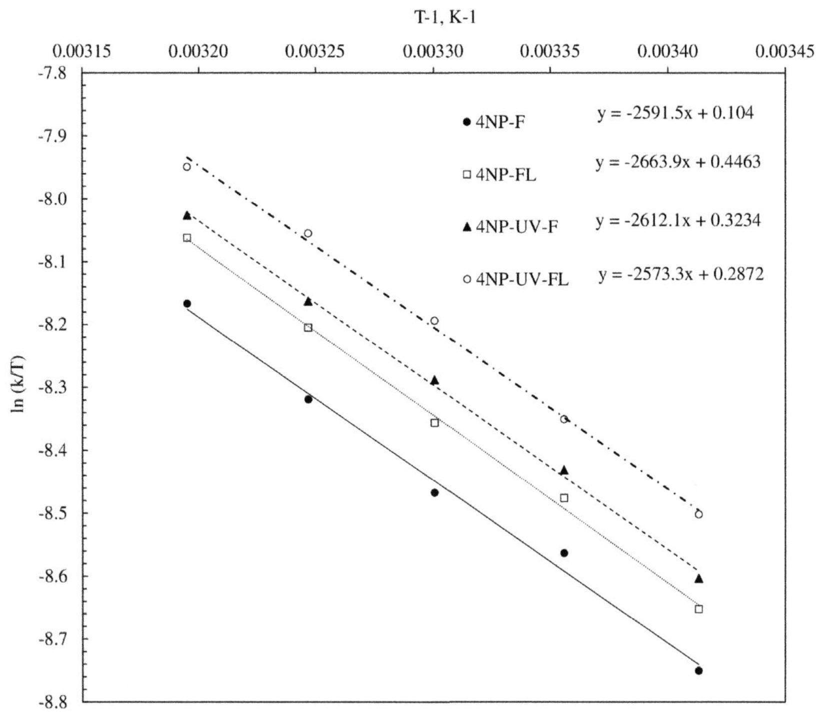

FIGURE 1.9 Arrhenius plot for 4-NP-Fenton processes.

the curve fitting constant. The analysis focuses on examining the impact of each individual independent variable on the degradation rate of the response variable. The values obtained through empirical observations and those derived from mathematical models are validated through the calculation of their respective mean values and standard errors of estimation (Table 1.7).

The exact mathematical exponential equations formed using the model are as under:

$$(Z_1) = 0.1670_*\left(\prod_1\right)^{-0.1017} *\left(\prod_2\right)^{0.1052} *\left(\prod_3\right)^{0.0307} *\left(\prod_4\right)^{0.2745} *\left(\prod_5\right)^{0.0598} *\left(\prod_6\right)^{0.3295}$$

Time for Degradation rate (Z_1) =

$$(De) = 0.1670_*[\mathbf{pH}]^{-0.1017}\left[\mathbf{Fe/H_2O}\right]^{0.1052}[\theta]^{0.0307}[(\mathbf{H_2}O)(Re)]^{0.2745}\left[\mathbf{IC/H_2O}\right]^{0.0598}$$
[(St)(Ti)]0.3295

Interpretation of the model:

The interpretation of the model is being presented in relation to several aspects, namely:

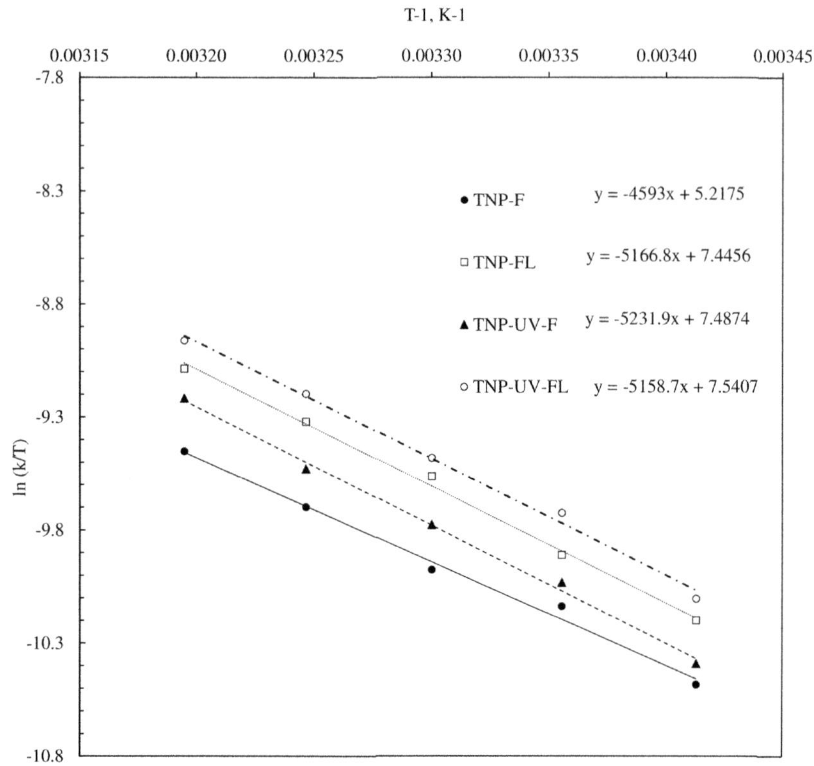

FIGURE 1.10 Arrhenius plot for TNP-Fenton processes.

(a) Interpretation of curve fitting constant (K_1) for Z_1:

The value of the curve fitting constant in this model for (Z_1) is 0.1670. This collectively represents the combined effect of all extraneous variables such as pressure, physiological, and biomechanical demands of the purification process (a variable that we have not considered into experimental observations). Further, as it is positive, this indicates that these causes have an increasing influence on degradation (Z_1).

The following conclusion can be drawn:

1. The absolute index of $\prod_6$ is the highest, viz. 0.3295. Thus, the term related to the specification of stirrer speed and time involved is the most influential p term in this model. The value of this index is positive, indicating that the degradation activity (Z_1) is directly proportional to the term related to the specification of stirrer speed and time involved, that is, $\prod_6$, which indicates that more time is required for degradation (Alfano & Cassano, 2008).

TABLE 1.6 Energy, enthalpy, entropy, and Gibbs free energy of activation values.

No.	System	ln A	E_a, kJ/mol	$\Delta H^{\#}$, kJ/mol	$\Delta S^{\#}$, J/mol K^{-1}	$\Delta G^{\#}$, kJ/mol
1	$2NP + H_2O_2 + Fe^{2+}$	7.214	25.29	22.77	−193.4	80.4
	$2NP + H_2O_2 + Fe^{3+}$	6.729	23.78	21.26	−197.4	80.1
	$2NP + UV + H_2O_2 + Fe^{2+}$	6.921	24.23	21.72	−195.8	80.07
	$2NP + UV + H_2O_2 + Fe^{3+}$	6.838	23.74	26.76	−196.5	79.8
2	$4NP + H_2O_2 + Fe^{2+}$	6.817	24.06	21.55	−196.7	80.2
	$4NP + H_2O_2 + Fe^{3+}$	7.159	24.66	22.15	−193.8	79.9
	$4NP + UV + H_2O_2 + Fe^{2+}$	7.037	24.23	21.72	−195.0	79.8
	$4NP + UV + H_2O_2 + Fe^{3+}$	7.001	23.91	21.40	−195.2	79.6
3	$TNP + H_2O_2 + Fe^{2+}$	11.931	40.70	38.19	−154.2	84.1
	$TNP + H_2O_2 + Fe^{3+}$	14.159	45.47	42.96	−135.6	83.4
	$TNP + UV + H_2O_2 + Fe^{2+}$	14.201	46.01	43.50	−135.2	83.8
	$TNP + UV + H_2O_2 + Fe^{3+}$	14.254	45.41	42.89	−134.9	83.1

2. The absolute index of $\prod_3$ is the lowest, viz. 0.0307. Thus, the term related to temperature is the least influential p term in this model. The value of this index is positive, indicating that the degradation (Z_1) is directly proportional to the term related to temperature [$\prod_1$]. The temperature increases as $\prod_1$ increases.

   As it may increase the kinetic energy and stimulate molecules for more degradation.
3. The sequence of influence of other independent p terms present in this model is $\prod_4$, $\prod_5$, and $\prod_2$, having absolute indices of 0.2745, 0.0598, and 0.1052 in the reducing order, respectively.

The degradation activity (Z_1) is inversely proportional to the term related to pH [$\prod_1$], with the index being -0.1017. With the increase in pH, the degradation is reduced.

Similarly, the time of AOP activity (Z_1) is directly proportional to the term related to H_2O and Re [$\prod_4$] with an index of 0.2745, increasing the degradation with the increase in H_2O and Re capacity (Zaharia & Suteu, 2014).

TABLE 1.7 Comparison of experimental calculated values and mathematical model equation base values.

Sr. no.	Experimentally calculated values and mathematical model equation base values			
	Calculated from mathematical model	Experimental observation	Error = Experimental observation − calculated value	Percentage error
1	0.548	0.4437	−0.1047	−23.5992
2	0.689	0.6837	−0.0054	−0.7900
3	0.788	0.8613	0.0738	8.5649
4	0.866	0.9380	0.0721	7.6901
5	0.911	0.9720	0.0610	6.2775
6	0.610	0.5570	−0.0527	−9.4597
7	0.766	0.7737	0.0075	0.9750
8	0.876	0.9257	0.0500	5.4054
9	0.963	0.9700	0.0073	0.7532
10	1.013	0.9823	−0.0305	−3.1070
11	0.635	0.6620	0.0268	4.0500
12	0.798	0.9073	0.1092	12.0318
13	0.912	0.9823	0.0701	7.1342
14	1.003	0.9943	−0.0086	−0.8672

15	1.055	0.9967	−0.0585	−5.8744
16	0.652	0.6800	0.0275	4.0512
17	0.820	0.8670	0.0471	5.4375
18	0.937	0.9457	0.0086	0.9118
19	1.030	0.9807	−0.0495	−5.0524
20	1.084	0.9887	−0.0952	−9.6318
21	0.599	0.5207	−0.0787	−15.1090
22	0.753	0.8130	0.0599	7.3665
23	0.861	0.9213	0.0606	6.5749
24	0.946	0.9613	0.0150	1.5595
25	0.996	0.9760	−0.0197	−2.0133
26	0.622	0.5780	−0.0438	−7.5826
27	0.781	0.8670	0.0856	9.8760
28	0.893	0.9430	0.0499	5.2957
29	0.982	0.9737	−0.0082	−0.8413
30	1.033	0.9840	−0.0490	−4.9815
31	0.635	0.662	0.0268	4.0500
32	0.798	0.907333	0.1092	12.0318
33	0.912	0.982333	0.0701	7.1342

(Continued)

TABLE 1.7 (Continued)

Sr. no.	Experimentally calculated values and mathematical model equation base values			
	Calculated from mathematical model	Experimental observation	Error = Experimental observation − calculated value	Percentage error
34	1.003	0.994333	−0.0086	−0.8672
35	1.055	0.996667	−0.0585	−5.8744
36	0.647	0.661333	0.0147	2.2287
37	0.812	0.882	0.0695	7.8803
38	0.929	0.952333	0.0237	2.4889
39	1.021	0.984333	−0.0366	−3.7214
40	1.074	0.992	−0.0822	−8.2824
41	0.662	0.62	−0.0419	−6.7628
42	0.832	0.848667	0.0169	1.9915
43	0.951	0.946333	−0.0043	−0.4567
44	1.045	0.976333	−0.0688	−7.0514
45	1.100	0.981333	−0.1183	−12.0554
46	0.647	0.662	0.0149	2.2542
47	0.813	0.907333	0.0942	10.3855

48	0.929	0.982333	0.0530	5.3962
49	1.022	0.994333	−0.0274	−2.7549
50	1.075	0.996667	−0.0783	−7.8559
51	0.635	0.626333	−0.0089	−1.4139
52	0.798	0.874333	0.0762	8.7116
53	0.912	0.968	0.0557	5.7591
54	1.003	0.985667	−0.0173	−1.7541
55	1.055	0.994	−0.0612	−6.1584
56	0.617	0.547333	−0.0695	−12.7055
57	0.775	0.752333	−0.0228	−3.0329
58	0.886	0.893	0.0071	0.7897
59	0.974	0.957333	−0.0167	−1.7449
60	1.025	0.979	−0.0458	−4.6772

1.7 Conclusion

The activation energy refers to the minimum quantity of energy necessary for the creation of an activated complex. An increase in temperature results in an acceleration of the reaction rate (Galvan et al., 2013). According to Levenspiel (Wiley, Hepburn, & Levenspiel, 1999), reactions with activation energy values falling within the range of 40–400 kJ/mol are classified as having low and high activation energy, respectively. The activation energies for both the Fenton and Fenton-like processes for 2-nitrophenol (2NP) and 4-nitrophenol (4NP) are approximately 25 kJ/mol, while the activation energy for TNP is approximately 45 kJ/mol. These values corroborate the slow degradation of TNP as compared to 2NP and 4NP. Further, these activation energy values for Fenton and Fenton-like systems are much lower as compared to the activation energy of 170 kJ/mol for noncatalytic oxidation of phenol in aqueous solution below 463K (Vicente, Rosal, & Díaz, 2002). This observation suggests that the Fenton and Fenton-like processes exhibit significantly enhanced reaction rates at room temperature and proceed along a reaction pathway characterized by a notably low activation energy (Ahmed et al., 2021).

Positive $\Delta H\#$ values are indicative of the necessity for an external energy source to elevate the energy level of reactants to facilitate the formation of a transition state (Shuchi, Suhan, Humayun, Haque, & Islam, 2021). The temperature rise has been observed to have a beneficial impact on the oxidative degradation of pollutants. This is primarily attributed to the higher likelihood of activated complex formation and the abundance of available resources, which contribute to an accelerated degradation rate at elevated temperatures. The study conducted by the researchers (Sun et al., 2009) demonstrated that the presence of OH radicals significantly increased the rate of pollutant degradation. These findings (Cardozo, 2011) demonstrate promising capabilities in the treatment of water and wastewater, resulting in improved efficiency in the degradation process. The rate of degradation (Z1) exhibits a direct proportionality to the terms associated with IC and H_2O [p5]. This relationship is characterized by an index of 0.0598, suggesting that an optimal initial concentration can significantly enhance the rate of degradation.

Similarly, the time of degradation rate activity (Z_1) is directly proportional to the term related to the Fe and H_2O [$\prod_2$]. As the Fe increases, degradation increases with an index of 0.1052.

To demonstrate the industrial applicability of AOPs, it is necessary to conduct comprehensive investigations aimed at identifying more economically viable AOPs at the pilot plant scale as opposed to the batch scale. Due to the intricate interplay between pollutants and their respective reaction pathways, the generalizability of findings from one pollutant to another is challenging. Consequently, further investigations employing authentic effluents are imperative. The treatment of actual wastewater necessitates the implementation of an integrated process that incorporates thermodynamics and reaction kinetic studies, as these factors play an important role in the estimation of costs.

Furthermore, the utilization of mechanistic and mathematical modeling techniques can be employed to enhance and establish the correlation between parental compounds and reactants utilized within treatment systems. Moreover, hybrid AOPs may potentially exhibit enhanced efficacy in the elimination of refractory compounds as compared to individual AOPs. Typically, the integration of various AOPs serves to augment the production of free radicals, thereby leading to an elevation in the rate of chemical reactions.

In summary, it can be inferred that AOPs have the potential to emerge as a formidable technique, provided that the challenges associated with system design and optimization are effectively addressed and resolved. Furthermore, it is imperative to thoroughly investigate and assess all the variables that contribute to the improvement of degradation efficiency.

References

Agrawal, Pratibha S., Belkhode, Pramod N., & Rokhum, Samuel Lalthazuala (2022). *Modelling and simulation to predict the performance of the diesel blends. Sustainability in Biofuel Production Technology* (pp. 209–266). Wiley. Available from 10.1002/9781119888864.ch10.

Ahmed, N., Vione, D., Rivoira, L., Carena, L., Castiglioni, M., & Bruzzoniti, M. C. (2021). A review on the degradation of pollutants by fenton-like systems based on zero-valent iron and persulfate: Effects of reduction potentials, ph, and anions occurring in waste waters. *Molecules*, *26*(15). Available from https://doi.org/10.3390/molecules26154584. Available from: https://www.mdpi.com/1420-3049/26/15/4584/pdf.

Alfano, Orlando M., & Cassano, Alberto E. (2008). Photoreactor modeling: Applications to advanced oxidation processes. *International Journal of Chemical Reactor Engineering*, *6*(1). Available from https://doi.org/10.2202/1542-6580.1617.

Ballard, T., Palada, H., Griffin, M., & Neal, A. (2021). An integrated approach to testing dynamic, multilevel theory: Using computational models to connect theory, model, and data. *Organizational Research Methods*, *24*(2), 251–284. Available from https://doi.org/10.1177/1094428119881209. Available from: http://www.sagepub.com/ejournals.

Belkhode, P., Sakhale, C., & Bejalwar, A. (2020). Evaluation of the experimental data to determine the performance of a solar chimney power plant. *Materials Today: Proceedings*, *27*, 102–106. Available from https://doi.org/10.1016/j.matpr.2019.09.006, https://www.sciencedirect.com/journal/materials-today-proceedings.

Belkhode, P. N. (2019a). Optimum choice of the front suspension of an automobile. *Journal of Engineering Sciences*, *6*(1), e21–e24. Available from https://doi.org/10.21272/jes.2019.6(1).e4.

Belkhode, P. N. (2019b). Development of mathematical model and artificial neural network simulation to predict the performance of manual loading operation of underground mines. *Journal of Materials Research and Technology*, *8*(2), 2309–2315. Available from https://doi.org/10.1016/j.jmrt.2019.04.015. Available from: http://www.elsevier.com/journals/journal-of-materials-research-and-technology/2238-7854.

Belkhode, P. N., Ganvir, V. N., Shelare, S. D., Shende, A., & Maheshwary, P. (2022). Experimental investigation on treated transformer oil (TTO) and its diesel blends in the diesel engine. *Energy Harvesting and Systems*, *9*(1), 75–81. Available from https://doi.org/10.1515/ehs-2021-0032. Available from: http://www.degruyter.com/view/j/ehs.

Belkhode, P. N., Ganvir, V. N., Shende, A. C., & Shelare, S. D. (2021). Utilization of waste transformer oil as a fuel in diesel engine. *Materials Today: Proceedings*, *49*, 262–268. Available from https://doi.org/10.1016/j.matpr.2021.02.008, https://www.sciencedirect.com/journal/materials-today-proceedings.

Belkhode, Pramod, Modak, J. P., Vidyasagar, V., & Agrawal, Pratibha (2021). *Analysis of performance of the model* (pp. 95–102). Informa UK Limited. Available from 10.1201/9781003132127-9.

Bisaria, K., Sinha, S., Singh, R., & Iqbal, H. M. N. (2021). Recent advances in structural modifications of photo-catalysts for organic pollutants degradation – A comprehensive review. *Chemosphere*, *284*. Available from https://doi.org/10.1016/j.chemosphere.2021.131263. Available from: http://www.elsevier.com/locate/chemosphere.

Cardozo, M. R. R. (2011). Advanced oxidation of chemicals of emerging concern: Modeling and experimental simulation. *Dissertation Abstracts International*.

Chairam, S., Konkamdee, W., & Parakhun, R. (2017). Starch-supported gold nanoparticles and their use in 4-nitrophenol reduction Starch-supported gold nanoparticles in 4-nitrophenol reduction. *Journal of Saudi Chemical Society*, *21*(6), 656–663. Available from https://doi.org/10.1016/j.jscs.2015.11.001. Available from: http://www.sciencedirect.com/science/journal/13196103.

Charamba, L. V. C., Santana, R. M. D. R., Do Nascimento, G. E., Charamba, B. V. C., De Moura, M. C., Coelho, L. C. B. B., … Napoleão, D. C. (2018). Application of the advanced oxidative process on the degradation of the green leaf and purple açaí food dyes with kinetic monitoring and artificial neural network modelling. *Water Science and Technology*, *78*(5), 1094–1103. Available from https://doi.org/10.2166/wst.2018.391. Available from: https://watermark.silverchair.com/wst078051094.pdf.

Copete-Pertuz, L. S., Serna-Galvis, E. A., Plácido, J., Torres-Palma, R. A., & Mora-Martínez, A. L. (2021). Coupling chemical oxidation processes and Leptosphaerulina sp. mycoremediation to enhance the removal of recalcitrant organic pollutants in aqueous systems. *Science of the Total Environment*, *772*. Available from https://doi.org/10.1016/j.scitotenv.2021.145449. Available from: http://www.elsevier.com/locate/scitotenv.

Deshpande, B. D., Agrawal, P. S., & Yenkie, M. K. N. (2019a). Nanoparticles aided aop for degradation of p-nitro benzoic acid. *Materials Today: Proceedings*, *32*, 519–523. Available from https://doi.org/10.1016/j.matpr.2020.02.924, https://www.sciencedirect.com/journal/materials-today-proceedings.

Deshpande, B. D., Agrawal, P. S., & Yenkie, M. K. N. (2019b). AOP as a degradative tool for oxidation of 4-hydroxybenzoic acid. *AIP Conference Proceedings*, *2104*, 15517616. Available from https://doi.org/10.1063/1.5100402, http://scitation.aip.org/content/aip/proceeding/aipcp.

Deshpande, B. D., Agrawal, P. S., & Yenkie, M. K. N. (2019c). Advanced oxidative degradation of benzoic acid and 4-nitro benzoic acid- A comparative study. *AIP Conference Proceedings*, *2142*. Available from https://doi.org/10.1063/1.5122650, 15517616 American Institute of Physics Inc. India, http://scitation.aip.org/content/aip/proceeding/aipcp.

Deshpande, B. D., Agrawal, P. S., Yenkie, M. K. N., & Dhoble, S. J. (2020). Prospective of nanotechnology in degradation of waste water: A new challenges. *Nano-Structures and Nano-Objects*, *22*. Available from https://doi.org/10.1016/j.nanoso.2020.100442. Available from: http://www.journals.elsevier.com/nano-structures-and-nano-objects/.

Deshpande, B.D., Yenkie, M.K. N., & Agrawal, P.S. (2018). Advanced oxidative degradation of benzoic acid and 4-hydroxy benzoic acid in aqueous phase: A comparative study. In: *Environmental Division 2018 - Core Programming Area at the 2018 AIChE Annual Meeting* 9781510876248 6-9 AIChE India.

Dhutekar, P., Mehta, G., Modak, J., Shelare, S., & Belkhode, P. (2021). Establishment of mathematical model for minimization of human energy in a plastic moulding operation. *Materials Today: Proceedings*, *47*, 4502–4507. Available from https://doi.org/10.1016/j.matpr.2021.05.330, Elsevier Ltd India, https://www.sciencedirect.com/journal/materials-today-proceedings.

Fu, L., Li, J., Wang, G., Luan, Y., & Dai, W. (2021). Adsorption behavior of organic pollutants on microplastics. *Ecotoxicology and Environmental Safety*, *217*. Available from https://doi.org/10.1016/j.ecoenv.2021.112207. Available from: http://www.elsevier.com/inca/publications/store/6/2/2/8/1/9/index.htt.

Galvan, D., Orives, J. R., Coppo, R. L., Silva, E. T., Angilelli, K. G., & Borsato, D. (2013). Determination of the kinetics and thermodynamics parameters of biodiesel oxidation reaction obtained from an optimized mixture of vegetable oil and animal fat. *Energy and Fuels*, *27*(11), 6866–6871. Available from https://doi.org/10.1021/ef401927x.

Williams, G. (2013). Architecting tacit information in conceptual data models for requirements process improvement. Thesis, King's College London.

Gondane, S. M., Belkhode, P. N., Joshi, M. P., Maheshwary, P. B., & Modak, J. P. (2022). *Experimental data-based model of fracture of adhesive joint for a link of mechanism made from bamboo* (pp. 735–745). Springer Science and Business Media LLC. Available from 10.1007/978-981-19-3716-3_60.

He, W., Zhu, Y., Zeng, G., Zhang, Y., Wang, Y., Zhang, M., ... Tang, W. (2020). Efficient removal of perfluorooctanoic acid by persulfate advanced oxidative degradation: Inherent roles of iron-porphyrin and persistent free radicals. *Chemical Engineering Journal*, *392*. Available from https://doi.org/10.1016/j.cej.2019.123640. Available from: http://www.elsevier.com/inca/publications/store/6/0/1/2/7/3/index.htt.

Kalaimurugan, D., Sivasankar, P., Durairaj, K., Lakshmanamoorthy, M., Ali Alharbi, S., Al Yousef, S. A., ... Venkatesan, S. (2021). Novel strategy for biodegradation of 4-nitrophenol by the immobilized cells of Pseudomonas sp. YPS3 with Acacia gum. *Saudi Journal of Biological Sciences*, *28*(1), 833–839. Available from https://doi.org/10.1016/j.sjbs.2020.11.019. Available from: http://www.ksu.edu.sa/sites/Colleges/CollegeofScinces/BotanyDepartment/SJBS/Default.aspx.

Khoshnamvand, M., Hao, Z., Huo, C., & Liu, J. (2020). Photocatalytic degradation of 4-nitrophenol pollutant and in vitro antioxidant assay of gold nanoparticles synthesized from Apium graveolens leaf and stem extracts. *International Journal of Environmental Science and Technology*, *17*(4), 2433–2442. Available from https://doi.org/10.1007/s13762-020-02632-0. Available from: http://www.springerlink.com/content/1735-1472.

Kulaksız, E., Kayan, B., Gözmen, B., Kalderis, D., Oturan, N., & Oturan, M. A. (2022). Comparative degradation of 5-fluorouracil in aqueous solution by using H2O2-modified subcritical water, photocatalytic oxidation and electro-Fenton processes. *Environmental Research*, *204*. Available from https://doi.org/10.1016/j.envres.2021.111898. Available from: http://www.elsevier.com/inca/publications/store/6/2/2/8/2/1/index.htt.

Liu, W. K., Karniadakis, G., Tang, S., & Yvonnet, J. (2019). A computational mechanics special issue on: Data-driven modeling and simulation—theory, methods, and applications. *Computational Mechanics*, *64*(2), 275–277. Available from https://doi.org/10.1007/s00466-019-01741-z. Available from: http://www.springer.com/journal/466.

Loures, Carla C. A., Alcântara, Marco A. K., Izário Filho, H. élcio J., Teixeira, Antonio C. S. C., Silva, Flávio T., Paiva, Teresa C. B., & Samanamud, Gisella R. L. (2013). Advanced oxidative degradation processes: Fundamentals and applications. *International Review of Chemical Engineering (IRECHE)*, *5*(2), 102. Available from https://doi.org/10.15866/ireche.v5i2.6909.

Ma, Jie, Ma, Yao, Rong, Xun, Song, Quanwei, Wu, Baichun, Lan, Xingying, ... Zhang, Ping (2021). Persulfate-based controlled release beads for in situ chemical oxidation of common organic pollutants. *Journal of Environmental Chemical Engineering*, *9*(5)105627. Available from https://doi.org/10.1016/j.jece.2021.105627.

Milling, P.M. (1995). In: *1995 Proceedings of the 1995 International System Dynamics Conference: System Dynamics '95 Managing and Understanding Techonological Innovations.*

Neelam, D., Meyerstein, J., Adhikary, A., Burg, D., Shamir, Y., & Albo. (2020). Zero-valent iron nanoparticles entrapped in SiO2 sol-gel matrices: A catalyst for the reduction of several pollutants. *Catalysis Communications, 133*. Available from https://doi.org/10.1016/j.catcom.2019.105819. Available from: http://www.elsevier/locate/catcom.

Qiao, X. X., Yu, K., Xu, J. Y., Cai, Y. L., Li, Y. F., Cao, H. L., & Lü, J. (2021). Engineered nanoscale schwertmannites as Fenton–like catalysts for highly efficient degradation of nitrophenols. *Applied Surface Science, 548*. Available from https://doi.org/10.1016/j.apsusc.2021.149248. Available from: http://www.journals.elsevier.com/applied-surface-science/.

Saheed, I. O., Oh, W. D., & Suah, F. B. M. (2021). Chitosan modifications for adsorption of pollutants – A review. *Journal of Hazardous Materials, 408*. Available from https://doi.org/10.1016/j.jhazmat.2020.124889. Available from: http://www.elsevier.com/locate/jhazmat.

Serrà, A., Artal, R., Pozo, M., Garcia-Amorós, J., & Gómez, E. (2020). Simple environmentally-friendly reduction of 4-nitrophenol. *Catalysts, 10*(4). Available from https://doi.org/10.3390/catal10040458. Available from: https://www.mdpi.com/2073-4344/10/4/458/pdf.

Shi, S., Han, X., Liu, J., Lan, X., Feng, J., Li, Y., ... Wang, J. (2021). Photothermal-boosted effect of binary Cu[sbnd]Fe bimetallic magnetic MOF heterojunction for high-performance photo-Fenton degradation of organic pollutants. *Science of the Total Environment, 795*. Available from https://doi.org/10.1016/j.scitotenv.2021.148883. Available from: http://www.elsevier.com/locate/scitotenv.

Shuchi, S. B., Suhan, M. B. K., Humayun, S. B., Haque, M. E., & Islam, M. S. (2021). Heat-activated potassium persulfate treatment of Sudan Black B dye: Degradation kinetic and thermodynamic studies. *Journal of Water Process Engineering, 39*. Available from https://doi.org/10.1016/j.jwpe.2020.101690. Available from: http://www.journals.elsevier.com/journal-of-water-process-engineering/.

Singh, N., & Balomajumder, C. (2021). Phytoremediation potential of water hyacinth (Eichhornia crassipes) for phenol and cyanide elimination from synthetic/simulated wastewater. *Applied Water Science, 11*(8). Available from https://doi.org/10.1007/s13201-021-01472-8. Available from: http://www.springer.com/earth + sciences + and + geography/hydrogeology/journal/13201.

Sokol, A., Ratkiewicz, A., Tomaszewska, I., & Karpinska, J. (2021). Kinetics and mechanistic studies of photochemical and oxidative stability of galaxolide. *Water (Switzerland), 13*(13). Available from https://doi.org/10.3390/w13131813. Available from: https://www.mdpi.com/2073-4441/13/13/1813/pdf.

Sonde, V. M., Belkhode, P. N., & Sakhale, C. N. (2014). Approach to formulate mathematical model for wood chipper cutter using HPFM concept. *International Conference On Multidisciplinary Research and Practices*, 509–511.

Sun, S. P., Li, C. J., Sun, J. H., Shi, S. H., Fan, M. H., & Zhou, Q. (2009). Decolorization of an azo dye Orange G in aqueous solution by Fenton oxidation process: Effect of system parameters and kinetic study. *Journal of Hazardous Materials, 161*(2-3), 1052–1057. Available from https://doi.org/10.1016/j.jhazmat.2008.04.080.

Tranchant, M., Serrà, A., Gunderson, C., Bertero, E., García-Amorós, J., Gómez, E., ... Philippe, L. (2020). Efficient and green electrochemical synthesis of 4-aminophenol using porous Au micropillars. *Applied Catalysis A: General, 602*. Available from https://doi.org/10.1016/j.apcata.2020.117698. Available from: http://www.sciencedirect.com/science/journal/0926860X.

Vicente, J., Rosal, R., & Díaz, M. (2002). Noncatalytic oxidation of phenol in aqueous solutions. *Industrial and Engineering Chemistry Research, 41*(1), 46–51. Available from https://doi.org/10.1021/ie010130w. Available from: http://pubs.acs.org/journal/iecred.

Waghmare, S. N., Sakhale, C. N., Tembhurkar, C. K., & Shelare, S. D. (2020). *Assessment of average resistive torque for human-powered stirrup making process, . Advances in Intelligent Systems and Computing* (1025, pp. 845–853). . Available from 10.1007/978-981-32-9515-5_79, http://www.springer.com/series/11156.

Wiley, J., Hepburn, K., & Levenspiel, O. (1999). *Chemical Reaction Engineering* (3rd edn). New York: Wiley.

Yan, N., Li, M., Liu, Y., Liu, F., & Brusseau, M. L. (2017). Kinetic and thermodynamic studies of chlorinated organic compound degradation by siderite-activated peroxide and persulfate. *Water, Air, and Soil Pollution, 228*(12). Available from https://doi.org/10.1007/s11270-017-3631-y. Available from: http://www.kluweronline.com/issn/0049-6979/.

Yang, H., Yang, L., Zhu, X., Huang, J. B., Guo, Y., Sheng, F., ... Shi, J. G. (2021). Research progress on promoting the chemical oxidation degradation of pollutants by organics. *Journal of Ecology and Rural Environment, 37*(9), 1137–1146. Available from https://doi.org/10.19741/j.issn.1673-4831.2020.0798. Available from: http://www.ere.ac.cn/EN/article/downloadArticleFile.do?attachType = PDF&id = 12435.

Zaharia, C., & Suteu, D. (2014). A preliminary modelling and optimisation study of a homogenous advanced oxidation process applied for an industrial Coloured effluent. *Journal of Environmental Protection and Ecology, 15*(4), 1680–1689. Available from: http://www.jepe-journal.info/journal-content.

Zhao, X., Li, J., Li, X., Huo, P., & Shi, W. (2021). Design of metal-organic frameworks (MOFs)-based photocatalyst for solar fuel production and photo-degradation of pollutants. *Chinese Journal of Catalysis, 42*(6), 872–903. Available from https://doi.org/10.1016/S1872-2067(20)63715-9. Available from: http://www.elsevier.com.

Chapter 2

Application of geospatial techniques in estimating the impact of Fani cyclone on land surface conditions of Puri district of Odisha, India

Seema Rani[1] and Rajesh Kumar[2]
[1]Department of Geography, Institute of Science, Banaras Hindu University, Varanasi, India,
[2]Department of Geography, Sikkim University, Gangtok, India

2.1 Introduction

Tropical cyclones are considered the most damaging natural disasters, associated with a very high intensity of rainfall, wind speed, and storm surge. Storms, encompassing tropical cyclones and hurricanes, ranked as the second leading cause of fatalities, claiming the lives of 233,000 people and impacting an estimated 726 million people globally between 1998 and 2017 (Ortiz et al., 2023; World Health Organization, 2022a). Within the last three decades, the percentage of the global population residing along cyclone-exposed coastlines has surged by 192%, consequently elevating the potential risks associated with mortality and morbidity in the event of a tropical cyclone (Klemas, 2009; World Health Organization, 2022a). A previous study evaluated the short-term mortality risks associated with tropical cyclones from 1980 to 2019, revealing a global estimate of 97,430 excess deaths per decade following cyclone exposure, with East Asia and South Asia registering the highest mortality rates (Huang et al., 2023). These cyclones can change the landscape of the coastal ecosystem. A significant adverse impact of tropical cyclones has been found on global sectors, particularly agriculture, trade, and tourism, spanning from 1970 to 2015 (Kunze, 2021).

Approximately 85 tropical storms emerge annually across the warm tropical expanses of the world's oceans (World Health Organization, 2022b). Among these storms, 72% originate in the northern hemisphere, with the

Water Resources Monitoring, Management, and Sustainability. DOI: https://doi.org/10.1016/B978-0-443-23665-5.00002-8

remaining 28% developing in the southern hemisphere. From 1990 to 2021, the occurrence of ephemeral named storms and events of rapid intensification (50 + knots per day) resulting in widespread damage has notably escalated, largely attributed to growing population densities and the increased valuation of coastal assets (Klotzbach et al., 2022). A significant upswing was observed in the intensification rates of tropical cyclones on a global scale between 1982 and 2017, attributing a positive correlation to anthropogenic influences (Bhatia et al., 2022). Moreover, they highlighted a more conducive thermodynamic environment surrounding these cyclones for intensification, with climate models affirming that anthropogenic warming substantially amplifies the likelihood of such changes (Bhatia et al., 2022). Due to climate change, the intensity and frequency of tropical cyclones have increased over time (Knutson et al., 2010). Nevertheless, a decreasing pattern is found in the annual frequency of tropical cyclones on both a global and regional scale throughout the twentieth century (Chand et al., 2022).

The North Indian Ocean region contributes to nearly 8% of all tropical cyclones worldwide, with approximately 75% of these cyclones originating specifically within the Bay of Bengal basin (Pandey & Liou, 2020). In the Indian Ocean (Arabian Sea and Bay of Bengal), a total of 1284 cyclones occurred during 1891–2018 (India Meteorological Department, 2020). About 94% of these cyclones have their origin in the Bay of Bengal basin, indicating the vulnerability of the area. Approximately 50% of the cyclones in the Bay of Bengal basin occurred between June and September. It has been perceived that the cyclonic activities in the Indian Ocean have amplified and become intense over the years (Balaji, Chakraborty, & Mandal, 2018; Hoarau, Bernard, & Chalonge, 2012). Besides other factors, increasing cyclonic activities are linked with ongoing climate change (Mishra, 2014). The coastal districts of Odisha, namely, Ganjam, Puri, Jagatsinghapura, Kendrapara, Bhadrak, Baleshwar, Khorda, Gajapati, Nayagarh, Cuttack, Dhenkanal, Keonjhar, Mayurbhanj, and Jajpur, are prone to tropical cyclones, storm surges, and rainfall-induced riverine flooding (Mohapatra, Mandal, Bandyopadhyay, Tyagi, & Mohanty, 2012). During 1971–2000, the coastal areas of Odisha witnessed 16 (moderate to severe) cyclonic events (Chittibabu et al., 2004). They also reported that 72 flooding events were directly related to cyclones during the 19th century, while the 20th century witnessed only 56 flooding events (Chittibabu et al., 2004). However, an increase in the intensity of cyclones and floods was reported during the late 20th century (Bahinipati, 2014). An escalating trend is found in the intensity of global tropical cyclones during the period 1982–2009, particularly noting substantial increases in the southern Indian Ocean (Kossin, Knapp, Olander, & Velden, 2020). A surge is highlighted in both the count and duration of very severe cyclonic storms (wind speed >64 knots) in the North Indian Ocean from 1997 to 2014 (Balaji et al. 2018). However, a slight decline is projected in overall tropical cyclone genesis frequency in the North Indian Ocean, coupled with increased genesis frequency in the Arabian Sea and a decrease in the

Bay of Bengal (Bell, Chand, Tory, Ye, & Turville, 2020). Likewise, other studies reported a notable rise in cyclonic storm occurrences and durations over the Arabian Sea, juxtaposed with a decrease in the Bay of Bengal (Deshpande et al., 2021; Singh, Panda, Sahoo, & Mohapatra, 2019). The lengthening durations of Arabian Sea cyclones were attributed to heightened mid-level humidity, increased column-averaged moist static energy (950−150 hPa), correlated significantly with rising sea surface temperatures, and tropical cyclone heat potential in the basin (Deshpande et al., 2021; Singh et al., 2019).

Geospatial technology is considered to have a greater potential for assessing the damage induced by tropical cyclones at different spatial scales and also enhancing the decision support system to manage the coastal ecosystem efficiently (Hoque, Pradhan, Ahmed, & Roy, 2019). Another study did a comprehensive analysis of cyclone disaster management studies using remote sensing and spatial analysis, emphasizing their concentration in Asia (55%) and North and Central America (40%), predominantly after 2004 (Hoque, Phinn, & Roelfsema, 2017). These studies primarily focused on preparedness (34%) and prevention/reduction (32%) phases. These days, policymakers, planners, and scientists are using remote sensing data from both active and passive sensors to map inundation or extract water pixels (Kumar, Singh, Gautam, & Pandey, 2018; Manjusree, Bhatt, Begum, Rao, & Bhanumurthy, 2015). Synthetic Aperture Radar (SAR) operates in all weather conditions and during both day and night. It penetrates dense cloud cover, unlike optical passive sensors. Therefore, these data (Indian Remote Sensing Satellite RISAT-1, RADARSAT 1−2, TerraSAR-X, COSMO-SkyMed, and Sentinel 1A-B) are extensively used in flood studies (Ban, Kwon, Shin, Ryu, & Hong, 2017; Bhatt et al., 2017; Greifeneder, Wagner, Sabel, & Naeimi, 2014). Recent studies on the Sentinel 1 A-B SAR data utilize images of VV polarization for the mapping of flood-inundated areas (Ban et al., 2017; Clement, Kilsby, & Moore, 2018; Kumar, 2018; Twele, Cao, Plank, & Martinis, 2016). The detection of inundated areas and water pixels from SAR data is mainly based on Ostu's approach and universal threshold range, supervised classification, gray-level co-occurrence matrix, segmentation and object-oriented image analysis techniques, active contour model, band ratio and change detection, fuzzy rules, and visual interpretation (Bhatt et al., 2017; Borghys, Yvinec, Perneel, Pizurica, & Philips, 2006; Giustarini et al., 2013; Greifeneder et al., 2014; Horritt, 1999; Mason, Schumann, Neal, Garcia-Pintado, & Bates, 2012; Oberstadler, Hönsch, & Huth, 1997; Pulvirenti, Chini, Pierdicca, & Boni, 2016; Schlaffer, Matgen, Hollaus, & Wagner, 2015; Song, Sohn, & Park, 2007). A smooth water surface causes a specular reflection of incident energy. Therefore, the backscattering values (in dB) of water bodies are negative. The quick mapping of flooded land use and land cover (LULC) can be done using passive remote sensing images of pre- and postinundation events (Kumar, 2016; Wang, 2004). Furthermore, a reduction in Normalized Difference Vegetation Index (NDVI) values between pre- and post-inundation events can be used as a proxy for damage to crops and natural vegetation (Kumar & Acharya, 2016).

On April 27, 2019, the Fani cyclone originated over the southeast Bay of Bengal, while its landfall occurred around Puri city (Odisha state) in India on May 3, 2019. In Odisha, this cyclone ranked among the top three most severe cyclones in the past 150 years, prompting the safe evacuation of approximately 1.34 million people before it made landfall (Chatterjee, 2020; UNICEF, 2019). The government of Odisha has mentioned the damage to infrastructure caused by the cyclonic storm (The United Nations, the World Bank, and the Asian Development Bank., 2019). However, they have not taken into account the damage caused by cyclone-induced floods on the land surface conditions such as built-up areas, crops, and natural vegetation in the district, which are very important resources for human livelihood and the ecosystem. Hence, the objective of this study is to assess the flood extent caused by the Fani cyclone and its impacts on the land surface conditions in the Puri district of the state of Odisha because this district was the landfall site of the cyclone.

2.2 Study area

The study area falls in the Puri district of the state of Odisha along the eastern coast of India (Fig. 2.1), which has a geographical area of about 3479 km^2 and a total population of 1,698,730 (Census of India CoI, 2011). Major geomorphic units of the district are coastal plains with saline marshy tracts, gently sloping alluvial plains, and undulating hard-rocky areas (CGWB, 2013). The largest brackish water lake of Chilika is located in the district. A subtropical climate prevails in the Puri district, with a hot and dry summer and a mild winter. Almost 75% of the total rainfall occurs in the southwest monsoon season (June–September) (CGWB, 2013). The average annual rainfall in the district is 1449 mm (CGWB, 2013). Paddy is the principal crop in the district. The district has been affected by 10 cyclonic activities since 1967 (Odisha District Disaster Management Authority DDMA, 2018).

2.3 Data and methods

Daily rainfall data (in mm) for all the districts (from April 27 to May 6, 2019) in Odisha have been collected from the Odisha Rainfall Monitoring System (2019). The VV polarization image of Sentinel 1A-B has been obtained to extract the inundated pixels for the pre- and post-Fani cyclone events (Copernicus, 2020). The Sentinel 1 SAR uses the C band at 5.405 GHz frequency (ESA, 2018). Moreover, the Sentinel 2 A multispectral instrument (MSI) images have also been used to obtain different LULC classes in the Puri district. Estimation of flood depth in the Puri district is done using the Shuttle Radar Topography Mission digital elevation model data at

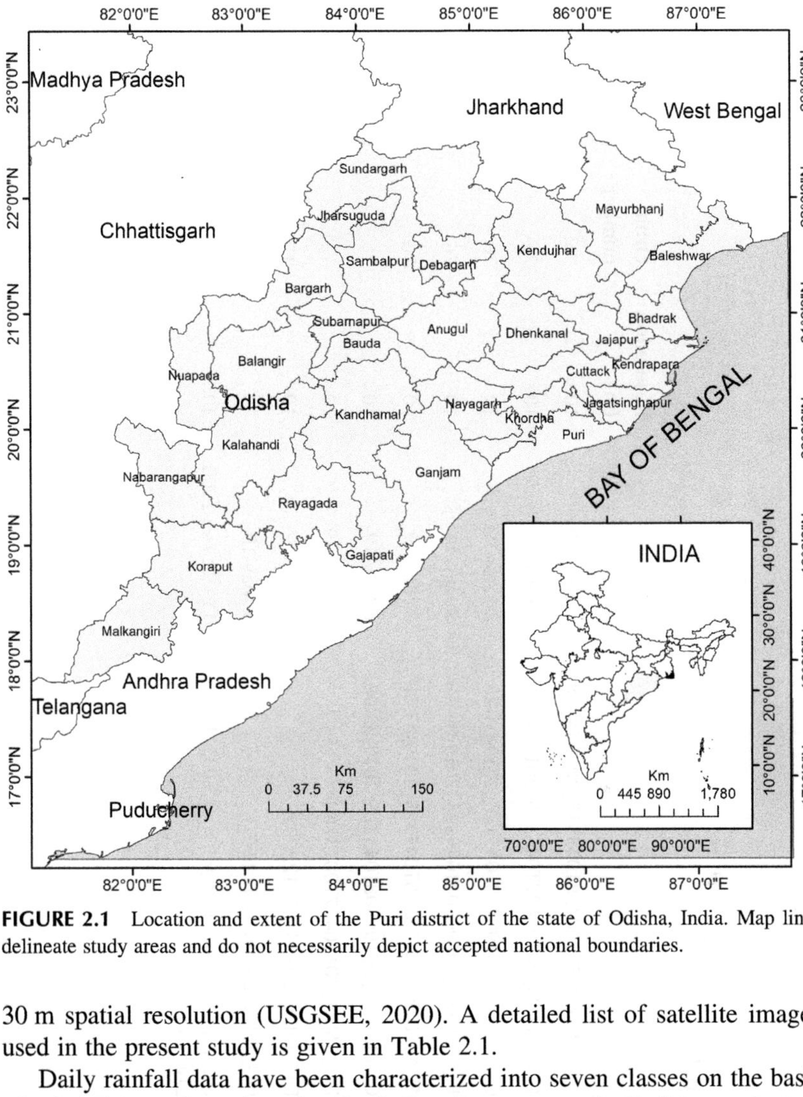

FIGURE 2.1 Location and extent of the Puri district of the state of Odisha, India. Map lines delineate study areas and do not necessarily depict accepted national boundaries.

30 m spatial resolution (USGSEE, 2020). A detailed list of satellite images used in the present study is given in Table 2.1.

Daily rainfall data have been characterized into seven classes on the basis of rainfall intensity, given by the Indian Department (India Meteorological Department (IMD), 2020) (Fig. 2.2). Rainfall deviation is also computed and plotted by making five categories, provided by the IMD (India Meteorological Department, IMD, 2020). The spatial distribution of rainfall was prepared by the choropleth technique in a GIS environment.

2.3.1 Flooded area estimation

The subset of Sentinel 1A-B images has been done to obtain the area of interest. Then, pre-processing (radiometric calibration, speckle filter, terrain correction,

TABLE 2.1 List of satellite data.

Satellite	Data	Date	Resolution (m)	Orbit number	Orbit direction	Band used
Sentinel-2A	L2A_T45QUB_A020093_20190428T045318_2019-04-28	28-Apr-19	10	76	Descending	B2, B3, B4, B8
Sentinel-2A	L2A_T45QUC_A020093_20190428T045318_2019-04-28	28-Apr-19	10	76	Descending	B2, B3, B4, B8
Sentinel-2A	L2A_T45QVC_A020093_20190428T045318_2019-04-28	28-Apr-19	10	76	Descending	B2, B3, B4, B8
Sentinel-2A	L2A_T45QUB_A020236_20190508T045325_2019-05-08	8-May-19	10	76	Descending	B2, B3, B4, B8
Sentinel-2A	L2A_T45QUC_A020236_20190508T045325_2019-05-08	8-May-19	10	76	Descending	B2, B3, B4, B8
Sentinel-2A	L2A_T45QVC_A020236_20190508T045325_2019-05-08	8-May-19	10	76	Descending	B2, B3, B4, B8
S1A	S1A_IW_GRDH_1SDV_20190427T001326	27-Apr-19	10	121	Descending	VV
S1A	S1B_IW_GRDH_1SDV_20190505T122702	5-May-19	10	158	Ascending	VV
S1B	S1B_IW_GRDH_1SDV_20190505T122731	5-May-19	10	158	Ascending	VV
SRTM DEM	n19_e085_1arc_v3; n19_e086_1arc_v3; n20_e085_1arc_v3 & n20_e086_1arc_v3	11–22 Feb-2000	30	N.A.	N.A.	X-Band Synthetic Aperture Radar (X-SAR)

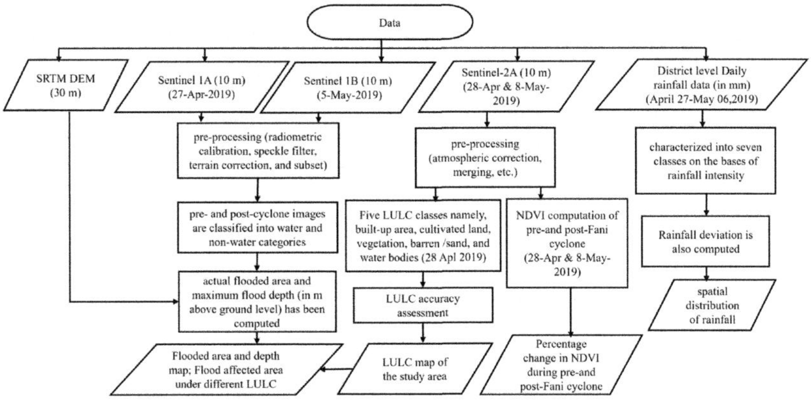

FIGURE 2.2 Methodology of the present study.

and subset) of the images was performed using the sentinel application platform (SNAP ver. 7.0). The backscatter values of the images are converted into physical values (dB). The study analyzed the histogram of the filtered backscatter coefficient to differentiate water pixels from non-water pixels. Low backscatter values signify water pixels, unlike non-water pixels. Subsequently, the threshold value of the VV polarization of Sentinel SAR data ranges from -5 to -15 dB. This threshold range has been decided based on the statistics of training sites in the flooded areas. Further, these threshold ranges have been used to segment the image using band arithmetic in the SNAP tool. The obtained pre- and postcyclone images are classified into water and non-water categories. A mosaic of post-classified images of the study area is prepared. The actual flooded area has been obtained by subtracting the pre-Fani cyclone water bodies from the post-Fani cyclone water bodies (e.g., Kumar, 2016). Subsequently, the volume and surface area of the actual flooded areas have been computed above different reference planes of polygons. The maximum flood depth (in m above ground level) has been computed using Eq. (2.1).

$$\text{Maximum flood depth} = \text{volume of polygon}/\text{surface area of polygon} \quad (2.1)$$

2.3.2 Land use and land cover and normalized difference vegetation index estimation

For classifying Sentinel 2 MSI images, an atmospheric correction has been done by the dark object subtraction method (Chavez, 1988). Then, all the corrected images are merged to prepare a subset of the study area. Five LULC classes, namely, built-up area, cultivated land, vegetation, barren/sand, and water bodies, have been recognized by the NRSC National Remote Sensing Centre & ISRO Indian Space Research Organization (2011). LULC classification has been done by developing a decision tree by spectral

TABLE 2.2 Accuracy assessment results of the classified LULC (land use and land cover) of the study area.

LULC classes	Built-up area	Cultivated land	Vegetation	Barren/ sand	Water bodies	Total
			(%)			
Built-up area	8	7	0	2	1	18
Cultivated land	13	80	9	2	1	105
Vegetation	0	2	40	0	1	43
Barren/ sand	0	3	0	62	1	66
Water bodies	2	1	2	1	64	70
Total	23	93	51	67	68	302

Overall accuracy = (254/302) 84.10%

Kappa coefficient = 0.7919

separability analysis. The error matrix (Congalton, 1991) along with the Kappa (K) coefficient (Cohen, 1960) was applied to estimate the accuracy of classified LULC. Overall accuracy and kappa coefficient of LULC classes are 84% and 0.79, respectively (Table 2.2).

The NDVI computation of pre- and post-Fani cyclones is computed using the Sentinel 2 MSI. The formula for NDVI computation has been given in Eq. (2.2).

$$\mathrm{NDVI} = (\mathrm{band8} - \mathrm{band4})/(\mathrm{band8} + \mathrm{band4}) \tag{2.2}$$

Thereafter, zonal statistics are computed by overlaying the LULC map on the flooded area map, LULC over the NDVI map, and flooded-LULC over the NDVI map. The percentage change in NDVI between pre- and post-Fani events is computed using Eq. (2.3).

$$\%\mathrm{change\ in\ NDVI} = (\mathrm{NDVI}_{\mathrm{Postevent}} - \mathrm{NDVI}_{\mathrm{Preevent}})/\mathrm{NDVI}_{\mathrm{Preevent}} \times 100 \tag{2.3}$$

2.4 Results and discussion

2.4.1 Spatial pattern of rainfall

On May 3, 2019, the Fani cyclone made landfall near Puri in the Odisha state of India. It has brought heavy downpours along with the high-speed wind. Coastal districts of Odisha received extremely heavy (Ganjam with 419 mm and Gajapati with 148.1 mm) and very heavy rainfall (Khordha with 72.7 mm, Puri with 78.37 mm, and Kandhamal with 88.8 mm) on May 3 (Fig. 2.3) with a wind speed of 180 kilometers per hour (kmph). On May 4, 2019, the highest rainfall was received by Khordha (132 mm), followed by Dhenkanal (110 mm), Nayagarh (107 mm), Cuttack (95 mm), and Mayurbhanj (81 mm).

Overall, the state received a mean rainfall of 1057 mm on May 3 and 1190 mm on May 4, 2019, due to cyclonic activities. Rainfall has declined drastically to less than 35 mm after the onward movement of cyclone Fani. Generally, rainfall deviation from normal was scanty, and there were no rain categories in most of the districts of Odisha on April 27 and May 3, 2019. The Fani cyclone produced normal to excess rainfall with a rainfall deviation of more than 20% in most of the districts, particularly on the way of the cyclone (Fig. 2.4). Interestingly, coastal districts received no rain after the dissemination of the cyclone Fani. Krushnaprasad, Satyabadi, Nimapara, Pipili, and Brahmagiri blocks of the Puri district had cumulative rainfall of more than 121 mm on May 3and 4, 2019 because of the landfall of the cyclone. It is obvious that the extremely heavy to very heavy rainfall led to inundation and damage to natural vegetation, crops (especially paddy), and infrastructure, as pointed out in international reports (The United Nations, the World Bank, & the Asian Development Bank, 2019).

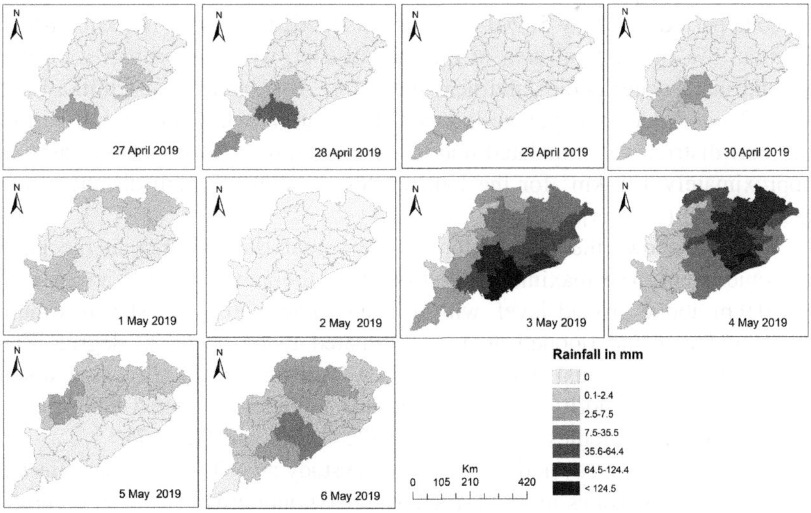

FIGURE 2.3 District-wise distribution of rainfall intensity during pre- and post-Fani cyclone. Please check the online version to view the color image of the figure. Map lines delineate study areas and do not necessarily depict accepted national boundaries.

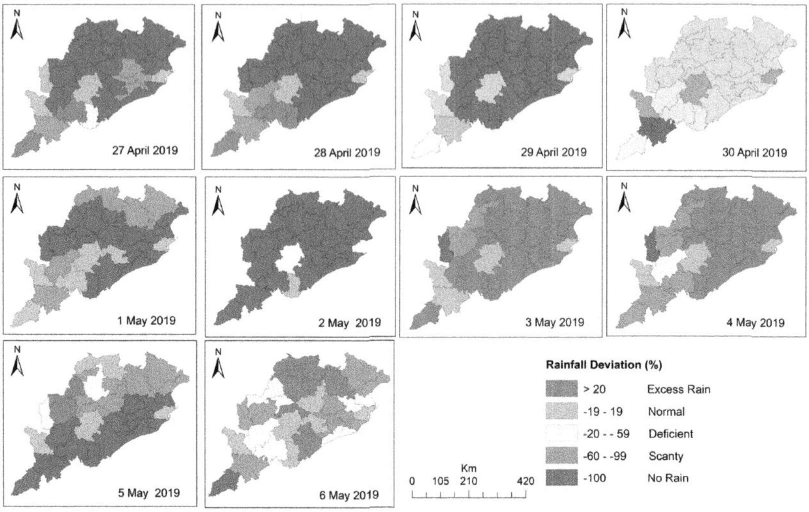

FIGURE 2.4 The district-wise distribution of rainfall deviation from normal rainfall during pre- and post-Fani cyclone. Map lines delineate study areas and do not necessarily depict accepted national boundaries. Please check the online version to view the color image of the figure.

2.4.2 Flooded area and depth

Cyclone-induced storm surges and heavy downpours have caused floods in the Puri district. The area of demarcated water bodies was about 670 km^2 in the pre-cyclone period (Fig. 2.5). It increased to about 774 km^2 in the post-cyclone period. On May 5, 2019, the actual flooded area was 102 km^2, which is about 3% of the total area of the Puri district, which is primarily located around main water bodies and near the coast. Also, areas near Chilika Lake were flooded due to a cyclone-induced storm surge, while the central part of the Puri district was inundated due to excess rainfall. An inundation area of approximately 128 km^2 for the Fani cyclone within a 75-km distance from the coastline (Chatterjee, 2020).

Fig. 2.6 shows the distribution of the actual flooded area in the post-cyclonic event. The maximum flood depth in the actual flooded areas is up to 30.9 m above ground level, with a mean value of 0.14 m. The maximum slope of the actual flooded areas is estimated to be up to 22 degree. The mean slope of the actual flooded areas is 0.74 degree. The LULC assessment reflects that the majority of the district area is under cultivated land (36%), followed by water bodies and vegetation (Table 2.3 and Fig. 2.7).

The built-up area covered about 6% of the study area. The flood has affected all the LULC categories in the district. The maximum effect has been observed on water bodies (35%), followed by barren/sand (32%), and cultivated land (19%) (Table 2.4). In financial terms, agricultural crops accounted for approximately 83% of the overall damage value in the study area (The United Nations, the World Bank, & the Asian Development Bank, 2019).

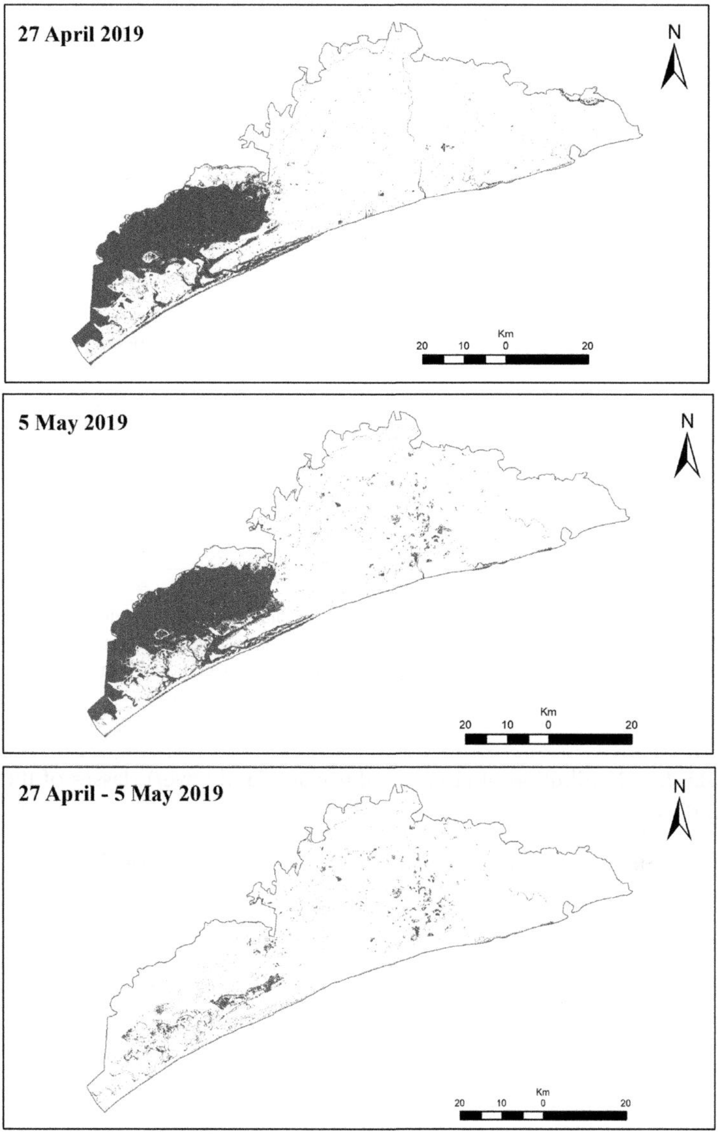

FIGURE 2.5 Spatial distribution of water bodies of the study area during the pre- and post-Fani cyclone event. Map lines delineate study areas and do not necessarily depict accepted national boundaries.

The impact of the cyclone on the lake, vegetation, and coast has been shown in Fig. 2.8. It shows the severe flooding and loss of vegetation along the coast of the district (e.g., Rajhans Island, Chilika Lake sea mouth, and Kusabhandra River). The flow of seawater to the Chilika Lake

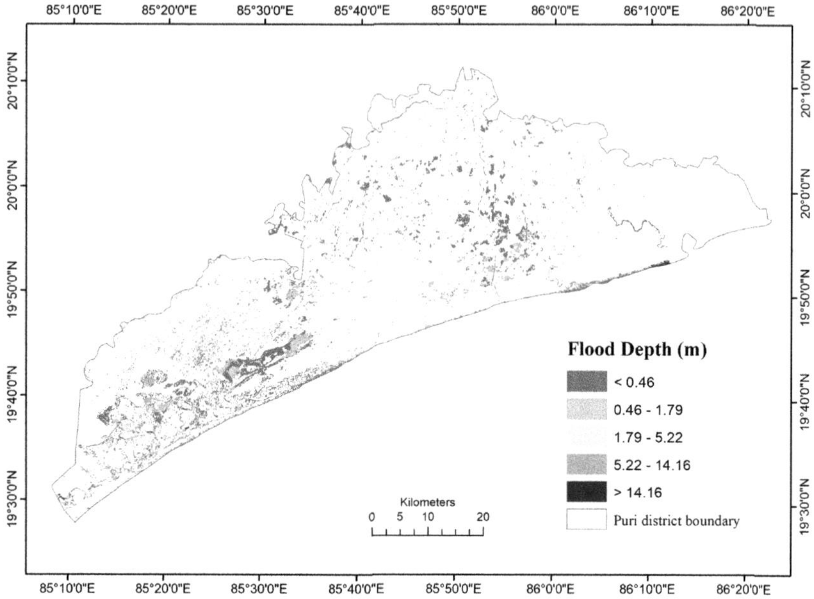

FIGURE 2.6 Maximum flood depth (in m, above ground level) in the actual flooded area of the Puri district. Map lines delineate study areas and do not necessarily depict accepted national boundaries. Please check the online version to view the color image of the figure.

TABLE 2.3 Distribution of LULC (land use and land cover) classes of the Puri district.

LULC classes	Area (km^2)	Area (%)
Built-up area	195	5.7
Cultivated land	1250	36.6
Vegetation	720	21.1
Barren/sand	466	13.6
Water bodies	782	22.9
	3413	100

due to cyclonic storms affects its hydrological parameters and can increase the water salinity. An increase is also mentioned in silicate concentrations in the Chilika Lake due to a severe cyclonic storm *"Phailin"* along the coast of Odisha in 2013 (Muduli, 2017). The Fani cyclone has also impacted the shoreline, as mentioned by Acharyya, Mishra, and Kar (2020), who studied the ecological impact of the cyclone along the coast

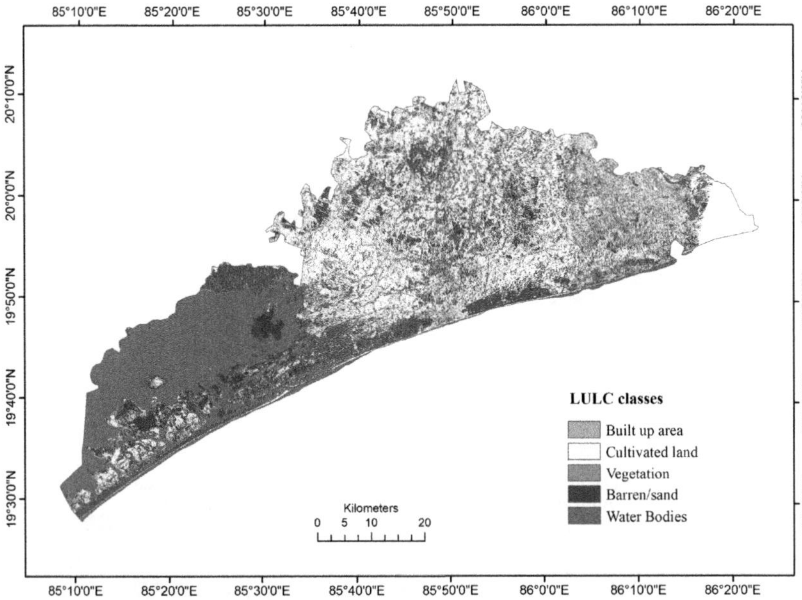

FIGURE 2.7 Distribution of land use and land cover of the Puri District on April 28, 2019. Please check the online version to view the color image of the figure. Map lines delineate study areas and do not necessarily depict accepted national boundaries.

TABLE 2.4 Flood affected area under different LULC (land use and land cover) classes in the Puri district.

LULC classes	Area (km^2)	Area (%)
Built-up area	8	7.40
Cultivated land	20	19.88
Vegetation	5	4.99
Barren/sand	33	32.14
Water bodies	36	35.59
	102	100

of the state and its impacts on the shoreline, which in turn can impact the lagoon ecosystem.

Fani cyclone leads to seawater intrusion into the lakes and rivers near the coast. The major effect on water bodies is because of their proximity to the coast (e.g., Chilika Lake). It is a matter of concern as it has damaged the fisheries occupation, which is also the main source of earnings for the population. Damage to cultivated land and vegetation in the district has spoiled

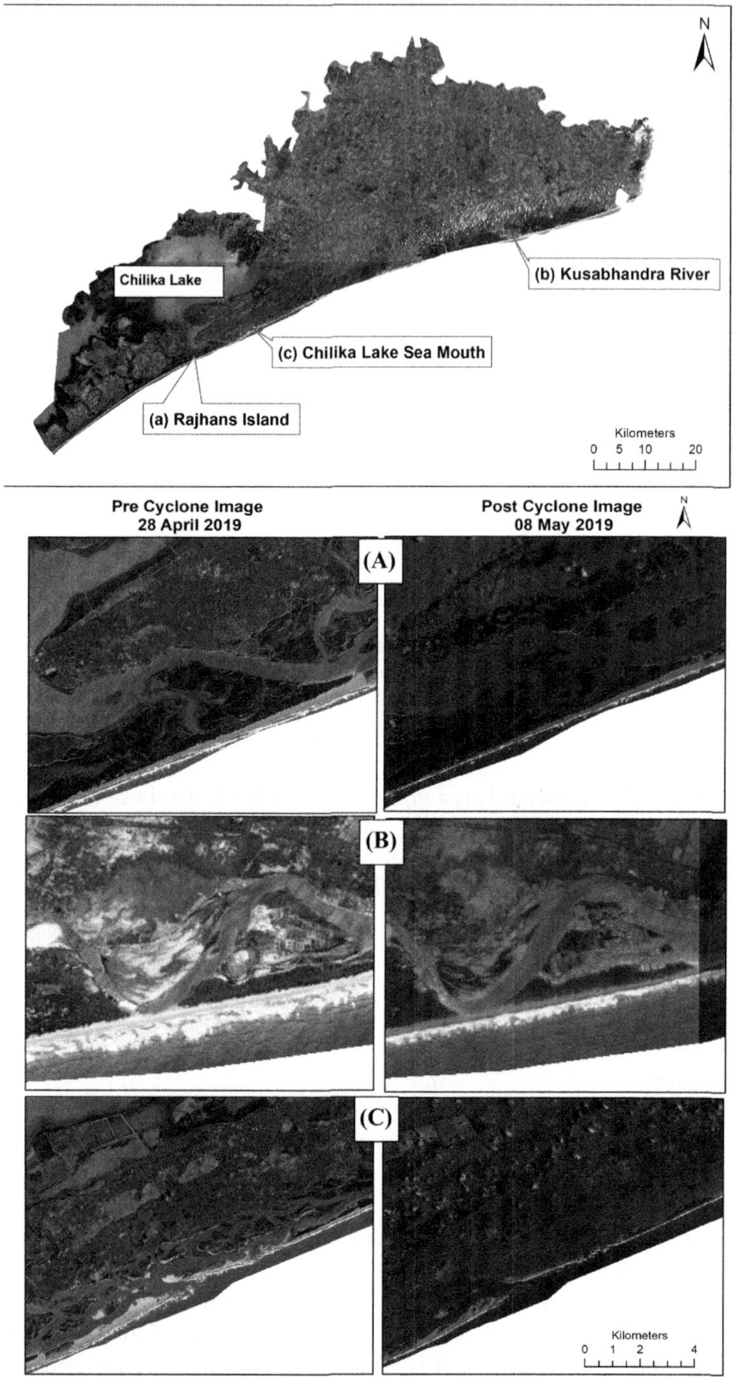

FIGURE 2.8 Impact of the Fani Cyclone along the coast of the Puri district. Expansion of water bodies and loss of vegetation cover has been observed in the post-cyclone period compared to the precyclone land surface conditions (a, b, and c).

one season crop which has influenced the livelihood of the population. People have fruit plantations and crops on the field, which are uprooted by the high-velocity wind. In the Puri district, 11 blocks and 1772 villages were affected by the cyclone, which has influenced the population of 1,968,228 (The United Nations, the World Bank, and the Asian Development Bank., 2019). According to the report (Government of Odisha, 2019), a total of 113,774 houses (pucca and kutcha) has been fully/severely damaged and a total agricultural area of 603 km^2 (Okra, Tomato Brinjal, Banana, Mango, Key Lime, Betel vine, Turmeric, Chilli Oil, Palm, Cashew, and Coconut) has been affected due to flooding and high-velocity wind. Loss of cropland is an important concern because agriculture is the primary activity of the state and about 62% of its population depends on this for their livelihood (Government of Odisha, 2019).

2.4.3 Change in normalized difference vegetation index values as a proxy for damage assessment

The study has shown the effect of cyclone-induced flooding on LULC conditions. However, land surface conditions are also impacted by the high-velocity winds, which were observed at about 180 kmph. It has damaged crops and plantations like bananas and coconuts. Therefore, NDVI is used as a proxy to further study the impact of the cyclone on vegetation outside and inside the flooded areas. The study has taken pre- and postcyclone satellite images to estimate NDVI. Normally, the NDVI values vary from -1 to $+1$. A high negative NDVI value (-1) shows water bodies. Rocky areas, sand, and snow correspond to -0.1 to 0.1 NDVI values. The NDVI values for scrub and grassland range between 0.2 and 0.4, while the high positive values correspond to temperate and tropical rainforests. The mean NDVI values of pre- and post-Fani cyclones in the Puri district have been shown in Fig. 2.9.

Major changes are found in NDVI near the coast, though the impact is also visible in the interior parts of the district. In the Puri district, the overall percentage reduction in mean NDVI values of vegetation cover is about 42%, which is the highest, followed by barren/sand (-37%) and cultivated land (-31%) during the pre- and post-Fani cyclone (Table 2.5). In the flooded areas, the percentage reduction in mean NDVI values shows severe destruction to natural vegetation (-113%), followed by cultivated land and barren/sand (Table 2.5). It shows that the high-velocity wind and flooding have impacted the land cover conditions.

A reduction in NDVI values shows that the vegetation density is inversely related to the cyclonic storm and resultant flooding in the Puri district (e.g., Dodangeh et al., 2020; Hu & Smith, 2018; Kumar, 2016; Tehrany, Pradhan, Mansor, & Ahmad, 2015). Coastal conservation needs such a database for better planning and management of the coastal ecosystem, as

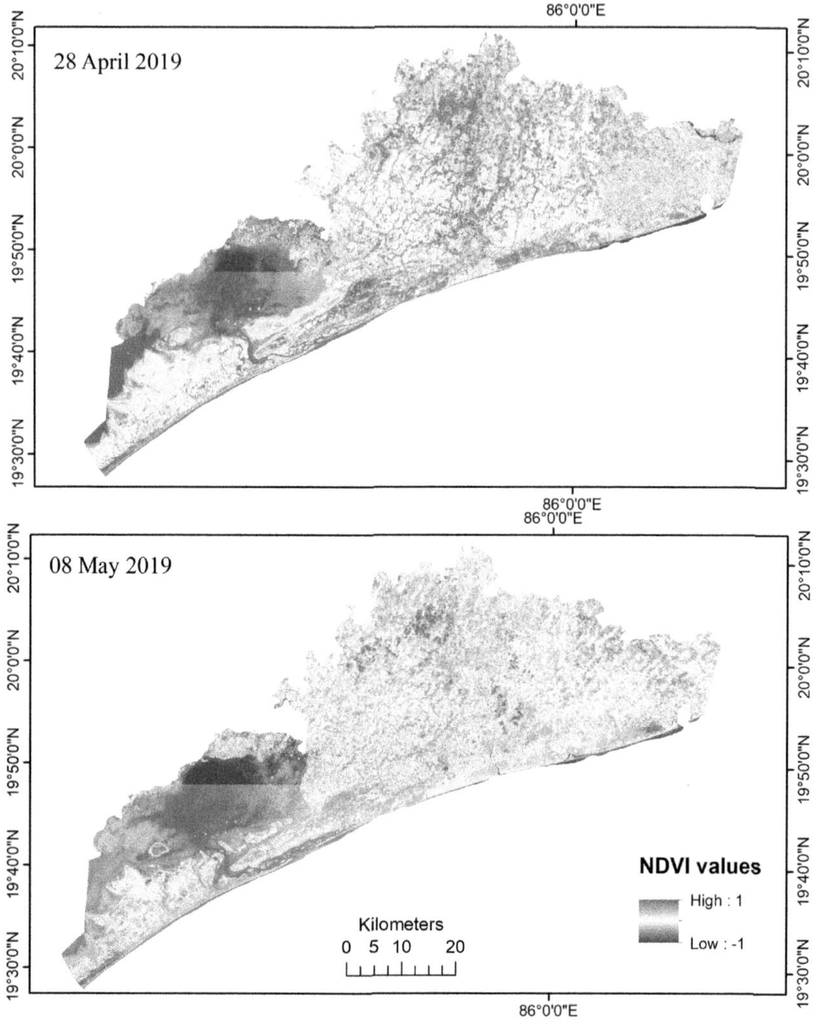

FIGURE 2.9 Spatial distribution of normalized difference vegetation index across the Puri district during pre- and post-Fani cyclone. Map lines delineate study areas and do not necessarily depict accepted national boundaries. Please check the online version to view the color image of the figure.

population density has been increasing along the coast of India and around the world as well. Tropical coasts are more vulnerable to cyclonic activities. The east coast of India has been experiencing frequent cyclonic activities over the years (India Meteorological Department, 2020). Hence, such an assessment of severe cyclones helps to understand their nature and extent of damage.

TABLE 2.5 Percentage change in mean NDVI (Normalized Difference Vegetation Index) values in the study area.

LULC classes	April 28, 2019 Mean NDVI	May 8, 2019 Mean NDVI	% Change in NDVI
Entire Puri district			
Cultivated land	0.36	0.25	−31
Vegetation	0.71	0.41	−42
Barren/sand	0.26	0.16	−37
Actual flooded area of the Puri district			
Cultivated land	0.32	−0.01	−103.4
Vegetation	0.68	−0.09	−113.2
Barren/sand	0.24	−0.01	−103.3

2.5 Conclusion

The landfall of Cyclone Fani near Odisha's Puri district resulted in a storm surge and exceptionally heavy rainfall exceeding 244.4 mm. This cyclone period witnessed a rainfall increase of over 20% across most parts of the district. The combined impact of the cyclone-induced storm surge and excessive rainfall led to flooding, covering 3% of the district's total area, notably concentrated around Chilika Lake. In certain regions of Puri district, the floodwaters reached depths of up to 14 meters. The flooding predominantly affected water bodies (35%), followed by barren/sand areas (32%) and cultivated land (19%). Apart from flooding, the high-velocity winds and excessive rain caused substantial damage to natural vegetation and crops, significantly impacting local livelihoods. The post-Fani cyclone NDVI values sharply declined by -113% compared to pre-Fani cyclone levels for natural vegetation, crops, and barren/sand areas, underscoring the severity of the damage. The potential harm to crops and natural vegetation remains a significant concern during cyclonic events due to the heavy reliance of a sizable population on agriculture within the affected region. Furthermore, the cyclone's influence extended to the hydrology of freshwater bodies due to seawater intrusion. These changes raise concerns among authorities as they disrupt the ecological balance of aquatic systems and have implications for the local fishery. Therefore, it is essential to comprehend the full scope of the severe cyclone's influence on the hydrological conditions within the region. The present study would be enormously important for the researchers and policymakers working in the coastal regions around the globe to protect the built-up areas, natural vegetation, and cropped land from cyclonic activities and resultant inundation.

Acknowledgments

The authors are grateful to the European Space Agency (ESA) for providing Sentinel data in the public domain. The authors also acknowledge the United States Geological Survey (USGS) for providing elevation data. The first author also expresses sincere thanks to the Banaras Hindu University, Varanasi, Uttar Pradesh (India), for providing a seed grant (No. R/Dev/D/IoE/Equipment/Seed Grant- II/2022–23/62810) under the Institute of Eminence (IoE).

References

Acharyya, T., Mishra, M., & Kar, D. (2020). Rapid impact assessment of extremely severe cyclonic storm Fani on morpho-dynamics & ecology of Chilika Lake, Odisha, India. *Journal of Coastal Conservation*, *24*(3). Available from https://doi.org/10.1007/s11852-020-00754-8, https://www.springer.com/journal/11852.

Bahinipati, C. S. (2014). Assessment of vulnerability to cyclones and floods in Odisha, India: A district-level analysis. *Current Science*, *107*(12), 1997–2007. Available from http://www.currentscience.ac.in/Volumes/107/12/1997.pdf.

Balaji, M., Chakraborty, A., & Mandal, M. (2018). Changes in tropical cyclone activity in north Indian Ocean during satellite era (1981–2014). *International Journal of Climatology*, *38*(6), 2819–2837. Available from https://doi.org/10.1002/joc.5463, http://onlinelibrary.wiley.com/journal/10.1002/(ISSN)1097-0088.

Ban, H. J., Kwon, Y. J., Shin, H., Ryu, H. S., & Hong, S. (2017). Flood monitoring using satellite-based RGB composite imagery and refractive index retrieval in visible and near-infrared bands. *Remote Sensing*, *9*(4). Available from https://doi.org/10.3390/rs9040313. Available from, http://www.mdpi.com/2072-4292/9/4/313/pdf.

Bell, S. S., Chand, S. S., Tory, K. J., Ye, H., & Turville, C. (2020). North Indian Ocean tropical cyclone activity in CMIP5 experiments: Future projections using a model-independent detection and tracking scheme. *International Journal of Climatology*, *40*(15), 6492–6505. Available from https://doi.org/10.1002/joc.6594. Available from, https://onlinelibrary.wiley.com/doi/abs/10.1002/joc.6594.

Bhatia, K., Baker, A., Yang, W., Vecchi, G., Knutson, T., Murakami, H., ... Whitlock, C. (2022). A potential explanation for the global increase in tropical cyclone rapid intensification. *Nature Communications*, *13*(1), 6626. Available from https://doi.org/10.1038/s41467-022-34321-6. Available from, https://www.nature.com/articles/s41467-022-34321-6.

Bhatt, C. M., Rao, G. S., Farooq, M., Manjusree, P., Shukla, A., Sharma, S. V. S. P., ... Dadhwal, V. K. (2017). Satellite-based assessment of the catastrophic Jhelum floods of September 2014, Jammu & Kashmir, India. *Geomatics, Natural Hazards and Risk*, *8*(2), 309–327. Available from https://doi.org/10.1080/19475705.2016.1218943, http://www.tandfonline.com/toc/tgnh20/current.

Borghys, D., Yvinec, Y., Perneel, C., Pizurica, A., & Philips, W. (2006). Supervised feature-based classification of multi-channel SAR images. *Pattern Recognition Letters*, *27*(4), 252–258. Available from https://doi.org/10.1016/j.patrec.2005.08.006.

Census of India (CoI) (2011). *District census handbook, Puri, Odisha*. Available from http://www.censusindia.gov.in/2011census/dchb/DCHB_A/21/2118_PART_A_DCHB_PURI.pdf.

CGWB. (2013). *Ground water booklet: Puri district, Orissa*. Central Ground Water Board, Ministry of Water Resources, Govt of India, 5–6. Available from https://cgwb.gov.in/District_Profile/Orissa/Puri.pdf.

Chand, S. S., Walsh, K. J. E., Camargo, S. J., Kossin, J. P., Tory, K. J., Wehner, M. F., ... Murakami, H. (2022). Declining tropical cyclone frequency under global warming. *Nature Climate Change*, *12*(7), 655–661. Available from https://doi.org/10.1038/s41558-022-01388-4, https://doi.org/10.1038/s41558-022-01388-4.

Chatterjee, S. (2020). Analytical Study of North Indian Oceanic Cyclonic Disturbances with special reference to extremely severe cyclonic storm fani: Meteorological variability, India's preparedness with terrible aftermath. *Natural Hazards and Earth System Sciences Discussions*, 1–24. Available from https://doi.org/10.5194/nhess-2020-287, https://nhess.copernicus.org/preprints/nhess-2020-287/.

Chavez, P. S. (1988). An improved dark-object subtraction technique for atmospheric scattering correction of multispectral data. *Remote Sensing of Environment*, *24*(3), 459–479. Available from https://doi.org/10.1016/0034-4257(88)90019-3.

Chittibabu, P., Dube, S. K., Macnabb, J. B., Murty, T. S., Rao, A. D., Mohanty, U. C., & Sinha, P. C. (2004). Mitigation of flooding and cyclone hazard in Orissa, India. *Natural Hazards*, *31*(2), 455–485. Available from https://doi.org/10.1023/B:NHAZ.0000023362.26409.22.

Clement, M. A., Kilsby, C. G., & Moore, P. (2018). Multi-temporal synthetic aperture radar flood mapping using change detection. *Journal of Flood Risk Management*, *11*(2), 152–168. Available from https://doi.org/10.1111/jfr3.12303, http://onlinelibrary.wiley.com/journal/10.1111/(ISSN)1753-318X.

Cohen, J. (1960). A coefficient of agreement for nominal scales. *Educational and Psychological Measurement*, *20*(1), 37–46. Available from https://doi.org/10.1177/001316446002000104.

Congalton, R. G. (1991). A review of assessing the accuracy of classifications of remotely sensed data. *Remote Sensing of Environment*, *37*(1), 35–46. Available from https://doi.org/10.1016/0034-4257(91)90048-B.

Copernicus. (2020). *Sentinels Scientific Data Hub*. Available from https://scihub.copernicus.eu/dhus/#/home.

Deshpande, M., Kumar Singh, V., Ganadhi, M. K., Roxy, M. K., Emmanuel, R., & Kumar, U. (2021). Changing status of tropical cyclones over the north Indian Ocean. *Climate Dynamics*, *57*(11), 3545–3567. Available from https://doi.org/10.1007/s00382-021-05880-z, https://doi.org/10.1007/s00382-021-05880-z.

Dodangeh, E., Choubin, B., Eigdir, A. N., Nabipour, N., Panahi, M., Shamshirband, S., & Mosavi, A. (2020). Integrated machine learning methods with resampling algorithms for flood susceptibility prediction. *Science of the Total Environment*, *705*. Available from https://doi.org/10.1016/j.scitotenv.2019.135983, http://www.elsevier.com/locate/scitotenv.

ESA. (2018). *Sentinel-1 mission details*. From https://earth.esa.int/web/guest/missions/esa-operationaleo-missions/sentinel-1.

Giustarini, L., Hostache, R., Matgen, P., Schumann, G. J. P., Bates, P. D., & Mason, D. C. (2013). A change detection approach to flood mapping in Urban areas using TerraSAR-X. *IEEE Transactions on Geoscience and Remote Sensing*, *51*(4), 2417–2430. Available from https://doi.org/10.1109/TGRS.2012.2210901.

Government of Odisha. (2019). *Memorandum. Extremely severe cyclonic storm "Fani," 3rd May 2019*. Special Relief Commissioner, Revenue & Disaster Management Department, Government of Odisha, Unpublished content.

Greifeneder, F., Wagner, W., Sabel, D., & Naeimi, V. (2014). Suitability of SAR imagery for automatic flood mapping in the Lower Mekong Basin. *International Journal of Remote Sensing*, *35*(8), 2857–2874. Available from https://doi.org/10.1080/01431161.2014.890299, https://www.tandfonline.com/loi/tres20.

Hoarau, K., Bernard, J., & Chalonge, L. (2012). Intense tropical cyclone activities in the northern Indian Ocean. *International Journal of Climatology, 32*(13), 1935–1945. Available from https://doi.org/10.1002/joc.2406.

Hoque, M. A.-A., Phinn, S., & Roelfsema, C. (2017). A systematic review of tropical cyclone disaster management research using remote sensing and spatial analysis. *Ocean & Coastal Management, 146*, 109–120. Available from https://doi.org/10.1016/j.ocecoaman.2017.07.001, https://www.sciencedirect.com/science/article/pii/S0964569117302806.

Hoque, M. A.-A., Pradhan, B., Ahmed, N., & Roy, S. (2019). Tropical cyclone risk assessment using geospatial techniques for the eastern coastal region of Bangladesh. *Science of The Total Environment, 692*, 10–22. Available from https://doi.org/10.1016/j.scitotenv.2019.07.132, https://www.sciencedirect.com/science/article/pii/S0048969719332450.

Horritt, M. S. (1999). A statistical active contour model for SAR image segmentation. *Image and Vision Computing, 17*(3–4), 213–224. Available from https://doi.org/10.1016/s0262-8856(98)00101-2, https://www.journals.elsevier.com/image-and-vision-computing.

Hu, T., & Smith, R. B. (2018). The impact of Hurricane Maria on the vegetation of Dominica and Puerto Rico using multispectral remote sensing. *Remote Sensing, 10*(6). Available from https://doi.org/10.3390/rs10060827, https://res.mdpi.com.

Huang, W., Li, S., Vogt, T., Xu, R., Tong, S., Molina, T., ... Guo, Y. (2023). Global short-term mortality risk and burden associated with tropical cyclones from 1980 to 2019: A multi-country time-series study. *The Lancet Planetary Health, 7*(8), e694–e705. Available from https://doi.org/10.1016/S2542-5196(23)00143-2, https://www.thelancet.com/journals/lanplh/article/PIIS2542-5196(23)00143-2/fulltext.

India Meteorological Department (IMD). (2020). *Terminologies and Glossary*. Available from http://imd.gov.in/section/nhac/termglossary.pdf.

India Meteorological Department (2020). *Cyclone E Atlas. Tracks of cyclones and depressions over North Indian Ocean, 1891–2018*. Available from http://www.rmcchennaieatlas.tn.nic.in.

Klemas, V. V. (2009). The role of remote sensing in predicting and determining coastal storm impacts. *Journal of Coastal Research, 25*(6), 1264–1275. Available from https://doi.org/10.2112/08-1146.1.

Klotzbach, P. J., Wood, K. M., Schreck III, C. J., Bowen, S. G., Patricola, C. M., & Bell, M. M. (2022). Trends in Global Tropical Cyclone Activity: 1990–2021. *Geophysical Research Letters, 49*(6), e2021GL095774. Available from https://doi.org/10.1029/2021GL095774, https://onlinelibrary.wiley.com/doi/abs/10.1029/2021GL095774.

Knutson, T. R., McBride, J. L., Chan, J., Emanuel, K., Holland, G., Landsea, C., ... Sugi, M. (2010). Tropical cyclones and climate change. *Nature Geoscience, 3*(3), 157–163. Available from https://doi.org/10.1038/ngeo779.

Kossin, J. P., Knapp, K. R., Olander, T. L., & Velden, C. S. (2020). Global increase in major tropical cyclone exceedance probability over the past four decades. *Proceedings of the National Academy of Sciences, 117*(22), 11975–11980. Available from https://doi.org/10.1073/pnas.1920849117, https://www.pnas.org/doi/10.1073/pnas.1920849117.

Kumar, R. (2016). Flood hazard assessment of 2014 floods in Sonawari sub-district of Bandipore district (Jammu & Kashmir): An application of geoinformatics. *Remote Sensing Applications: Society and Environment, 4*, 188–203. Available from https://doi.org/10.1016/j.rsase.2016.10.002, http://www.journals.elsevier.com/remote-sensing-applications-society-and-environment/.

Kumar, R. (2018). *Flood inundation and hazard mapping of 2017 floods in the Rapti River basin using Sentinel-1A synthetic aperture radar images Applications and Challenges of*

Geospatial Technology: Potential and Future Trends (pp. 77–98). India: Springer International Publishing. Available from http://www.springer.com/in/book/9783319998817, 10.1007/978-3-319-99882-4-6.

Kumar, R., & Acharya, P. (2016). Flood hazard and risk assessment of 2014 floods in Kashmir Valley: A space-based multisensor approach. *Natural Hazards*, *84*(1), 437–464. Available from https://doi.org/10.1007/s11069-016-2428-4, http://www.wkap.nl/journalhome.htm/0921-030X.

Kumar, R., Singh, R., Gautam, H., & Pandey, M. K. (2018). Flood hazard assessment of August 20, 2016 floods in Satna District, Madhya Pradesh, India. *Remote Sensing Applications: Society and Environment*, *11*, 104–118. Available from https://doi.org/10.1016/j.rsase.2018.06.001, http://www.journals.elsevier.com/remote-sensing-applications-society-and-environment/.

Kunze, S. (2021). Unraveling the effects of tropical cyclones on economic sectors worldwide: Direct and indirect impacts. *Environmental and Resource Economics*, *78*(4), 545–569. Available from https://doi.org/10.1007/s10640-021-00541-5. https://doi.org/10.1007/s10640-021-00541-5.

Manjusree, P., Bhatt, C. M., Begum, A., Rao, G. S., & Bhanumurthy, V. (2015). A decadal historical satellite data analysis for flood hazard evaluation: A case study of Bihar (North India). *Singapore Journal of Tropical Geography*, *36*(3), 308–323. Available from https://doi.org/10.1111/sjtg.12126India, http://onlinelibrary.wiley.com/journal/10.1111/(ISSN)1467-9493.

Mason, D. C., Schumann, G. J. P., Neal, J. C., Garcia-Pintado, J., & Bates, P. D. (2012). Automatic near real-time selection of flood water levels from high resolution Synthetic Aperture Radar images for assimilation into hydraulic models: A case study. *Remote Sensing of Environment*, *124*, 705–716. Available from https://doi.org/10.1016/j.rse.2012.06.017.

Mishra, A. (2014). Temperature rise and trend of cyclones over the Eastern Coastal Region of India. *Journal of Earth Science & Climatic Change*, *05*(09). Available from https://doi.org/10.4172/2157-7617.1000227.

Mohapatra, M., Mandal, G. S., Bandyopadhyay, B. K., Tyagi, A., & Mohanty, U. C. (2012). Classification of cyclone hazard prone districts of India. *Natural Hazards*, *63*(3), 1601–1620. Available from https://doi.org/10.1007/s11069-011-9891-8.

Muduli, P. R. (2017). The impact of tropical cyclone 'Phailin' on the hydrology of Chilika Lagoon, India. *International Journal of Environmental Sciences & Natural Resources*, *4*(2). Available from https://doi.org/10.19080/IJESNR.2017.04.555632.

NRSC (National Remote Sensing Centre) and ISRO (Indian Space Research Organization). (2011). *Manual on preparation of geospatial layers using high resolution (Cartosat-1Pan + LISS-IV mx) Orthorectified satellite imagery*. Space based information support for decentralized planning (SIS-DP), remote sensing and GIS applications area National Remote Sensing Centre, Indian space research organisation (ISRO), Department of Space, government of India, Hyderabad.

Oberstadler, R., Hönsch, H., & Huth, D. (1997). Assessment of the mapping capabilities of ERS-1 SAR data for flood mapping: A case study in Germany. *Hydrological Processes*, *11* (10), 1415–1425. Available from. Available from http://onlinelibrary.wiley.com/journal/10.1002/(ISSN)1099-1085.

Odisha District Disaster Management Authority (DDMA). (2018). *District Disaster Management Plan (2017-18) Puri, Odisha*. Available from https://cdn.s3waas.gov.in/s3e836d813fd184325132f-ca8edcdfb40e/uploads/2018/05/2018051650.pdf.

Odisha Rainfall Monitoring System. (2019). *Odisha_rainfall_district wise 2019*. Available from http://rainfall.nic.in/PubRainChartDtl.asp.

Ortiz, A. M. D., Chua, P. L. C., Salvador, D., Dyngeland, C., Albao, J. D. G., & Abesamis, R. A. (2023). Impacts of tropical cyclones on food security, health and biodiversity. *Bulletin of the World Health Organization, 101*(2), 152–154. Available from https://doi.org/10.2471/BLT.22.288838.

Pandey, R. S., & Liou, Y.-A. (2020). Decadal behaviors of tropical storm tracks in the North West Pacific Ocean. *Atmospheric Research, 246*105143. Available from https://doi.org/10.1016/j.atmosres.2020.105143, https://www.sciencedirect.com/science/article/pii/S0169809520310796.

Pulvirenti, L., Chini, M., Pierdicca, N., & Boni, G. (2016). Use of SAR data for detecting floodwater in urban and agricultural areas: The role of the interferometric coherence. *IEEE Transactions on Geoscience and Remote Sensing, 54*(3), 1532–1544. Available from https://doi.org/10.1109/TGRS.2015.2482001.

Schlaffer, S., Matgen, P., Hollaus, M., & Wagner, W. (2015). Flood detection from multi-temporal SAR data using harmonic analysis and change detection. *International Journal of Applied Earth Observation and Geoinformation, 38*, 15–24. Available from https://doi.org/10.1016/j.jag.2014.12.001.

Singh, K., Panda, J., Sahoo, M., & Mohapatra, M. (2019). Variability in tropical cyclone climatology over North Indian Ocean during the Period 1891 to 2015. *Asia-Pacific Journal of Atmospheric Sciences, 55*(2), 269–287. Available from https://doi.org/10.1007/s13143-018-0069-0. https://doi.org/10.1007/s13143-018-0069-0.

Song, Y. S., Sohn, H. G., & Park, C. H. (2007). Efficient water area classification using radarsat-1 SAR imagery in a high relief mountainous environment. *Photogrammetric Engineering and Remote Sensing, 73*(3), 285–296. Available from https://doi.org/10.14358/PERS.73.3.285. Available from, http://www.asprs.org/Photogrammetric-Engineering-and-Remote-Sensing/PE-RS-Journals.html.

Tehrany, M. S., Pradhan, B., Mansor, S., & Ahmad, N. (2015). Flood susceptibility assessment using GIS-based support vector machine model with different kernel types. *Catena, 125*, 91–101. Available from https://doi.org/10.1016/j.catena.2014.10.017, http://www.elsevier.com/inca/publications/store/5/2/4/6/0/9.

The United Nations, the World Bank, and the Asian Development Bank. (2019). *Cyclone fani: damage, loss and needs assessment: Odisha (India)*. Available from https://www.ilo.org/wcmsp5/groups/public/-ed_emp/documents/publication/wcms_732468.pdf.

Twele, A., Cao, W., Plank, S., & Martinis, S. (2016). Sentinel-1-based flood mapping: A fully automated processing chain. *International Journal of Remote Sensing, 37*(13), 2990–3004. Available from https://doi.org/10.1080/01431161.2016.1192304, https://www.tandfonline.com/loi/tres20.

UNICEF. (2019). *India: Cyclone Fani*. Available from https://www.unicef.org/appeals/files/UNICEF_India_Humanitarian_SitRep_Cyclone_Fani_12_May_2019.pdf.

USGSEE. (2020). *EarthExplorer web Portal*. Available from http://earthexplorer.usgs.gov/.

Wang, Y. (2004). Using Landsat 7 TM data acquired days after a flood event to delineate the maximum flood extent on a coastal floodplain. *International Journal of Remote Sensing, 25*(5), 959–974. Available from https://doi.org/10.1080/0143116031000150022.

World Health Organization. (2022a). Tropical cyclones. Available from https://www.who.int/health-topics/tropical-cyclones.

World Health Organization. (2022b). *Tropical Cyclones*. Available from https://public-old.wmo.int/en/our-mandate/focus-areas/natural-hazards-and-disaster-risk-reduction/tropical-cyclones.

Chapter 3

Sensitivity of cryosphere to climate change in Western Himalaya: a study from Dras Basin

Suhail A. Lone and Gh Jeelani
Department of Earth Sciences, University of Kashmir, Srinagar, India

3.1 Introduction

The Himalaya, also known as the water towers of Asia, hosts the world's richest snow and glacier cover after the Arctic and Antarctic polar regions. A reliable amount of water supplied by this cryospheric cover is extremely important in defining the agro-economic and political stability of the region (Scott et al., 2019; Jeelani et al., 2020; Lone et al., 2023). Apart from its role in domestic, agriculture, and hydropower generation, this cryospheric cover plays an important role in regulating the atmospheric-meteorological phenomena of the globe in general and the Himalayas in particular (Bolch et al., 2012; Somers, 2019). The health of this white mass is highly dependent on the two largest weather systems, viz., Western Disturbance and Indian Summer Monsoons (Immerzeel et al., 2015; Jeelani & Lone, 2019; Lone et al., 2023). Himalayan snow and glacier cover are highly sensitive to climate change. The most immediate impact of this rise in temperature will adversely affect the glacier receding rate in the Himalayas because of its sensitivity to changing climate (Lone et al., 2017). Studies have suggested that the rate of retreat has increased during the past few decades due to a rise in temperature by 0.4% per year (Bolch et al., 2012; Lone et al., 2017). The studies have also reported that ~65%–70% of annual stream water discharge and groundwater recharge is sustained by water derived from snow and glacier melt (Lone et al., 2021; Lone et al., 2022) Therefore variability in precipitation and receding of snow/glaciers can have adverse effects on the regional hydrology of the basin (Lone et al., 2023).

Over the past two to three decades, the Himalayan region has become a focal point for numerous scientific investigations aimed at assessing how

Water Resources Monitoring, Management, and Sustainability. DOI: https://doi.org/10.1016/B978-0-443-23665-5.00003-X

high-altitude glaciers are reacting to the impacts of recent climate change (Lone & Jeelani, 2024). These studies have consistently underscored the profound influence of the region's unique topography and climatic conditions as significant drivers behind the glaciers' retreat (Rao et al., 2016; Lone & Jeelani, 2024). To delve into the intricate dynamics of glacier response, researchers have extensively employed remote sensing-based methodologies (Rao et al., 2016; Lone & Jeelani, 2024 and references their in). These encompass a range of techniques, including on-screen delineation, supervised and unsupervised classification, and image rationing. Remote sensing offers a valuable tool through the provision of multi-sensor and multi-temporal satellite images (Shekhar et al., 2010; Rao et al., 2016). These images play a crucial role in the mapping, monitoring, and systematic assessment of glacial extent and changes, providing the advantage of a synoptic view over vast areas. The utility of satellite imagery becomes particularly evident in challenging terrains like the rugged Himalayas, complementing field-based glaciological measurements. Recognizing the dynamic nature of mountainous glaciers and their responses to diverse climatic variables, it becomes apparent that regular and meticulous glacier monitoring is of paramount importance. Such monitoring not only serves as a foundation for sound economic and water resource management policies but also caters to the specific needs of local and regional contexts. Hence, the main objective of our present study is to evaluate the recession rate of Machoi glacier from the Dras Basin under recent climatic change scenarios using geospatial data. Monitoring glacier recession holds the potential to provide invaluable insights into the long-term response of Himalayan glaciers to changing climates. Furthermore, this information can be leveraged to reconstruct past climatic conditions and ascertain the equilibrium line altitude. The study is highly significant in contributing to policies related to water resource management and ecosystem restoration across the Himalayas. This research is a critical step in addressing the challenges posed by climate change and its cascading effects on vital water resources and ecological systems in mountainous regions.

3.2 Study area

Gateway of Ladakh, Dras sub-basin is one of the micro-watersheds of the Indus Basin, situated within the geographic coordinates of latitude 34° 25′ 50″ N and longitude 75° 45′ 85″ E (Fig. 3.1). The altitude of the Dras basin varies from ~2680 m above sea level (asl) to more than ~5500 m asl.

The area of the basin is 3465 km^2 and it is classified as one of the coldest places in the world. The Dras basin experiences a continental Mediterranean climate (Köppen classification system), receiving a mean annual precipitation of 550 mm from December to March, mostly in the form of snow (Fig. 3.2). The mean annual temperature of Dras is 5.2°C, with maximum (23°C) and minimum (−23°C) observed in August and December,

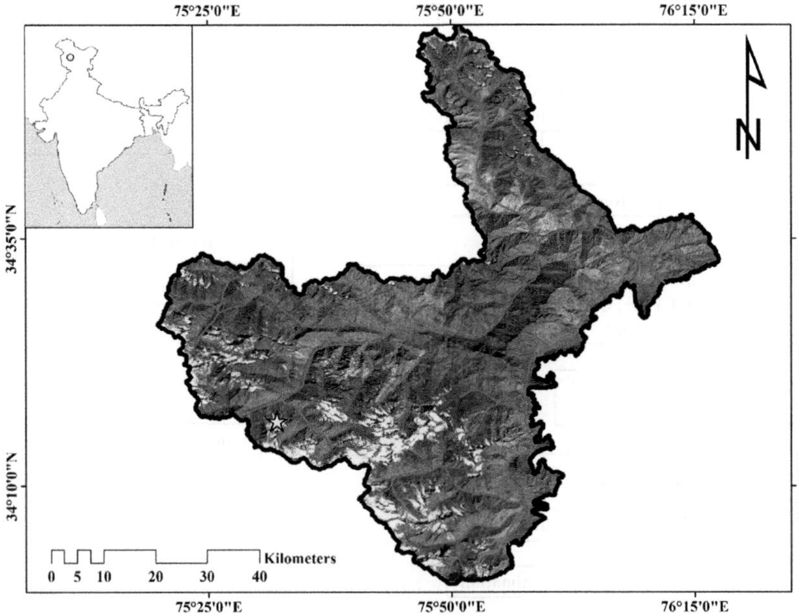

FIGURE 3.1 Study area map showing the location of the Dras basin with respect to India and Ladakh. The Landsat TM Image of October 2022 was used as a base map. The blue line on the map represents the drainage network and the white mass depicts the glacier cover. The star symbol on the map depicts the location of Machoi Glacier. Map lines delineate study areas and do not necessarily depict accepted national boundaries.

respectively. Westerlies and southwest monsoons are two major weather systems controlling the meteorology of the basin. The winter-accumulated snowfall is one of the major sources of water sustaining the hydrology of the basin. Apart from seasonal snow, the basin is bestowed by numerous glaciers, among which Machoi is a major glacier with a present aerial extent of 5 km^2. The glacier lies at an altitude of ~4000–5500 m asl. The glacier is one of the major sources of perennial flow to the Dras River, playing a substantial role in sustaining the local and regional hydrology of the basin.

3.3 Materials and methods

In the present study, we embarked on a comprehensive study to explore how the Dras glacier in the Himalayas has responded to changing temperature and precipitation patterns over a substantial time frame, spanning from 1901 to 2022. The primary focus of our study was centered on assessing alterations in the glacier's snout position and its overall area over the past few decades. To gather the necessary data for our investigation, we diligently acquired time-series datasets of temperature and precipitation from reliable sources. These datasets provided critical insights into the climatic conditions

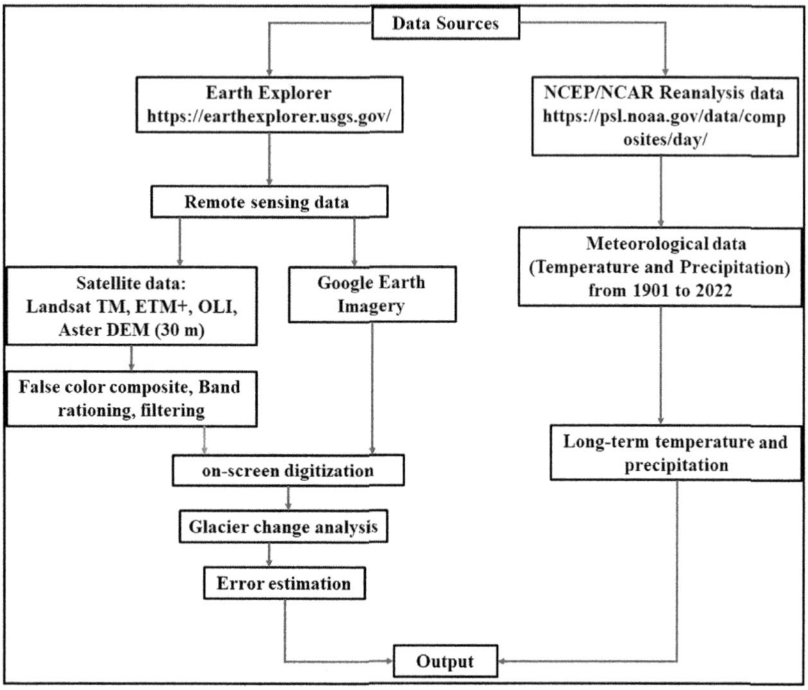

FIGURE 3.2 Methodology flow chart depicting the datasets and methods used in this study.

that have prevailed in the Himalayan region throughout the 20th and early 21st centuries. Concurrently, we harnessed remote sensing satellite data to obtain a comprehensive view of the Dras glacier's evolving state. This data was instrumental in achieving our research objectives. To access the required satellite imagery, we accessed Landsat data from the years 1980 to 2022, utilizing various sensor resolutions. These images were sourced from the Earth Explorer website, managed by the United States Geological Survey (https://earthexplorer.usgs.gov/). To ensure the accuracy and reliability of our findings, we cross-verified the results obtained from satellite data with imagery available on Google Earth and data collected through field validation.

Selecting the most suitable satellite images for glacier mapping demanded meticulous consideration of various factors. These factors included the ablation period, snow cover, and cloud conditions, all of which significantly impact the quality and precision of the results. To pinpoint the snout position of the glacier within the satellite data, we employed a well-established methodology (Fig. 3.2). This involved identifying terminal moraines, which are distinct features indicative of glacier termini (Lone, 2016). Subsequently, we engaged in on-screen digitization to accurately document the changing snout positions. This meticulous process was carried out at an average scale of 1:150,000 using Arc-GIS 10.7.1 software, ensuring a high level of precision in our analysis.

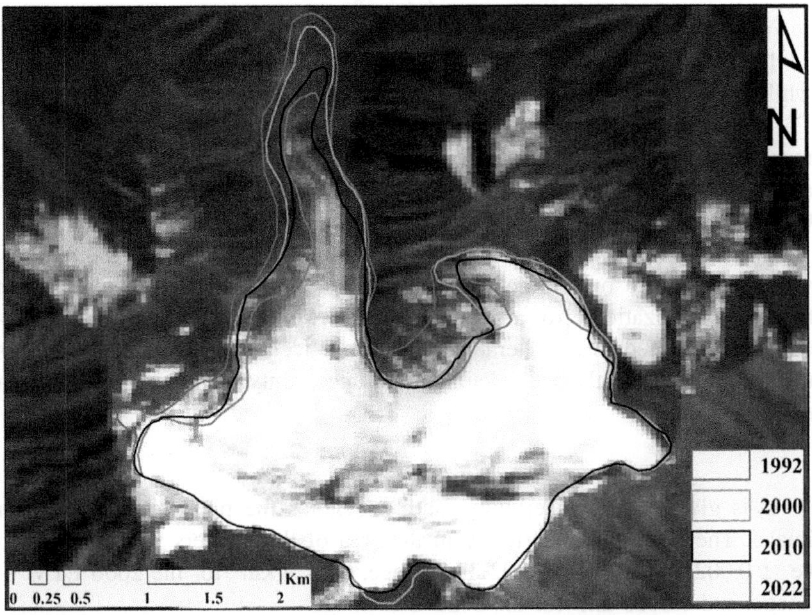

FIGURE 3.3 The plot depicting the time-series areal change of Machoi glacier in the Dras basin. It can be observed from the layout that the area of Machoi glacier has decreased significantly from 1980 to 2022.

Additionally, we applied various image processing techniques, such as filtering, false color composites, and band ratio analysis, to effectively distinguish glacial features (Fig. 3.3). These methods enhanced our ability to precisely delineate the present snout position and measure changes in glacier area over time. In tandem with our remote sensing analysis, we complemented our findings with long-term meteorological data, specifically temperature and precipitation records. These meteorological parameters were sourced from the National Ocean and Atmospheric Administration (NOAA), providing crucial context for our glacier response analysis.

3.3.1 Uncertainty estimations

Differences in snow cover, cloud cover, shadows, and image resolution introduce variations in the precision of our measurements. The inaccuracies observed in estimating glacier length and demarcating glacier areas can be attributed to the process of image co-registration. Consequently, conducting a thorough error estimation becomes imperative to ascertain the degree of accuracy and the significance of our research outcomes. In the present study, the snout change was evaluated using the formula:

$$U = \sqrt{a^2 + b^2 + \sigma}$$

In our analysis, we denote "a" and "b" as the resolutions of the images, while "σ" represents the uncertainty associated with image georeferencing. To assess image georeferencing error, a visual evaluation was conducted, resulting in measurements of 8, 7, and 0 m for the 1992 TM, 2000 ETM + , and 2022 OLI images when georeferenced to the base image of Landsat OLI 2022.

To determine the uncertainties associated with glacier area estimations, we employed the buffer method. The size of the buffer region was set at twice the digitization error around each glacier, in accordance with the methodology proposed in the literature (Lone and Jeelani (2024) and references their in). Subsequently, we calculated the uncertainty measurements for each glacier, given as:

$$U_{\text{area}} = 2\text{UV}$$

U is glacier area uncertainty, and V denotes the pixel resolution of an image. Therefore the uncertainty in the area of the glaciers was assessed to be $\pm 0.0028\ \text{km}^2$ for the 1992 TM image, $0.0027\ \text{km}^2$ for the 2000 ETM + image, and $0.0025\ \text{km}^2$ for the 2022 OLI image.

3.4 Results

Our primary data source for this analysis was reanalysis data obtained from NOAA. When examining the long-term temperature data spanning from 1901 to 2022, we observed a notable overall increase in temperature. Specifically, the data revealed an average temperature rise of approximately 0.166°C during this period. However, this temperature increase did not follow a steady upward trajectory. Instead, it exhibited distinct temporal trends. Initially, there was a modest temperature rise of 0.02°C per year during two periods, from 1901 to 1930 and from 1974 to 1990. This was followed by a slight decline in temperature of 0.0002°C per year from 1931 to 1973. However, in the more recent period from 1991 to 2022, we observed a substantial increase in temperature, registering a significant 0.10°C per year. The precipitation data also displayed variability over time. While the average annual precipitation exhibited a decrease of approximately 0.032 mm per year, there was an intriguing observation. Between 1910 and 1930, there was actually an increase in average precipitation of 0.8 mm per year. Beyond these long-term trends, our analysis also uncovered seasonal variations in temperature and precipitation in the western Himalayas. In terms of temperature, we noted a slight increase in the summer season, amounting to 0.006°C, whereas a more significant temperature increase of 0.08°C was observed in the winter season. Regarding precipitation, we observed a decrease in winter precipitation by 0.18 mm per year. In contrast, the other seasons showed varying trends, with spring witnessing an increase of 0.21 mm per year, summer experiencing a 0.09 mm per year increase, and autumn exhibiting a 0.18 mm per year increase.

The analysis of this extensive remote sensing dataset revealed compelling evidence of glacier retreat. From 1992 to 2022, the area of the Machoi glacier dwindled from 5.90 to 4.56 km^2 (Fig. 3.3). In 1992, the total area of the glacier was 5.90 km^2, which was reduced to 5.38 km^2 in 2000, 4.96 km^2 in 2010, and 4.56 km^2 in 2022. This staggering deglaciation equates to a loss of 1.34 km^2 of glacier coverage within the Dras basin over the course of 30 years.

3.5 Discussions

3.5.1 Long-term temperature and precipitation

Our investigation focused on the analysis of extensive statistical data pertaining to critical atmospheric variables, specifically temperature and precipitation, within the western Himalayas. This dataset offered invaluable insights into the evolving climatic conditions of the region over an extended period of time. The examination of seasonal trends, coupled with the examination of long-term temperature and precipitation patterns, highlights the intricate and dynamic nature of the climate in the western Himalayas. The prolonged meteorological dataset revealed an overarching trend of increasing temperatures and decreasing precipitation, particularly noticeable from 1980 onward. This escalation is of significant concern, given the heightened sensitivity of cryospheric cover in the mountainous catchments of the basin and surrounding areas. The interplay of diverse factors, encompassing both natural and anthropogenic influences, contributes to these observed fluctuations. Comprehending these climatic trends holds paramount importance for regional planning, resource management, and adapting to the changing climate conditions in this vulnerable and crucial region. The alarming temperature rise, coupled with a reduction in precipitation, underscores the urgency for informed decision-making and strategic interventions to address the challenges posed by these climatic shifts. It is imperative to foster sustainable practices and policies that consider the delicate balance of ecosystems and ensure the resilience of communities in the face of evolving climatic conditions. Similar conclusions have been put forward by studies carried out in other parts (Shekhar et al., 2010; Lone et al., 2019).

3.5.2 Glacier recession

Our in-depth investigation aimed to elucidate the temporal dynamics of the Machoi glacier, a prominent glacier situated within the Dras basin (Fig. 3.3). The outcomes of our research underscore a notable reduction in the overall extent of the Machoi glacier over this three-decade observational period. The observed decline in glacier extent aligns with patterns identified in other studies focused on Himalayan glaciers (Lone & Jeelani, 2024). This parallel trend emphasizes the widespread nature of glacier changes in the region and raises important questions about the underlying factors contributing to these transformations. The extended temporal scope of our data allows us to capture nuanced dynamics and trends, providing a more comprehensive understanding of the glacier's behavior over time. The findings not only contribute to the broader scientific knowledge of glacial dynamics in the Himalayas but also hold significance for local and global contexts. Documenting and comprehending the changes in the Machoi glacier's extent is pivotal for assessing the impacts of climate change on this specific glacier and, by extension, the hydrological systems and communities dependent on it. This research contributes valuable insights to ongoing discussions about the vulnerability of Himalayan glaciers and the broader implications for water resources and environmental stability in the region. However, it is crucial to note that the rate of glacier retreat is not uniform with other glaciers in the region. Variations in size, elevation, slope, aspect, and physiography of the individual basins all play a significant role in determining the pace of glacier recession. Several studies have proposed that the response of glaciers to receding trends is closely linked to glacier thickness, which is, in turn, influenced by the size of the glacier itself. Generally, smaller glaciers tend to retreat at a faster rate compared to their larger counterparts. Recent scientific investigations have reinforced this notion, emphasizing that changing climate effects are more pronounced in smaller glaciers than in larger ones. Furthermore, the continuous temperature rise can exacerbate the rate of melting in smaller glaciers, making them particularly vulnerable to climate change impacts. Another critical factor influencing glacier recession is the aspect (the direction a slope faces) and the elevation of the glacier snout. In our analysis, we divided the sub-basins into various sun-facing classes to evaluate the aspect effect. It became evident that south- and northeast-facing glaciers are retreating more rapidly than those facing north. This aspect-driven difference in retreat rates underscores the complex interplay between glacier orientation and response to changing environmental conditions.

Additionally, we considered the elevation effect on glacier recession. Our observations revealed that glaciers in the Liddar basin, situated at altitudes ranging from 3500 to 5000 m asl, have experienced more significant recession compared to glaciers in the Suru basin, which are located at higher altitudes, ranging from 4200 to 6000 m asl. This elevation-related disparity highlights the nuanced relationship between glacier altitude and its vulnerability to the forces of climate change. In conclusion, our study paints a detailed picture of glacier dynamics in the Dras basin over a 30-year span. It underscores the complexity of glacier responses to changing climatic conditions, influenced by factors such as size, aspect, elevation, and temperature trends. Understanding these intricate dynamics is crucial for both scientific comprehension and effective climate change mitigation and adaptation strategies in the Himalayan region.

3.5.3 Future hydrological regimes of the region

The climate, hydrology, and overall ecosystem sustainability of Southeast Asia are intricately linked to the vital meltwaters originating from the Himalayan cryosphere. The snow and glaciers of the Himalayas are pivotal components of the region's hydrological cycle, as emphasized by several studies (Gruber et al., 2017; Lone et al., 2022, 2023; Jeelani et al., 2024). These water resources play a fundamental role in supporting the livelihoods of communities residing in the major Himalayan river basins. However, the adverse impacts of a warming climate can disrupt the usual patterns of cryospheric advance and retreat, consequently disturbing both aquatic and terrestrial ecosystems (Immerzeel et al., 2015). The variations observed in long-term meteorological parameters indicate a direct connection between alterations in the hydrological patterns of Himalayan river basins and historical and ongoing climatic shifts. It is anticipated that these hydrological patterns may undergo further changes due to the continual rise in winter temperatures and a decrease in precipitation, particularly in the western Himalayas. In this region, both surface and sub-surface hydrology are primarily controlled by meltwaters (Lone et al., 2023; Nepal et al., 2021; Shah & Lone, 2019). The observed and projected variability in meteorological parameters underscores the vulnerability of the Himalayan cryosphere to the adverse effects of climate change (You et al., 2020; Jeelani et al., 2024). It is worth noting that the Indus River Basin, which is home to the world's largest irrigated agriculture system, heavily relies on these cryospheric waters. Consequently, these unprecedented externalities pose a significant threat to public water supply and the agro-economy of the region, particularly affecting mountainous communities and potentially leading to instability. The interconnectedness of these factors underscores the urgent need for proactive measures to mitigate the impacts of climate change on the Himalayan cryosphere and the downstream regions that depend on it for their sustenance and well-being.

3.6 Conclusion

In this study, we employed long-term optical remote sensing data and meteorological datasets to comprehensively evaluate and gain insights into how the Machoi glacier has responded to recent shifts in climate. Our analysis of temperature and precipitation data over the Himalayan region, spanning from 1901 to 2022, revealed some noteworthy trends. Specifically, there has been a gradual warming, with temperatures increasing at a rate of 0.2°C per decade, coupled with a concerning decline in precipitation, decreasing by 0.032 millimeters per year over the same period. One of the most important findings of our study is the significant reduction in the extent of the Machoi glacier during the observation period. We found that the total glaciated area of the Machoi glacier has shrunk from 5.90 km^2 in

1992 to 4.56 km^2 in 2022, exhibiting a deglaciation of about 1.34 km^2 in the last 30 years. The implications of these findings are far-reaching, particularly in the context of water resource management in the Himalayan region. The dwindling size of the Machoi glacier underscores the urgent need for proactive measures and informed policy decisions to address the challenges posed by climate change. This glacier's decline has direct consequences for downstream water resources and communities that rely on it for sustenance and various activities. In essence, our study offers crucial insights and data that can serve as a foundation for crafting water resource management policies.

Acknowledgment

The authors are highly thankful to the United States Geological Survey (USGS) and National Oceanic and Atmospheric Administration (NOAA) for open access to Satellite and meteorological datasets.

References

Bolch, T., Kulkarni, A., Kääb, A., Huggel, C., Paul, F., Cogley, J. G., ... Stoffel, M. (2012). The state and fate of Himalayan Glaciers. *Science (New York, N.Y.)*, *336*(6079), 310–314. Available from https://doi.org/10.1126/science.1215828.

Gruber, S., Fleiner, R., Guegan, E., Panday, P., Schmid, M. O., Stumm, D., ... Zhao, L. (2017). Review article: Inferring permafrost and permafrost thaw in the mountains of the Hindu Kush Himalaya region. *Cryosphere*, *11*(1), 81–99. Available from https://doi.org/10.5194/tc-11-81-2017, http://www.the-cryosphere.net/volumes_and_issues.html.

Immerzeel, W. W., Wanders, N., Lutz, A. F., Shea, J. M., & Bierkens, M. F. P. (2015). Reconciling high-altitude precipitation in the upper Indus basin with glacier mass balances and runoff. *Hydrology and Earth System Sciences*, *19*(11), 4673–4687. Available from https://doi.org/10.5194/hess-19-4673-2015.

Jeelani, G., Hassan, W., Padhya, V., Deshpande, R.D., Dimri, A.P., & Lone, S.A. (2024). Significant role of permafrost in regional hydrology of the Upper Indus Basin, India. *Science of The Total Environment*, 170863, 919. Available from: https://doi.org/10.1016/j.scitotenv.2024.170863.

Jeelani, G., & Lone, S. A. (2019). Establishing the effect of aridity on the stable isotopes (O and D) of precipitation in cold desert, Ladakh. *International Conference on Water Resources and Arid Environments*, *8*(2), 161–168.

Jeelani, G., Lone, S. A., Un Nisa, A., Mukherjee, A., & Deshpande, R. D. (2020). Sources and processes of groundwater arsenic mobilization in upper Jhelum basin, western Himalayas. *Journal of Hydrology*, 591. Available from https://doi.org/10.1016/j.jhydrol.2020.125292.

Lone, S. (2016). Mapping the Glacial-Geomorphological Landforms in East Liddar Valley, NW Himalaya Kashmir India. *Journal of Geography & Natural Disasters*, *s6*. Available from https://doi.org/10.4172/2167-0587.S6-004.

Lone, S. A., Jeelani, G., Deshpande, R. D., Mukherjee, A., Jasechko, S., & Lone, A. (2021). Meltwaters dominate groundwater recharge in cold arid desert of Upper Indus River Basin (UIRB), western Himalayas. *Science of The Total Environment*, *786*. Available from https://doi.org/10.1016/j.scitotenv.2021.147514.

Lone, S. A., Jeelani, G., Deshpande, R. D., Sultan Bhat, M., & Padhya, V. (2023). Assessing the hydrological controls on spatio-temporal patterns of streamwater in glacierized mountainous Upper Indus River Basin (UIRB), western Himalayas. *Journal of Hydrology, 619*. Available from https://doi.org/10.1016/j.jhydrol.2023.129310.

Lone, S. A., Jeelani, G., Padhya, V., & Deshpande, R. D. (2022). Identifying and estimating the sources of river flow in the cold arid desert environment of Upper Indus River Basin (UIRB), western Himalayas. *Science of The Total Environment, 832*. Available from https://doi.org/10.1016/j.scitotenv.2022.154964.

Lone, S. A, & Jeelani, G. H. (2024). *Evaluating the potential impact of climate change on glacier dynamics in Western Himalayas, India. Climate Crisis: Adaptive Approaches and Sustainability* (pp. 499–510). Springer Nature.

Lone, S. A., Jeelani, G., Deshpande, R. D., & Mukherjee, A. (2019). Stable isotope (δ 18 O and δD) dynamics of precipitation in a high altitude Himalayan cold desert and its surroundings in Indus river basin, Ladakh. *Atmospheric Research, 221*, 46–57. Available from https://doi.org/10.1016/j.atmosres.2019.01.025, http://www.elsevier.com/inca/publications/store/5/0/3/3/2/3/index.htt.

Lone, S. A., Jeelani, G., Deshpande, R. D., & Shah, R. A. (2017). Evaluating the sensitivity of glacier to climate by using stable water isotopes and remote sensing. *Environmental Earth Sciences, 76*(17). Available from https://doi.org/10.1007/s12665-017-6937-6, http://www.springerlink.com/content/121380/.

Nepal, S., Khatiwada, K. R., Pradhananga, S., Kralisch, S., Samyn, D., Bromand, M. T., ... Chevallier, P. (2021). Future snow projections in a small basin of the Western Himalaya. *Science of the Total Environment*, 795. Available from https://doi.org/10.1016/j.scitotenv.2021.148587.

Rao, V. B., Franchito, S. H., Pedroso Gerólamo, R. O., Giarolla, E., Venkata Siva Ramakrishna, S. S., Srinivasa Rao, B. R., & Naidu, C. V. (2016). Himalayan warming and climate change in India. *American Journal of Climate Change, 05*(04), 558–574. Available from https://doi.org/10.4236/ajcc.2016.54038.

Scott, C. A., Zhang, F., Mukherji, A., Immerzeel, W., Mustafa, D., & Bharati, L. (2019). *Water in the Hindu Kush HimalayaThe Hindu Kush Himalaya Assessment: Mountains, Climate Change, Sustainability and People* (pp. 257–299). United States: Springer International Publishing, United States Springer International Publishing. Available from https://link.springer.com/book/10.1007/978-3-319-92288-1, 10.1007/978-3-319-92288-1-8.

Shah, R. A., & Lone, S. A. (2019). Hydrogeomorphological mapping using geospatial techniques for assessing the groundwater potential of Rambiara river basin, western Himalayas. *Applied Water Science, 9*(3). Available from https://doi.org/10.1007/s13201-019-0941-9, https://www.springer.com/journal/13201.

Shekhar, M. S., Chand, H., Kumar, S., Srinivasan, K., & Ganju, A. (2010). Climate-change studies in the western Himalaya. *Annals of Glaciology, 51*(54), 105–112. Available from https://doi.org/10.3189/172756410791386508.

Somers,L.D. (2019). *Mountain water resources, groundwater and climate change in the Peruvian Andes.*

You, Q., Chen, D., Wu, F., Pepin, N., Cai, Z., Ahrens, B., ... AghaKouchak, A. (2020). Elevation dependent warming over the Tibetan Plateau: Patterns, mechanisms and perspectives. *Earth-Science Reviews, 210*. Available from https://doi.org/10.1016/j.earscirev.2020.103349.

Chapter 4

Geostatistics-based geospatial modeling and hydro-geochemical characterization of groundwater quality: a case study in Jharia Coalfield Region, India

Binay Prakash Panigrahy[1], Girija Shankar Behera[2], Rahul Kumar Singh[3], Suren Nayak[4] and Nirasindhu Desinayak[5]

[1]Department of Environmental Science & Engineering, IIT (ISM), Dhanbad, Jharkhand, India, [2]Department of Applied Geology, IIT (ISM), Dhanbad, Jharkhand, India, [3]Geological Survey of India, Ranchi, Jharkhand, India, [4]Department of Geology, Utkal University, Bhubaneswar, Odisha, India, [5]Department of Geology, Ravenshaw University, Cuttack, Odisha, India

4.1 Introduction

Groundwater resources are inherently dynamic and subject to various influences, including the expansion of irrigation, industrialization, urbanization, and mining activities (Chatterjee, Tarafder, & Paul, 2010). Surface and groundwater frequently experience pollution and contamination in areas replete with coal mining operations. These issues are exacerbated by various factors, including the dumping of overburden (OB), the disposal of refuse in excavated regions, industrial waste, mine water, acid mine drainage, and tailing ponds. The water quality in the Jharia Coalfield, renowned for housing the premier coking coal in the nation, has been significantly impacted by extensive exploration and exploitation (Pandey, Kumar, Singh, & Mohalik, 2016). Unfortunately, the water bodies in this region bear the brunt of these operations. Residents in proximity to coal mining areas in this region are encountering significant difficulties in accessing clean drinking water. Therefore, there is a pressing need for effective water resource management and conservation, especially given the deterioration of water quality that has transpired in the past within this area.

Water Resources Monitoring, Management, and Sustainability. DOI: https://doi.org/10.1016/B978-0-443-23665-5.00004-1

The Jharia Coalfield area features numerous standing water bodies situated in proximity to collieries and industrial zones. Over time, land use studies reveal a diminishing trend in the coverage of water bodies within this coalfield. Specifically, the area occupied by water bodies has decreased from 7.3% in 1925 to approximately 3% in 1987 (Singh, 1987). More recent data indicates a further reduction, with water bodies covering only about 2.9% of the Jharia coalfield area. The coalfield boasts an extensive network of streams, with the Damodar River serving as the primary drainage system for the region.

The Water Quality Index (WQI) is a method used to evaluate and rate the overall quality of water by considering the combined impact of individual water quality parameters (Singh, Tiwari, & Mahato, 2013). It involves a mathematical formula that condenses a large set of water quality data into a single numerical value. WQI is designed to give planners a straightforward and easily understandable measure that will assist them in comprehending the quality of water as well as the prospective uses of that water body. To determine whether water is suitable for human consumption, the WQI is determined by comparing the water quality data obtained from the analysis of samples to the prescribed standards for drinking water as specified in BIS 2003 (IS:10500). For this, the drinking water standards established by ISI in India are utilized, as well as the associated WQI categories.

A geostatistical technique is a statistical tool commonly employed by researchers to analyze the spatial variation of groundwater characteristic features (Adhikary, Chandrasekharan, Chakraborty, & Kamble, 2010; Delhomme, 1979; Teikeu et al., 2016). This approach is particularly valuable in hydrogeology and environmental science for understanding the distribution and variability of properties and parameters related to groundwater. In this geostatistical investigation, a significant aspect involves the creation of Kriging estimation maps specifically tailored to various groundwater quality parameters, viz., pH, electrical conductivity (EC), TDS, F^-, Cl^-, HCO_3^-, SO_4^{2-}, NO_3^-, Mg^{2+}, Na^+, K^+, Ca^{2+}, and TH. These maps contribute to evaluating the effectiveness and reliability of the geostatistical techniques applied. Additionally, considerable emphasis has been placed on refining the accuracy of variogram models, which are then superimposed on the experimental variograms. This iterative process aims to enhance the precision of the variogram parameters, which are subsequently subjected to validation procedures. The development of Kriging estimation maps is a pivotal step in this study, as it allows for a visual representation of the spatial distribution patterns of different groundwater quality parameters. By employing geostatistical methods, these maps provide insights into the spatial trends and variations in the studied groundwater attributes. Kriging, a spatial interpolation technique, leverages the spatial relationships between sample points to generate predictions across the study area.

Calculating the WQI involves three steps. During the first stage, each of the 10 parameters (pH, EC, TDS, F^-, Cl^-, HCO_3^-, SO_4^{2-}, NO_3^-, Mg^{2+}, Na^+, K^+, Ca^{2+}, and TH) is allocated a weight (wi) according to its relative importance in establishing the overall suitability of water for drinking. The parameters TDS, F^-, Cl^-, NO_3-, and SO_4^{2-} are given a weight of 5, reflecting their significant role in assessing water quality (Vasanthavigar et al., 2010). Conversely, HCO_3^- is assigned the lowest weight of 1. Weights (wi) ranging from 1 to 5 are allocated to parameters such as Mg^{2+}, Ca^{2+}, Na^+, and TH based on their individual contributions in determining water quality.

4.2 Study area

The proposed study will be conducted within the confines of the Jharia Coalfields area, encompassing the geographical coordinates of longitude 86°11′ to 86°27′ E and latitude 23°39′ to 23° 48′ N, spanning a total area of 450 km^2. It falls within the Survey of India topographic map numbers 73 I/1, 73 I/2, 73 I/5, and 73 I/6 on a 1:50,000 scale. In the Jharia coalfield, the groundwater quality parameters are predominantly influenced by the existing topography and geomorphic agents. The Damodar River plays a pivotal role in shaping the drainage patterns of the Jharia coalfield. Flowing along the southern boundary of the coalfield, the Damodar River is guided by the Great Boundary Fault.

The Jharia Coalfield is located within a basin of sedimentary sequences, forming a distinctive sickle-shaped contour, and is surrounded by Precambrian metamorphic rocks. The study area is a component of the Permo-Carboniferous Gondwana basin, characterized by lower Gondwana sedimentary formations overlaying Precambrian metamorphic rocks. Within this region, the groundwater conditions in various litho-stratigraphic units can be categorized into three primary divisions: weathered formations, fractured formations, and abandoned waterlogged regions.

4.3 Geology of the study area

The Jharia coalfield is located in a major sedimentary basin, which is part of Gondwana and is located in the Damodar Valley of eastern India (Fox, 1930; Mehta & Murthy, 1957). According to Chandra (Chandra, 1992), the coalfield is formed like an oval and has an aerial extent of around 458 km^2. The geological formation of the region is distinguished by a prominent coal basin encompassed by regions, including crystalline rocks. The basement rock consists of ancient rock formations from the Dharwar and post-Dharwar periods. On top of this, the lower Gondwana group of sedimentary strata has formed, which includes coal seams and pockets of sandstone. The region is notable for its significant deposits of lower Gondwana coal, which are located within these fields. The Jharia coalfield basin is mostly oriented along a west-northwest to east-southeast axis, with a gentle slope toward the west. This is indicated by the dipping of the Raniganj strata in the direction of the Jamunia River.

Fig. 4.1 depicts the sedimentary sequence that overlays the ancient gneissic basement. This sequence starts with the Talchir Formation, marked by glaciogenic sediments, followed by the successive deposition of fluvial and fluviolacustrine sediments belonging to the Barakar, Barren Measures, and Raniganj Formations. The Barren Measures Formation, which has an estimated thickness of 750 meters, is clearly evident in the center region of the basin. The boundaries of the region are defined by regional faults that separate it from the Barakar Formation in the north and east. The Barren Measures Formation lies beside the Archean basement at the southern boundary, with divergence caused by the southern boundary fault. The formation consists of alternating parts dominated by shale and sandstone, which originated from river flow. Even the thickest layer of shale, specifically referred to as the Tetengabad Shale, shows clear indications of regular influx from rivers. The frequent shifts between river and oceanic sediment types indicate the existence of a lacustrine environment.

The climate of a specific area is affected by the average combination of several climatic factors, including rainfall, temperature, humidity, sunlight, wind, pressure, and cloud cover, among others. Mining activities not only significantly affect the distribution of various vegetation types but are also

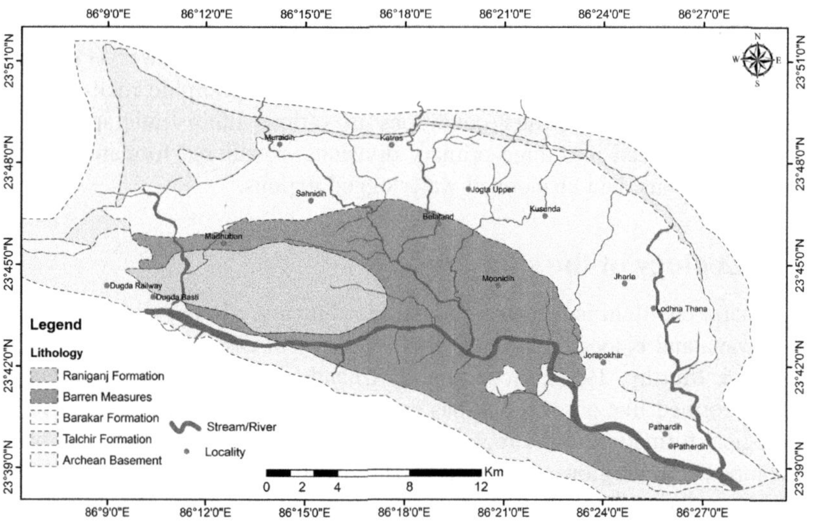

FIGURE 4.1 Geological map of Jharia. Map lines delineate study areas and do not necessarily depict accepted national boundaries. *Modified after From Paul, S., & Chatterjee, R. (2011). Mapping of cleats and fractures as an indicator of in-situ stress orientation, Jharia coalfield, India. International Journal of Coal Geology, 88(2–3), 113–122. https://doi.org/10.1016/j.coal.2011.09.006.*

believed to have detrimental effects on groundwater quality, either through direct or indirect exacerbation or redistribution. During a single year, the Jharia Coalfield is subject to the occurrence of four distinct seasons. The winter season begins at the end of November and continues until the beginning of March; the summer season begins at the end of March and continues until May; the rainy season begins in June and continues until September or early October; and the southwest monsoon season begins in October and continues until November.

4.4 Materials and methods

A total of twenty-nine (29) sampling locations (dug wells) were selected for groundwater quality from different mining areas of Jharia Coalfield. The groundwater of the Jharia coal field is neutral to slightly alkaline in nature. Organizing groundwater quality data in a structured database is essential for effective analysis. This typically involves collecting, cleaning, and formatting the data to make it suitable for statistical and geostatistical techniques. Statistical methods have been used to describe and summarize the data. Descriptive statistics, such as mean, standard deviation, variance, skewness and kurtosis have been analysed for understanding of the water quality parameters.

Geostatistics focuses on the spatial distribution and correlation of data points. Geostatistical methods like ordinary Kriging (OK) have been employed to interpolate values between sampled locations and create spatial distribution maps, providing insights into the spatial variability of water quality parameters across the Jharia Coalfield region.

The groundwater quality zones are delineated on maps and developed by geostatistical estimation employing the OK technique for all cations and anions. Then, spatial distribution maps are developed, providing insights into the spatial variability and helping to identify areas that may require attention in terms of water quality management or remediation efforts.

4.4.1 Geostatistical study

In this study, the spatial distribution of key hydrogeochemical parameters associated with groundwater quality was determined using OK and semi-variography analysis. These approaches are widely recognized and have been extensively employed in hydrogeological studies.

Semi-variograms are a key component of geostatistical analysis. They are used to model the spatial autocorrelation, or variability, in the dataset. Semi-variogram modeling involves fitting a mathematical function to the experimental semi-variogram derived from the data. The semi-variogram model provides insights into the range, sill, and nugget effect, which are essential parameters for understanding the spatial structure of the dataset. The semi-variogram function (h) can be expressed mathematically as follows: Z(xi) represents the value of a sample taken at location xi, and Z(xi + h) represents the value at a distance of "h" from the xi point. Semi-variograms are generated by establishing a lag interval, which represents a constantly increasing distance between a sample value and the remaining values (Armstrong, 1998; Isaak, 1989). The semi-variance function (h) can be expressed mathematically as follows:

$$\gamma(h) = 1/2\ N[Z(x_i) - Z(x_i + h)]^2$$

OK is a geostatistical interpolation method that provides optimal estimates of values at unsampled locations based on the weighted average of neighboring sample points. It takes into account the spatial correlation or variability of the data. OK is particularly suitable for situations where data points are unevenly spaced and exhibit spatial autocorrelation, as is often the case in groundwater quality studies. By analyzing the spatial arrangement of sample points and their variabilities, OK produces estimation maps that offer a detailed representation of the parameter's spatial distribution.

Kriging is a geostatistical method that uses the parameters derived from a semi-variogram to estimate the values of a regionalized variable. Let G* be the kriged estimate of the grade G average for the samples with g1, g2, g3,... gn values. Assume that each value is assigned a weight of a1, a2, a3,... an so that $G^* = \sum ai\ gi$ and $\sum ai = 1$. As a result, the estimation obtains objectivity; for a large number of estimated values, the mean error is zero and the estimated variance is at its minimum. The formula for the Kriging variance is k2 = (gi − G*). 2. The ideal solution of the Kriging system uses a function known as the Lagrange multiplier (λ) to minimize the Kriging variance. Kriging for point estimates is referred to as "point kriging," while Kriging for estimating a block of surface is referred to as "block kriging." The geostatistical modeling study is particularly effective for groundwater quality studies because it considers both the available data points and the spatial relationships between them. This approach allows for the creation of spatial distribution maps that not only provide estimates of groundwater quality parameters at unsampled locations but also incorporate the underlying spatial variability and autocorrelation. As a result, these techniques are valuable tools for assessing and managing groundwater resources and quality.

4.5 Result and discussion

4.5.1 Statistical analysis

A statistical analysis has been carried out to evaluate the water assessment of the groundwater in the higher catchments of the Jharia coalfield, emphasizing its hydrogeochemical properties. The water sample has undergone analysis for EC, pH, TDS, alkalinity, acidity, major cations (Ca^{2+}, Mg^{2+}, Na^{+}, and K^+), and major anions (Cl^-, F^-, SO_4^{2-}, NO_3^-, and HCO_3^-). The objective of this analysis is to measure and determine the suitability of the groundwater for domestic use by comparing its various parameters with the groundwater standards specified in IS: 10500 (World Health Organization, 2017). Representative twenty-nine samples from groundwater are collected from different dug wells in mining areas. Statistical analysis of the parameters involves estimating the mean, standard deviation, variance, skewness, and kurtosis. pH value ranges from 6.50 to 8.30, which means 7.62 is in the permissible limit; EC ranges from 630 to 1590 (μS/cm), whereas in some locations it occurs beyond the permissible limit presented in Table 4.1. Sulfate was the predominant anion among the main ones, accounting for an average of 54% of the total anions. Bicarbonate is the second most prevalent negatively charged ion, accounting for approximately 37% of the total anions. The presence of chlorides as ions was less significant, contributing to only 8% of the total anions. Nitrate is the least prevalent anion among all the anions. The relative abundance of anions on the surface was determined to be in the following order: $SO_4^{2-} > HCO_3^- > Cl^- > NO_3^-$. Calcium was the most common cation, accounting for approximately 46% of all cations. Sodium and magnesium ions have a subordinate significance, constituting around 25% and 23% of the overall cations, respectively. Potassium had the lowest level of dominance among the cations and represented 6% of the overall cation count. The cation abundance followed the order: $Ca^{2+} > Na^+ > Mg^{2+} > K^+$.

Plotting the analytical results on the Piper (1944) trilinear diagram provides an additional evaluation of the geochemical characteristics of dissolved ions in water as well as their relationship to each other. The Piper diagram is a valuable tool for assessing the geochemical characteristics of the groundwater in the studied area. The Piper diagram is a deliberately designed diagram that comprises two triangles positioned at the lower left and lower right corners. These triangles depict the proportionate distribution of cations and anions. In between these triangles, there is a diamond-shaped field that reflects the composition of water in terms of both cations and anions, as shown in Fig. 4.2. The data plot on the trilinear diagram indicates that most water samples are not concentrated in any particular zone, but a smaller number of samples are found in the Ca^{2+} and Mg^{2+} zones in the cation facies and the HCO_3^- and SO_4^{2-} zone in the anion facies. The groundwater of the Jharia coalfield was analyzed, and the results showed that alkaline earth elements ($Ca^{2+} + Mg^{2+}$) were more prevalent than alkalies ($Na^+ + K^+$). Additionally, strong acids ($SO_4^{2-} + Cl^-$) were found to be more abundant than weak acids (HCO_3^-), and there was a presence of non-carbonate hardness. The majority of samples from the Jharia coalfield exhibit an abundance of alkaline earth elements ($Ca^{2+} + Mg^{2+}$) over alkalies ($Na^+ + K^+$). Approximately 31% of the samples exhibit intermediate chemical characteristics, as indicated by their displayed points falling inside the

TABLE 4.1 Descriptive statistics of groundwater quality.

Groundwater quality (29 sample locations)	Min	Max	Mean	Standard deviation	Variance	Skewness	Kurtosis	Permissible limit
pH	6.50	8.30	7.62	0.37	0.14	0.54	1.64	7.00–8.50
EC (μS/cm)	630.00	1590.00	1080.13	274.70	75,460.55	0.28	− 1.01	750
TDS (mg/L)	532.00	1377.00	900.79	233.97	54,742.67	0.15	− 0.80	500
F − (mg/L)	0.46	1.87	1.06	0.41	0.17	0.55	− 0.94	1.00
Cl^- (mg/L)	54.00	254.00	116.25	45.57	2077.13	1.09	1.43	250
HCO_3^- (mg/L)	192.10	451.00	266.24	65.29	4263.73	1.45	1.91	200
SO_4^{2-} (mg/L)	89.70	397.00	231.87	100.80	10,162.14	0.01	− 1.22	200
NO_3^- (mg/L)	2.10	46.00	12.64	10.27	105.60	1.74	3.28	50
Ca^{2+} (mg/L)	35.10	205.00	105.65	43.42	1886.04	0.66	− 0.04	75
Mg^{2+} (mg/L)	37.60	137.70	68.53	27.11	735.18	1.38	1.32	30
Na^+ (mg/L)	27.90	189.00	78.98	40.62	1650.27	1.21	1.01	200
K^+ (mg/L)	3.90	38.90	16.48	9.71	94.35	0.82	− 0.15	100
TH (mg/L)	294.00	917.00	545.82	173.40	300,070.58	0.58	− 0.58	500

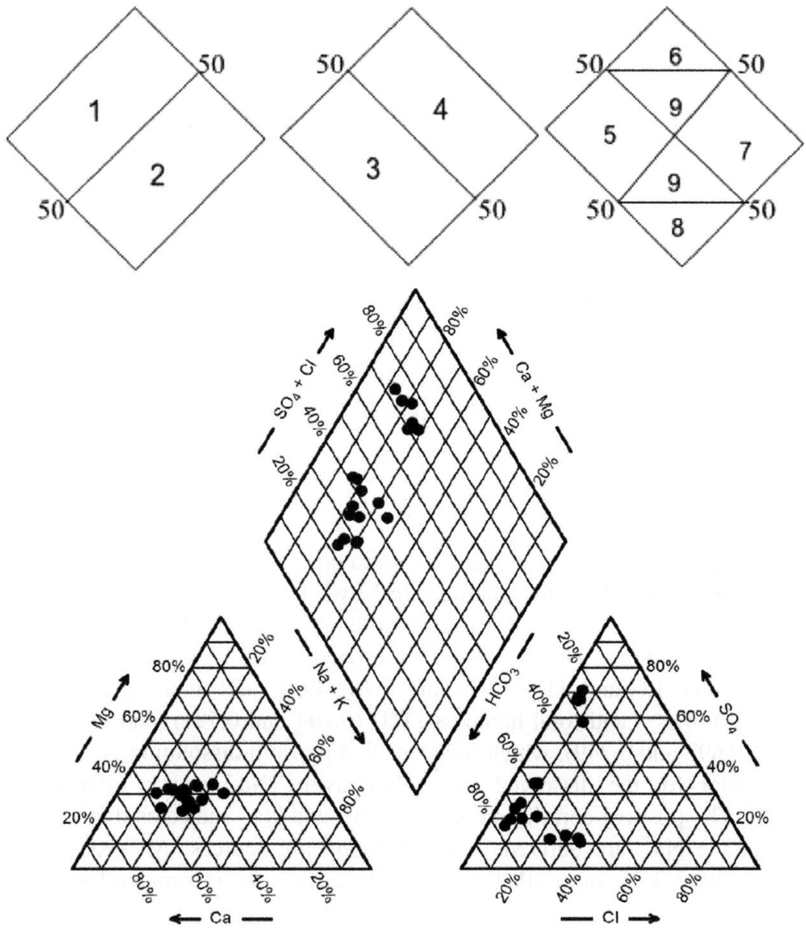

FIGURE 4.2 Hydrogeochemical facies of groundwater.

central region of the diamond-shaped field. Alkalies surpass alkaline earth by a mere 7%. The present study relies on the facies mapping approach to identify two primary types of water in the study area: Ca–Mg SO_4^{2-} and Ca–Mg-HCO_3. These types are determined based on the concentration of key ions, except for a few samples of Na–HCO_3. Classification of the water diagram tabulated in Table 4.2.

4.5.2 Statistical correlation of groundwater

Table 4.3 shows the Pearson correlation of the groundwater in the study area. Correlation analysis quantifies the degree of association between selected variables. A correlation value closer to +1 or −1 indicates a strong

TABLE 4.2 Hydrochemical facies.

Fields	Characteristics of groundwater
1	Alkaline earth ($Ca^+ + Mg^+$) exceed alkalies ($Na^+ + K^+$)
2	Alkalies exceed alkaline earth
3	Weak acids (CO3 + HCO3) exceed strong acids ($SO_4^{2-} + Cl^-$)
4	Strong acids exceed weak acid
5	Carbonate hardness (secondary alkalinity)
6	Non-carbonate hardness (secondary salinity)
7	Non-carbonate alkali (primary salinity)
8	Carbonate alkali (primary alkalinity)
9	No one cation-anion pair

linear relationship between the two variables. This investigation aims to determine the nature of the relationship between the measures of groundwater water quality. There is a strong correlation between TDS and EC with Mg^{2+}, Ca^{2+}, SO_4^{2-}, HCO_3^-, and TH, respectively. There is a good correlation between EC and TDS (0.98 and 0.96), Ca^{2+} and total hardness (0.78 and 0.79), Mg^{2+} and total hardness (TH) (0.794 and 0.791), and SO_4^{2-} and Ca^{2+} (0.697 and 0.749), which indicates that gypsum dissolution is the major contributor for the dissolved ions, respectively, Na^+ and k^+ (0.664 and 0.700). The positive correlation between Cl^- and Na^+ indicates that Cl^- and part of the Na^+ are derived from anthropogenic sources. All groundwater samples have TDS values that significantly surpass the recommended threshold of 500 mg/L.

4.5.3 Geostatistical analysis

The spherical semi-variogram model is one of the most commonly used variogram models. It is a differentiable variogram, which means it is smooth and continuous. This model assumes that the spatial correlation between data points levels off and becomes constant after a certain distance. The best-fitted theoretical spherical semi-variogram models have been generated by selecting trial and error parameters. The nugget, continuity, and range of best-fitted spherical models have been given in Table 4.4. The range (m) of parameters varies from 5100 to 7700 m. The nugget-to-sill ratio is less than 25%, which indicates that there is relatively little measurement error or microscale variability and that the data exhibit strong spatial dependence (Bodrud-Doza et al., 2016; Shi et al., 2007). This suggests that nearby data points are highly correlated. The nugget-to-sill ratio of EC, pH, TDS, Cl^-,

TABLE 4.3 Correlation matrix of groundwater parameter of the study area.

	pH	EC	TDS	F^-	Cl^-	HCO_3^-	SO_4^{2-}	NO_3^-	Ca^{2+}	Mg^{2+}	Na^+	K^+	TH
pH	1	0.143	0.167	0.135	0.286	0.400*	-0.091	-0.096	0.030	0.228	0.004	0.256	0.165
EC		1	0.981**	0.895**	0.911**	0.405*	0.809**	0.558**	0.763**	0.549**	0.527**	0.719**	0.831**
TDS			1	0.856**	0.904**	0.443*	0.811**	0.543**	0.755**	0.567**	0.554**	0.740**	0.837**
F^-				1	0.846**	0.261	0.738**	0.388*	0.716**	0.450*	0.445*	0.610**	0.737**
Cl^-					1	0.372*	0.679**	0.465*	0.664**	0.605**	0.438*	0.694**	0.804**
HCO_3^-						1	-0.112	0.198	0.164	0.247	0.312	0.530**	0.262
SO_4^{2-}							1	0.407*	0.697**	0.541**	0.355	0.417*	0.783**
NO_3^-								1	0.510**	0.171	0.316	0.379*	0.430*
Ca^{2+}									1	0.243	0.162	0.433*	0.782**
Mg^{2+}										1	-0.121	0.238	0.794**
Na^+											1	0.664**	0.024
K^+												1	0.424*
TH													1

**A moderate correlation between elements in groundwater.*
***A good correlation having a value greater than 0.5.*

TABLE 4.4 Summary of semi-variogram analysis.

Parameters	Nugget (C_0)	Continuity (C)	Range (m)	Sill ($C_0 + C$)	(C_0/ $C_0 + C$)
pH	0.048	0.0996	6400	0.14	0.32
EC	37,896	36,702.00	7300	74,598.00	0.50
TDS (mg/L)	7300	25,000.00	7100	56,400.00	0.55
F^-	0.085	0.064	7300	0.15	0.57
Cl^- (mg/L)	1061.00	904.00	7100	1965.00	0.54
HCO_3^- (mg/L)	1814.609	2407.09	6330	4221.70	0.43
SO_4^{2-} (mg/L)	6890.00	2760.00	7100	9650.00	0.71
NO_3^- (mg/L)	47.21	41.08	7100	88.29	0.53
Ca^{2+} (mg/L)	404.53	1377.47	7100	1782.00	0.22
Mg^{2+} (mg/L)	386.037	422.718	5100	808.76	0.47
Na^+ (mg/L)	523.82	1099.87	7150	1623.69	0.32
K^+ (mg/L)	64.70	29.30	7700	94.00	0.68
TH (mg/L)	16,966.00	13,330.00	5570	30,296.00	0.56

F^-, HCO_3^-, SO_4^{2-}, NO_3^-, Mg^{2+}, Na^+, K^+, and TH fall within the range of 25%–75%, which suggests that there is a balance between microscale variability and larger-scale spatial structures in the data. This is considered moderate spatial dependence, indicating that both short- and long-range correlations are present in the data. Mathematical spherical semi-variograms with fitted models of pH, EC, TDS, F^-, Cl^-, and HCO_3^- are displayed in Fig. 4.3, and SO_4^{2-}, NO_3^-, Ca^{2+}, Mg^{2+}, Na^+, K^+, and TH are shown in Fig. 4.4. Spherical models fitted to the experimental semi-variograms of all cations and anions are numerically expressed in the following equations:

$$\gamma(h) = 0.048 + 0.0996\left[(1.5) \times (h/6400) - 0.5 \times (h/6400)^3\right], \text{for pH} \quad (4.1)$$

$$\gamma(h) = 37896 + 36702\left[(1.5) \times (h/7300) - 0.5 \times (h/7300)^3\right], \text{for EC} \quad (4.2)$$

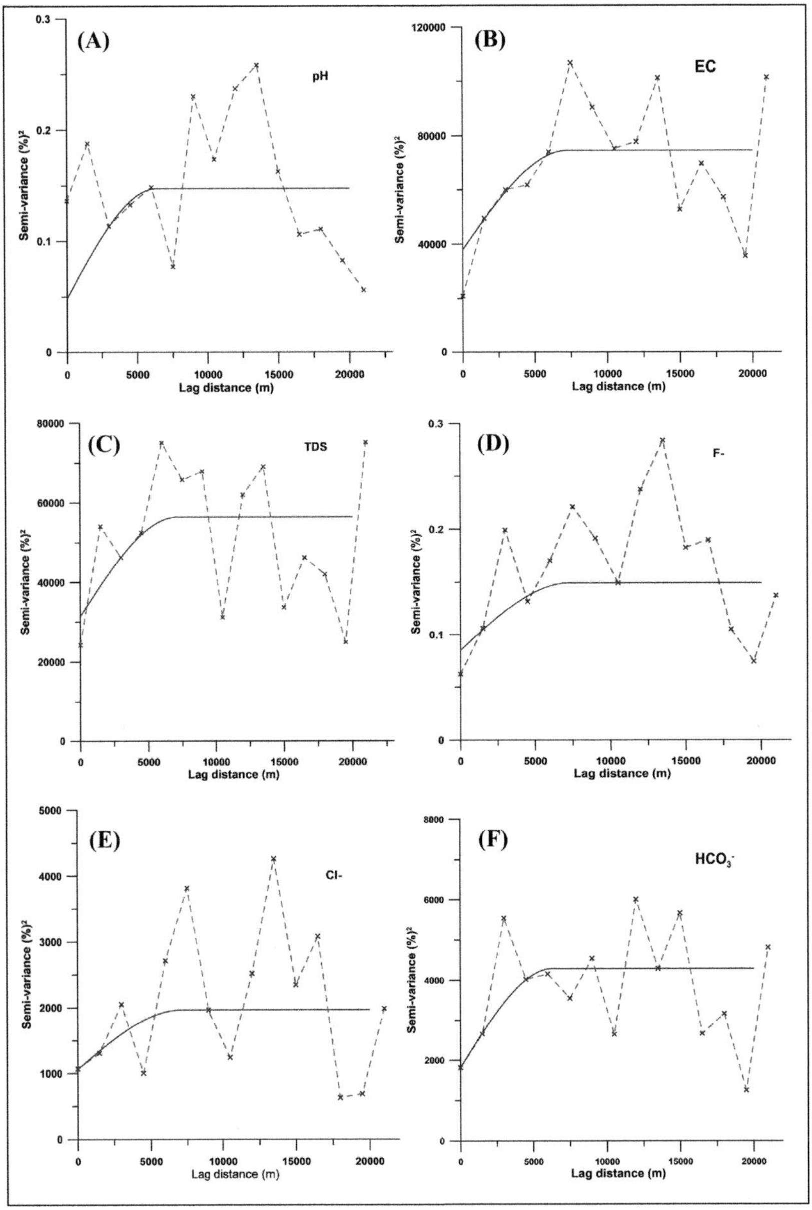

FIGURE 4.3 Spherical semi-variogram model of cations and anions. (A – F) Semi-variogram model of pH, EC, TDS, F^-, Cl^-, HCO_3^-.

$$\gamma(h) = 7300 + 25000\left[(1.5) \times (h/7100) - 0.5 \times (h/7100)^3\right], \text{for TDS} \quad (4.3)$$

$$\gamma(h) = 0.085 + 0.064\left[(1.5) \times (h/7300) - 0.5 \times (h/7300)^3\right], \text{for } F^- \quad (4.4)$$

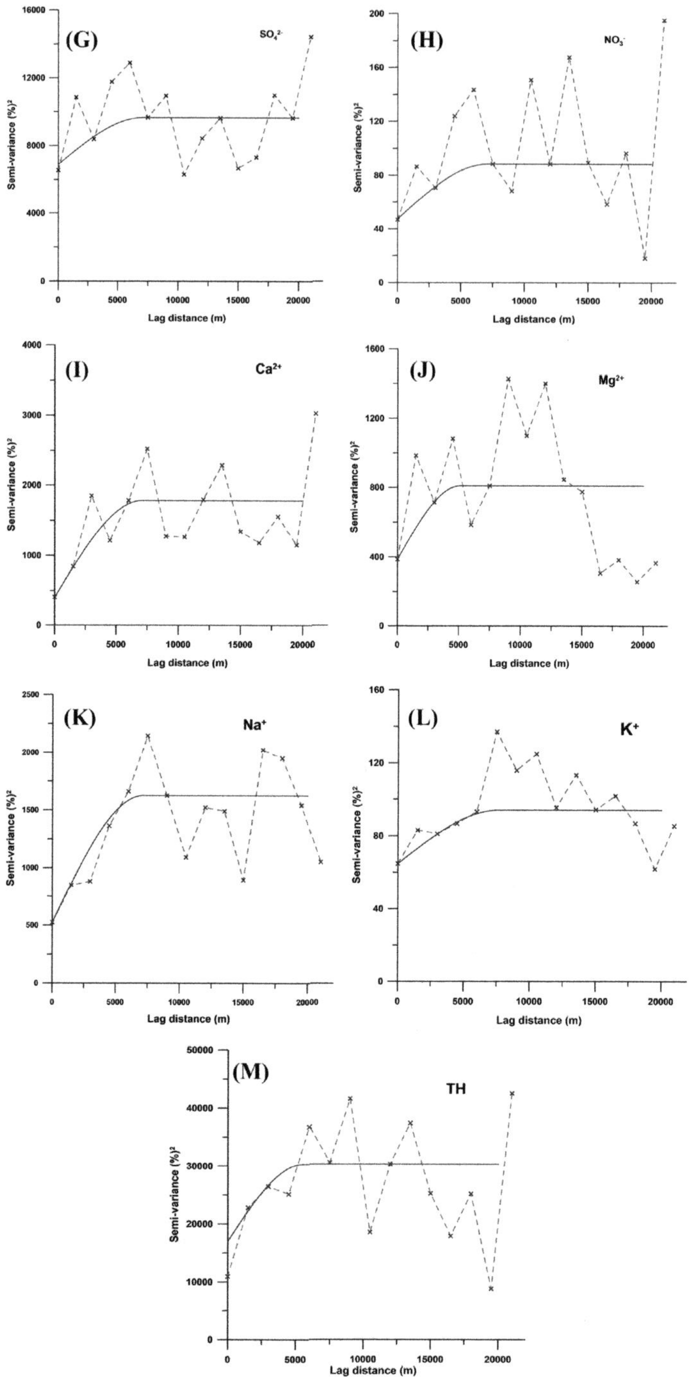

FIGURE 4.4 Spherical semi-variogram model of cations and anions. (G – M) Semi-variogram model of SO_4^{2-}, NO_3^-, Ca^{2+}, Mg^{2+}, Na^+, K^+, TH.

$$\gamma(h) = 1061.00 + 904.00\left[(1.5)\times\left(h/7100\right) - 0.5\times\left(h/7100\right)^3\right], \text{for } Cl^- \tag{4.5}$$

$$\gamma(h) = 1814.60 + 2407.09\left[(1.5)\times\left(h/6330\right) - 0.5\times\left(h/6330\right)^3\right], \text{for } HCO_3^- \tag{4.6}$$

$$\gamma(h) = 6890 + 2760\left[(1.5)\times\left(h/7100\right) - 0.5\times\left(h/7100\right)^3\right], \text{for } SO_4^{2-} \tag{4.7}$$

$$\gamma(h) = 47.21 + 41.08\left[(1.5)\times\left(h/7100\right) - 0.5\times\left(h/7100\right)^3\right], \text{for } NO_3^- \tag{4.8}$$

$$\gamma(h) = 404.53 + 1377.47\left[(1.5)\times\left(h/7100\right) - 0.5\times\left(h/7100\right)^3\right], \text{for } Ca^{2+} \tag{4.9}$$

$$\gamma(h) = 386.03 + 422.71\left[(1.5)\times\left(h/5100\right) - 0.5\times\left(h/5100\right)^3\right], \text{for } Mg^{2+} \tag{4.10}$$

$$\gamma(h) = 523.82 + 1099.87\left[(1.5)\times\left(h/7150\right) - 0.5\times\left(h/7150\right)^3\right], \text{for } Na^+ \tag{4.11}$$

$$\gamma(h) = 64.70 + 29.30\left[(1.5)\times\left(h/7700\right) - 0.5\times\left(h/7700\right)^3\right], \text{for } K^+ \tag{4.12}$$

$$\gamma(h) = 16966.00 + 13330.00\left[(1.5)\times\left(h/5570\right) - 0.5\times\left(h/5570\right)^3\right], \text{for TH} \tag{4.13}$$

For the estimation of different groundwater quality parameters, OK has been employed. The spatial distribution maps of different cations and anions are generated by employing OK, showing variation in different regions of the mining region. The method considers both the data values at nearby points and their spatial correlation. It provides the best linearly unbiased estimates for the unsampled points. The generated maps can be interpreted to identify areas with different levels of cations and anions. These maps can help in understanding the spatial patterns and variations in groundwater quality within the mining region. Spatial distribution maps of pH, EC, TDS, F^-, Cl^-, HCO_3^-, SO_4^{2-}, NO_3^-, Mg^{2+}, and TH are displayed in the Figs. 4.5–4.15. As shown in Fig. 4.5, the spatial distribution map pH is not a major issue in the coalfield region. It is within the permissible limit of WHO standards. It varies from 7.02 to 7.77. In the kriged spatial distribution, it is observed that EC ranges from 830.04 to 1328.90. Towards the south of Katras and north of Belatand high EC value of 1260.44 to 1328.90 are shown

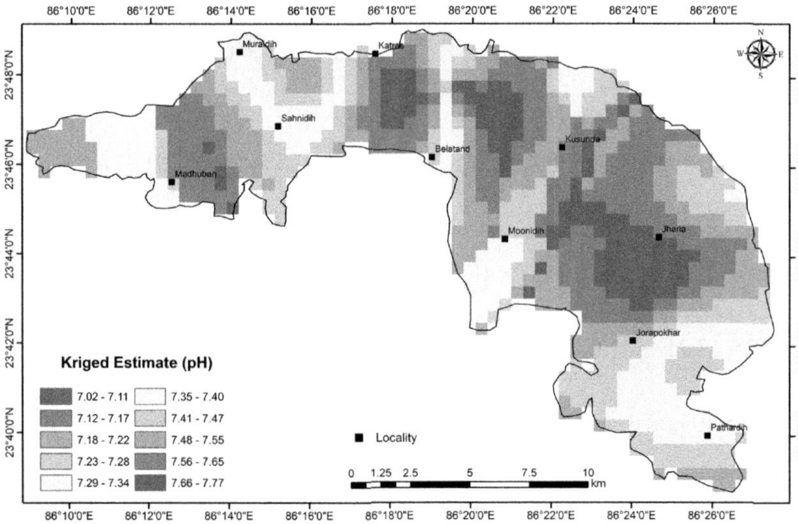

FIGURE 4.5 Kriged estimate map of pH. Map lines delineate study areas and do not necessarily depict accepted national boundaries.

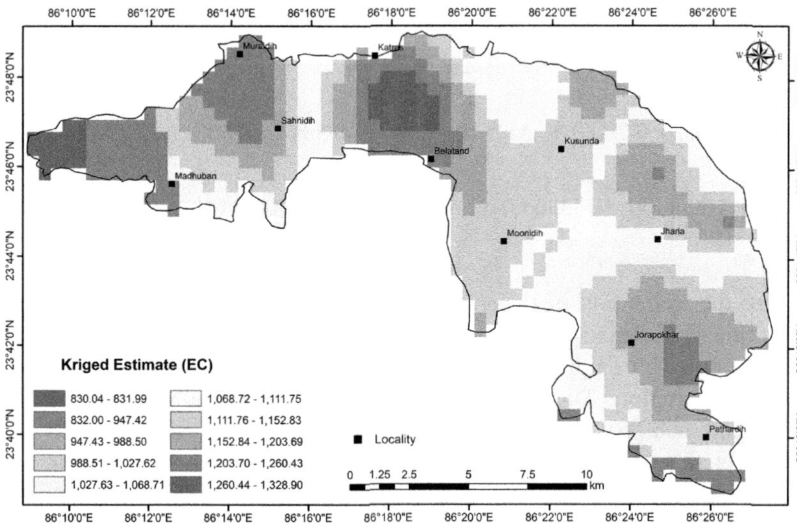

FIGURE 4.6 kriged estimate map of EC. Map lines delineate study areas and do not necessarily depict accepted national boundaries.

in Fig. 4.6. This high EC may be attributed to extensive mining and excavation activities in those areas, which can introduce various minerals and ions into the groundwater. TDS is estimated to be high in the Katras, Moondih, and Belatand regions. The value ranges from 893.36 to 1105.08, whereas a slightly lower value is observed in the N-E and S-E parts of the Jharia

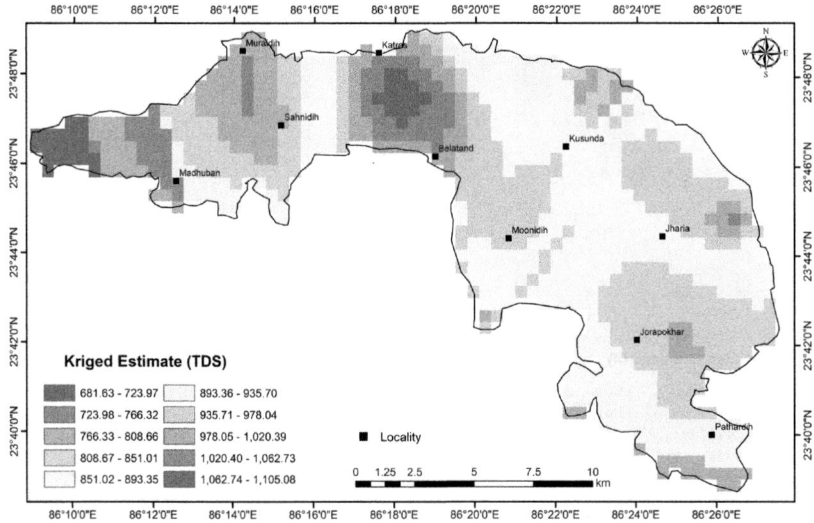

FIGURE 4.7 Kriged estimate map of TDS. Map lines delineate study areas and do not necessarily depict accepted national boundaries.

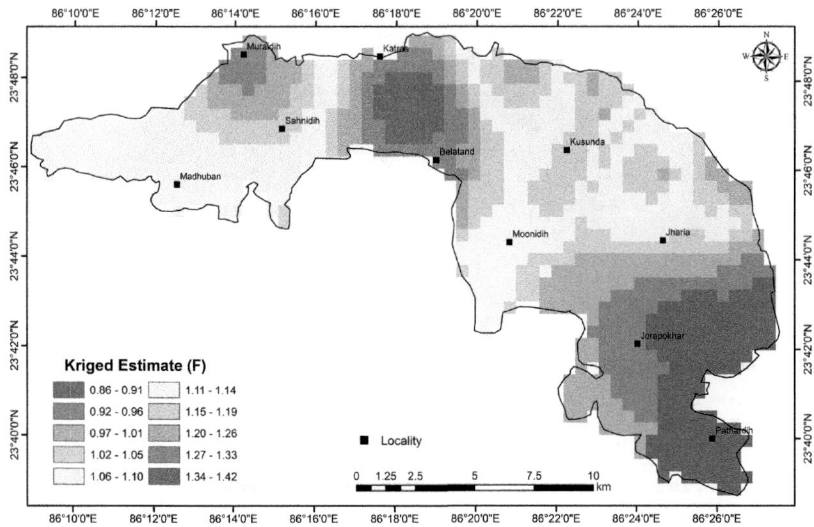

FIGURE 4.8 Kriged estimate map of F^-. Map lines delineate study areas and do not necessarily depict accepted national boundaries.

(Fig. 4.7). Higher TDS in the mentioned areas may also be linked with mining and associated activities. Fluoride (F^-) content is evaluated high in some middle parts of the Jharia near Katras and Belatand and is observed low toward south-east regions (Fig. 4.8). The chloride (Cl^-) content is a less dominant and effective parameter in this zone; it ranges from 90.63 to

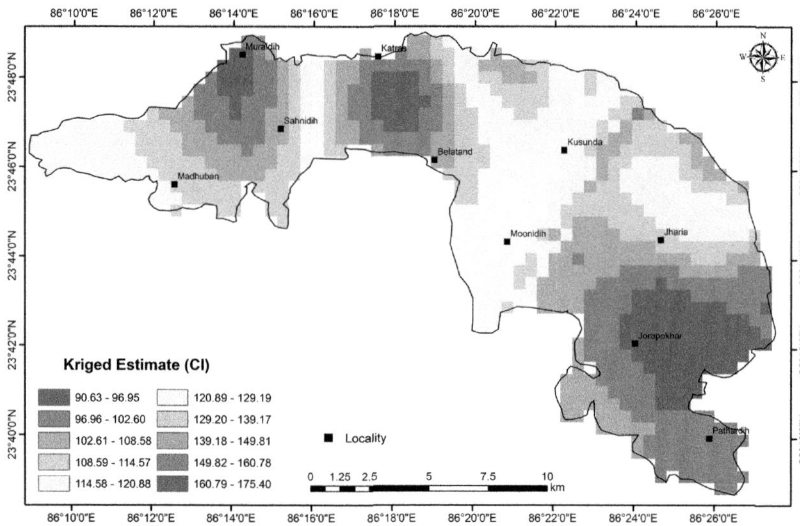

FIGURE 4.9 Kriged estimate map of Cl^-. Map lines delineate study areas and do not necessarily depict accepted national boundaries.

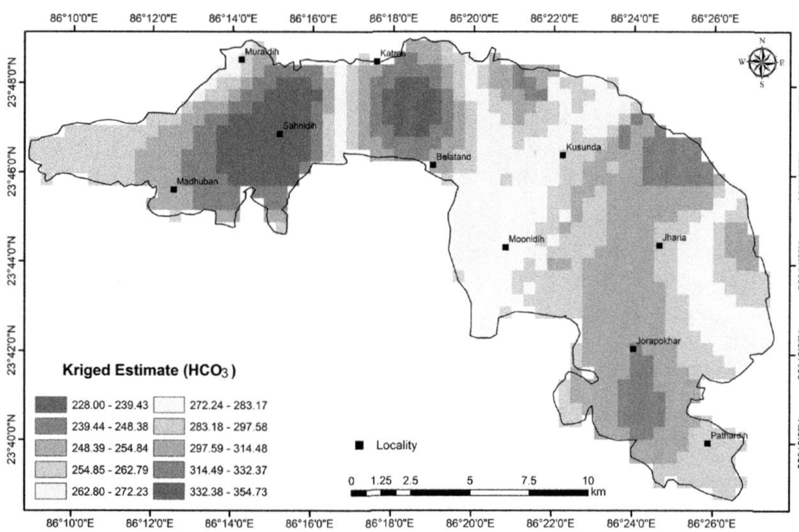

FIGURE 4.10 Kriged estimate map of HCO_3^-. Map lines delineate study areas and do not necessarily depict accepted national boundaries.

175.40 mg/L (Fig. 4.9). HCO_3^- ranges from 228.00 to 354.73 throughout the study area, and it is estimated to be higher nearby Katras, Belatand, Kusunda, and toward the east of Jharia, as shown in the kriged estimated map, Fig. 4.10. The presence of bicarbonate ions (HCO_3^-) in coalfield groundwater can be caused by a number of hydrogeological and geological

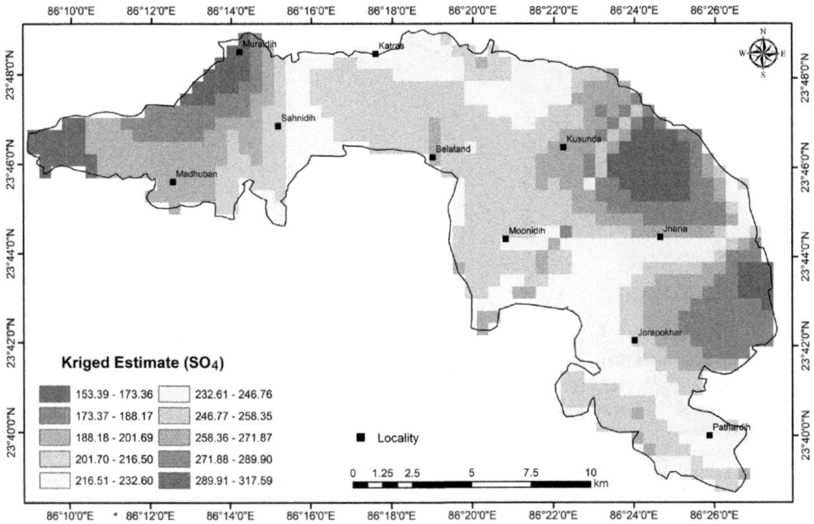

FIGURE 4.11 Kriged estimate map of SO_4^{2-}. Map lines delineate study areas and do not necessarily depict accepted national boundaries.

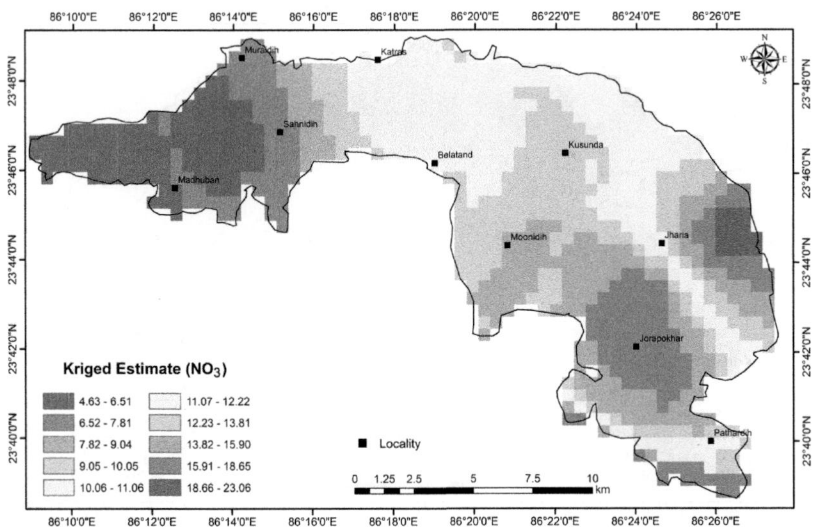

FIGURE 4.12 Kriged estimate map of NO_3^-. Map lines delineate study areas and do not necessarily depict accepted national boundaries.

phenomena. Bicarbonate, a frequent anion in natural waters, may occur from the reaction of carbon dioxide (CO_2) with water or from the dissolution of minerals, especially carbonate minerals. There could be several reasons why groundwater in coalfield areas contains bicarbonate ions. SO_4^{2-} ranges from 153.39 to 317.59. According to estimates, oxidative weathering of sulfide-

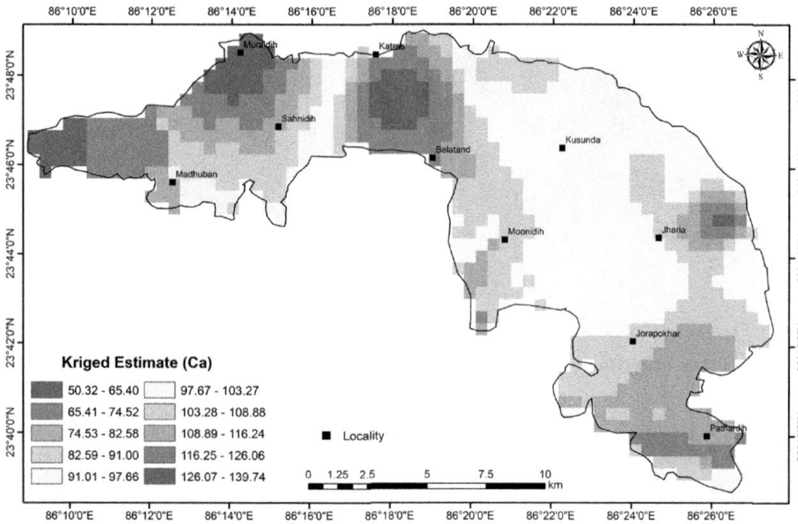

FIGURE 4.13 Kriged estimate map of Ca^{2+}. Map lines delineate study areas and do not necessarily depict accepted national boundaries.

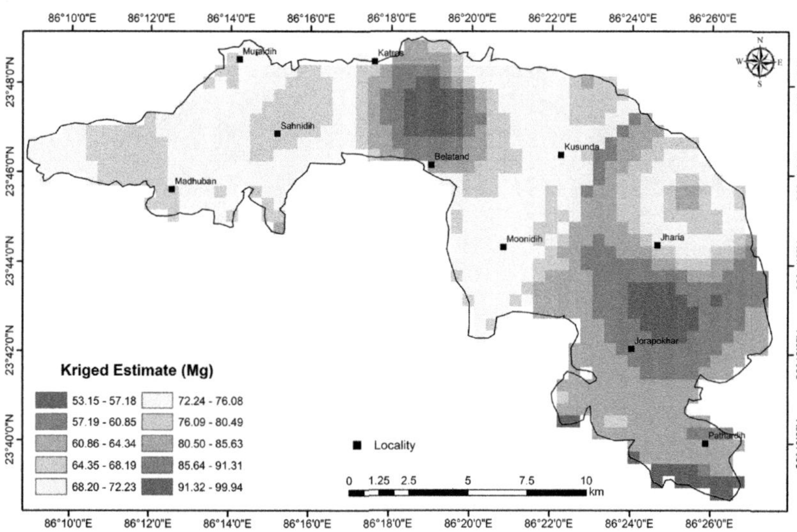

FIGURE 4.14 Kriged estimate map of Mg^{2+}. Map lines delineate study areas and do not necessarily depict accepted national boundaries.

bearing minerals like pyrite (FeS_2) may be the cause of the substantial range of 289.91−317.59 observed toward the N-E side of Jharia, as displayed in Fig. 4.11. The estimated NO_3^- is less effective in mining regions of Jharia and surrounding areas, which range from 4.63 to 23.06 (Fig. 4.12). Kriged

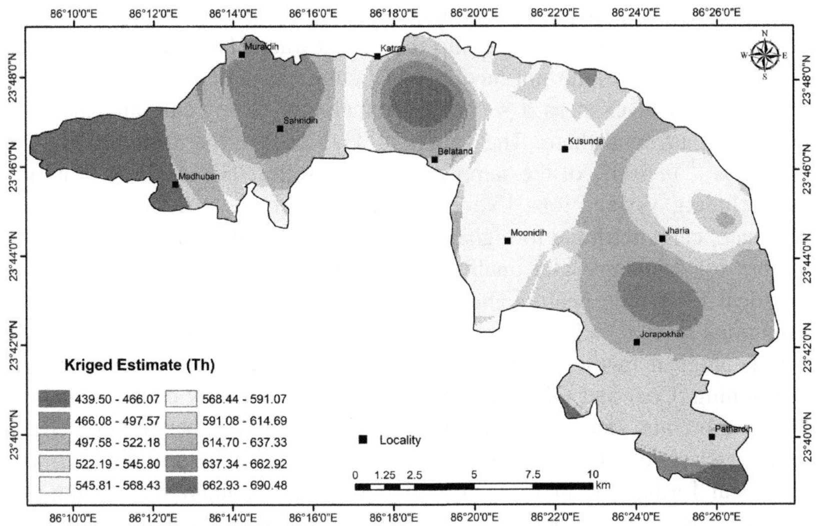

FIGURE 4.15 Kriged estimate map of TH. Map lines delineate study areas and do not necessarily depict accepted national boundaries.

estimated Ca^{2+} is observed to be very high in and surrounding regions of Katras and Belantand villages. It also ranges from 97.67 to 126.06 mg/L in the middle and south-eastern parts of Jharia, as displayed in Fig. 4.13. Mg^{2+} is also observed high between Katras and Belatand village. It ranges from 53.15 to 99.94. Higher levels of Mg^{2+} could be indicative of certain geological formations or weathering of minerals that release magnesium ions into the groundwater (Fig. 4.14). Among major cations, both Na^+ and K^+ concentrations are relatively low, which can be indicative of freshwater or groundwater that has not been significantly impacted by saline or brackish water sources. Na^+ ranges from 46.12 to 127.90, whereas K^+ varies from 7.21 to 21.91. TH ranges from 439.50 to 690.48 toward the northern part of the study area (Fig. 4.15). Estimated TH in water is a measure of the concentration of divalent metallic cations, primarily calcium (Ca^{2+}) and magnesium (Mg^{2+}), that are present in the form of salts.

4.6 Conclusion

Geostatistics-based Geospatial modeling and hydro-geochemical characterization of groundwater water quality study has been carried out to study the variation in groundwater quality in the Jharia mining region. Water samples had been collected in the Jharia coalfield at several locations. Twenty-nine groundwater samples had been collected from various areas inside the coalfield. The nature of the groundwater samples ranges from slightly acidic to alkaline. The high EC value identified near Katras and Belatand region may be attributed to extensive mining and excavation activities, which can introduce various minerals and ions into the groundwater. High SO_4^{2-} estimated in some regions, despite not having much sulfur in the coalfield area, pyrite (FeS_2) is said to exist as a secondary mineral in these coals and the sediments that surround them. The values of TDS exceed the desirable limit of 500 mg L^{-1} in 100% of the samples of groundwater. The leaching of minerals, including various ions like calcium, magnesium, sodium, sulfate, and carbonate, can contribute to higher TDS levels. High fluoride levels can be a concern for drinking water quality, and understanding its spatial distribution is critical for public health. The observation of higher calcium ($Ca^{2+)}$ concentrations in coal field regions can be attributed to various geological and hydrogeological factors, as well as anthropogenic activities associated with coal mining. Coal and OB layers in coal field regions may contain calcium-bearing minerals. During the mining process, interactions between water and these materials can lead to the release of calcium into the groundwater. The geological formations in the coalfield region may contain magnesium-rich minerals, and the weathering or dissolution of these minerals can release magnesium into the groundwater. Natural waters frequently include the bicarbonate ion (HCO^{3-}), which can have both beneficial and detrimental effects depending on the circumstances. Some key considerations regarding the impact of bicarbonate in various scenarios include water alkalinity, mineral precipitation/dissolution, and the impact on drinking water quality. A variety of geological and hydrogeological elements, as well as human activities associated with coal mining, can have an impact on the TH of the water in coalfield locations.

By employing OK and geostatistical methods, it makes informed decisions based on the spatial characteristics of groundwater quality in the Jharia coalfield. Overall, kriged spatial distribution maps provide a valuable foundation for understanding groundwater quality in the Jharia coalfield and taking appropriate measures to manage and protect this vital resource. The spatial distribution maps offer valuable insights into the distribution of water quality parameters in the study area. They can guide decision-making processes related to groundwater resource management, environmental protection, and remediation efforts. This approach helps in understanding the heterogeneity of water quality and can be instrumental in addressing environmental and public health concerns associated with mining and related activities in the region.

References

Adhikary, P. P., Chandrasekharan, H., Chakraborty, D., & Kamble, K. (2010). Assessment of groundwater pollution in West Delhi, India using geostatistical approach. *Environmental Monitoring and Assessment*, *167*(1–4), 599–615. Available from https://doi.org/10.1007/s10661-009-1076-5.

Armstrong, A. M. (1998). *Basic linear geostatistics*. Springer.

Bodrud-Doza, M., Islam, A. R. M. T., Ahmed, F., Das, S., Saha, N., & Rahman, M. S. (2016). Characterization of groundwater quality using water evaluation indices, multivariate statistics and geostatistics in central Bangladesh. *Water Science*, *30*(1), 19–40. Available from https://doi.org/10.1016/j.wsj.2016.05.001, https://www.tandfonline.com/toc/twas20/current.

Chandra, D. (1992). *Jharia coalfields*. Bangalore: Geological Society of India.

Chatterjee, R., Tarafder, G., & Paul, S. (2010). Groundwater quality assessment of Dhanbad district, Jharkhand, India. *Bulletin of Engineering Geology and the Environment*, *69*(1), 137–141. Available from https://doi.org/10.1007/s10064-009-0234-x.

Delhomme, J. P. (1979). Spatial variability and uncertainty in groundwater flow parameters: A geostatistical approach. *Water Resources Research*, *15*(2), 269–280. Available from https://doi.org/10.1029/WR015i002p00269.

Fox, C. S. (1930). The Jharia coalfield. *Memoir Geological Survey of India*, *56*.

Isaak, M. (1989). Srivastava *An introduction to applied geostatistics*. Oxford University Press.

Mehta, D. R., & Murthy, B. R. (1957). A revision of the geology and coal resources of Jharia Coalfield. Memoir Geological Survey of India. GSI. 84.

Pandey, J., Kumar, D., Singh, V. K., & Mohalik, N. K. (2016). Environmental and socio-economic impacts of fire in Jharia coalfield, Jharkhand, India: An appraisal. *Current Science*, *110*(9), 1639–1650. Available from https://doi.org/10.18520/cs/v110/i9/1639-1650, http://www.currentscience.ac.in/Volumes/110/09/1639.pdf.

Shi, Jiachun, Wang, Haizhen, Xu, Jianming, Wu, Jianjun, Liu, Xingmei, Zhu, Haiping, & Yu, Chunlan (2007). Spatial distribution of heavy metals in soils: A case study of Changxing, China. *Environmental Geology*, *52*(1), 1–10. Available from https://doi.org/10.1007/s00254-006-0443-6.

Singh, P. K., Tiwari, A. K., & Mahato, M. K. (2013). Qualitative assessment of surface water of West Bokaro coalfield, Jharkhand by using water quality index method. *International Journal of ChemTech Research*, *5*(5), 2351–2356. Available from http://www.sphinxsai.com/2013/JulySept13/chPDF/CT = 42%282351-2356%29JS13.pdf.

Singh, T. N. (1987). *Geo environment of Indian coalfields* (1st edn.). Sharda Prakasan Dhanbad.

Teikeu, W. A., Meli'i, J. L., Njandjock Nouck, P., Tabod, T. C., Enyegue, F., Nyam, A., & Aretouyap, Z. (2016). Assessment of groundwater quality in Yaoundé area, Cameroon, using geostatistical and statistical approaches. *Environmental Earth Sciences*, *75*(1), 1–15. Available from https://doi.org/10.1007/s12665-015-4779-7, http://www.springerlink.com/content/121380/.

Vasanthavigar, M., Srinivasamoorthy, K., Vijayaragavan, K., Rajiv Ganthi, R., Chidambaram, S., Anandhan, P., … Vasudevan, S. (2010). Application of water quality index for groundwater quality assessment: Thirumanimuttar sub-basin, Tamilnadu, India. *Environmental Monitoring and Assessment*, *171*(1–4), 595–609. Available from https://doi.org/10.1007/s10661-009-1302-1.

World Health Organization. (2017). Guidelines for drinking-water quality, 4th edn, World Health Organization.

Chapter 5

Flood susceptibility mapping of Malda District using analytical hierarchy process and remote sensing-based applications

Sarbajit Chowdhury, Jatisankar Bandyopadhyay and Bijay Halder
Department of Remote Sensing and GIS, Vidyasagar University, Midnapore, West Bengal, India

5.1 Introduction

One of the most common natural calamities to strike the world is flooding and related climate actions. When rivers overflow into adjacent land, either as a result of heavy rainfall or poor river management, floods happen (Dong et al., 2016; Hu et al., 2018). Heavy rainfall that triggers flash floods results in extensive loss of life and property. In addition, it degrades drinking water quality, propagates illnesses, and harms the ecosystem (Pakhale, Gondkar, & Saudagar, 2019). Several thousand people lost their lives in some of the deadliest floods that have happened in China (1935, 1931, and 1887), Guatemala (1949), Bangladesh (1974), Venezuela (1999), Iran (1954), India (2013), Japan (1953), Peru (1941), and many other places. Information systems and current technologies allow for proper monitoring of these natural disasters. Floods can provide certain advantages, such as distributing fish to small water bodies and bringing rich soil to agricultural areas, but they also cause terrible damage to the areas around them (Barbier, Georgiou, Enchelmeyer, & Reed, 2013). The tainted water from these floods contributed to the spread of cholera, typhoid fever, leptospirosis, hepatitis A, malaria, and dengue.

West Bengal's main issue is flooding; the Malda district, like other areas in the state, experienced substantial losses as a result of the flooding (Ghosh & Kar, 2018; Mehebub, Raihan, Nuhul, & Haroon, 2015). The Malda district is susceptible to flooding due to its geographic location and geomorphology. This district is situated where the Rajmahal Hills allow the Ganga River to enter Bengal. At the end of the Pleistocene Ice Age, the powerful Ganga-Brahmaputra delivered vast amounts of rain and snowmelt water with high

Water Resources Monitoring, Management, and Sustainability. DOI: https://doi.org/10.1016/B978-0-443-23665-5.00005-3

loads that were deposited here, creating the depositional fan that is home to Malda. Numerous rivers drain the area, but they are slow to flow because of the gradient drop and are full of silt, which causes them to overflow and cause floods. Malda has three distinct physiographic units: barind, which is composed of lateritic or older alluvial soil with remnants of an older flood plain, covering relatively elevated ground and being less prone to floods (Das & Samanta, 2022; Ghosh & Kar, 2018; Mehebub et al., 2015). Another anthropogenic activity that has exacerbated flood problems in and around the district is the construction of the Farakka Barrage on the River Ganga, downstream of Malda. Next is Tal, a marshy area that is mostly composed of older alluvium and bog areas that are located in pockets inside flood plains along inland drainages that are declining and are subject to flooding throughout the rainy season. The Tal merges with the following physiographic unit, the Diara, as it gradually slopes southwest (Das & Samanta, 2022). As its name suggests, the "Diara," which is surrounded by rivers, is a flat, well-drained area of younger alluvium that is less than 27 m above mean sea level (MSL). Every year, floods enhance alluvium deposits, making them extremely fertile and appropriate for agriculture and allied livelihoods. Tal and Diara are prone to flooding, but they are both densely populated areas. In addition to the floods, agricultural land flows into the river beds as a result of the gangs' ever-eroding banks. Furthermore, when there is flooding, the roadways become submerged, creating issues for the transportation system. Furthermore, the issues with drinking water and electricity still exist (Yaseen et al., 2018).

Three common methods are used to anticipate floods: the frequency of flooding can be analyzed statistically to determine the recurrence interval for any given year and stream discharge (without explicitly characterizing the flood area); storm progress can be tracked to predict short-term flood events; and flood hazard maps can be used to determine the extent of flooded areas based on estimated discharges for various return periods (Tehrany et al., 2017; Aydin & Iban, 2023). Hazard maps, however, simply show where hazards are likely to occur; they do not show how they may influence human lives. To address this vulnerability, vulnerability variables such as infrastructure and population must also be taken into account (Balica, Wright, & van der Meulen, 2012; Mollah, 2016). The geographical information system (GIS) is employed in this work to provide accessibility for the general public. Criterion weights in Multi-criteria Decision Management (MCDM) can be determined in a variety of ways, including pairwise comparison, rating, and ranking (Saaty, 1977, 2006, 2008). One of the most popular MCDM techniques is the Analytic Hierarchy Process (AHP), which was developed by Saaty and is frequently used to address water resource-related decision-making challenges (Feizizadeh, Jankowski, & Blaschke, 2014; Kowalski, McDaniel, & Olek, 2010). The relative relevance of each factor to the others is ascertained by pairwise comparison. All that is needed for this method to generate susceptibility maps is spatial information.

Floods are among the worst natural catastrophes; they destroy a great deal of property and communication infrastructure, resulting in a great deal of livestock and human casualties, as well as the loss of farmlands and agricultural products. Because floods are multifaceted and dynamic occurrences, the extent of flooded areas has been primarily explored using GIS and remote sensing (RS) data (Bui et al., 2018; Tehrany, Pradhan, & Jebur, 2013). Monitoring floods in close to real time is necessary to lessen their effects. It has been discovered that measures linked to LULC changes, such as environmental degradation and climatic change, can guarantee improved monitoring capacity in addition to cumulative flood hazard delineation (Barbier et al., 2013; Grimaldi, Xu, Li, Pauwels, & Walker, 2020). To accomplish an assessment in almost real-time, pixel-based flood analysis needs a significant amount of time and processing power. Flood vulnerability measures the likelihood of property and human loss in flood-prone communities by combining sociological data and the extent of inundation (Balica et al., 2012; Mollah, 2016). In industrialized nations with dense populations, flood exposure, and vulnerability are currently mapped using hydrodynamic inundation models in conjunction with high-resolution population distribution data. Malda is among the areas most susceptible to flooding. Every year during the monsoon, it experiences severe flooding that results in irreversible damage to both property and human lives. In Malda, around 42% of the total land area is vulnerable to flooding. Maps showing areas that are vulnerable to flooding are important tools for managing and reducing flood risk (Yaseen et al., 2020). Planning tools for evaluating flood hazards and landscape vulnerability can be derived from these maps, which can offer precise geographical data regarding the extent of floods. Using RS and GIS techniques, flood inundation mapping serves as a naturally helpful tool for tracking the extent of floods and evaluating their effects on land cover and use. Globally, a lot of people utilize the Landsat Thematic Mapper to find the extent of floods. By superimposing the flood inundation map in a GIS environment, surface elevation data (ASTER DEM) can be efficiently used for mapping flood zonation.

5.2 Study area

Malda District is one of the 23 districts that make up the Indian state of West Bengal. It is located between 88°28'10" East and 87°45'50" East longitude and between 25°32'08" North and 24°40'20" North latitude. The Murshidabad district lies to the south; Uttar Dinajpur and Dakshin Dinajpur districts to the north; the 165.5-km international boundary with Bangladesh to the east; and Purnea and Santhal Paragonas in Jharkhand and Bihar to the west. The district's entire size in 1935 was 1987 square miles, according to a report on settlement operations and the survey. India was 2004 square miles in size before it was divided, but it is currently 1441.6 square miles (3733 km^2). There are two subdivisions in the district: Chanchal and Malda Sadar.

Six community development blocks make up Chanchal: Harishchandrapur I, Harishchandrapur II, Ratua I, Ratua II, Chanchal I, and Chanchal II. The Old Malda and English Bazar municipalities, as well as nine community development blocks (English Bazar, Gazole, Habibpur, Kaliachak–I, Kaliachak–II, Kaliachak–III, Manickchak, Old Malda, and Bamangola), make up the Malda Sadar subdivision. Headquarters for the district are in English Bazar. This district contains 2 municipalities, 146 gram panchayats, 12 police stations, 15 development blocks, and 3701 villages (Fig. 5.1). One of the characteristics that can be used to explain the type and nature of surface water bodies in the district—essentially the "bils"—is the district's climate. Temperature, humidity, and precipitation all have major roles in how much water is used during the summer and how much is replenished during the monsoon and post-monsoon seasons. Naturally, rainfall has a little greater influence on the amount and quality of water that recharges the bils. The district supplies the replenishing quantity and quality components of the district's bils due to its infamous flooding history.

The Ganga River system has surface hydrological conditions that are highly active in the genuine sense, with several of its tributaries having longitudinal alignments that originate from the Himalayas and its foothills in the north. Furthermore, as was already noted, the Malda district's basin lies between Raajmahal and the Garo Hills. This configuration presents no

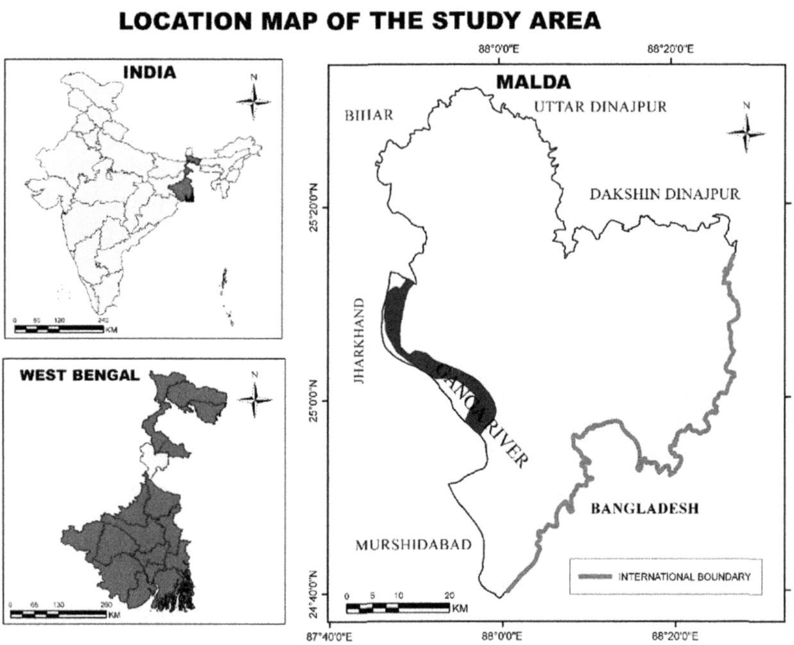

FIGURE 5.1 Location map of the study area. Map lines delineate study areas and do not necessarily depict accepted national boundaries.

structural (geological) or relief (geomorphological) barriers to the surface hydrological features' free play area stage. First, the territory is traversed by several perennial rivers, including the Ganga, which makes the hydrological activity obvious. Secondly, there is compelling evidence of a changing river channel, and third, floods continue to occur nearly every day throughout the entire region. It follows that among the physical indicators of the district's hydrological characteristics, surface hydrology elements in particular have left behind clear, noticeable, and long-lasting traces. This notable feature has also left its mark on the district's ethno-sociocultural mosaic of human life, in addition to penetrating some of the district's explicated physical and geographical features.

5.3 Materials and method

5.3.1 Data used

To create several thematic maps and estimate the flood susceptibility zone, GIS is crucial to this study. Various types of data are gathered from various sources for the current investigation (Table 5.1). These are primarily thematic maps, numerical data, and other things that support the study of the project activity. The precipitation map is derived from rainfall data, while the land use and land cover classification map is derived from the Sentinel 2 (10*10 meter resolution) image. For the compilation of maps showing slope, elevation, drainage density, and topographic wetness index (TWI), ASTER DEM (30*30 m resolution) is utilized. LANDSAT 8 with a resolution of

TABLE 5.1 Details about the datasets.

Sl no.	Attribute	Source
1	Precipitation	Indian Metrological Department (IMD) [http://www.imdpune.gov.in]
2	Land use and land cover	SENTINEL 2, USGS (earthexplorer.usga.gov.in)
3	Slope, elevation, TWI, drainage density	ASTER DEM, USGS (earthexplorer.usga.gov.in)
4	Normalized Difference Vegetation Index	LANDSAT 8 USGS (earthexplorer.usga.gov.in)
5	Soil	National Bureau of Soil Survey and Land Use Planning (NBSS and LUP), {Scale- 1:500000} [esdac.jrc.ec.europa.eu]
6	Population	Census of India (censusindia.gov.in)

30*30 m to prepare a Normalized Difference Vegetation Index (NDVI) map, utilize this image. Regional soil groups are displayed on soil maps. To present distances from roads and rivers on maps, data from road and drainage shape files is gathered from open street maps. Population statistics are used to represent demographics.

5.3.2 Flood susceptibility analysis

Several studies were cited to choose the main standards for evaluating floods. Ten factors were found to be crucial in producing floods, and as a result, they were the main emphasis for identifying flood-prone areas (Ha et al., 2021; Tehrany, Pradhan, & Jebur, 2014; Aydin & Iban, 2023). These fundamental standards were grouped into five main categories according to shared characteristics and consistency. These categories are as follows: (I) Hydrological standard: precipitation and density of river network; (II) Morphometric standard: elevation, slope, and distance from rivers; (III) Permeability standard: soil type and TWI; (IV) LULC and NDVI are the land cover dynamics criteria; and (V) anthropogenic interference, which is the distance from highways. Every criterion was specified and preprocessed using vector and raster datasets. After they were ranked and rated, their weights were estimated using the AHP. Using multi-criteria analysis and the weighted linear combination approach of AHP, a final flood susceptibility map was produced. A diagrammatic flow chart that depicts the study's methodology is provided (Feizizadeh et al., 2014; Rajasekhar, Gadhiraju, Kadam, & Bhagat, 2020; Yalew, van Griensven, Mul, & van der Zaag, 2016) (b). The WGS84/UTM/Zone 45 North coordinate system was used in the ArcGIS 10.8 software environment to assess all of the primary criteria and sub-criteria influencing floods in the study area. Each of the different parameters' ranges was categorized into five susceptibility categories according to how likely it was to flood (5: very high, 4: high, 3: moderate, 2: low, and 1: very low). To determine the susceptibility of a major criterion, the susceptibility of each criterion was assessed and aggregated. Based on the following data, the primary and secondary criteria were chosen and estimated. Each component that went into creating the flood susceptibility map is explained individually (Fig. 5.2).

5.3.3 Criteria

5.3.3.1 Topographic wetness index

The Topographic wetness index (TWI) is a crucial component in the mapping of flood susceptibility. This study can yield data on the hydrogeology, geomorphology, and even biological processes occurring in the basin, allowing for a qualitative and quantitative description of the region (Fig. 5.3). The slope and the upstream contributing area per unit width orthogonal to the flow direction determine the index (Bui et al., 2018). Following the application of a certain

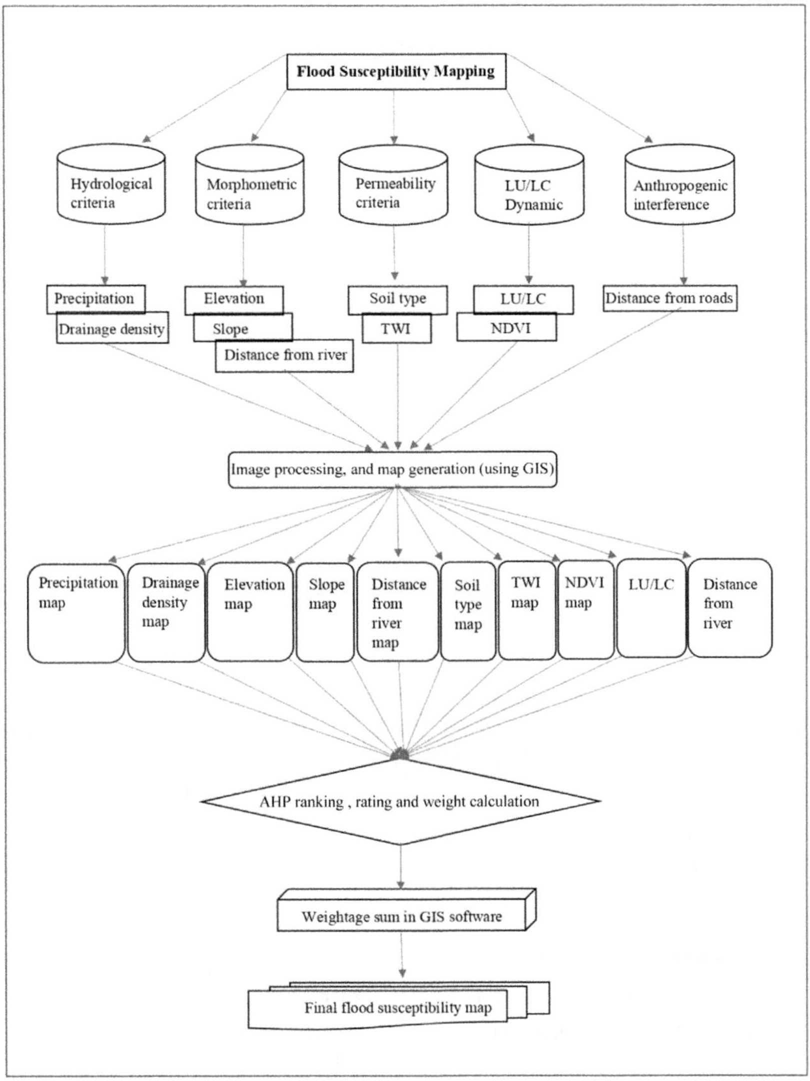

FIGURE 5.2 Applied methodology of the study area .

formula and mathematical method, it is observed that the TWI has a maximum value of 24 in the vicinity of the river and a minimum value of 4.6 in the most remote portions of the river. The TWI describes the spatial distribution of topographic features that characterize soil moisture, groundwater depth, and soil wetness (TWI). Areas that are conducive to water accumulation and high runoff have high TWI ratings. High TWI values indicate more saturated soil with poor drainage, which could result in flooding.

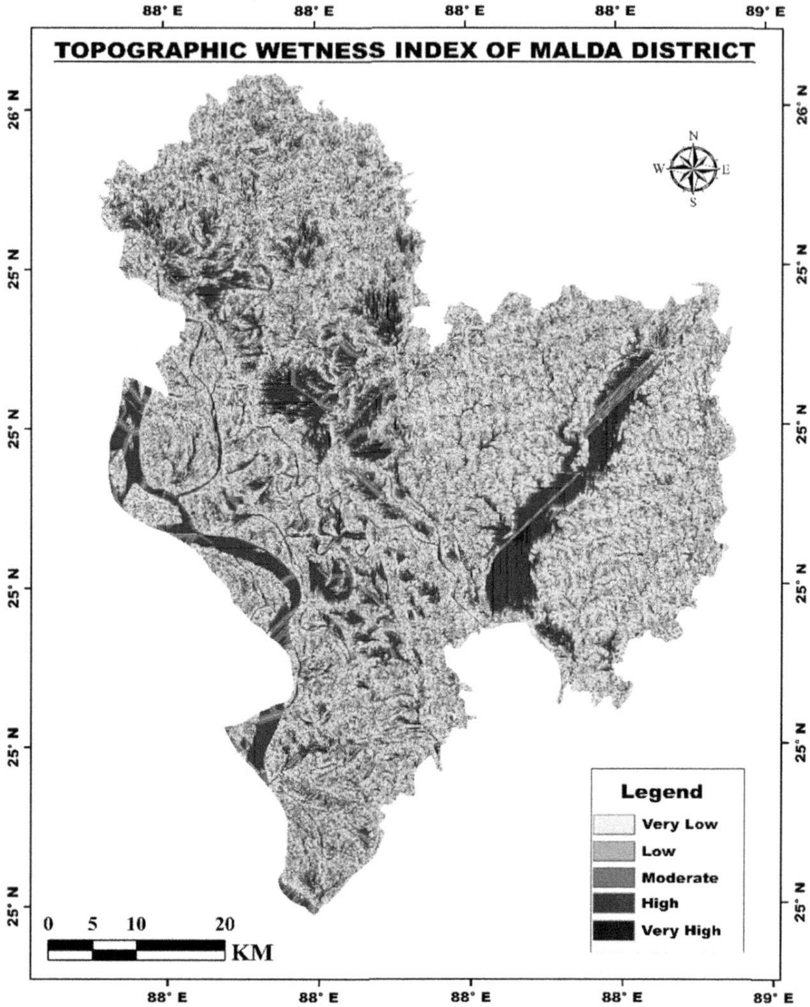

FIGURE 5.3 TWI map of the study area. *TWI*, topographic wetness index. Map lines delineate study areas and do not necessarily depict accepted national boundaries.

5.3.3.2 Elevation

Elevation significantly and effectively influences one's susceptibility to flooding. The local relief and the highest and lowest points within the landscape above mean sea level are indicated by elevation, also known as altitude (Hosseiny, 2021). Variations in altitude or elevation can occasionally be used to determine the climate, and these changes in the climate can bring about variances in the types of plants and soil, among other things (Fig. 5.4). Since low-lying locations are where many rivers join, the likelihood of flooding increases with elevation. Conversely, higher-elevation regions are less likely to

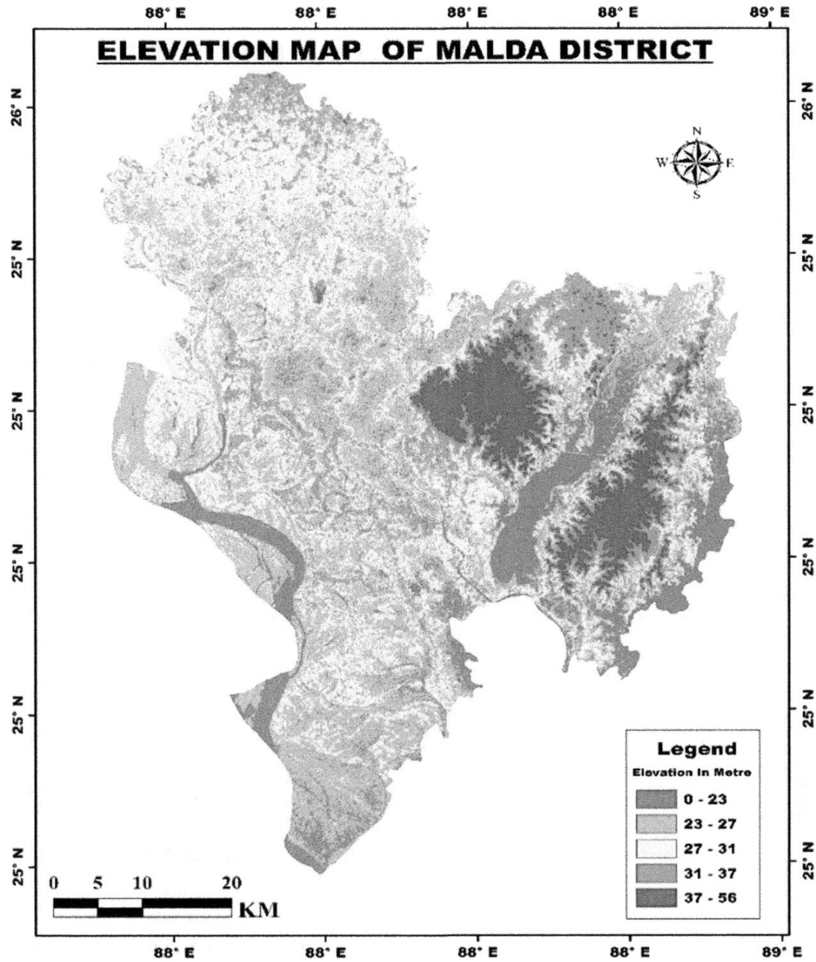

FIGURE 5.4 Elevation map of the study area. Map lines delineate study areas and do not necessarily depict accepted national boundaries.

experience flooding. It is clear from the five classes into which the elevation chart of the area is split (0–23; 23–27; 27–31; 31–37; and 37–56 m) that the smallest elevation is 23 m and the greatest elevation is 56 m.

5.3.3.3 Slope

Another significant factor that has a direct bearing on flooding is slope. The rise or decrease in the ground surface is called a slope. An oblique direction refers to a line that deviates from a horizontal line or direction and occasionally an inclination from one line or surface to another (Fig. 5.5). Rainwater infiltration and surface runoff are influenced by slopes. Because of the low

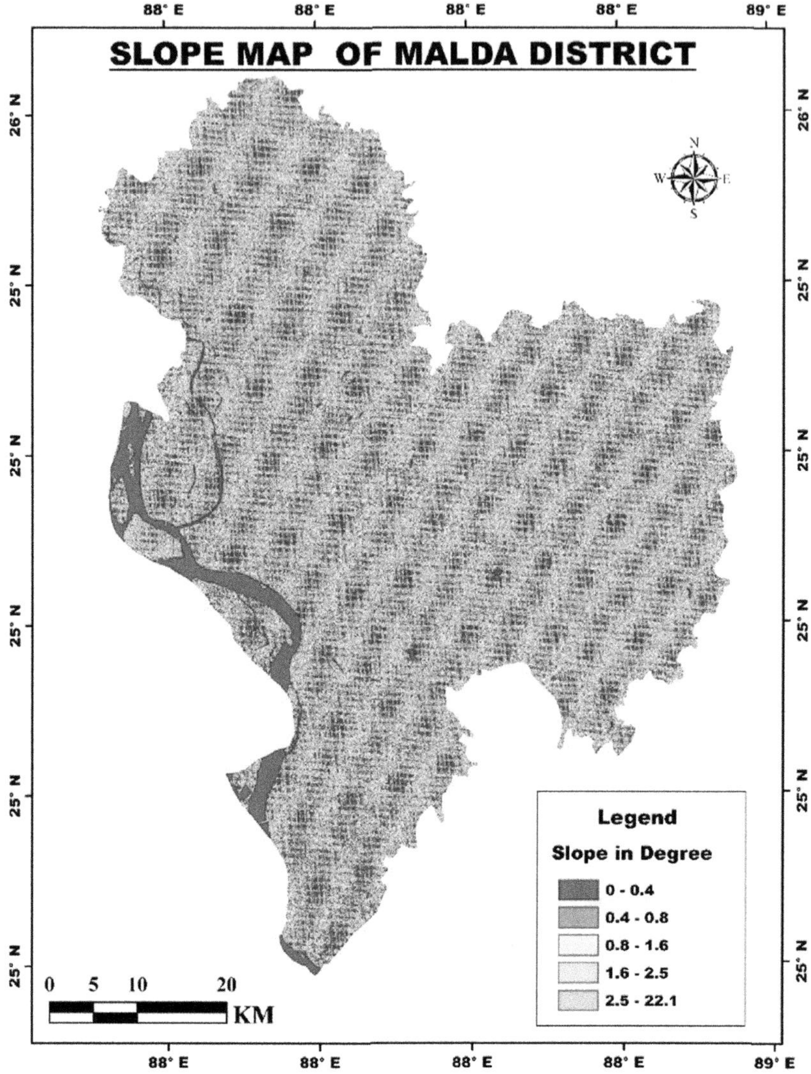

FIGURE 5.5 Slope map of the study area. Map lines delineate study areas and do not necessarily depict accepted national boundaries.

surface runoff velocity, areas with low slopes flood more quickly (Myronidis, Ioannou, Fotakis, & Dörflinger, 2018). Conversely, high-slope regions have high runoff velocities that are unsuitable for floods. Five classes, ranging from 0% to 22%, are depicted on the Malda district slope map, presented in percentage terms. Malda district locations that have a slope of less than 2% are less likely to flood, and places that have an average slope of more than 15% are more likely to flood.

5.3.3.4 Precipitation

Inundation is mostly caused by precipitation or rainfall in Maldha district. Following a downpour or series of downpours, streams may extend over their beds. On the other hand, not all locations in a locality receive the same quantity of rainfall. Consequently, regions receiving the highest amounts of rainfall will be deemed more vulnerable to flooding (Fig. 5.6). The Malda district's rainfall distribution chart is split into 5 classes, with total amounts

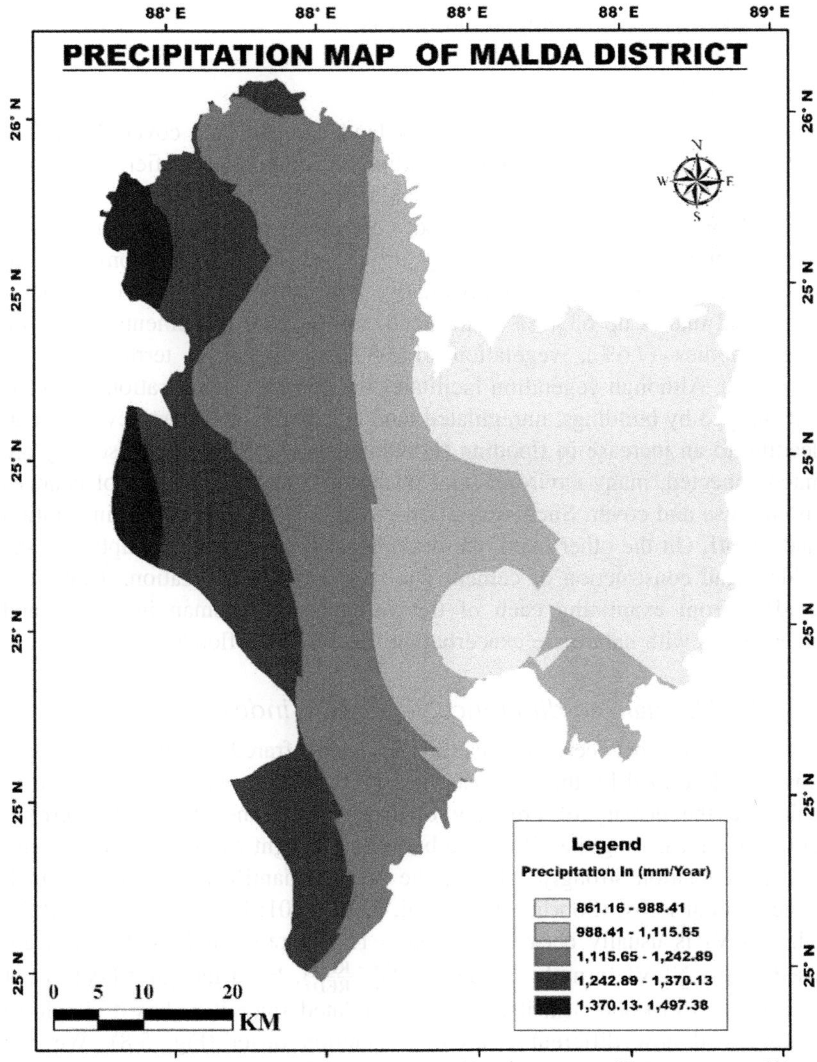

FIGURE 5.6 Precipitation of the study area. Map lines delineate study areas and do not necessarily depict accepted national boundaries.

ranging from 850 to 1500 mm (Vormoor, Lawrence, Heistermann, & Bronstert, 2015). The southern part of the Malda district (Old Malda, English Bazar, Manikchak, Habibpur, Kaliachak-1, Kaliachak-2, and Kaliachak-3 blocks) receives rainfall less than 1400 mm, while the blocks of Harishchandrapur-1, Harishchandrapur-2, Chanchal-1, and Chanchal-2 receive rainfall more than 1400 mm. It is possible to detect heavy and low rainfall in certain locations since rainfall is not distributed equally everywhere. The rainfall map makes it evident that the western portion of the Malda district received more rainfall than the eastern portion, indicating a strong likelihood of flooding in those areas.

5.3.3.5 LULC

Flooding is influenced significantly by land use and land cover. Using the ERDAS IMAGINE environment's maximum likelihood classifier, supervised classification was completed. The Malda district's land use and cover are divided into five categories: bare land, vegetation, water bodies, settlements, and agricultural land. Areas falling under various classifications of land cover and use are discussed individually. Within the Malda district, agricultural land makes up 63.48% of the land, followed by settlements (15.89%), water bodies (7.6%), vegetation (11.08%), and barren terrain (1.95%) (Fig. 5.7). Although vegetation facilitates the process of infiltration, runoff is encouraged by buildings, unregulated land use, and other urban development, leading to an increase in flooding (Fontanelli et al., 2022). Because they are interconnected, many environmental phenomena alter as a result of changes in land use and cover. Such vegetation cover is closely related to infiltration and runoff. On the other hand, the drainage system is being disrupted by the widespread construction of cement due to growing urbanization. Thus, it is evident from examining each of the variables that human ignorance and interactions with nature are exacerbating the effects of floods.

5.3.3.6 Normalized difference vegetation index

The difference between the visible and near-infrared reflectance of plant cover is described by the dimensionless NDVI index, which can be used to calculate the density of greenery on a piece of land (Weier & Herring, 2000). By measuring the difference between red light and near-infrared light, which vegetation strongly reflects, the NDVI quantifies vegetation (which vegetation absorbs) (Sobrino, Raissouni, & Li, 2001; Fontanelli et al., 2022). The NDVI is usually between -1 and +1. The red and NIR channels are used in the NDVI formula. $\left(\text{NDVI} = \frac{\text{NIR} - \text{RED}}{\text{NIR} + \text{RED}}\right)$. Even though NDVI is not directly tied to flood risk, it is indirectly related since it makes it simple to comprehend how different types of vegetation differ (Fig. 5.8). We can quickly comprehend the variations in a region's vegetation and water bodies by calculating its NDVI. The Malda District's NDVI was calculated using

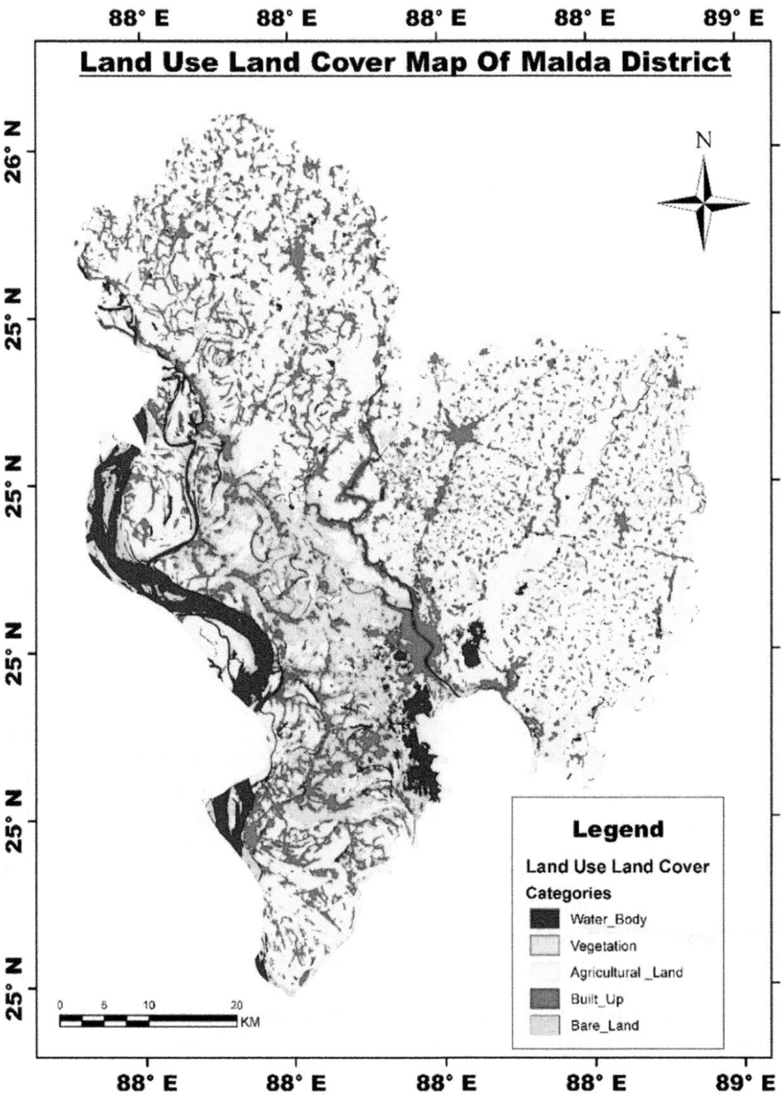

FIGURE 5.7 LULC map of the study area. Map lines delineate study areas and do not necessarily depict accepted national boundaries.

specific mathematical formulas, and the results show that the values -0.2 to -0.3 represent deep water bodies, -0.3–0.1 represent shallow water bodies, 0.1–0.2 represent sandbars or bare land, 0.2–0.3 represent shrubs, and 0.3–0.6 represent healthy vegetation. The places that are most vulnerable to flooding are shown here by values that are less than zero.

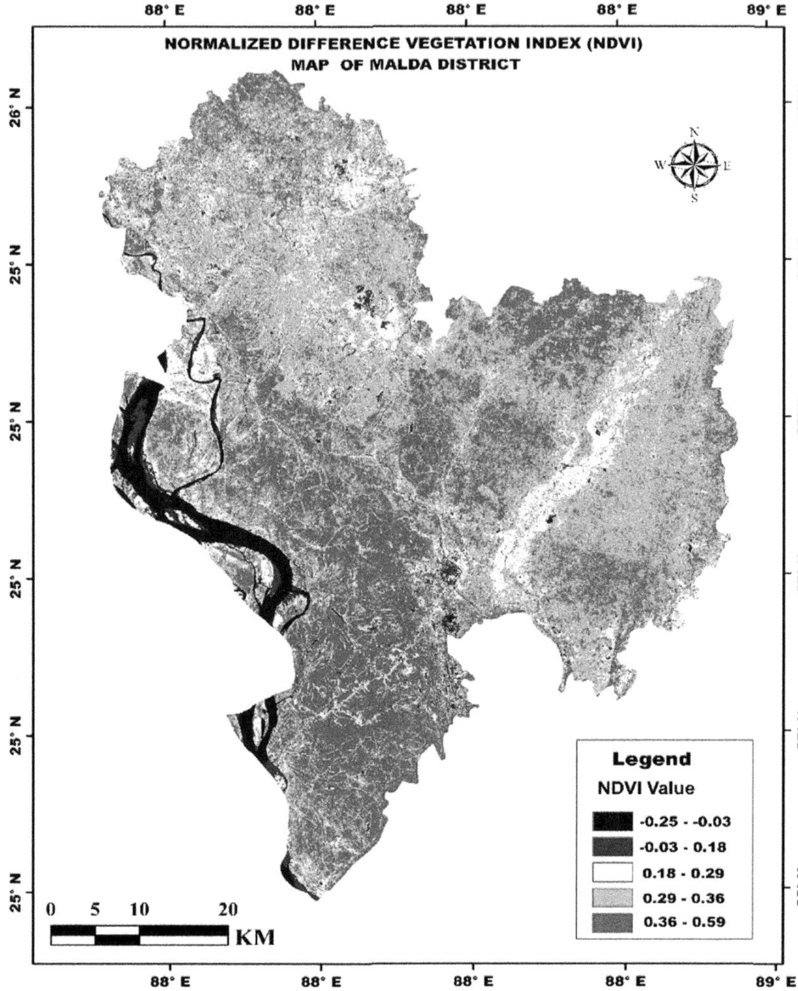

FIGURE 5.8 NDVI map of the study area. *NDVI*, Normalized Difference Vegetation Index. Map lines delineate study areas and do not necessarily depict accepted national boundaries.

5.3.3.7 Distance from river

One river-generated characteristic that emphasizes the vulnerability of the channel to flooding is the distance from the river. A river map's distance from the streams' effects on the flooding process has been developed. There are five categories for distance from a river. Flood susceptibility is largely determined by distance from rivers; a closer distance from a river suggests a lower risk of flooding, while a closer distance from a river indicates a higher risk of flooding (Fig. 5.9). There are five primary groups on the area's distance map: 0–10, 10–20, 20–40, 40–60, and 60–100 m. With a score of 5,

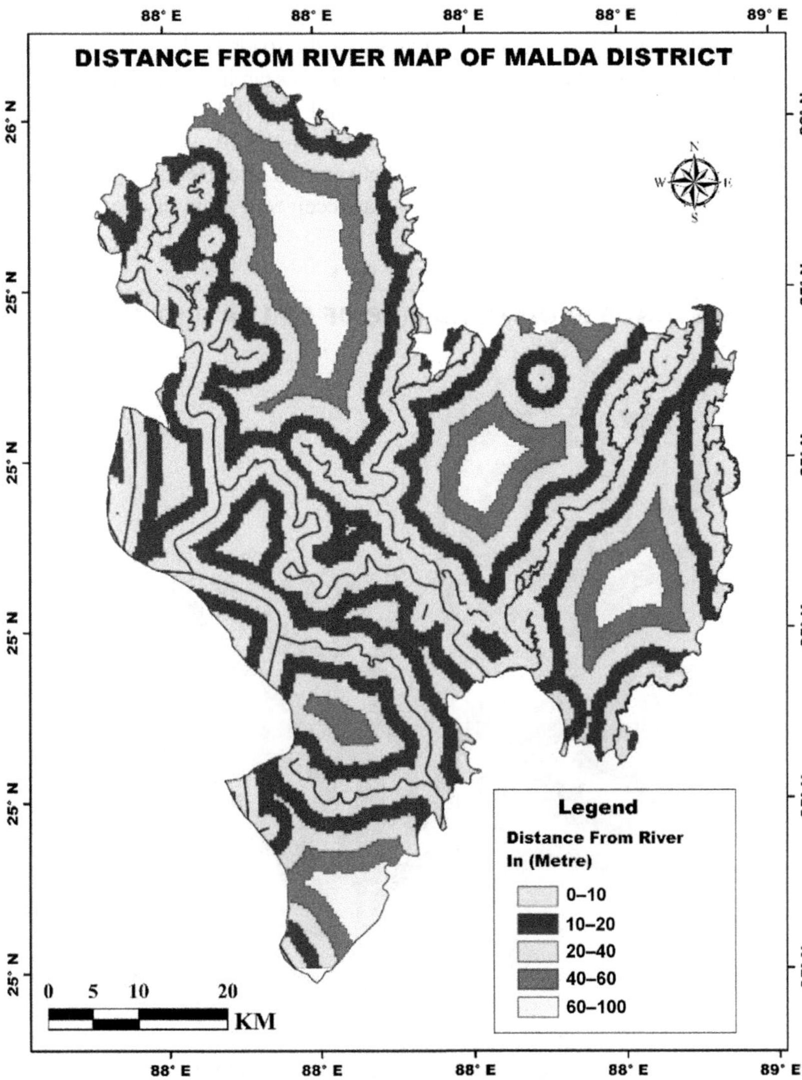

FIGURE 5.9 Distance from river map of the study area. Map lines delineate study areas and do not necessarily depict accepted national boundaries.

the distance from the river class with the lowest values (0–10 m) has a significant danger of flooding. The opposite is true for those with high river distance values (60–100 m). A rating of one has been given to the latter class.

5.3.3.8 Distance from road

One of the most crucial variables used to map flood risk is the distance from a road. It is connected to the flood indirectly, even though it is not directly related.

The risk of flooding decreases with increasing distance from the submerged area and increases with increasing proximity of the roads to the submerged area (Fig. 5.10). Based on the Euclidean distance, the research area is separated into five classes: 0–10, 10–20, 20–40, 40–60, and 60–100 m. The roads with a score of 5 are those that are closest to the water (0–10 m), as they are the most vulnerable to floods. Those with values between 60 and 100 m from the road have the opposite impact. The latter class has received a score of one.

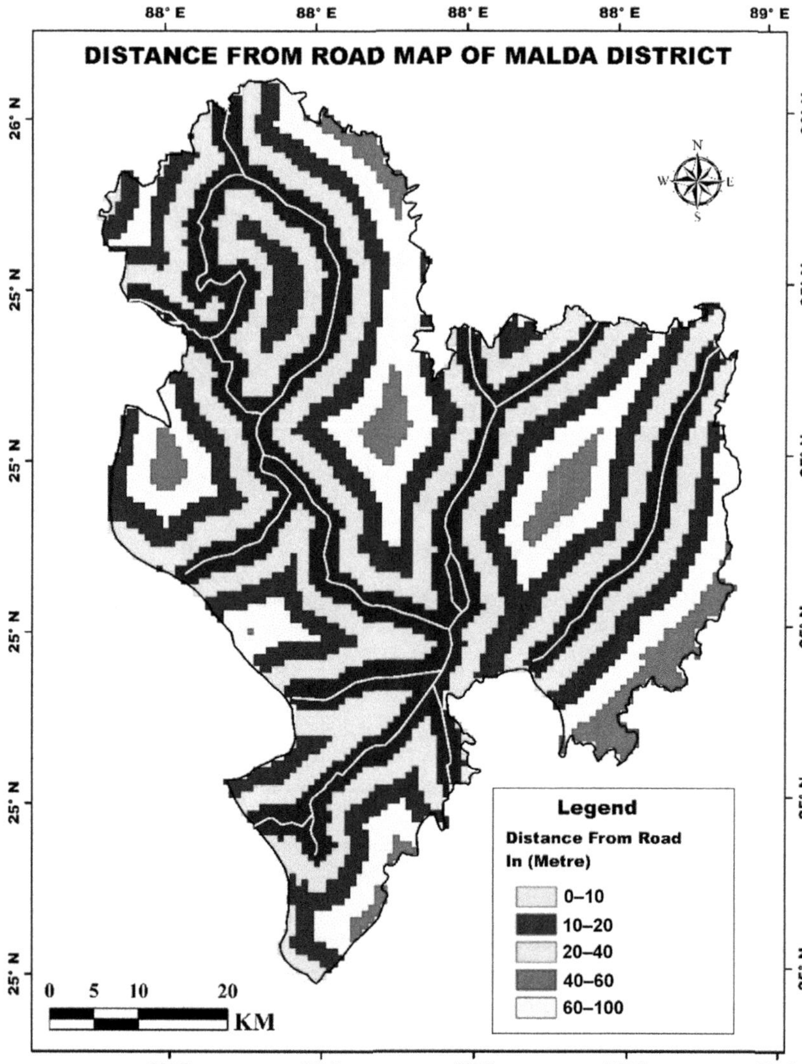

FIGURE 5.10 Distance from river map of the study area. Map lines delineate study areas and do not necessarily depict accepted national boundaries.

5.3.3.9 *Drainage density*

The most significant element influencing flood susceptibility is drainage density. One measure used to characterize the physical characteristics of a drainage basin is drainage density. The following equation, which represents drainage density, is defined as the channel in a drainage basin divided by the entire area: $\mathbf{d} = \frac{\sum L}{\mathbf{Abasin}}$. The drainage basin's physical features and climate both affect drainage density (Fig. 5.11). A watershed's runoff is influenced by the underlying rock type and soil permeability, or the difficulty of

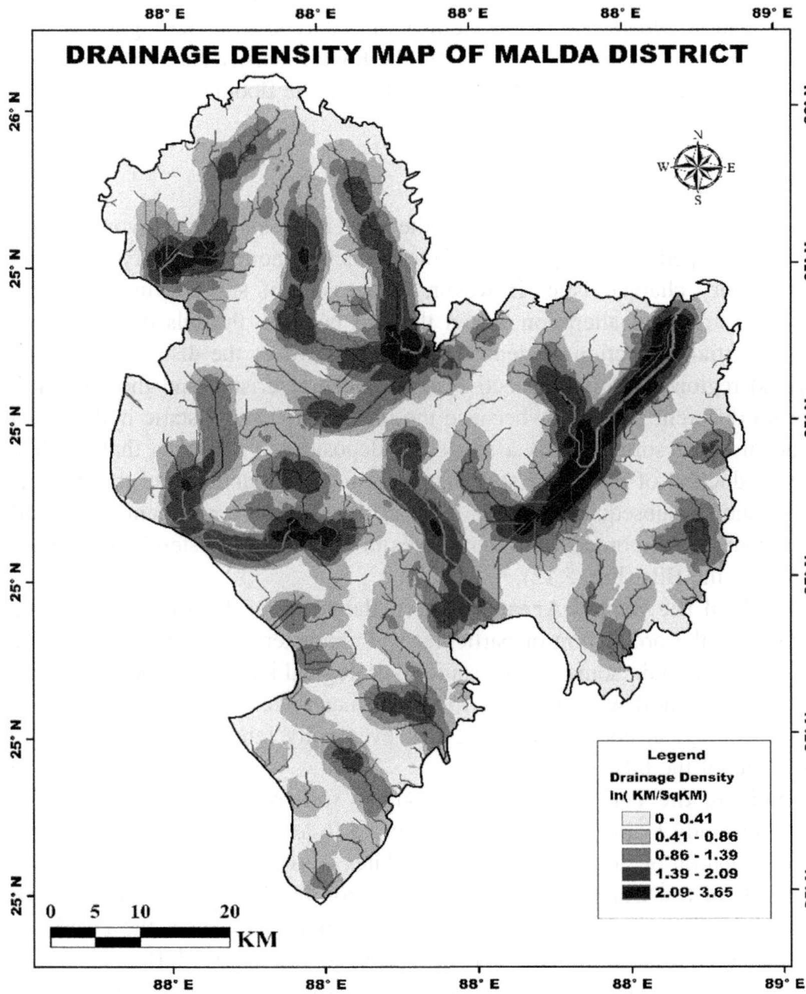

FIGURE 5.11 Drainage density map of the study area. Map lines delineate study areas and do not necessarily depict accepted national boundaries.

infiltration. Exposed bedrock or impermeable land will increase surface water runoff, which will result in more frequent streams. If the other features of the drainage basin remain the same, rugged or highly elevated areas will also have a higher drainage density than other drainage basins (Ghute & Babar, 2021). Drainage density relates to the hydrograph's shape and indicates the permeability and infiltration of a drainage basin. A high ratio is correlated with high drainage densities. It seems that an area's drainage density would have a major influence on the likelihood of flooding: the higher the density, the greater the likelihood of flooding. Five classes have been added to the drainage density map, which was previously displayed. The values in these classes range from 0.41 to 3.65 km/km^2. These values are highest in confluences and the vicinity of watercourses. The highest density class has received a rating of 5, while the lowest density class has received a grade of 1, based on its notable contribution to the flooding process.

5.3.3.10 Soil types

As far as the thematic spreads of the current research endeavor are concerned, the pedological qualities of the district do not have a very significant explanatory characteristic. However, as a result of surface runoff and the ensuing structural alteration across the soil horizon, the bils or water body areas partially function as depositional zones. All of the district, except the Barind region, has a strong active alluvial characteristic, and the beel areas are no different. However, because the bils' areas have a static hydrological state of some sort, there is a scope for depositional activities that result in the production of particular soil characteristics. It is anticipated that the surface runoff's subsequent addition of more modern soil constituents will result in a pedological composition that mixes in with the ecosystem's abiotic elements in the Bils (Fig. 5.12).

Soils of the Barind Area: However, the scope of depositional processes leading to the production of particular soil characteristics occurs because the bils' areas contain some sort of static hydrological state. It is anticipated that the surface runoff will introduce fresh soil materials, resulting in the creation of a pedological composition that will eventually integrate with the abiotic elements of the bils ecosystem.

Soils of the Tal Area: The Tal region's soils range in texture from clay loam to sandy loam. These soils are referred to as "Do-ash" light loams. It is a combination of sand and clay and is a later alluvial structure.

Soils of the Diara Area: The Diara region has the richest and relatively new soils. These soils have a clay and sand blend. However, the District of Malda has 12 categories of soil, according to the National Bureau of Soil Survey (NBSS) and Land Use Planning (LUP) chart of soil occurrences in West Bengal.

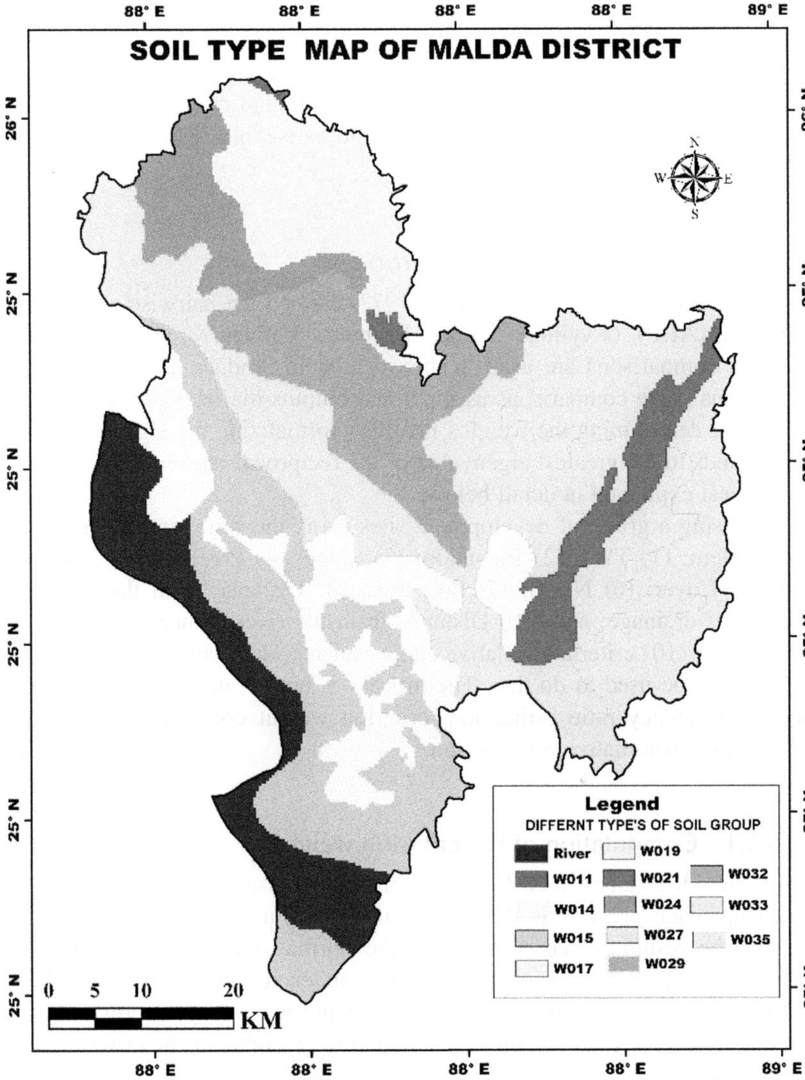

FIGURE 5.12 **Soil types map of the study area.** Map lines delineate study areas and do not necessarily depict accepted national boundaries.

5.3.4 Assessing flood susceptibility through analytical hierarchy process modeling approaches

Using a set of choices, users, and planners can statistically determine a scale of preference using the AHP multi-perspective, multi-objective decision-making paradigm. Due to its intrinsic capacity to evaluate several options based on a range of criteria to potentially identify the optimal alternative,

AHP has gained popularity as a viable tool for assessing complicated real-world problems (Ghosh & Kar, 2018; Radwan, Alazba, & Mossad, 2019; Yalew et al., 2016). For their ultimate implementation, these options may be thoroughly investigated in the future. By using spatial multi-criteria decision analysis, the user can determine which alternative best fits a given set of evaluation criteria.

5.3.4.1 Pairwise comparison method

In the framework of the AHP, Saaty (1977) created the pairwise comparison approach (AHP) (Ayyildiz, 2022; Hadipour, Vafaie, & Deilami, 2020). Pairwise comparisons are used in this method to produce a ratio matrix. It receives pairwise comparisons as input and outputs the relative weights. The process of determining the weights involves normalizing the eigenvector that corresponds to the greatest eigenvalue of the reciprocal ratio matrix. The process is best explained in detail below.

Assessing a group of development sites using the following 10 criteria is the problem: (1) TWI; (2) Elevation; (3) Slope; (4) Precipitation; (5) Land Use and Cover; (6) NDVI; (7) Soil type; (8) Distance from the road; (9) Density of drainage; and (10) Distance from the river. It necessitates determining the 10 criteria's relative importance. The pairwise comparison method can be used to do this. The process is broken down into three main steps: consistency ratio estimation, criterion weight computation, and pairwise comparison matrix construction.

5.3.4.1.1 Computation of the criterion weight

In this step, the values in each column of the pairwise comparison matrix are added up; each element is divided by the total number of columns in the matrix (the resulting matrix is called the normalized pairwise comparison matrix); and the average of the elements in each row of the normalized matrix is calculated by dividing the sum of the normalized scores for each row by 10 (the number of criteria). The relative weights of the criteria under comparison are approximated by these averages.

By employing this technique, the weights are understood to represent the mean of all feasible comparisons between the criteria. The criterion weights for TWI, elevation, slope, precipitation, land use and land cover (LULC), NDVI, distance from the river, distance from the road, drainage density, and soil type are, as can be seen, 0.13, 0.12, 0.10, 0.13, 0.07, 0.06, 0.13, 0.06, 0.09, and 0.12. This indicates that the most significant criteria are TWI, elevation, slope, precipitation, distance from the river, and soil type, followed by NDVI, LULC, and distance from the road.

5.3.4.1.2 Estimation of the consistency ratio

We check the consistency of our comparisons at this stage. The following procedures are involved: The weights for the first criterion (TWI) and the second weight (elevation) are multiplied times the first and second columns of the original pairwise comparison matrix, respectively. This process is repeated for all factors, and the weighted sum vector is then determined by dividing the weighted sum vector by the criterion weights that were previously determined. Finally, the weighted sum vector is summed over the rows (Tables 5.2;5.3;5.4;5.5;5.6; and 5.7).

5.4 Results and discussion

The range of results for each criterion in each of the five classes was used to produce the flood susceptibility map. The sub-criteria categorized under each criterion were used to create the combined flood susceptibility map for the five criteria. The final flood zoning map was further developed using the major criteria-based flood susceptibility maps. After building PCM, the relative relevance of each criterion was calculated using the AHP. To ascertain the relative importance and derive a rating score, the values of every row and every column are compared. The final flood susceptibility map was produced with a lambda (λ) max value of 11.26 and a CR value of 0.09 (<0.1, validated) after the weighting sum of all causative factors were applied. The highest flood susceptibility zones were identified based on the river and soil type criteria, as shown by the weights of 0.13, 0.12, 0.10, 0.13, 0.13, and 0.12 assigned using the MCA technique in AHP (Fig. 5.13).

Density of drainage and land use, as indicated by the weights of 0.09 and 0.07, the land cover criterion has the second-highest weight. The remaining NDVI criteria with weights of 0.06 and 0.05, distance from the road is also the least significant factor. However, each weight is shown on the graph, and each criterion has the potential to independently increase flood susceptibility at any particular time. Using data from previous flood episodes and secondary topography information, the AHP reclassification technique created the flood susceptibility maps for ten criteria. Here, we can observe that places close to rivers are more vulnerable to flooding, while areas farther from rivers are less vulnerable, according to susceptibility maps based on primary criteria that were also assessed. Because of the extremely low slope angle, the areas along the Ganga River are particularly prone to flooding each year. Following analysis, the Malda district's flood susceptibility map was found to be classified into five classes: very low, low, moderate, high, and very high. According to this study, 12.87% of the region is classified as extremely high, 26.61% as high, and 29.61% as moderate. 9.2% of the region is classified as extremely low, and 21.47% as low.

TABLE 5.2 Pairwise comparison method.

Criteria	Pairwise comparison of the evaluation criteria									
	TWI	Elevation	Slope	Precipitation	LULC	Normalized Difference Vegetation Index	Distance from river	Distance from road	Drainage density	Soil types
Topographic Wetness Index	1.00	1.00	1.00	1.00	3.00	5.00	1.00	3.00	1.00	1.00
Elevation	1.00	1.00	1.00	1.00	2.00	3.00	1.00	3.00	1.00	1.00
Slope	1.00	1.00	1.00	1.00	3.00	1.00	0.50	1.00	1.00	1.00
Precipitation	1.00	1.00	1.00	1.00	3.00	2.00	2.00	3.00	1.00	1.00
LULC	0.33	1.00	0.33	0.33	1.00	1.00	0.33	3.00	1.00	1.00
NDVI	0.20	0.33	1.00	0.50	1.00	1.00	0.20	1.00	1.00	1.00
Distance from river	1.00	1.00	2.00	0.50	3.00	5.00	1.00	3.00	1.00	1.00
Distance from road	0.33	0.33	1.00	0.33	0.33	1.00	0.33	1.00	1.00	1.00
Drainage density	1.00	0.50	1.00	1.00	1.00	2.00	1.00	1.00	1.00	1.00
Soil types	2.00	1.00	4.00	1.00	1.00	1.00	1.00	1.00	1.00	1.00
Total	8.86	8.16	13.33	7.66	18.33	22.00	8.36	20.00	10.00	10.00

TABLE 5.3 Normalization of each criterion.

Criteria	Normalization of each score									
	TWI	Elevation	Slope	Precipitation	LULC	Normalized Difference Vegetation Index	Distance from river	Distance from road	Drainage density	Soil types
Topographic Wetness Index	0.11	0.12	0.08	0.13	0.16	0.22	0.12	0.15	0.10	0.10
Elevation	0.11	0.12	0.08	0.13	0.12	0.14	0.12	0.15	0.10	0.10
Slope	0.11	0.12	0.08	0.13	0.16	0.05	0.06	0.05	0.10	0.10
Precipitation	0.11	0.12	0.08	0.13	0.16	0.09	0.24	0.15	0.10	0.10
LULC	0.04	0.12	0.03	0.04	0.05	0.05	0.04	0.15	0.10	0.10
NDVI	0.02	0.04	0.08	0.07	0.05	0.05	0.02	0.05	0.10	0.10
Distance from river	0.11	0.12	0.13	0.07	0.18	0.21	0.12	0.15	0.10	0.10
Distance from road	0.04	0.06	0.08	0.04	0.02	0.05	0.04	0.05	0.10	0.10
Drainage density	0.11	0.06	0.08	0.13	0.05	0.09	0.12	0.05	0.10	0.10
Soil types	0.24	0.12	0.28	0.13	0.05	0.05	0.12	0.05	0.10	0.10
Total	1.00	1.00	1.00	1.00	1.00	1.00	1.00	1.00	1.00	1.00

TABLE 5.4 Arithmetic values of each criterion.

Criteria	Arithmetic values	Weight
Topographic Wetness Index	(0.11 + 0.12 + 0.08 + 0.13 + 0.16 + 0.23 + 0.12 + 0.15 + 0.10 + 0.10)/10	0.13
Elevation	(0.11 + 0.12 + 0.08 + 0.13 + 0.11 + 0.14 + 0.12 + 0.15 + 0.10 + 0.10)/10	0.12
Slope	(0.11 + 0.12 + 0.08 + 0.13 + 0.16 + 0.05 + 0.06 + 0.05 + 0.10 + 0.10)/10	0.10
Precipitation	(0.11 + 0.12 + 0.08 + 0.13 + 0.16 + 0.09 + 0.24 + 0.15 + 0.10 + 0.10)/10	0.13
LULC	(0.04 + 0.12 + 0.03 + 0.04 + 0.05 + 0.05 + 0.04 + 0.15 + 0.10 + 0.10)/10	0.07
Normalized Difference Vegetation Index	(0.02 + 0.04 + 0.08 + 0.07 + 0.05 + 0.05 + 0.02 + 0.05 + 0.10 + 0.10)/10	0.06
Distance from river	(0.11 + 0.12 + 0.15 + 0.07 + 0.16 + 0.23 + 0.12 + 0.15 + 0.10 + 0.10)/10	0.13
Distance from road	(0.04 + 0.04 + 0.08 + 0.04 + 0.02 + 0.05 + 0.04 + 0.05 + 0.10 + 0.10)/10	0.06
drainage density	(0.11 + 0.06 + 0.08 + 0.13 + 0.05 + 0.09 + 0.12 + 0.05 + 0.10 + 0.10)/10	0.09
Soil types	(0.23 + 0.12 + 0.30 + 0.13 + 0.05 + 0.05 + 0.12 + 0.05 + 0.10 + 0.10)10	0.12
Total		1.00

TABLE 5.5 Estimation of the consistency ratio.

Criteria	Step 1	Step 2
Topographic wetness index	(0.13*1 + 0.12*1 + 0.10*1 + 0.13*1 + 0.07*1 + 0.06*5 + 0.13*1 + 0.06*3 + 0.09*1 + 0.12*1) = 1.15	1.15/0.13 = 11.61
Elevation	(0.13*1 + 0.12*1 + 0.10*1 + 0.13*1 + 0.07*2 + 0.06*3 + 0.13*1 + 0.06*3 + 0.09*1 + 0.12*1) = 1.32	1.32/0.12 = 11.00
Slope	(0.13*1 + 0.12*1 + 0.10*1 + 0.13*1 + 0.07*3 + 0.06*1 + 0.13*0.5 + 0.06*1 + 0.09*1 + 0.12*1) = 1.08	1.08/0.10 = 10.8
Precipitation	(0.13*1 + 0.12*1 + 0.10*1 + 0.13*1 + 0.07*3 + 0.06*2 + 0.13*2 + 0.06*3 + 0.09*1 + 0.12*1) = 1.46	1.46/0.13 = 11.23
LULC	(0.13*0.33 + 0.12*1 + 0.10*0.33 + 0.13*0.33 + 0.07*1 + 0.06*1 + 0.13*0.33 + 0.06*3 + 0.09*1 + 0.12*1) = 0.80	0.80/0.07 = 11.42
Normalized Difference Vegetation Index	(0.13*0.2 + 0.12*0.33 + 0.10*1 + 0.13*0.5 + 0.07*1 + 0.06*1 + 0.13*0.2 + 0.06*1 + 0.09 + 1 + 0.12*1) = 0.66	0.66/0.06 = 11.00
Distance from River	(0.13*1 + 0.12*1 + 0.10*2 + 0.13*0.05 + 0.07*3 + 0.06*5 + 0.13*1 + 0.06*5 + 0.09*1 + 0.12*1) = 1.54	1.54/0.13 = 11.84
Distance from Road	(0.13*0.33 + 0.12*0.33 + 0.10*1 + 0.13*0.33 + 0.07*0.33 + 0.06*1 + 0.13*0.33 + 0.06*1 + 0.09*1 + 0.12*1) = 0.62	0.62/0.06 = 10.33
Drainage Density	(0.13*1 + 0.12*0.50 + 0.10*1 + 0.13*1 + 0.07*1 + 0.06*2 + 0.13*1 + 0.06*1 + 0.09*1 + 0.12*1) = 1.01	1.01/0.09 = 11.22
Soil Types	(0.13*2 + 0.12*1 + 0.10*4 + 0.13*1 + 0.07*1 + 0.06*1 + 0.13*1 + 0.06*1 + 0.09*1 + 0.12*1) = 1.44	1.44/0.12 = 12.00

TABLE 5.6 Random Consistency Index (RI) (Saaty 1980, 2000).

n	1	2	3	4	5	6	7	8	9	10	11	12	13	14	15
RI	0	0	0.58	0.9	0.12	1.24	1.32	1.41	1.45	1.49	1.51	1.48	1.56	1.57	1.59

TABLE 5.7 Sub criteria values for flood susceptibility mapping.

Sl no.	Flood criteria	Class	Susceptibility level	Class rating
1	Topographic wetness index	4.64–8.63	Very Low	1
		8.63–10.97	Low	2
		10.97–13.08	Moderate	3
		13.08–15.95	High	4
		15.95–23.87	Very High	5
2	Elevation	0–23	Very High	5
		23–27	High	4
		27–31	Moderate	3
		31–37	Low	2
		37–56	Very Low	1
3	Slope	0–0.4	Very High	5
		0.4–0.8	High	4
		0.8–1.6	Moderate	3
		1.6–2.5	Low	2
		2.5–22.1	Very Low	1
4	Precipitation	861.16–988.41	Very Low	1
		988.41–1115.65	Low	2
		1115.65–1242.89	Moderate	3
		1242.89–1370.13	High	4
		1370.13–1497.38	Very High	5
5	LULC	Water Body	Very High	5
		Agricultural Land	High	4
		Built up	Moderate	3
		Barren Land	Low	2
		Vegetation	Very Low	1
6	Normalized Difference Vegetation Index	(-)0.25 - (-)0.03	Very High	5
		(-)0.03–0.18	High	4
		0.18–0.29	Moderate	3
		0.29–0.36	Low	2
		0.36–0.59	Very Low	1

(Continued)

TABLE 5.7 (Continued)

Sl no.	Flood criteria	Class	Susceptibility level	Class rating
7	Distance from River	0–10	Very High	5
		10–20	High	4
		20–40	Moderate	3
		40–60	Low	2
		60–100	Very Low	1
8	Distance from Road	0–10	Very High	5
		10–20	High	4
		20–40	Moderate	3
		40–60	Low	2
		60–100	Very Low	1
9	Drainage Density	0–0.41	Very Low	1
		0.41–0.86	Low	2
		0.86–1.39	Moderate	3
		1.39–2.09	High	4
		2.09–3.65	Very High	5
10	Soil Types	Sandy Clay Loam	Low	2
		Loam	Moderate	3
		Sandy Loam	Very High	5

The Malda district is highly vulnerable to floods because of its geographic location and geomorphology, particularly in the low-lying Tal and Diara regions. The Tal and Diara regions of Malda sustain a higher population density than the average density of the state, despite yearly floods that hurt people's lives and means of subsistence. This indicates that the local population has adapted to living with floods because they have become more resilient to flood dangers, which has decreased their vulnerability. The current research argues that the inner villages of Nurpur and Ratua have not developed as effective coping and adaptation mechanisms to floods as Maniknagar and Gopalpur, which are situated near two major flood-producing rivers, the Ganga in Diara and Fulhar in Tal.

As a result of their increased resilience to flood threats, which lowers their vulnerability, the local population has demonstrated that they have learned to live with floods. According to the current report, Maniknagar and Gopalpur villages—

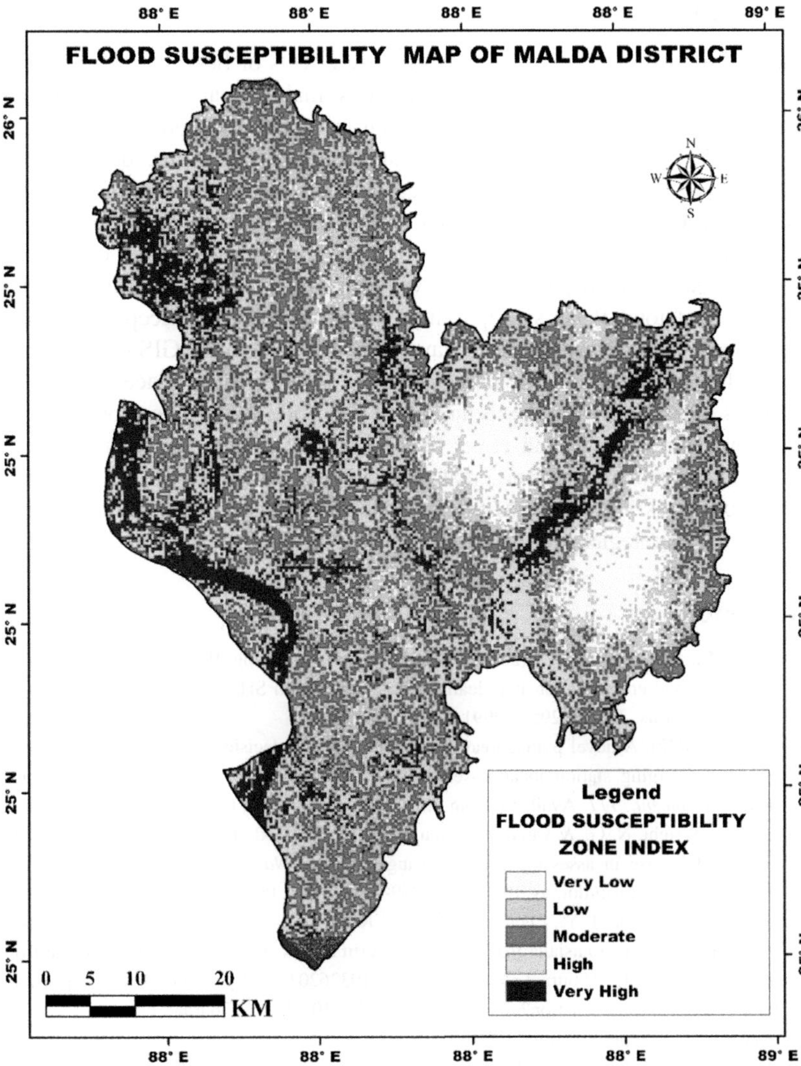

FIGURE 5.13 **Flood susceptibility mapping of the Malda district.** Map lines delineate study areas and do not necessarily depict accepted national boundaries.

which are situated near two of the main rivers that cause flooding, the Ganga in Diara and Fulhar in Tal—have developed more resilient coping mechanisms for flooding than have the interior villages of Nurpur and Ratua. Consistent with our findings, researchers have found that soil, terrain, and rainfall all have a major impact on the likelihood of floods. Modern techniques, including the frequency ratio, support vector machine model, and naïve Bayes method of alternate decision trees and random forests, might help improve flood susceptibility mapping.

5.5 Conclusion

One of the most effective ways to lessen the effects of flood hazards and help decision-makers, planners, and stakeholders maintain appropriate oversight over the areas most vulnerable to flooding is to map the flood disaster-prone zones. This ensures appropriate and sustainable socioeconomic development. An effective substitute method for determining the flood-prone areas in the Malda district of West Bengal, India, is the RS GEE-based multi-parametric AHP-GIS technique. Based on their proximity and relative amount of involvement in creating floods, 10 variables were classified into 5 key criteria for the evaluation of total flood susceptibility. To create a flood susceptibility map, every flood that affected the raster layer was analyzed in the GIS environment. The hydrologic criterion was ranked highest for identifying places vulnerable to floods using Saaty's AHP technique. It was followed by the morphometric criterion, permeability, land cover dynamics, and anthropogenic interference. To minimize vulnerability and create a society that is resilient to disasters, this article examines and evaluates the level of preparedness, coping mechanisms, and adaptive strategies of individuals at the household level.

References

Aydin, H. E., & Iban, M. C. (2023). Predicting and analyzing flood susceptibility using boosting-based ensemble machine learning algorithms with SHapley Additive exPlanations. *Natural Hazards*, *116*(3), 2957–2991.

Ayyildiz, E. (2022). A novel pythagorean fuzzy multi-criteria decision-making methodology for e-scooter charging station location-selection. *Transportation Research Part D: Transport and Environment.*, *111*. Available from https://doi.org/10.1016/j.trd.2022.103459, 13619209.

Balica, S. F., Wright, N. G., & van der Meulen, F. (2012). A flood vulnerability index for coastal cities and its use in assessing climate change impacts. *Natural Hazards*, *64*(1), 73–105. Available from https://doi.org/10.1007/s11069-012-0234-1, 0921-030X.

Barbier, E. B., Georgiou, I. Y., Enchelmeyer, B., & Reed, D. J. (2013). The value of wetlands in protecting southeast louisiana from hurricane storm surges. *PLoS One*, *8*(3). Available from https://doi.org/10.1371/journal.pone.0058715, 19326203, http://www.plosone.org/article/fetchObjectAttachment.action?uri = info%3Adoi%2F10.1371%2Fjournal.pone.0058715&representation = PDF.

Bui, D. T., Panahi, M., Shahabi, H., Singh, V. P., Shirzadi, A., Chapi, K., … Ahmad, B. B. (2018). Novel hybrid evolutionary algorithms for spatial prediction of floods. *Nature Publishing Group, Viet Nam Scientific Reports*, *8*(1). Available from https://doi.org/10.1038/s41598-018-33755-7, 20452322, http://www.nature.com/srep/index.html.

Das, R., & Samanta, G. (2022). Impact of floods and river-bank erosion on the riverine people in Manikchak Block of Malda District. *Development and Sustainability*, 1–23.

Dong, J., Xiao, X., Menarguez, M. A., Zhang, G., Qin, Y., Thau, D., … Moore, B. (2016). Mapping paddy rice planting area in northeastern Asia with Landsat 8 images, phenology-based algorithm and Google Earth Engine. *Remote Sensing of Environment*, *185*, 142–154. Available from https://doi.org/10.1016/j.rse.2016.02.016, 00344257, http://www.elsevier.com/inca/publications/store/5/0/5/7/3/3.

Feizizadeh, B., Jankowski, P., & Blaschke, T. (2014). A GIS based spatially-explicit sensitivity and uncertainty analysis approach for multi-criteria decision analysis. *Computers & Geosciences*, *64*, 81–95. Available from https://doi.org/10.1016/j.cageo.2013.11.009, 00983004.

Fontanelli, G., Lapini, A., Santurri, L., Pettinato, S., Santi, E., Ramat, G., ... Paloscia, S. (2022). Early-season crop mapping on an agricultural area in Italy using X-band dual-polarization SAR satellite data and convolutional neural networks. *IEEE Journal of Selected Topics in Applied Earth Observations and Remote Sensing*, *15*, 6789–6803.

Ghosh, A., & Kar, S. K. (2018). Application of analytical hierarchy process (AHP) for flood risk assessment: A case study in Malda district of West Bengal, India. *Natural Hazards*, *94*(1), 349–368. Available from https://doi.org/10.1007/s11069-018-3392-y, 15730840, http://www.wkap.nl/journalhome.htm/0921-030X.

Ghute, B. B., & Babar, S. M. (2021). An approach to mapping groundwater recharge potential zones using geospatial techniques in Kayadhu River Basin, Maharashtra. *Indian Journal of Agricultural Research*, *55*(1), 23–32.

Grimaldi, S., Xu, J., Li, Y., Pauwels, V. R. N., & Walker, J. P. (2020). Flood mapping under vegetation using single SAR acquisitions. *Remote Sensing of Environment.*, *237*. Available from https://doi.org/10.1016/j.rse.2019.111582, 00344257.

Ha, H., Luu, C., Bui, Q. D., Pham, D. H., Hoang, T., Nguyen, V. P., ... Pham, B. T. (2021). Flash flood susceptibility prediction mapping for a road network using hybrid machine learning models. *Natural Hazards*, *109*(1), 1247–1270. Available from https://doi.org/10.1007/s11069-021-04877-5, 15730840, http://www.wkap.nl/journalhome.htm/0921-030X.

Hadipour, V., Vafaie, F., & Deilami, K. (2020). Coastal flooding risk assessment using a GIS-based spatial multi-criteria decision analysis approach. *Water*, *12*(9). Available from https://doi.org/10.3390/w12092379, 2073–4441.

Hosseiny, H. (2021). A deep learning model for predicting river flood depth and extent. *Environmental Modelling & Software*, *145*. Available from https://doi.org/10.1016/j.envsoft.2021.105186, 13648152.

Hu, C., Wu, Q., Li, H., Jian, S., Li, N., & Lou, Z. (2018). Deep learning with a long short-term memory networks approach for rainfall-runoff simulation. *MDPI AG, China Water (Switzerland)*, *10*(11). Available from https://doi.org/10.3390/w10111543, 20734441, https://www.mdpi.com/2073-4441/10/11/1543/pdf.

Kowalski, K.J., McDaniel, R.S., & OlekJ. (2010). Identification of laboratory technique to optimize Superpave HMA surface friction characteristics, Purdue University.

Mehebub, S., Raihan, A., Nuhul, H., & Haroon, S. (2015). Assessing flood inundation extent and landscape vulnerability to flood using geospatial technology: A study of Malda district of West Bengal, India. *Forum Geografic*, *XIV*(2), 156–163. Available from https://doi.org/10.5775/fg.2067-4635.2015.144.d, 1583-1523.

Mollah, S. (2016). Assessment of flood vulnerability at village level for Kandi block of Murshidabad district, West Bengal. *Indian Academy of Sciences, India Current Science*, *110*(1), 81–86. Available from https://doi.org/10.18520/cs/v110/i1/81-98, 00113891, http://www.currentscience.ac.in/Volumes/110/01/0081.pdf.

Myronidis, D., Ioannou, K., Fotakis, D., & Dörflinger, G. (2018). Streamflow and hydrological drought trend analysis and forecasting in cyprus. *Water Resources Management*, *32*(5), 1759–1776. Available from https://doi.org/10.1007/s11269-018-1902-z, 0920-4741.

Pakhale, N. V., Gondkar, S. B., & Saudagar, R. B. (2019). Effervescent floating drug delivery system: A review. *Journal of Drug Delivery and Therapeutics*, *9*(3–s), 836–838.

Radwan, F., Alazba, A. A., & Mossad, A. (2019). Flood risk assessment and mapping using AHP in arid and semiarid regions. *Acta Geophysica*, *67*(1), 215–229. Available from https://doi.org/10.1007/s11600-018-0233-z, 1895-6572.

Rajasekhar, M., Gadhiraju, S. R., Kadam, A., & Bhagat, V. (2020). Identification of groundwater recharge-based potential rainwater harvesting sites for sustainable development of a semiarid region of southern India using geospatial, AHP, and SCS-CN approach. *Arabian Journal of Geosciences*, *13*(2). Available from https://doi.org/10.1007/s12517-019-4996-6, 18667538, http://www.springer.com/geosciences/journal/12517?cm_mmc = AD-_-enews-_-PSE1892-_-0.

Sobrino, J. A., Raissouni, N., & Li, Z. L. (2001). A comparative study of land surface emissivity retrieval from NOAA data. *Remote Sensing of Environment*, *75*(2), 256–266. Available from https://doi.org/10.1016/S0034-4257(00)00171-1, 00344257.

Tehrany, M. S., Pradhan, B., & Jebur, M. N. (2014). Flood susceptibility mapping using a novel ensemble weights-of-evidence and support vector machine models in GIS. *Journal of Hydrology*, *512*, 332–343. Available from https://doi.org/10.1016/j.jhydrol.2014.03.008, 00221694.

Tehrany, M. S., Pradhan, B., & Jebur, M. N. (2013). Spatial prediction of flood susceptible areas using rule based decision tree (DT) and a novel ensemble bivariate and multivariate statistical models in GIS. *Journal of Hydrology*, *504*, 69–79. Available from https://doi.org/10.1016/j.jhydrol.2013.09.034, 00221694.

Tehrany, M. S., Shabani, F., Jebur, M. N., Hong, H., Chen, W., & Xie, X. (2017). GIS-based spatial prediction of flood prone areas using standalone frequency ratio, logistic regression, weight of evidence and their ensemble techniques. *Geomatics, Natural Hazards and Risk*, *8* (2), 1538–1561. Available from https://doi.org/10.1080/19475705.2017.1362038, 1947-5705.

Vormoor, K., Lawrence, D., Heistermann, M., & Bronstert, A. (2015). Climate change impacts on the seasonality and generation processes of floods – Projections and uncertainties for catchments with mixed snowmelt/rainfall regimes. *Hydrology and Earth System Sciences*, *19*(2), 913–931. Available from https://doi.org/10.5194/hess-19-913-2015, 1607-7938.

Weier, J., & Herring, D. (2000). Measuring vegetation (ndvi & evi). *NASA Earth Observatory*, *20*(2).

Yalew, S. G., van Griensven, A., Mul, M. L., & van der Zaag, P. (2016). Land suitability analysis for agriculture in the Abbay basin using remote sensing, GIS and AHP techniques, Springer Science and Business Media Deutschland GmbH*Netherlands Modeling Earth Systems and Environment*, *2*(2). Available from https://doi.org/10.1007/s40808-016-0167-x, Springer.com/journal/40808.

Yaseen, Z. M., Fu, M., Wang, C., Mohtar, W. H. M. W., Deo, R. C., & El-shafie, A. (2018). Application of the Hybrid artificial neural network coupled with rolling mechanism and grey model algorithms for streamflow forecasting over multiple time horizons, Springer Netherlands, Viet Nam*Water Resources Management*, *32*(5), 1883–1899. Available from https://doi.org/10.1007/s11269-018-1909-5, http://www.wkap.nl/journalhome.htm/0920-4741.

Yaseen, Z. M., Naganna, S. R., Sa'adi, Z., Samui, P., Ghorbani, M. A., Salih, S. Q., & Shahid, S. (2020). Hourly river flow forecasting: Application of emotional neural network versus multiple machine learning paradigms. *Water Resources Management*, *34*(3), 1075–1091. Available from https://doi.org/10.1007/s11269-020-02484-w, 15731650, http://www.wkap.nl/journalhome.htm/0920-4741.

Chapter 6

Hydrological analysis and assessment of inland flood risk using geospatial data

Cathy Koloa and Sailesh Samanta
School of Surveying and Land Studies, The Papua New Guinea University of Technology, Lae, Papua New Guinea

6.1 Introduction

Papua New Guinea (PNG) is situated close to the equator and falls within the moist tropical rainforest zone with high to very high rainfall. According to the World Bank's Natural Disaster Hotspot study, the country is ranked 54th among countries most exposed to multiple hazards, such as inland and coastal flooding, besides earthquakes, volcanoes, and several others (UNDP, 2016). Climate change is expected to cause changes in flooding patterns, including increased rainfall and changes in anthropogenic influence (Antea Group, 2017; Mercer, Kelman, Suchet-Pearson, & Lloyd, 2009). Inland floods in tropical regions like PNG are particularly severe due to factors like large catchment areas, high precipitation, and slow response times (Eccles, Zhang, & Hamilton, 2019). A specific type of inland flooding occurs in the Sepik River Basin, characterized by gradually rising and receding water levels (Ibbitt, Thompson, & Turner, 2005). Over the past years, there has been constant news of flooding events along the Sepik River, which has disturbed the lives of communities living along the Sepik River (Silverman, 2018; The National, 2019). It was reported that the huge impact of the floods on the communities living along the river caused deaths and affected about 11,500 people without proper sanitation and access to clean and safe drinking water (ABC News, 2013).

Thus, to mitigate the effects of floods, flood risk management systems and early warning systems need to be implemented. In 2018, a piloted flood monitoring and early warning system was established in the Bumbu River of Morobe Province, signifying the positive milestone of early warning system (EWS) in PNG being replicated in other major river systems (UNDRR, 2019).

Water Resources Monitoring, Management, and Sustainability. DOI: https://doi.org/10.1016/B978-0-443-23665-5.00006-5

Therefore, flood risk mapping and vulnerability assessment for the Sepik River basin are important for providing useful information to plan for flood management and emergency response planning (prioritization of interventions). Such an approach will improve land-use planning and limit development in flood-prone areas, identify and protect wetlands, forests, or other ecosystems that could provide flood protection benefits, and provide a visual representation of flood risks for awareness-raising in local communities. The use of flood modeling techniques has been used by researchers around the world to determine flood risk. These include models based on hydrologic, hydraulic, numerical, rainfall-runoff, remote sensing, Geographic Information Systems (GIS), artificial intelligence and machine learning, and multiple-criteria decision analysis (Kumar, Sharma, Caloiero, Mehta, & Singh, 2023). Multi-criteria Decision Analysis (MCDA) within a GIS environment has proven to be very effective for making well-informed decisions when confronted with complex and conflicting decisions (Kumar & Jha, 2023; Nachappa et al., 2020). MCDA has been widely used for disaster management, especially floods, to minimize the risk of flood damage or to reduce the effects of a flood event (Agustina, Putra, & Susanti, 2023; Arabameri, Rezaei, Cerdà, Conoscenti, & Kalantari, 2019; Cabrera & Lee, 2019; de Brito, Almoradie, & Evers, 2019; Ganji, Gharechelou, Ahmadi, & Johnson, 2022; Khosravi et al., 2019; Samanta, Koloa, Pal, & Palsamanta, 2016; Sekac, Jana, & Pal, 2015; Varo, Sekac, & Jana, 2020).

Analytical hierarchical process (AHP) and analytic network process are the commonly used MCDA methods for pairwise comparisons (Kumar et al., 2023; Nachappa et al., 2020). Literature shows the flexibility of the AHP process in determining weights, where pairwise comparisons are used to find variables that are most superior compared to other variables with predetermined criteria previously (Chen, Yeh, & Yu, 2011; Kazakis, Kougias, & Patsialis, 2015; Nachappa et al., 2020; Rahmati, Zeinivand, & Besharat, 2016; Yilmaz, 2022). Therefore, this study aimed to identify and analyze flood-prone areas within the Sepik River Basin using an MCDA and GIS. The weighted overlay approach was used in this study to develop the final flood susceptibility map, which was validated using historical flood datasets. At the end of this study, it was anticipated that the research output would be focused on providing information about the potential flooding zones and creating risk maps for local governments, non-governmental organisations (NGOs), and planners to plan out flood mitigation plans and early warning systems.

6.2 Rationale of the study

The rationale for this study is rooted in the pressing need to address the severe and recurrent flooding issues within the Sepik River Basin in PNG. As a tropical rainforest zone with high rainfall and a location close to the equator, the region is particularly vulnerable to flooding. Additionally, climate change is expected to exacerbate flooding patterns in the region, further increasing the risks faced by local communities. The purpose of this study is to assess flood susceptibility and risk in the Sepik River Basin, which has experienced significant flooding events in recent years, resulting in fatalities, displacements, and a host of socioeconomic hardships. There is no flood-warning system in PNG, including the Sepik River Basin (Antea Group, 2015; UNDRR, 2019), emphasizing the importance of flood risk mapping. These assessments are crucial in informing land-use planning and raising awareness about flood risks among local communities. Therefore, this study intends to bridge this critical gap by employing flood modeling techniques, specifically MCDA, within a GIS environment. To reduce communities' vulnerabilities to devastating flood events within the Sepik River Basin, the research output will provide critical information to local governments, nongovernmental organizations, and urban planners so that they can develop flood mitigation plans and establish effective early warning systems.

6.3 Materials and methods

6.3.1 Geographical setting of the study area

The hydrological analysis, namely flood susceptibility and its risk assessment, was performed within the Sepik River Basin. The Sepik basin lies within the latitudinal and longitudinal extensions of 3° 7′ 01.62″S, 140° 58′ 41.35″E and 5° 51′ 12.01″S, 144° 40′ 45.75″E (Fig. 6.1). The Sepik River flows across extensive lowland plains in the northwest of PNG, and its basin runs parallel to the Central Ranges. In its middle and lower reaches, the flat nature of the terrain causes the Sepik River to meander through broad and swampy flood plains. There are no major earthworks, such as dams and channel diversions.

The Sepik River basin was found to be an 11th-order drainage basin according to software calculations from the Digital Elevation Model (DEM). For visibility purposes, only the 5th to 11th-order channels are presented in Fig. 6.2.

The drainage pattern for this basin was observed to be dendrite. This pattern specifies the similarity in texture and the absence of structural control. The high bifurcation values were found in the lower stream orders, indicating a greater structural complexity, which shows more overland flow and a high number of channels contributing to the mainstream (Table 6.1).

6.3.2 Methodology

Inland flood modeling was conducted using AHP to assess flood risk in the entire Sepik River Basin. Different mandatory input parameters, namely elevation, slope, rainfall, soil drainage, hydrological soil group, distance from drainage, landform and land use, and land cover, were used in the ranking

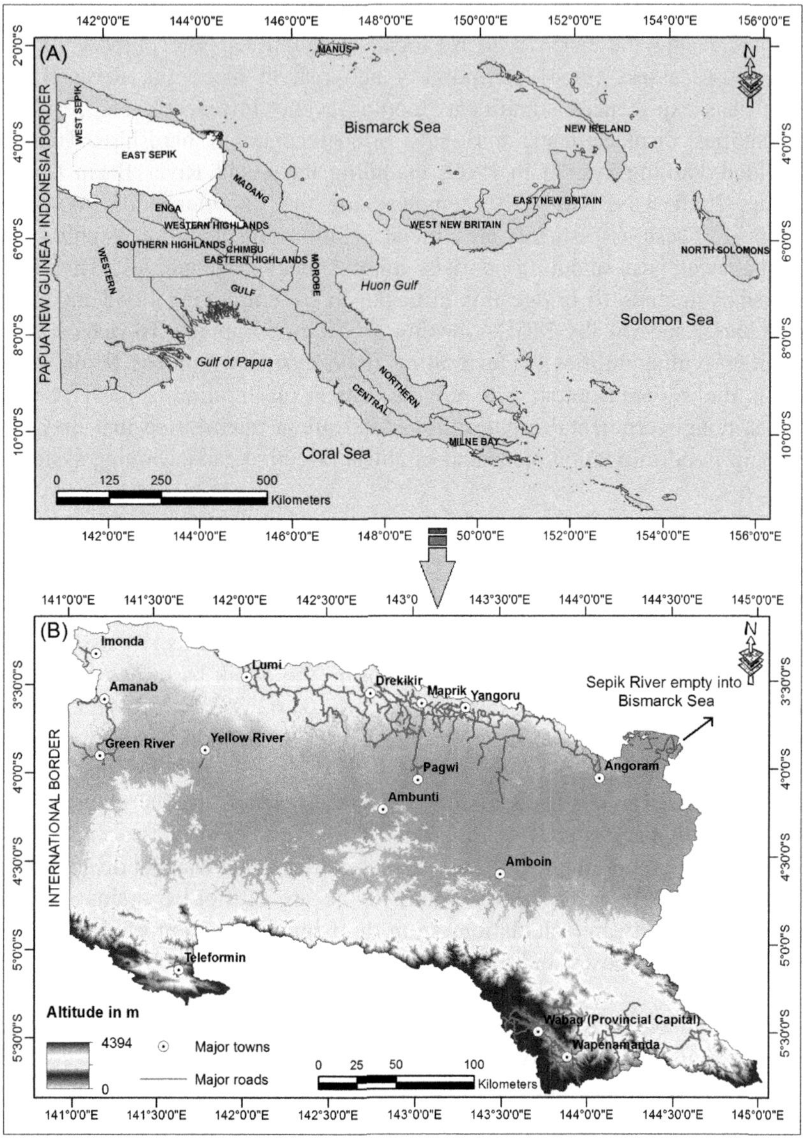

FIGURE 6.1 Geographical location of the study area (A) Papua New Guinea and (B) the Sepik River basin. Map lines delineate study areas and do not necessarily depict accepted national boundaries.

and weight model for inland flood risk mapping. The Advanced Space-borne Thermal Emission and Reflection Radiometer provides a high-resolution (30-m spatial resolution) Global Digital Elevation Model (GDEM) (Abrams et al., 2015; Abrams, 2000). This elevation data was used to prepare the

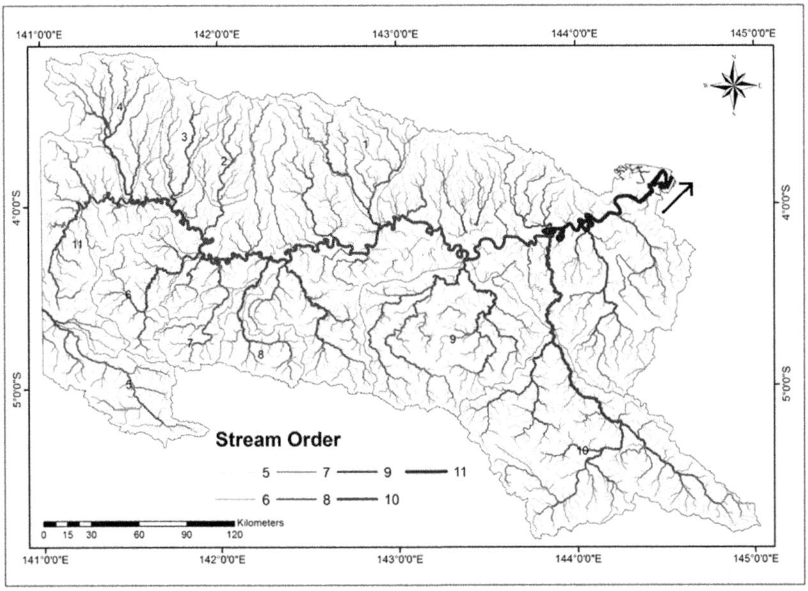

FIGURE 6.2 Stream order characteristics of the Sepik River basin. Map lines delineate study areas and do not necessarily depict accepted national boundaries.

elevation and slope databases. Optical standard false color combination (FCC) bands of the enhanced thematic mapper plus (ETM +) and Landsat-7 satellite imagery were used to produce a land use/land cover database of the study area. The only available national metadata database, the PNG Resource Information System (PNGRIS), was used to derive rainfall characteristics, soil drainage pattern, hydrological characteristics of the soil, distance from drainage, and the landform characteristics of the study area.

All the subclasses of the selected parameters were ranked based on their preferences for flooding. The ranking criterion was adopted from Saaty's AHP theory (Saaty, 1987). The fundamental scale is 1–5, where 1 is less favorable and 5 is highly favorable. The weights were assigned to each parameter to identify which criterion has a more favorable influence on flooding and which has less.

6.4 Results and discussion

Supervised classification was performed on the individual geometrically rectified scenes, which were mosaicked later on, and the catchment was finally extracted. From the tonal values of the original FCC imageries, land cover classes were selected, and a signature editor was created to supervise the classification. Seven (7) land use land cover classes were extracted, namely, agriculture and crops, forest, grassland and shrubs, wetland, water area, bare land/disturbed land, and

TABLE 6.1 Linear parameters for Sepik River basin based on 90 m DEM data.

Stream order	Number of streams (Nu)	Stream length (Lu) (km)	Mean stream length (Lum)	Bifurcation ratio (Rb)	Stream length ratio (RL)	Mean (stream length ratio)
5	3534	12478	4		0.5	2.13
6	823	6201	8	4.29	0.24	2.36
7	201	3575	18	4.09	0.26	1.94
8	52	1792	34	3.87	0.29	1.44
9	15	746	50	3.47	0.27	3.74
10	4	745	186	4	0.25	0.78
11	1	145	145	4	0	0

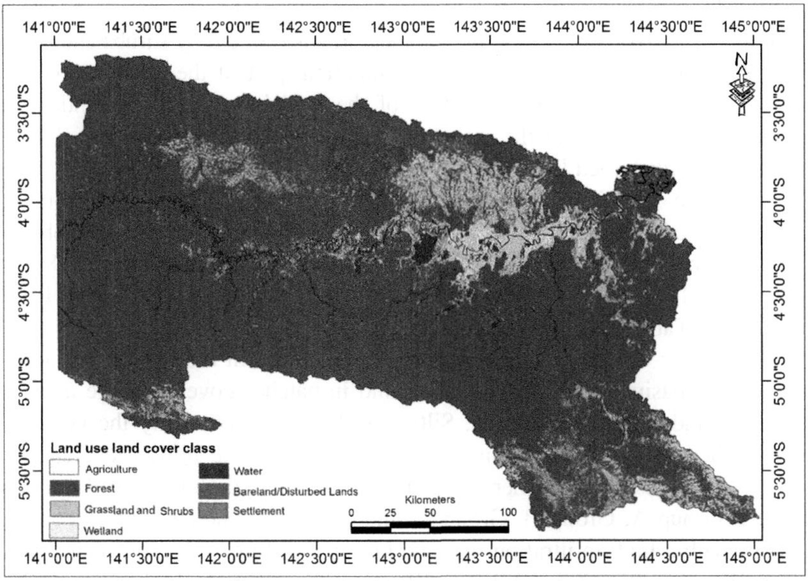

FIGURE 6.3 Land use/land cover of the Sepik River basin, generated from Landsat-7, ETM + . Map lines delineate study areas and do not necessarily depict accepted national boundaries.

TABLE 6.2 Statistics for different land use/land cover classes in the study area.

Value	Land use/Land cover	Area (ha)	Percentage %
1	Agriculture and crops	18653	0.24
2	Forest	6491763	84.88
3	Grassland and shrubs	853795	11.16
4	Wetland	84725	1.11
5	Water	134574	1.76
6	Bare land/disturbed lands	64845	0.85
7	Settlement	214	0.003
Total area		7648569	100

settlement/built-up (Fig. 6.3). Post-classification verification was done to evaluate the accuracy of the classification through field observations and extracting the reference data from recent Google Earth imagery. The overall accuracy of the classification was 93%, which is more than the acceptable rate (Manandhar, Odehi, & Ancevt, 2009). As presented in Table 6.2, 84.88% of the study area is covered by forest, covering both the northern and southern parts of the study area. About 11.16% of the land in the study area is covered by shrubs and grassland. Agriculture activity is also a very important part of the basin and covers 0.24% of the land. Less than 0.003% of the land is used for residential and industrial purposes, while the water class covers 1.76%, wetlands are 1.11%, and bare land/disturbed land is 0.85%.

There are 12 soil texture classes found, along with a few other non-soil classes, based on the PNGRIS (PNGRIS, 2009) and Geobook (Geobook, 2009) databases of PNG. Major soil texture classes are sandy clay (30.89%), silty clay (18.88%), and silty clay loam (14.70%) within the Sepik River basin area (Table 6.3).

Sandy clay can be observed covering large areas in the south- and northwest of the basin, while silty clay is found in patches covering large areas in the north and south of the basin. Silty clay loam covers mostly the northeast and southern parts of the basin (Fig. 6.4). Based on the infiltration rate, the soil texture was thus further classified into the hydrological soil groups, namely, Group A, Group B, Group C, and Group D, where Group A consists of soils with a high infiltration rate and Group D refers to the soil with a poor infiltration rate (Fig. 6.5). A low infiltration rate refers to higher runoff and

TABLE 6.3 Soil texture classes of the Sepik River basin.

Sl. no.	Soil texture	Area in ha	Area in %
1	Clay	703,521.5	9.2
2	Silty clay	1,444,379	18.88
3	Silty clay loam	1,124,601	14.7
4	Sit	380,401.2	4.97
5	Loamy	42,873.03	0.56
6	Clay loam	13,492.89	0.18
7	Silt loam	176,811.3	2.31
8	Sandy loam	453,999.2	5.94
9	Sandy clay loam	178,450.3	2.33
10	Sandy clay	2,362,495	30.89
11	Loamy sand	2.25	0.0
13	Peat	765,973.4	10.01
15	Gravel	1568.97	0.02
Total area		7,648,569	100

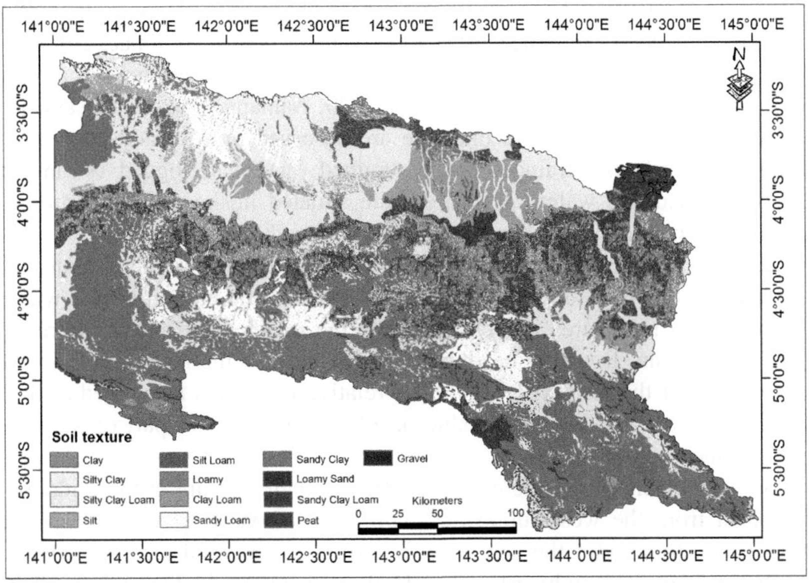

FIGURE 6.4 Spatial distribution of soil texture in the Sepik River basin. Map lines delineate study areas and do not necessarily depict accepted national boundaries.

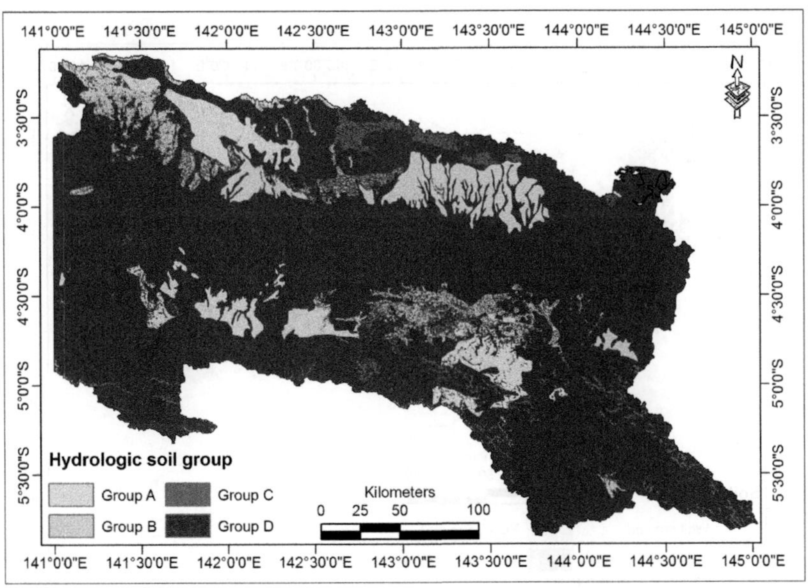

FIGURE 6.5 Spatial distribution of hydrological soil group in the Sepik River basin. Map lines delineate study areas and do not necessarily depict accepted national boundaries.

higher erosion (Quinton & Catt, 2004). The higher runoff refers to the higher potential of a flood (Zaharia, Costache, Prăvălie, & Ioana-Toroimac, 2017).

Soil drainage also refers to the probability of flood occurrences in a basin area (Chowdhuri, Pal, & Chakrabortty, 2020; Samanta, Pal, & Palsamanta, 2018). Four (4) types of soil drainage are observed in the study area: well-drained, imperfectly drained, poorly drained, and waterlogged (swampy). Poorly drained and waterlogged areas are situated in the middle portion of the basin, which areas have more potential to flood (Fig. 6.6).

The absolute relief with a maximum altitude of 4394 is found along the southern part of the study area, while almost $^2/_3$ of the entire basin is within the minimum relief area, which is mainly the middle part and the northern part of the basin, from the headwaters in the west to the river mouth in the eastern part of the basin (Fig. 6.7). The relative relief was calculated as having a maximum relative relief value of 723 in the southern part of the river basin in the headwaters.

Slope analysis shows the characteristics of flatter terrain in the middle of the basin from the west, along the main valley toward the east, where the river basin drains into the sea. Steeper terrain is observed on both sides of the basin from the southwest end right through to the southeast part of the basin. The north-west of the basin also covers some steeper slopes. The maximum slope of 89.72 degrees is seen in the South, South West, and South East parts of the River Basin (Fig. 6.8).

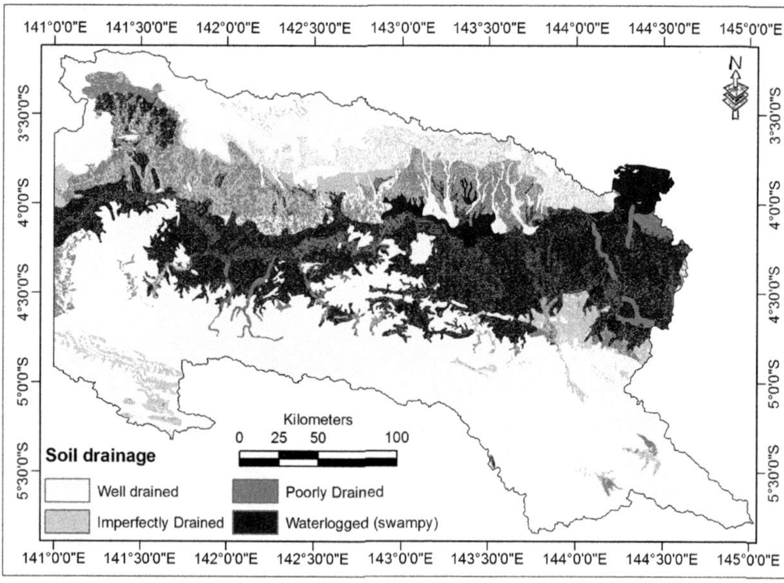

FIGURE 6.6 Soil drainage characteristics of the Sepik River basin. Map lines delineate study areas and do not necessarily depict accepted national boundaries.

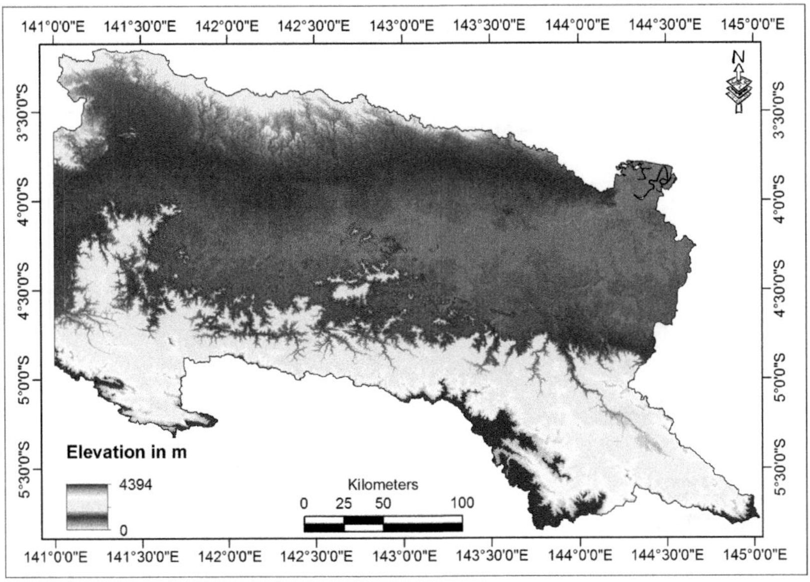

FIGURE 6.7 The altitude variation of the Sepik River basin. Map lines delineate study areas and do not necessarily depict accepted national boundaries.

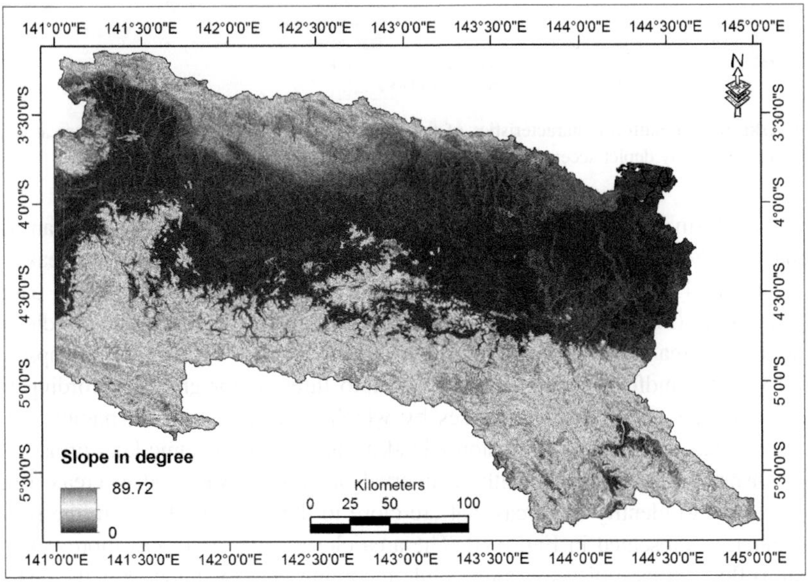

FIGURE 6.8 DEM-based slope in degree characteristics in the Sepik River basin. Map lines delineate study areas and do not necessarily depict accepted national boundaries.

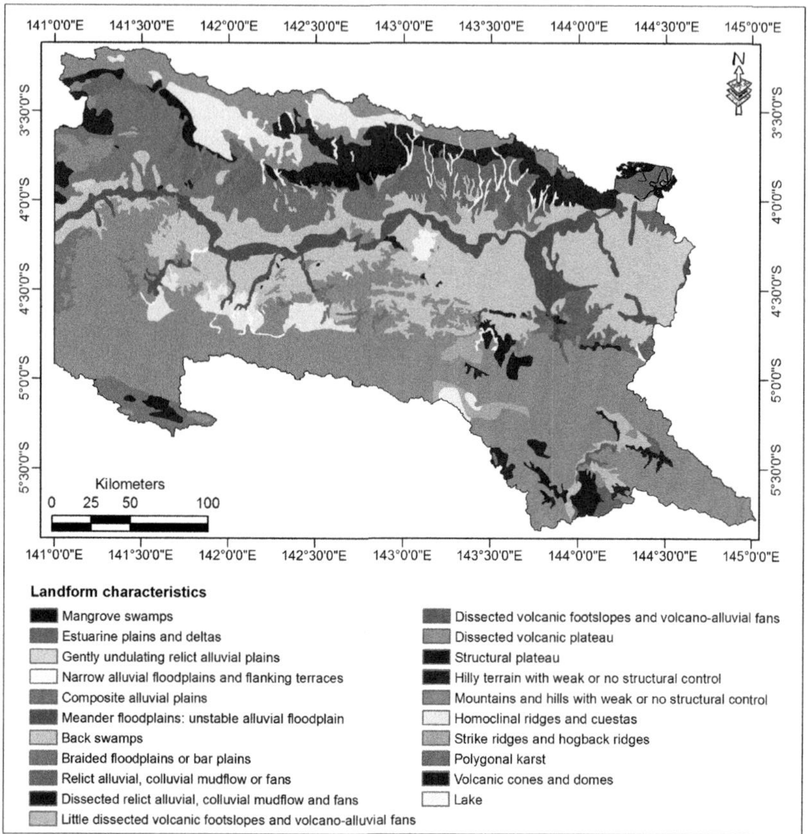

FIGURE 6.9 Landform characteristics of the study area. Map lines delineate study areas and do not necessarily depict accepted national boundaries.

Landform refers to the recurring pattern of topography (MacMillan & Shary, 2009). Most of the area is covered by mountains and hills with weak or no structural control, which are mostly found in the northern and southern parts of the basin (Fig. 6.9). Meander flood plains and back swamps in the middle of the basin area have more potential to flood (Croke, Fryirs, & Thompson, 2013). The landform types were divided into three major groups according to the dominant geomorphic processes by which they were formed, namely: (1) depositional landforms, (2) erosional landforms, and (3) volcanic landforms.

Distance from the river with an interval of 1000 m was used to create five (5) classes to identify the areas that are close to the river and far from the river channel, as presented in Fig. 6.10. The area closer to the river has more potential to flood in case of heavy rain, while incoming water volume exits the channel capacity (Bonasia & Ceragene, 2021). The rainfall database was prepared with five distinguished classes, as represented in Fig. 6.11. The average annual rainfall ranges from 1850 to 9000 mm. The higher rainfall occurs in the southwest mountain terrain than in the coastal area in the north (Fig. 6.11).

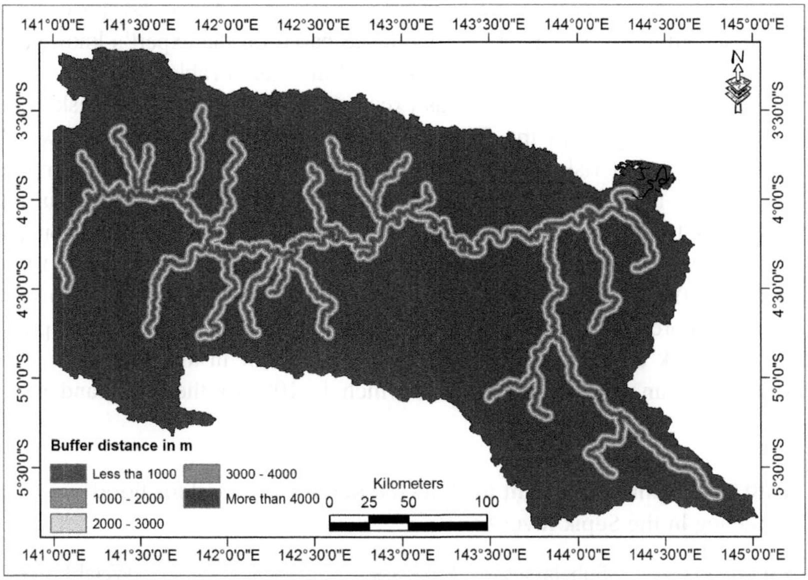

FIGURE 6.10 The buffer distance from drainage characteristics of the study area. Map lines delineate study areas and do not necessarily depict accepted national boundaries.

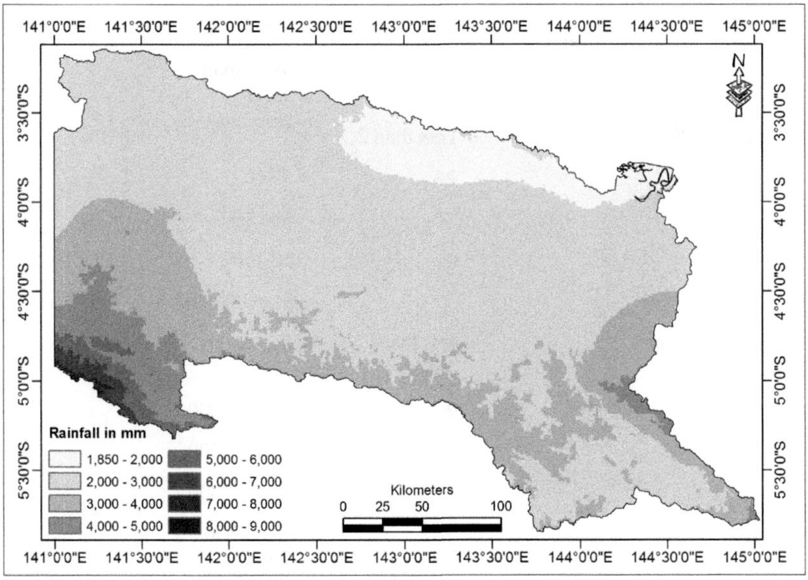

FIGURE 6.11 Annual average rainfall characteristics of the Sepik River basin. Map lines delineate study areas and do not necessarily depict accepted national boundaries.

Weights were assigned to all eight (8) parameters, and their subclasses were ranked according to their relative importance based on the expert's knowledge and results from other associated research and literature (Table 6.4).

A weighted overlay analysis was executed to identify the flood risk areas of the lowland region of the Sepik River Basin during high storm flows. A spatial distribution risk map of the inland flood is presented according to flood extent, along the lower basin area of the Sepik River. The resulting output was further reclassified into five (5) suitability groups based on the risk index value, namely: (1) low risk (index value of less than 30), (2) low risk (30–40), (3) medium risk (40–45), (4) high risk (45–55), and (5) very high risk (more than 55). The blue color pixel in Fig. 6.12 indicates the probable flood risk area where the land slope is very low in the study area. The result shows an area of 765,092 ha, which is 10% of the total land areas

TABLE 6.4 Rank and weight for selected parameters used for flood modeling in the Sepik River basin.

Parameters	Subclasses number	Subclass name	Rank	Weight	
Elevation (in m)	1	Up to 2000	5		
		2	2000–2500	2	4
		3	2500–3000		3
		4	3000–3500		2
		5	More than 3500		1
Slope (Degree)	1	Less than 2	5		
		2	5–10	3	4
		3	10–20		3
		4	20–30		2
		5	More than 30		1
Rainfall (in mm)	1	less than 3500	1		
		2	3500–4000	1	2
		3	4000–4500		3
		4	4500–5000		4
		5	more than 5000		5

(*Continued*)

TABLE 6.4 (Continued)

Parameters	Subclasses number	Subclass name	Rank	Weight	
Soil drainage	1	Well drained	1		
		2	Imperfectly Drained	2	2
		3	Poorly Drained		3
		4	Waterlogged (swampy)		4
Hydrological soil group	1	Group A	1		
		2	Group B	1	2
		3	Group C		3
		4	Group D		4
Distance from drainage (m)	1	Up to 1000 m	5		
		2	2000 m	4	4
		3	3000 m		3
		4	4000 m		2
		5	More than 4000 m		1
Landforms	1	Depositional Landform	3	5	
		2	Volcanic Landform		2
		3	Erosional Landform		1
Land use and land cover	1	Forest	1		
		2	Grassland	1	4
		3	Agricultural Land		3
		4	Urban Built-Up		2
		5	Barren Land		4
		6	Water		5
		7	Shrubs		3

within the basin, are marked as high-risk flood zones. A total population of 54,947 of 224 villages falls within the very high-risk zone and is more vulnerable to the hazardous impact of inland floods (Fig. 6.12 and Table 6.5). The result shows several existing infrastructures within the very high-risk category (5), including 54 primary schools, nine airports/airstrips, three towns, seven health centers, 42 aid posts, and 4 bridges, as derived from existing datasets of PNGRIS (PNGRIS, 2009) (Table 6.6).

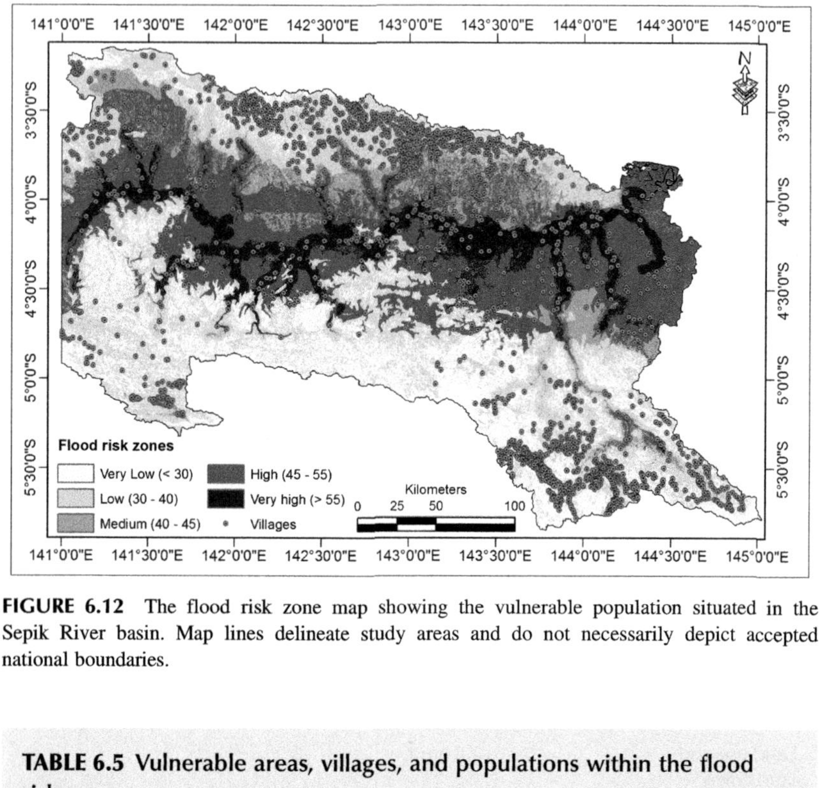

FIGURE 6.12 The flood risk zone map showing the vulnerable population situated in the Sepik River basin. Map lines delineate study areas and do not necessarily depict accepted national boundaries.

TABLE 6.5 Vulnerable areas, villages, and populations within the flood risk zone.

Zones	Flood risk	Area (ha)	%	Villages	Population
1	Very Low (Up to 30)	1,367,314	17.9	135	30,912
2	Low (30–40)	2,650,796	34.7	1418	389,226
3	Medium (40–45)	708,632	9.3	298	89,528
4	High (45–55)	2,156,735	28.2	403	86214
5	Very High (More than 55)	765,092	10.0	224	54,947

TABLE 6.6 Vulnerable area and infrastructural elements for estimated flood depth.

Flood Risk	High schools	Primary school	Airports/Airstrips	Towns	Health Center	Aid posts	Bridges
Very low	2	20	5	0	3	27	11
Low	11	281	31	7	57	249	116
Medium	8	82	23	3	16	63	64
High	5	94	39	4	19	84	38
Very high	0	54	9	3	7	42	4

6.5 Conclusions

A hydrological disaster (flood) is a powerful, critical change that occurs at unexpected times in the spread or movement of water on the surface of an overflow of an expanse of water in which land is inundated by water. It is not considered a hazard until land resources, population in villages and cities, infrastructure, or inhabited areas are inundated by water. This study represents all aspects of flood risk zone identification and risk assessment. Validation is a must to assess the level of acceptability of each model used for flood risk mapping (Kwag, Gupta, & Dinh, 2018). The validation test was performed after overlaying historical flood points on the resulting flood susceptibility database. These flood points were derived using fishnet analysis (Samanta et al. 2018) using the existing inundation database of PNGRIS (PNGRIS, 2009). Very high and high-risk zones were considered positive for potential flooding zones, while on the other hand, moderately low and very low susceptibility zones were ignored for flooding. The analysis (Table 6.7) represents 46 flood points within the medium flood-risk zone, 360 flood points within the high flood-risk zone, and 137 flood points under the very high-risk zone (Fig. 6.13). The analysis further estimates that 497 (360 + 137) historical flood points are situated within the potential flooding zone out of 543, which leads to 91.69% accuracy and validates the model and methodology used in this study. The methodology used in this study is quite user-friendly and can be used by disaster management authorities. These types of risk maps can be developed for any other river basins in Papua New Guinea to manage flood hazards in the context of regional and local knowledge, using data gathered by site-specific gauges and other relevant technology.

TABLE 6.7 Historical flood points within the flood risk zone.

Risk zones	Flood risk	Historical flood points under flood risk	%
1	Very Low (Up to 30)	0	0
2	Low (30–40)	0	0
3	Medium (40–45)	46	8.5
4	High (45–55)	360	66.3
5	Very High (More than 55)	137	25.2
Total		543	100

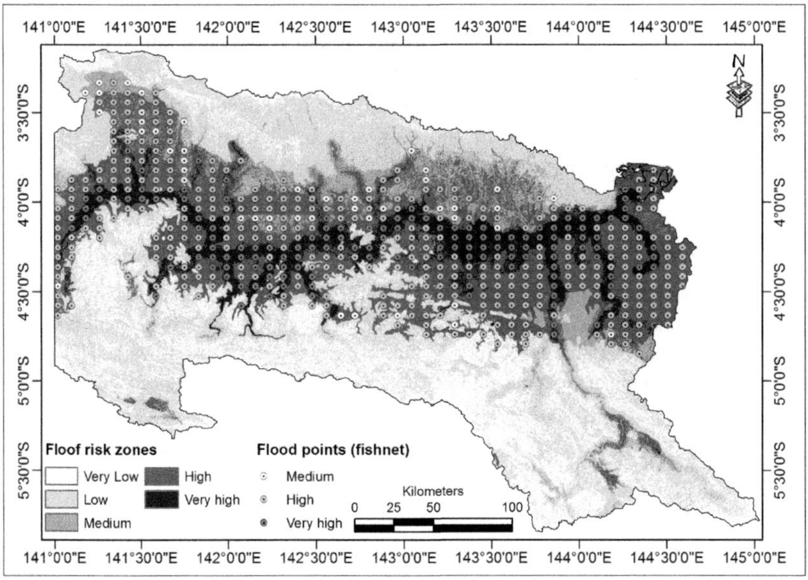

FIGURE 6.13 Historical flood points within flood risk zones in the Sepik River basin. Map lines delineate study areas and do not necessarily depict accepted national boundaries.

6.5.1 Recommendations

The findings from this research have observed how the different basin characteristics and other geography and hydrological parameters can influence inland flooding within a basin. Thus, hydrologic assessments have proven to be very helpful in examining inland flood risk behavior. The approach used in this research was based on logical conditions and reasoning and can be considered in other river basins for identifying and mapping vulnerable areas for efficient planning and management. Indeed, the present work makes an initial attempt at disaster mapping, particularly flood hazard modeling over a coastal basin area, but it also evokes a lot of scope for further studies. This study recommends further improvements in land use/land cover characterization, as well as using coastal topography, bathymetric data built from higher-resolution satellite data, and updated rainfall data. Emphasis might be placed on constructing the model selection for different independent variables. Variables such as soil erosion, surface runoff, coastal land use, El-Niño, and prevailing global and local wind patterns might be considered for further improvements.

6.5.2 Limitations of the study

The major limitation of this research was related to the size of the Sepik basin, which is very large in spatial terms. The portion of the basin that falls under another country (Indonesia) was excluded from this research as some information was not available in those areas. The national databases of Papua New Guinea, PNGRIS (PNGRIS, 2009), and Geobook (Geobook, 2009) are quite old (14 years old) and may be questionable, but there are no other options to gather some of the data sets used in this research, like soil texture, soil drainage, historical flood/inundation, and rainfall.

Acknowledgment

The authors are grateful to the Department of Surveying and Land Studies as well as The Papua New Guinea University of Technology for providing the opportunity and providing a digital image interpretation laboratory facility to carry out the research work. The authors are also grateful to the anonymous referees for their valuable comments and suggestions.

Funding sources

The research was not funded.

Declarations on conflict of interest

The authors declare that there is no conflict of interest for the publication of this article.

Data availability

Data related to the manuscript are kept electronically and will be made available on reasonable request.

References

ABC News. (2013). Unpublished content PNG floods at crisis point. Available from https://www.abc.net.au/news/2013-05-07/an-png-floods-at-crisis-point/4674414.

Abrams, M. (2000). The Advanced Spaceborne Thermal Emission and Reflection Radiometer (ASTER): Data products for the high spatial resolution imager on NASA's Terra platform. *International Journal of Remote Sensing*, *21*(5), 847–859. Available from https://doi.org/10.1080/014311600210326.

Abrams, M., Tsu, H., Hulley, G., Iwao, K., Pieri, D., Cudahy, T., ... Kargel, J. (2015). The Advanced Spaceborne Thermal Emission and Reflection Radiometer (ASTER) after fifteen years: Review of global products. *International Journal of Applied Earth Observation and Geoinformation*, *38*, 292–301. Available from https://doi.org/10.1016/j.jag.2015.01.013.

Agustina, R. D., Putra, R. P., & Susanti, S. (2023). Mapping Greater Bandung flood susceptibility based on multi-criteria decision analysis (MCDA) using AHP method. *Environmental Earth Sciences*, *82*(15). Available from https://doi.org/10.1007/s12665-023-11062-3, https://www.springer.com/journal/12665.

Antea Group. (2017). Climate risk, Vulnerability and Risk Assessment in the New Ireland Province in Papua New Guinea. Ref. No. PNG/AF/VNA/2014. Available from https://info.undp.org/docs/pdc/Documents/PNG/Report_Climate%20Risk%20Vulnerability%20Assessment_New%20Ireland.pdf.

Antea Group. (2015). An Assessment of Early Warning Systems for Inland and Coastal Flooding in Papua New Guinea. Ref. No. PNG/AF/EWS/2014. Available from https://info.undp.org/docs/pdc/Documents/PNG/Report_Assessment%20of%20Early%20Warning%20System%20%20in%20PNG.pdf.

Arabameri, A., Rezaei, K., Cerdà, A., Conoscenti, C., & Kalantari, Z. (2019). A comparison of statistical methods and multi-criteria decision making to map flood hazard susceptibility in Northern Iran. *Science of the Total Environment*, *660*, 443–458. Available from https://doi.org/10.1016/j.scitotenv.2019.01.021, http://www.elsevier.com/locate/scitotenv.

Bonasia, R., & Ceragene, M. (2021). Hydraulic numerical simulations of La Sabana River Floodplain, Mexico, as a Tool for a Flood Terrain Response Analysis. *Water*, *13*(24), 3516. Available from https://doi.org/10.3390/w13243516.

Cabrera., & Lee. (2019). Flood-prone area assessment using GIS-based multi-criteria analysis: A case study in Davao Oriental, Philippines. *Water*, *11*(11), 2203. Available from https://doi.org/10.3390/w11112203.

Chen, Y. R., Yeh, C. H., & Yu, B. (2011). Integrated application of the analytic hierarchy process and the geographic information system for flood risk assessment and flood plain management in Taiwan. *Natural Hazards*, *59*(3), 1261–1276. Available from https://doi.org/10.1007/s11069-011-9831-7.

Chowdhuri, I., Pal, S. C., & Chakrabortty, R. (2020). Flood susceptibility mapping by ensemble evidential belief function and binomial logistic regression model on river basin of eastern India. *Advances in Space Research*, *65*(5), 1466–1489. Available from https://doi.org/10.1016/j.asr.2019.12.003, http://www.journals.elsevier.com/advances-in-space-research/.

Croke, J., Fryirs, K., & Thompson, C. (2013). Channel-floodplain connectivity during an extreme flood event: Implications for sediment erosion, deposition, and delivery. *Earth Surface Processes and Landforms*, *38*(12), 1444–1456. Available from https://doi.org/10.1002/esp.3430.

de Brito, M. M., Almoradie, A., & Evers, M. (2019). Spatially-explicit sensitivity and uncertainty analysis in a MCDA-based flood vulnerability model. *International Journal of Geographical Information Science*, *33*(9), 1788–1806. Available from https://doi.org/10.1080/13658816.2019.1599125, http://www.tandf.co.uk/journals/titles/13658816.asp.

Eccles, R., Zhang, H., & Hamilton, D. (2019). A review of the effects of climate change on riverine flooding in subtropical and tropical regions. *Journal of Water and Climate Change*, *10*(4), 687–707. Available from https://doi.org/10.2166/wcc.2019.175, https://watermark.silverchair.com/jwc0100687.pdf?.

Ganji, K., Gharechelou, S., Ahmadi, A., & Johnson, B. A. (2022). Riverine flood vulnerability assessment and zoning using geospatial data and MCDA method in Aq'Qala. *International Journal of Disaster Risk Reduction*, *82*. Available from https://doi.org/10.1016/j.ijdrr.2022.103345, http://www.journals.elsevier.com/international-journal-of-disaster-risk-reduction/.

Geobook. (2009). The Remote Sensing Centre of University of Papua New Guinea, Port Moresby, Papua New Guinea Unpublished content Provincial GIS-based Planning Tools.

Ibbitt, R., Thompson, C., & Turner, R. (2005). Skill assessment of a linked precipitation-runoff flood forecasting system. *Journal of Hydrology New Zealand*, *44*(2), 91–104.

Kazakis, N., Kougias, I., & Patsialis, T. (2015). Assessment of flood hazard areas at a regional scale using an index-based approach and Analytical Hierarchy Process: Application in Rhodope-Evros region, Greece. *Science of the Total Environment*, *538*, 555–563. Available from https://doi.org/10.1016/j.scitotenv.2015.08.055, http://www.elsevier.com/locate/scitotenv.

Khosravi, K., Shahabi, H., Pham, B. T., Adamowski, J., Shirzadi, A., Pradhan, B., ... Prakash, I. (2019). A comparative assessment of flood susceptibility modeling using Multi-Criteria Decision-Making Analysis and Machine Learning Methods. *Journal of Hydrology*, *573*,

311–323. Available from https://doi.org/10.1016/j.jhydrol.2019.03.073, http://www.elsevier.com/inca/publications/store/5/0/3/3/4/3.

Kumar, N., & Jha, R. (2023). GIS-based flood risk mapping: The Case Study of Kosi River basin, Bihar, India. *Engineering, Technology and Applied Science Research, 13*(1), 9830–9836. Available from https://doi.org/10.48084/etasr.5377, https://etasr.com/index.php/ETASR/article/download/5377/2952.

Kumar, V., Sharma, K. V., Caloiero, T., Mehta, D. J., & Singh, K. (2023). Comprehensive overview of flood modeling approaches: A review of recent advances. *Hydrology, 10*(7). Available from https://doi.org/10.3390/hydrology10070141, http://www.mdpi.com/journal/hydrology.

Kwag, S., Gupta, A., & Dinh, N. (2018). Probabilistic risk assessment based model validation method using Bayesian network. *Reliability Engineering and System Safety, 169*, 380–393. Available from https://doi.org/10.1016/j.ress.2017.09.013.

MacMillan, R. A., & Shary, P. A. (2009). Landforms and landform elements in geomorphometry. *Developments in Soil Science, 33*. Available from https://doi.org/10.1016/S0166-2481(08)00009-3, http://www.sciencedirect.com.

Manandhar, R., Odehi, I. O. A., & Ancevt, T. (2009). Improving the accuracy of land use and land cover classification of landsat data using post-classification enhancement. *Remote Sensing, 1*(3), 330–344. Available from https://doi.org/10.3390/rs1030330, http://www.mdpi.com/2072-4292/1/3/330/pdf.

Mercer, J., Kelman, I., Suchet-Pearson, S., & Lloyd, K. (2009). Integrating indigenous and scientific knowledge bases for disaster risk reduction in Papua New Guinea. *Geografiska Annaler, Series B: Human Geography, 91*(2), 157–183. Available from https://doi.org/10.1111/j.1468-0467.2009.00312.x.

Nachappa, T. G., Piralilou, S. T., Gholamnia, K., Ghorbanzadeh, O., Rahmati, O., & Blaschke, T. (2020). Flood susceptibility mapping with machine learning, multi-criteria decision analysis and ensemble using Dempster Shafer Theory. *Journal of Hydrology., 590.*

PNGRIS. (2009). University of Papua New Guinea The Land-Use Section, Science and Technology Branch, Department of Agriculture and Livestock, Boroko, Papua New Guinea Unpublished content Papua New Guinea Resource Information System.

Quinton, J. N., & Catt, J. A. (2004). The effects of minimal tillage and contour cultivation on surface runoff, soil loss and crop yield in the long-term Woburn Erosion Reference Experiment on sandy soil at Woburn, England. *Soil Use and Management, 20*(3), 343–349. Available from https://doi.org/10.1079/SUM2004267.

Rahmati, O., Zeinivand, H., & Besharat, M. (2016). Flood hazard zoning in Yasooj region, Iran, using GIS and multi-criteria decision analysis. *Geomatics, Natural Hazards and Risk, 7*(3), 1000–1017. Available from https://doi.org/10.1080/19475705.2015.1045043, http://www.tandfonline.com/toc/tgnh20/current.

Saaty, R. W. (1987). The analytic hierarchy process-what it is and how it is used. *Mathematical Modelling, 9*(3-5), 161–176. Available from https://doi.org/10.1016/0270-0255(87)90473-8.

Samanta, S., Koloa, C., Pal, D. K., & Palsamanta, B. (2016). Flood risk analysis in lower part of Markham River based on Multi-Criteria Decision Approach (MCDA). *Hydrology, 3*(3), 29. Available from https://doi.org/10.3390/hydrology3030029.

Samanta, S., Pal, D. K., & Palsamanta, B. (2018). Flood susceptibility analysis through remote sensing, GIS and frequency ratio model. *Applied Water Science, 8*(2). Available from https://doi.org/10.1007/s13201-018-0710-1, https://www.springer.com/journal/13201.

Sekac, T., Jana, S. K., & Pal, D. K. (2015). A remote sensing and GIS approach to assessing multiple environmental factors leading to delineation of flood hazard risk zone in the Busu River catchment, Morobe Province. *Melanes J Geomat Prop Stud., 1*, 40–55.

Silverman, E. K. (2018). *The Sepik River, Papua New Guinea: Nourishing tradition and modern catastrophe* (pp. 187–221). ANU Press. Available from 10.22459/IR.06.2018.08.

The National. (2019). The National Unpublished content Rising water levels put Sepik River communities at risk. Available from https://www.thenational.com.pg/rising-water-levels-put-sepik-river-communities-at-risk/.

UNDP. (2016). UNDP United Nations Development Program Unpublished content Papua New Guinea to develop information systems for disasters. Available from https://www.preventionweb.net/quick/27836.

UNDRR. (2019). Disaster Risk Reduction in Papua New Guinea: Status Report United Nations Office for Disaster Risk Reduction (UNDRR), Regional Office for Asia and the Pacific Unpublished content Disaster Risk Reduction in Papua New Guinea: Status Report.

Varo, J., Sekac, T., & Jana, S. K. (2020). Flood hazard micro zonation from a geomatic perspective on Vitilevu Island, Fiji. *International Journal of Geoinformatics*, *16*(3), 37–47. Available from http://file:///C:/Users/304401/Downloads/1599-1889-1-PB.pdf.

Yilmaz, O. S. (2022). Flood hazard susceptibility areas mapping using Analytical Hierarchical Process (AHP), Frequency Ratio (FR) and AHP-FR ensemble based on Geographic Information Systems (GIS): A case study for Kastamonu, Türkiye. *Acta Geophysica*, *70*(6), 2747–2769. Available from https://doi.org/10.1007/s11600-022-00882-9, https://www.springer.com/journal/11600.

Zaharia, L., Costache, R., Prăvălie, R., & Ioana-Toroimac, G. (2017). Mapping flood and flooding potential indices: A methodological approach to identifying areas susceptible to flood and flooding risk. Case study: The Prahova catchment (Romania). *Frontiers of Earth Science*, *11*(2), 229–247. Available from https://doi.org/10.1007/s11707-017-0636-1, http://www.springer.com/west/home?SGWID = 4-102-70-173677829-0&changeHeader = true.

Chapter 7

Impact of soil salinity on the increasing trends of Aqua farming in the coastal blocks of Purba Medinipur District: a Geospatial Approach

Suman Das[1], Jatisankar Bandyopadhyay[1], Nirupam Acharyya[1], Mallika Jana[1] and Saroj Maity[2]
[1]*Department of Remote Sensing and GIS, Vidyasagar University, Midnapore, West Bengal, India,*
[2]*Indian Space Research Organization, Space Applications Center, Ahmedabad, Gujarat, India*

7.1 Introduction

The global issue of climate change has emerged as a widespread environmental problem (Kawser, Nath, & Hoque, 2022; Paul & Rashid, 2016). The dynamic relationship between land and sea is a prominent feature in coastal areas, which face particular hazards and coastal residents worldwide are growing increasingly vulnerable to natural disasters (Thomas & LLpez, 2015). According to the World Bank Group report, 6−12 million people might be impacted by coastal flooding by 2070−2100, depending on the emission pathway, and climate change would cause a 3.5% decline in national income by 2050 (Nguyen, Liou, Tran, Hoang, & Nguyen, 2020). Several studies consider sea-level rise to be the most important factor influencing changes in coastal ecosystems (Kawser *et al.*, 2022; Mukhopadhyay, Sarkar, Jat, Sharma, & Bolan, 2021; Nath et al., 2023; Sarkar, Islam, & Majumder, 2021). Deltaic areas are particularly vulnerable to climate change due to sea level rise, cyclones, waterlogging, increased salinity, and land subsidence, which are affecting the coastal areas, including soil productivity, water quality, Agriculture lands, freshwater ponds, decreases in wetland and groundwater table, and an increase in soil and water salinity from saltwater intrusion (Kawser *et al.*, 2022).

Water Resources Monitoring, Management, and Sustainability. DOI: https://doi.org/10.1016/B978-0-443-23665-5.00007-7

Soil salinity is influenced by constant rainfall, evaporation rates, and seawater intrusion. The daily existence and economic viability of coastal communities and regions have been consistently affected by soil salinity (Mukhopadhyay et al., 2021; Yin et al., 2016; Yu et al., 2014). Soil salinity contributes significantly to soil degradation phenomena such as increased soil erosion, soil dispersion, and biochemical characteristics. According to Nguyen et al. (2020), when salts are dissolved in water and accumulated in the soil, salinity intrusion expands, affecting agricultural production, the environment, and the economy. As a result of the increase in soil salinity, the coastal area has seen notable changes in the land use land cover (LULC) classes. Cyclones, including the most destructive cyclone Alia in 2009, have caused the salinization of coastal landscapes, including agriculture lands, sweet water ponds, canal beds, and riverbeds over the coastal regions of West Bengal and Odisha (Gayathri, Murty, Bhaskaran, & Srinivasa Kumar, 2016). The Bay of Bengal is especially vulnerable to the effects of climate change because of several coastal factors, such as waves, tides, wind, saltwater intrusion, storm surge, sea-level rise, and human involvement (Muhammad Abdullah et al., 2021; Nath et al., 2023). Aside from natural processes, anthropogenic activities contribute to increased soil salinity in coastal areas. In India, population expansion has been rapid in recent decades, and as a result, land alteration has occurred, threatening the coastal ecology as well as long-term growth.

Lower crop rates are the result of saltwater intrusion, which has a significant influence on agricultural production worldwide and along India's coasts (Hossen, Yabar, & Faruque, 2022). As a result, the socioeconomic development of West Bengal's coastal areas has greatly benefited from aquaculture farming. Aquaculture farming began to progressively expand commercially due to the strengthening demand on the global market, the higher economic return from small investments, and the approval of government policies. Consequently, the shrimp-farming zone of West Bengal was converted into the coastal areas of South 24 Parganas and Purba Medinipur. This makes aquaculture farming associated with more detrimental effects on the environment, such as pollution, disease outbreaks, sedimentation, salinization, decreasing rice production, shrinking mangroves, and even losses in biodiversity (Akber, Islam, Ahmed, Rahman, & Rahman, 2017). As a result, aquaculture farming has been a major source of land use conflict in this research area.

Satellite-based remote sensing and Geographic Information Systems (GIS) are widely applied for assessing, mapping, and measuring soil salinity using the time series Landsat imageries in various studies (Abbas, 2023; Abuelgasim & Ammad, 2019; Fourati, Bouaziz, Benzina, & Bouaziz, 2015; Hassan, Ahmed, Islam, Alam, & Xie, 2021; Meng, Zhou, Zhang, & Bi, 2016). Several researchers have used the Normalized Difference Vegetation Index (NDVI) for vegetation change and soil salinity estimation (Fourati et al., 2015; Thiam et al., 2021). The Normalized Difference Salinity Index

(NDSI) has been applied to determine the soil salinity using Landsat imageries (Abbas, 2023; Hassan et al., 2021; Sahana et al., 2020). For salinity analysis, a previous study used the Enhanced Vegetation Index (Nguyen et al., 2020). Several studies used the salinity indices (SI-1, SI-2, SI-3, and SI-4) for analyzing soil salinity (Hassan et al., 2021; Nguyen et al., 2020). For LULC change analysis, several studies have used the Maximum Likelihood (ML) classification (Devi, Deka, Aneesh, Srinivas, & Nair, 2022; Muttitanon & Tripathi, 2005; Nath et al., 2023), Support Vector Machine (SVM) (Rahaman et al., 2022), and Random Forest (RF) (Rodriguez-Galiano, Ghimire, Rogan, Chica-Olmo, & Rigol-Sanchez, 2012; Van Beijma, Comber, & Lamb, 2014) classification approach. Quantification of LULC classes can help to acquire valuable information about land alteration and possible natural disaster-related activities. The LULC is under pressure from both natural disasters and rapid population growth. This leads to significant environmental effects such as changes in the shoreline, vegetation cover, intrusion of salt water, depletion of biodiversity, and deterioration of the soil over coastal blocks in the Purba Medinipur district. In this context, agriculture land was transformed into aquaculture farming, which has decreased crop production. For this reason, this study can be helpful for future investigation, awareness, planning, and management purposes.

In this present study, the RS and GIS methods are extensively used for the LULC classification, for identifying the land area alteration, and for monitoring the changes and environmental aspects. The lower part or major coastal blocks of the Purba Medinipur district were mapped using the Landsat 5 TM and Landsat 8 OLI/TIRS datasets for various years. The specific objectives were to (1) estimate the soil salinity assessment using NDSI, (2) analyze the change of salinity-prone areas using NDSI, (3) determine the dynamics of the waterbody over the study area using NDWI, (4) detect vegetation cover change throughout the period using NDVI, and (5) assess the driving factors for LULC change detection over the last 30 years. Land use planners, environmentalists, decision-makers, and other interested parties are expected to utilize this work to develop ecologically sound planning strategies and policies based on an adequate understanding of the dynamics of LULC change.

7.2 Rationale of the study

In this study, the Purba Medinipur district's coastal blocks' soil salinity was measured using geospatial techniques. The geospatial-based salinity index such as NDSI used in this paper to calculate the salinity of the soil. The present level of soil salinization in the coastal region of West Bengal can be evaluated by this study. Simultaneously, the ecological balance in saline areas can be constructed based on the variational characteristics of soil salinity to provide decision support for the utilization and management of cultivated land resources, as well as the protection and improvement of soil fertility. Administrators, policymakers, local planners, and researchers were also accommodated in this study for their research and development in this area. This study may be applicable in the otherpart of coastal area in India and abroad.

7.3 Materials and methods

7.3.1 Study area

The present study is situated in the East Midnapore district, which was a part of the Kanthi Coastal Plain of the Bay of Bengal. The study area comprises Nandigram-I, Nandigram-II in the upper portion (northern), Khejuri-I, Khejuri-II, Deshpran, Contai-III in the middle portion, Contai-I, and Contai Municipality in the lower portion of the study area. Geographically, this area is located between 22°5’56.82"N to 21°41’23.29"N latitude and 87°37’1.15"E to 88° 2’26.29"E longitude. Tropical climatic conditions along with the tropical cyclone were observed in this region. Over the study area, an average monthly temperature range of 22°C–34°C was observed. The entire area is influenced by the southwest monsoon. Maximum monthly rainfall of 280 mm was observed in the monsoon season (June and July), and the lowest monthly rainfall of 1 mm was observed in December. The study area’s dynamic features, which include its varied topography, municipalities, and community development blocks, provide important insights into the difficulties and effects of soil salinity in this part of East Midnapore (Fig. 7.1).

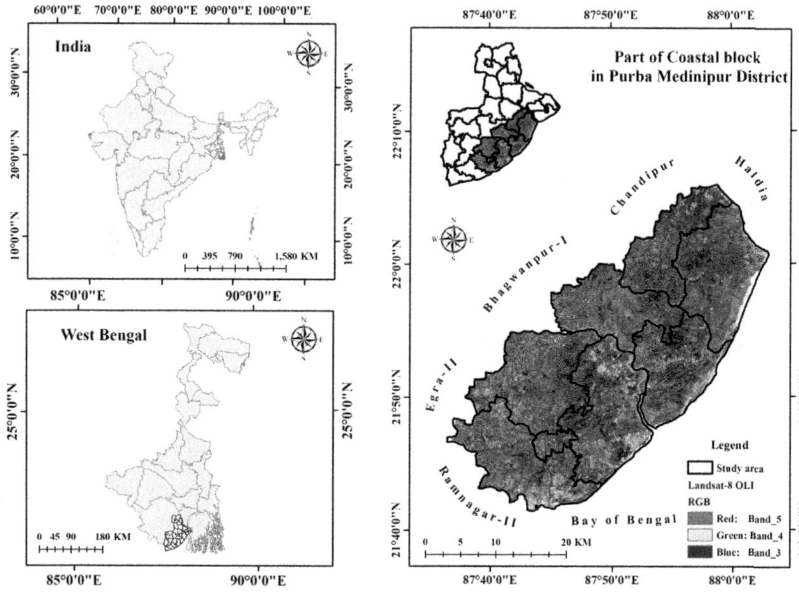

FIGURE 7.1 Location map of the study area. Map lines delineate study areas and do not necessarily depict accepted national boundaries.

7.3.2 Data used

For the present study, Landsat satellite images with no cloud cover were downloaded from USGS Earth Explorer (https://earthexplorer.usgs.gov/) for the years 1990, 2000, 2010, and 2020. Table 7.1 displays the specifics of satellite images. The complete methodology flowchart is shown in Fig. 7.2 and discussed in the following sections. For the years 1990–2010, Landsat 5 TM satellite images were used, and the satellite image for the rest of the year was selected from Landsat 8 OLI/TIRS images. Furthermore, these images are preprocessed before being used for soil salinity assessment and LULC change analysis.

7.3.3 Data preprocessing

All the Landsat data imageries in the winter season were acquired and preprocessed by different image-processing software. The satellite data have undergone multiple procedures, including radiometric correction, atmospheric correction, subsetting, and layer stacking. At first, the radiometric correction was applied over blue, red, NIR, and SWIR bands using Arc GIS 10.4 software. Then, the Fast Line-of-Sight Atmospheric Analysis of Spectral Hypercubes atmospheric correction was used for all the images by the ENVI 5.2 software. In addition, Arc GIS 10.4.1 georeferenced these corrected satellite images to the WGS 1984 Universal Transverse Mercator 45 N zone projection system. Finally, the corrected images were used for extracting the geospatial indicators and LULC analysis.

7.3.3.1 Calculation of Normalized Differnce Salinity Index (NDSI)

The NDSI is an indicator of soil salinity. The red (R) and near infra-red (NIR) bands are used to calculate the NDSI. Several studies (Nguyen et al., 2020; Sahana et al., 2020) used the NDSI for soil salinity assessment. The equation of the NDSI is given below:

TABLE 7.1 Specification of satellite.

Satellite images and sensor	Acquired date and year	Resolution	Path/Row	Source
Landsat 5 TM	30-01-1990	30 × 30 m	138/045	https://earthexplorer.usgs.gov/
Landsat 5 TM	26-01-2000	30 × 30 m	138/045	
Landsat 5 TM	21-01-2010	30 × 30 m	138/045	
Landsat 8 OLI/TIRS	17-01-2020	30 × 30 m	138/045	

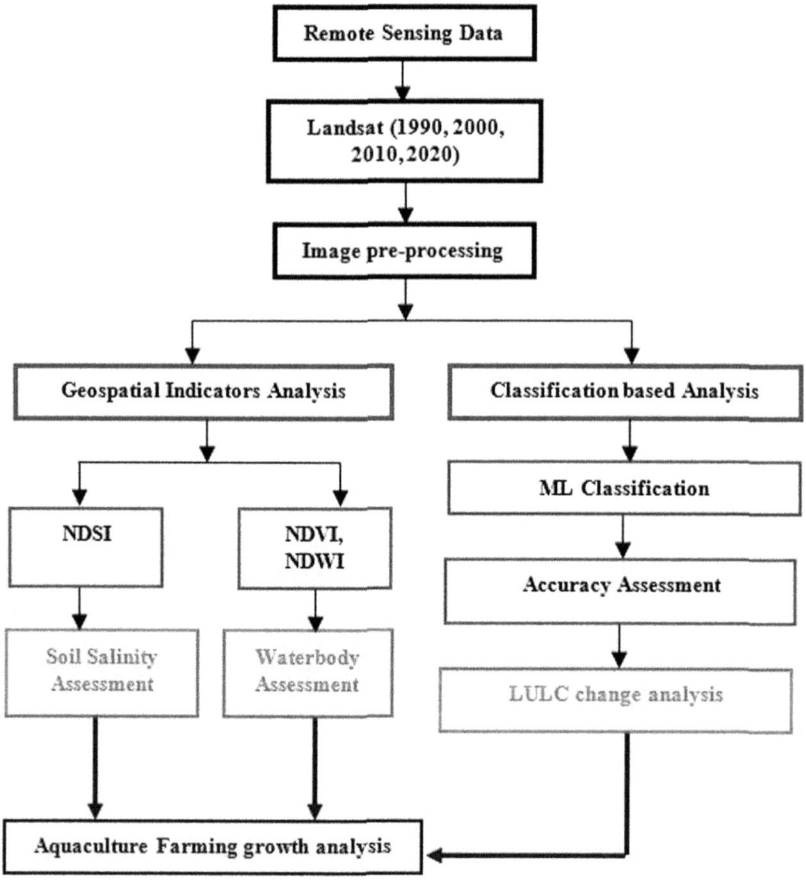

FIGURE 7.2 Methodological flowchart of the study.

$$\text{NDSI} = \frac{\text{R} - \text{NIR}}{\text{R} + \text{NIR}}$$

where R is the red band and NIR is the near-infrared band. The NDSI values lie between -1 and $+1$ and values approaching $+1$ are associated with high soil salinity (Gerardo & de Lima, 2022).

7.3.3.2 Calculation of NDVI

The NDVI is the most commonly used index for the vegetation cover on the earth's surface. Using the NIR and red bands, the NDVI was used to extract the greenness of vegetation. The NDVI value varies from $+1$ to -1, where values close to $+1$ represent healthy vegetation. On the other hand, values near -1 show the water body area over the study area. The equation of the NDVI is given below:

$$\mathrm{NDVI} = \frac{\mathrm{NIR} - \mathrm{RED}}{\mathrm{NIR} + \mathrm{RED}}$$

7.3.3.3 Calculation of Normalized Difference Water Index (NDWI)

McFeeters developed the NDWI in 1996 (McFeeters, 1996). The NIR band and green band of the NDWI were used to define the waterbody area. The NDWI formula is provided below:

$$\mathrm{NDWI} = \frac{\mathrm{GREEN} - \mathrm{NIR}}{\mathrm{GREEN} - \mathrm{NIR}}$$

NDWI values range from -1 to $+1$. The NDWI value is higher in the lower portion of the study area than that for the water body, and thus, the enlarged water area may have an influence on the satellite-observed amount of waterbody throughout the study area.

7.3.3.4 Maximum Likelihood (ML) supervised classification method

ML supervised classification was applied to the Landsat data from 1990 to 2020 using Erdas Imagine 2014. Many studies have adopted different classification techniques for LULC mapping, such as the RF classification used by Zhao, An, Xiong, Ou, & Jiang (2021), SVM by Shih, Stow, & Tsai (2019), and iterative Self-Organizing Data Analysis clustering by Datta & Deb (2012) and Akter, Gazi, & Mia (2021), for their respective study areas. The spatial heterogeneity, or homogeneity of pixels, played a significant role in the classification approaches mentioned. The ML classification was widely used, and a statistical decision criterion that helps in the categorization of overlapping signatures and assigns pixels to the class with the highest probability is a benefit of the ML approach (Willie, Pillay, Zhou, & Orimoloye, 2019). In this study, five LULC classes were classified. These are agricultural land, vegetation, waterbodies, settlements, and bare land or sand, which are described in Table 7.2.

7.3.3.5 Accuracy assessment

After the classification, an accuracy assessment was calculated by the Erdas Imagine 2014 software. Accuracy assessment was the most important factor after classification because this technique represents the accuracy of the classification result (Cohen, 1968). This method was used to compare the ground truth data with the classified one. The ground truth data were collected from field surveys and Google Earth Pro. In accuracy assessment, there are four types of statistics calculated: overall accuracy (OA), user's accuracy (UA), producer's accuracy (PA), and Kappa coefficient (k) for all the LULC classes over the study area (Saha, Bandopadhyay, Kumar, & Mitra, 2020;

TABLE 7.2 Description of LULC (land use and land cover) classes.

LULC Classes	Descriptions
Agriculture land	Farmland that can be cultivated normally
Vegetation	The area under parks, plantations, and natural vegetation
Waterbody	Natural ponds, aquaculture ponds, rivers, and lake
Settlement	Rural residential area, Urban area, industrial, transport, and tourism spot
Bare land or Sand	Area with no vegetation, uncultivated agricultural lands, mud flat, and beach area

Sharma & Vashishtha, 2023). The equation for each accuracy statistic is given below:

Overall accuracy (OA)

$$\text{Over all accuracy} = \frac{\text{Total number of corrected samples}}{\text{Total number of all samples}} \times 100$$

The classification accuracy within each class was computed through UA and PA assessments using the following equation:

Commission error (CE)

$$\text{CE} = \left(\frac{\sum \text{off diagonal elemenmt of row}}{\text{Column total}}\right) \times 100$$

So, the calculation of **UA** is

$$\text{UA}(\%) = 100\% - \text{error of commission}(\%)$$

Now, calculate the **omission error (OE) for PA**.

$$\text{OE} = \left(\frac{\sum \text{off diagonal elemenmt of column}}{\text{Column total}}\right) \times 100$$

So, the **calculation of PA** is

$$\text{PA}(\%) = 100\% - \text{error of ommission }(\%)$$

After that, k was calculated for each classified image. The purpose of calculating k was to measure the inter-observer agreement for classified items. The k was calculated following this equation:

$$\text{K} = \frac{(\text{total} \times \text{sum of correct}) - \text{sum of all the (row total} \times \text{column total)}}{(\text{total squared} - \text{sum of all the (row total} \times \text{column total))}}$$

TABLE 7.3 Scale of Kappa coefficient.

Sl. No.	Value of K	Strength of agreement
1	<0.20	Poor
2	0.21–0.40	Fair
3	0.41–0.60	Moderate
4	0.61–0.80	Good
5	0.81–1.00	Very Good

The k is always less than or equal to 1. The K values range from 0.70 to 0.85, which is considered very high accuracy, and >0.85 is considered outstanding accuracy of the classified map (Monserud & Leemans, 1992). The scale of K is given in Table 7.3.

7.4 Results and discussion

7.4.1 Spatio-temporal variation of soil salinity based on NDSI

Geospatial indicators of SI such as SI 1, SI 2, SI 3, SI 4, SI 5, SI 6, SI 7, SI 8, and NDSI were used for delineating soil salinity based on the multispectral images (Landsat-4–5 TM, 7 ETM+, 8 OLI/TIRS, and Sentinel-2A/B). In this study, soil salinity maps were generated using the NDSI index. In the year 1990, the salinity area occupied a small amount in the south and southwestern parts of the study area (Fig. 7.3). The orange color shows the salinity area, the yellow color indicates the bare area, the green color highlights the agricultural land, and the blue color shows the vegetation cover over the study area. The high NDSI values were 0.31 (1990), 0.41 (2000), 0.37 (2010), and 0.40 (2020), which are given below in Fig. 7.3. Natural hazards (cyclonic activity and flood-related work), saltwater intrusion, sea level rise, and human modification over the study area increased the saline area in 2020. The spatial distribution of the high saline zone was found in the southern part, south-eastern part, and northern part of the study area, where aquaculture farming or salt beds exist. The Pichaboni River in the southern part and the Rasulpur River are situated in the middle portion (south-eastern) of this study area, and these two rivers meet the Bay of Bengal. So, the riverine saline water plays an important role in soil salinity fluctuation (Fig. 7.4).

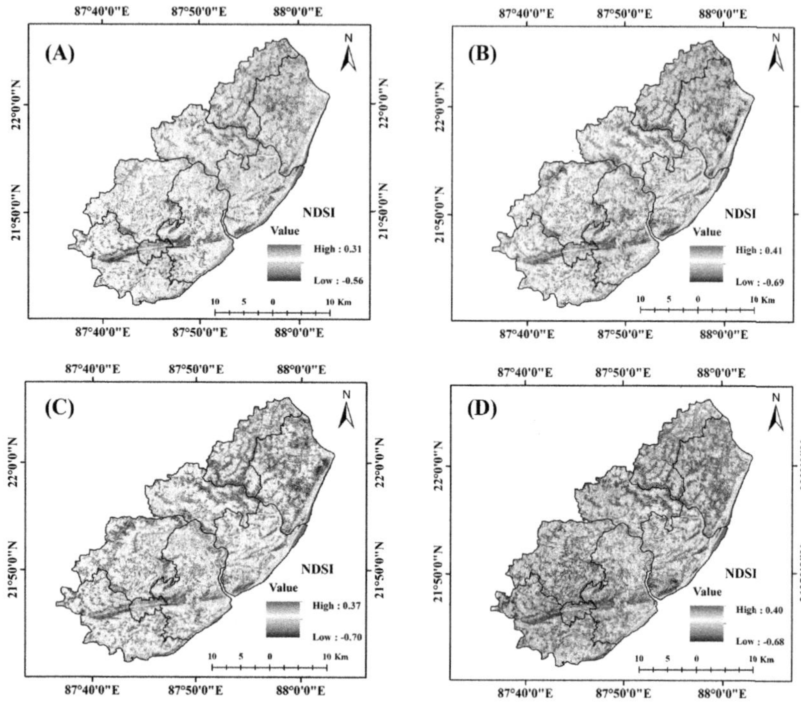

FIGURE 7.3 Shows the salinity map (A) NDSI in 1990, (B) NDSI in 2000, (C) NDSI in 2010 (D) NDSI in 2020. *NDSI*, Normalized Difference Salinity Index. Map lines delineate study areas and do not necessarily depict accepted national boundaries.

7.4.2 Spatio-temporal variation of aquaculture farming based on NDWI

The NDWI index was derived from Landsat images from 1990, 2000, 2010, and 2020. The NDWI values in 1990 varied from −0.48 to 0.40 but the values in 2000 ranged from −0.63 to 0.47. The positive value shows that the waterbody area was increased in between 1990–2000. On the other hand, the values ranged from −0.59 to 0.51 in 2010 and −0.56 to 0.46 in 2020. The NDWI map is represented in Fig. 7.5. The waterbody area was increased in the period of 2010–2020. Throughout the years, a significant alteration in waterbody area was highlighted and depicted in the Fig. 7.6. The results observed that the positive value of NDWI in 1990 was 0.40 and 0.46 in 2020. Thus, our result indicate that huge change was occured over the study area after 1990.

7.4.3 Spatio-temporal variation of vegetation based on NDVI

According to Fig. 7.7, the study area's NDVI values range from −0.31 to 0.56 in 1990, −0.41 to 0.69 in 2000, −0.36 to 0.69 in 2010, and −0.39 to

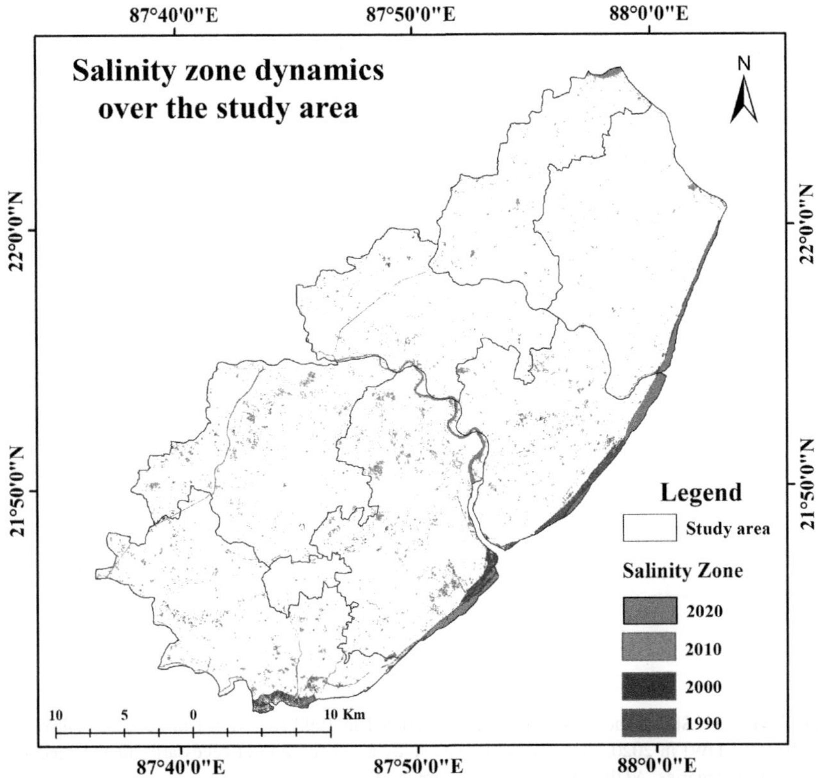

FIGURE 7.4 Shows soil salinity change map in 1990, 2000, 2010, and 2020 over the study area. Map lines delineate study areas and do not necessarily depict accepted national boundaries.

0.68 in 2020. The moderate NDVI value indicates that there is a moderate amount of vegetation in the coastal blocks of the Purba Medinipur district. Due to saltwater intrusion, soil salinity might have severely affected the growth of the vegetation, agricultural land, and also fisheries, and from the NDVI map, the soil of the study area is brutally affected by the salinity. The negative value of NDVI is represented by the waterbody of this study area. The waterbody was gradually increased due to the reduction of agricultural production rate due to conversion of agriculture land to aquaculture land in coastal areas with the rise of soil salinity (Fig. 7.8).

7.4.4 Spatio-temporal changes of the land use land cover classes between 1990, 2000, 2010, and 2020

Five types of LULC classes were classified in this study area; agricultural land, vegetation, waterbody, bare land/sand, and settlement. The Landsat 4/5TM and Landsat 8 OLI/TIRS data were applied to detect the LULC classes

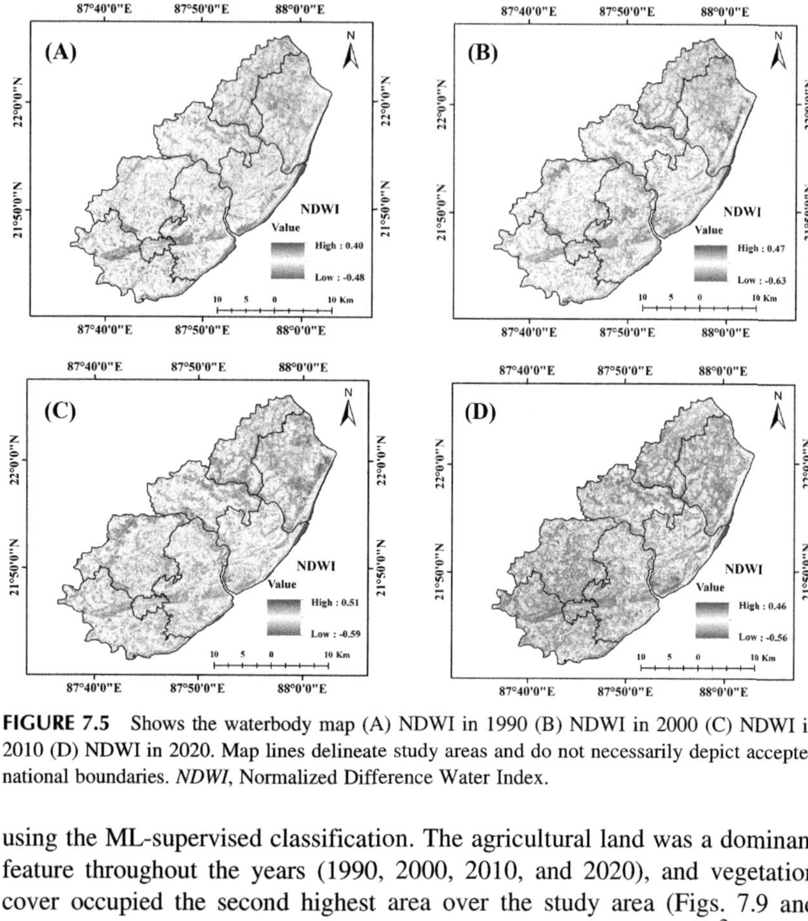

FIGURE 7.5 Shows the waterbody map (A) NDWI in 1990 (B) NDWI in 2000 (C) NDWI in 2010 (D) NDWI in 2020. Map lines delineate study areas and do not necessarily depict accepted national boundaries. *NDWI*, Normalized Difference Water Index.

using the ML-supervised classification. The agricultural land was a dominant feature throughout the years (1990, 2000, 2010, and 2020), and vegetation cover occupied the second highest area over the study area (Figs. 7.9 and 7.10). The agricultural land areas were estimated at 807.21 km^2 (1990), 695.74 km^2 (2000), 761.73 km^2 (2010), and 689.68 km^2 (2020) over the study area. Conversely, Table 7.4 and Fig. 7.8 show that the vegetation covers were 113.78 km^2 (1990), 177.37 km^2 (2000), 168.44 km^2 (2010), and 171.49 km^2 (2020). The agricultural areas were gradually decreased due to the natural hazards and man-made land transformations in the study area. The natural vegetation gently increased due to the land alteration and the awareness of cyclone mitigation. Some of the vegetation cover represents active agricultural land due to the crop calendar change and several crop practices. The most important feature of this study area was the water body. The waterbody areas in four years were 40.00 km^2 (1990), 90.23 km^2 (2000), 33.00 km^2 (2010), and 69.10 km^2 (2020) (Table 7.4) over the study area. A maximum portion of waterbody areas were located in the Deshapran and Khejuri-I blocks. The agricultural land was converted into aquaculture farming, which is why the waterbody area was gradually increased due to the environmental disturbances in the study area. The demand for

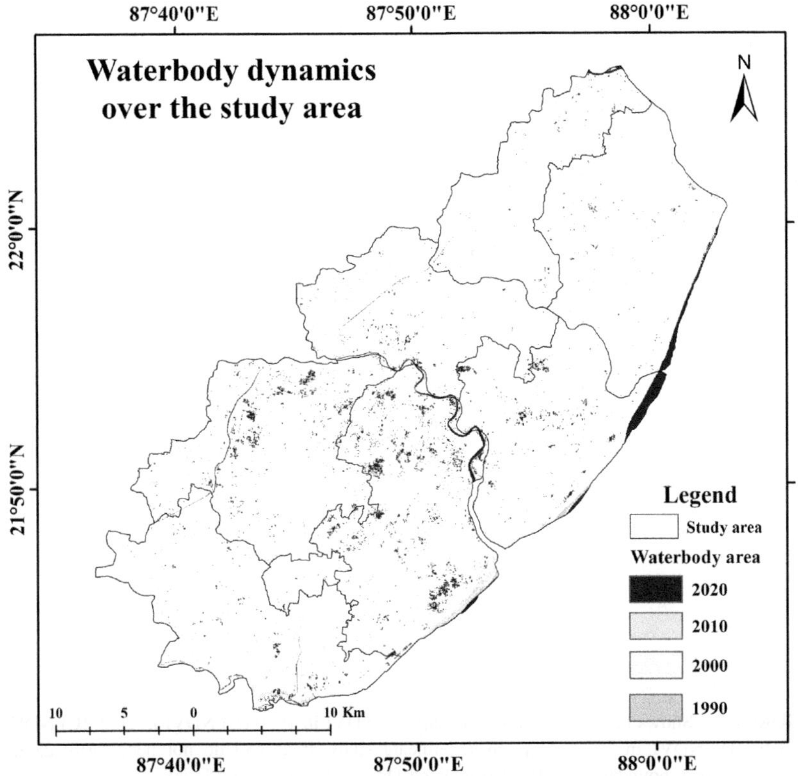

FIGURE 7.6 Shows waterbody change map in 1990, 2000, 2010, and 2020 over the study area. Map lines delineate study areas and do not necessarily depict accepted national boundaries.

aquaculture products such as shrimp and crabs was highly profitable in the international market. So, the local people converted their agricultural land to aquaculture farming to develop their economic stability.

The bare land or sandy areas fluctuated due to the shoreline change and also the anthropogenic activity in the study area. The bare land and sandy areas were assessed at 9.50 km^2 (1990), 2.93 km^2 (2000), 3.50 km^2 (2010), and 11.74 km^2 (2020) over the study area. This area was found in the lower portion of the study area (Table 7.4). The settlement areas were estimated at 40.78 km^2 (1990), 77.96 km^2 (2000), 44.54 km^2 (2010), and 62.98 km^2 (2020) over the study area. The settlement area increased due to population pressure, land alteration, and the development of tourism in this study area. The main reasons for land alternation were highlighted, including the occurrence of several cyclones such as "Aila" in 2009, "Fani" in 2019, "Bulbul" in 2019, and "Amphan" in 2020. The cyclonic condition of saltwater intrusion over the study area destroyed the ecological balance. The infiltration of salt water had an impact on the soil properties, micronutrients in soil and water, groundwater, and plant health (Fig. 7.8).

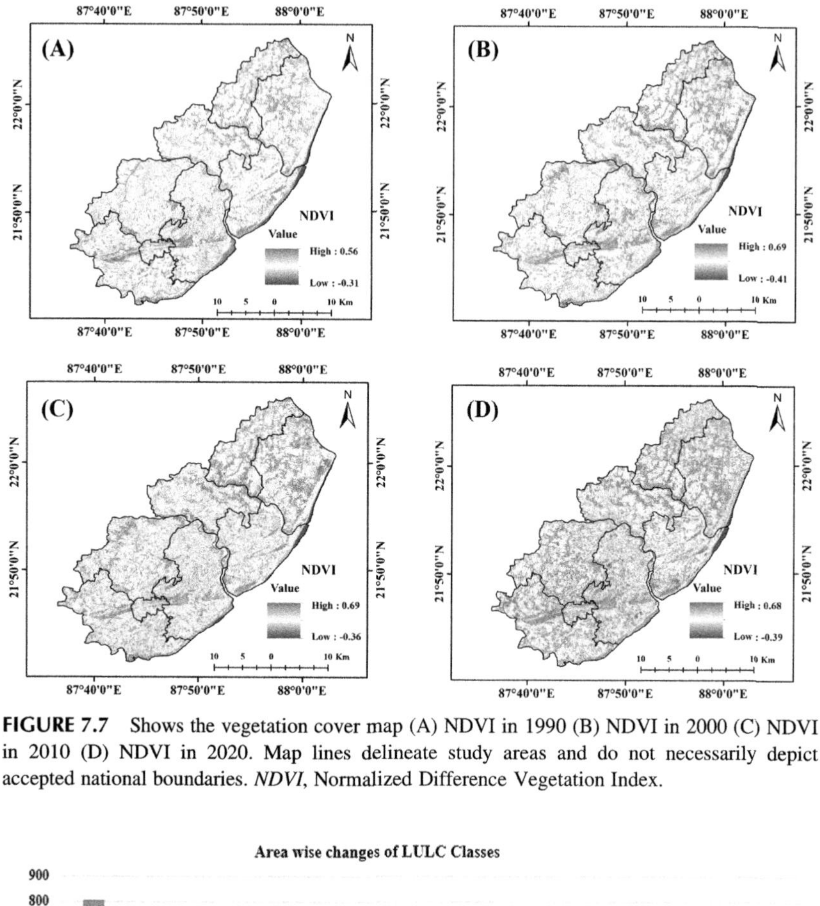

FIGURE 7.7 Shows the vegetation cover map (A) NDVI in 1990 (B) NDVI in 2000 (C) NDVI in 2010 (D) NDVI in 2020. Map lines delineate study areas and do not necessarily depict accepted national boundaries. *NDVI*, Normalized Difference Vegetation Index.

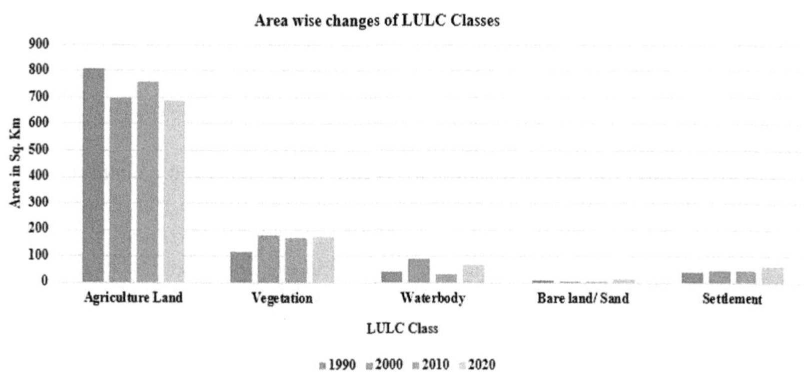

FIGURE 7.8 Area-wise distribution of LULC classes. *LULC*, land use and land cover.

7.4.5 Accuracy assessment of the land use land cover classification map

Google Earth images and field survey referencing points were used to determine the accuracy assessment, and the LULC classification maps in 1990, 2000,

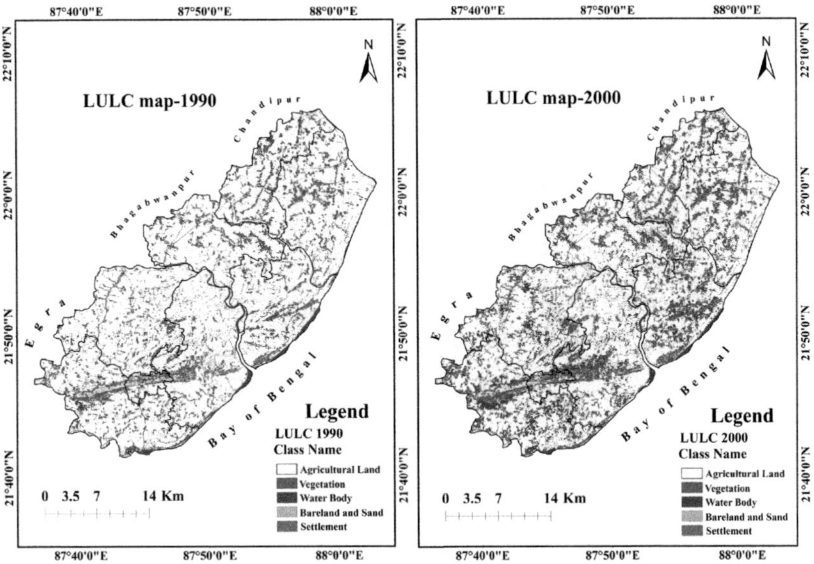

FIGURE 7.9 Shows the LULC map of the study area in 1990 and 2000. Map lines delineate study areas and do not necessarily depict accepted national boundaries. *LULC*, land use and land cover.

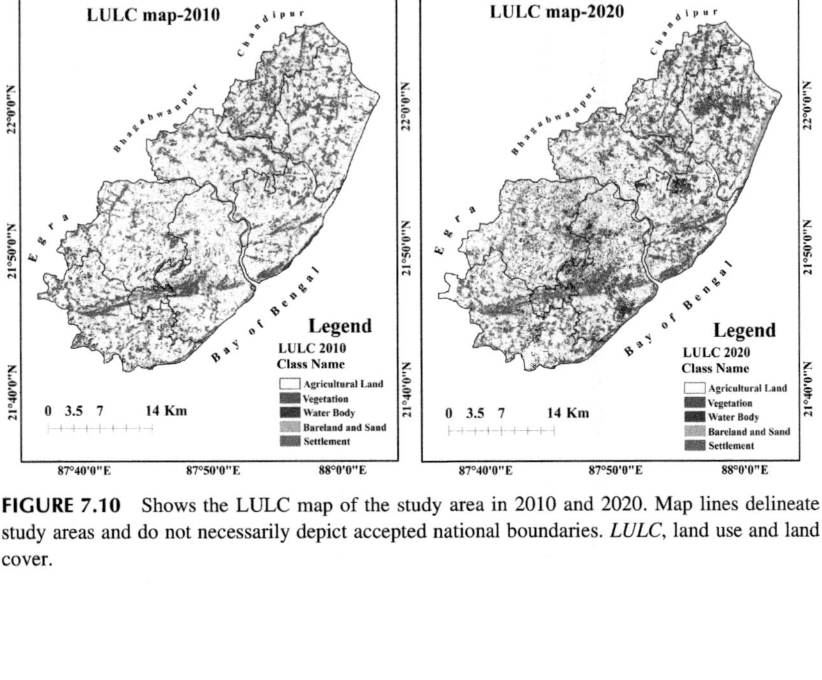

FIGURE 7.10 Shows the LULC map of the study area in 2010 and 2020. Map lines delineate study areas and do not necessarily depict accepted national boundaries. *LULC*, land use and land cover.

TABLE 7.4 Area-wise change estimation of LULC (land use and land cover) classes.

LULC Classes	Area in km^2				Area in percentage (%)			
	1990	**2000**	**2010**	**2020**	**1990**	**2000**	**2010**	**2020**
Agricultural Land	807.21	695.74	761.73	689.68	79.82	68.80	75.33	68.20
Vegetation	113.78	177.37	168.44	171.49	11.25	17.54	16.66	16.96
Waterbody	40.00	90.23	33.00	69.10	3.96	8.92	3.26	7.45
Bare land/Sand	9.50	2.93	3.50	11.74	0.94	0.29	0.35	1.16
Settlement	40.78	44.96	44.54	62.98	4.03	4.45	4.40	6.23
Total	1011.27	1011.23	1011.21	1011.19	100	100	100	100

2010, and 2020 were derived from Landsat 4/5 TM and 8 OLI/TIRS of this study area. Overall accuracy was estimated at 97% (1990) in Table 7.5, 97% (2000) in Table 7.6, 95.51% (2010) in Table 7.7, and 93.55% (2020) in Table 7.8, respectively . On the other hand, the Kappa coefficients were found to be 0.94 (1990), 0.96 (2000), 0.94 (2010), and 0.91 (2020) over the classification maps of the study area. The accuracy assessment results revealed that the waterbody area was showing the best classification result over the study area.

7.4.6 Analysis of land transformation from 1990–2020 over the study area

The LULC change detection or land transformation occurred in different periods. Table 7.9 shows the loss and gain of land areas in agricultural land, vegetation, waterbody, bare land/sand, and settlement areas over the study area. The LULC analysis disclosed that agricultural land and bare land/sand areas were reduced to −117.53 and −2.24 km^2, respectively, between 1990 and 2020 (Figs. 7.11 and 7.15). As a result, the waterbody and settlement areas were gained over the study area. In this same time period, vegetation, waterbodies, and settlement areas increased by 57.71, 29.1, and 22.2 km^2, respectively (Fig. 7.15). In comparison to the first time period between 1990 and 2000, vegetation cover decreased by −63.59 km^2 (Fig. 7.12). However, in this same study period, the bare land/sand area was decreased to −6.57 km^2 over the study area (Fig. 7.12). On the other hand, the settlement increased between 1990 and 2000. In the case of the second study time period (2000–10), the waterbody and settlement areas were reduced by −57.23 and −0.42 km^2, respectively (Fig. 7.13) due to effect of Aila cyclone. In comparison to the third study period (2010–20) (Fig. 7.14), the settlement area grew by 18.44 km^2 over the study period due to anthropogenic activity such as torism, population presuure, etc.. The waterbody area significantly increased by 36.1 km^2 between 2010 and 2020 (Figs. 7.11–7.15) due to conversion of agriculture land to aquaculture land.

7.4.7 Important factors for rises of soil salinity in this study area

Climate change has many adverse environmental impacts on soil salinity. Several natural disasters like the Aila, Fani, Bulbul, and Amphan have affected the flood-prone areas and forced them to convert the land area. Saltwater intrusion is a major contributor to rising soil salinity in the study area. Saltwater plays an important role in decreasing the soil fertility rate and changing the soil electrical conductivity, pH, and other soil physicochemical parameters. This change in soil properties has affected the agricultural production rate directly. In contrast to the LULC change, agricultural land areas were gradually converted to aquaculture farming in this study area because fish cultivation is more profitable for the local people. This fish cultivation continuously reformed the water quality, which affected plant health,

TABLE 7.5 Accuracy assessment and Kappa co-efficient estimation in 1990.

Land use and land cover Classes	Ground truthing/reference					Row Total	CE	UA (%)
	Agricultural land	Vegetation	Waterbody	Bare land/ Sand	Settlement			
Agricultural land	35	1	0	0	0	36	2.78%	97.22%
Vegetation	2	32	0	0	0	34	5.88%	94.12%
Waterbody	0	0	50	2	0	52	3.85%	96.15%
Bare land/Sand	0	0	0	25	1	26	3.85%	96.15%
Settlement	1	0	0	1	30	32	6.25%	93.75%
Total	38	33	50	28	31	180		
OE	7.89%	3.03%	0.00%	10.71%	3.23%			
PA	92.11%	96.97%	100.00%	89.29%	96.77%			
OA	97.00%							
Kappa coefficient	0.94378429							

CE, Commission Error; UA, User's Accuracy; OE, Omission Error; PA, Producer Accuracy; OA, Overall Accuracy.

TABLE 7.6 Accuracy assessment and kappa co-efficient estimation in 2000.

Land use and land cover Classes	Ground truthing/reference					Row total	CE	UA (%)
	Agricultural land	Vegetation	Waterbody	Bare land/ Sand	Settlement			
Agricultural land	40	1	0	0	0	41	2.78%	97.22%
Vegetation	1	31	0	0	0	32	5.88%	94.12%
Waterbody	0	0	65	2	0	66	3.85%	96.15%
Bare land/Sand	0	0	0	28	1	29	3.85%	96.15%
Settlement	1	0	0	1	30	32	6.25%	93.75%
Total	42	32	65	30	31	200		
OE	4.76%	3.13%	0.00%	6.67%	3.23%			
PA	95.24%	96.88%	100.00%	93.33%	96.77%			
OA	97.00%							
Kappa coefficient	0.961417272							

CE, Commission Error; UA, User's Accuracy; OE, Omission Error; PA, Producer Accuracy; OA, Overall Accuracy.

TABLE 7.7 Accuracy assessment and kappa co-efficient estimation in 2010.

land use and land cover classes	Ground truthing/reference					Row total	CE	UA (%)
	Agricultural land	Vegetation	Waterbody	Bare land/Sand	Settlement			
Agricultural land	45	1	0	0	0	46	2.17%	97.83%
Vegetation	2	20	0	0	0	22	9.09%	90.91%
Waterbody	0	0	50	2	0	52	3.85%	96.15%
Bare land/Sand	0	0	0	25	1	26	3.85%	96.15%
Settlement	1	0	0	1	30	32	6.25%	93.75%
Total	48	21	50	28	31	178		
OE	6.25%	4.76%	0.00%	10.71%	3.23%			
PA	93.75%	95.24%	100.00%	89.29%	96.77%			
OA	95.51%							
Kappa coefficient	0.94239							

CE, Commission Error; UA, User's Accuracy; OE, Omission Error; PA, Producer Accuracy; OA, Overall Accuracy.

TABLE 7.8 Accuracy assessment and kappa co-efficient estimation in 2020.

Land use and land cover classes	Ground truthing/reference					Row Total	CE	UA (%)
	Agricultural land	Vegetation	Waterbody	Bare land/Sand	Settlement			
Agricultural land	50	1	0	0	0	51	1.96%	98.04%
Vegetation	2	20	0	0	0	22	9.09%	90.91%
Waterbody	0	0	30	2	0	32	6.25%	93.75%
Bare land/Sand	0	0	0	25	1	26	3.85%	96.15%
Settlement	3	0	0	1	20	24	16.67%	83.33%
Total	55	21	30	28	21	155		
OE	9.09%	4.76%	0.00%	10.71%	4.76%			
PA	90.91%	95.24%	100.00%	89.29%	95.24%			
OA	93.55%							
Kappa coefficient	0.91653							

CE, Commission Error; UA, User's Accuracy; OE, Omission Error; PA, Producer Accuracy; OA, Overall Accuracy.

TABLE 7.9 Loss and gain analysis of LULC classes in different periods over the study area.

Class name	Change detection assessment of LULC classes (area in km^2)			
	(1990–2000)	(2000–2010)	(2010–2020)	(1990–2020)
Agricultural Land	−111.47	65.99	−72.05	−117.53
Vegetation	−63.59	8.93	3.05	57.71
Waterbody	50.23	−57.23	36.1	29.1
Bare land/Sand	−6.57	0.57	8.24	−2.24
Settlement	4.18	−0.42	18.44	22.2

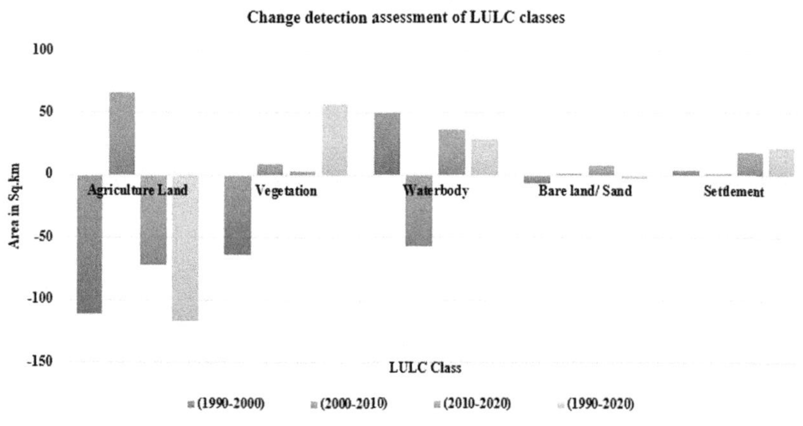

FIGURE 7.11 It shows the change detection analysis of LULC classes in different years over. *LULC*, land use and land cover.

groundwater, and agricultural land. Basically, the agricultural land in coastal blocks is closely associated with soil salinization.

7.5 Limitations of the study

The Landsat 5 TM and 8 OLI/TIRS datasets were used for calculating the soil salinity based on NDSI, aquaculture farming change analysis (NDVI and NDWI), and the LULC change in the selected coastal blocks of the Purba Medinipur district. The soil salinity and aquaculture farming changes were estimated using four multi-temporal datasets (1990, 2000, 2010, and 2020). The disadvantage of the Landsat data is the difficulty of region identification due to pixel mixing and resolution issues. On the other hand, the other

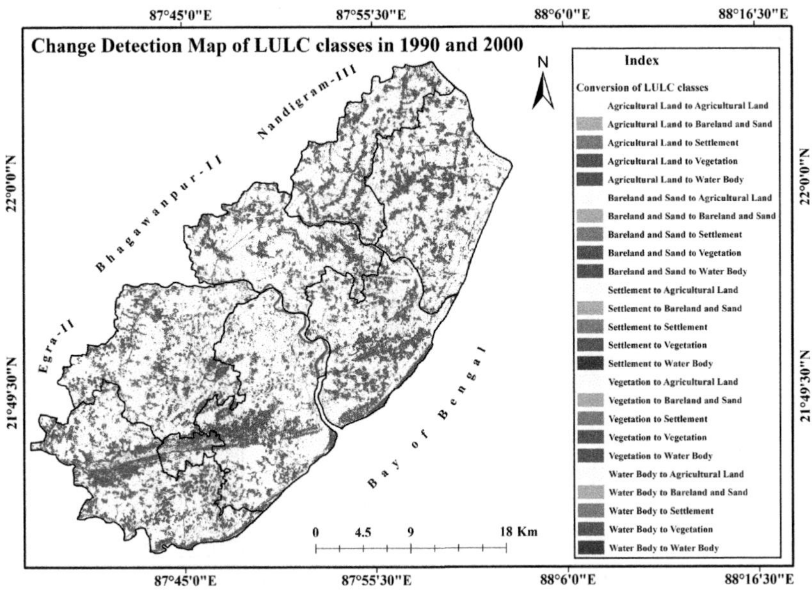

FIGURE 7.12 Change detection map of the study area between 1990 and 2000. Map lines delineate study areas and do not necessarily depict accepted national boundaries.

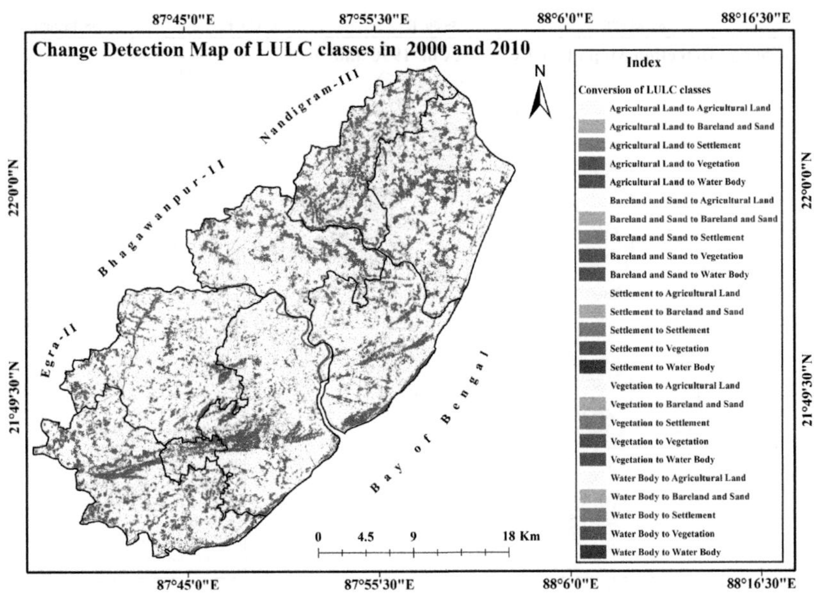

FIGURE 7.13 Change detection map of the study area between 2000 and 2010. Map lines delineate study areas and do not necessarily depict accepted national boundaries.

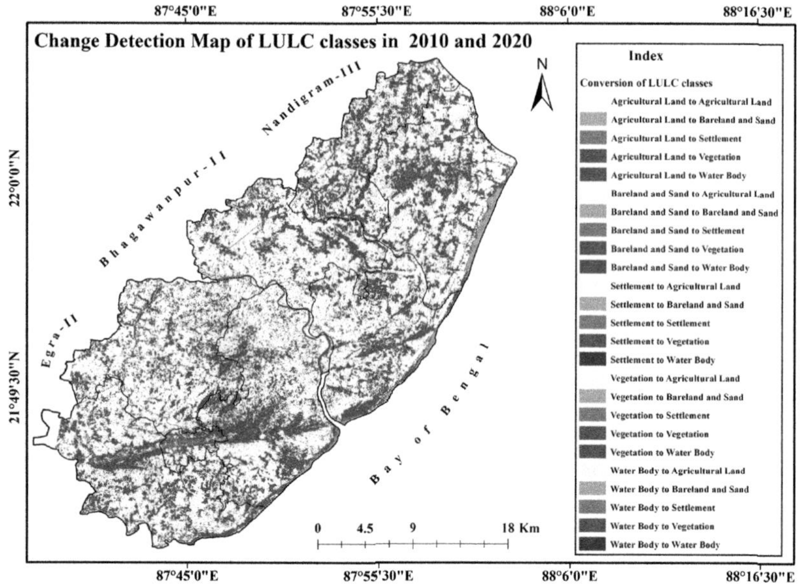

FIGURE 7.14 Change detection map of the study area between 2010 and 2020. Map lines delineate study areas and do not necessarily depict accepted national boundaries.

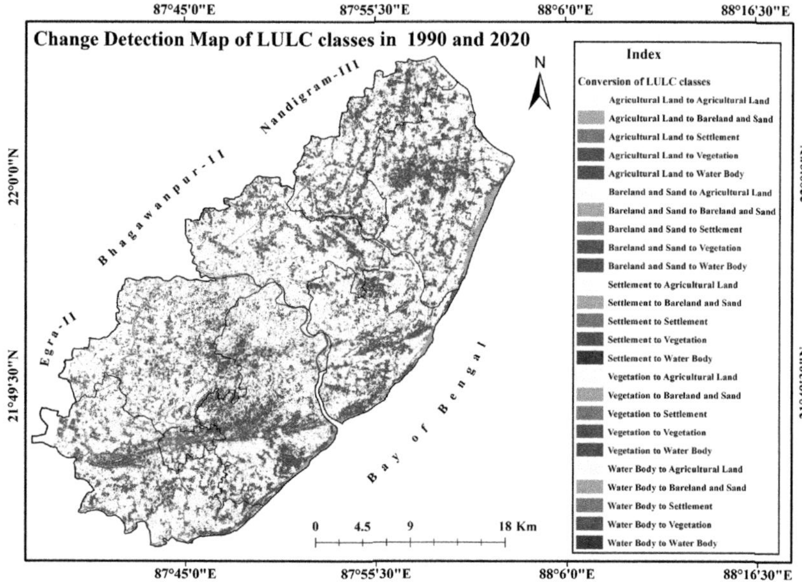

FIGURE 7.15 Change detection map of the study area between 1990 and 2020. Map lines delineate study areas and do not necessarily depict accepted national boundaries.

limitations of these images were pixel identification and cloud coverage. A geospatial indicator such as NDSI has some limitation for the assessment of soil salinity. Additionally, the MLS-supervised classification for estimation of LULC classes did not give 100% classification accuracy. As a result, it is assumed that the salinity indices have also some limitation to determine the soil salinity using the satellite data in the coastal area of Purba Medinipur district.

7.6 Recommendations

The following are the main research areas that require focus in the near future on a global or regional scale to maintain salt-affected soil for sustainable development in the coastal area:

The techniques of stopping or controlling saline water intrusion by shrimp farming, the use of natural fertilizers and antibiotics, agroforestry, tree-based farming, and water management practices with the introduction of salt-tolerant varieties of different crops can all be used to increase land reclamation and cropping intensity.

Growing saline water shrimp in a freshwater pond can be an efficient way to minimize soil and water salinity. This shrimp cultivation approach has the potential to mitigate the negative effects of present shrimp farming.

As sea level rise is expected to increase in the future, it is recommended that the region's soil salinity pattern be studied further in the next few decades.

7.7 Conclusions

In this work, the important coastal blocks of the Purba Medinipur district were subjected to LULC change detection analysis and soil salinity estimation using multi-temporal remote sensing pictures. The coastal areas of West Bengal have had significant impacts from climate change. Climate change-related soil salinity predominates in rural coastal areas and has a substantial detrimental impact on crop growth and development, ultimately harming agricultural crop productivity and food security. Because clean drinking water is scarce, it also has a negative influence on public health. Based on the LULC map, it can be inferred that around half of the study area is covered by agricultural land, indicating that agriculture is the primary source of income for the local population. Therefore, salinity-affected land area identification and monitoring are crucial for creating adequate land-use planning for agricultural crop production as well as for addressing the shortage of fresh drinking water. Because the locations, intensities, and patterns of areas affected by salt vary over time, more research is required. However, salinity is usually determined by combining field measurements with laboratory analysis, which can be challenging, costly, and time-consuming to do on a large scale. The goal of this research was to use some methods, equations, and salinity indicators that were used in this studies which can be applied in other parts of the world to assess the soil salinity study. The agricultute, aquaculture, settlement, vegetation and livelihood of the people close to this coastal area face a serious risk of flooding due to climate change and sea level rise. Therefore, everyone involved must understand the threat that lies ahead. It is crucial that all stakeholder involved in this area should aware the looming threat and the government should take the necessary steps to regulate aquaculture farming in this area.

Acknowledgments

The authors are grateful for the help of Vidyasagar University, West Bengal, and Space Applications Center, Ahmedabad. We also thank the USGS Earth Explorer data portal for providing the Landsat satellite image.

References

Abbas, J. A. A. (2023). Soil salinity assessment by using spectral salinity indices in Al-Sweira project middle of the Iraqi alluvial plain. *International Journal of Environmental Science and Technology*, *20*(10), 10847–10860. Available from https://doi.org/10.1007/s13762-022-04733-4, 17352630, https://www.springer.com/journal/13762.

Abuelgasim, A., & Ammad, R. (2019). Mapping soil salinity in arid and semi-arid regions using Landsat 8 OLI satellite data. *Remote Sensing Applications: Society and Environment*, *13*, 415–425. Available from https://doi.org/10.1016/j.rsase.2018.12.010, 23529385, http://www.journals.elsevier.com/remote-sensing-applications-society-and-environment/.

Akber, M. A., Islam, M. A., Ahmed, M., Rahman, M. M., & Rahman, M. R. (2017). Changes of shrimp farming in southwest coastal Bangladesh. *Aquaculture International*, *25*(5), 1883–1899. Available from https://doi.org/10.1007/s10499-017-0159-5, 1573143X, http://www.wkap.nl/journalhome.htm/0967-6120.

Akter, T., Gazi, M. Y., & Mia, M. B. (2021). Assessment of land cover dynamics, land surface temperature, and heat Island growth in northwestern Bangladesh using satellite imagery. *Bangladesh Environmental Processes*, *8*(2), 661–690. Available from https://doi.org/10.1007/s40710-020-00491-y, 21987505, http://www.springer.com/earth + sciences + and + geography/environmental + science + %26 + engineering/journal/40710.

Cohen, J. (1968). Weighted kappa: Nominal scale agreement provision for scaled disagreement or partial credit. *Psychological Bulletin*, *70*(4), 213–220. Available from https://doi.org/10.1037/h0026256, 00332909.

Datta, D., & Deb, S. (2012). Analysis of coastal land use/land cover changes in the Indian Sunderbans using remotely sensed data. *Geo-Spatial Information Science*, *15*(4), 241–250. Available from https://doi.org/10.1080/10095020.2012.714104, 10095020, http://www.tandfonline.com/loi/tgsi20.

Devi, A. B., Deka, D., Aneesh, T. D., Srinivas, R., & Nair, A. M. (2022). Predictive modelling of land use land cover dynamics for a tropical coastal urban city in Kerala, India. *Arabian Journal of Geosciences*, *15*(5). Available from https://doi.org/10.1007/s12517-022-09735-7, 1866–7511.

Fourati, H. T., Bouaziz, M., Benzina, M., & Bouaziz, S. (2015). Modeling of soil salinity within a semi-arid region using spectral analysis. *Arabian Journal of Geosciences*, *8*(12), 11175–11182. Available from https://doi.org/10.1007/s12517-015-2004-3, 1866–7511.

Gayathri, R., Murty, P. L. N., Bhaskaran, P. K., & Srinivasa Kumar, T. (2016). A numerical study of hypothetical storm surge and coastal inundation for AILA cyclone in the Bay of Bengal. *Environmental Fluid Mechanics*, *16*(2), 429–452. Available from https://doi.org/10.1007/s10652-015-9434-z, 15731510, http://www.wkap.nl.

Gerardo, R., & de Lima, I. P. (2022). Sentinel-2 satellite imagery-based assessment of soil salinity in irrigated rice fields in Portugal. *Agriculture*, *12*(9), 1490. Available from https://doi.org/10.3390/agriculture12091490, 2077-0472.

Hassan, R., Ahmed, Z., Islam, M. T., Alam, R., & Xie, Z. (2021). Soil salinity detection using salinity indices from Landsat 8 Satellite Image at Rampal, Bangladesh. *Bangladesh Remote Sensing in Earth Systems Sciences*, *4*(1–2). Available from https://doi.org/10.1007/s41976-020-00041-y, 25208209, https://www.springer.com/earth + sciences + and + geography/geography/journal/41976.

Hossen, B., Yabar, H., & Faruque, M. J. (2022). Exploring the potential of soil salinity assessment through remote sensing and GIS: Case study in the coastal rural areas of Bangladesh. *MDPI, Japan Land*, *11*(10). Available from https://doi.org/10.3390/land11101784, 2073445X, http://www.mdpi.com/journal/land/.

Kawser, U., Nath, B., & Hoque, A. (2022). Observing the influences of climatic and environmental variability over soil salinity changes in the Noakhali Coastal Regions of Bangladesh using geospatial and statistical techniques. *Environmental Challenges*, *6*, 100429. Available from https://doi.org/10.1016/j.envc.2021.100429, 26670100.

McFeeters, S. K. (1996). The use of the Normalized Difference Water Index (NDWI) in the delineation of open water features. *International Journal of Remote Sensing*, *17*(7), 1425–1432. Available from https://doi.org/10.1080/01431169608948714, 13665901.

Meng, L., Zhou, S., Zhang, H., & Bi, X. (2016). Estimating soil salinity in different landscapes of the Yellow River Delta through Landsat OLI/TIRS and ETM + Data. *Journal of Coastal Conservation*, *20*(4), 271–279. Available from https://doi.org/10.1007/s11852-016-0437-9, 18747841, http://www.springerlink.com/content/1400-0350.

Monserud, R. A., & Leemans, R. (1992). Comparing global vegetation maps with the Kappa statistic. *Ecological Modelling.*, *62*(4), 275–293. Available from https://doi.org/10.1016/0304-3800(92)90003-W, 03043800.

Muhammad Abdullah, H., Ahmed, S. M., Khan, B. M., Mohana, N. T., Ahamed, T., & Islam, I. (2021). Agriculture and fisheries production in a regional blending and dynamic fresh and saline water systems in the coastal area of Bangladesh. *Environmental Challenges*, *4*. Available from https://doi.org/10.1016/j.envc.2021.100089, 26670100, https://www.journals.elsevier.com/environmental-challenges.

Mukhopadhyay, R., Sarkar, B., Jat, H. S., Sharma, P. C., & Bolan, N. S. (2021). Soil salinity under climate change: Challenges for sustainable agriculture and food security. *Journal of Environmental Management*, *280*. Available from https://doi.org/10.1016/j.jenvman.2020.111736, 10958630, https://www.sciencedirect.com/journal/journal-of-environmental-management.

Muttitanon, W., & Tripathi, N. K. (2005). Land use/land cover changes in the coastal zone of Ban Don Bay, Thailand using Landsat 5 TM data. *Thailand International Journal of Remote Sensing*, *26*(11), 2311–2323. Available from https://doi.org/10.1080/0143116051233132666, 13665901, https://www.tandfonline.com/loi/tres20.

Nath, A., Koley, B., Choudhury, T., Saraswati, S., Ray, B. C., Um, J.-S., & Sharma, A. (2023). Assessing coastal land-use and land-cover change dynamics using geospatial techniques. *Sustainability*, *15*(9), 7398. Available from https://doi.org/10.3390/su15097398, 2071-1050.

Nguyen, K. A., Liou, Y. A., Tran, H. P., Hoang, P. P., & Nguyen, T. H. (2020). Soil salinity assessment by using near-infrared channel and Vegetation Soil Salinity Index derived from Landsat 8 OLI data: A case study in the Tra Vinh Province, Mekong Delta, Vietnam. *Progress in Earth and Planetary Science*, *7*(1). Available from https://doi.org/10.1186/s40645-019-0311-0, 21974284, https://progearthplanetsci.springeropen.com/.

Paul, B. K., & Rashid, H. (2016). *Climatic hazards in coastal Bangladesh: Non-structural and structural solutions* (pp. 1–327). United States: Elsevier Inc.. Available from http://www.sciencedirect.com/science/book/9780128052761.

Rahaman, Z. A., Kafy, A. A., Faisal, A. A., Al Rakib, A., Jahir, D. M. A., Fattah, M. A., ... Rahman, M. T. (2022). Predicting microscale land use/land cover changes using cellular automata algorithm on the northwest coast of Peninsular Malaysia. *Malaysia Earth Systems and Environment*, *6*(4), 817–835. Available from https://doi.org/10.1007/s41748-022-00318-w, 25099434, https://www.springer.com/journal/41748.

Rodriguez-Galiano, V. F., Ghimire, B., Rogan, J., Chica-Olmo, M., & Rigol-Sanchez, J. P. (2012). An assessment of the effectiveness of a random forest classifier for land-cover classification. *ISPRS Journal of Photogrammetry and Remote Sensing*, *67*(1), 93–104. Available from https://doi.org/10.1016/j.isprsjprs.2011.11.002, 09242716.

Saha, P., Bandopadhyay, S., Kumar, C., & Mitra, C. (2020). Multi-approach synergic investigation between land surface temperature and land-use land-cover. *Journal of Earth System Science*, *129*(1). Available from https://doi.org/10.1007/s12040-020-1342-z, 0973774X, https://rd.springer.com/journal/12040.

Sahana, M., Rehman, S., Patel, P. P., Dou, J., Hong, H., & Sajjad, H. (2020). Assessing the degree of soil salinity in the Indian Sundarban Biosphere Reserve using measured soil electrical conductivity and remote sensing data–derived salinity indices. *United Kingdom Arabian Journal of Geosciences*, *13*(24). Available from https://doi.org/10.1007/s12517-020-06310-w, 18667538, http://www.springer.com/geosciences/journal/12517?cm_mmc = AD-_-enews-_-PSE1892-_-0.

Sarkar, B., Islam, A., & Majumder, A. (2021). Seawater intrusion into groundwater and its impact on irrigation and agriculture: Evidence from the coastal region of West Bengal, India. *Regional Studies in Marine Science*, *44*, 101751. Available from https://doi.org/10.1016/j.rsma.2021.101751, 23524855.

Sharma, A., & Vashishtha, D. (2023). Spatio-temporal assessment of land use land cover changes and their impact on variations of land surface temperature in Aligarh Municipality. *Journal of the Indian Society of Remote Sensing*, *51*(4), 799–827. Available from https://doi.org/10.1007/s12524-022-01652-2, 09743006, https://www.springer.com/journal/12524.

Shih, Hc, Stow, D. A., & Tsai, Y. H. (2019). Guidance on and comparison of machine learning classifiers for Landsat-based land cover and land use mapping. *International Journal of Remote Sensing*, *40*(4), 1248–1274. Available from https://doi.org/10.1080/01431161.2018.1524179, 13665901, https://www.tandfonline.com/loi/tres20.

Thiam, S., Villamor, G. B., Faye, L. C., Sène, J. H. B., Diwediga, B., & Kyei-Baffour, N. (2021). Monitoring land use and soil salinity changes in coastal landscape: A case study from Senegal. *Ghana Environmental Monitoring and Assessment*, *193*(5). Available from https://doi.org/10.1007/s10661-021-08958-7, 15732959, https://link.springer.com/journal/10661.

Thomas, V., & LLpez, R. (2015). Global increase in climate-related disasters. *SSRN Electronic Journal*. Available from https://doi.org/10.2139/ssrn.2709331, 1556–5068.

Van Beijma, S., Comber, A., & Lamb, A. (2014). Random forest classification of salt marsh vegetation habitats using quad-polarimetric airborne SAR, elevation and optical RS data. *Remote Sensing of Environment*, *149*, 118–129. Available from https://doi.org/10.1016/j.rse.2014.04.010, 00344257, http://www.elsevier.com/inca/publications/store/5/0/5/7/3/3.

Willie, Y. A., Pillay, R., Zhou, L., & Orimoloye, I. R. (2019). Monitoring spatial pattern of land surface thermal characteristics and urban growth: A case study of King Williams using remote sensing and GIS. *Earth Science Informatics*, *12*(4), 447–464. Available from https://doi.org/10.1007/s12145-019-00391-2, 18650481, http://www.springer.com/geosciences/computer + %26 + mathematical + applications/journal/12145.

Yin, A., Zhang, M., Gao, C., Yang, X., Xu, Y., Wu, P., & Zhang, H. (2016). Salinity evolution of coastal soils following reclamation and intensive usage, Eastern China. *Environmental Earth Sciences*, *75*(18). Available from https://doi.org/10.1007/s12665-016-6095-2, 1866–6280.

Yu, J., Li, Y., Han, G., Zhou, D., Fu, Y., Guan, B., ... Wang, J. (2014). The spatial distribution characteristics of soil salinity in coastal zone of the Yellow River Delta. *Environmental Earth Sciences*, *72*(2), 589–599. Available from https://doi.org/10.1007/s12665-013-2980-0, 18666299, http://www.springerlink.com/content/121380/.

Zhao, Y., An, R., Xiong, N., Ou, D., & Jiang, C. (2021). Spatio-temporal land-use/land-cover change dynamics in Coastal Plains in Hangzhou Bay Area, China from 2009 to 2020 Using Google Earth Engine. *Land.*, *10*(11), 1149. Available from https://doi.org/10.3390/land10111149, 2073-445X.

Chapter 8

Unveiling the dynamic seasons of the Vaal River: a multitemporal monitoring and analysis on a cloudbased geospatial platform

Idowu Racheal Bodunrin and Ednah Kwamboka Onyari
Department of Civil Engineering, College of Science, Engineering and Technology, University of South Africa, Florida, Johannesburg, South Africa

8.1 Introduction

Water resource monitoring is one of the most common measures when it comes to water availability, management, and sustainability. Monitoring the changes in water resources is imperative in any given ecosystem, particularly in drought-affected regions where the impacts are exaggerated. According to the Global Risks Report (2020), the water crisis, including water shortages, water pollution, and other water issues, is ranked fifth out of the top 10 risks in terms of impact. The scarcity of water, where its demands surpass its availability, is a determining factor of water scarcity, and it directly affects the health and well-being of urban citizens, urban environmental quality, and socio-economic development (McDonald et al., 2014).

There has been a critical global challenge in ensuring a safe, equitable, and sustainable supply of water (Bischoff-Mattson et al., 2020). There has been a shift in the patterns of water availability, variability, and distribution due to a change in global climate (Milly, Dunne, & Vecchia, 2005). The effects of water scarcity vary across different sectors, including household welfare, employment, and household consumption, which are borne mostly by low-income households (Briand, Reynaud, Viroleau, Markantonis, & Branciforti, 2023).

Anthropogenic factors, such as water withdrawals, mining activities, and natural effects like seasonal variations, climate change, and evapotranspiration (ET), have impacted water availability, usually resulting in exacerbated water

Water Resources Monitoring, Management, and Sustainability. DOI: https://doi.org/10.1016/B978-0-443-23665-5.00008-9

scarcity. A rapid increase in urbanization typically necessitates strategies for water provision to cater for agricultural, domestic, and industrial uses, leading to a substantial increase in water demands. On a global scale, water is predominantly utilized by showering, toilets, bathing, and washing machines (Roshan and Kumar, 2020). According to Koop, Van Dorssen, & Brouwer (2019), countries have been predicted to encounter a 40% water shortage by 2030, possibly compromising the realization of reaching the United Nations Sustainable Development Goals (SDGs), particularly SDG11 (Sustainable Cities and Communities) and SDG6 (Clean Water and Sanitation), established in 2015.

Satellite observation makes it possible for spatial and temporal analysis of surface water dynamics (Gujrati & Jha, 2018). Remotely sensed satellites, which acquire high-frequency data over a large scale, possess visible, infrared, and micro wavelengths that are very effective in monitoring water changes. Due to spatial and temporal coverage, remote sensing (RS) technology has been used to reflect the spatial distribution and dynamic changes in water quality components (Kallio et al., 2001; Malahlela, 2019; Shi, Zhang, Zhu, Qin, & Pan, 2018; Song et al., 2011).

Google Earth Engine (GEE), which is an open application programming interface (API), provides free computational resources and satellite data on a cloud computational platform, thereby lessening the user's requirement for computational resources and data availability (Gorelick et al., 2017). GEE offers a cloud platform for big data processing through freely available satellite data, including those of Landsat-8, Sentinel-1, and Sentinel-2 (Shelestov, Lavreniuk, Kussul, Novikov, & Skakun, 2017). The synergetic process of combining satellite data and GEE cloud-based platforms makes it feasible to regularly monitor the seasonal dynamics of surface water for water conservation and sustainability.

South Africa, located within the sub-Saharan part of Africa, is facing severe water scarcity due to sporadic water levels. It is characterized by low and variable annual rainfall along with high natural evaporation levels, making it the 30th driest country in the world (2013). The country possesses a well-developed water resource infrastructure; however, it is fast approaching the full utilization of its surface water yields in certain areas, and in others, yields have already been exceeded, making suitable sites for new dams increasingly unfeasible (2013). Climate change has pronounced effects, particularly in semiarid regions, as it influences precipitation patterns, temperature, and weather outcomes. Its influence in South Africa is expected to witness the strongest decrease in precipitation in Sub-Saharan Africa with concurrent risks of drought (Serdeczny et al., 2017).

Out of the 223 river ecosystem forms in South Africa, 60% are in danger of extinction, and 25% of those are classified as critically endangered (2013). A recent South African scuffle with water scarcity was the 2018-day zero threat in Cape Town, where water provision was expected to be depleted. There is uniqueness in each city in the details resulting in any given scarcity crisis; however, many cities share similar attributes: growing population, demographic, or economic changes that modify patterns of demand, dependence on scarce infrastructure, fluctuating rainfall variability, public demands for equity or effectiveness, a struggle involving urban and rural consumptive uses, or a necessity to maintain environmental flows (Godinez Madrigal, van der Zaag, and van Cauwenbergh, 2018; Millington, 2018). The National Development Plan (2030) stated clearly the interconnection between food, fuel, and water, specifically in the context of climate change and how they impact one another in the long term. To ensure that the country operates optimally both now and in the future, regular monitoring of the water resources holds great significance.

The nation has quite several water resources, and among the valuable rivers within the country is the Vaal River. The Vaal Water Management Area (WMA) is perceived to be one of the most important WMAs in South Africa because it houses the highest concentration of urban, industrial, agricultural, mining, and power generation developments in South Africa (Department of Water and Sanitation, 2016). An increase in the demand for water due to the increase in population, economic growth, and the need for development puts pressure on the supply capacity of the Vaal River (Remilekun, Thando, Nerhene, & Archer, 2021). Climate change has attributed further challenges in the Vaal River catchment, which supports the country's economic hub, more than 30% of the population, and about 70% of its maize production (Remilekun et al., 2021) and water demand goes beyond the existing yield in the Vaal River catchment (Ilunga, 2017).

The seasonal water monitoring of the Vaal River is very crucial, particularly because it falls within a region that has been categorized as semiarid/drought regions. Semiarid regions are regions known to exhibit water scarcity due to limited precipitation and high rates of evaporation, making water scarcity a major concern. The Vaal River is particularly important within the region as it supplies water over a wide area. The major water uses of the Vaal WMA incorporate industrial, mining sectors, power generation, commercial agriculture (comprising stock watering, small and large irrigation schemes, dryland farming, and forestry), nature conservation, and urban and rural human settlements (Department of Water and Sanitation, 2016).

Depletion in the volume of water within the region, influenced by the interplay of anthropogenic activities, climatic variation, and hydrological processes, means that most sectors, along with the quality of life for residents, may experience adverse effects, resulting in economic disruption. Considering the circumstances observed in semiarid regions, it is crucial to emphasize the importance of robust monitoring to ensure water availability, effective management, and long-term sustainability of the Vaal water resource.

The study aims to analyze the seasonal variation of the Vaal River from 2013 to 2023. As a result, the objectives of the study are as follows: (1) to utilize geospatial platform technology to create digital maps to illustrate the distribution of water in the Vaal River; (2) to carry out a seasonal variability assessment of the Vaal River; and (3) to provide insights and recommendations for policymakers and water resource managers based on the results of the study. In the analyses, the study will also provide an average time series for both the dry and wet seasons along the Vaal River axis. By analyzing the seasonal variation in the water levels, the study would provide valuable information that could assist with detecting long-term trends and changes in the hydrological regime of the Vaal River. Engaging in seasonal monitoring of the Vaal River would assist in evaluating how climate change variability affects this important water resource. The study will also assist with the planning and management of water resources, particularly during water shortage periods, to avoid potential drought events and water challenges that may arise in the future.

The application of RS techniques and the GEE offer opportunities to conduct comprehensive seasonal water monitoring. These geospatial tools employ the use of satellite imagery captured at 30 m resolution to identify spatiotemporal patterns in the dynamic Vaal River. This will ensure that the water that contributes to meeting the demands of a nation of about 60.6 million people is adequately monitored.

8.2 Limitations of the study

Satellite and remotely sensed data have been very beneficial for obtaining temporal data over a wide span of geographical space. Nevertheless, for inland water, such as small lakes or narrow rivers, there are substantial challenges in acquiring remotely sensed data with fine spectral, spatial, and temporal resolution concurrently. For this study, the availability and quality of the RS data, especially for the Vaal region, posed some limitations. This is common with studies using RS images for their studies. The weather conditions that result in cloud and fog contamination impact greatly the output of remotely sensed data obtained from satellites.

8.3 Materials and methods

8.3.1 Study area

The Vaal River is South Africa's second-longest river, flowing for about 1,210 kilometers along four out of the nine provinces in South Africa. It is located between latitudes 26° 19'- 29° 4' S and longitude 23° 38'- 30° 11'E (Fig. 8.1). With its source in Ermelo, in the Mpumalanga Province, west of

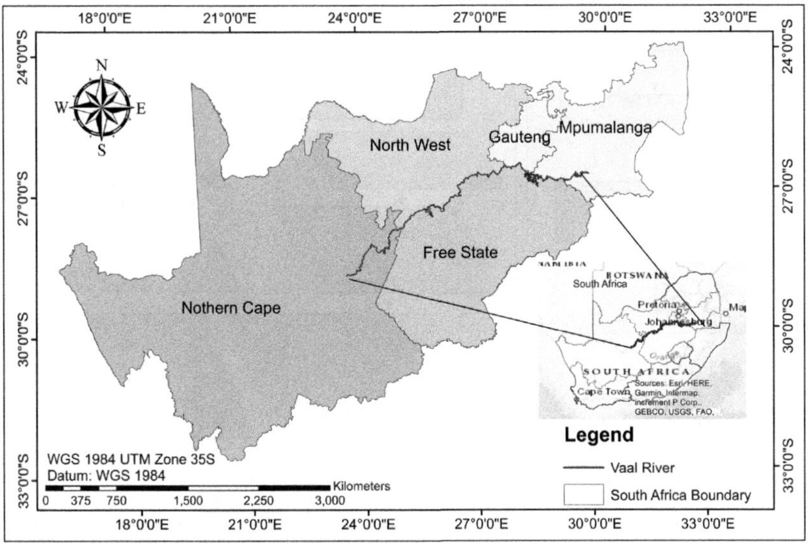

FIGURE 8.1 Location of the Vaal River in the context of South Africa. Map lines delineate study areas and do not necessarily depict accepted national boundaries.

Swaziland, it flows westward toward Gauteng, the Free State and Northwest Province, and into the Northern Cape. It then joins the Orange River, which flows into the Atlantic Ocean (Department of Water and Sanitation, 2016). The Vaal River system provides water resource services to roughly 60% of the national economy and provides additional water to more than 45% of South Africa's population—approximately 20 million people (Department of Water Affairs, 2013).

The region is typically identified as a semiarid region. South Africa, where the Vaal is located, is characterized by four seasons, i.e., summer (December, January, and February), autumn (March, April, and May), winter (June, July, and August), and spring (September, October, and November). The dry season occurs during the winter months, and the wet season occurs during the summer months, with rainfall peaking mostly in January and then into February. The climate around the Vaal River is a sub-tropical dry savanna, with a mean annual evaporation of approximately 1300 mm, greatly surpassing the rainfall of approximately 600 mm (Jury, 2016). Precipitation within the area has spatial variability and ranges from 800 to 1000 mm in the east to around 300 mm in the west (Department of Water Affairs and Sanitation, 2015), with a mean annual temperature of 15°C (Department of Water Affairs and Sanitation, 2015). The final action plan map has been prepared and incorporated after incorporating the parameters required to meet the objectives of the study (Fig. 8.2).

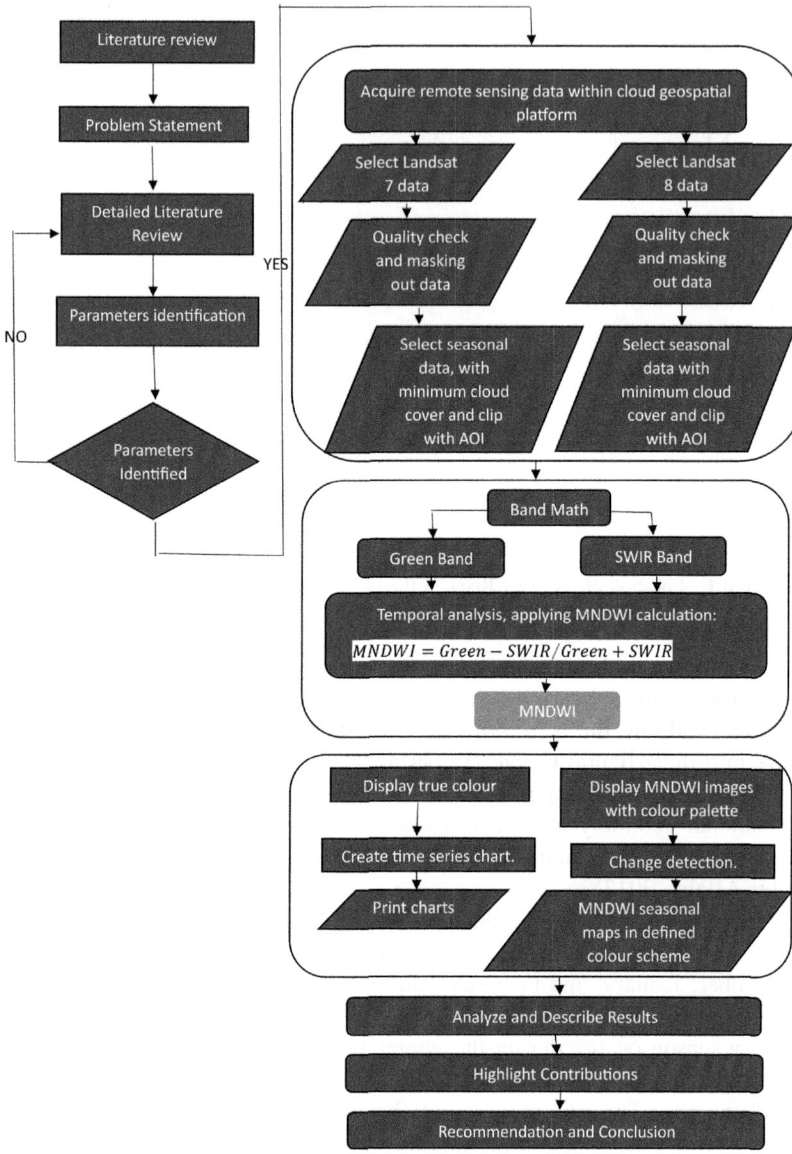

FIGURE 8.2 The Methodological Flowchart for the study.

8.3.2 Data

RS images acquired by the Landsat-7 and Operational Land Imager (OLI) sensor Landsat 8 satellite were used to monitor the Vaal River area between 2013 and 2023. Landsat 7 acquires images of bands (1–8), some in 15, 30 and 60 m spatial resolution. The Landsat-8 OLI acquires images of bands

1–11 in 30 m spatial resolution and band 8 in 15 m spatial resolution. The United States Geological Services Landsat-7 and Landsat 8 Top of Atmosphere collections were downloaded within the GEE platform. The GEE is a comprehensive platform that combines geospatial data like shapefiles and satellite imagery from various sources like MODIS, Landsat, and various computational tools, including AI algorithms, to enable advanced geospatial analysis and visualization. The study specifically employed images captured within the peak of the dry season (July–August) and wet season (January–February), with a focus on particularly selecting images that had a very low cloud cover—precisely less than 30%.

Water index (WI) estimations are very useful in delineating the existence of surface water resources. The Normalized Difference Water Index (NDWI) proposed by McFeeters (1996) has been efficient for RS of inland water content from space (Gao, 1996). The idea behind the NDWI method is to suppress non-water features while enhancing water features using the green band and the near-infrared band. According to Xu (2006), it was detected that the application of NDWI in water regions around built-up environments did not achieve the goal as water information was mixed with built-up noise, resulting in built-up land features having positive values in the NDWI image. As a result, modifications were made to the NDWI to remediate its shortcomings.

The Modified Normalized Difference Water Index (MNDWI) is a WI developed by Xu in 2006. According to Xu (2006), MNDWI is a more efficient way to extract water body features than NDWI. It uses the green (GREEN) and the short-wave infrared bands. This modification improved the enhancement of open water bodies, and it has been tested by several researchers:

$$\text{MNDWI} = \text{Green} - \text{SWIR}/\text{Green} + \text{SWIR} \tag{8.1}$$

8.4 Results

8.4.1 Spatiotemporal dynamics of water index across the dry season

The Vaal area as presented in Figs. 8.3 and 8.4 generally indicates the dominating presence of extreme water stress, despite indicating the presence of water (Vaal River); hence, it is expected to struggle with water supply at different intensities across seasons depending on its output. Fig. 8.3 represents the spatiotemporal dynamics of the WI based on the MNDWI for the dry season for random years between 2013 and 2023 along the Vaal River.

In 2013, the presence of the Vaal River with a WI value between 0.3 and 1 (abundant water) was evident, particularly with its concentration in two different spots within the Upper Vaal. As the river flows from east to west, another concentration of the river can also be seen at the tail end of the Middle Vaal, which also has a WI ranging between 0.3 and 1. Comparing

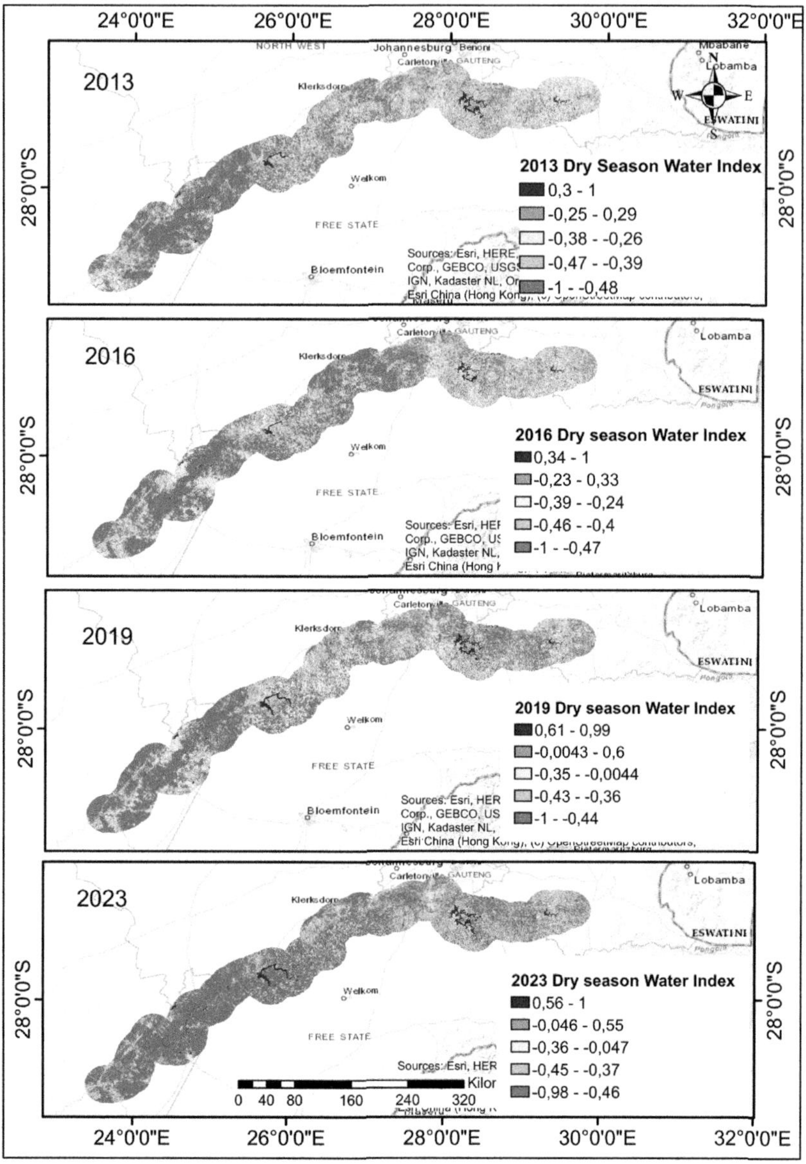

FIGURE 8.3 MNDWI for the dry season. Map lines delineate study areas and do not necessarily depict accepted national boundaries. *MNDWI*, Modified Normalized Difference Water Index.

this size to that seen in 2016, there was a reduction in area despite its presence. The results from 2019 showed an increase in the water concentration. This increase was bigger than it was in 2013 and 2016. Looking very similar in size to what was observed in 2019, but slightly bigger, 2023 also depicted

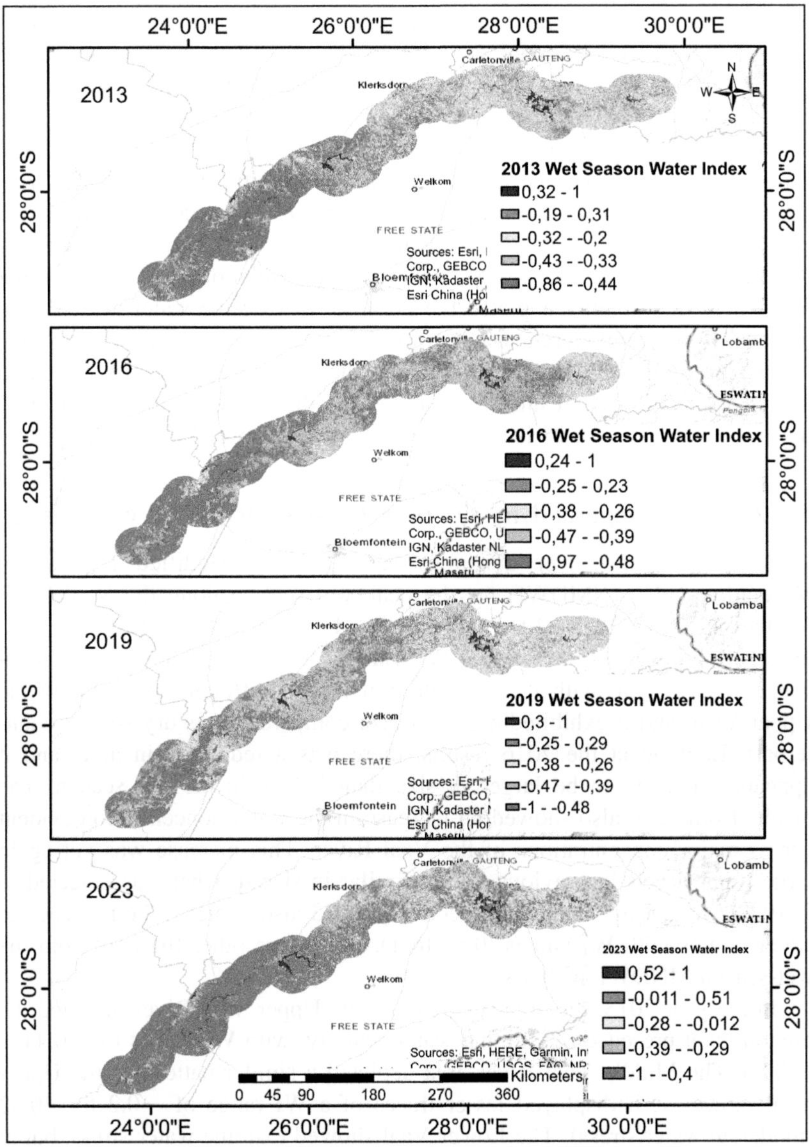

FIGURE 8.4 MNDWI for the wet season. Map lines delineate study areas and do not necessarily depict accepted national boundaries. *MNDWI*, Modified Normalized Difference Water Index.

a WI value ranging from 0.56 to 1, signifying the presence of a water body, among other information.

Other findings from Fig. 8.3 also revealed that in 2013, the Upper and Middle extents of the Vaal River had an almost uniform WI value ranging from −0.47 to −0.39 (severe water stress), with a few patches indicating the lowest WI values from −0.48 to −1, indicating extreme water scarcity, otherwise known as extreme drought. This corresponds to the 2016 results observed in the Upper Vaal, as seen in Fig. 8.3. The 2013 result of the Lower Vaal showed that the region was mostly dominated by a WI ranging between −0.48 and −1 (extreme water scarcity), with patches of WI values between −0.26 and −0.38 (moderate water stress). The results from 2016 showed a heavier presence of extreme water scarcity, which spread across the Middle Vaal to the Lower Vaal. More traces of moderate water stress (−0.24 to −0.39) can also be seen in the Lower Vaal region compared to the previous year. The Upper and Middle Vaal results of 2019 had a similar WI, where they mainly depicted the presence of extreme water scarcity and patches of severe/moderate water stress. The 2023 results revealed extreme water scarcity generally. This was seen in the Upper Vaal, and this increased appreciably as it shifted into the Middle, and the concentration of WI ranging from −0.48 to −0.98 is greatly evident in the Lower Vaal.

8.4.2 Spatiotemporal dynamics of water index across the wet season

The spatiotemporal river dynamics based on the Modified Normalized Vegetation Index (NMDWI) for the peak wet season along the Vaal River for the period of 2013–2023 were assessed in Fig. 8.4. In 2013, the existence of the Vaal River with a WI value between 0.32 and 1 (abundant water) was evident with its concentration in two different spots within the Upper Vaal, and it exhibited a greater size compared to its dry-season counterpart. Looking at the 2016 results, there was a reduction in area, and it appears yet again to be larger in size than it was in the dry season. The results from 2019 also showed an increase in the water concentration evident across major concentrations of the Vaal River. This increase was yet again larger than it was in previous years. Similar in size to what was detected in 2019, yet larger and smaller, respectively, 2023 also depicted the presence of water bodies with WI values (0.52 to 1). Below are other findings from the wet-season MNDWI analyses.

In the year 2013, the area surrounding the Upper Vaal River depicted the predominant presence of extreme water scarcity, with WI values of −0.44 to −0.86. The Middle Vaal River area depicted a similar pattern as the Upper Vaal; however, it displayed lesser traces of a WI range of −0.2 to −0.32 (moderate water stress). The Lower Vaal showed extreme water stress, but at a lesser intensity when compared to the 2013 dry season. The result from 2016 for the Upper Vaal depicted a WI value indicating moderate water stress (−0.39 to −0.47). This is similar to what was observed in the Middle Vaal, although with more patches indicating moderate water stress. The Lower Vaal showed conditions slightly more intense than those observed in 2013, indicating extreme water stress. In 2019, the water index in the Upper Vaal took a turn as it predominantly indicated moderate water stress, possessing a WI value of −0.26 to −0.38. The 2023 Upper Vaal was dominated by a WI value indicating moderate water stress, but its case had the least wet scenario out of the selected years analyzed. The Middle Vaal was seen to have a dominant WI value of −0.4 to −1. This year has the most pronouncedly least WI value out of the selected years of study. Generally, WI values indicating declining wet conditions were usually observed from the Upper, Middle to Lower Vaal (i.e., from their source in Ermelo until they meet with the Orange River).

A further decline in wet volumes at the Vaal River was evident in 2019 along the entire Upper, Middle, and Lower Vaals. This was also evident in the larger blue color that signified moderate water stress, the presence of the Vaal River, and other WI values. In terms of the general topography around the Vaal, this year depicted the wettest out of the randomly selected years in the decade considered for the analysis. This was succeeded by the much dryer year of 2023. The dynamics had reverted to a similar situation as was obtained in 2016, but in an even more severe condition. Extreme water stress took over the greatest portion, as it recorded MNDWI values of −0.4 to −1 for over 50% of the Vaal. Thus, the results revealed that 2016 and 2023 had the least wet conditions for the wet season along the areas surrounding the Vaal River. However, the spatial distribution of the MNDWI values of all the randomly selected years within the 10-year period in 2023 proved to be the least wet year for river catchment.

8.4.3 Average modified normalized difference water index time series for dry and wet seasons

Figs. 8.5 and 8.6 represent the time series chart for the respective wet and dry seasons over randomly selected years between the 2013 and 2023 periods. The dry season result revealed that there was a steady increase in the average MNDWI values of about 0.945 in 2013, increasing to about 0.968 in

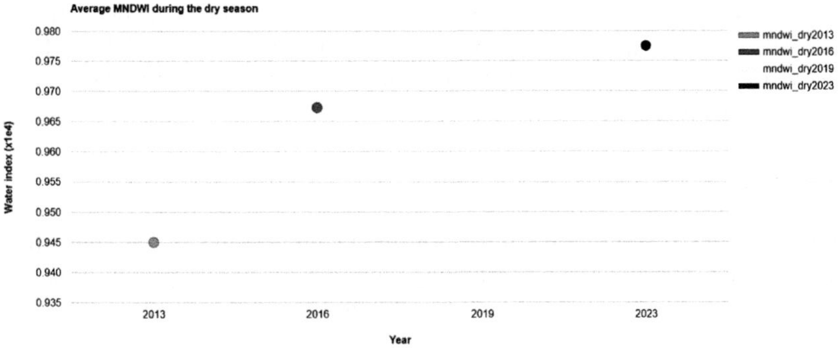

FIGURE 8.5 Average MNDWI for dry seasons from 2013 to 2023. *MNDWI*, Modified Normalized Difference Water Index.

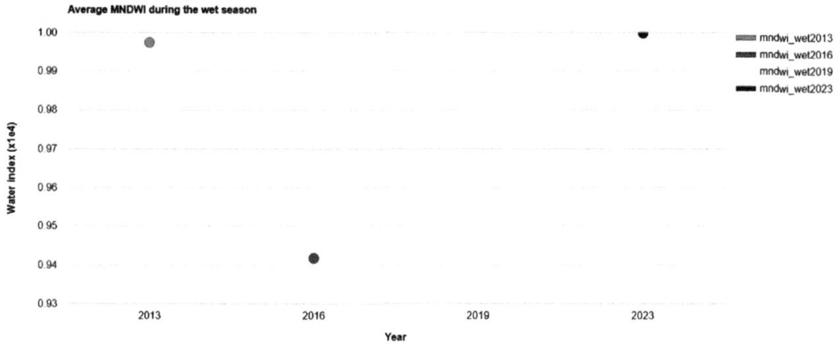

FIGURE 8.6 Average MNDWI for wet seasons from 2013 to 2023. *MNDWI*, Modified Normalized Difference Water Index.

2016, peaking at about 0.977 in 2019, and slightly increasing to 0.978 in 2023. Generally, the average MNDWI over time can be seen to have observed a steady increase until it remained almost constant. As seen in Fig. 8.6, the 2013 wet season results revealed a very high WI average of about 0.997. The year 2016 showed a sharp drop to 0.942, which appeared to be the lowest in the years analyzed. The value picked up again in 2019 to 0.961 and then reached the value of 1 in 2023. Results revealed that the average MNDWI was slightly higher than in the previous year, reaching a WI average of approximately 1 in 2023 and 2013, respectively.

8.4.4 Area of the Vaal River during both dry and wet seasons

The chart in Fig. 8.7 shows the seasonal area of the Vaal River for the randomly selected years over a decade. It can be observed that in all the years studied, the area of the Vaal during the dry season was smaller than in the wet season. The 2013 dry season area had an area of 553,289,783 km^2, and in the wet season, it was 597,992,968 km^2. In 2016, the area was further reduced to its smallest of 378,473,184 km^2. The area showed appreciation in the years 2019 and 2023. The dry area was about 696,045,182 km^2. From the years analyzed, the area of the Vaal River in the 2013 and 2023 wet seasons was the highest.

8.5 Discussion

Water availability and variability are highly unpredictable, and this study demonstrates the effectiveness of using cloud-based geospatial platforms and water indices in appraising the water variability across seasons in the semi-arid region of Vaal in South Africa.

The spatial distribution of seasonal MNDWI values calculated in this study indicates a contrasting WI in the last decade. The results of the dry

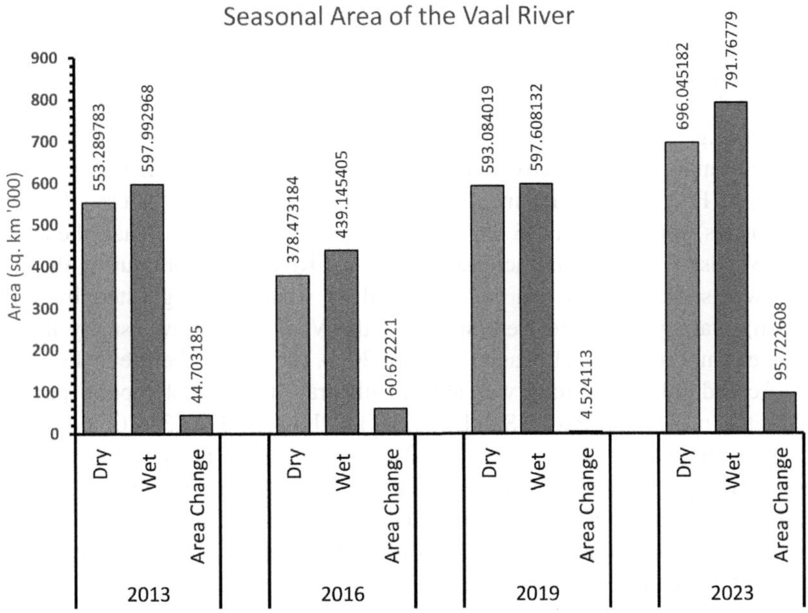

FIGURE 8.7 Area of Change of the Vaal River during both dry and wet seasons.

season water study revealed that Vaal River's average WI has progressively experienced a temporal rise in the pattern that shows an appreciation in average dry season NMDWI from 2013 till 2019 before remaining almost constant in 2023 (Fig. 8.5). The average MNDWI for the wet season showed a high WI in 2013 and then a sharp fall in 2016. This has earlier been documented by Botai, Botai, & Adeola (2018), who identified the drop in precipitation pattern in the 2016/2017 hydrological year, which contributed to the drought experienced in several provinces across South Africa. The average WI later appreciated in 2019, and then it indicated its highest in 2023.

The results of this study revealed that the average wet WI for both 2016 and 2019 was about 0.942 and 0.96, respectively, as against their average dry season WI, which showed that the WI values were 0.961 and 0.979 for the years 2016 and 2019. Pérez-Ruiz, Vivoni, & Sala (2022) reported that in arid and semiarid landscapes, and usually during the wet season, soil infiltration and loss of channel transmission lead to water stored in the subsurface for subsequent dry seasons. During the wet season, larger rainfall occurrences coupled with higher radiation and atmospheric water demand generally resulted in an increase in loss of water through ET and from overflow occurring through channels (e.g., McKenna & Sala, 2018; Méndez-Barroso, Vivoni, Watts, & Rodríguez, 2009; Pérez-Ruiz et al., 2021; Pierini, Vivoni, Robles-Morua, Scott, & Nearing, 2014; Templeton et al., 2014; Vivoni, Rodríguez, & Watts, 2010). On the other hand, the lower concentration of precipitation during the winter (dry season), when radiation and atmospheric water demands are much lower, on average favors the downward percolation of water through deeper soil layers, with minimal ET and surplus production, nevertheless supporting ecosystem productivity and water use within the dry season (e.g., Anderson-Teixeira, Delong, Fox, Brese, & Litvak, 2011; Scott & Biederman, 2019; Wilcox, Dowhower, Teague, & Thurow, 2006).

From Fig. 8.4, despite majorly depicting extreme water stress (drought tendencies) around about 25 km radius away from the Lower Vaal, the average WI also suggested an increase in the Vaal River area both during the dry and wet seasons. This can be attributed to what was suggested by Jury (2016), stating that there were spikes in the Vaal River flow usually in the wet season, one of which was in January 2010, three years before 2013.

Accordingly, this study was able to illustrate the spatiotemporal pattern of water in the Vaal River, South Africa, for the period 2013–2023. Similar to other studies, Remilekun et al. (2021) also suggested the variability of the Vaal River water, and Niranjannaik et al. (2022) identified similar groundwater variability in a semiarid river basin. Water variations in the Vaal can be attributed to climatic factors such as El Nino and La Nina events, influencing the weather and thus affecting rainfall intensities.

8.6 Challenges and solutions

Challenges encountered during the period of the study ranged from data quality issues and availability to cloud cover and atmospheric effects, resulting in the presence of scan lines, especially for the 2013 dry and wet images. Other challenges were around data integration and various data formats as different software, in addition to the GEE, was utilized to produce results. Proffering solutions to these challenges required a combination of technical skills and interdisciplinary collaboration, among other things.

8.7 Recommendations

This research has showcased the great potential of utilizing GEE as a powerful tool for conducting cloud-based geospatial analyses and for addressing complex hydrological and environmental challenges. To ensure the reliability of the analysis conducted within the GEE, updates and privacy guidelines as adopted in the study should be adhered to. Data quality checks, quality assurance, and control protocols are also to be ensured. Also, quality control procedures should be implemented for satellite images and data obtainable from the platform. Codes developed and used within the platform should be compatible with the platform's API when carrying out future studies.

To ensure long-term monitoring and sustainability of the Vaal River's dynamic seasons using the GEE, collaborations and partnerships should be explored with local agencies or organizations like the Department of Water and Sanitation and Rand Water. This collaboration will contribute to a sustained understanding of the Vaal River's dynamics, ensuring continuous monitoring of the river beyond the duration stipulated in this analysis. Several measures were taken to ensure the quality, reliability, and ethical conduct of this research. The authors recommend that the sources of the data used to monitor the Vaal River are relevant and up-to-date and clear information about the goals of future studies and potential benefits should be provided.

Sustainable water management policies should be made and enforced that incorporate insights from the findings of this research to guide decision-making, especially during seasons when there is likely to be a low water supply in the river. There should be public awareness of the significance of the river, its ecosystem, and the carrying out of sustainable practices and water conservation measures to ensure continuous availability of water during all seasons. There should be local community engagement and empowerment to participate in the monitoring and conservation of the river and its ecosystem and the preservation of the biodiversity around the Vaal River should be advocated for. Communication mediums should be established with the community, updating them about the findings from this study and involving them in decision-making process that would ensure sustainable solutions for the Vaal River and the region it covers.

The findings obtained from this study should be built upon by researchers and organizations. Additional seasonal water reservation plans that would cater for seasonal water fluctuation and water demands should be considered. This will assist in drought preparedness plans for identified dynamic seasons and establish early warning systems and response strategies to any extreme weather events.

The authors recommend that policymakers, institutions, and researchers continue to harness the capabilities of GEE in future studies, especially in monitoring water resources for efficient management and sustainability. The GEE possesses a user-friendly interface that has a vast archive of RS data from different sources and displays effective data processing capabilities for diverse applications, as conducted in this study. GEE can play a crucial role in assisting regional and global projects on issues related to water monitoring, conservation, and sustainability. The authors recommend that governments, organizations such as Rand Water, and institutions like the Department of Water and Sanitation, among others, explore GEE to leverage its capabilities for addressing the pressing demand for water and the challenges its supply has faced due to the effects of climate change. Recommendations were developed based on the findings of this research to promote sustainable water management practices for the Vaal River.

8.8 Conclusion

The study was aimed at evaluating the seasonal spatial-temporal characteristics of the Vaal River in the last decade and providing insight into the dynamic attributes of the Vaal River in South Africa. The Landsat 7 and Landsat 8 OLI RS data were used to calculate MNDWI values at two timescales (i.e., dry and wet seasons). From the findings obtained, the WI along the Vaal exhibits noticeable spatial-temporal contrast. Several reports, as shown in the study, suggest that Vaal and other areas within the semiarid region of South Africa were faced with a very low to moderate WI, indicating a moderate to severe drought. As documented in several global and domestic reports and in this study, millions of people face water-related issues due to shortages and poor management. Due to the importance of water, the absence of water often poses steady and sporadic fears for the functioning of ecosystems and the lives of humans. The spatiotemporal dynamics of the Vaal River based on the MNDWI showed that the worst affected areas along the Vaal included the middle and lower Vaal, with MNDWI values indicating a depreciation in values surrounding the Vaal River.

Climate change, which has worsened the precipitation deficit in some years more than others, is more likely to impact the water availability of the river, especially in a semiarid region. This situation may have been further intensified by a low level of sustainable and management strategies in many parts of the country. As such, the recurrent pattern of depressed water supply will further lead to an even greater water crisis, resulting in the probable shortage of food, the loss of livestock, and, in the worst-case scenarios, the loss of human lives due to dehydration.

The study demonstrates the effectiveness of using geospatial and RS indices rather than conventional methods in appraising the water availability across seasons in a semiarid Vaal region in South Africa. However, the importance of RS and cloud-based geospatial analysis, as presented in this study, suggests that proactive decision support measures can be developed and implemented to initiate early warning systems that will assist with monitoring water volume changes and the sustainability of water supply. This would improve the ongoing efforts within this region to tackle the challenges of water supply that have resulted from the changing climate.

Acknowledgments

The authors would like to acknowledge the Department of Civil Engineering, College of Science, Engineering and Technology (CSET) of the University of South Africa (UNISA) for providing the necessary infrastructure and research facilities. Furthermore, we acknowledge the financial support provided by the University of South Africa (UNISA). Their funding made it possible for the authors to carry out this research and disseminate our findings. The authors also acknowledge Enoch Anning, who assisted with most of the GEE analysis.

References

Anderson-Teixeira, K. J., Delong, J. P., Fox, A. M., Brese, D. A., & Litvak, M. E. (2011). Differential responses of production and respiration to temperature and moisture drive the carbon balance across a climatic gradient in New Mexico. *Global Change Biology*, *17*(1), 410–424. Available from https://doi.org/10.1111/j.1365-2486.2010.02269.x.

Bischoff-Mattson, Z., Maree, G., Vogel, C., Lynch, A., Olivier, D., & Terblanche, D. (2020). Shape of a water crisis: Practitioner perspectives on urban water scarcity and 'Day Zero' in South Africa. *Water Policy*, *22*(2), 193–210. Available from https://doi.org/10.2166/wp.20200.233. Available from, https://watermark.silverchair.com/022020193.pdf?token = AQECAHi208BE49Ooan9kkhW_Ercy7Dm3ZL_9Cf3qfKAc485ysgAAA2Uwgg NhBgkqhkiG9w0BBwagggNSMIIDTgIBADCCA0cGCSqGSIb3DQEHATAeBglghkg BZQMEAS4wEQQMy3AThSAcfgz.

Botai, C. M., Botai, J. O., & Adeola, A. M. (2018). Spatial distribution of temporal precipitation contrasts in South Africa. *Journal of Science*, *114*(7-8). Available from https://doi.org/10.17159/sajs.2018/20170391. Available from:, https://www.sajs.co.za/article/view/4853/6992.

Briand, A., Reynaud, A., Viroleau, F., Markantonis, V., & Branciforti, G. (2023). Assessing the Macroeconomic Effects of Water Scarcity in South Africa using a CGE Model. *Environmental Modeling and Assessment*, *28*(2), 259–272. Available from https://doi.org/10.1007/s10666-023-09883-4. Available from, https://www.springer.com/journal/10666.

Department of Water Affairs (2013). National Water Resource Strategy (2nd ed.).Water Resources, 201. Available from http://www.dwa.gov.za/documents/Other/StrategicPlan/NWRS2-Final-email-version.pdf. http://www.dwa.gov.za/documents/Other/StrategicPlan/NWRS2-Final-email-version.pdf.

Gao, B. (1996). NDWI—A normalized difference water index for remote sensing of vegetation liquid water from space. *Remote Sensing of Environment.*, *58*(3). Available from https://doi.org/10.1016/S0034-4257(96, 67–3.

Godinez Madrigal, Jonatan, van der Zaag, Pieter, & van Cauwenbergh, Nora (2018). A half-baked solution: Drivers of water crises in Mexico. *Proceedings of the International Association of Hydrological Sciences*, *376*, 57–62. Available from https://doi.org/10.5194/piahs-376-57-2018.

Gorelick, N., Hancher, M., Dixon, M., Ilyushchenko, S., Thau, D., & Moore, R. (2017). Google Earth Engine: Planetary-scale geospatial analysis for everyone. *Remote Sensing of Environment*, *202*, 18–27. Available from https://doi.org/10.1016/j.rse.2017.060.031. Available from, http://www.elsevier.com/inca/publications/store/5/0/5/7/3/3.

Gujrati, A., & Jha,V.B. (2018). Surface Water Dynamics of Inland Water Bodies of India Using Google Earth Engine. *ISPRS Annals of the Photogrammetry, Remote Sensing and Spatial Information Sciences*, IV-5, 467–472. http://www.isprs.org/publications/annals.aspx.

Ilunga, M. (2017). Assessing catchment resilience usingentropy associated with mean annual runofffor the Upper Vaal catchment in South Africa. *Entropy*, *19*(5). Available from https://doi.org/10.3390/e19050147. Available from, http://www.mdpi.com/1099-4300/19/5/147/pdf.

Jury, M. R. (2016). Climate influences on Vaal River flow. *South African Water Research Commission, Puerto Rico Water SA*, *42*(2), 232–242. Available from https://doi.org/10.4314/wsa.v42i2.07. Available from, http://www.wrc.org.za/Knowledge%20Hub%20Documents/Water%20SA%20Journals/Manuscripts/2016/02%20April%202016/3153.pdf.

Kallio, K., Kutser, T., Hannonen, T., Koponen, S., Pulliainen, J., Vepsäläinen, J., & Pyhälahti, T. (2001). Retrieval of water quality from airborne imaging spectrometry of various lake types in different seasons. *Science of the Total Environment*, *268*(1-3), 59–77. Available from https://doi.org/10.1016/S0048-9697(00)00685-9.

Koop, S. H. A., Van Dorssen, A. J., & Brouwer, S. (2019). Enhancing domestic water conservation behaviour: A review of empirical studies on influencing tactics. *Journal of Environmental Management*, *247*, 867–876. Available from https://doi.org/10.1016/j.jenvman.2019.060.126. Available from:, https://www.sciencedirect.com/journal/journal-of-environmental-management.

Malahlela, O.E. (2019). Spatio-temporal assessment of inland surface water quality using remote sensing data in the wake of changing climate.In: *IOP conference series: earth and environmental science*. https://iopscience.iop.org/journal/1755-1315.

McDonald, R. I., Weber, K., Padowski, J., Flörke, M., Schneider, C., Green, P. A., ... Montgomery, M. (2014). Water on an urban planet: Urbanization and the reach of urban water infrastructure. *Global Environmental Change*, *27*(1), 96–105. Available from https://doi.org/10.1016/j.gloenvcha.2014.040.022. Available from, http://www.elsevier.com/inca/publications/store/3/0/4/2/5.

McFeeters, S. K. (1996). The use of the Normalized Difference Water Index (NDWI) in the delineation of open water features. *International Journal of Remote Sensing*, *17*(7), 1425–1432. Available from https://doi.org/10.1080/01431169608948714.

McKenna, O. P., & Sala, O. E. (2018). Groundwater recharge in desert playas: Current rates and future effects of climate change. *Environmental Research Letters*, *13*(1). Available from https://doi.org/10.1088/1748-9326/aa9eb6. Available from, http://iopscience.iop.org/article/10.1088/1748-9326/aa9eb6/pdf.

Millington, N. (2018). Producing water scarcity in São Paulo, Brazil: The 2014-2015 water crisis and the binding politics of infrastructure. *Political Geography*, *65*, 26–34. Available from https://doi.org/10.1016/j.polgeo.2018.040.007. Available from, http://www.elsevier.com/inca/publications/store/3/0/4/6/5/index.htt.

Milly, P. C. D., Dunne, K. A., & Vecchia, A. V. (2005). Global pattern of trends in streamflow and water availability in a changing climate. *Nature*, *438*(7066), 347–350. Available from https://doi.org/10.1038/nature04312. Available from, http://www.nature.com/nature/index.html.

Méndez-Barroso, L. A., Vivoni, E. R., Watts, C. J., & Rodríguez, J. C. (2009). Seasonal and interannual relations between precipitation, surface soil moisture and vegetation dynamics in the North American monsoon region. *Journal of Hydrology*, *377*(1-2), 59–70. Available from https://doi.org/10.1016/j.jhydrol.2009.080.009.

Niranjannaik, M., Kumar, A., Beg, Z., Singh, A., Swarnkar, S., & Gaurav, K. (2022). Groundwater variability in a Semi-Arid River Basin, Central India. *Hydrology*, *9*(12). Available from https://doi.org/10.3390/hydrology9120222. Available from:, http://www.mdpi.com/journal/hydrology.

Pierini, N. A., Vivoni, E. R., Robles-Morua, A., Scott, R. L., & Nearing, M. A. (2014). Using observations and a distributed hydrologic model to explore runoff thresholds linked with mesquite encroachment in the Sonoran Desert. *Water Resources Research*, *50*(10), 8191–8215. Available from https://doi.org/10.1002/2014WR015781UnitedStates. Available from, http://agupubs.onlinelibrary.wiley.com/hub/journal/10.1002/(ISSN)1944-7973/.

Pérez-Ruiz, E. R., Vivoni, E. R., Yépez, E. A., Rodríguez, J. C., Gochis, D. J., Robles-Morua, A., ... Adams, D. K. (2021). Landscape controls on water-energy-carbon fluxes across different ecosystems during the North American Monsoon. *Journal of Geophysical Research: Biogeosciences*, *126*(5). Available from https://doi.org/10.1029/2020JG005809. Available from:, https://agupubs.onlinelibrary.wiley.com/loi/21698961.

Pérez-Ruiz, E. R., Vivoni, E. R., & Sala, O. E. (2022). Seasonal carryover of water and effects on carbon dynamics in a dryland ecosystem. *Ecosphere*, *13*(7). Available from https://doi.

org/10.1002/ecs2.4189. Available from, http://esajournals.onlinelibrary.wiley.com/hub/journal/10.1002/(ISSN)2150-8925/.

Remilekun, A. T., Thando, N., Nerhene, D., & Archer, E. (2021). Integrated assessment of the influence of climate change on current and future intra-annual water availability in the Vaal River catchment. *Journal of Water and Climate Change*, *12*(2), 533–551. Available from https://doi.org/10.2166/wcc.2020.269, https://watermark.silverchair.com/jwc0120533.pdf?.

Roshan, Ajmal, & Kumar, Manish (2020). Water end-use estimation can support the urban water crisis management: A critical review. *Journal of Environmental Management*, *268*, 110663. Available from https://doi.org/10.1016/j.jenvman.2020.110663.

Scott, R. L., & Biederman, J. A. (2019). Critical zone water balance over 13 years in a Semiarid Savanna. *Water Resources Research*, *55*(1), 574–588. Available from https://doi.org/10.1029/2018WR023477. Available from, http://agupubs.onlinelibrary.wiley.com/hub/journal/10.1002/(ISSN)1944-7973/.

Serdeczny, O., Adams, S., Baarsch, F., Coumou, D., Robinson, A., Hare, W., ... Reinhardt, J. (2017). Climate change impacts in Sub-Saharan Africa: From physical changes to their social repercussions. *Regional Environmental Change*, *17*(6), 1585–1600. Available from https://doi.org/10.1007/s10113-015-0910-2. Available from, http://springerlink.metapress.com/app/home/journal.asp?wasp = 64cr5a4mwldrxj984xaw&referrer = parent&backto = browsepublicationsresults,451,542.

Shelestov, A., Lavreniuk, M., Kussul, N., Novikov, A., & Skakun, S. (2017). Exploring Google earth engine platform for big data processing: Classification of multi-temporal satellite imagery for crop mapping. *Frontiers in Earth Science*, *5*, 1–10. Available from https://doi.org/10.3389/feart.2017.00017. Available from, https://www.frontiersin.org/journals/earth-science.

Shi, K., Zhang, Y., Zhu, G., Qin, B., & Pan, D. (2018). Deteriorating water clarity in shallow waters: Evidence from long term MODIS and in-situ observations. *International Journal of Applied Earth Observation and Geoinformation*, *68*, 287–297. Available from https://doi.org/10.1016/j.jag.2017.120.015. Available from:, http://www.elsevier.com/locate/jag.

Song, K., Wang, Z., Blackwell, J., Zhang, B., Li, F., Zhang, Y., & Jiang, G. (2011). Water quality monitoring using Landsat Themate Mapper data with empirical algorithms in Chagan Lake, China. *Journal of Applied Remote Sensing*, *5*(1). Available from https://doi.org/10.1117/1.3559497. Available from:, http://www.spie.org/x3636.xml.

Templeton, R. C., Vivoni, E. R., Méndez-Barroso, L. A., Pierini, N. A., Anderson, C. A., Rango, A., ... Scott, R. L. (2014). High-resolution characterization of a semiarid watershed: Implications on evapotranspiration estimates. *Journal of Hydrology*, *509*, 306–319. Available from https://doi.org/10.1016/j.jhydrol.2013.110.047.

Vivoni, E. R., Rodríguez, J. C., & Watts, C. J. (2010). On the spatiotemporal variability of soil moisture and evapotranspiration in a mountainous basin within the North American monsoon region. *Water Resources Research*, *46*(2). Available from https://doi.org/10.1029/2009WR008240.

Wilcox, B. P., Dowhower, S. L., Teague, W. R., & Thurow, T. L. (2006). Long-term water balance in a semiarid shrubland. *Rangeland Ecology and Management*, *59*(6), 600–606. Available from https://doi.org/10.2111/06-014R3.1. Available from, http://www.bioone.org/loi/rama.

Xu, H. (2006). Modification of normalised difference water index (NDWI) to enhance open water features in remotely sensed imagery. *International Journal of Remote Sensing*, *27*(14), 3025–3033. Available from https://doi.org/10.1080/01431160600589179. Available from:, https://www.tandfonline.com/loi/tres20.

Chapter 9

An appraisal of groundwater potentiality of Kharkai watershed, India

Partha Modak, Susanta Mandi, Pradyut Bera, Padma Ruhidas and Mrinal Mandal
Department of Geography, Sidho-Kanho-Birsha University, Purulia, West Bengal, India

9.1 Introduction

Globally, the crisis related to freshwater availability has been witnessed owing to changes in the atmosphere and increasing pollution. Moreover, ever-increasing water demand from each corner of society significantly raises havoc pressure on readily available groundwater resources (Gautam, Pande, Kothari, Kumar Singh, & Agrawal, 2023). Groundwater has always played a crucial role in the process of human development. A significant proportion of the gross domestic product is contributed by the groundwater (Ghosh, Mandal, Banerjee, & Karmakar, 2020). In the case of India, groundwater contributes around 50% of irrigated agriculture (Ghosh, Mandal, Banerjee, et al., 2020; Mandal et al., 2023; Nag & Das, 2017). The rapid population growth, industrial advancement, expansion of urban settlements, changing weather and climate, and erratic nature of downpours exert a huge impact on groundwater resources (Thapa, Gupta, Gupta, Reddy, & Kaur, 2018). The present scenario of groundwater exploitation points toward a treacherous future. Therefore, it is important to assess the availability of groundwater resources and their management for sustainable utilization (Ghosh, Mandal, Banerjee, et al., 2020; Ghosh, Mandal, Karmakar, Banerjee, & Mandal, 2020; Mandal et al., 2023; Thapa et al., 2018). Groundwater, being the most vivacious source of drinking water, also serves various other purposes, such as domestic use, irrigation, fishing (fishery pond), industrial activities, and the service sector. The use of groundwater for different purposes is mainly high in areas of hard rock terrain with a greater volume of surface runoff (Khadri & Pande, 2016). In the hilly region, the availability of groundwater is an important concern (Singh, Dahiphale, Yadav, & Singh, 2018), which

Water Resources Monitoring, Management, and Sustainability. DOI: https://doi.org/10.1016/B978-0-443-23665-5.00009-0

primarily depends on several hydrogeological factors (Das, 2017). The arid and semi-arid regions of the globe, characterized by diverse physiographic conditions, experience severe challenges due to water shortages (Etikala, Golla, Li, & Renati, 2019). Thus, groundwater assessment is an important task that requires continuous monitoring and proper mapping to ensure its sustainable utilization (Gautam et al., 2023; Ghosh, Mandal, Banerjee, et al., 2020; Ghosh, Mandal, Karmakar, et al., 2020; Karmakar, Banerjee, Mandal, & Ghosh, 2021; Mandal et al., 2023). In India, groundwater contributes 85%, 62%, and 45% of the total water requirements of agriculture, irrigation, and urban areas, respectively (Abijith, Singh, Jennifer, Saranya, & Parthasarathy, 2020; CGWB, 2013). Even around 30% and more than 90% of the metropolitan and rural populations, respectively, depend on groundwater for drinking and domestic purposes (Parthasarathy & Deka, 2021). The total annual groundwater recharge in India is estimated at 436.15 bcm, whereas the annual extractable groundwater resources are 397.62 bcm (CGWB, 2022). The annual availability of per capita water <1700 m^3 and <1000 m^3 is considered water stress and water scarcity conditions. In India, the availability of average annual per capita water was 1816 and 1545 m^3 in 2001 and 2011, respectively. Based on these figures, India belongs to the category of water-stressed conditions. In the future, the situation may deteriorate more drastically due to rapid population growth (Ministry of Jal Shakti, Government of India, 2023).

Over the years, researchers from all over the world have engaged themselves to evaluate groundwater potential zones (GWPZ) mainly of hard rock terrains in arid and semi-arid regions. To have a proper understanding of the groundwater potentiality of any region, hydrogeological factors are required to be considered. Each factor has a certain role in the process of groundwater development. To enumerate the influence of each factor, several researchers across the world have been employing different multi-criteria decision making (MCDM) models coupled with remote sensing (RS) and geographical information systems (GIS) to identify the GWPZ. Several studies related to groundwater assessment have already been carried out using RS, GIS, and MCDM techniques (Al-Abadi, Shahid, Ghalib, & Handhal, 2017; Das & Pardeshi, 2018; Ghosh, Mandal, Banerjee, et al., 2020; Ghosh, Mandal, Karmakar, et al., 2020; Kaliraj, Chandrasekar, & Magesh, 2013; Karmakar et al., 2021; Mandal et al., 2023; Pinto, Shrestha, Babel, & Ninsawat, 2017; Rajasekhar, Bramaiah, Deepthi, Amaravathi, & Raju, 2019; Sashikkumar, Selvam, Kalyanasundaram, & Johnny, 2017; Selvam et al., 2014; Tiwari et al., 2017). The multi-influencing factors (MIF), one of the widely accepted and used MCDM techniques, along with RS and GIS techniques, are applied to assess the GWPZ. The MIF technique is also a reliable and cost-effective instrument for understanding suitable groundwater sites. The best side of this technique is to identify the influence of each factor on the hydrogeological arrangement of any terrain (Das, Gupta, & Ghosh, 2017; Etikala et al., 2019; Magesh, Chandrasekar, &

Soundranayagam, 2012; Nasir, Khan, Zahid, & Khan, 2018; Thapa et al., 2018). Moreover, the MIF is the most elementary and comprehensive method, and it requires sufficient knowledge of the area of investigation along with the relative importance of individual parameters in determining and influencing groundwater availability (Mandal, Ranjan, & Dharanirajan, 2021). The present study aims to evaluate the GWPZ of the Kharkai watershed in eastern India, employing the MIF technique along with geospatial technology.

9.2 Study area

The river Kharkai originates from the Bamanghati-Kolhan upland area of Odisha and Jharkhand states and flows over the Singhbhum plain and Archean terrain of southern Singhbhum (Mukherjee & Patel, 2022). The Kharkai river mainly flows from the northern slope of Darbarmela Hill and the western slope of Tungru Hill in the Mayurbhanj district, Odisha (Das & Guchait, 2016). The river merges with the Subarnarekha River in the Adityapur area of Jamshedpur, Jharkhand (Das & Guchait, 2016; Mukherjee & Patel, 2022). The major tributaries of the river are Kandria, Nusa, Barhai, Karanjia, and Sanjai (Das & Guchait, 2016). The length of the Kharkai river is 136 km, with a watershed area of 6256.15 km^2. The geographical extensions of the river watershed are 85°20′ E to 86°20′ E and 22°5′ N to 22°55′ N (Fig. 9.1).

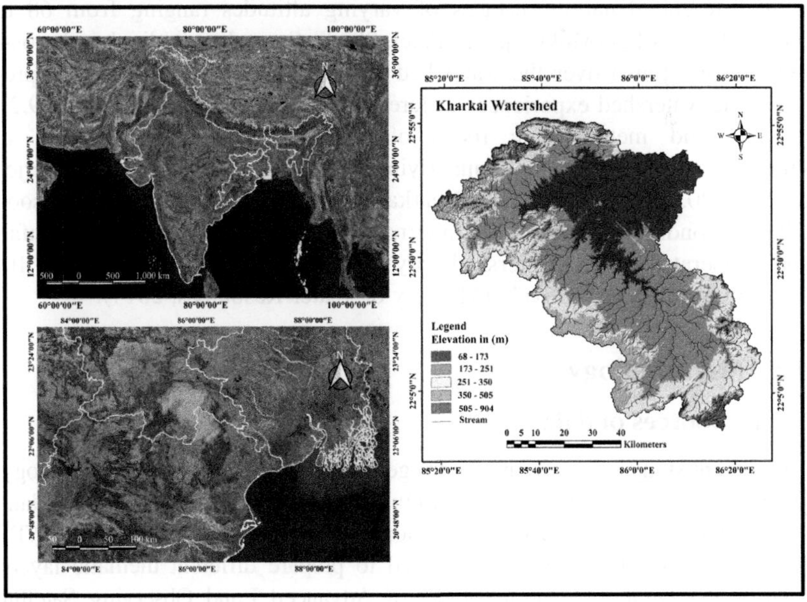

FIGURE 9.1 Location map of the study area. Map lines delineate study areas and do not necessarily depict accepted national boundaries.

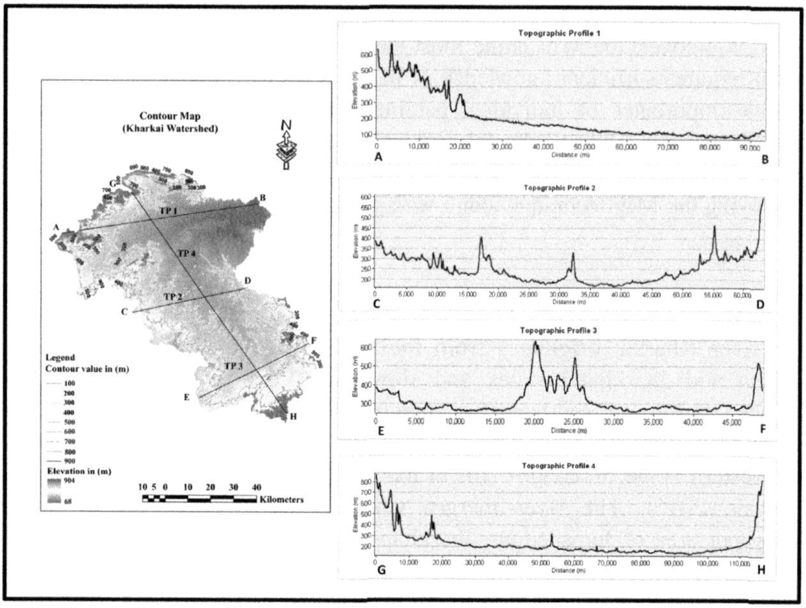

FIGURE 9.2 Longitudinal and cross profiles of the Kharkai watershed. Map lines delineate study areas and do not necessarily depict accepted national boundaries.

Geomorphologically, the watershed is a part of the Subarnarekha basin with different planation surfaces of varying altitudes ranging from 68 to 904 m (Fig. 20.12) (Mukherjee & Patel, 2022). Some longitudinal and cross-sections are drawn over the watershed to understand the topographical profiles. The watershed experiences different topographical variations (Fig. 9.2). Igneous and metamorphic rocks of low- to high-grade Archean to Proterozoic ages are the main underlying rocks of the study area (Mukherjee & Patel, 2022) (Fig. 20.5). The Kharkai watershed enjoys tropical monsoon climatic conditions, followed by alternate wet and dry seasons. The rainfall and temperature of the watershed vary from 1158 to 1702 mm (Fig. 20.10) and 10°C–42°C, respectively (Ministry of Water Resources, 2014).

9.3 Methodology

9.3.1 Sources of data

The present study employs nine hydro-geological parameters, such as lithology, geomorphology, lineament density (LD), soil type, drainage density (DD), land use and land cover (LULC), rainfall, slope, and elevation, to assess the GWPZ (Fig. 9.3). Several datasets are acquired to prepare different thematic layers, such as the Digital Elevation Model of Advanced Land Observing Satellite Polarimetric Phased Array L-band Synthetic Aperture Radar with a spatial

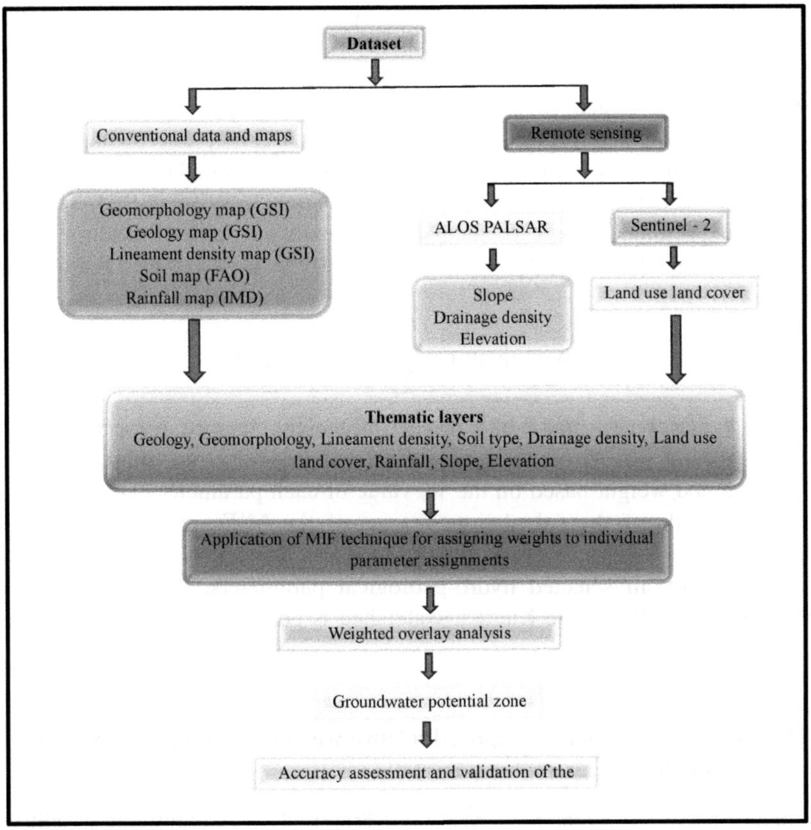

FIGURE 9.3 Flow chart of the methodology.

resolution of 12.5 m for elevation, slope, and DD maps in 2015; the Esri land cover Sentinel-2 satellite image of 2021 with a spatial resolution of 10 m for the LULC map; yearly gridded rainfall (0.25 × 0.25 degree) NetCDF data file downloaded from Indian Meteorological Department (IMD), Pune for rainfall map; geomorphological map and lithological map of Geological Survey of India on a 1:250,000 and 1:50,000 scale, respectively; LD map downloaded from https://bhuvan.nrsc.gov.in/; and soil map on a 1:5,000,000 scale obtained from https://data.apps.fao.org/ of Food and Agriculture Organization (FAO) of the United Nations in RS and GIS software (ArcGIS-10.3.1, QGIS-3.12.3, and Golden software Grapher-12).

9.3.2 Multi influencing factor

Each hydrogeological factor has its own distinctive role in the process of groundwater development and also influences the level of importance of other factors. In brief, all the factors are mutually interdependent. To assign weight to each parameter based on the relative strength of other factors, the MIF technique is applied in this study (Ghosh, Mandal, Banerjee, et al., 2020; Ghosh, Mandal, Karmakar, et al., 2020; Kaliraj et al., 2013; Magesh et al., 2012; Siddi Raju, Sudarsana Raju, & Rajasekhar, 2019; Thapa et al., 2018). The major determining factors are assigned a weight of 1, while 0.5 weight is allocated to the minor influencing parameters (Table 9.1). The following expression is used to calculate the weight of each influencing factor (Ghosh, Mandal, Banerjee, et al., 2020; Magesh et al., 2012; Siddi Raju et al., 2019):

$$\text{Weight}(W_i) = \frac{(X+Y)}{\sum(X+Y)} \times 100 \tag{9.1}$$

where X and Y are major and minor influencing factors, respectively. W_i represents the weight of each parameter. Subsequently, the sub-classes are also assigned weight based on the W_i value of each parameter (Table 9.1).

The details of the calculation process of the MIF are available in the study done by Ghosh, Mandal, and Banerjee et al. (2020). After computing the weight of all selected hydro-geological parameters, the final thematic map of the GWPZ of the Kharkai watershed is prepared, integrating all nine

TABLE 9.1 Major, minor, proposed relative weight, and proposed weight of each influential parameter using MIF.

Influential factors	Major effects (X)	Minor effects (Y)	Proposed relative weight (X + Y)	Proposed weight of influential parameter (P_i)
Geology	4	0	4	17
Geomorphology	3	0.5	3.5	15
Lineament density	3	0.5	3.5	15
Soil type	2	0	2	8
Drainage density	1	0.5	1.5	6
LULC	2	1.5	3.5	15
Rainfall	1.5	0.5	2	8
Slope	1.5	0.5	2	8
Elevation	2	0	2	8
Total			$\sum 24$	$\sum 100$

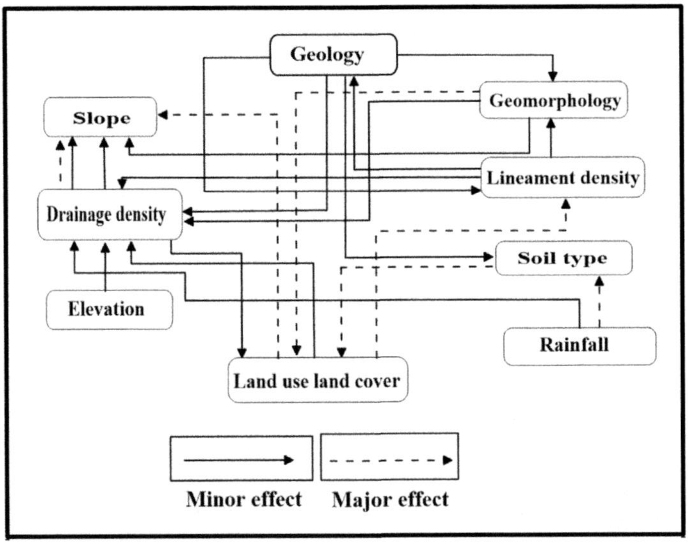

FIGURE 9.4 Major and minor hydrogeological factors and their interrelationship.

thematic layers. In this process, the Universal Transverse Mercator Projection is adopted by selecting the World Geodetic System 84 and 45 N zones and following the weighted overlay method in the Arc GIS software 10.3.1. A schematic diagram related to the major and minor effects of factors on one another is represented in Fig. 9.4.

9.4 Results

9.4.1 Lithology

The quantity and quality of groundwater are highly influenced by the sub-surface geological structure, and the permeability and porosity of rocks also determine the rate of groundwater recharge (Das & Mukhopadhyay, 2020; Mandal et al., 2021; Murmu, Lal, Sonker, & Singh, 2019; Rahmati, Pourghasemi, & Melesse, 2016). There are several lithological units and groups from the Archean to the Proterozoic ages observed in the study area. Half of the watershed (50.16%) is mainly composed of granite from the Chakradharpur granite gneiss group from the Archean age, followed by shale and phyllite (32.41%) of the Kolhan group from the Palaeoproterozoic to Mesoproterozoic, and hornblende schist, chlorite schist, and metabasic rocks of the Gorumahisani group from the Palaeoproterozoic (Table 9.2).

Besides these, mica schist, phyllite, carbon phyllite, hornblende schist, and epidiorite (6.41%) of the Singhbhum group and volcanic agglomerates (3.86%) of the Dalma group from the Palaeoproterozoic also exist in the Kharkai watershed (Fig. 9.5).

TABLE 9.2 Lithological units of the watershed.

Lithological units	Group name	Age	Area (%)
Sandstone and conglomerate	Kolhan	Palaeoproterozoic—Mesoproterozoic	0.61
Volcanic agglomerates	Dalma	Palaeoproterozoic	3.86
Chert, jaspery, and quartzite	Older metamorphics	Archean	1.77
Hematite, magnetite, and quartzite	Dalma	Palaeoproterozoic	0.14
Mica schist, phyllite, and carbon phyllite	Singhbhum	Palaeoproterozoic	3.07
Hornblende schist and epidiorite	Singhbhum	Palaeoproterozoic	3.34
Shale and phyllite	Kolhan	Palaeoproterozoic—Mesoproterozoic	32.41
Hornblende schist, chlorite schist, and metabasic rocks	Gorumahisani	Palaeoproterozoic	4.64
Granite	Chakradharpur granite gneiss	Archean	50.16

9.4.2 Geomorphology

The geomorphological variation in any region plays a significant role in the occurrence of groundwater (Ghosh, Mandal, Banerjee, et al., 2020; Ghosh, Mandal, Karmakar, et al., 2020; Karmakar et al., 2021; Mandal et al., 2023). Different landform features help in identifying and assessing the groundwater prospect zone (Etikala et al., 2019; Nair Nithya, Srinivas, Magesh, & Kaliraj, 2019; Senapati & Das, 2022). Based on the geomorphological units, the watershed is categorized into eight groups (Fig. 9.6).

The dominant geomorphological unit is found to be a pediment pediplain complex with a share of 70.10%, followed by moderately dissected hills and valleys (25.01%). The proportions of low dissected hills and valleys and highly dissected hills and valleys are 1.28% and 1.11%, respectively, while dams and reservoirs account for 1.89% (Table 9.3).

9.4.3 Lineament density

The porosity and permeability of hard rocks have a positive correlation with LD as they help to augment groundwater prospects (Fagbohun, 2018; Ghosh,

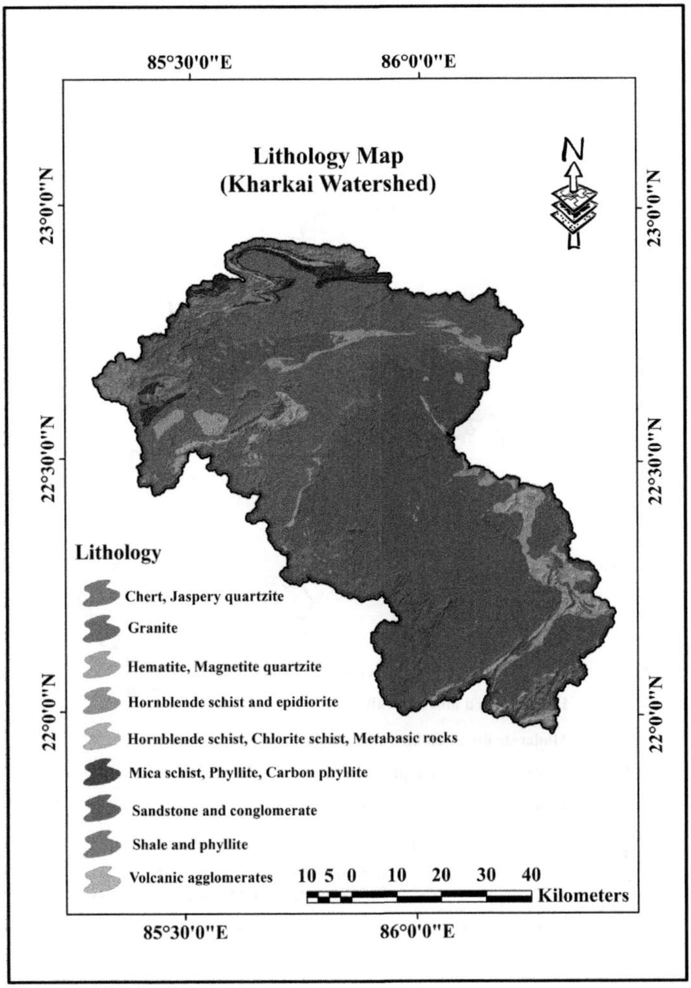

FIGURE 9.5 Lithology map of the Kharkai watershed. Map lines delineate study areas and do not necessarily depict accepted national boundaries.

Mandal, Banerjee, et al., 2020; Ghosh, Mandal, Karmakar, et al., 2020; Karmakar et al., 2021; Mandal et al., 2023). Since fracturing and faulting are the basic elements of lineament, the groundwater potentiality is greatly influenced by this secondary porosity (Magesh, Chandrasekar, & Soundranayagam, 2012; Rajaveni, Brindha, & Elango, 2017; Travaglia & Ammar, 1998). Around 33% area of the watershed experiences LD $\leq$4.03 km/km^2, whereas about 57% of the area consists of LD from $\geq$4.03 to $\leq$14.24 km/km^2. The higher proportion (10.78%) of LD $\geq$14.24 to $\leq$22.06 km/km^2 is enjoyed by the study of interest (Table 9.3,Fig. 9.7).

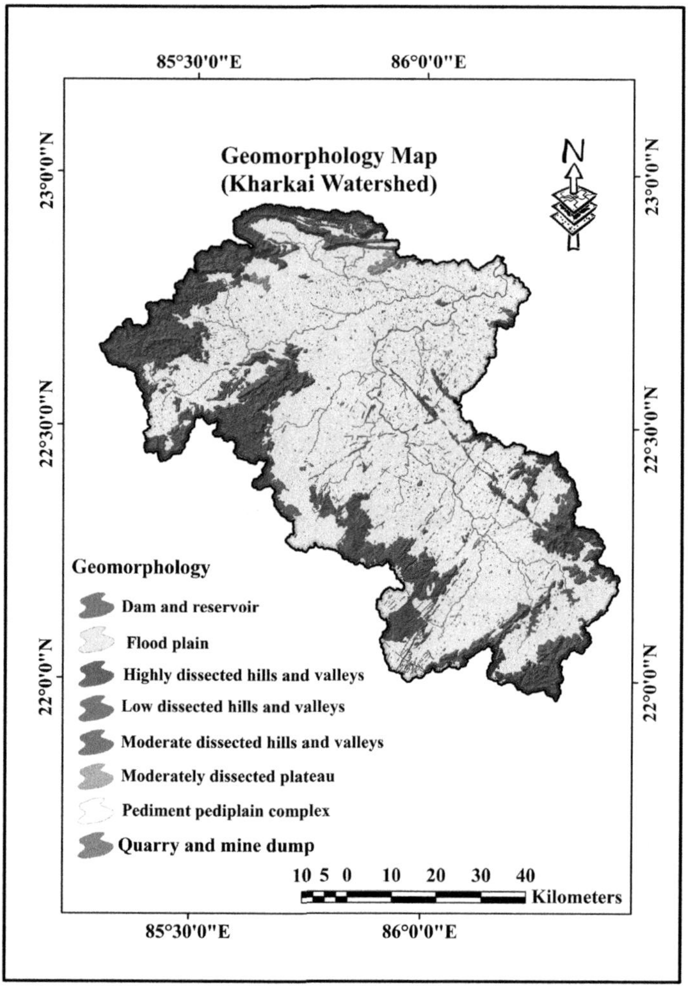

FIGURE 9.6 Geomorphology map of Kharkai watershed. Map lines delineate study areas and do not necessarily depict accepted national boundaries.

9.4.4 Drainage density

In the process of shaping the earth's landscape, DD plays a significant role, and it also influences runoff potential and groundwater development (Ghosh, Mandal, Banerjee, et al., 2020; Ghosh, Mandal, Karmakar, et al., 2020; Karmakar et al., 2021; Mandal et al., 2023). The DD, being the numerical measurement, often indicates the rate of infiltration of surface water (Magesh et al., 2012; Mandal et al., 2023). Areas with lower and higher DD values generally experience greater and lesser groundwater recharge, respectively (Arefin, 2020a, 2020b; Mandal et al., 2023). The Kharkai watershed is divided into five distinctive classes based on DD (Fig. 9.8).

TABLE 9.3 Classification of factors, subclasses, and weight influencing the groundwater potential zone.

Influential factors	Sub_classes	Area (km^2)	Area (%)	Proposed weight for each influential factor	Proposed weight for sub-classes
Lithology	Sandstone and conglomerate	38.03	0.61	17	17
	Volcanic agglomerates	134.54	3.86		15
	Chert, jaspery, and quartzite	110.65	1.77		13
	Hematite, magnetite, and quartzite	8.82	0.14		11
	Mica schist, phyllite, and carbon phyllite	192.24	3.07		9
	Hornblende schist and epidiorite	208.66	3.34		7
	Shale and phyllite	2027.65	32.41		5
	Hornblende schist, chlorite schist, and metabasic rocks	290.41	4.64		3
	Granite	3138.26	50.16		1
Geomorphology	Dams and reservoirs	118.17	1.89	15	15
	Flood plain	5.42	0.09		13
	Pediment pediplain complex	4385.56	70.10		11
	Low-dissected hills and valleys	79.90	1.28		9
	Moderately dissected plateau	29.57	0.47		7
	Moderately dissected hills and valleys	1564.67	25.01		5
	Quarry and mine dump	3.24	0.05		3
	Highly dissected hills and valleys	69.64	1.11		1

(Continued)

TABLE 9.3 (Continued)

Influential factors	Sub_classes	Area (km^2)	Area (%)	Proposed weight for each influential factor	Proposed weight for sub-classes
Lineament density	22.06–32.14	177.70	2.84	15	15
	14.24–22.06	496.67	7.94		12
	8.70–14.24	1488.36	23.79		9
	4.03–8.70	2055.98	32.86		6
	0–4.03	2037.45	32.57		3
Soil type	Charomic vertisols	2635.89	42.12	8	8
	Lithosols	2311.78	36.95		5
	Dystric nitosols	1309.28	20.93		2
Drainage density	0.01–0.17	782.32	12.50	6	6
	0.17–0.30	1496.37	23.92		5
	0.30–0.40	1849.85	29.57		4
	0.40–0.51	1451.14	23.20		3
	0.51–0.77	676.47	10.81		2
LULC	Water body	63.67	1.02	15	15
	Agricultural land	3060.90	48.93		13
	Vegetation	1931.38	30.87		11
	Scrub land	901.34	14.41		9
	Wet-land	0.87	0.01		7
	Bare land	0.83	0.01		5
	Built area	297.16	4.75		3

Rainfall	1561–1702	307.01	4.91	8	8
	1444–1561	790.20	12.63		6
	1350–1444	1142.46	18.26		4
	1258–1350	2966.54	47.42		2
	1158–1258	1049.95	16.78		1
Slope	0–2	2087.47	33.37	8	8
	2–8	2794.15	44.66		6
	8–15	700.56	11.20		4
	15–23	457.42	7.31		2
	23–75	216.56	3.46		1
Elevation	68–173	1468.30	23.47	8	8
	173–251	2356.63	37.67		6
	251–350	1325.41	21.19		4
	350–505	851.72	13.61		2
	505–904	254.10	4.06		1

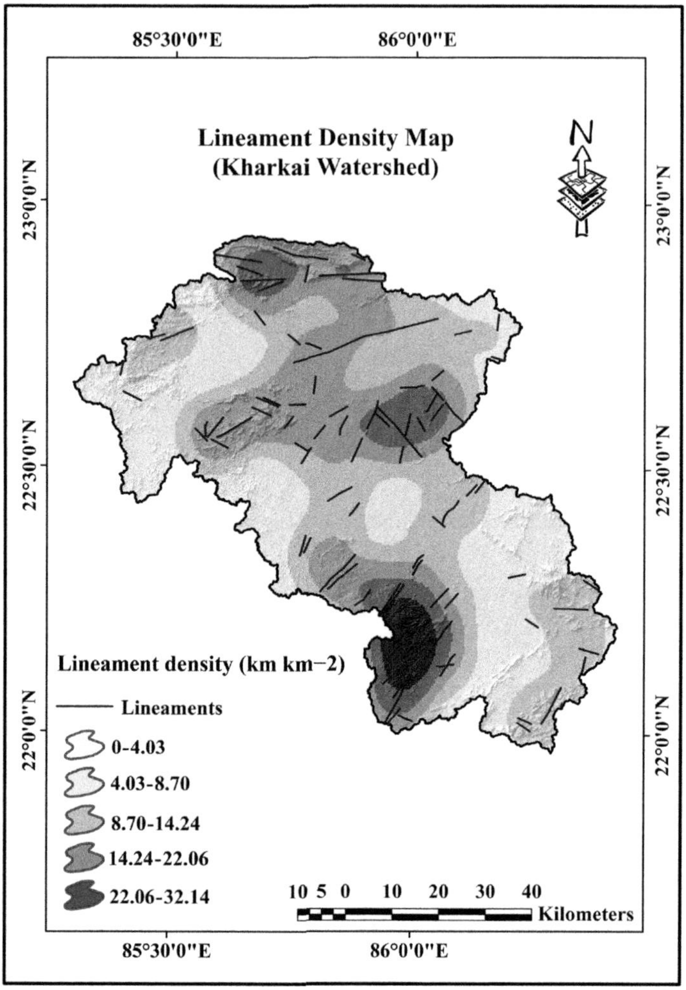

FIGURE 9.7 Lineament density map of the Kharkai watershed. Map lines delineate study areas and do not necessarily depict accepted national boundaries.

Around 66% of the watershed enjoys DD ≤ 0.40 km/km^2, while higher DD (≥ 0.40 to ≤ 0.77 km/km^2) is experienced by the rest of the study area (Table 9.3).

9.4.5 Land use land cover

The prospect of groundwater potentiality is also attributed to the LULC, as it impacts the rate of overland flow, evapotranspiration, and soil erosion (Karmakar et al., 2021; Saravanan, Jacinth, Singh, Saranya, & Sivaranjani, 2019; Wang, Yang, Chen, & Jumpei, 2005). During the monsoon period, the run-off rate and vegetation cover are enhanced meaningfully, and there is a high

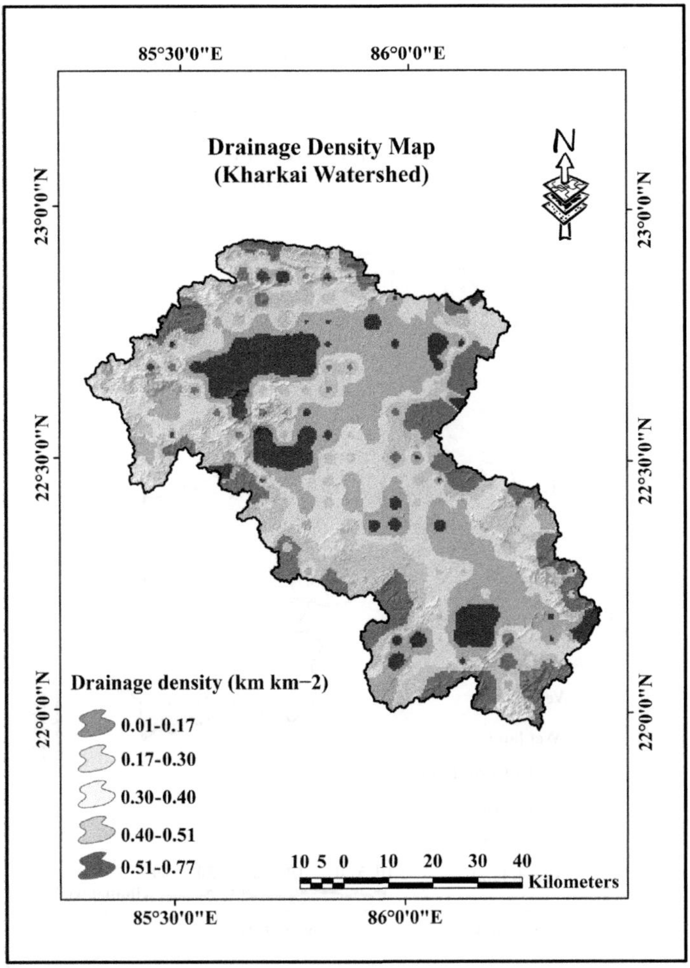

FIGURE 9.8 Drainage density map of the Kharkai watershed. Map lines delineate study areas and do not necessarily depict accepted national boundaries.

probability of groundwater recharge (Das et al., 2017; Ghosh, Mandal, Banerjee, et al., 2020; Ghosh, Mandal, Karmakar, et al., 2020; Mandal et al., 2023). A total of seven types of LULC are observed in the watershed (Fig. 9.9).

Almost half of the watershed (48.93%) is used for agricultural practices, followed by vegetation cover (30.87%). Around 15% of the study area is classified as scrubland, and only 4.75% of the area is found to be built-up (Table 9.3).

9.4.6 Rainfall

Rainfall is the fundamental element of groundwater recharge. The rate of water infiltration into the ground and the rise and fall of the groundwater

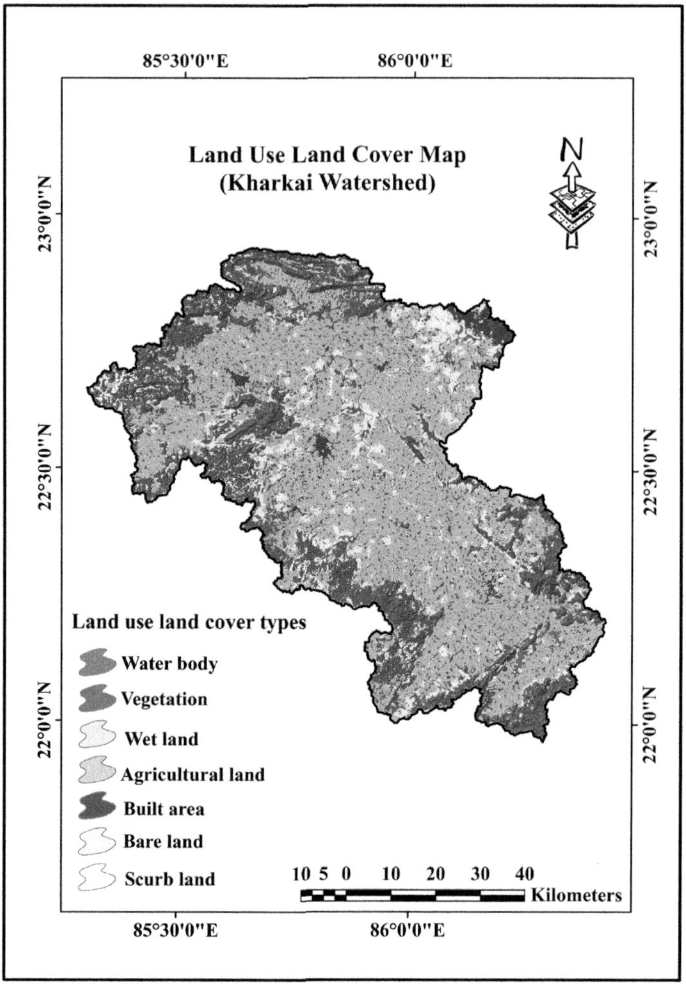

FIGURE 9.9 Land use land cover map of Kharkai watershed. Map lines delineate study areas and do not necessarily depict accepted national boundaries.

table markedly depend on the amount of rainfall (Etikala et al., 2019; Karmakar et al., 2021; Mandal et al., 2023; Thapa et al., 2018). The watershed mainly enjoys rainfall from July to September due to the southwest monsoon. The proportion of rainfall received by the study area ranges between 1158 and 1702 mm (Fig. 9.10), which is quite high (IMD 2022). A significant proportion of rainfall ($\geq$1158 to $\leq$1350 mm) is experienced by 64.20% of the watershed, whereas only 35.80% of the study area enjoys rainfall $\geq$1350 to $\leq$1702 mm (Table 9.3).

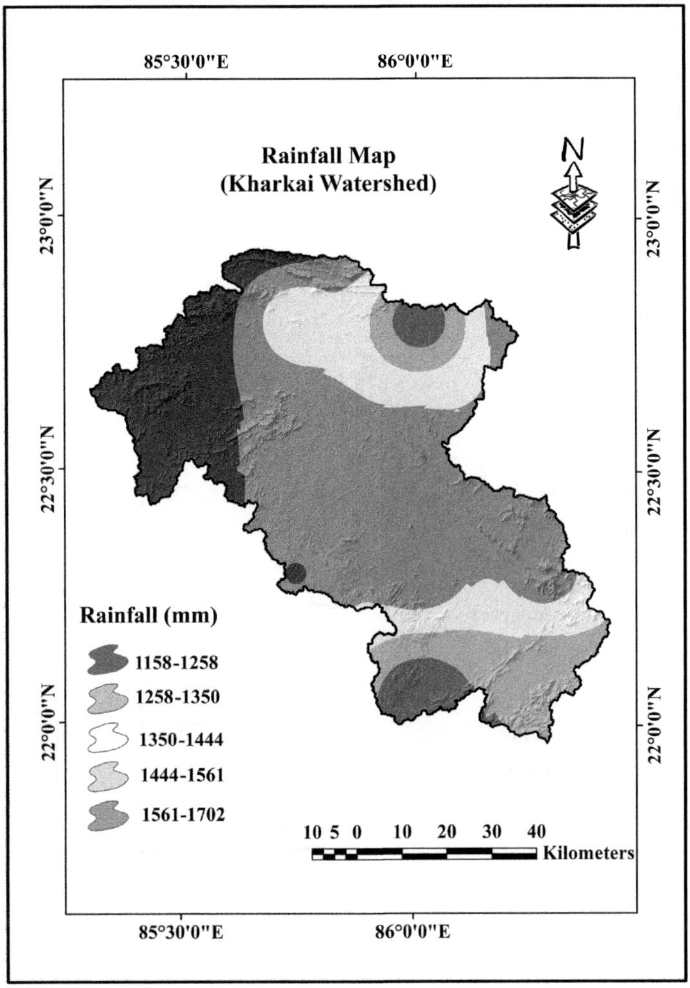

FIGURE 9.10 Rainfall map of the Kharkai watershed. Map lines delineate study areas and do not necessarily depict accepted national boundaries.

9.4.7 Slope

There is an inverse relationship between slope and groundwater recharge (Gautam et al., 2023; Pande et al., 2022). Any region with a higher degree of slope often indicates a lower possibility of water infiltration into the ground and vice versa (Gautam et al., 2023; Mandal et al., 2023). In the present study, the watershed is divided into five categories based on the magnitude of the slope, which varies from <2 up to 75° (Fig. 9.11).

Around 78% of the study of interest possesses a slope ≤8°, while 10.77% of the area experiences a slope >15° to ≤75° (Table 9.3).

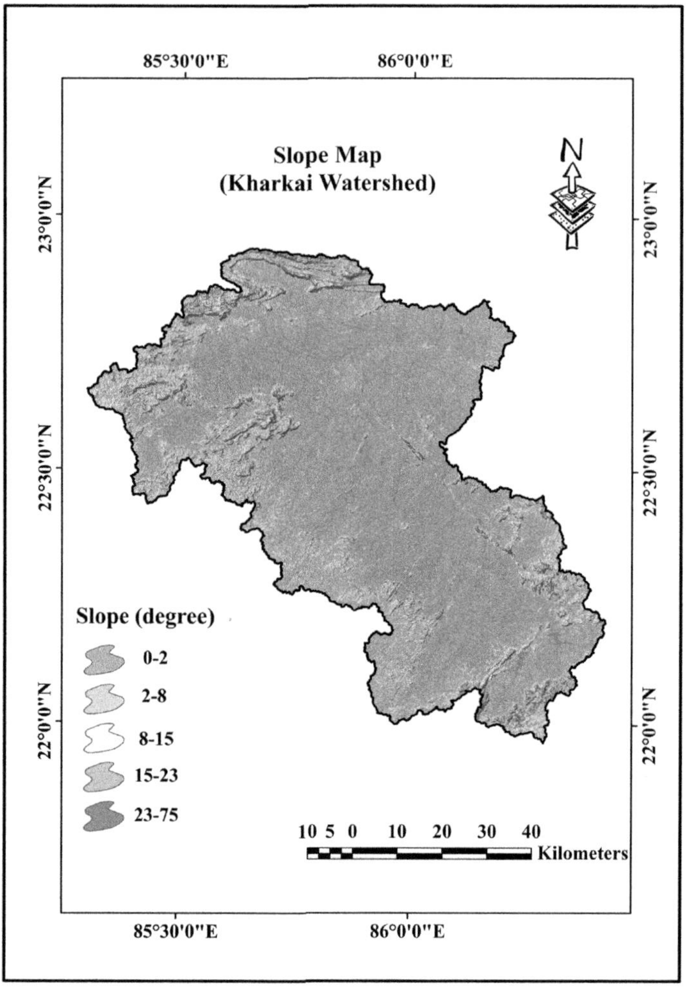

FIGURE 9.11 Slope map of the Kharkai watershed. Map lines delineate study areas and do not necessarily depict accepted national boundaries.

9.4.8 Elevation

The roughness of the terrain may be assessed from the elevation map. With the increasing rate of elevation, the run-off amount increases, and the proportion of infiltration decreases. It simply refers to the negative relationship between elevation and groundwater recharge. Thus, regions with lower elevations, generally plain areas, are more appropriate for groundwater prospects (Patra, Mishra, & Mahapatra, 2018; Senapati & Das, 2022; Thapa et al., 2018). The elevation of the watershed varies between 68 and 904 m (Fig. 9.12).

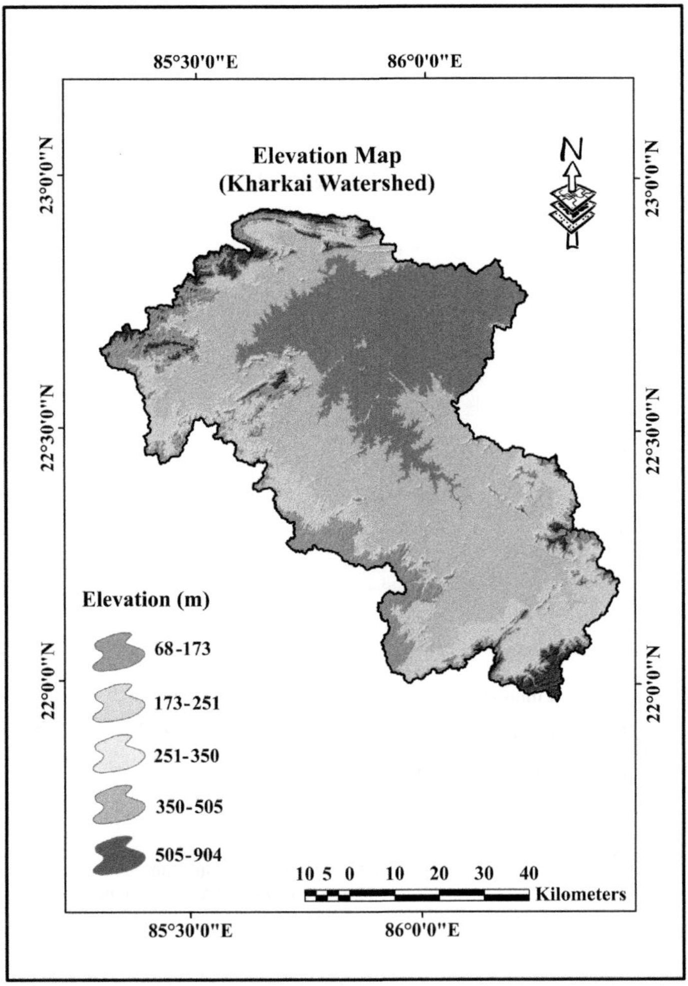

FIGURE 9.12 Elevation map of Kharkai watershed. Map lines delineate study areas and do not necessarily depict accepted national boundaries.

The majority of the study area (61.14%) falls under the elevation of 68–251 m, while 34.80% of the area comes under the elevation of 251–505 m. The rest of the portion (4.06%) experiences elevations >505 to ≤904 m (Table 9.3).

9.4.9 Soil type

Different properties of soil determine the rate of overland flow and water infiltration into the soil (Mandal et al., 2023; Modak, Mandal, Mandi, &

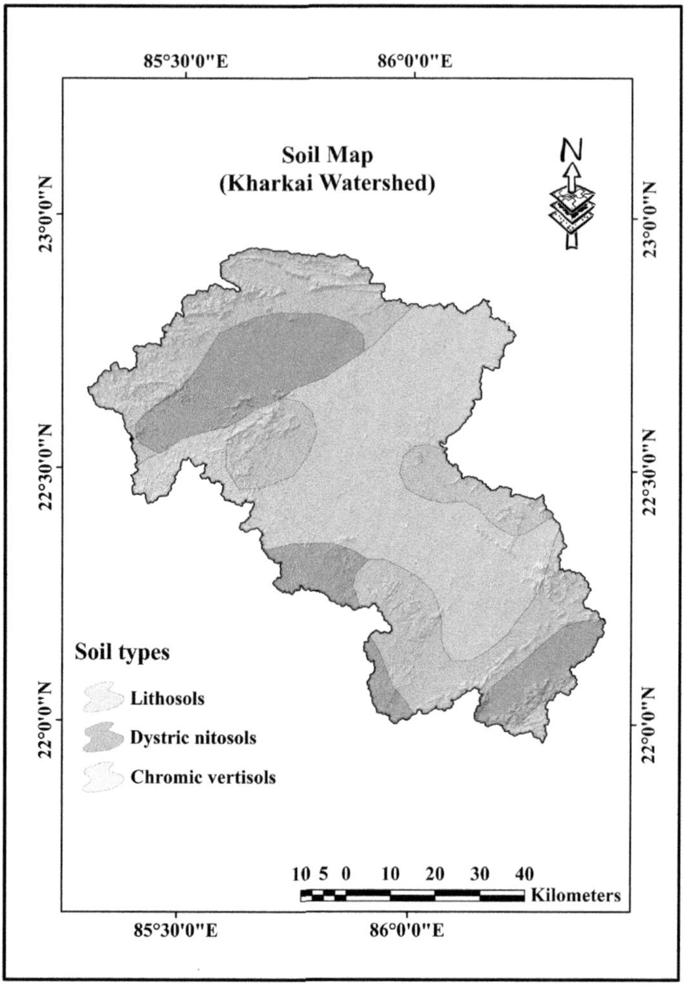

FIGURE 9.13 Soil map of the Kharkai watershed. Map lines delineate study areas and do not necessarily depict accepted national boundaries.

Ghosh, 2022). The movement of groundwater in the unsaturated zone is greatly influenced by the soil (Ghosh, Mandal, Karmakar, et al., 2020; Mandal et al., 2023). The less groundwater recharge, the finer the soil texture, and vice versa. In contrast, the sandy soil texture comparatively produces less runoff potential and a higher infiltration rate (Senapati & Das, 2022). Three distinctive soil types are observed in the watershed (Fig. 9.13).

The chromic vertisols comprise 42.12% of the study area, followed by lithosols (36.95%) and dystric nitosols (20.93%) (Table 9.3).

9.4.10 Groundwater potential zones

The hydrogeological parameters adopted in this study to assess the groundwater prospect are assigned weight considering their major and minor impacts on each other. The final map of the GWPZ of the Kharkai watershed is produced by amalgamating all the layers of parameters in the ArcGIS software using weighted overlay analysis with the help of RS and GIS techniques coupled with the MIF model. The GWPZ map is categorized into five distinct classes (Fig. 9.14). Around 42% (2615.09 km^2) of the watershed experiences low to very low groundwater prospects, while 24.42% (1527.77 km^2) of the

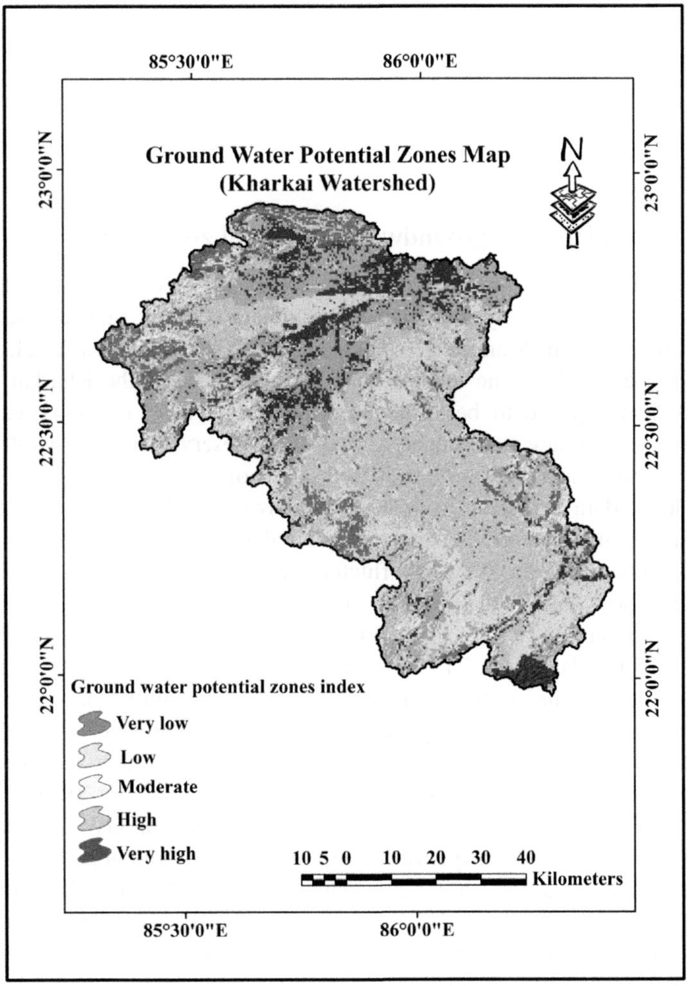

FIGURE 9.14 Groundwater potential zones map of the Kharkai watershed. Map lines delineate study areas and do not necessarily depict accepted national boundaries.

TABLE 9.4 Distribution of groundwater potential zone.

GWPZ	Area (km^2)	Area (%)	No. of reference dug well
Very Low	534.54	8.54	1
Low	2080.55	33.26	1
Moderate	1527.77	24.42	6
High	1418.37	22.67	12
Very High	694.92	11.11	13
Total	**6256.15**	**100**	**33**

area bears a moderate category of groundwater possibility. In contrast, an area of 2113.29 km^2 (33.78%) enjoys high to very high groundwater potentiality (Table 9.4).

9.4.11 Validation of groundwater potential zones map

There is no use of any model without its accuracy assessment, as it enhances the level of acceptability among the researchers (Ghosh, Mandal, Banerjee, et al., 2020; Ghosh, Mandal, Karmakar, et al., 2020; Karmakar et al., 2021; Mandal et al., 2023). The derived map of the GWPZ of the Kharkai watershed is thus required to be validated with ground level reference data. To perform the accuracy assessment, a total of 33 observatories dug wells existing across the watershed are considered based on the availability of groundwater level data accessed from https://indiawris.gov.in/wris/ in 2022. The fluctuation data of groundwater level are calculated from the pre monsoon and post monsoon data. Now, the fluctuation data are classified into five distinct categories using the natural break classification method (Jenks) in ArcGIS software: very high (<1.14 mbgl), high (1.14–1.71 mbgl), moderate (1.71–2.75 mbgl), low (2.75–4.33 mbgl), and very low (>4.33 mbgl). The reference dug wells of those areas with high fluctuations in groundwater levels are recognized as having a low prospect of groundwater, and vice versa (Table 9.5). The accuracy level of the GWPZ map is now validated with reference data using the Receiver Operating Characteristics (ROC) curve, which is one of the important validating techniques of GWPZ (Ashwini, Verma, Sriharsha, Chourasiya, & Singh, 2023; Das & Mukhopadhyay, 2020; Kumar, Singh, & Singh, 2023; Senapati & Das, 2021, 2022; Yadav et al., 2023). The match between GWPZ and reference data is carried out using binary numbers, where 1 represents agreement and 0 refers to disagreement in SPSS 2023 software (Table 9.5).

TABLE 9.5 Geographical distribution of dug wells and fluctuation data of groundwater level.

S. no. of dug well	Location of dug well		Dug well site	Groundwater fluctuation (mbgl)	Reference category	Map category	Binary code
	Latitude	Longitude					
1	22.709730°	85.537992°	Karaikela	2.60	Moderate	Moderate	1
2	22.547282°	85.802502°	Chaibasa	0.46	Very high	Very high	1
3	22.676479°	85.625457°	Chakradharpur	2.46	Moderate	Moderate	1
4	22.411224°	85.765722°	Jhinkpani (Hat Gamhariya)	0.72	Very high	High	0
5	22.413918°	85.760668°	Jhinkpani	0.02	Very high	Very high	1
6	22.623683°	85.722224°	Khuntpani	1.13	Very high	Very high	1
7	22.570897°	85.460764°	Sonua	2.15	Moderate	Moderate	1
8	22.464973°	85.913521°	Kokcho	1.11	Very high	High	0
9	22.383593°	85.631522°	Tonto	1.30	High	High	1
10	22.596536°	85.984016°	Jamdih	0.56	Very high	Very high	1
11	22.702330°	85.927895°	Saraikela	1.35	High	High	1
12	22.808319°	86.100821°	Adityapur (Gamharia)	1.52	High	High	1
13	22.782874°	85.825504°	Kharsawan	2.72	Moderate	High	0
14	22.407923°	86.125050°	Bahalda Road (Kona)	1.20	Moderate	Moderate	1
15	22.308745°	86.136992°	Dandabose	1.44	High	High	1
16	22.326463°	86.126846°	Gambharia	0.53	Very high	Low	0

(*Continued*)

TABLE 9.5 (Continued)

S. no. of dug well	Location of dug well		Dug well site	Groundwater fluctuation (mbgl)	Reference category	Map category	Binary code
	Latitude	Longitude					
17	22.414700°	86.129802°	Poilakunda	1.35	High	High	1
18	22.229891°	86.133197°	Budhamara	3.60	Low	Low	1
19	22.234673°	86.214449°	Bademtolia	0.48	Very high	Moderate	0
20	22.088173°	86.111278°	Badampahad-1	1.63	High	High	1
21	22.244866°	86.268503°	Ambadiha	0.52	Very high	High	0
22	22.127854°	86.299459°	Bhatachhatra	1.29	High	High	1
23	22.164499°	86.402395°	Bisoi	6.15	Very low	Low	0
23	22.212896°	86.300848°	Karanjei-Bijatola Chhak	0.28	Very high	Very high	1
25	22.080625°	86.236513°	Manda	1.25	High	High	1
26	22.362093°	86.031926°	Moranda	1.03	Very high	Moderate	0
27	22.267234°	86.137655°	Champrai	1.57	High	High	1
28	22.304628°	86.198195°	Niranjan	2.28	Moderate	Moderate	1
29	22.137058°	86.177612°	Purunapani	1.05	Very high	High	0
30	22.146815°	86.096049°	Kherna	1.37	High	High	1
31	22.207286°	86.126350°	Naujara	0.61	Very high	Moderate	0
32	22.422068°	86.060174°	Nischintapur	1.26	High	High	1
33	22.505183°	86.072738°	Tiring	2.08	Moderate	Moderate	1

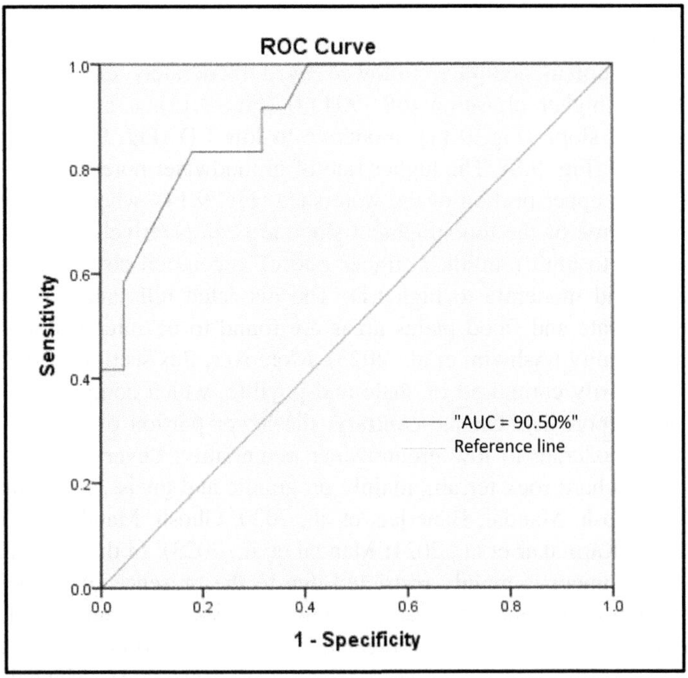

FIGURE 9.15 ROC curve for groundwater potential zones map of the watershed. *ROC*, Receiver Operating Characteristics.

The value of area under curve (AUC), which is the relationship between forecast accuracy and AUC, generally ranges from 0.5 to 1.0, where 1 depicts perfect accuracy and 0.5 means ambiguity in the model. The AUC value varying from 0.9 to 1 indicates excellent agreement with the model (Arabameri, Pradhan, Pourghasemi, & Rezaei, 2018; Ashwini et al., 2023; Kumar et al., 2023; Pande et al., 2021; Senapati & Das, 2021, 2022; Yadav et al., 2023). It was computed in the present study, and the value of the AUC is 0.905, which is excellent in terms of the result (Fig. 9.14). Finally, the accuracy level of the GWPZ map is found to be 90.50%, which is quite impressive (Fig. 9.15).

9.5 Discussion

The present study assesses the GWPZ of the Kharkai watershed employing the MIF model, the RS technique, and the GIS technique. Based on the map of GWPZ, which is prepared considering a number of hydrogeological parameters, it is observed that around 58% (around 3641 km^2) of the watershed enjoys moderate to very high groundwater potentiality. The watershed is mainly characterized by granite, shale, and phyllite, followed by hornblende schist, chlorite schist, metabasic rocks, volcanic agglomerates, hornblende schist, and epidiorite (Fig. 9.5 and Table 9.3). The study area is part of a pediment pediplain complex, followed by a moderately dissected plateau (Fig. 9.6), a higher elevation (68−904 m) (Fig. 9.12), a higher degree of topographical slope (Fig. 9.11), moderate to low LD (Fig. 9.7), and low to moderate DD (Fig. 9.8). The higher rate of groundwater potentiality is found mainly in the upper portion of the watershed (Fig. 9.14), where the elevation level and degree of the topographical slope are comparatively low, followed by moderate to high rainfall, a higher rate of vegetation cover and agricultural land, and moderate to high LD. The dissected hills and valleys from low to moderate and flood plains areas are found to be suitable for groundwater potentiality (Ashwini et al., 2023). Moreover, this section of the watershed is primarily composed of shale and phyllite, which comparatively bear higher water porosity. On the contrary, the lower portion of the study area encounters moderate to low groundwater potentiality. Several earlier works conducted on hard rock terrain, mainly on granite and gneiss, found this type of result (Ghosh, Mandal, Banerjee, et al., 2020; Ghosh, Mandal, Karmakar, et al., 2020; Karmakar et al., 2021; Mandal et al., 2023). In this area, groundwater development is mainly restricted due to the presence of nonperennial rivers with lower floodplain areas, varying degrees of topographical slope, undulating topography, less permeable rock, and shortterm highintensity rainfall during the monsoon season (Karmakar et al., 2021; Mandal et al., 2023), which is subject to surface runoff (Ghosh, Mandal, Banerjee, et al., 2020; Ghosh, Mandal, Karmakar, et al., 2020). Despite this, the secondary porosity of the rocks is an important parameter of groundwater recharge during the rainy season (Mandal et al., 2023).

9.6 Conclusion

The assessment of groundwater prospect zones of the Kharkai watershed is conducted in this study employing RS, GIS, and MIF techniques, considering nine hydrogeological parameters. The lithological map reveals that around 50% of the study area is composed of granite from the Chakradharpur granite gneiss group, followed by shale and phyllite (32.41%) of the Kolhan group. The derived GWPZ map reveals that around 33.78% (2113.29 km^2) of the watershed enjoys high to very high groundwater prospects, followed by moderate groundwater availability of 24.42% (1527.77 km^2). In contrast, around 42% (2615.09 km^2) of the area possesses low to very low groundwater prospects. The upper portion of the watershed, primarily composed of shale and phyllite of the Kolhan group (Palaeoproterozoic-Mesoproterozoic), where groundwater potentiality is found to be moderate to very high, is characterized by moderate to high rainfall, comparatively lower elevation and degree of topographical slope, moderate to high LD, and greater vegetation cover and agricultural land. In the case of moderate to very low groundwater possibility, it is mainly observed in the areas of Archean granite (Chakradharpur granite gneiss group), mica schist, phyllite, carbon phyllite, hornblende schist, and epidiorite of the Singhbhum group (Palaeoproterozoic), where the primary porosity of rock is comparatively low, followed by moderate to high altitude. The validation of the GWPZ map is conducted considering fluctuation data of the groundwater level of 33 dug wells distributed across the watershed. The accuracy level of the GWPZ map is computed by applying the ROC curve, and the level of accuracy is calculated to be 90.50%, which is quite good for the validity of the model. This study could have been more precise if the geophysical resistivity survey had been done. The present research also opens the future directions of new studies for indepth evaluation of groundwater prospects. Interested researchers may take up further work on the identification of groundwater recharge zones, nature and storage of aquifers, rate of recharge and exploitation, groundwater quality, and contamination.

References

Abijith, Devanantham, Saravanan, Subbarayan, Singh, Leelambar, Jennifer, Jesudasan Jacinth, Saranya, Thiyagarajan, & Parthasarathy, K. S. S. (2020). GIS-based multi-criteria analysis for identification of potential groundwater recharge zones - a case study from Ponnaniyaru watershed, Tamil Nadu, India. *HydroResearch.*, *3*, 1−14. Available from https://doi.org/10.1016/j.hydres.2020.02.002.

Al-Abadi, A. M., Shahid, S., Ghalib, H. B., & Handhal, A. M. (2017). A GIS-based integrated fuzzy logic and analytic hierarchy process model for assessing water-harvesting zones in Northeastern Maysan Governorate, Iraq. *Arabian Journal for Science and Engineering*, *42* (6), 2487−2499. Available from https://doi.org/10.1007/s13369-017-2487-1, https://link.springer.com/journal/13369.

Arabameri, A., Pradhan, B., Pourghasemi, H. R., & Rezaei, K. (2018). Identification of erosion-prone areas using different multi-criteria decision-making techniques and gis. *Geomatics, Natural Hazards and Risk*, *9*(1), 1129−1155. Available from https://doi.org/10.1080/19475705.2018.1513084, http://www.tandfonline.com/toc/tgnh20/current.

Arefin, R. (2020a). Groundwater potential zone identification using an analytic hierarchy process in Dhaka City, Bangladesh. *Environmental Earth Sciences*, *79*(11). Available from https://doi.org/10.1007/s12665-020-09024-0, https://link.springer.com/journal/12665.

Arefin, R. (2020b). Groundwater potential zone identification at Plio-Pleistocene elevated tract, Bangladesh: AHP-GIS and remote sensing approach. *Groundwater for Sustainable Development.*, *10*, 100340. Available from https://doi.org/10.1016/j.gsd.2020.100340.

Ashwini, K., Verma, R. K., Sriharsha, S., Chourasiya, S., & Singh, A. (2023). Delineation of groundwater potential zone for sustainable water resources management using remote sensing-GIS and analytic hierarchy approach in the state of Jharkhand, India. *Groundwater for Sustainable Development.*, *21*, 100908. Available from https://doi.org/10.1016/j.gsd.2023.100908.

CGWB. (2013). Ground Water Pollution by Industrial Clusters, Bhu-Jal News. Ministry of Water Resources, River Development and Ganga Rejuvenation, Bhujal Bhawan, Unpublished content. Available from http://cgwb.gov.in/bhujal-news-0.

CGWB. (2022). Aquifer mapping and management plan of Purulia district West Bengal, Department of Water Resources, River Development & Ganga Rejuvenation Ministry of Jal Shakti, Government of India. Rejuvenation Ministry of Jal Shakti, Government of India Unpublished content. Available from https://cgwb.gov.in/AQM/NAQUIM_REPORT/WEST-BENGAL/Purulia/20Final/20West/20Bengal.pdf.

Das, G. K., & Guchait, R. (2016). Modeling of risk of soil erosion in Kharkai Watershed using RUSLE and TRMM Data: A geospatial approach. *International Journal of Science and Research (IJSR)*, *5*(10), 1–10. Available from https://www.ijsr.net/archive/v5i10/ART20161984.pdf.

Das, N., & Mukhopadhyay, S. (2020). Application of multi-criteria decision making technique for the assessment of groundwater potential zones: A study on Birbhum district, West Bengal, India. *Environment, Development and Sustainability*, *22*(2), 931–955. Available from https://doi.org/10.1007/s10668-018-0227-7, http://www.wkap.nl/journalhome.htm/1387-585X.

Das, S. (2017). Delineation of groundwater potential zone in hard rock terrain in Gangajalghati block, Bankura district, India using remote sensing and GIS techniques. *Modeling Earth Systems and Environment.*, *3*(4), 1589–1599. Available from https://doi.org/10.1007/s40808-017-0396-7, http://springer.com/journal/40808.

Das, S., Gupta, A., & Ghosh, S. (2017). Exploring groundwater potential zones using MIF technique in semi-arid region: A case study of Hingoli district, Maharashtra. *Spatial Information Research*, *25*(6), 749–756. Available from https://doi.org/10.1007/s41324-017-0144-0, http://springer.com/journal/41324.

Das, S., & Pardeshi, S. D. (2018). Integration of different influencing factors in GIS to delineate groundwater potential areas using IF and FR techniques: A study of Pravara basin, Maharashtra, India. *Applied Water Science*, *8*(7). Available from https://doi.org/10.1007/s13201-018-0848-x, https://www.springer.com/journal/13201.

Etikala, Balaji, Golla, Veeraswamy, Li, Peiyue, & Renati, SiddiRaju (2019). Deciphering groundwater potential zones using MIF technique and GIS: A study from Tirupati area, Chittoor District, Andhra Pradesh, India. *HydroResearch*, *1*, 1–7. Available from https://doi.org/10.1016/j.hydres.2019.04.001.

Fagbohun, B. J. (2018). Integrating GIS and multi-influencing factor technique for delineation of potential groundwater recharge zones in parts of Ilesha schist belt, southwestern Nigeria. *Environmental Earth Sciences*, *77*(3). Available from https://doi.org/10.1007/s12665-018-7229-5, http://www.springerlink.com/content/121380/.

Gautam, V. K., Pande, C. B., Kothari, M., Kumar Singh, P., & Agrawal, A. (2023). Exploration of groundwater potential zones mapping for hard rock region in the Jakham river basin using geospatial techniques and aquifer parameters. *Advances in Space Research*, *71*(6), 2892–2908. Available from https://doi.org/10.1016/j.asr.2022.11.022, http://www.journals.elsevier.com/advances-in-space-research/.

Ghosh, D., Mandal, M., Banerjee, M., & Karmakar, M. (2020). Impact of hydro-geological environment on availability of groundwater using analytical hierarchy process (AHP) and geospatial techniques: A study from the upper Kangsabati river basin. *Groundwater for Sustainable Development*, *11*, 100419. Available from https://doi.org/10.1016/j.gsd.2020.100419, https://www.sciencedirect.com/science/article/abs/pii/S2352801X19301146.

Ghosh, D., Mandal, M., Karmakar, M., Banerjee, M., & Mandal, D. (2020). Application of geospatial technology for delineating groundwater potential zones in the Gandheswari watershed, West Bengal. *SustainableWater Resources Management*, *6*(1), 14. Available from https://doi.org/10.1007/s40899-020-00372-0, http://springer.com/journal/40899.

Kaliraj, S., Chandrasekar, N., & Magesh, N. S. (2013). Identification of potential groundwater recharge zones in Vaigai upper basin, Tamil Nadu, using GIS-based analytical hierarchical process (AHP) technique. *Arabian Journal of Geosciences, 7*(4), 1385–1401. Available from https://doi.org/10.1007/s12517-013-0849-x, http://www.springer.com/geosciences/journal/12517?cm_mmc = AD-_-enews-_-PSE1892-_-0.

Karmakar, M., Banerjee, M., Mandal, M., & Ghosh, D. (2021). *Application of AHP for groundwater potential zones mapping in Plateau Fringe Terrain: Study from western province of West BengalGroundwater and Society: Applications of Geospatial Technology* (pp. 189–219). India: Springer International Publishing. Available from https://link.springer.com/book/10.1007/978-3-030-64136-8, 10.1007/978-3-030-64136-8_9.

Khadri, S. F. R., & Pande, C. (2016). Ground water flow modeling for calibrating steady state using MODFLOW software: A case study of Mahesh River basin, India. *Modeling Earth Systems and Environment, 2*(1). Available from https://doi.org/10.1007/s40808-015-0049-7, http://springer.com/journal/40808.

Kumar, M., Singh, P., & Singh, P. (2023). Machine learning and GIS-RS-based algorithms for mapping the groundwater potentiality in the Bundelkhand region, India. *Ecological Informatics., 74*, 101980. Available from https://doi.org/10.1016/j.ecoinf.2023.101980.

Magesh, N. S., Chandrasekar, N., & Soundranayagam, J. P. (2012). Delineation of groundwater potential zones in Theni district, Tamil Nadu, using remote sensing, GIS and MIF techniques. *Geoscience Frontiers, 3*(2), 189–196. Available from https://doi.org/10.1016/j.gsf.2011.10.007, https://www.sciencedirect.com/journal/geoscience-frontiers.

Mandal, K. K., Ranjan, A., & Dharanirajan, K. (2021). Delineation of Groundwater Potential Zones (GWPZ) of Port Blair, Andaman Islands, India, using Multi Influencing Factors (MIF) method and geospatial techniques. *Remote Sensing Applications: Society and Environment, 24*, 100631. Available from https://doi.org/10.1016/j.rsase.2021.100631.

Mandal, M., Mandi, S., Modak, P., Ghosh, B., Panda, B., Mukherjee, A., … Senapati, T. (2023). Groundwater quality enumeration and health risk in the extended part of Chhotanagpur granite gneiss complex of India. *Environmental Science and Pollution Research.* Available from https://doi.org/10.1007/s11356-023-26397-4, https://www.springer.com/journal/11356.

Ministry of Jal Shakti Government of India. (2023). 2023 Department of Water Resources, River Development & Ganga Rejuvenation New Delhi Annual Report 2022-23, Department of Water Resources, River Development & Ganga Rejuvenation. Available from https://jalshakti.gov.in/.

Ministry of Water Resources. (2014). Ministry of Water Resources, Government of India. 'Subarnarekha Basin. Available from https://indiawris.gov.in/wiki/doku.php?id = subarnarekha

Modak, P., Mandal, M., Mandi, S., & Ghosh, B. (2022). Gully erosion vulnerability modelling, estimation of soil loss and assessment of gully morphology: A study from cratonic part of eastern India. *Environmental Science and Pollution Research, 30*(55), 116656–116687. Available from https://doi.org/10.1007/s11356-022-22118-5, https://www.springer.com/journal/11356.

Mukherjee, Jayesh, & Patel, Priyank Pravin (2022). *Landscape characterization using geomorphometric parameters for a small Sub-Humid River Basin of the Chota Nagpur Plateau, Eastern India* (pp. 127–152). Springer Science and Business Media LLC. Available from 10.1007/978-3-030-79634-1_6.

Murmu, Preetilata, Kumar, Mukesh, Lal, Deepak, Sonker, Irjesh, & Singh, Sudhir Kumar (2019). Delineation of groundwater potential zones using geospatial techniques and analytical hierarchy process in Dumka district, Jharkhand, India. *Groundwater for Sustainable Development, 9*, 100239. Available from https://doi.org/10.1016/j.gsd.2019.100239.

Nag, S. K., & Das, S. (2017). Assessment of groundwater quality from Bankura I and II Blocks, Bankura District, West Bengal, India. *Applied Water Science*, *7*(6), 2787–2802. Available from https://doi.org/10.1007/s13201-017-0530-8, https://www.springer.com/journal/13201.

Nair Nithya, C., Srinivas, Y., Magesh, N. S., & Kaliraj, S. (2019). Assessment of groundwater potential zones in Chittar basin, Southern India using GIS based AHP technique. *Remote Sensing Applications: Society and Environment*, *15*, 100248. Available from https://doi.org/10.1016/j.rsase.2019.100248.

Nasir, M. J., Khan, S., Zahid, H., & Khan, A. (2018). Delineation of groundwater potential zones using GIS and multi influence factor (MIF) techniques: A study of district Swat, Khyber Pakhtunkhwa, Pakistan. *Environmental Earth Sciences*, *77*(10). Available from https://doi.org/10.1007/s12665-018-7522-3, http://www.springerlink.com/content/121380/.

Pande, C. B., Moharir, K. N., Panneerselvam, B., Singh, S. K., Elbeltagi, A., Pham, Q. B., ... Rajesh, J. (2021). Delineation of groundwater potential zones for sustainable development and planning using analytical hierarchy process (AHP), and MIF techniques. *Applied Water Science.*, *11*(12), 186. Available from https://doi.org/10.1007/s13201-021-01522-1.

Pande, C. B., Moharir, K. N., Singh, S. K., Elbeltagi, A., Pham, Q. B., Panneerselvam, B., ... Kouadri, S. (2022). Groundwater flow modeling in the basaltic hard rock area of Maharashtra, India. *Applied Water Science*, *12*(1). Available from https://doi.org/10.1007/s13201-021-01525-y, http://www.springer.com/earth + sciences + and + geography/hydrogeology/journal/13201.

Parthasarathy, K. S. S., & Deka, P. C. (2021). Remote sensing and GIS application in assessment of coastal vulnerability and shoreline changes: A review. *ISH Journal of Hydraulic Engineering.*, *27*(1), 588–600. Available from https://doi.org/10.1080/09715010.2019.1603086, http://www.tandfonline.com/tish.

Patra, S., Mishra, P., & Mahapatra, S. C. (2018). Delineation of groundwater potential zone for sustainable development: A case study from Ganga Alluvial Plain covering Hooghly district of India using remote sensing, geographic information system and analytic hierarchy process. *Journal of Cleaner Production*, *172*, 2485–2502. Available from https://doi.org/10.1016/j.jclepro.2017.11.161.

Pinto, D., Shrestha, S., Babel, M. S., & Ninsawat, S. (2017). Delineation of groundwater potential zones in the Comoro watershed, Timor Leste using GIS, remote sensing and analytic hierarchy process (AHP) technique. *Applied Water Science*, *7*(1), 503–519. Available from https://doi.org/10.1007/s13201-015-0270-6, https://www.springer.com/journal/13201.

Rahmati, O., Pourghasemi, H. R., & Melesse, A. M. (2016). Application of GIS-based data driven random forest and maximum entropy models for groundwater potential mapping: A case study at Mehran Region, Iran. *Catena.*, *137*, 360–372. Available from https://doi.org/10.1016/j.catena.2015.10.010, http://www.elsevier.com/inca/publications/store/5/2/4/6/0/9.

Rajasekhar, Raju, Bramaiah., Deepthi., Amaravathi, Y., & Raju, Siddi (2019). Delineation of groundwater potential zones of Semi-Arid region of YSR Kadapa District Andhra Pradesh India using RS GIS and analytic hierarchy process. *Remote Sensing of Land*, *2*, 76–86. Available from https://doi.org/10.21523/gcj1.18020201.

Rajaveni, S. P., Brindha, K., & Elango, L. (2017). Geological and geomorphological controls on groundwater occurrence in a hard rock region. *Applied Water Science*, *7*(3), 1377–1389. Available from https://doi.org/10.1007/s13201-015-0327-6, https://www.springer.com/journal/13201.

Saravanan, S., Jacinth, J. J., Singh, L., Saranya, T., & Sivaranjani, S. (2019). *Impact of land-use change on soil erosion in the Coonoor watershed, Nilgiris mountain range, Tamil Nadu, India*

Advances in Science, Technology and Innovation (pp. 109–111). India: SpringerNature. Available from https://www.springer.com/series/15883, 10.1007/978-3-030-01440-7_26.

Sashikkumar, M. C., Selvam, S., Kalyanasundaram, V. L., & Johnny, J. C. (2017). GIS based groundwater modeling study to assess the effect of artificial recharge: A case study from Kodaganar river basin, Dindigul district, Tamil Nadu. *Journal of the Geological Society of India*, *89*(1), 57–64. Available from https://doi.org/10.1007/s12594-017-0558-2, http://www.springerlink.com/content/121451/.

Selvam, S., Magesh, N. S., Sivasubramanian, P., Soundranayagam, J. P., Manimaran, G., & Seshunarayana, T. (2014). Deciphering of groundwater potential zones in Tuticorin, Tamil Nadu, using remote sensing and GIS techniques. *Journal of the Geological Society of India*, *84*(5), 597–608. Available from https://doi.org/10.1007/s12594-014-0167-2, http://www.springerlink.com/content/121451/.

Senapati, U., & Das, T. K. (2021). Assessment of basin-scale groundwater potentiality mapping in drought-prone upper Dwarakeshwar River basin, West Bengal, India, using GIS-based AHP techniques. *Arabian Journal of Geosciences*, *14*(11). Available from https://doi.org/10.1007/s12517-021-07316-8, http://www.springer.com/geosciences/journal/12517?cm_mmc = AD-_-enews-_-PSE1892-_-0.

Senapati, U., & Das, T. K. (2022). GIS-based comparative assessment of groundwater potential zone using MIF and AHP techniques in Cooch Behar district, West Bengal. *Applied Water Science*, *12*(3). Available from https://doi.org/10.1007/s13201-021-01509-y, http://www.springer.com/earth + sciences + and + geography/hydrogeology/journal/13201.

Siddi Raju, R., Sudarsana Raju, G., & Rajasekhar, M. (2019). Identification of groundwater potential zones in Mandavi River basin, Andhra Pradesh, India using remote sensing, GIS and MIF techniques. *HydroResearch*, *2*, 1–11. Available from https://doi.org/10.1016/j.hydres.2019.09.001.

Singh, P. K., Dahiphale, Praveen, Yadav, K. K., & Singh, Manjeet (2018). *Delineation of groundwater potential zones in Jaisamand Basin of Udaipur District* (pp. 3–20). Springer Nature. Available from 10.1007/978-981-10-5789-2_1.

Thapa, R., Gupta, S., Gupta, A., Reddy, D. V., & Kaur, H. (2018). Use of geospatial technology for delineating groundwater potential zones with an emphasis on water-table analysis in Dwarka River basin, Birbhum, India. *Hydrogeology Journal*, *26*(3), 899–922. Available from https://doi.org/10.1007/s10040-017-1683-0, http://www.springerlink.com/content/102028/.

Tiwari, A. K., Lavy, M., Amanzio, G., De Maio, M., Singh, P. K., & Mahato, M. K. (2017). Identification of artificial groundwater recharging zone using a GIS-based fuzzy logic approach: A case study in a coal mine area of the Damodar Valley, India. *Applied Water Science*, *7*(8), 4513–4524. Available from https://doi.org/10.1007/s13201-017-0603-8, https://www.springer.com/journal/13201.

Travaglia, C., & Ammar, O. (1998). Groundwater exploration by satellite remote sensing in the Syrian Arab Republic. Rome. *FAO*, *6*, 33.

Wang, G., Yang, L., Chen, L., & Jumpei, K. (2005). Impacts of land use changes on groundwater resources in the Heihe River Basin. *Journal of Geographical Sciences*, *15*(4), 405–414. Available from https://doi.org/10.1007/BF02892147.

Yadav, B., Malav, L. C., Jangir, A., Kharia, S. K., Singh, S. V., Yeasin, Md, … Yadav, K. K. (2023). Application of analytical hierarchical process, multi-influencing factor, and geospatial techniques for groundwater potential zonation in a semi-arid region of western India. *Journal of Contaminant Hydrology*, *253*, 104122. Available from https://doi.org/10.1016/j.jconhyd.2022.104122.

Chapter 10

Modeling and prediction of surface runoff for the humid areas: a geospatial analysis

Surendra Singh and H.J. Syiemlieh
Department of Geography, North-Eastern Hill University, Shillong, India

10.1 Introduction

Measurement of surface runoff is always a debatable issue with the increasing use of digital tools and computer-based techniques in assessing and managing water resources. A conventional approach to surface runoff assessment has long been practiced to gauge the stream discharge at the mouth of the watershed, considering the variability of affecting factors such as climate and catchment factors like size and shape of the area, topography, vegetation, and many more (Ward, 1968). A contemporary holistic approach to the assessment and prediction of surface runoff is desirably appropriate to be integrated to establish the "cause-effect relationship" of hydrological processes with physical factors of land terrain (composed of land slope, drainage pattern, soil types, and land use systems) (Sharma, Sundryal, Rai, & Krishna, 1998). A focus on integrated approach to runoff assessment through the application of appropriate tools and techniques can provide a better understanding of the variable hydrological conditions and processes prevalent in different locations. The current work carried out in the humid area of India raises several inescapable questions requiring the following expectations:

1. What are the hydrologic characteristics of the arid areas vis-a-vis the most humid areas in India?
2. Do hydrologic responses in humid areas need to be evaluated carefully?
3. Is physically-based distributed runoff prediction technique appropriate for humid tropical environments?

With tropical and monsoonal hydrological regimes in India, there are two distinctive water circulation conditions, as the isohyet of 1000 mm divides the country into two typical hydrological divisions (Fig. 10.1) introduced below:

Water Resources Monitoring, Management, and Sustainability. DOI: https://doi.org/10.1016/B978-0-443-23665-5.00010-7

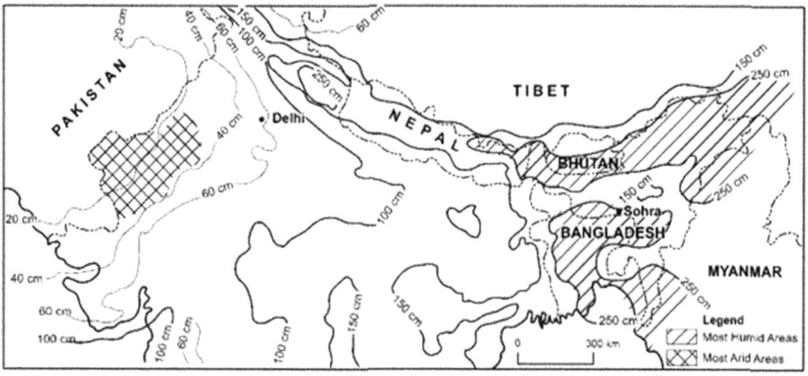

FIGURE 10.1 Distinctive Hydrologic Regimes shown with Annual Average Rainfall of Northern India. Map lines delineate study areas and do not necessarily depict accepted national boundaries. *From Survey of India (1987): School Atlas (II Edn), Government of India, Plate at p. 28.*

1. In the most humid conditions of North East India, high annual average rainfall (> 2000 mm), the existence of intensive soil leaching, weak aquifers, low groundwater recharge, low infiltration capacity, and low water retention capacity with gravelly soils (Pawel, 2010) happen, which allows sub-surface water to be added to runoff. As a result, the saturated areas vary quickly in their area extent, especially during heavy rainfall (Singh & Syiemlieh, 2010). On the other hand, increasing evaporative demand in the atmosphere (because of high temperatures >35°C) controls evaporation, which is a significant factor in runoff (Mishra, Sarkar, Dutta, & Panigrahy, 2008; Singh, 2007).
2. The extremely deficient soil water conditions of the arid regions of Rajasthan and Gujarat, where very little rainfall (around 300 mm annually) occurs, are associated with low rainfall intensity and high temperatures (above 45°C in the driest June month). In this situation, a high evaporative demand condition is created. Further, thick sandy soils (varying from 1 m to more than 5 m) associated with high porosity and barren (scanty) vegetative cover consequently result in high infiltration capacity, leading to high-capacity absorption of rainwater with high porosity (k > 100 cm/day). It may have a soil retention capacity of 800 mm of water depth. The hydrologic conditions of these areas are deep aquifers, flash floods, no full recharge of soils, and, thus, no discharge through sub-surface flow.

From the above examples, the areas of surplus soil water (case-1) are most dependent on "surplus-excess processes" of runoff generation, while the conditions of arid regions (case-2) follow the Hortonian "infiltration-excess processes" of overland flow only when precipitation exceeds infiltration. Heavy rainfall with variable intensity of rainstorms in north-east India alters the runoff processes (Singh & Syiemlieh, 2010). To assess whether a physical-based distributed runoff approach is suitable to evaluate geo-hydrological parameters for flood forecasting systems in surplus moisture conditions, a small watershed of about 874 ha in size was selected to evaluate the hydrologic responses of runoff.

10.2 Experimental watershed and the geo-hydrologic characteristics

10.2.1 The experimental watershed

The Umpher watershed, which is located on the margins of the foothills of the northern slopes of the Meghalaya plateau, was selected as an experimental watershed for testing the validity of the geo-hydrological conditions of the humid regions of north-east India. The Umpher watershed lies in low-laying foothills between the Meghalaya plateau in its south and the central Brahmaputra plains in the north. It is in close vicinity of Guwahati town, the regional capital of the north-east, near Byrnihat settlement (Fig. 10.2). With the consultation of a group of geo-hydrologists and the Government of Meghalaya agencies, an appropriate site for this watershed was selected to conduct geomorphic and hydrologic experiments in the north-east region of India, located at the cross of 26.07° N latitude and 91.85° E longitude. This slightly oblong watershed covers an area of about 874 ha, encompassing undulating hillocks at an elevation of about 500 m.a.s.l. and a stream flowing in a north-easterly direction (Fig. 10.3).

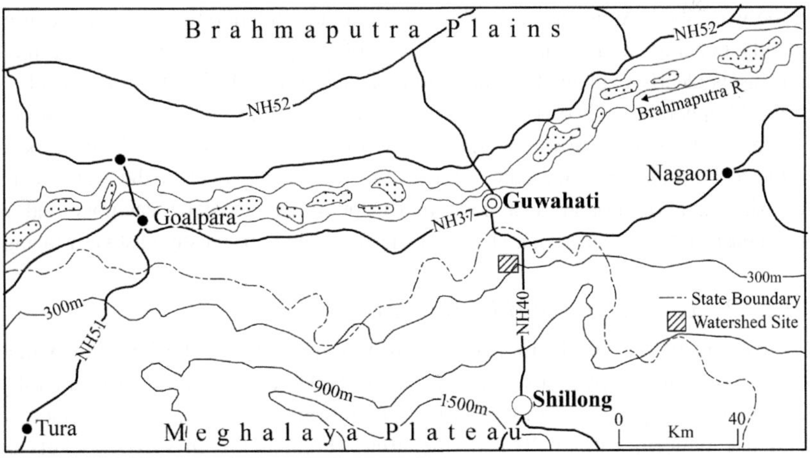

FIGURE 10.2 Location of Umpher watershed in its geographical surroundings. Map lines delineate study areas and do not necessarily depict accepted national boundaries.

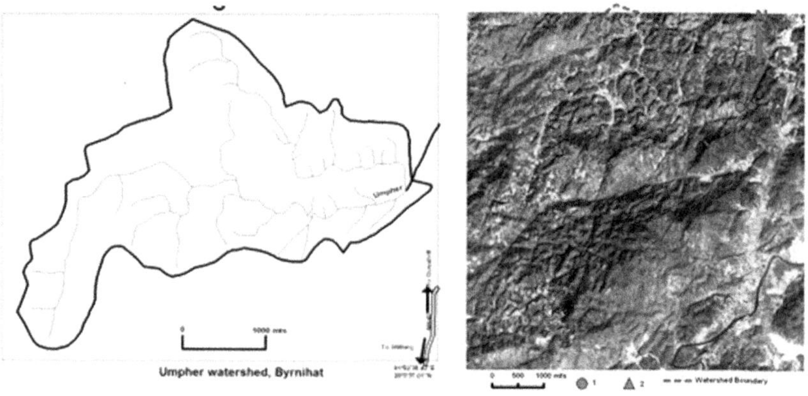

FIGURE 10.3 Surroundings of Umpher experimental watershed; the location of **1**- Autometic Weather Station, and **2**- Autometic Water Level Recorder. Map lines delineate study areas and do not necessarily depict accepted national boundaries.
N.B.: The dark color of the image shows the dense forests, the light-dark shows the grasses, and the white shade represents the establishments, settlements, roads, and quarries. Source: Cartosat-II Image, NRSA Hyderabad.

10.2.2 Land surface and weather conditions

Evaluation of physically-based distributed parameters of land surface and weather conditions is necessary to map them and ascertain their links with the runoff prediction process. Three land surface factors with intensity and pattern of precipitation in the watershed have been spatially characterized as follows.

10.2.2.1 Slope steepness

The steepness of the slope is directly related to the flow and retention of rainfall in the watershed. Areas of gentle slopes have a large retention capacity and consequently generate wetter surfaces than areas of steep slopes. The area-slope relationship is clearly explained by the digital elevation model (DEM). A few pockets in the Umpher watershed (Byrnihat) have been classified under very steep slopes ($>45^{\circ}$; Fig. 10.4A). However, most areas of gentle slopes ($<20\%$) have been found to be under wet rice cultivation in the watershed. Most of the watershed's recorded area of around 65% falls under the moderate slope category and is south-east-oriented; hence, the slopes of the watershed are mostly eastward facing.

10.2.2.2 Land cover and land use pattern

On account of the humid monsoonal climatic conditions, the Umpher watershed is dominated by dense mixed forests (40% area), which are generally located on steep slopes and hilltops. Plantations occupy a second place in land use patterns (about 20% of the area) (Fig.10.4B). In some pockets of fertile lands situated either on the flat hilltops or on the gentle slopes of valleys, paddy cultivation and other crops were found to dominate, and this practice reduces the infiltration rate and checks the surface runoff.

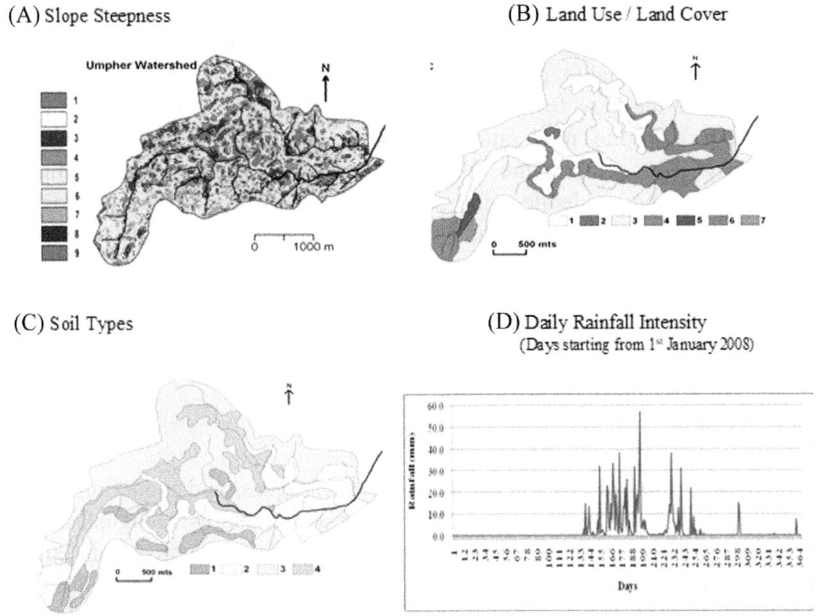

FIGURE 10.4 Spatial pattern of slope steepness, soil, and land use with daily rainfall. Map lines delineate study areas and do not necessarily depict accepted national boundaries. (A) Surface slope (%): 1 = flat (<2%), 2 = very gentle (2–4), 3 = gentle (4–10), 4 = moderate (10–20), 5 = moderately steep (20–35), 6 = steep (35–60), 7 = very steep (60–100), 8 = most steep (100–175), 9 = extremely steep (above 175%); (B) land use/land cover categories: 1 = paddy field, 2 = settlement, 3 = open mixed forest, 4 = quarry, 5 = nursery, 6 = plantation, 7 = slash and burn; (C) soil types: 1 = highly porous sandy, 2 = sandy loam, 3 = loam clay, 4 = shallow impermeable (low porosity); (D) rainfall Intensity: high rainfall intensity in monsoonal season (time scale at horizontal axis starts from January 1, 2008).

10.2.2.3 The soils

This land surface factor predominates in generating and changing the areal extent of saturation excess since, through the soil column, the soil porosity, and aquifer transmissivity determine the infiltration capacity, which is the key element of soil moisture availability (Schmocker-Fackel, 2004: p7). Aquifer transmissivity (i.e., called hydraulic conductivity) is evaluated by measuring two levels of groundwater between two places. Considering Darcy's law of hydraulic conductivity (k), the hydraulic gradient (G_h) that is the ratio of head loss (difference of groundwater level between two places), that is, $\Delta h = (h_1 - h_2)$, and the distance of water level difference between them (d) is expressed as $G_h = (\Delta h/d)$ (Brassington, 1980). The role of the aquifer is necessary to understand soil moisture retention capacity.

10.2.2.4 Weather conditions and rainfall pattern

The experimental watershed is under the influence of the humid monsoon weather of the central Brahmaputra valley, where a major share of annual rainfall precipitates in the summer season. As a result, a large share of surface runoff dominates during summers (May–September) due to intensive rainfall with a variable pattern. However, depleting soil moisture storage contributes to the base flow on which stream discharge is dependent in the dry winters. Despite moderate atmospheric air pressure (993–1000 mb), moderately high temperature (35°C and above), high relative humidity (80%–90%), and moderate wind speed (5–7 m/s) during the summers, a larger amount of daily rainfall precipitates in the form of rainstorms (Fig. 10.4D). This determines the fluctuation in stream discharge and the variation in the expansion/contraction of saturated areas in the watershed.

10.3 Methods and materials

10.3.1 The methodological issues

Keeping aside the discussion on the tools, techniques, and modeling developed mostly for dry hydrologic regimes such as lumped hydrological modes with the use of piezometers and stream flow (Pelletier & Andreassian, 2022), the focus is on the hydrologic response in the high-humid areas using a "physically-based distributed runoff mechanism," through which the effects of many land surface parameters of the catchment may be included in the analysis. This technique is the development of semi-lumped physically-based runoff modeling that assesses the hydrologic responses in terms of dividing the watershed (spatial domain) into sub-watershed area units and then totaling the responses of the physical elements of each unit (Beven, 2001; Crawford & Linsley, 1966; Freeze & Harlan, 1969). It estimates stream discharge considering the smaller spatial units of the watershed, which are sufficiently variable in their land surfaces. Keeping in mind the site-variability of the physical characteristics of the land surface, Beven & Kirkby (1979) modified this technique to understand how land surface conditions respond to runoff generation and to identify the location of wet areas. In fact, the evaluation of the variable pattern of precipitation intensity in relation to hydrologic responses linked with the spatially stable factors of the land surface is a necessary dimension of this model (Beven & Kirkby, 1979; Beven, 2001).

Hydrologic responses are predominately dependent on water flow conditions that are governed by hydrologic conductivity (i.e., sometimes water permeability or aquifer transmissivity). Available soil moisture storage plus incremental precipitation intensity determines the surface (overland) flow, while depletion of soil moisture storage is responsible for sub-surface flow that regulates stream discharge even in the dry season.

Revisiting the idea of the inclusion of surface conditions for water delivery systems, the Soil Conservation Services, Department of Agriculture, and US Government (USDA-SCS) developed a concept of curve numbers (CNs) in which rainfall-runoff relationships were determined by CNs for consideration of two major factors of the land surface, namely, land use pattern and soil types, called the "Hydrologic Response Unit," or HRU. The proportional areas of different HRUs of the watershed are weighted to CN for computation of maximum retention of three different precipitation intensity conditions, such as antecedent soil moisture condition (AMC), termed AMC I (dry), AMC-II (normal), and AMC-III (wet) (Das, 2002; Singh & Syiemlieh, 2010; USDA-SCS, 1972). Runoff prediction through the SCS-CN method is greatly controlled by loss of precipitation through land use and soils to calculate the potential maximum retention (or soil moisture deficit) known as "effective average moisture storage, S_e". Soil and Water Analysis Tool (SWAT) has been developed by using the same method to assess surface runoff (Arnold, William, Srinivasan, & King, 1996). However, it works at the sub-watershed level (so-called semi-lumped). The physically-based distributed runoff model is further developed by preparing routines for the physical characteristics of the catchment, namely, the analyses of hill slope topography, shape and size of catchment, soil moisture characteristics, and kinetic overland flow that are known as the "Variable Source Area (VSA) method," which is compiled using GIS grid-based data for land surface features (Beven, Warren, & Zauni, 1980: 124-127).

A close scrutiny of the SCS-CN method confirms the assemblage of internal mechanisms where maximum soil retention (effective soil storage) controls the runoff in each sub-area unit of the watershed, while the VSA method adds geomorphic factors (slope and direction) to the SCS-CN mechanism to identify the location of saturated surface in each grid (cell). The former is "area-specific" and the latter is "location-specific," but both methods are physically-based distributed runoff. The VSA method explains the dynamics (expansion or contraction) of the saturated area. This method satisfactorily integrates the three components of the geo-hydrologic process for runoff assessment, namely, the preparation of the topographic index of the land surface to measure the speed and direction of surface water flow, the measurement of available moisture storage that controls the sub-surface flow, and the observation of precipitation intensity, the most dynamic element of the atmospheric system, to identify the location and extent of saturation excess areas on the landscape (Beven & Kirkby, 1979; Schneiderman et al., 2007). So far as scrutiny of current tools and techniques of runoff assessment like SCS-CN-based SWAT modeling is concerned, this antecedent soil moisture-based technique needs to rely on map algebra for strengthening and applying the Spatial Decision Support System (SDSS). The exploration and application of an improved version of the runoff generation mechanism using the VSA approach can be realized with the inclusion of several land surface variables expressing topo-features, in addition to the existing variables, coupled with the surface wetness algorithm to predict runoff more accurately.

10.3.2 What is variable source area (VSA) mechanism and modeling?

The description of the development of structural componential relationships and dynamics of runoff assessment is necessary and can be done in the following manner:

10.3.2.1 Preparation of wetness index (WI, λ)

In a gridded watershed, cells receiving accumulated flow from flat land were considered wet with full saturation, while cells located on the steep slopes became dry as rainwater flowed downstream (Arnold & William, 1995). The effects of land surface parameters, namely, the surface gradient, accumulation of water flow, saturated hydraulic conductivity, and soil depth, are captured by a single integrated index called the "Wetness Index" (Beven & Kirkby, 1979; Lyon, Seibert, Lembo, & Steenhuis, 2006; Lyon, Walter, Gerard-Marchant, & Steenhuis, 2004). The functional relationship between these parameters is expressed as follows:

$$\mathrm{WI} = \ln\left[\alpha/(\tan\beta^{*}\mathrm{k}^{*}\mathrm{d})\right], \tag{10.1}$$

where α indicates flow accumulation that is the flow contributing catchment area in m^2 (number of cells drained to outlet to feed flow to the neighboring cells); it is simple to measure the number of upstream elements multiplied by the grid cell area while creating a flow accumulation topology (details given below under paragraph 10.3.2.1.1); the term tan β shows local surface gradient at unit scale (m/m), d refers to soil depth (cm), and k means the saturated hydraulic conductivity (cm/day).

It is to be noted that it is the physical index that shows the combined effect of the physical characteristics of the land surface within the watershed. Since the physical parameters of the land surface are assumed to be stable but spatially vary, the wetness index mapping reveals the factorial effects of physical parameters to determine the location and extent of saturated areas. This index controls the dynamics of precipitation acting upon the wetness map. There are two methodological issues pertaining to the topology of flow accumulation where the location of saturated areas is being identified and, second, the understanding of the mechanism of hydrologic responses in humid areas where precipitation controls the dynamics of location and extent of saturation-excess areas within the watershed.

10.3.2.1.1 Creation of flow accumulation topology

The first issue is the search for the steepest downhill path of flow that forms the drainage pattern, leading to the accumulation and discharge of rainwater in each cell covered by the gridded surface. Cell-by-cell discrete operation in GIS is extended by considering the state of a cell that does not have only the vertical operation of overlaying layers but also the lateral interactions with its neighboring cells (Burrough & McDonnell, 1998). To analyze the flow accumulation process, the "square window" is the most widely used method of spatial filtering, a technique to estimate slope and aspect for determining the steepest downhill flow paths. A window of the smallest size of 3*3 cells is possible to analyze the drainage direction as it is based on eight neighboring cells representing eight directions (E, W, N, S, NE, NW, SE, and SW) to measure the effect of directions and their gradients toward the central cell. Following the spatial filtering procedure that starts from a discretized sequence of operations to the involvement of using a continuous surface for passing a square window, it creates a flow accumulation topology. The spatial-filtering-based slope-weighted algorithms and stream-tube algorithms were prepared for the measurement of the dynamics of the flow process (Moore, 1996: 143–148). The flow accumulation algorithm is built into the ILWIS toolbox, enabling the identification of the direction of flow on a gridded surface, which is used to prepare the wetness index in the present case.

10.3.2.1.2 Computation of saturation excess areas

The second is the procedural issue concerning modeling the saturation-excess runoff and establishing the triangular relationship between precipitation, surface features (particularly the slopes and soils), and runoff. In the condition when precipitation intensity exceeds soil storage capacity in the cells of the gridded surface due to shallowness and poor retention capacity of the soil, the soil becomes fully saturated to produce overland flow (Schneiderman et al., 2007). Such areas of the watershed are counted under the condition of saturation excess. The mechanism of saturated excess areas is concerned with a few componential relationships:

1. Direct runoff (Q_d) is defined as the areas of overland flow where scanty vegetation with undulating slope topography exists.
2. The three precipitation retention factors of land surface: Initial abstraction (I_a), ponding (F_a), and soil moisture storage (S) are important and necessary to understand the generation of runoff. For example, at the time of initial abstraction before ponding, it does not yield runoff because the total amount of precipitation is initially abstracted ($P < I_a$). Furthermore, the stage of ponding starts after the initial abstraction. In this condition, the overland flow does not appear until precipitation exceeds these conditions. After initial abstraction and ponding, the precipitated water is added to the surface soil for continuous recharge of soil moisture storage, S_t, until it becomes fully recharged, which is the stage of maximum retention capacity (S_m) under conditions ($S_m > S_t$).
3. The effective available storage (S_e) is the difference between the amount of soil storage minus the initial abstraction, as $S_e = (S - I_a)$.

4. Effective precipitation (P_e) is the function of initial abstraction and evapotranspiration (ET) components ($P_e = P - I_a - ET$) especially in tropical areas where the evaporative demand of the atmosphere significantly reduces the amount of runoff (Mishra et al., 2008).

The SCS-CN method developed the equation by establishing a relationship between direct runoff, effective precipitation, and effective available soil storage as follows:

$$Q_d = \left[(P - I_a)^2/(P + S - I_a)\right] \text{for } P \geq I_a; \text{otherwise } Q_d = 0 \text{ for } P \leq I_a, \quad (10.2)$$

It is obvious that in this equation, I_a includes the ET value, especially the calculation of Q_d for the most humid areas. In the key concept of establishing a relationship between effective rainfall, $P_e = (P - I_a)$, and soil moisture storage, the USDA-SCS (1972) developed the CN technique, assuming that the depth of direct runoff per unit of effective rainfall (Q_d/P_e) is proportional to the detained rainwater for soil recharge per unit of the maximum retention capacity of soil moisture storage (F_a/S_m), which is expressed as:

$$\left(Q_d/P_e\right) \approx \left(F_a/S_m\right). \quad (10.3)$$

It may be written as:

$$\left(Q_d/P_e\right) = \lambda\left(F_a/S_m\right) \quad (10.4)$$

where the proportionality coefficient λ determines the wetness intensity. This equation explains that increasing the maximum soil retention capacity (S_m) diminishes the runoff intensity. Since parameter λ has a negative relationship with F_a, the lower its values, the greater the wetness intensity, and vice versa. Thus, a non-linear negative relationship between direct runoff Q_d and soil moisture storage S_m can be observed (Chow, Maidment, & Mays, 1988: 147–155).

10.3.2.2 The fractional depth of runoff

It is the dynamic part of the model to obtain an important parameter that identifies the saturated areas. The relationship between rainfall and runoff is established by Eq. (10.2), which is modified for the dynamism of precipitation intensity by differentiating the rainfall-runoff equation, notated as A_f. It transforms as:

$$A_f = dQ_d/dP_e = 1 - \left[S^2/(P_e + S)^2\right], \quad (10.5)$$

The mechanism of the VSA module provides improved results in saturated areas to predict runoff accurately in a catchment (Frankenberger, Brooks, Walter, Walter, & Steenhuis, 1999). However, several improved deterministic models based on semi-lumped techniques overpredict the results of peak runoff, especially for the bigger size catchments of the monsoonal environment (Babel, Najim, & Loof, 2004).

10.3.3 Materials collected

Carrying out the creation of a spatial database under GIS to map topo-features, land cover/land use pattern and soil types, images from CARTOSAT-II (spatial resolution 2.5 m) supplied by the National Remote Sensing Agency (now Corporation) NRSC Hyderabad, and a topo-survey of watershed were conducted for detailing DEM and land use features at R.F. 1/5000 scale during March–April, 2008. Soil samples from various sites with surface features were collected, analyzed, and mapped.

For measurement of weather conditions in the watershed, the hourly weather statistics were generated by installing an automatic weather station at Byrnihat on top of a flat ridge located in the central part of the watershed site. The accuracy of hourly rainfall (mm) is counted as 5.0% of the full range (i.e., 0.2 mm).

In addition, by installing an automatic water level recorder at the mouth of the stream in the experimental watershed and preparing a concrete structure of rectangular weirs 2.78 m wide, the hourly flow rate was measured. Later on, it was converted into a stream discharge rate by applying the rating curve equation expressed as:

$$\text{Flow rate} = 2.78*1.83*(1.0 - 0.2\text{h})*(\text{h})^{1.5}, \tag{10.6}$$

where h = head of water (m) and constant 2.78 is the length of the weirs (in m).

Despite regular monitoring of hydrometric instruments and recording for more than 2 years (August 2007 to December 2009), there were data gaps in different time-slots of the recorded time-schedule caused by lightning, especially during the pre-monsoon season. Sometimes, other mechanical problems occur, leading to disruptions in data recording. However, hourly statistics of several significant events were captured to use them for the development of the physically-based distributed model and its validation.

10.4 Discussion and analysis

10.4.1 Hydrograph analysis: seasonal variation in flow discharge

The flow discharge regime of tropical monsoon areas such as north-east India largely depends on the rainfall pattern (Lyon et al., 2004). However, the causes of discharge variation are different under topographic as well as land use and land cover variations. Due to the location of the catchment's topographic influence, the sub-surface water regulates stream discharge, especially in dry winters. There will be an abundant downward flow of

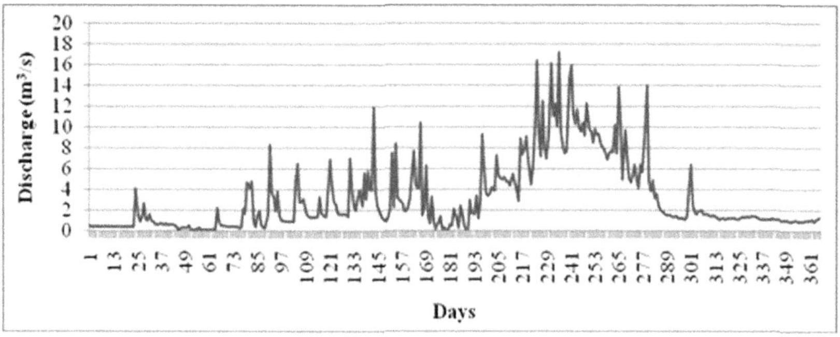

FIGURE 10.5 Daily variation in discharge rate in watershed for 366 days (starting from January 1, 2008, to December 31, 2008) (N.B.: figures at vertical scale shows 10 times values of actual discharge).

underground water where stream discharge is regulated by a robust aquifer. The abundant contribution and continuous base flow help in smoothening the discharge (Fig. 10.5). However, a maximum discharge rate of 1.738 m^3/s with a significant annual variation (CV = 109.35%) was recorded at this station.

Adopting a conventional water-budget approach to the hourly weather data collected for specified rainstorms of 29 hours starting at 18:00 hours (IST) on September 21, 2009, till 18:00 (IST) on September 23, 2009, with dry antecedent moisture condition (AMC-I), the hydrologic characteristics of the watershed were analyzed. An amount of 40.26 mm of runoff was recorded when 47.6 mm of rain was precipitated during the storm. In fact, an amount of 7.4 mm of rainfall was computed as loss during the rainfall event (i.e., 15.5%), which is comparatively lesser than the average loss (30%–40%) computed in the north-eastern region of the country. However, a water yield of 8.3 $m^3/hr/ha$ was recorded in the watershed. Peak flow received after 15.3 hours of rising it in the stream with a rising trend at an average rate of 0.2612m^3/s per hour. To evaluate the impulse response function, it is assumed that the runoff yield system is working continuously in its linear trend over time; this means that the difference between the input (precipitation) and output of the system (direct runoff) over time equals the difference in soil storage, expressed as $I(t)-Q(t) = dS/dt$. The impulse duration is recorded long enough in the watershed (that is, 34 hours), where the average event response is 0.334 $m^3/s/cm$ with variability of about 75.4% during the event duration (Table 10.1).

10.4.2 The Wetness Index (WI = λ)

The creation of input data files in ArcGIS to map the model parameters is a prime task to use raster-based map algebraic functions as required for

TABLE 10.1 Hydrologic characteristics of rainstorm event (duration 49 h).

Description	Umpher watershed
1. Size of watershed (ha)	874.1
2. Antecedent moisture conditions (AMC)	I
3. Hydrologic event duration(h)	49
4. Total surface water (RO) produced(m^3)	351,926.25
5. Average surface water produced per hour (m^3/h)	7182.168
6. Average surface water yield produced per hr/ha (m^3/h/ha)	8.2166
7. Total amount of storage in channels (m^3)	75,277.2
8. Share of storage of total volume per cm of Pnet (%)	21.39
9. Total precipitation (mm)	47.60
10. Total loss (mm)	7.40
11. Share of loss to total (%)	15.54
12. P net (P-loss) = (ΣQdt/A) (mm)	40.260
13. Water travel time (hrs)	12.11
14. Peak flow (m^3/s)	4.474
15. Average discharge (m^3/s)	1.995
16. Minimum discharge (m^3/s)	0.81603
17. Standard Deviation of discharge (m^3/s)	1.0027
18. Co-efficient of variation (%)	50.260
Impulse response	
19. Maximum intensity of response (m^3/s/cm)	0.8629
20. Increasing rate of response up to its maximum level (m^3/s/cm)	0.07177
21. Average response of even (m^3/s/cm)t	0.3340

satisfying Eq. (10.1). With the use of the wetness index tool available in the ILWIS 3.6 open-source version, the land surface features are parameterized to compute flow accumulation for each cell of 20 m*20 m size on the gridded surface. To observe the role of soil types on wet surfaces, the parameters of soil transmissivity are used in two ways: unconstrained soil variability (to use soil parameters in their variable values) and constrained soil factors (considering the soil effect as spatially homogeneous at an average

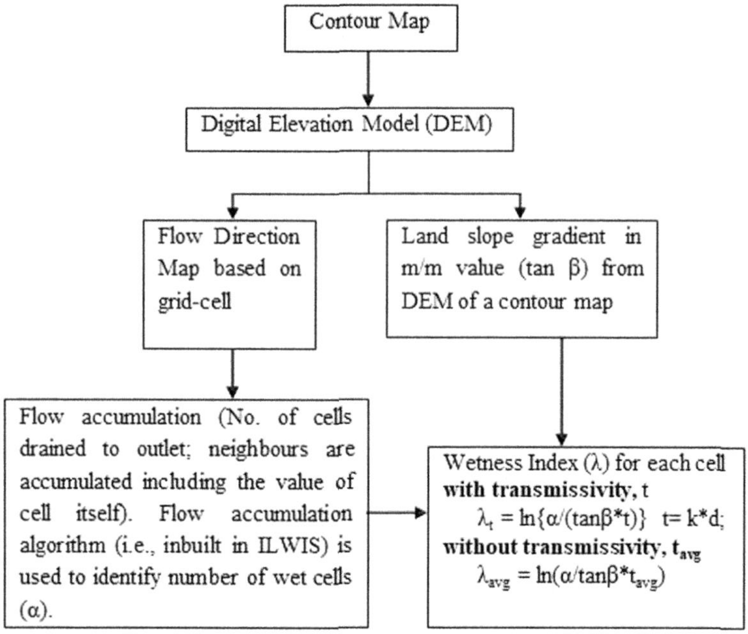

FIGURE 10.6 Development of Wetness Index (λ) for identification of wet areas (Beven and Kirkby, 1979).

value of soil transmissivity, t_{avg}). Fig. 10.6 illustrates the procedural steps to measure the wetness index by counting the number of cells under each category for both the unconstrained and constrained soil moisture effects on wet areas (Fig. 10.7 and Table 10.2).

The location of saturated area (A_S) is identified by equating the incremental fraction contributing runoff depth (A_f defined in Eq. 10.5) with the fractional contributing area (Ac/A) as $A_f = (Ac/A)$. We obtain a critical value of λ that determines the number of wet cells within the watershed. The soils start becoming saturated to release excessive water at an initial value of λ, so the extent of saturated area expands fast in the watershed and vice versa. As a result, the graph of the cumulative fractional contributing area with λ (ogive) shows the "S" type of relationship.

Fig. 10.8A illustrates the effect of soil moisture (unconstrained) on wet surface features, while Fig. 10.8B shows constrained soil moisture effects when a t_{avg} value of 4500 cm^3/day is used as a spatially constant. A comparison of these two maps would provide a sound base for determining area differences in the effect of soils on the saturation excess process. The values of available soil moisture (unconstrained) that deviated from the t_{avg} (constrained) explain the direct impact of soils on the location and extent of wet areas. Categorizing the distribution of wetness into ten classes, most of the areas of the watershed fall under moderately wet to moderately dry classes, varying from 9.00 to 13.00 λ_t values. The spatially variable soil moisture (unconstrained condition) shows that the share of wet areas is higher because of moist soils, displaying wetness index values below 10 (see Col. 1 of Table 10.2).

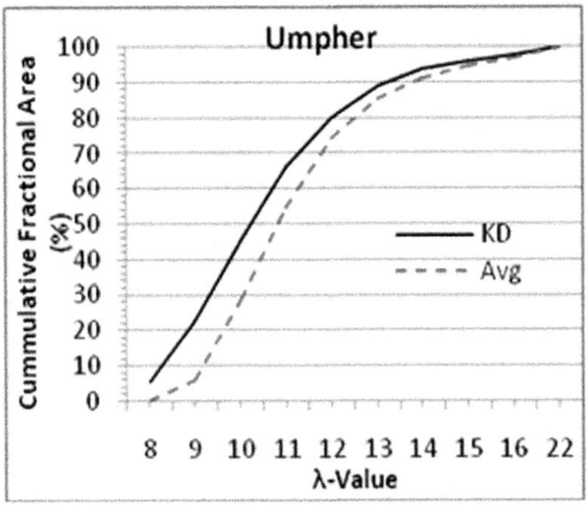

FIGURE 10.7 Cumulative wet areas (%) subject to unconstrained (KD) and constrained (t_{Avg}) Soil moisture conditions.

TABLE 10.2 Percentage share of contributing areas in various wetness classes in different watersheds.

Wetness class (λ value)	Umpher watershed		
	λ_t (% area)	λ_{avg} (% area)	Deviation ($\lambda_t - \lambda_{avg}$)
Extremely wet (Below 8)	5.48	0.25	+
Very wet (8–9)	17.14	5.72	+
Moderately wet (9–10)	22.78	22.71	+
Wet (10–11)	20.55	26.14	–
Dry (11–12)	14.23	19.71	–
Moderately dry (12–13)	8.76	11.05	–
Very dry (13–14)	4.67	5.6	–
Very very dry (14–15)	2.39	3.38	–
Extremely dry (15–16)	1.65	2.24	–
Driest (Above 16)	2.35	3.19	–

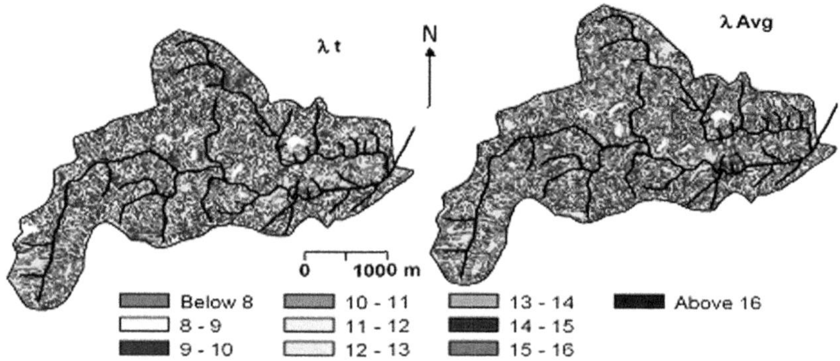

FIGURE 10.8 Pattern of Wetness Index for unconstrained (λ_t) and constrained (λ_{avg}) soil moisture conditions (white patches show the clouds). Map lines delineate study areas and do not necessarily depict accepted national boundaries. Note: See Table 10.2 for the name of wet area categories shown on the map.

An observation regarding the effect of soil moisture on saturation shown graphically (Fig. 10.7) makes it clear that the porosity and depth of soils are significant factors that directly affect soil moisture storage. A continuous infiltration due to the porous soils increases the share of wet areas in the watershed (also see col-3, Table 10.2), especially in the flat valley lands (Fig. 10.8).

The location of the saturated area and runoff depth of each cell (q) based on the excess saturated overland flow that is produced is called effective precipitation minus maximum retention soil moisture as:

$$q = (Pe - Se) \text{ for } Pe > Se, \text{ otherwise } q = 0. \tag{10.7}$$

A total quantity of discharge (Q) is then the multiplication factor of runoff depth with the area of saturated cells as:

$$Q = \sum (q_i * ac_i), \quad i = 1, 2, 3, \ldots, n \text{ cells falling within saturated area.} \tag{10.8}$$

10.5 The model application

10.5.1 Model calibration and validation

The calibration of the model's parameters is based on the three rainstorm patterns collected hourly and the hydrologic statistics of five major rainstorms that occurred in the monsoon season of 2008 (Table 10.3). Despite some short-duration rainstorms of 3, 6, and 4 hours, the event-wise observed parametric values like maximum retention soil moisture (Se), which is computed at 15 mm for this watershed as per its soil conditions, and the initial abstraction before ponding that does not produce any runoff (I_a). It also includes the amount of ET. Its value is drawn from the amount of precipitation until the beginning of runoff. The three rainstorms' average value of I_a is used to calibrate the model. It is counted as 4.5 mm.

The validation phase is based on the observed and predicted volume of discharge of the last two rainstorms with their continuous recording of 12 hours (Table 10.3). Despite limited sample data size, the results can provide support to be a good testing ground of the model. A regression-based validation tool developed by Nash & Sutcliffe, (1970) was applied to evaluate the output performance. This tool measures the efficiency coefficient (E) following an unexplained variation of the hydrologic event, expressed as:

$$E = 1 - \left[\sum (Q_o - Q_p)^2 / \sum (Q_o - Qo_{mean})^2\right]. \tag{10.9}$$

If E = 1.00, it shows perfect agreement between observed Q_o and predicted Q_p values. Increasing unexplained variation diminishes the efficiency of the runoff model, and vice versa. In the present case, E is computed at

TABLE 10.3 Rainstorm pattern and peak discharge.

Days	Duration (in h)	Precipitation amount (mm)	Precipitation intensity (mm/h)	Peak discharge rate (Qo in m^3/s)
22 May 2008	3 (2:00 am–4:00 am)	6.8	2.1	7.7
2 June 2008	6(1:00 am–6:00 am)	18.4	3.0	8.3
10 June 2008	4(7:00 pm–10:00 pm)	13.2	3.3	10.5
16 June 2008	7(4:00 pm–10:00 pm)	20.6	2.94	12.3
29–30 June 2008	10(9:00 pm–7:00 am)	2.4	0.24	6.7

Source: From Singh, S., & Syimlieh, H. J. (2008): Hourly weather statistics of the central Meghalaya plateau (June 2007–June 2008), Monograph No. GE/NRDMS/DST/SS/8, Department of Geography, North Eastern Hill University Shilling.

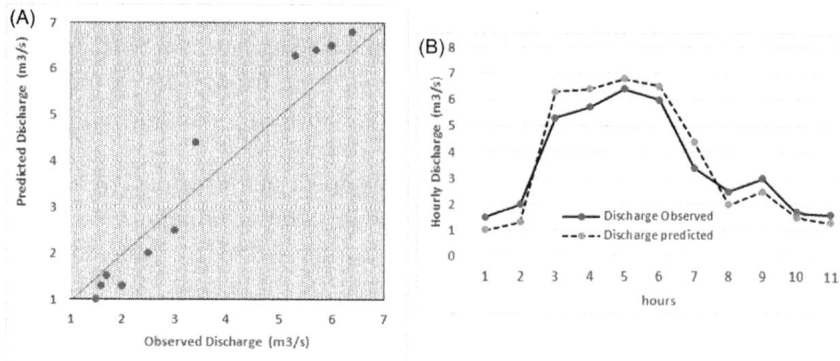

FIGURE 10.9 (A) Scattered Q_o - Q_p Plot; (B) hourly trend of Q_o and Q_p.

0.93578. It means model performance is overall good to predict the runoff in this watershed (Fig. 10.9A). However, it predicts higher runoff in the peak hours of effective precipitation (Fig. 10.9B).

10.5.2 Evaluation and spatial simulation

Precipitation-based spatial simulations were conducted to calibrate the VSA model at Se = 15.00 mm and I_a = 4.5 mm for the watershed computing PET with the use of the T-M procedure of soil water availability for wet soil conditions (AMC-III). Note that the simulations were made to consider unconstrained soil moisture conditions to include its spatially variable characteristics as depicted in Fig. 10.4C. The location and extent of saturated areas within the watershed were simulated at different intensities of effective precipitation and explained precisely.

While the intensity of effective precipitation increases substantially, the VSA method performs satisfactorily as an increasing volume of runoff appears to be a significant expansion of wet areas in the watershed. The precipitation intensity below 5.00 mm does not generate surface runoff due to the effective infiltration and evaporative demand of the atmosphere. Effective precipitation of 1.00 mm saturates about 9.5% of hilltop areas in the watershed. With the increase in effective precipitation intensity from 3.00 to 5.00 mm, the saturation excess processes are accelerated fast to add saturated areas of about 5.00%, especially toward hill slopes and later, in the valley flats (Fig. 10.10C). However, during the most intensive effective precipitation (above 7.0%), though saturation area increases (Fig. 10.10 and Table 10.4), it occurs at a diminishing rate.

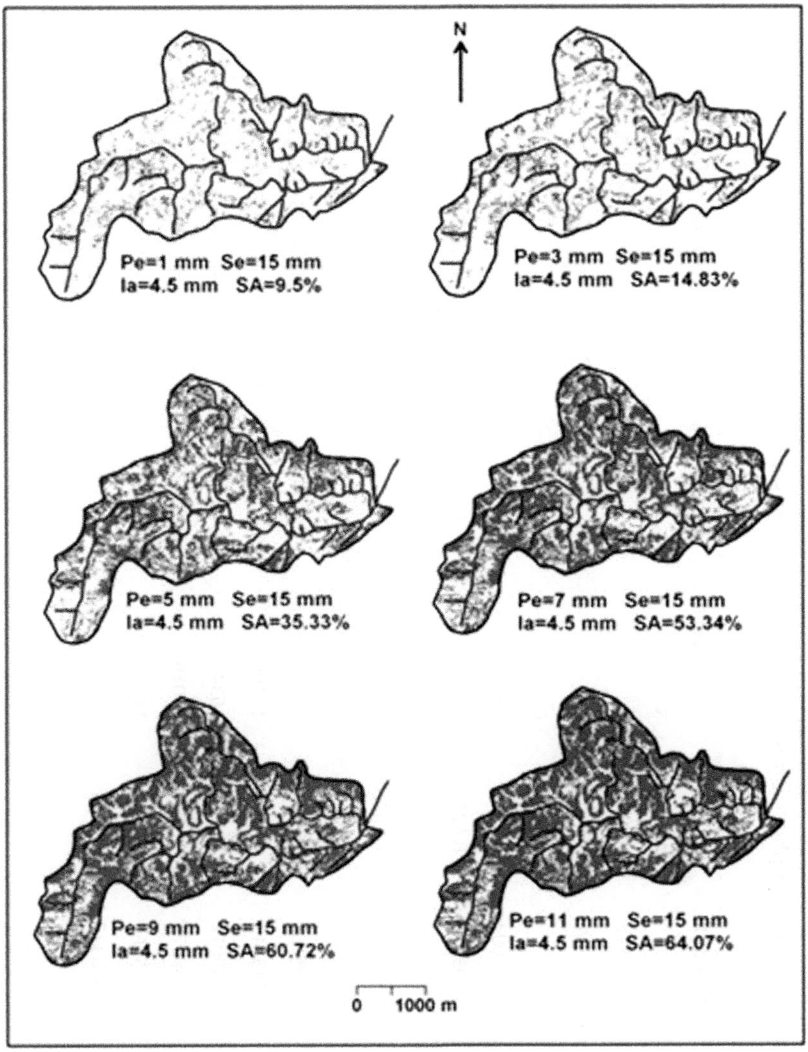

FIGURE 10.10 Pattern of saturated areas at different effective rainfall intensities. Map lines delineate study areas and do not necessarily depict accepted national boundaries.

TABLE 10.4 Simulation of saturated area in umpher watershed with uncontrolled soil conditions (Ia = 4.5 mm, Se = 15.00 mm).

Rainfall (mm)	Pe (mm)	Af (mm)	λt	No. of cells	Saturated area (ha)	Saturated fractional area
1	**2**	**3**	**4**	**5**	**6**	**7**
0.55	-4	-0.85951	0	0	0	0
2.55	-2	-0.33136	0	0	0	0
4.55	0	0	0	0	0	0
5.55	1	0.121094	8.314	2077	83.0811	0.09504633
7.55	3	0.305556	8.621	3242	120.968	0.14835831
9.55	5	0.437523	9.801	8833	350.332	0.40421004
11.55	7	0.535124	10.408	11657	460.628	0.53344011
13.55	9	0.609375	10.716	13268	530.072	0.60716165
15.55	11	0.667161	10.951	14002	560.008	0.64075049
17.55	13	0.713012	11.403	15599	620.396	0.71383137
19.55	15	0.750012	11.651	16346	650.384	0.74801514
21.55	17	0.780273	11.906	17019	680.076	0.77881249
23.55	19	0.805363	12.100	17248	680.992	0.78929184
24.55	20	0.816327	12.121	17484	690.936	0.80009152

10.6 Conclusion

A physically-based distributed runoff model provides a rich foundation and great flexibility with the inclusion of a number of land surface parameters in the process of identifying the areas of saturation excess within a watershed. This approach is useful for SDSS, providing ways to predict runoff, particularly when hydrological statistics of humid areas are available and, if not, generated weather and base-flow statistics can be applied to existing models. The current VSA sub-model, say a specific module to identify saturated areas subject to land surface parameters (Eq. 10.1), has been linked with other modules to express the transfers occurring in saturated areas into predicted runoff depth (Eqs. 10.5 and 10.8). This model provides accurate runoff prediction using spatial simulation capabilities, which is an important feature. It has been possible to predict the precipitation effect and also to assess the impact of land factors, particularly the accumulated water flow, surface gradient, and soil moisture, on runoff depth. This is achievable due to the fact that the model offers a combination of two modules, the VSA sub-model, and the transfer model. Physically-based distributed runoff modeling has been used as a successful tool to simulate precipitation effects on runoff intensity and flood forecasting in a watershed, particularly within the humid environments of tropical and monsoonal lands around the world.

Acknowledgments

We sincerely acknowledge the financial assistance of the Ministry of Science and Technology, New Delhi provided under an individual Project (ES/11/833/2004) under an NRDMS Program grant (DST). This is a modified version of the part of this Project Report entitled "Runoff Processes in Extremely Humid Areas of the Central Meghalaya Plateau." We express our gratitude to the Director, Regional Soil Laboratory of the National Bureau of Soil Survey and Land Use Planning (NBSS&LUP), Jorhat (Assam) for soil sample analysis for the project and also to James M. Lyngdoh, Department of Geography, North Eastern Hill University, Shillong (India) for cartographic assistance.

References

Arnold, J. G., & William, J. R. (1995). In SWRRB (Ed.), A watershed scale model for soil and water resources management, water resource publication (pp. 847–908). Highland Ranch Co.

Arnold, J. G., William, J. R., Srinivasan, R., & King, K. W. (1996). *Soil and Water Assessment Tool (SWAT): User Manual*. USDA: Agriculture research sercives. Temple: Grassland Soil and Water Research Laboratory.

Babel, M. S., Najim, M. M. M., & Loof, R. (2004). Assessment of agricultural nonpoint source model for a watershed in tropical environment. *Journal of Environmental Engineering, 130*(9), 1032–1041. Available from https://doi.org/10.1061/(ASCE)0733-9372(2004)130:9(1032).

Beven, K. J., & Kirkby, M. J. (1979). A physically based, variable contributing area model of basin hydrology. *Hydrological Sciences Bulletin, 24*(1), 43–69. Available from https://doi.org/10.1080/02626667909491834.

Beven, K. J. (2001). Rainfall-runoff mdelling- the primer (pp. 124–127). John Willey.

Beven, K. J., Warren, R., & Zauni, J. (1980). SHE: Towards a methodology for physically based distributed forecasting in hydrology (pp. 133–137). Wallingford: IAHS Publications, No 129.

Brassington, R. (1980). *Field hydrology* (II, pp. 66–85). John Wiely and Sons.

Burrough, P. A., & McDonnell, R. A. (1998). *Principles of geographical information systems* (pp. 183–197). Oxford University Press.

Chow, V. T., Maidment, D. R., & Mays, L. W. (1988). Applied hydrology. New York: Mc Graw-Hill Book Co. Inc.

Crawford, N. H., & Linsley, R. K. (1966). Digital simulation in hydrology: Stanford watershed model IV; trchnical report No 39. USA: Stanford University.

Das, G. (2002). Hydrology and soil conservation engineering (pp. 70–78). New Delhi: Printice Hall of India.

Frankenberger, J. R., Brooks, E. S., Walter, M. T., Walter, M. F., & Steenhuis, T. S. (1999). A GIS- based variable source hydrology model. *Hydrological Processess, 13*, 805–822.

Freeze, R. A., & Harlan, R. L. (1969). Blueprint for a physically based digitally simulated hydrologic responce model. *Journal of Hydrology, 9*, 237–258.

Lyon, S. W., Seibert, J., Lembo, A. J., & Steenhuis, T. S. (2006). Geostatistical investigation into the temporal evaluation of spatial structure in a shallow water table. *Hydrology and Earth System Sciences, 10*, 113–125.

Lyon, S. W., Walter, T. M., Gerard-Marchant, P., & Steenhuis, T. S. (2004). Using a topographic index to distribute variable source area runoff predicted with the SCS curve-number equation. *Hydrological Processes, 18*(15), 2757–2771. Available from https://doi.org/10.1002/hyp.1494.

Mishra, S. K., Sarkar, R., Dutta, S., & Panigrahy, S. (2008). A physically hydrological model for paddy agriculture dominated hill watershed in tropical region. *Journal of Hydrology, 357*, 389–404.

Moore, I. D. (1996). Hydrological modelling and GIS. In GIS and environmental modelling: Processes and research issues, (pp. 143–148). Fort Collins, Colorado: GIS World Books.

Nash, J. E., & Sutcliffe, J. V. (1970). River flow forecasting through conceptual model: Part-I, A discussion of principles. *Journal of Hydrology (Amsterdam), 10*(3), 282–290.

Pawel, P. (2010). *Soil resources and soil degradation.* Land degradation and ecorestoration in the extremely wet monsoon environment of Cherrapunji area (pp. 61–68). New Delhi: Bookwell Publications.

Pelletier, A., & Andreassian, V. (2022). On contributing a lumped hydrological model with both piezometry and stream flow: Results of a large sample evaluation. *HESS, 26*(10), 2733–2758.

Schmocker-Fackel, P. (2004). A method to delineate processes in a catchment and its implications for runoff simulations (Ph.D. dissertation) (p. 186.) Zurich: Swiss Federal Institute of Technology.

Schneiderman, E. M., Steenhuis, T. S., Thongs, D. J., Easton, Z. M., Zion, M. S., Neal, A. L., & Walter, M. T. (2007). Incorporating variable source area hydrology into a curve-number-based watershed model. *Hydrological Processes, 21*(25), 3420–3430. Available from https://doi.org/10.1002/hyp.6556.

Sharma, E., Sundryal, R. C., Rai, S. C., & Krishna, A. P. (1998). *Watershed: A functional unit of management for sustainable development*. Modern trends in ecology and environment (pp. 171–185). Amesterdom: Backhuya Publications.

Singh, S. (2007). Rainfall-runoff conditions in Meghalaya plateau, research monograph No GE/UGC/SS/5. Shillong: North-Eastern Hill University.

Singh, S., & Syiemlieh, H. J. (2010). *Runoff Processes in Extremely Humid Areas of the Central Meghalaya Plateau. Research Monograph No.: GE/NRDMS/DST/10*. Shillong: Department of Geography, North-Eastern Hill University.

USDA-SCS. (1972). *Hydrology in soil conservation service*. National engineering hand book. Washington DC: U S Department of Agriculture.

Ward, R. C. (1968). Principles of hydrology. *Soil Science*, *106*(4), 326. Available from https://doi.org/10.1097/00010694-196810000-00023.

Chapter 11

Application of GIS-based analytical hierarchy process (AHP) for the identification of groundwater potential zone in Dibrugarh District, Assam, India

Praduyt Dey[1] and Arpana Handique[2]

[1]*Department of Geography, Rajiv Gandhi University, Doimukh, Arunachal Pradesh, India,*
[2]*Center for Studies in Geography, Dibrugarh University, Dibrugarh, Assam, India*

11.1 Introduction

Groundwater is considered a crucial resource for the progress of any region. It serves as a primary source of drinking water supply for its population and facilitates the growth of agricultural practices in that region (Namous et al., 2021). With the increasing demand for freshwater in developing nations, groundwater is considered a valuable asset and a high-priority natural resource (Todd & Mays, 2004). Around the world, groundwater is a significant source of drinking water as well as for industrial activities, mining, and agricultural irrigation. As per the United States Geological Survey (USGS), around 30.1% of the freshwater on the earth is accumulated as groundwater, whereas surface water, including lakes, streams, rivers, and water from other freshwater sources, makes up only 1.2%. Despite the widespread availability of groundwater resources, it is important to use them sustainably and scientifically (Evans, Jones, Williams, Ames, & Nelson, 2020). As per the estimate proposed by Shao, Huq, Cai, Altan, & Li (2020), around 36% of the total groundwater resources of the world are used for domestic purposes, around 27% are used by the industrial sector, and nearly 42% are utilized by the agricultural sector. About 2.5 billion people around the world are estimated to rely on groundwater for drinking water (Achu, Thomas, & Raghunath, 2020). In India, 86% of the drinking water requirements and above 60% of the water

Water Resources Monitoring, Management, and Sustainability. DOI: https://doi.org/10.1016/B978-0-443-23665-5.00011-9

required for irrigation are met by groundwater (Kolanuvada, Ponpandian, & Sankar, 2019). Unregulated extraction of groundwater has significantly decreased the groundwater level (GWL) in several parts of India (Kumar, Chandniha, Lohani, Nema, & Krishan, 2018). Thus, it is crucial to identify areas where groundwater sources have ample potential to meet water requirements and maintain sustainability (Saravanan, Saranya, Abijith, Jacinth, & Singh, 2021).

Groundwater availability is determined by numerous factors, such as geological arrangement, lineament density, geomorphology, porosity, weathering processes, drainage characteristics, land use and land cover (LULC) practices, and precipitation (Mukherjee, Singh, & Mukherjee, 2012). The dominance of these factors in determining groundwater availability varies from one region to another because of the variation in natural and environmental conditions. Prudent methods of identifying groundwater potential or resources need high-end instruments and methodologies (Rahmati, Nazari Samani, Mahdavi, Pourghasemi, & Zeinivand, 2015).

Remote sensing and GIS have become widely used techniques that are adequately used for the identification of groundwater potential zones in various administrative divisions and watersheds (Das & Pardeshi, 2018). Satellite-based remote sensing techniques have been providing valuable data that is applied to understand factors like geology, geomorphological characteristics, land use/cover, drainage patterns, and lineaments on which the storage and flow of groundwater depend (Aggarwal, Saravanan, Jennifer, & Abijith, 2019). In recent years, many researchers have adopted different methodologies for delineating groundwater potential zones, namely, AHP and geospatial techniques (Das, Mukherjee, Bhattacharyya, Patel, & Banerjee, 2022; Kom, Gurugnanam, & Sunitha, 2022; Muthu & Sudalaimuthu, 2021), fuzzy logic methods (Şener, Şener, & Davraz, 2018), and machine learning techniques (Bai, Liu, & Liu, 2022). In the present study, the Analytical Hierarchy Process (AHP) and geospatial techniques were used to identify and assess the groundwater potential zones of the study area. The application of AHP techniques is considered a powerful tool in delineating groundwater regimes as it deals with complex decision-making through pairwise comparisons. Later, the findings can also be checked for consistency assessment to avoid biases in the decision-making process (Fatti, 1989).

Although the Dibrugarh district of Assam is well-drained, highly populated, and receives significant annual rainfall, the groundwater potential estimation has not been done before. Therefore, geospatial techniques were used along with the AHP to obtain high-accuracy groundwater potential results. A total of ten predictive factors were used, including geomorphological characteristics, geological structures, lineament density, drainage density, slope, LULC, soil, rainfall, evapotranspiration, and topographic wetness index (TWI). The main aim of this research is to identify the groundwater potential zones in the Dibrugarh district of Assam using multi-criteria decision-making techniques (AHP) in combination with remote sensing and GIS to instigate sustainable management of groundwater resources.

11.2 Materials and methods

11.2.1 Study Area

Dibrugarh is located between 27°5'38″ to 27°42'03″ N latitude and 94°33'46″ E to 95°29'08″ E longitude in Assam, covering an approximate area of 3381 km^2 (Fig. 11.1). It is located in the upper part of the Brahmaputra Valley. Its northern boundary is demarcated by the Dhemaji and Lakhimpur districts; the Tirap district of Arunachal Pradesh lies to its south; and Tinsukia and Sivasagar districts lie to its east and west, respectively. Physiographically, the region exhibits a wide range of features, including extensive flood plains, swamps, wetlands, and some isolated hills. The Brahmaputra River flows along the northern border of the district, and the Burhi Dihing River, a major south-bank tributary of the Brahmaputra, flows through the district from south-east to north-west, covering a significant part of the region and contributing to the picturesque landscape with bamboo fields, tea plantations, and agriculture fields. Dibrugarh district of Assam experiences a humid subtropical climate characterized by wet summers and comparatively dry winters (Bora, Das, Bhuyan, & Hazarika, 2022).

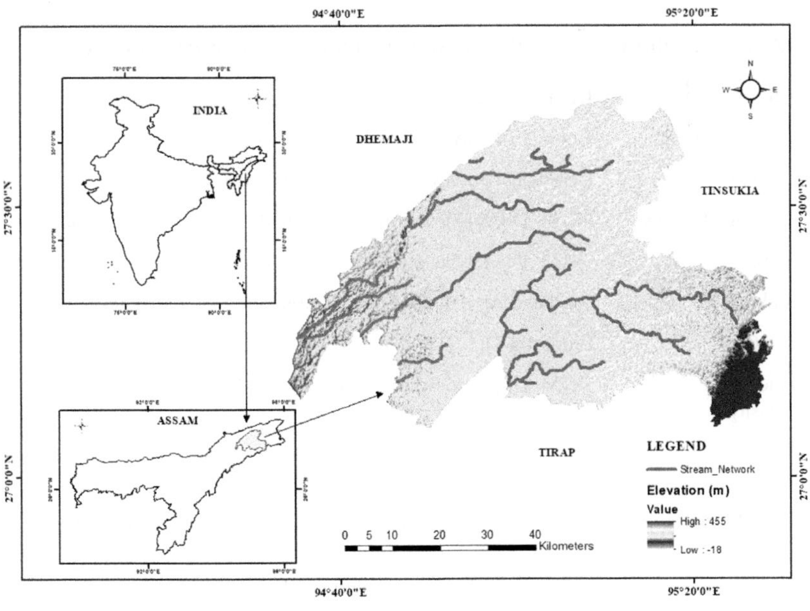

FIGURE 11.1 Study area Location Map. Map lines delineate study areas and do not necessarily depict accepted national boundaries.

11.2.2 Data acquisition

The data used to prepare the thematic layers is extracted from different sources (Table 11.1). The data source includes the SRTM DEM with a 30 m resolution downloaded from the USGS, the soil map obtained from the Food and Agriculture Organization of the United States Soil dataset portal, the geomorphology and geology map collected from the Geological Survey of India (*Geological Survey of India*, 2023), the Landsat OLI 8 (Operational Land Imager) satellite data acquired on March 22, 2023, downloaded from Survey (2023), the rainfall data for 2021–22 collected from the Climatic Research Unit high-resolution gridded time series dataset at 0.5° resolution, and the evapotranspiration data (MODIS/TerraNet Evapotranspiration 8 Day L4 Global 500 m S/N Grid) derived from The Level-1 and Atmosphere Archive & Distribution System Distributed Active Archive Center *LAADS DAAC* (2023). The data used for model validation was derived from the Central Ground Water Control Board through the India Water Resources Information System, and personal field data was collected from various locations in the study area.

11.2.3 Preparation of the geohydrological control factor layers

Understanding the impact of geohydrological control factors is crucial for assessing groundwater potential and susceptibility and developing effective water resource management strategies (Fetter, 2014). The process of

TABLE 11.1 Illustration of the causative factors and their data sources.

Factors	Data sources
Geomorphology	Geological Survey of India (1:250,000)
Geology	Geological Survey of India (scale 1: 2,000,000)
Drainage Density	SRTM- Digital Elevation Model (DEM) 30 × 30 M resolution
Slope	SRTM- Digital Elevation Model (DEM) 30 × 30 M resolution
Soil	Food and Agriculture Organization of the United States (FAO)
LULC	Landsat 8 OLI
Lineament Density	SRTM- Digital Elevation Model (DEM) 30 × 30 M
Rainfall	Climatic Research Unit (CRU) high-resolution gridded time series dataset at 0.5° resolution (2021–22)
Evapotranspiration	MODIS/TerraNet Evapotranspiration 8 Day L4 Global 500 m S/N Grid
TWI	SRTM- Digital Elevation Model (DEM) 30 × 30 M resolution

generating thematic layers to represent groundwater potential zones includes analyzing satellite data, converting conventional maps from various secondary sources into digital format, and applying statistical and spatial models within the ArcGIS environment.

The initial step involves digitizing the geomorphological and geological maps to create the thematic layer. Subsequently, slope, drainage density, and TWI raster layers are generated using digital processing techniques in ArcGIS 10.8 software, utilizing the Digital Elevation Model (SRTM DEM). The lineament density map was generated through a manual digitization method using SRTM DEM, which involves opening a new layer in ArcGIS, selecting a line as a digitizing tool, generating hill shades of varying degrees to represent the terrain, tracing the linear features in the hill shade layer, and finally calculating the lineament density by determining the number of lineaments present within specific areas or grid cells. The LULC layer was generated using Landsat 8 OLI using a supervised classification process, and the maximum likelihood classification algorithm was employed. The classification procedure involved selecting the training area sequentially and generating a signature file based on vector or polygonal representations. The soil layer was obtained through the integration of a soil map of the world (Esri shapefile format) with the study area layer file. Subsequently, the annual rainfall layer was generated using the Inverse Distance Weighting (IDW) interpolation method, which allowed for the estimation of rainfall based on available data points. The IDW algorithm was applied in ArcGIS 10.8 software to spatially interpolate the rainfall values across the study area. Finally, the Evapotranspiration layer was produced using MODIS/TerraNet Evapotranspiration data, adding a spatial boundary layer (study area) and visualization in ArcGIS software.

11.2.4 Analytical hierarchy process and weight assignment

The AHP is an extensive decision-making process that combines an expert's judgment with empirical data (Şener, Sener, & Karagüzel, 2011). The data is accumulated, identified, and evaluated for its importance in decision-making. Decisions are made by building a hierarchical comparison matrix and assigning ratings based on the degree of influence using Saaty's scale ranging from 1 to 9 (Table 11.2). The selected factors were compared with each other based on their influence on groundwater potential (Table 11.3). Weights are determined through expert choice and judgment and then normalized using Saaty's AHP method (Saaty, 1980) to mitigate bias in the weight assignment of thematic layers (Table 11.4). Once the weights are calculated, each criterion and sub-criterion are ranked based on their assigned weights. The layers with relatively higher weights are considered to have a greater influence on the groundwater potential (Table 11.5).

TABLE 11.2 Suggested scale for pair-wise comparison matrix using AHP (Saaty, 1980).

Scales	Degree of preferences	Descriptions
1	Equally	Both factors are equally important
3	Moderately	Experiences and judgment slightly tend to be a factor
5	Strongly	Experiences and judgment strongly tend to be a factor
7	Very Strongly	A factor is strongly favored over another, and its dominance is shown in practice
9	Extremely	The evidence of favoring one factor over another is of the highest degree possible of an affirmation
2, 4, 6, and 8	Intermediate values	Used to represent compromises between the preferences in weights 1, 3, 5, 7, and 9

Inconsistency can arise in the methodology adopted for performing AHP due to the subjective nature of pairwise comparisons, which rely on personal judgments (Arunbose, Srinivas, Rajkumar, Nithya, & Kaliraj, 2021). This deviation in judgments may introduce fluctuations in the degree of inconsistency during the decision-making process. The consistency of the comparison matrix was assessed through the Consistency Index (CI) and consistency ratio (CR) calculated using Eqs. (11.1) and (11.2), respectively (Saaty, 1980). These measures were used to quantify the level of consistency in the pairwise comparisons made within the matrix:

$$\text{Lambda}(\lambda)\text{max} = (107.94/10) = 10.79$$

$$\text{Consistency Index} = \frac{\text{Maximum principal Eigenvalue}(\lambda)\text{-No.offactor}(n)}{\text{Number offactor-1}}$$

$$= (10.79 - 10/10\text{-}1) = 0.09 \tag{11.1}$$

where n is the number of factors; λ is the principal eigenvalue:

$$\text{Consistency Ratio(CR)}\frac{\text{Consistency Index(CI)}}{\text{Random Consistency Index(RCI)}} \tag{11.2}$$

$$= (0.09/1.49) = 0.06$$

The random CI is derived from Saaty's standard ratio index (Table 11.6). A CR value equal to or below 0.10 is considered acceptable, indicating that the pairwise comparisons within the analysis are reasonably consistent (Saaty, 1980). With a CR value of 0.06, the pairwise comparisons within the analysis

TABLE 11.3 Pairwise comparison matrix assigned for identifying groundwater potential zone using AHP.

Factors	Geom.	Geol.	Slope	LD	DD	RF	ET	TWI	Soil	LULC
Geomorphology	1	2	2	3	3	2	4	5	5	2
Geology	0.5	1	2	3	2	2	3	3	4	2
Slope	0.5	0.5	1	4	2	0.5	4	4	2	3
Lineament Density	0.33	0.33	0.25	1	0.33	0.5	4	3	0.33	2
Drainage Density	0.33	0.5	0.5	2	1	0.33	2	2	0.5	2
Rainfall	0.5	0.5	2	3	3	1	4	2	2	3
Evapotranspiration	0.25	0.33	0.25	0.25	0.5	0.25	1	0.5	0.33	0.5
TWI	0.2	0.33	0.25	0.33	0.5	0.5	2	1	0.33	0.5
Soil	0.2	0.25	0.5	3	2	0.5	3	3	1	2
LULC	0.5	0.5	0.33	0.5	0.5	0.33	2	2	0.5	1

TABLE 11.4 Normalized pairwise comparison matrix and computation of criterion weights.

Factors	Geom.	Geol.	Slope	LD	DD	RF	ET	TWI	Soil	LULC	Weight
Geomorphology	0.23	0.33	0.22	0.15	0.20	0.25	0.14	0.20	0.31	0.11	0.21
Geology	0.12	0.16	0.22	0.15	0.13	0.25	0.10	0.12	0.25	0.11	0.16
Slope	0.12	0.08	0.11	0.20	0.13	0.06	0.14	0.16	0.13	0.17	0.13
Lineament Density	0.08	0.05	0.03	0.05	0.02	0.06	0.14	0.12	0.02	0.11	0.07
Drainage Density	0.08	0.08	0.06	0.10	0.07	0.04	0.07	0.08	0.03	0.11	0.07
Rainfall	0.12	0.08	0.22	0.15	0.20	0.13	0.14	0.08	0.13	0.17	0.14
Evapotranspiration	0.06	0.05	0.03	0.01	0.03	0.03	0.03	0.02	0.02	0.03	0.03
TWI	0.05	0.05	0.03	0.02	0.03	0.06	0.07	0.04	0.02	0.03	0.04
Soil	0.05	0.04	0.06	0.15	0.13	0.06	0.10	0.12	0.06	0.11	0.09
LULC	0.12	0.08	0.04	0.02	0.03	0.04	0.07	0.08	0.03	0.06	0.06

TABLE 11.5 Assigned weights and ranks of all thematic layers through AHP.

Factor	Classes	Weight	Influence (%)	Rank	Consistency ratio
Geomorphology	Alluvial Plain	0.21	21	4	0.083
	Dam and Reservoir			5	
	Flood Plain			4	
	Highly Dissected Hills and Valleys			1	
	Pediment Pediplain Complex			3	
	Waterbodies Other			5	
	Waterbodies River			5	
Geology	Tipam Sandstone Fm.	0.16	16	4	0.085
	Undiff.Fluvial/Eolian/Coastal and Glacial Sediments			5	
	Disang Fm.			3	
Slope	0–2	0.13	13	5	0.083
	2–4			4	
	4–6			3	
	6–8			2	
	>8			1	

(Continued)

TABLE 11.5 (Continued)

Factor	Classes	Weight	Influence (%)	Rank	Consistency ratio
Lineament Density	Very Low	0.07	7	1	0.077
	Low			2	
	Moderate			3	
	High			4	
	Very High			5	
Drainage Density	Very Low	0.07	7	5	0.081
	Low			4	
	Moderate			3	
	High			2	
	Very High			1	
Rainfall	3971–4075	0.14	14	1	0.083
	4075–4120			2	
	4120–4173			3	
	4173–4233			4	
	4233–4305			5	

Evapotranspiration	136–202	0.03	3	5	0.078
	202–268			4	
	268–346			3	
	346–544			2	
	>544			1	
TWI	<7.3	0.04	4	1	0.078
	7.3–11.55			2	
	11.55–15.72			3	
	15.72–19.89			4	
	>19.89			5	
Soil	Sandy Clay Loam (Af)	0.09	9	4	0.083
	Sandy Clay Loam (Ao)			5	
	Clay Loam (Ao)			2	
LULC	River Bodies	0.06	6	5	0.073
	Sand deposits			5	
	Forest			5	
	Agriculture			4	
	Plantation			4	
	Settlement/Built-up areas			1	

TABLE 11.6 Random consistency index (RCI) (Saaty, 1980).

1	2	3	4	5	6	7	8	9	10	11	12
0.00	0.00	0.58	0.90	1.12	1.24	1.32	1.41	1.45	1.49	1.51	1.48

exhibit a high level of consistency, meeting the acceptable threshold. Thus, the groundwater potential zone map (GWPZ) of the Dibrugarh district was created using Eq. (11.3) in ArcGIS 10.8 through spatial analyst tools (Aykut, 2021):

$$\text{GWPZ} = GMw \times GMr + Gw \times Gr + SLw \times SLr + LDw \times LDr + DDw \times DDr + Rw \times Rr + ETw \times ETr + TWIw \times TWIr + Sw \times Sr + LULCw \times LULCr \tag{11.3}$$

where GM is geomorphology, G is geology, SL is the slope, LD denotes lineament density, DD is drainage density, R signifies rainfall, ET represents evapotranspiration, TWI is topographic wetness index, S represents soil, and LULC indicates land use and land cover. The suffix w represents the assigned weight, while r indicates the rank of each thematic layer, respectively. The GWPZ map was categorized into four groups: high, moderate, and poor, representing different levels or classes based on the assessment outcome. The methodology adopted in the study is depicted in Fig. 11.2.

11.3 Results

11.3.1 Geomorphology

Geomorphology is the study of the landform features on the earth's surface and their evolution processes (Ajay Kumar, Mondal, & Ahmed, 2020), which controls the hydrological and hydrogeological processes, specifically runoff, seepage of water, and the groundwater recharge process (Abijith et al., 2020). The study area displays diverse geomorphological features (Fig. 11.3), including an alluvial plain of about 63% (2121.29 km^2), a flood plain of around 16% (546.12 km^2), highly dissected hills and valleys about 2% (57.56 km^2), a pediment pediplain complex of nearly 2% (75.97 km^2), and waterbodies of around 17% (579.35 km^2). The floodplain and water bodies region were considered highly significant for the Groundwater Potential Zone (GWPZ) due to its high infiltration capacity, and the highly dissected hills and valleys were considered least significant due to high surface runoff and a lower recharge area (Table 11.5).

11.3.2 Geology

The geological settings, to a large extent, determine the availability and recharge of groundwater (Kaur, Rishi, Singh, & Nath Thakur, 2020). The characteristics of porosity and permeability differ depending on the type of rock formation. Generally, rock formations with higher hardness allow limited water infiltration. In this research, the significance of geology was emphasized alongside geomorphology due to its permeability and capacity for infiltration. The geological structures found in the Dibrugarh district are displayed in Fig. 11.4. About 96% (3254.68 km^2) of the study area is

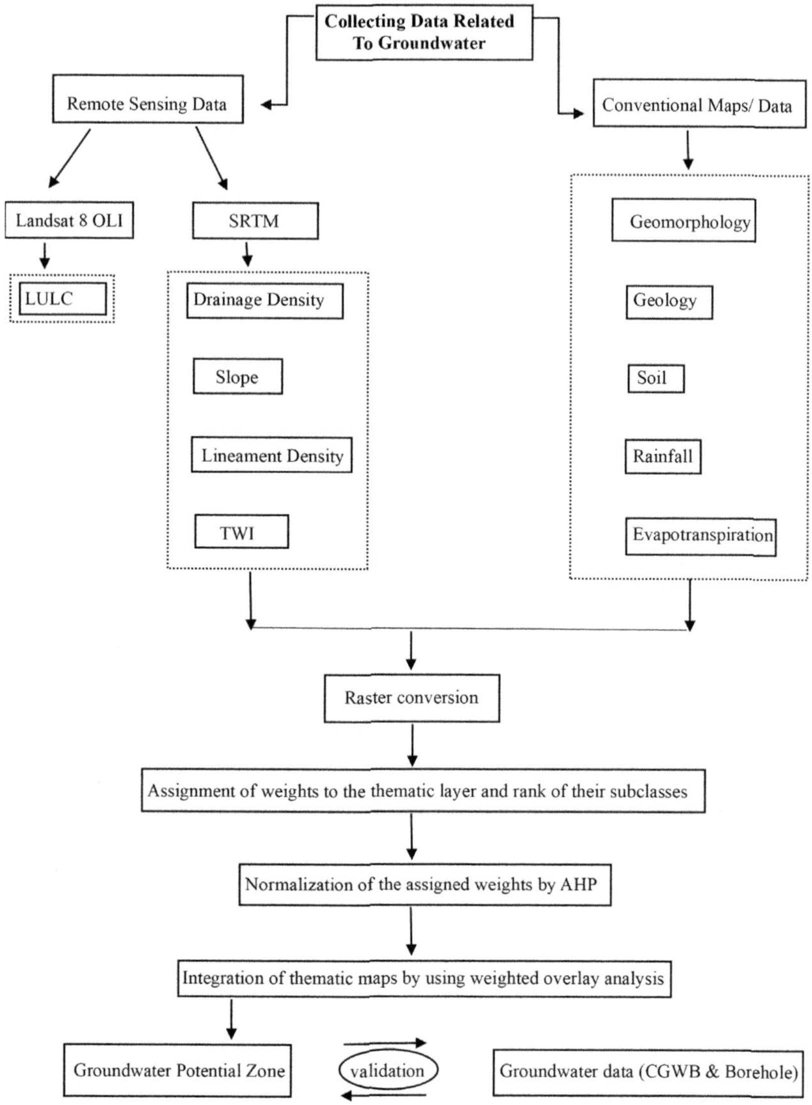

FIGURE 11.2 Flow chart of the methodology adopted in the delineation of the Groundwater Potential Zone.

dominated by fluvial sediments as the entire region is drained by the Brahmaputra and its tributaries, followed by tipam sandstone formation, which covers nearly 4% (125.56 km^2), and the remaining portion is covered by disang formation, which covers 0.001% (0.04 km^2). Fluvially originating sediments generally possess high groundwater potential due to their good permeability and porosity. They provide favorable conditions for the storage

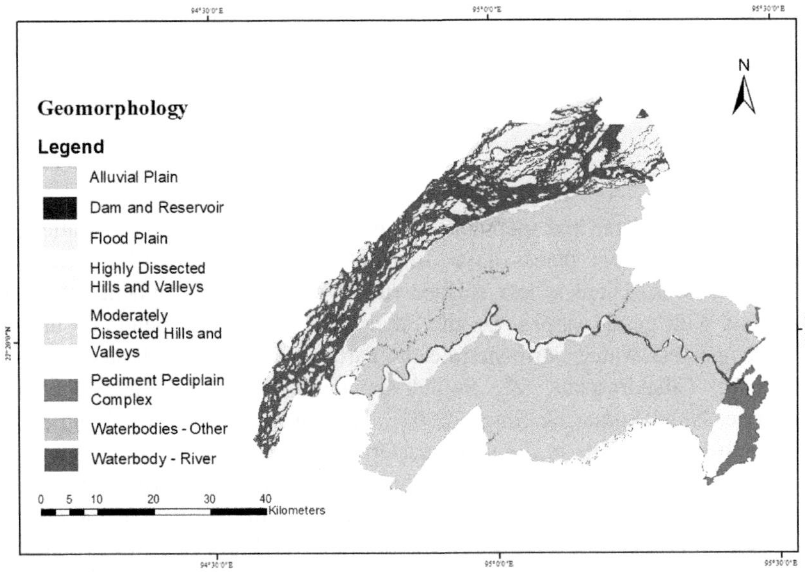

FIGURE 11.3 Geomorphological Map. Map lines delineate study areas and do not necessarily depict accepted national boundaries. *From Geological Survey of India.*

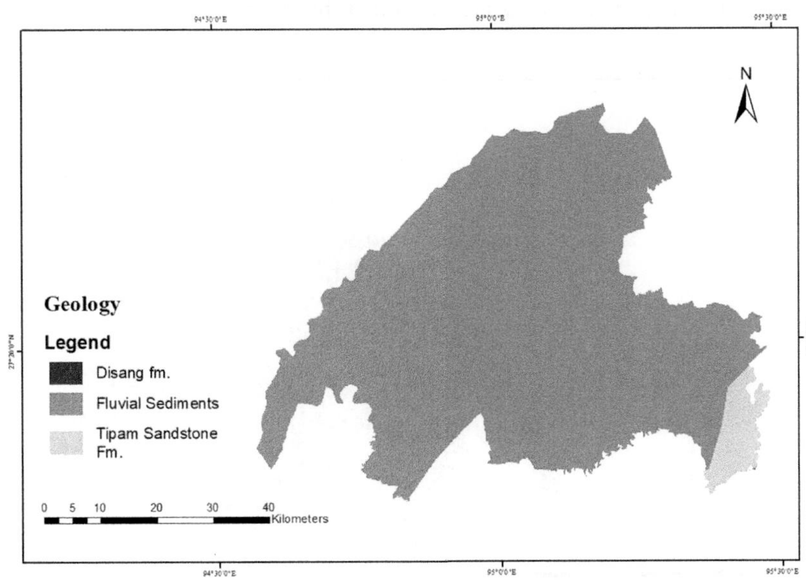

FIGURE 11.4 Geological Map of the study area. Map lines delineate study areas and do not necessarily depict accepted national boundaries. *From Geological Survey of India.*

and movement of groundwater and, therefore, are given the highest weightage, whereas disang formation, due to its lower permeability and porosity, was given the least weightage (Table 11.5).

11.3.3 Slope

The amount of water that seeps into the ground depends on the inclination of the terrain. Different places have different rates of water infiltration from surface sources. There is less surface runoff and more water penetration in locations with gentle slopes. On the other hand, there is tremendous surface runoff, and less water can penetrate the earth in locations with steep slopes (Magesh, Chandrasekar, & Soundranayagam, 2012; Prasad, Mondal, Banerjee, Nandakumar, & Singh, 2008). The study area is a low-lying plain area with isolated hills in the south-eastern portion ranging from 0 degree to 50 degree slope angle, which has been reclassified into five categories (Fig. 11.5). Areas with the lowest slope, ranging from 0–2 degree and covering 1833 km^2 (54.4%), were given the highest weightage due to low surface runoff and a higher infiltration rate. The rest of the area with varied slope angles, that is, 2–4 degree covering 875 km^2 (26%), 4–6 degree covering 343 km^2 (10.5%), 6–8 degree covering 152 km^2 (4.5%), and more than 8 degree covering 165 km^2 (4.8%), is assigned weight accordingly. The subclasses and the assigned ranks are listed in Table 11.5.

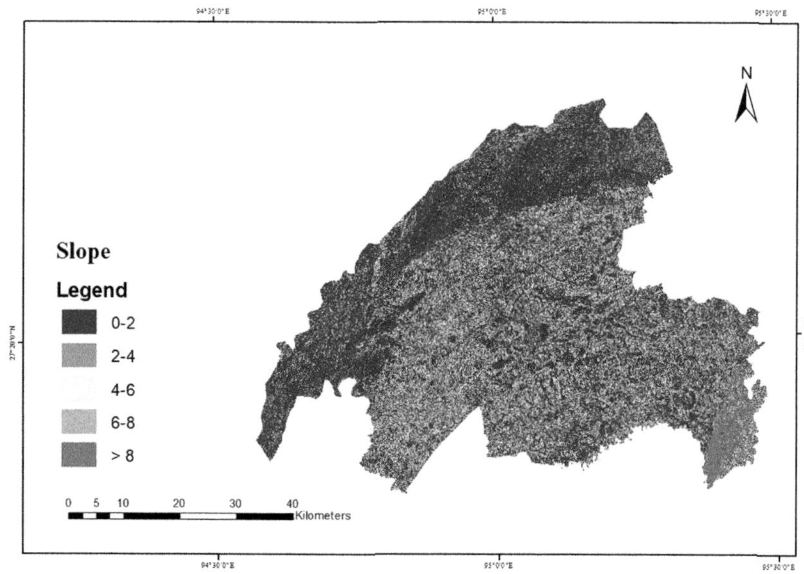

FIGURE 11.5 Slope Map of the study area. Map lines delineate study areas and do not necessarily depict accepted national boundaries.

11.3.4 Lineament density

Lineaments are linear features on the earth's surface that indicate rock joints, faults, and fractures that are observable through satellite imagery. These are basically the zones of high porosity and permeability and are believed to be important in groundwater studies (Goitsemang, Das, Raul, Subudhi, & Panigrahi, 2020). In regions characterized by solid rock formations, the dynamics of groundwater movement and storage are influenced directly or indirectly by the existence of faults and areas with fractured rock. In areas where lineaments traverse drainage networks in a parallel manner, the potential for higher groundwater yields is observed compared to other regions. Consequently, lineaments play an important role in understanding groundwater flow and storage patterns, and they help in identifying groundwater zones during hydrogeological investigations (Rao, Chakradhar, and Srinivas, 2001). The prepared lineament density map was classified as follows: very low about 97% (3296.95 km^2), low about 0.7% (24.5 km^2), moderate nearly 0.8% (26.76 km^2), high roughly 0.4% (12.3 km^2), and very high about 0.5% (18.42 km^2). The study area has relatively low lineament density, with the smallest geographical coverage of regions with extremely high lineament densities in the southwest (Fig. 11.6).

11.3.5 Drainage density

The drainage density in an area holds significance in the identification of groundwater storage and sediment transportation (Sahoo, Dhar, Kar, & Ram, 2017).

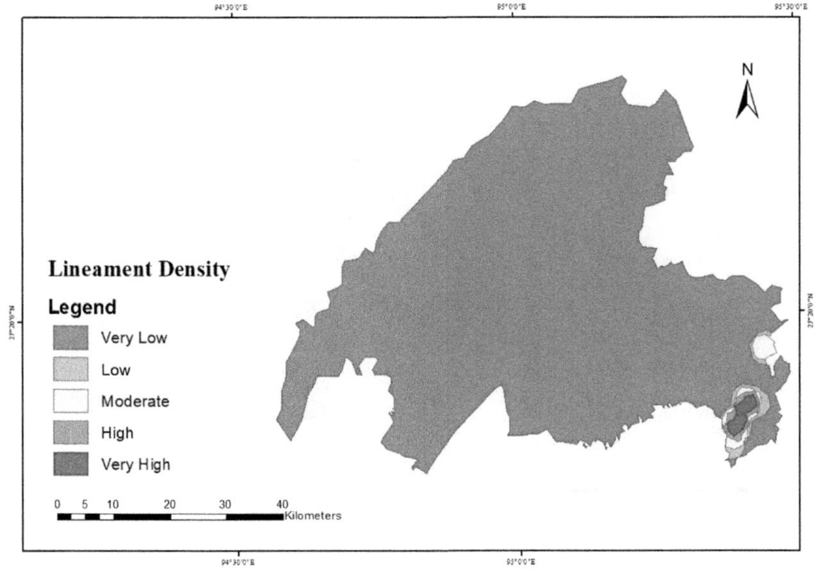

FIGURE 11.6 Lineament Density Map. Map lines delineate study areas and do not necessarily depict accepted national boundaries.

The spacing between stream channels and the surface material is revealed by drainage density, which is related to soil permeability (Singha et al., 2019). The drainage density map has been reclassified and was divided into five categories, namely, very low (1993.83 km^2), low (381.81 km^2), medium (541.1 km^2), high (329.54 km^2), and very high (132.67 km^2) (Fig. 11.7). The very low drainage density among other classes covers approximately 52% of the region. These regions have a very low drainage density, thus a high infiltration rate, and are weighed relatively higher in comparison to other classes (Table 11.5).

11.3.6 Rainfall

The amount of rainfall received in a particular area determines the recharge of groundwater. Regions with higher rainfall have greater chances of groundwater recharge, and vice versa (Machiwal, Jha, and Mal, 2011). The study area lies in the subtropical climatic zone affected by southwest and northeast monsoonal rain and thus receives a very high amount of annual rainfall ranging from 3971 to 4305 mm (Fig. 11.8). The region with the highest rainfall concentration had a higher amount of water source for infiltration and groundwater recharge, which in turn is also influenced by the geological settings, soil type, and permeability. The areas with 4233 to 4305 mm (11.33%) of rainfall were given the highest weightage, whereas the areas with 3971 to 4075 mm (25.39%) of rainfall were given the lowest weightage (Table 11.5).

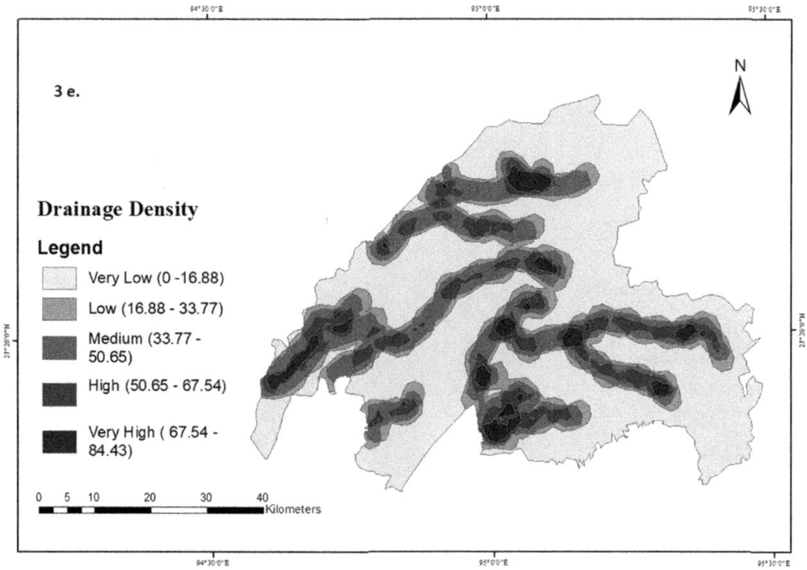

FIGURE 11.7 Drainage Density of the study area. Map lines delineate study areas and do not necessarily depict accepted national boundaries.

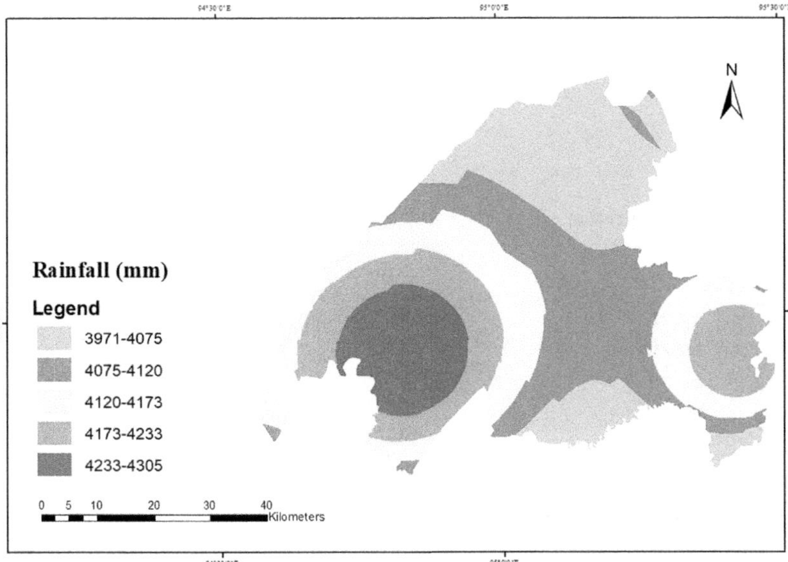

FIGURE 11.8 Rainfall map of the study area. Map lines delineate study areas and do not necessarily depict accepted national boundaries. *From Climatic Research Unit (CRU).*

11.3.7 Evapotranspiration

Evapotranspiration results in the widespread occurrence of shallow water tables (Jiang et al., 2018). The overall water balance of any particular region is greatly influenced by evapotranspiration since it affects the amount of water available for groundwater recharge. Low GWLs and lower groundwater potential can result from high evapotranspiration rates. Contrary to this, regions with lower evapotranspiration may have higher groundwater potential as more water can infiltrate and replenish the aquifers. A large part of the study area (Fig. 11.9), about 3252.19 km^2 (96.5%), experiences a very high rate of evapotranspiration (more than 544 $kg/m^2/8$ day) and is thus assigned the lowest weightage, whereas the areas that receive very low evapotranspiration (136–202 $kg/m^2/8$ day) were given the highest weightage (Table 11.5).

11.3.8 Topographic wetness index

TWI is a widely used method to assess the influence of topographic factors on hydrological processes that include the seepage of surface water for groundwater recharge (Sørensen, Zinko, & Seibert, 2006). The groundwater aquifer is properly recharged in areas with a high rate of flow accumulation (Makonyo & Msabi, 2021). Thus, the groundwater potential will be high in areas with high TWI values. Fig. 11.10 illustrates the TWI map, where

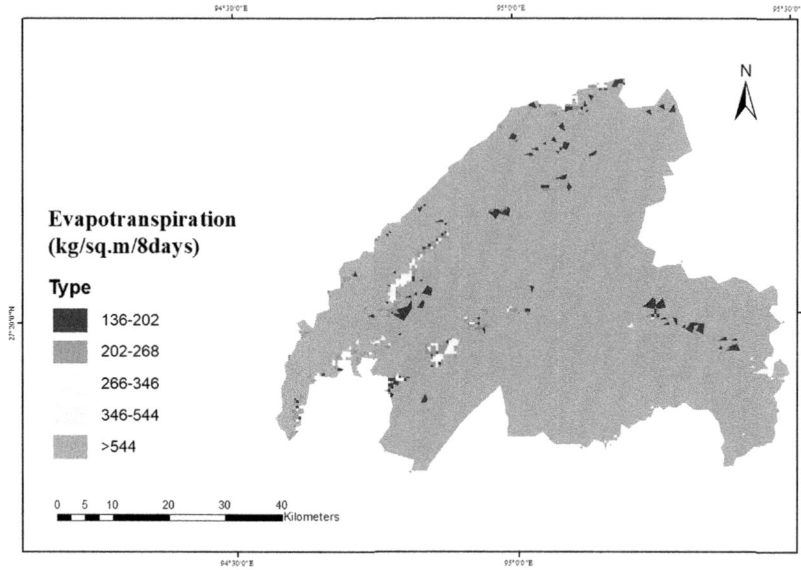

FIGURE 11.9 Evapotranspiration Map of the study area. Map lines delineate study areas and do not necessarily depict accepted national boundaries. *From LAADS DAAC.*

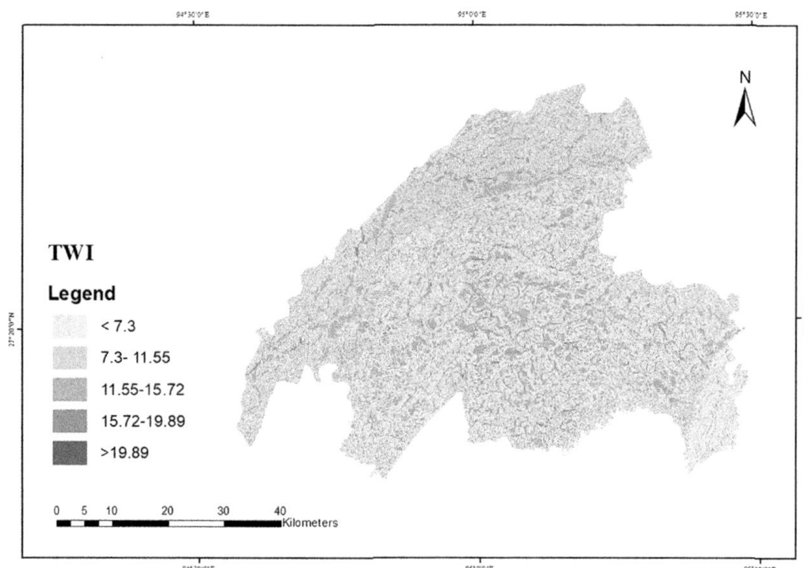

FIGURE 11.10 Topographic Wetness Index Map. Map lines delineate study areas and do not necessarily depict accepted national boundaries.

values are reclassified into five classes: very low (less than 7), low (7.3–11.55), moderate (11.55–15.72), high (15.72–19.89), and very high (more than 19.89). Since a major portion of the study area is dominated by very low (34%) and low (43.32%) TWI values, these areas were assigned lower weightage and the higher weightage was assigned to regions with very high (0.29%) TWI values (Table 11.5).

11.3.9 Soil type

Soils play a very important role in facilitating the movement of groundwater in the unsaturated zone (Machiwal, Jha, and Mal, 2011). The infiltration rates of surface water are greatly determined by the soil profile of an area (Tolche, 2021). The availability of mixed-textured soil types, like fine sand or sandy loam, is advantageous for groundwater recharge (Costache, Hong, & Wang, 2019). The study area basically consists of three different types of soil (Fig. 11.11), namely, clay loam (Ao) covering 46.1 km^2 area (1.3%), sandy clay loam (Af) covering 2075.03 km^2 area (61.6%), and sandy clay loam (Ao) covering 1257.9 km^2 area (37.3%). The main difference between sandy clay loam (Af) and sandy clay loam (Ao) types of soil is that Af soil has a higher clay content, making it more cohesive than Ao soil. Thus, the sandy clay loam (Ao) soil type was given the highest weightage, followed by Af, and the least weightage was given to clay loam (Ao) while assessing their role in water infiltration and groundwater recharge (Table 11.5).

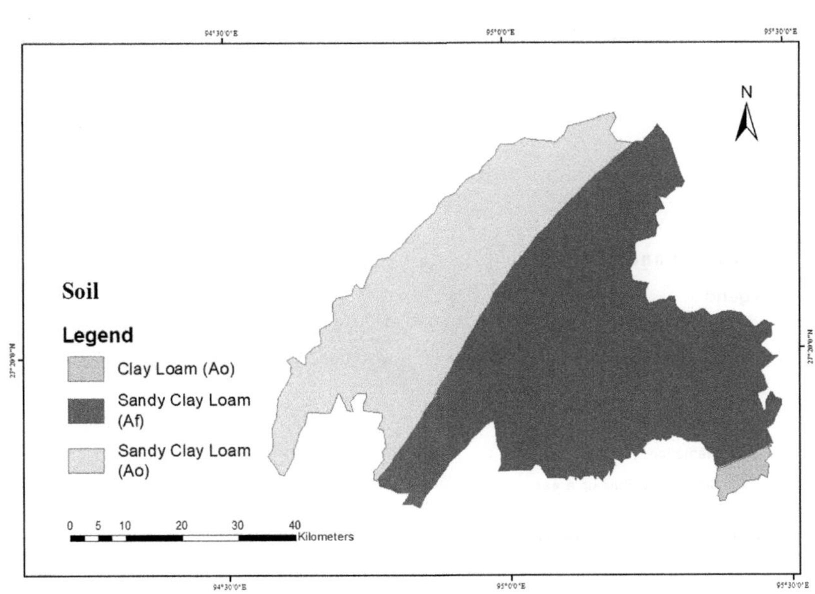

FIGURE 11.11 Soil Type Map. Map lines delineate study areas and do not necessarily depict accepted national boundaries. *From Food and Agriculture Organization (FAO).*

11.3.10 Land use and land cover

The land cover types and land use practices play a substantial role in the hydrologic components of any region, such as surface water flow, water seepage, and evapotranspiration (Leduc, Favreau, & Schroeter, 2001). The LULC of the study area consists of river bodies covering 239.51 km^2 (7.1%), sand deposits covering 657.01 km^2 (19.5%), forest areas covering 238.74 km^2 (7.08%), agricultural land covering 967.31 km^2 (28.7%), plantations covering 731.8 km^2 (21.7%), and settlement and built-up areas covering 545.54 km^2 (16.1%) (Fig. 11.12). Water bodies, sand deposits, and forest areas were given higher weightage as they contribute significantly to groundwater recharge, whereas settlements and built-up areas were given the least weightage as they increase surface runoff and prevent infiltration of surface water (Table 11.5).

11.3.11 Identifying groundwater potential zones

AHP techniques were utilized to generate the weights of 10 different thematic layers, including geomorphology, geology, slope, lineament density, drainage density, rainfall, evapotranspiration, TWI, soil, and LULC. Tables 11.2 and 11.3 show the pairwise comparison matrix and the normalized weight obtained by AHP, respectively. Once the weights were generated, the groundwater potential zone was identified by using Eq. (11.3), which solely depends on the weights of each factor and the ranking of the

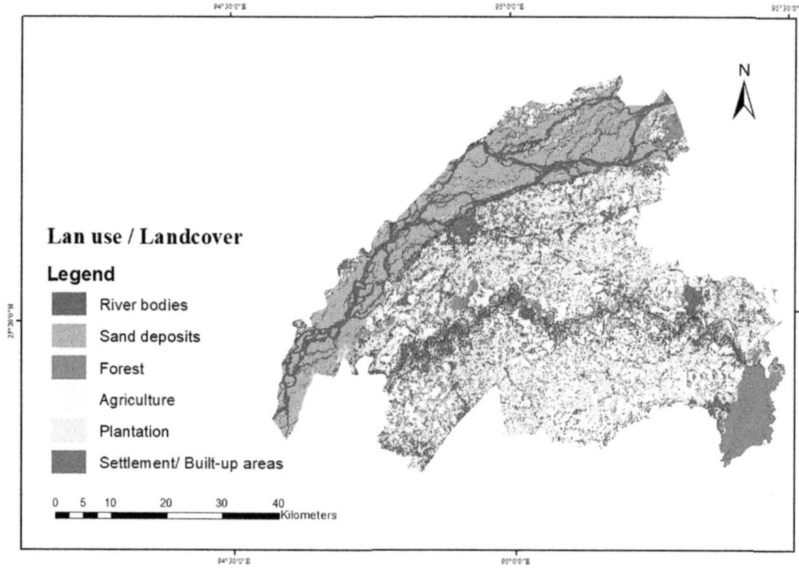

FIGURE 11.12 Land use and Land cover classes. Map lines delineate study areas and do not necessarily depict accepted national boundaries.

sub-classes of the thematic layers depicted in Table 11.5. The resulting groundwater potential index obtained from weighted overlay analysis was classified into four classes: poor covering 16.43 km^2 (0.5%), moderate covering 850.25 km^2 (25.58%), good covering 2456.02 km^2 (73.89%), and very good accounting for 0.83 km^2 (0.02%) groundwater potential areas. In this study, the CR is 0.06, which signifies that the assigned weights and ranks are satisfactory.

11.4 Discussions

The GWPZ map (Fig. 11.13) reveals that areas with "good" groundwater potential, which have the largest share in terms of area coverage, are primarily located in the region associated with alluvial and flood plains dominated by fluvially originated sediments that are beneficial for the percolation of surface water, with a high to very high drainage density stimulating the process of groundwater recharge. The slope angle is categorized as a very low to low slope (0–4 degrees), facilitating the percolation of rainwater as a lower slope angle allows water to remain stagnant for a considerable period of time. Moreover, the region is dominated by agriculture and plantation activities, where irrigation facilities are given less importance, thereby allowing water retention and groundwater recharge. Here, the relatively higher rate of evapotranspiration is compensated by high rainfall (more than 4000 mm annually), which is evident from a moderate to very high TWI

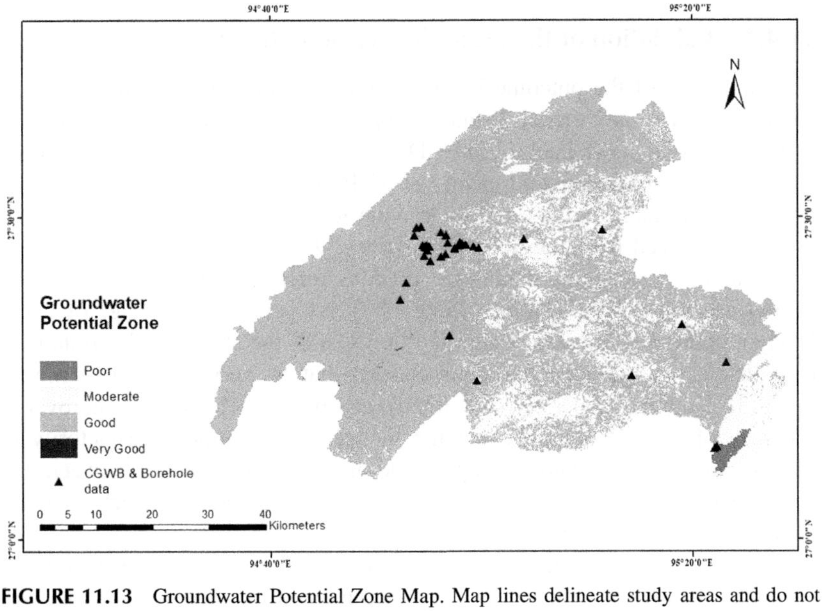

FIGURE 11.13 Groundwater Potential Zone Map. Map lines delineate study areas and do not necessarily depict accepted national boundaries.

value. Furthermore, areas categorized as "moderate" groundwater potential zones are seen in the areas with moderate drainage density, viz., the region drained by the Brahmaputra, Burhi Dihing, and other tributaries and wetlands in the central part, as well as the south-eastern part of the study area. Areas with moderate drainage density indicate comparatively lesser amounts of surface water that can infiltrate and be stored in the subsurface as groundwater. The south-eastern part of the area is characterized by highly dissected hills and valleys, mainly the foothills zone of the Patkai range with a relatively higher slope angle (more than 8 degrees), where most of the surface water drains off with high velocity and water retention is not possible, thus making the region unsuitable for groundwater infiltration and recharge. The small, isolated pocket of poor groundwater potential zone is noticed in the southern part of the study area, mainly the areas of human settlement near Jeypore Reserve Forest, where artificial structures obstruct water infiltration and restrict groundwater recharge. Areas with very good groundwater potential occupy the least area and are located mainly in the north-western part of the study area, mostly covered by river bodies and wetlands.

The assessment of groundwater potential zones in the Dibrugarh district provided useful information into the significance of different thematic layers and their individual classes in the delineation of the areas with high groundwater potential. The results of the study indicated that approximately 99% of the study area exhibited good and moderate groundwater potential zones owing to various geo-hydrological factors affecting the availability of groundwater sources and their gradual recharge.

11.4.1 Validation of the groundwater potential zone

The validation of the outcome is an essential step in evaluating the accuracy and credibility of any model. Without proper validation, models lack scientific significance (Chung & Fabbri, 2003; Das, 2019). For validation of the groundwater potential map prepared using the AHP, the GWL data, including seven observation wells from the Central Groundwater Commission (CGWB) and 36 boreholes derived from the household survey, were utilized. Hence, a total of 43 locations' water-level data were used as reference data to compute the overall accuracy of the GWPZ (Table 11.7). As per the ideal range of water level and preferred zonation suggested by CGWB, the study area's groundwater potential was divided into four classes, that is, very good (0–3 mbgl), good (3–6 mbgl), moderate (6–9 mbgl), and poor (9–12 mbgl). The overall accuracy of 76.74% signifies that the model has performed satisfactorily in groundwater zonation, and a Kappa value of 0.62 (which ranges between −1 and 1, where 1 suggests perfect agreement and −1 suggests perfect disagreement) indicates near-perfect agreement.

Also, to evaluate the fairness of the model, the Receiver Operating Characteristics (ROC) curve, a widely used method for model validation, was

TABLE 11.7 Location, water level, and statement of agreement with groundwater potential zone map.

Sl. no.	Location name	Longitude	Latitude	Source	Water level (mbgl)	Reference class	Map class	Agreement
1	Azarguri gaon	94.9500	27.3167	CGWB	3.21	G	G	Agree
2	Bamunbari	94.9931	27.2461	CGWB	4.51	G	G	Agree
3	Barbaruah	94.8819	27.3986	CGWB	4.25	G	G	Agree
4	Chabua	95.1917	27.4815	CGWB	4.39	G	M	Disagree
5	Dikom	95.0681	27.4667	CGWB	4.69	G	M	Disagree
6	Jaipur naharani	95.2381	27.255	CGWB	2.27	VG	M	Disagree
7	Lapetkata	94.8733	27.3722	CGWB	4.22	G	G	Agree
8	Panoi Path	94.913	27.4541	Borehole	4.37	G	G	Agree
9	Natun Nirmali Gaon	94.9128	27.4547	Borehole	5.26	G	G	Agree
10	Bylane No. 11	94.9483	27.4605	Borehole	6.09	M	G	Partially Agree
11	Duliajan	95.3167	27.3333	Borehole	5.18	G	M	Disagree
12	Joypur	95.3876	27.2751	Borehole	7.62	M	G	Disagree
13	Panoi Path-2	94.9118	27.4542	Borehole	4.24	G	G	Agree
14	Natun Nirmali Gaon-2	94.9138	27.456	Borehole	5.89	G	G	Agree
15	Natun Nirmali Gaon-3	94.915	27.4571	Borehole	5.7	G	G	Agree

(*Continued*)

TABLE 11.7 (Continued)

Sl. no.	Location name	Longitude	Latitude	Source	Water level (mbgl)	Reference class	Map class	Agreement
16	Gorokhabari	94.9088	27.4568	Borehole	5.93	G	G	Agree
17	Namghar Path	94.9197	27.4552	Borehole	4.32	G	G	Agree
18	Natun Nirmali Gaon-4	94.9168	27.4554	Borehole	4.56	G	G	Agree
19	Mancotta (Tepor Gaon)	94.9157	27.4552	Borehole	4.96	G	G	Agree
20	Mancotta (Lakshmi Nagar)	94.9153	27.4482	Borehole	5.13	G	G	Agree
21	Railway Colony	94.9105	27.4405	Borehole	5.62	G	M	Disagree
22	Chiring Gaon	94.921	27.4323	Borehole	5.98	M	G	Agree
23	Taor Gaon	94.9372	27.4392	Borehole	6.08	M	M	Agree
24	Dibrugarh MB and OG Part	94.9447	27.4433	Borehole	6.45	M	M	Agree
25	Boiragimoth Kachari Gaon	94.9582	27.451	Borehole	5.98	G	G	Agree
26	Boiragimoth Kachari Gaon-2	94.9592	27.4528	Borehole	5.32	G	M	Disagree
27	Chaulkhowa Grant Gaon	94.9654	27.456	Borehole	5.33	G	G	Agree
28	Bokel Tea Estate	94.9756	27.4571	Borehole	5.16	G	G	Agree

29	Bokel Tea Estate-2	94.9761	27.4574	Borehole	5.65	G	M	Disagree
30	Lahowal	94.9969	27.4529	Borehole	4.23	G	M	Disagree
31	Bashbari Gaon	94.9882	27.4543	Borehole	5.12	G	G	Agree
32	Hiloidhari Chandoi Gaon	94.9705	27.4596	Borehole	4.87	G	G	Agree
33	Hiloidhari Chandoi Gaon-2	94.967	27.4621	Borehole	4.66	G	G	Agree
34	Chamoguri Bangali Gaon	94.9455	27.4729	Borehole	5.31	G	G	Agree
35	Chamoguri Bangali Gaon-2	94.9455	27.4729	Borehole	5.45	G	G	Agree
36	Podum Nagar-2	94.9369	27.477	Borehole	4.89	G	G	Agree
37	Phool Bagan Tinali	94.9064	27.4857	Borehole	4.66	G	G	Agree
38	Phool Bagan Tinali-2	94.9064	27.4857	Borehole	4.65	G	G	Agree
39	A.T. Road Panchali	94.8997	27.4841	Borehole	3.33	G	G	Agree
40	A.T. Road Panchali-2	94.8997	27.4841	Borehole	3.24	G	G	Agree
41	Amolapatty	94.8955	27.4726	Borehole	3.12	G	G	Agree
42	Dilli Ghat	95.3678	27.1411	Borehole	7.2	M	M	Agree
43	Joypur Colliery L.P. School	95.3715	27.1435	Borehole	8.26	M	M	Agree

VP, Very Poor; P, Poor; M, Moderate; G, Good; VG, Very Good.

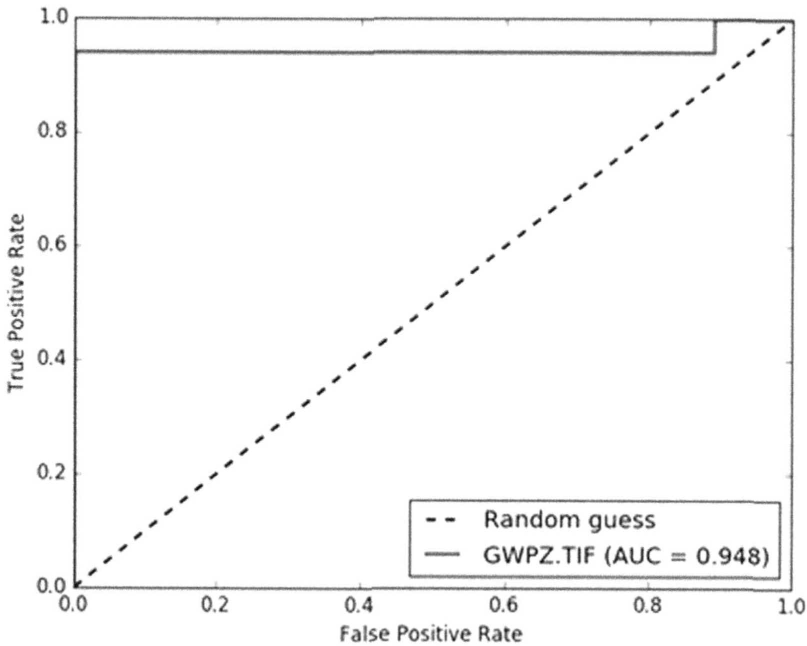

FIGURE 11.14 **ROC-AUC.** Receiver operating characteristics-area under curve.

utilized to establish a relationship between the existing observation well data collected from the CGWB of India, the borehole data, and the groundwater potential index map. The accuracy of the model, as calculated by the area under the curve (AUC), was ascertained to be 0.948 (94.8% global success rate) (Fig. 11.14), which is assumed to be very good. Hence, it can be stated that the approach employed in this research exhibited favorable accuracy in mapping groundwater potential zones, exceeding the threshold of 70%.

11.5 Conclusions

The groundwater potential zone in the Dibrugarh district of Assam was identified with the application of AHP and geospatial techniques. It is evident from the review of the literature that so far no such studies have been done in the region; therefore, a proper assessment of the groundwater resources is a prime requisite. The region experiences a sufficient amount of annual rainfall and has suitable alluvial soil; therefore, most of the people of the region are either dependent on agriculture or plantations as their main economic activity, which requires groundwater. The study revealed that the region has very good to poor groundwater potential, wherein poor covers 16.43 km^2 (0.5%), moderate covers 850.25 km^2 (25.58%), good covers 2456.02 km^2 (73.89%), and very good accounts for 0.83 km^2 (0.02%) groundwater potential areas. The CR of the AHP-enabled comparison matrix was calculated to be 0.06, which implies that the weights and ranks were assigned correctly and are acceptable. The overall accuracy of 76.74% and Kappa coefficient value of 0.62 suggest that the prediction of groundwater potential in the study area is remarkable, and this is further supported by a high ROC-AUC value of 0.948, indicating a strong performance of the predictive model in differentiating positive and negative instances. It was found that a large portion of the region has good groundwater potential, whereas the moderate groundwater potential zones are identified as scattered across the central portion, northeastern, eastern, and southern parts of the study area. The good groundwater potential in the district restricts the need for artificial recharge. Water-exhausting cropping practices, intensive pumping of groundwater, and changes in rainfall intensity have started exerting pressure on the groundwater resources in the region. Since the Dibrugarh district of Assam is among the rapidly developing industrial cities of India, the increasing demand for groundwater resources has become a critical aspect to be focused on. The present research work provides ideas for the application of integrated techniques that can reduce time, are cost-effective, and are very efficient for sustainable groundwater management processes.

References

Abijith, D., Saravanan, S., Singh, L., Jacinth Jennifer, J., Saranya, T., & Parthasarathy, K. S. S. (2020). GIS-based multi-criteria analysis for identification of potential groundwater recharge zones - a case study from Ponnaniyaru watershed, Tamil Nadu, India. *HydroResearch*, *3*, 1–14. Available from https://doi.org/10.1016/j.hydres.2020.020.002.

Achu, A. L., Thomas, J., & Raghunath, R. (2020). 3 a tropical river basin using remote sensing, GIS, and analytical hierarchy process (AHP). *International Journal of Energy and Water Resources.*, *1*. Available from https://doi.org/10.1016/J.GSD.2020.100365.

Aggarwal, M., Saravanan, S., Jennifer, J. J., & Abijith, D. (2019). *Delineation of groundwater potential zones for hard rock region in Karnataka using AHP and GIS Advances in Science, Technology and Innovation* (pp. 315–317). India: Springer Nature. Available from https://www.springer.com/series/15883, 10.1007/978-3-030-01440-7_71.

Ajay Kumar, V., Mondal, N. C., & Ahmed, S. (2020). Identification of groundwater potential zones using RS, GIS and AHP techniques: A case study in a part of deccan volcanic province (DVP), Maharashtra, India. *Journal of the Indian Society of Remote Sensing*, *48*(3), 497–511. Available from https://doi.org/10.1007/s12524-019-01086-3.

Arunbose, S., Srinivas, Y., Rajkumar, S., Nithya, C. N., & Kaliraj, S. (2021). Remote sensing, GIS and AHP techniques based investigation of groundwater potential zones in the Karumeniyar river basin, Tamil Nadu, southern India. *Groundwater for Sustainable Development*, *14*. Available from https://doi.org/10.1016/j.gsd.2021.100586.

Aykut, T. (2021). Determination of groundwater potential zones using Geographical Information Systems (GIS) and Analytic Hierarchy Process (AHP) between Edirne-Kalkansogut (northwestern Turkey). *Groundwater for Sustainable Development*, *12*. Available from https://doi.org/10.1016/j.gsd.2021.100545.

Bai, Z., Liu, Q., & Liu, Y. (2022). Groundwater potential mapping in Hubei Region of China using machine learning, ensemble learning, deep learning and AutoML methods. *Natural Resources Research, 31*(5), 2549–2569. Available from https://doi.org/10.1007/s11053-022-10100-4.

Bora, S. L., Das, J., Bhuyan, K., & Hazarika, P. J. (2022). *Flood susceptibility mapping using GIS and multi-criteria decision analysis in Dibrugarh District of Assam, North-East India* (pp. 65–79). Springer Science and Business Media LLC. Available from 10.1007/978-3-031-15377-8_4.

Chung, C. J. F., & Fabbri, A. G. (2003). Validation of spatial prediction models for landslide hazard mapping. *Natural Hazards., 30*(3), 451–472. Available from https://doi.org/10.1023/B:NHAZ.0000007172.62651.2b.

Costache, R., Hong, H., & Wang, Y. (2019). Identification of torrential valleys using GIS and a novel hybrid integration of artificial intelligence, machine learning and bivariate statistics. *CATENA*, 183. Available from https://doi.org/10.1016/j.catena.2019.104179.

Das, S., & Pardeshi, S. D. (2018). Integration of different influencing factors in GIS to delineate groundwater potential areas using IF and FR techniques: A study of Pravara basin, Maharashtra, India. *Applied Water Science, 8*(7). Available from https://doi.org/10.1007/s13201-018-0848-x, https://www.springer.com/journal/13201.

Das, S., Mukherjee, J., Bhattacharyya, S., Patel, P. P., & Banerjee, A. (2022). Detection of groundwater potential zones using analytical hierarchical process (AHP) for a tropical river basin in the Western Ghats of India. *Environmental Earth Sciences, 81*(16). Available from https://doi.org/10.1007/s12665-022-10543-1, https://link.springer.com/journal/12665.

Das, S. (2019). Comparison among influencing factor, frequency ratio, and analytical hierarchy process techniques for groundwater potential zonation in Vaitarna basin, Maharashtra, India. *Groundwater for Sustainable Development, 8*, 617–629. Available from https://doi.org/10.1016/j.gsd.2019.030.003.

Evans, S. W., Jones, N. L., Williams, G. P., Ames, D. P., & Nelson, E. J. (2020). Groundwater level mapping tool: An open source web application for assessing groundwater sustainability. *Environmental Modelling & Software*, 131. Available from https://doi.org/10.1016/j.envsoft.2020.104782.

Fatti, L. P. (1989). Decision making for leaders: The analytical hierarchy process for decisions in a complex world. *European Journal of Operational Research, 42*(1), 107–109. Available from https://doi.org/10.1016/0377-2217(89)90066-0.

Fetter, C. W. (2014). Applied hydrogeology. *Essex Pearson Education cop.*

Geological Survey of India. (2023). Available from <https://bhukosh.gsi.gov.in/Bhukosh/MapViewer.aspx> 16.07.23.

Goitsemang, T., Das, D. M., Raul, S. K., Subudhi, C. R., & Panigrahi, B. (2020). Assessment of groundwater potential in the Kalahandi District of Odisha (India) using remote sensing, geographic information system and analytical hierarchy process. *Journal of the Indian Society of Remote Sensing, 48*(12), 1739–1753. Available from https://doi.org/10.1007/s12524-020-01188-3.

Jiang, X. W., Wan, L., Wang, X. S., Wang, D., Wang, H., Wang, J. Z., ... Zhao, K. Y. (2018). A multi-method study of regional groundwater circulation in the Ordos Plateau, NW China. *Hydrogeology Journal, 26*(5), 1657–1668. Available from https://doi.org/10.1007/s10040-018-1731-4, http://www.springerlink.com/content/102028/.

Kaur, L., Rishi, M. S., Singh, G., & Nath Thakur, S. (2020). *Groundwater potential assessment of an alluvial aquifer in Yamuna sub-basin (Panipat region) using remote sensing and GIS techniques in conjunction with analytical hierarchy process (AHP) and catastrophe theory (CT, . Ecological Indicators* (110). India: Elsevier B.V. Available from http://www.elsevier.com/locate/ecolind, https://doi.org/10.1016/j.ecolind.2019.105850.

Kolanuvada, S. R., Ponpandian, K. L., & Sankar, S. (2019). Multi-criteria-based approach for optimal siting of artificial recharge structures through hydrological modeling. *Journal of Geosciences*, *12*(6). Available from https://doi.org/10.1007/s12517-019-4351-y, http://www.springer.com/geosciences/journal/12517?cm_mmc = AD-_-enews-_-PSE1892-_-0.

Kom, K. P., Gurugnanam, B., & Sunitha, V. (2022). Delineation of groundwater potential zones using GIS and AHP techniques in Coimbatore district, South India. *International Journal of Energy and Water Resources*. Available from https://doi.org/10.1007/s42108-022-00188-y.

Kumar, P., Chandniha, S. K., Lohani, A. K., Nema, A. K., & Krishan, G. (2018). Trend analysis of groundwater level using non-parametric tests in Alluvial Aquifers of Uttar Pradesh, India. *Current World Environment*, *13*(1), 44–54. Available from https://doi.org/10.12944/cwe.13.1.05.

LAADS DAAC. (2023). Available from https://ladsweb.modaps.eosdis.nasa.gov/> 19.07.23.

Leduc, C., Favreau, G., & Schroeter, P. (2001). Long-term rise in a Sahelian water-table: The Continental Terminal in South-West Niger. *Journal of Hydrology*, *243*(1–2), 43–54. Available from https://doi.org/10.1016/s0022-1694(00)00403-0.

Machiwal, D., Jha, M. K., & Mal, B. C. (2011). Assessment of groundwater potential in a semi-arid Region of India Using Remote Sensing, GIS and MCDM Techniques. *Water Resources Management*, *25*(5), 1359–1386. Available from https://doi.org/10.1007/s11269-010-9749-y.

Magesh, N. S., Chandrasekar, N., & Soundranayagam, J. P. (2012). Delineation of groundwater potential zones in Theni district, Tamil Nadu, using remote sensing, GIS and MIF techniques. *Geoscience Frontiers*, *3*(2), 189–196. Available from https://doi.org/10.1016/j.gsf.2011.100.007, https://www.sciencedirect.com/journal/geoscience-frontiers.

Makonyo, M., & Msabi, M. M. (2021). Identification of groundwater potential recharge zones using GIS-based multi-criteria decision analysis: A case study of semi-arid midlands Manyara fractured aquifer, North-Eastern Tanzania. *Remote Sensing Applications: Society and Environment*, *23*. Available from https://doi.org/10.1016/j.rsase.2021.100544.

Mukherjee, P., Singh, C. K., & Mukherjee, S. (2012). Delineation of groundwater potential zones in Arid Region of India-A Remote Sensing and GIS Approach. *Water Resources Management*, *26*(9), 2643–2672. Available from https://doi.org/10.1007/s11269-012-0038-9.

Muthu, K., & Sudalaimuthu, K. (2021). Integration of remote sensing, GIS, and AHP in demarcating groundwater potential zones in Pattukottai Taluk, Tamilnadu, India. *Arabian Journal of Geosciences*, *14*(17). Available from https://doi.org/10.1007/s12517-021-08110-2, http://www.springer.com/geosciences/journal/12517?cm_mmc = AD-_-enews-_-PSE1892-_-0.

Namous, M., Hssaisoune, M., Pradhan, B., Lee, C.-W., Alamri, A., Elaloui, A., ... Tagma, T. (2021). Spatial prediction of groundwater potentiality in large Semi-Arid and Karstic Mountainous Region using machine learning models. *Water*, *13*(16). Available from https://doi.org/10.3390/w13162273.

Prasad, R. K., Mondal, N. C., Banerjee, P., Nandakumar, M. V., & Singh, V. S. (2008). Deciphering potential groundwater zone in hard rock through the application of GIS. *Environmental Geology*, *55*(3), 467–475. Available from https://doi.org/10.1007/s00254-007-0992-3.

Rahmati, O., Nazari Samani, A., Mahdavi, M., Pourghasemi, H. R., & Zeinivand, H. (2015). Groundwater potential mapping at Kurdistan region of Iran using analytic hierarchy process and GIS. *Journal of Geosciences*, *8*(9), 7059–7071. Available from https://doi.org/10.1007/s12517-014-1668-4, http://www.springer.com/geosciences/journal/12517?cm_mmc = AD-_-enews-_-PSE1892-_-0.

Rao, N. S., Chakradhar, G. K. J., & Srinivas, V. (2001). Identification of groundwater potential zones using remote sensing techniques in and Around Guntur Town, Andhra Pradesh, India. *Journal of the Indian Society of Remote Sensing*, *29*(1–2), 69–78. Available from https://doi.org/10.1007/bf02989916.

Saaty, T.L. (1980). The Analytic Hierarchy Process. McGraw Hill, Juc.

Sahoo, S., Dhar, A., Kar, A., & Ram, P. (2017). Grey analytic hierarchy process applied to effectiveness evaluation for groundwater potential zone delineation. *Geocarto International, 32* (11), 1188–1205. Available from https://doi.org/10.1080/10106049.2016.1195888.

Saravanan, S., Saranya, T., Abijith, D., Jacinth, J. J., & Singh, L. (2021). Delineation of groundwater potential zones for Arkavathi sub-watershed, Karnataka, India using remote sensing and GIS. *Environmental Challenges., 5*. Available from https://doi.org/10.1016/j.envc.2021.100380, https://www.journals.elsevier.com/environmental-challenges.

Shao, Z., Huq, M. E., Cai, B., Altan, O., & Li, Y. (2020). Integrated remote sensing and GIS approach using Fuzzy-AHP to delineate and identify groundwater potential zones in semi-arid Shanxi Province, China. *Environmental Modelling & Software*, 134. Available from https://doi.org/10.1016/j.envsoft.2020.104868.

Singha, S., Pasupuleti, S., Durbha, K. S., Singha, S. S., Singh, R., & Venkatesh, A. S. (2019). An analytical hierarchy process-based geospatial modeling for delineation of potential anthropogenic contamination zones of groundwater from Arang block of Raipur district, Chhattisgarh, Central India. *Environmental Earth Sciences, 78*(24). Available from https://doi.org/10.1007/s12665-019-8724-z, https://link.springer.com/journal/12665.

Sørensen, R., Zinko, U., & Seibert, J. (2006). On the calculation of the topographic wetness index: Evaluation of different methods based on field observations. *Hydrology and Earth System Sciences, 10*(1), 101–112. Available from https://doi.org/10.5194/hess-10-101-2006.

Şener, E., Şener, Ş., & Davraz, A. (2018). Groundwater potential mapping by combining fuzzy-analytic hierarchy process and GIS in Beyşehir Lake Basin, Turkey. *Journal of Geosciences, 11*(8). Available from https://doi.org/10.1007/s12517-018-3510-x, http://www.springer.com/geosciences/journal/12517?cm_mmc = AD-_-enews-_-PSE1892-_-0.

Şener, Ş., Sener, E., & Karagüzel, R. (2011). Solid waste disposal site selection with GIS and AHP methodology: A case study in Senirkent–Uluborlu (Isparta) Basin, Turkey. *Environmental Monitoring and Assessment, 173*(1–4), 533–554. Available from https://doi.org/10.1007/s10661-010-1403-x.

Todd, D. K., & Mays, L. W. (2004). *Groundwater hydrology*. John Wiley & Sons.

Tolche, A. D. (2021). Groundwater potential mapping using geospatial techniques: A case study of Dhungeta-Ramis sub-basin, Ethiopia. *Ethiopia Geology, Ecology, and Landscapes, 5*(1), 65–80. Available from https://doi.org/10.1080/24749508.2020.1728882, https://www.tandfonline.com/toc/tgel20/current.

U.S. Survey (2023). Landsat 8 Operational Land Imager (OLI) Data. U.S. Geological Survey.

Chapter 12

Emission of nitrous oxide in tidal-influenced mangrove ecosystem of Indian Sundarban

Nilanjan Das, Harisankar Ray, Soumyadip Pal and Sudipto Mandal
Ecology and Environmental Modelling Laboratory, Department of Environmental Science, The University of Burdwan, Purba Bardhaman, West Bengal, India

12.1 Introduction

The greenhouse gas nitrous oxide (N_2O) has a long lifetime in the atmosphere, lasting between 118 and 131 years. It has the potential to cause three hundred times greater global warming than CO_2 and is mainly responsible for the depletion of stratospheric ozone (Ming, de Richter, Shen, & Caillol, 2016). In 2000, a comprehensive survey of mangrove ecosystems revealed that the total surface area covered by mangroves in coastal areas, north to south of the equator, was 137,760 km^2 (Ahmed, Nazim, & Khan, 2022; Ruan, Yan, Zhang, Fan, & Yang, 2022). The estimated total annual global N_2O emission from mangroves for the year 2000 is 0.023 Tg N $year^{-1}$ based on this surface area (Mao, Ullah, Gorelick, Hannah, & Krause, 2021). N_2O can be produced by both nitrification and denitrification. Denitrifying bacteria use the $NO_3 \rightarrow NO_2 \rightarrow NO \rightarrow N_2O \rightarrow N_2$ metabolic cycle to progressively reduce NO_3^- to molecular nitrogen (N_2) in anoxic and suboxic settings (Yoon, Song, Phillips, Chang, & Song, 2019). Although N_2O is the source of the inherent high variability in several coastal biogeochemical cycles, the mechanisms and reasons governing coastal N_2O cycling remain poorly understood (Giles, Morley, Baggs, & Daniell, 2012).

A complex web of causes may have an impact on N_2O generation and consumption in mangrove ecosystems (Hershey, Nandan, Vasu, & Tait, 2021). The availability of ammonium (NH_4^+) and NO_x is associated with N_2O cycling (Zhu, Burger, Doane, & Horwath, 2013). Additionally, it is impacted by a number of local and environmental parameters, including biomass, species composition, root structure, and terrain, as well as environmental factors such as redox potential, temperature, groundwater exchange, and

Water Resources Monitoring, Management, and Sustainability. DOI: https://doi.org/10.1016/B978-0-443-23665-5.00012-0

sulfur cycling (Krause et al., 2011; Maher, Sippo, Tait, Holloway, & Santos, 2016). If not, sedimentary dissolved inorganic nitrogen (DIN) level increases, which is transferred into coastal oceans from river catchments, greatly enhancing N_2O fluxes from ecosystems made up of mangroves (Mandal, Ray, & Ghosh, 2012; Mandal, Ray, & Ghosh, 2013; Mao et al., 2021). The atmospheric concentration of N_2O is experiencing an annual increase of around 0.25% as a result of continuous increases in nutrient imports from anthropogenic sources, such as wastewater discharge, excessive fertilizer use, and increased agriculture (Wang et al., 2023; Zhang, Zhang, Xu, & Zhang, 2006). Because organic carbon acts as the electron donor required for denitrification, the C:N ratio affects both denitrification and N_2O emission (Lee et al., 2021). Lee et al. (2021) observed a rise in the C:N ratio and a decrease in N_2O emission under anaerobic circumstances.

Seasonal and temperature variations modify microbial activity, which in turn impacts N_2O production (Butterbach-Bahl & Dannenmann, 2011). Huang, Gao, Guo, Ma, & Liu (2013) conducted a study on the impact of seasonal variations on N_2O generation. The study found that emissions were higher (average 67%) during the summer and lower (average 2%) during the winter. According to a study by Lisa, Song, Tobias, & Hines (2015), both nitrification and denitrification rates rise with seasonal temperature. Higher N_2O generation results from increased faunal activity brought on by the higher seasonal temperature (Kuiper, de Deyn, Thakur, & Groenigen, 2013). Though less research has been done on mangrove habitats than on other coastal chemicals, it is believed that mangrove ecosystems have a significant role in atmospheric NH_2OH, a possible precursor of N_2O (Schreiber, Wunderlin, Udert, & Wells, 2012). According to Inamori et al. (2008), oxygen released by aquatic plants' roots accelerates the nitrification process. N_2O emission may be impacted by the oxygen that macrophytes release into the rhizosphere (Su et al., 2021). Varying plant species have varying effects on N_2O production.

A crucial and brief intermediary in the nitrogen cycle is NH_2OH (Soler-Jofra, Pérez, & Van Loosdrecht, 2021). Because of their high reactivity, they are essential for the removal of nitric oxide (NO) and N_2O from soil (Heil, Vereecken, & Brüggemann, 2016). Ozone depletion in the stratosphere is mostly caused by NH_2OH (Duan, Shen, Jiang, Yan, & Xiong, 2020; Korth, Kock, Arévalo-Martínez, & Bange, 2019). According to Qi et al. (2023), NH_2OH is a crucial nitrification intermediate that can support the synthesis of N_2O *via* both biotic and abiotic processes. There is still much to learn about the concentrations of NH_2OH and the related biogeochemical controls in estuary water and sediments. This is very reactive and has comparatively high rates of self-decomposition that are influenced by the pH, metals, and organic matter in the soil (Das et al., 2020; Wei et al., 2017).

Priya, Ranjan, Sappal, & Ramanathan (2017) provide a review of the dynamics of reactive nitrogen in Indian mangrove ecosystems. Gao et al. (2022)

suggest that investigating the mechanisms of N_2O production and reduction at different elevations in intertidal areas might enhance our comprehension dynamics of N_2O in wetlands along the coast. A study was conducted to investigate the mechanisms behind N_2O emission in the low, medium, and high intertidal regions of the Yangtze Estuary using a "natural-abundance isotope" method. Higher levels of N_2O intake were primarily responsible for lower levels of N_2O emissions in low-tide zones. While bacterial denitrification accounted for the bulk ($>84\%$) of N_2O generation in high tidal flats, both NH_2OH oxidation and fungal denitrification made a substantial contribution. Bacterial denitrification increased, and NH_2OH oxidation/fungal denitrification decreased as elevation decreased, contributing to the generation of N_2O. In research on mangroves in South India, Krithika, Purvaja, & Ramesh (2008) obtained a similar result of N_2O flux of 0.40–0.78 mmol/m^2/h.

Mangrove ecosystems, which range in latitude from 30 N to 38 S, are coastal wetlands characterized by woody plants that grow in a variety of conditions, from warm temperate to arid or wet tropical (Feller et al., 2010). Mangroves can be found in geomorphological settings ranging from marine islands to riverine systems, at tidal heights of less than one meter to more than four meters, and they can withstand oligotrophic or eutrophic environments (Feller et al., 2010). Although the seasonal dynamics of N_2O flux in mangrove ecosystem are not well understood. According to Murray, Erler, Rosentreter, Maher, & Eyre (2018), the forests of mangroves have the potential to either contribute to or absorb N_2O into the atmosphere. According to the literature that is currently available, the DIN in Indian mangroves changes periodically, peaking during the monsoon season (Mandal, Ray, & Ghosh, 2009). This results from lower production combined with increased N leaching and higher terrestrial runoff (Wang & Li, 2019). The examined mangroves in India have a N_2O flux ranging from 4.3 to 208.1 μg/m^2/h, with a greater concentration in areas with agricultural runoff and terrestrial sources. Anthropogenic activities that include aquaculture pond runoff, sewage discharge, agricultural runoff, the construction and growth of ports and cities along the coast, and the vicinity of fertilizer factories all have a significant impact on Indian estuaries and mangroves (Priya et al., 2017).

12.1.1 Study area

The Sundarbans is a tide-dominated tidal wetland that spans an area of 9600 km^2 and is situated near the deltaic plains of the Ganges and Brahmaputra rivers (Das, Mondal, Saha, Ghosh, & Mandal, 2023) (Fig. 12.1). Rivers in the Sundarbans transport and deposit a massive amount of nutrients along their banks, and this amount grows every day (Das, Chakrabortty, Pal, Mondal, & Mandal, 2023; Das, Mondal, & Mandal, 2022b), and here we attempt (1) to estimate N_2O emission and NH_2OH concentration through the enclosed box chamber method and spectrophotometric method, and (2) to

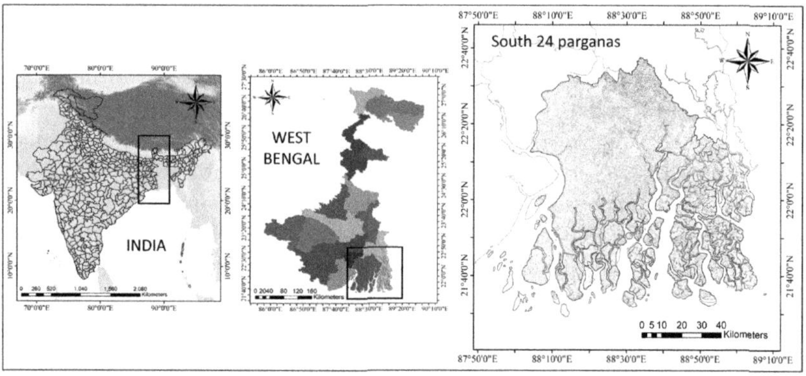

FIGURE 12.1 Geographical location of the mangrove-dominated major harbor sites under South 24 Parganas district of West Bengal in Eastern India. Map lines delineate study areas and do not necessarily depict accepted national boundaries.

identify the key sensitive environmental variable that influences and regulates the NH_2OH concentration and N_2O emissions.

12.2 In situ measurement of NH_2OH and N_2O in mangrove soil

The NH_2OH concentration of soil samples taken from ten different sites is measured using the spectrophotometric method developed by Deepa, Balasubramanian, & Nagaraja (2004).

At every sampling site, five different points were used to gather gas samples. Located within the sediment of the mangrove bed, a specially made, closed, dark container held the gas for two hours without interruption. Gas samples were collected from each site of the investigation. Three samples of gas were collected from every sampling point by a "Gas-tight syringe" (SUPELCO 509426). Then the samples were filled into a "Tedlar bag" (SUPELCO 24633) and transferred into a "Multiple Gas Analyzer" (HORIBA JAPAN, VA-VS, 5000), within a 24-hour period. To determine the N_2O flux of gas samples, measurement data were combined and recorded using the formula (Eq. 12.1) followed by Das, Chakrabortty, et al. (2023):

$$N_2O_{\text{flux}} = \frac{\frac{dG}{dt} \times V \times P}{(R \times T \times A)} \tag{12.1}$$

12.3 Spatial distribution of NH_2OH content and N_2O emission

Under the given conditions, N_2O emission and NH_2OH concentration were mostly observed in the southwest part of the Sundarban mangrove ecosystem, namely, along the Hooghly River and nearby streams. Contaminated wetlands have greater N_2O emission rates and nutrient concentrations than

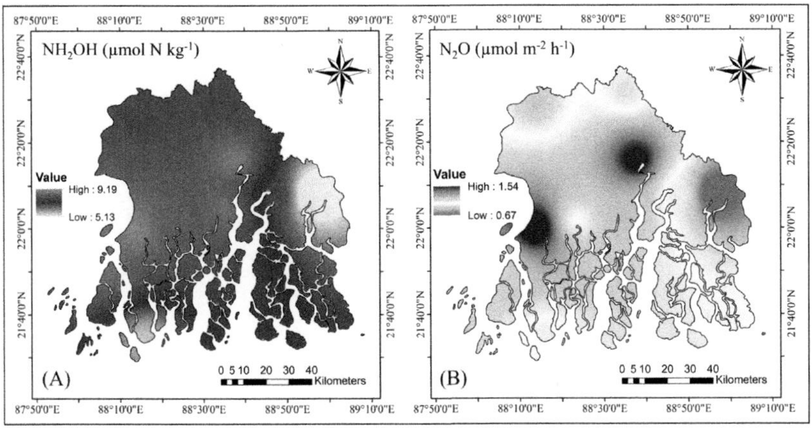

FIGURE 12.2 Spatial variation of estimated NH_2OH concentration (A), and observed N_2O emission (B).

unpolluted wetlands (Konnerup, Betancourt-Portela, Villamil, & Parra, 2014). There are a lot of factories around the Hooghly Estuary, and these factories have been dumping wastewater and household effluents into the estuary either directly or indirectly (Das, Mondal, & Mandal, 2022a; Das et al., 2022b). This waste material is used as a nutrient source by nitrifiers and denitrifiers (Das & Mandal, 2022; Mondal, Reichelt-Brushett, Jonathan, Sujitha, & Sarkar, 2018; Sarkar et al., 2017). Human activities, such as the discharge of residential and industrial wastewater, release high amounts of nutrients into the sediments of rivers, which may use nitrification and denitrification substrates (Das et al., 2022a; Lin et al., 2020). The severely depleted and contaminated administrative areas of South 24 Pargana around the Hooghly River, namely, Patharpratima, Namkhana, Kakadwip, Kulpi, Sagar, Basanti, Kultali, Mathurapur II, Canning I block, Gosaba, and the central-middle regions of Sundarban mangroves, exhibit greater concentrations of NH_2OH and N_2O emissions. The predominantly wooded surrounds of South 24 Parganas showed low NH_2OH concentrations and N_2O emissions, with the exception of the elevated and somewhat distributed blocks (Fig. 12.2).

12.4 Relation between NH_2OH content and N_2O emission

Regression analysis revealed a substantial positive relationship ($R^2 = 0.98$) between NH_2OH content and N_2O emission (Fig. 12.3). The two most significant intermediate products in nitrification are NO_2^- and NH_2OH. Through biotic and abiotic processes, NH_2OH and NO_2^- can contribute to the formation of N_2O, despite their short half-lives and unfavorable conditions (Zhao et al., 2022). The abiotic NO_2^- and NH_2OH reactions were eliminated, contributing to the creation of N_2O. The following describes the abiotic process

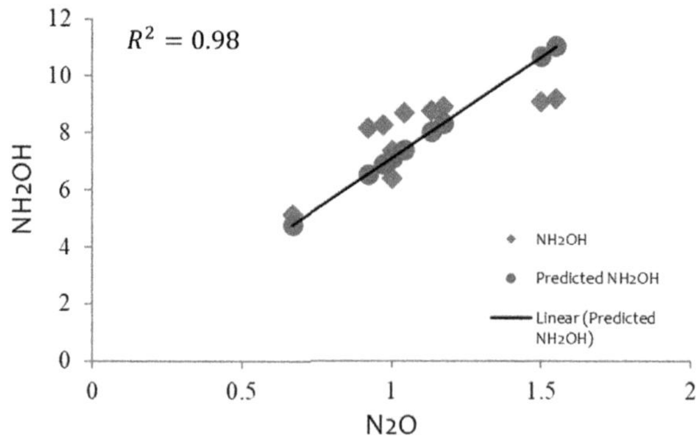

FIGURE 12.3 Linear regression analysis between NH_2OH concentration and N_2O emission.

of NH_2OH combined with iron (Fe) and manganese oxides (MnO) (Xu, Qi, Lin, & Li, 2022):

$$4Fe^{3+} + 2NH_2OH \rightarrow 4Fe^{2+} + N_2O + H_2O + 4H^+$$

$$2Mn^{4+} + 2NH_2OH \rightarrow 2Mn^{2+} + N_2O + H_2O + 4H^+$$

Nitrite (NO_2^-) produced by abiotic processes is typically combined with Fe^{2+} and Mn^{2+} to produce N_2O (Cojean, Moritz, Elizabeth, Bo, & Jakob, 2020):

$$4Fe^{2+} + 2NO^{2-} + 5H_2O \rightarrow 2Fe^{3+} + N_2O + 6H^+$$

$$2Mn^{2+} + 2NO^{2-} + 6H^+ \rightarrow 2Mn^{4+} + N_2O + 3H_2O$$

By using NH_2OH, NH_{4+} is oxidized to produce nitrite (NO_{2+}) and nitrate (NO_{3+}). Because NH_2OH is both a necessary nitrification intermediary and a possible player in the creation of soil N_2O, it seems to be a particularly intriguing compound (Russow, Stange, & Neue, 2009). When ammonia and NH_2OH are oxidized, several nitrifiers, such as *Nitrosomonas europaea* and *Alcaligenes faecalis*, can create N_2O (Wrage, Velthof, van Beusichem, & Oenema, 2001). In denitrifiers like *Pseudomonas sp.*, NH_2OH can also react with NO_2^- to create hybrid N_2O (Spott, Russow, & Stange, 2011). Large-scale N_2O emissions were also seen from sterilized soil following the addition of NH_2OH, suggesting that NH_2OH and other soil elements may also be important players in the chemical processes that lead to the generation of N_2O.

12.5 Factors controlling NH_2OH content and N_2O emission in mangrove soil

Six essential environmental factors linked to NH_2OH and N_2O emissions are shown in Fig. 12.4. Anticipated NH_2OH concentration and N_2O emission

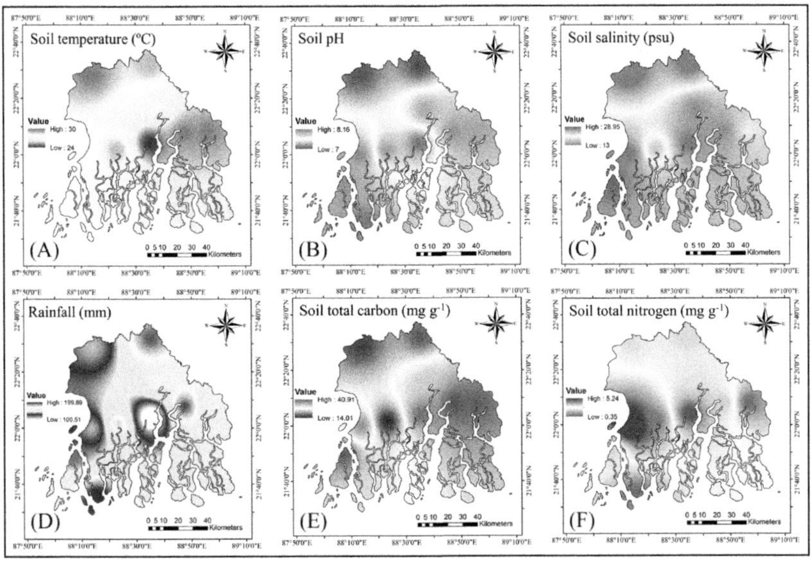

FIGURE 12.4 Spatial distribution of different environmental variables (A) soil temperature (ST), (B) soil pH (SpH), (C) soil salinity (SS), (D) rainfall (R), (E) soil total carbon (TC), and (F) soil total nitrogen (TN).

served as the dependent variables in an ANN model, which employed the ambient variables as covariates (inputs). The anticipated NH_2OH concentration and N_2O emission after the model run are shown to be somewhat significant (Fig. 12.5). In contrast, the results indicate that the most sensitive variables for NH_2OH concentration and N_2O emission in the mangrove environment were rainfall and soil salinity (Fig. 12.5). Principal component analysis revealed that component 2 has 26.36% variance with an eigenvalue of 2.61, while component 1 shows 57.60% variance with an eigenvalue of 6.17. A robust positive association of 0.85 was found between the total nitrogen content and component 1 (eigenvalue 6.17). A robust positive association of 0.85 was found between the total nitrogen content and component 1 (Fig. 12.6). Nonetheless, NH_2OH concentration and N_2O emissions exhibit a substantial positive link, according to the Pearson correlation finding (Fig. 12.7).

Nitrification, denitrification, dissimilatory reduction to NH_3, and even fungal pathways are among the several bacterial processes of N_2O generation that have been reported (Wrage-Mönnig et al., 2018; Yoon et al., 2019). The majority of the reviewed research addresses the processes of bacterial denitrification, which is the heterotrophic reduction of NO_3^- and NO_2^- to N_2O and N_2, and bacterial nitrification, which is the autotrophic oxidation of NH_4^+ to NO_3^- (Stein, 2014). The most abundant products in nitrification are usually NO_3^- and NO_2^-; in denitrification, the most abundant product is

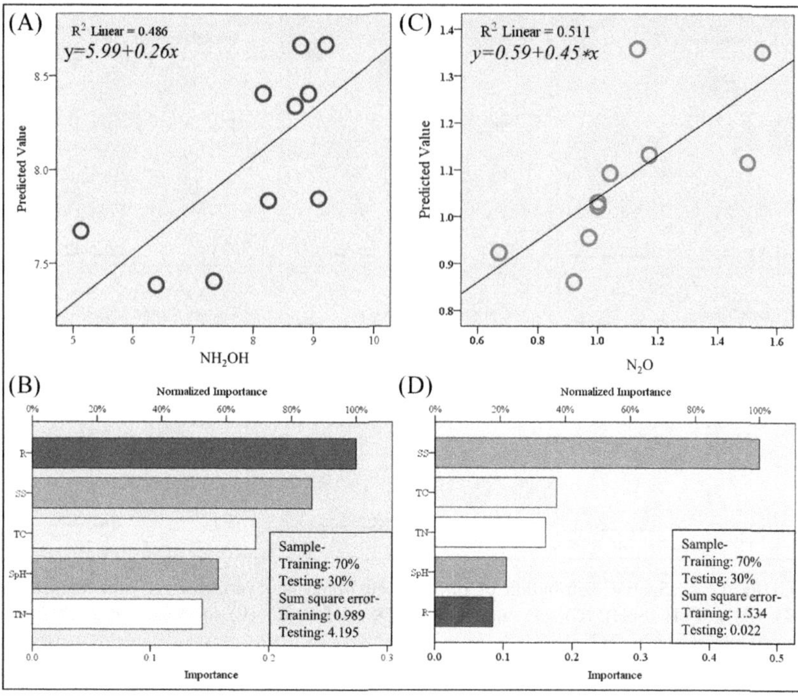

FIGURE 12.5 Regression between estimated and artificial neural network (ANN) model predicted values of NH_2OH concentration and N_2O emission (A and D). Important variable analysis between NH_2OH concentration and N_2O emission (B and C).

N_2 (Heil et al., 2016). When compared to denitrification, nitrification is assumed to yield less N_2O in estuarine conditions. The predominant method for producing N_2O in this system was found to be water-column nitrification (Yang et al., 2022).

The water content and sediment structure have an impact on the rate of N_2O consumption (Petersen, Schjønning, Thomsen, & Christensen, 2008). High water content is well known to prevent gas diffusion through the soil and limit sediment aeration, which prolongs the time that N_2O is present in the sediment and raises the possibility that N_2O will be consumed by denitrifiers (Saggar et al., 2013). Cavazos, Taillefert, Tang, & Glass (2018) state that the "abiotic NH_2OH pathway" is the primary mechanism responsible for producing N_2O. This is largely due to the ability of metal oxides to stimulate nitrification, which produces a significant amount of NH_2OH. This implies that the abiotic mechanisms involved in the formation of N_2O can be influenced by metal oxide concentrations. Since surface-bound Fe^{2+} and mineral classification may catalyze the NO_2^- reduction, most research findings suggest that "iron mineral classification" might influence the degree of "chemo

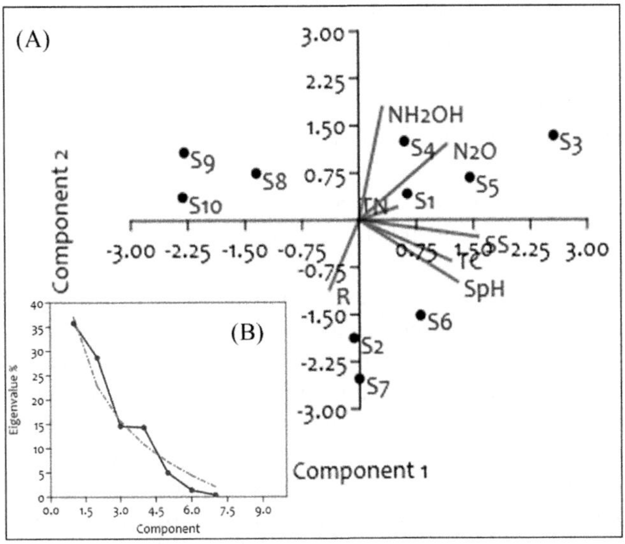

FIGURE 12.6 Biplot (A), and scree plot (B) of principle component analysis (PCA) between all the environmental variables, observed NH_2OH concentration and N_2O emission.

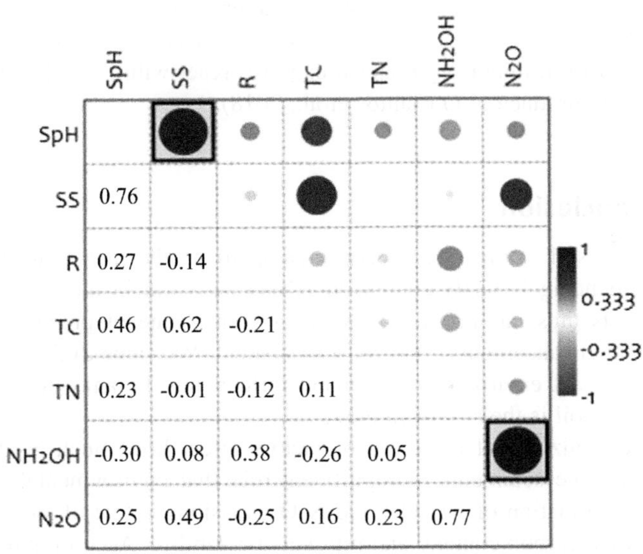

FIGURE 12.7 Correlation analysis between all the environmental variables, NH_2OH concentration, and N_2O emission.

denitrification" and "isotopic fractionation" effects (Xu et al., 2022). Moreover, different forms of iron have different N_2O production processes: Fe(III) can produce N_2O by oxidizing NH_2OH, whereas Fe^{2+} oxidation in conjunction with NO_2^- and NO_3^- can produce N_2O (Tischer et al., 2022). In environments, that are richer in NO_2, this process can produce larger quantities of N_2O due to the high synthesis of N_2O that results from chemo denitrification (NO_2^-). Consequently, NO_2^- may accumulate during nitrification, and the production of N_2O depends on the NO_2^- level (Song et al., 2023).

The fate and reaction activity is affected by the "abiotic NH_2OH" pathway and nitrite, which are reliant on pH (Su, Huang, Zhang, Wei, & Ng, 2023). pH $>$ 7 settings favor NO_2^- destiny and facilitate easy Fe^{2+} reaction (Choi & Oh, 2020). Abiotic processes can cause NH_2OH to produce N_2O in alkaline environments (Liu et al., 2017). The NH_2OH availability that is in charge of the significant N_2O generation may benefit from high nitrification (Zhao et al., 2022). Because of this, NH_2OH and NO_2^- accumulations are encouraged in estuarine wetlands, which increases the production of N_2O (Blackwell, Yamulki, & Bol, 2010). A recent study found that lower organic content and higher Mn^{4+} are advantageous for the oxidation of NH_2OH to N_2O, implying that more N_2O can be produced by the abiotic process of NH_2OH than by NO_2^- (Xu et al., 2022). More N_2O can be produced, and NO_2^- reduction is accelerated under hypoxic conditions. However, hypoxic conditions can also reduce NH_2OH for N_2O generation, suggesting that oxygen saturation plays a significant role in altering both NH_2OH and NO_2^- (Martin-Pozas et al., 2020). However, NO_2^- can also combine with Mn^{2+} and Cu^{2+} to form minor N_2O, and it can also react with phenolic chemicals and humus to produce N_2O (Santos et al., 2018).

12.6 Conclusion

Rainfall and soil salinity have been thought to be the most important elements influencing NH_2OH content and N_2O emission in the soils of mangrove forests thus far. From a lower to a greater rate of precipitation, we observed an enhancement of the NH_2OH content. We came to the conclusion that a significant environmental element influencing the amount of NH_2OH in mangrove soil is the rate of rainfall.

It is recognized that a variety of factors affect how much N_2O is produced in marine sediments. A significant role that environmental elements play in the formation of NH_2OH and N_2O emissions is that of soil temperature, pH, total carbon content, and nitrogen availability. According to statistical analysis, denitrification, organic carbon availability, and pore water nutrient concentrations were some of the key variables affecting the generation of N_2O in the current investigation. Although the relationship was positive, NH_2OH content and N_2O gas production in these sediments had a highly significant association, indicating that NH_2OH had an impact on N_2O gas production.

References

Ahmed, M., Nazim, K., & Khan, M. U. (2022). Mangrove ecosystem with changing climate: A review. *International Journal of Biology and Biotechnology, 19*(1), 77–88.

Blackwell, M. S. A., Yamulki, S., & Bol, R. (2010). Nitrous oxide production and denitrification rates in estuarine intertidal saltmarsh and managed realignment zones. *Estuarine, Coastal and Shelf Science, 87*(4), 591–600. Available from https://doi.org/10.1016/j.ecss.2010.02.017.

Butterbach-Bahl, K., & Dannenmann, M. (2011). Denitrification and associated soil N2O emissions due to agricultural activities in a changing climate. *Current Opinion in Environmental Sustainability, 3*(5), 389–395. Available from https://doi.org/10.1016/j.cosust.2011.08.004.

Cavazos, A. R., Taillefert, M., Tang, Y., & Glass, J. B. (2018). Kinetics of nitrous oxide production from hydroxylamine oxidation by birnessite in seawater. *Marine Chemistry, 202*, 49–57. Available from https://doi.org/10.1016/j.marchem.2018.03.002, https://www.sciencedirect.com/journal/marine-chemistry.

Choi, H., & Oh, S. (2020). Abiotic transient nitrite occurrences from nitrate reduction through goethite-mediated Fe(III)/Fe(II) cycle with labile organic materials and ammonia. *Water, 12* (4), 1202. Available from https://doi.org/10.3390/w12041202.

Cojean, A. N. Y., Moritz, F. L., Elizabeth, K. R., Bo, T., Jakob, Z., et al. (2020). Controls of H_2S, Fe^{2+}, and Mn^{2+} on microbial NO_3^--reducing processes in sediments of an Eutrophic Lake. *Frontiers in Microbiology, 11*, 1–17, 1158. Available from https://doi.org/10.3389/fmicb.2020.01158.

Das, N., & Mandal, S. (2022). Microbial populations regulate greenhouse gas emissions in Sundarban mangrove ecosystem, India. *Acta Ecologica Sinica, 42*(6), 641–652. Available from https://doi.org/10.1016/j.chnaes.2021.07.011, https://www.journals.elsevier.com/acta-ecologica-sinica/.

Das, N., Mondal, A., Gangopadhay, S., Banerjee, R., Batabyal, S., Ghosh, P. B., ... Mandal, S. (2020). Environmental variables regulating organic carbon dynamics of *Sundarban mangrove* ecosystem, India. *International Journal of Chemical and Environmental Sciences, 2* (1), 7–17. Available from https://doi.org/10.15864/ijcaes.2101.

Das, N., Mondal, A., & Mandal, S. (2022a). Polluted waters of the reclaimed islands of Indian *Sundarban promote* more greenhouse gas emissions from mangrove ecosystem. *Stochastic Environmental Research and Risk Assessment, 36*(5), 1277–1288. Available from https://doi.org/10.1007/s00477-021-02135-5, http://link.springer-ny.com/link/service/journals/00477/index.htm.

Das, N., Mondal, A., & Mandal, S. (2022b). Maximum entropy modelling for predicting the potential distribution of methanogens in Sundarban mangrove ecosystem, India. *Theoretical and Applied Ecology* (2), 42–47. Available from https://doi.org/10.25750/1995-4301-2022-2-042-047, http://envjournal.ru/ari/v2022/v2/22205.pdf.

Das, N., Chakrabortty, R., Pal, S. C., Mondal, A., & Mandal, S. (2023). A novel coupled framework for detecting hotspots of methane emission from the vulnerable Indian Sundarban mangrove ecosystem using data-driven models. *Science of The Total Environment, 866*, 161319. Available from https://doi.org/10.1016/j.scitotenv.2022.161319.

Das, N., Mondal, A., Saha, N. C., Ghosh, S., & Mandal, S. (2023). Decadal loss of aboveground biomass and subsequent potential CO_2 emission from the Sundarbans mangrove ecosystem, India. *Acta Ecologica Sinica, 43*(3), 452–458. Available from https://doi.org/10.1016/j.chnaes.2021.11.002, https://www.journals.elsevier.com/acta-ecologica-sinica/.

Deepa, B., Balasubramanian, N., & Nagaraja, K. S. (2004). Spectrophotometric determination of hydroxylamine and its derivatives in pharmaceuticals. *Chemical and Pharmaceutical*

Bulletin, 52(12), 1473–1475. Available from https://doi.org/10.1248/cpb.52.1473India, http://cpb.pharm.or.jp/cpb/200412/c12_1473.pdf.

Duan, P., Shen, H., Jiang, X., Yan, X., & Xiong, Z. (2020). The contributions of hydroxylamine and nitrite to NO and N_2O production in alkaline and acidic vegetable soils. *Journal of Soils and Sediments, 20*(7), 2903–2911. Available from https://doi.org/10.1007/s11368-020-02645-9, http://www.springerlink.com/content/1439-0108.

Feller, I. C., Lovelock, C. E., Berger, U., McKee, K. L., Joye, S. B., & Ball, M. C. (2010). Biocomplexity in mangrove ecosystems. *Annual Review of Marine Science, 2*(1), 395–417. Available from https://doi.org/10.1146/annurev.marine.010908.163809.

Gao, D., Hou, L., Liu, M., Zheng, Y., Yin, G., & Niu, Y. (2022). N_2O emission dynamics along an intertidal elevation gradient in a subtropical estuary: Importance of N_2O consumption. *Environmental Research, 205*, 112432. Available from https://doi.org/10.1016/j.envres.2021.112432.

Giles, M., Morley, N., Baggs, E. M., & Daniell, T. J. (2012). Soil nitrate reducing processes - Drivers, mechanisms for spatial variation, and significance for nitrous oxide production. *Frontiers in Microbiology, 3*. Available from https://doi.org/10.3389/fmicb.2012.00407, http://www.frontiersin.org/Journal/DownloadFile.ashx?pdf = 1&FileId = 18455&articleId = 30024&Version = 1&ContentTypeId = 21&FileName = fmicb-03-00407.pdf.

Heil, J., Vereecken, H., & Brüggemann, N. (2016). A review of chemical reactions of nitrification intermediates and their role in nitrogen cycling and nitrogen trace gas formation in soil. *European Journal of Soil Science, 67*(1), 23–39. Available from https://doi.org/10.1111/ejss.12306, http://www.blacksci.co.uk/~cgilib/jnlpage.bin?Journal = ejss &File = ejss&Page = aims.

Hershey, N. R., Nandan, S. B., Vasu, K. N., & Tait, D. R. (2021). Anthropogenic nutrient loads and season variability drive high atmospheric N_2O fluxes in a fragmented mangrove system. *Scientific Reports, 11*(1). Available from https://doi.org/10.1038/s41598-021-85847-6, http://www.nature.com/srep/index.html.

Huang, L., Gao, X., Guo, J., Ma, X., & Liu, M. (2013). A review on the mechanism and affecting factors of nitrous oxide emission in constructed wetlands. *Environmental Earth Sciences, 68*(8), 2171–2180. Available from https://doi.org/10.1007/s12665-012-1900-z, https://link.springer.com/journal/12665.

Inamori, R., Wang, Y., Yamamoto, T., Zhang, J., Kong, H., Xu, K., … Inamori, Y. (2008). Seasonal effect on N_2O formation in nitrification in constructed wetlands. *Chemosphere, 73*(7), 1071–1077. Available from https://doi.org/10.1016/j.chemosphere.2008.07.064.

Konnerup, D., Betancourt-Portela, J. M., Villamil, C., & Parra, J. P. (2014). Nitrous oxide and methane emissions from the restored mangrove ecosystem of the Ciénaga Grande de Santa Marta, Colombia. *Estuarine, Coastal and Shelf Science, 140*, 43–51. Available from https://doi.org/10.1016/j.ecss.2014.01.006.

Korth, F., Kock, A., Arévalo-Martínez, D. L., & Bange, H. W. (2019). Hydroxylamine as a potential indicator of nitrification in the Open Ocean. *Geophysical Research Letters, 46*(4), 2158–2166. Available from https://doi.org/10.1029/2018GL080466, http://agupubs.onlinelibrary.wiley.com/hub/journal/10.1002/(ISSN)1944-8007/.

Krause, S., Hannah, D. M., Fleckenstein, J. H., Heppell, C. M., Kaeser, D., Pickup, R., … Wood, P. J. (2011). Inter-disciplinary perspectives on processes in the hyporheic zone. *Ecohydrology, 4*(4), 481–499. Available from https://doi.org/10.1002/eco.176.

Krithika, K., Purvaja, R., & Ramesh, R. (2008). Fluxes of methane and nitrous oxide from an Indian mangrove. *Current Science, 94*(2), 218–224. Available from http://www.ias.ac.in/currsci/jan252008/218.pdf.

Kuiper, I., de Deyn, G. B., Thakur, M. P., & Groenigen, J. W. V. (2013). Soil invertebrate fauna affect N_2O emissions from soil. *Global Change Biology*, *19*(9), 2814–2825. Available from https://doi.org/10.1111/gcb.12232.

Lee, S. I., Park, H. J., Jeong, Y. J., Seo, B. S., Kwak, J. H., Yang, H. I., ... Choi, W. J. (2021). Biochar-induced reduction of N_2O emission from East Asian soils under aerobic conditions: Review and data analysis. *Environmental Pollution, 291*. Available from https://doi.org/10.1016/j.envpol.2021.118154, https://www.journals.elsevier.com/environmental-pollution.

Lin, J., Chen, N., Yuan, X., Tian, Q., Hu, A., & Zheng, Y. (2020). Impacts of human disturbance on the biogeochemical nitrogen cycle in a subtropical river system revealed by nitrifier and denitrifier genes. *Science of The Total Environment, 746*, 141139. Available from https://doi.org/10.1016/j.scitotenv.2020.141139.

Lisa, J. A., Song, B., Tobias, C. R., & Hines, D. E. (2015). Genetic and biogeochemical investigation of sedimentary nitrogen cycling communities responding to tidal and seasonal dynamics in Cape Fear River Estuary. *Estuarine, Coastal and Shelf Science, 167*, A313–A323. Available from https://doi.org/10.1016/j.ecss.2015.09.008, http://www.elsevier.com/inca/publications/store/6/2/2/8/2/3/index.htt.

Liu, S., Han, P., Hink, L., Prosser, J. I., Wagner, M., & Brüggemann, N. (2017). Abiotic conversion of extracellular NH_2OH contributes to N_2O emission during ammonia oxidation. *Environmental Science and Technology, 51*(22), 13122–13132. Available from https://doi.org/10.1021/acs.est.7b02360, http://pubs.acs.org/journal/esthag.

Maher, D. T., Sippo, J. Z., Tait, D. R., Holloway, C., & Santos, I. R. (2016). Pristine mangrove creek waters are a sink of nitrous oxide. *Scientific Reports, 6*. Available from https://doi.org/10.1038/srep25701, http://www.nature.com/srep/index.html.

Mandal, S., Ray, S., & Ghosh, P. B. (2012). Comparative study of mangrove litter nitrogen cycling to the adjacent estuary through modelling in pristine and reclaimed islands of Sundarban mangrove ecosystem, India. *Procedia Environmental Sciences, 13*, 340–362. Available from https://doi.org/10.1016/j.proenv.2012.01.033.

Mandal, S., Ray, S., & Ghosh, P. B. (2009). Modelling of the contribution of dissolved inorganic nitrogen (DIN) from litterfall of adjacent mangrove forest to Hooghly-Matla estuary, India. *Ecological Modelling, 220*(21), 2988–3000. Available from https://doi.org/10.1016/j.ecolmodel.2009.01.028.

Mandal, S., Ray, S., & Ghosh, P. B. (2013). Impact of mangrove litterfall on nitrogen dynamics of virgin and reclaimed islands of Sundarban mangrove ecosystem, India. *Ecological Modelling, 252*(1), 153–166. Available from https://doi.org/10.1016/j.ecolmodel.2012.06.038, http://www.elsevier.com/inca/publications/store/5/0/3/3/0/6.

Mao, F., Ullah, S., Gorelick, S. M., Hannah, D. M., & Krause, S. (2021). Increasing nutrient inputs risk a surge of nitrous oxide emissions from global mangrove ecosystems. *One Earth, 4*(5), 742–748. Available from https://doi.org/10.1016/j.oneear.2021.04.007, http://www.cell.com/one-earth.

Martin-Pozas, T., Sanchez-Moral, S., Cuezva, S., Jurado, V., Saiz-Jimenez, C., Perez-Lopez, R., ... Fernandez-Cortes, A. (2020). Biologically mediated release of endogenous N_2O and NO_2 gases in a hydrothermal, hypoxic subterranean environment. *Science of The Total Environment, 747*, 141218. Available from https://doi.org/10.1016/j.scitotenv.2020.141218.

Ming, T., de Richter, R., Shen, S., & Caillol, S. (2016). Fighting global warming by greenhouse gas removal: Destroying atmospheric nitrous oxide thanks to synergies between two breakthrough technologies. *Environmental Science and Pollution Research, 23*(7), 6119–6138. Available from https://doi.org/10.1007/s11356-016-6103-9, http://www.springerlink.com/content/0944-1344.

Mondal, P., Reichelt-Brushett, A. J., Jonathan, M. P., Sujitha, S. B., & Sarkar, S. K. (2018). Pollution evaluation of total and acid-leachable trace elements in surface sediments of Hooghly River Estuary and Sundarban Mangrove Wetland (India). *Environmental Science and Pollution Research, 25*(6), 5681–5699. Available from https://doi.org/10.1007/s11356-017-0915-0, http://www.springerlink.com/content/0944-1344.

Murray, R., Erler, D., Rosentreter, J., Maher, D., & Eyre, B. (2018). A seasonal source and sink of nitrous oxide in mangroves: Insights from concentration, isotope, and isotopomer measurements. *Geochimica et Cosmochimica Acta, 238*, 169–192. Available from https://doi.org/10.1016/j.gca.2018.07.003, http://www.journals.elsevier.com/geochimica-et-cosmochimica-acta/.

Petersen, S. O., Schjønning, P., Thomsen, I. K., & Christensen, B. T. (2008). Nitrous oxide evolution from structurally intact soil as influenced by tillage and soil water content. *Soil Biology and Biochemistry, 40*(4), 967–977. Available from https://doi.org/10.1016/j.soilbio.2007.11.017.

Priya, N., Ranjan, P., Sappal, S. M., & Ramanathan, A. L. (2017). *The Indian Nitrogen Assessment: Sources of reactive nitrogen, environmental and climate effects, management options, and policies. Reactive nitrogen dynamics in the mangroves of India* (pp. 335–359). India: Elsevier. Available from http://www.sciencedirect.com/science/book/9780128118368, https://doi.org/10.1016/B978-0-12-811836-8.00022-7.

Qi, M., Qian, W., Sardans, J., Li, Y., Peñuelas, J., Liu, M., … Li, X. (2023). Spatial and seasonal variability of hydroxylamine concentrations in a human-impacted estuary off Southeast China. *Journal of Geophysical Research: Biogeosciences, 128*(3). Available from https://doi.org/10.1029/2022JG007208, https://agupubs.onlinelibrary.wiley.com/loi/21698961.

Ruan, L., Yan, M., Zhang, L., Fan, X. S., & Yang, H. (2022). Spatial-temporal NDVI pattern of global mangroves: A growing trend during 2000–2018. *Science of The Total Environment, 844*, 157075. Available from https://doi.org/10.1016/j.scitotenv.2022.157075.

Russow, R., Stange, C. F., & Neue, H. U. (2009). Role of nitrite and nitric oxide in the processes of nitrification and denitrification in soil: Results from 15N tracer experiments. *Soil Biology and Biochemistry, 41*(4), 785–795. Available from https://doi.org/10.1016/j.soilbio.2009.01.017.

Saggar, S., Jha, N., Deslippe, J., Bolan, N. S., Luo, J., Giltrap, D. L., … Tillman, R. W. (2013). Denitrification and N_2O: N_2 production in temperate grasslands: Processes, measurements, modelling and mitigating negative impacts. *Science of the Total Environment, 465*, 173–195. Available from https://doi.org/10.1016/j.scitotenv.2012.11.050.

Santos, C., Goufo, P., Fonseca, J., Pereira, J. L. S., Ferreira, L., Coutinho, J., & Trindade, H. (2018). Effect of lignocellulosic and phenolic compounds on ammonia, nitric oxide and greenhouse gas emissions during composting. *Journal of Cleaner Production, 171*, 548–556. Available from https://doi.org/10.1016/j.jclepro.2017.10.050.

Sarkar, S. K., Mondal, P., Biswas, J. K., Kwon, E. E., Ok, Y. S., & Rinklebe, J. (2017). Trace elements in surface sediments of the Hooghly (Ganges) estuary: Distribution and contamination risk assessment. *Environmental Geochemistry and Health, 39*(6), 1245–1258. Available from https://doi.org/10.1007/s10653-017-9952-3, http://www.wkap.nl/journalhome.htm/0269-4042.

Schreiber, F., Wunderlin, P., Udert, K. M., & Wells, G. F. (2012). Nitric oxide and nitrous oxide turnover in natural and engineered microbial communities: Biological pathways, chemical reactions, and novel technologies. *Frontiers in Microbiology, 3*. Available from https://doi.org/10.3389/fmicb.2012.00372, http://www.frontiersin.org/Journal/DownloadFile.ashxpdf = 1&FileId = 12088&articleId = 28255&Version = 1&ContentTypeId = 21&FileName = fmicb-03-00372.pdf.

Soler-Jofra, A., Pérez, J., & Van Loosdrecht, M. C. M. (2021). Hydroxylamine and the nitrogen cycle: A review. *Water Research, 190*, 116723. Available from https://doi.org/10.1016/j.watres.2020.116723.

Song, Y., Wu, D., Ju, X., Dörsch, P., Wang, M., Wang, R., ... Yu, Y. (2023). Nitrite stimulates HONO and NOx but not N_2O emissions in Chinese agricultural soils during nitrification. *Science of The Total Environment, 902*, 166451. Available from https://doi.org/10.1016/j.scitotenv.2023.166451.

Spott, O., Russow, R., & Stange, C. F. (2011). Formation of hybrid N_2O and hybrid N_2 due to codenitrification: First review of a barely considered process of microbially mediated N-nitrosation. *Soil Biology and Biochemistry, 43*(10), 1995–2011. Available from https://doi.org/10.1016/j.soilbio.2011.06.014.

Stein, L. Y. (2014). *Nitrification and nitrifier denitrification. Heterotrophic* (pp. 95–114). American Society for Microbiology. Available from https://doi.org/10.1128/9781555817145.ch5.

Su, R., Huang, R., Zeng, J., Zhao, D., He, R., Yu, Z., & Wu, Q. L. (2021). Rhizosphere-associated nosZII microbial community of Phragmites australis and its influence on nitrous oxide emissions in two different regions. *Journal of Soils and Sediments, 21*(10), 3326–3341. Available from https://doi.org/10.1007/s11368-021-02967-2, http://www.springerlink.com/content/1439-0108.

Su, Q., Huang, S., Zhang, H., Wei, Z., & Ng, H. Y. (2023). Abiotic transformations of sulfamethoxazole by hydroxylamine, nitrite and nitric oxide during wastewater treatment: Kinetics, mechanisms and pH effects. *Journal of Hazardous Materials, 444*, 130328. Available from https://doi.org/10.1016/j.jhazmat.2022.130328.

Tischer, J., Zopfi, J., Frey, C., Magyar, P. M., Brand, A., Oswald, K., ... Lehmann, M. F. (2022). Isotopic signatures of biotic and abiotic N_2O production and consumption in the water column of meromictic, ferruginous Lake La Cruz (Spain). *Limnology and Oceanography, 67*(8), 1760–1775. Available from https://doi.org/10.1002/lno.12165, http://aslopubs.onlinelibrary.wiley.com/hub/journal/10.1002/(ISSN)1939-5590/.

Wang, Z. H., & Li, S. X. (2019). Nitrate N loss by leaching and surface runoff in agricultural land: A global issue (a review). *Advances in Agronomy, 156*. Available from https://doi.org/10.1016/bs.agron.2019.01.007, http://www.elsevier.com/wps/find/bookdescription.cws_home/675816/description#description.

Wang, J., Vilmin, L., Mogollón, J. M., Beusen, A. H. W., van Hoek, W. J., Liu, X., ... Bouwman, A. F. (2023). Inland waters increasingly produce and emit nitrous oxide. *Environmental Science and Technology, 57*(36), 13506–13519. Available from https://doi.org/10.1021/acs.est.3c04230, http://pubs.acs.org/journal/esthag.

Wei, J., Amelung, W., Lehndorff, E., Schloter, M., Vereecken, H., & Brüggemann, N. (2017). N_2O and NOx emissions by reactions of nitrite with soil organic matter of a Norway spruce forest. *Biogeochemistry, 132*(3), 325–342. Available from https://doi.org/10.1007/s10533-017-0306-0, http://www.wkap.nl/journalhome.htm/0168-2563.

Wrage-Mönnig, N., Horn, M. A., Well, R., Müller, C., Velthof, G., & Oenema, O. (2018). The role of nitrifier denitrification in the production of nitrous oxide revisited. *Soil Biology and Biochemistry, 123*, A3–A16. Available from https://doi.org/10.1016/j.soilbio.2018.03.020, http://www.elsevier.com/inca/publications/store/3/3/2.

Wrage, N., Velthof, G. L., van Beusichem, M. L., & Oenema, O. (2001). Role of nitrifier denitrification in the production of nitrous oxide. *Soil Biology and Biochemistry, 33*(12–13), 1723–1732. Available from https://doi.org/10.1016/s0038-0717(01)00096-7.

Xu, C., Qi, M., Lin, W., & Li, X. (2022). nitrous oxide from abiotic processes of hydroxylamine and nitrite in estuarine and coastal ecosystems: A review. *Journal of Marine Science and Engineering, 10*(5), 623. Available from https://doi.org/10.3390/jmse10050623.

Yang, J. Y. T., Hsu, T. C., Tan, E., Lee, K., Krom, M. D., Kang, S., ... Kao, S. J. (2022). Sedimentary processes dominate nitrous oxide production and emission in the hypoxic zone off the Changjiang River estuary. *Science of the Total Environment, 827*. Available from https://doi.org/10.1016/j.scitotenv.2022.154042, http://www.elsevier.com/locate/scitotenv.

Yoon, S., Song, B., Phillips, R. L., Chang, J., & Song, M. J. (2019). Ecological and physiological implications of nitrogen oxide reduction pathways on greenhouse gas emissions in agroecosystems. *FEMS Microbiology Ecology, 95*(6). Available from https://doi.org/10.1093/femsec/fiz066, http://femsec.oxfordjournals.org/.

Zhang, G., Zhang, J., Xu, J., & Zhang, F. (2006). Distributions, sources and atmospheric fluxes of nitrous oxide in Jiaozhou Bay. *Estuarine, Coastal and Shelf Science, 68*(3–4), 557–566. Available from https://doi.org/10.1016/j.ecss.2006.03.007.

Zhao, J., Zhao, J., Yang, W., Hu, B., Huang, T., Xie, S., ... Hou, W. (2022). Mechanisms of NO and N_2O production by enriched nitrifying sludge in a sequencing batch reactor: Effects of hydroxylamine. *Journal of Environmental Management, 316*, 115237. Available from https://doi.org/10.1016/j.jenvman.2022.115237.

Zhu, X., Burger, M., Doane, T. A., & Horwath, W. R. (2013). Ammonia oxidation pathways and nitrifier denitrification are significant sources of N_2O and NO under low oxygen availability. *Proceedings of the National Academy of Sciences, 110*(16), 6328–6333. Available from https://doi.org/10.1073/pnas.1219993110.

Section II

Water resources management

Chapter 13

Integrating geospatial approach with citizen science to combat marine plastic pollution

Wiame Waheed Emam and Kareem Mohamed Soliman
Department of Zoology, Faculty of Science, Ain Shams University, Cairo, Egypt

13.1 Introduction

The Earth's oceans, spanning more than 70% of the planet's surface, play a crucial role in sustaining life and regulating global climate and weather systems. Amid these crucial functions, the oceans face various challenges, with marine plastic pollution emerging as a particularly pervasive threat (Yu & Singh, 2023). Annually, an alarming 11 million tons of litter, primarily consisting of plastic (60%–95%), infiltrates the oceans, wreaking havoc on marine ecosystems and wildlife (UNEP, 2021; 2023).

Despite growing awareness, marine plastic pollution persists, exacerbated by the staggering global plastic production that reached 400 million tons in 2022 (Plastics Europe, 2023). If we continue with the prevailing business-as-usual approach, this figure will double by 2040. The need for a unified and comprehensive strategy to reduce plastic usage and prevent waste from entering waterways is undeniable.

Marine plastic pollution represents a complex global challenge that demands systemic solutions integrating social, economic, and environmental dimensions. In this pursuit, the DPSIR framework, denoting "Driver-Pressure-State-Impact-Response," emerges as a valuable analytical tool (Fet, 2023; Troian, Gomes, Tiecher, Berbel, & Gutiérrez-Martín, 2021; Tscherning, Helming, Krippner, Sieber, & Paloma, 2012). This framework unravels the socio-economic drivers behind plastic pollution, unveiling cascading pressures, conditions, and impacts on marine life and habitats. Understanding these interconnected factors can help identify effective policies, technologies, and stakeholder responses.

Innovative approaches are essential to tackling this challenge effectively. The integration of geospatial technology and citizen science offers promising avenues to address both upstream and downstream aspects of marine plastic

Water Resources Monitoring, Management, and Sustainability. DOI: https://doi.org/10.1016/B978-0-443-23665-5.00013-2

pollution. Geospatial technology, including geographic information systems (GIS), global positioning systems (GPS), and remote sensing, enables the collection, visualization, and analysis of extensive geospatial data (Emam & Soliman, 2020; 2022; Emam, Khalil, Nassif, & Soliman, 2019; Emam, El-Kafrawy, & Soliman, 2021; Mohamed et al., 2022; Mohamed et al., 2023). For instance, by analyzing and visualizing geospatial data on marine plastic pollution, researchers comprehensively understand plastic debris sources, transport pathways, and hotspots in marine environments (Coleby & Grist, 2019).

Additionally, citizen science, which involves the active engagement of the public in scientific research, data collection, and decision-making processes, has proven instrumental in various environmental studies (McKinley et al., 2017). Citizen science initiatives provide valuable insights into local-scale dynamics and enhance spatial coverage and temporal resolution. For example, through the involvement of citizens in data collection efforts, we can gather information on plastic pollution in coastal areas, rivers, and even remote regions that may be challenging to access by traditional means.

While the integration of geospatial technology and citizen science holds great promise, it is crucial to acknowledge potential counterarguments and limitations. For instance, there may be concerns about data accuracy and reliability when involving non-experts in data collection. Additionally, the accessibility and affordability of geospatial technology may pose challenges in certain regions or for specific communities. Recognizing these limitations, efforts must be made to address quality control, data validation, and inclusivity to ensure that the benefits of these approaches are maximized while minimizing potential drawbacks.

Accordingly, the objectives of this chapter are threefold: (1) to delve into each component of the Driver-Pressure-State-Impact-Response (DPSIR) framework to understand the issue of marine plastic pollution; (2) to explore the contributions of geospatial tools and citizen science in facilitating evidence-based responses; and (3) to present, for the first time, a comprehensive method protocol that integrates and leverages the combined capabilities of geospatial technology and citizen science. This protocol will encompass project planning, implementation, monitoring, and evaluation, aiming to improve data collection precision, encourage stakeholder collaboration, and facilitate the development of sustainable solutions to combat the pervasive issue of marine plastic pollution.

13.2 Marine plastic pollution: driver-pressure-state-impact-response analysis

This section explains how each component of the DPSIR framework corresponds to different stages in the cause-and-effect chain of marine plastic pollution:

13.2.1 Driving forces

Marine plastic pollution is driven by several interrelated socio-economic drivers (Fig. 13.1). The main socio-economic drivers that fuel marine plastic pollution include unsustainable patterns of plastic production, consumption, and waste management.

13.2.1.1 Plastic production and consumption

Plastic Production Trends: The plastic industry has witnessed exponential growth, with production quantities increasing from 1.5 million metric tons in 1950 to an astounding 400.3 million metric tons in 2022 (Fig. 13.2). Throughout this timeframe, the cumulative worldwide plastic production has surpassed 10.2 billion metric tons, with around 54% of this quantity being generated within the last two decades. While plastics offer many benefits to modern society through applications in construction, packaging, automotive, electronics, and healthcare sectors, rampant production far outpaces notions of sustainability. If we persist with the prevailing patterns in production, it is projected that the cumulative global plastic production in the next 28 years will surpass that of the previous 72 years by 1.5 times, reaching over 25 billion by 2050 (Geyer, Jambeck, and Lavender Law, 2017).

Plastic production is not evenly distributed across the globe. In 2022, Asia emerged as the leading producer, contributing a substantial amount of

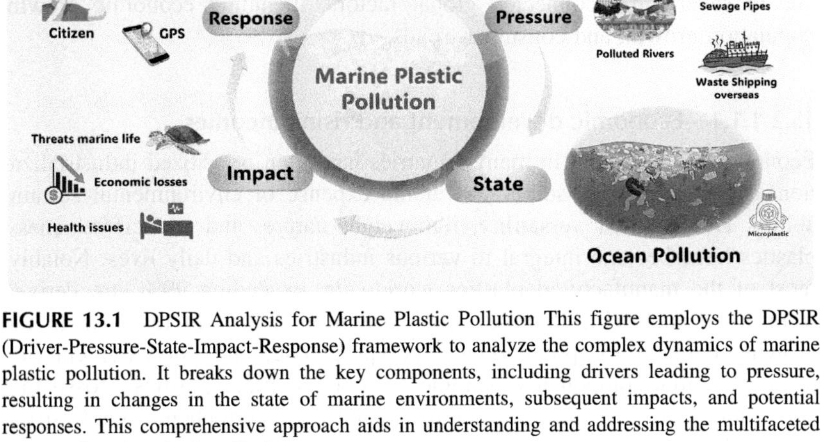

FIGURE 13.1 DPSIR Analysis for Marine Plastic Pollution This figure employs the DPSIR (Driver-Pressure-State-Impact-Response) framework to analyze the complex dynamics of marine plastic pollution. It breaks down the key components, including drivers leading to pressure, resulting in changes in the state of marine environments, subsequent impacts, and potential responses. This comprehensive approach aids in understanding and addressing the multifaceted aspects of marine plastic pollution.

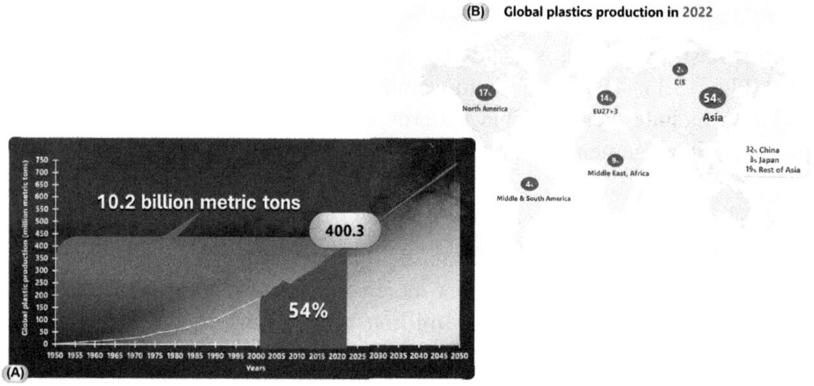

FIGURE 13.2 Global Plastic Production Trends (1950–2022) (A) Annual global plastic production (million metric tons) from 1950–2022. (B) The share of each continent in the global plastic production during 2022. *(A) From PlasticsEurope Market Research Group. https://plasticseurope.org/resources/infographics/.*

216.1 million metric tons, followed by 68 million metric tons in North America. The European Union, its 27 member states, and Norway, Iceland, and Liechtenstein collectively produced 56 million metric tons. Collectively, they contributed to approximately 85% of global production in 2022 (PlasticsEurope, 2023).

Plastic Consumption Trends: In 2022, the global plastic market size—the total volume of plastic products produced and sold globally—was valued at USD 609.01 billion (around 399 million metric tons of plastic consumption) and is expected to grow at a compound annual growth rate of 4.0% from 2023 to 2030 (Grand View Research, 2023).

This surge in the demand for plastics has reached unprecedented levels due to several interconnected global factors, including economic growth, population growth, and consumer trends.

13.2.1.1.1 Economic development and rising incomes

Economic development in many countries has often prioritized industrialization and consumption, sometimes at the expense of environmental sustainability. Due to their versatility, lightweight nature, and cost-effectiveness, plastics have become integral to various industries and daily lives. Notably, most of the manufactured plastics worldwide, exceeding 99%, are derived from fossil fuels (Li, Bataille, & Vogt-Schilb, 2023). In response to the growing demand for polyethylene, polypropylene, and polyvinyl chloride resin—essential components of numerous commonplace plastics—fossil fuel corporations are investing billions to expand their petrochemical infrastructure worldwide.

Once discretionary spending rises in developing countries, their populations aspire to consumer lifestyles associated with plastic-intensive goods. Single-use plastics are seen as symbols of modernity (Williams & Rangel-Buitrago, 2022).

13.2.1.1.2 Coastal population growth and urbanization

Population growth increases demand for consumer goods that often rely on plastic packaging. Coastal regions, where 40% of the world's population resides within 100 km of the coast (Kaniewski et al., 2021), are particularly vulnerable to plastic pollution. More people worldwide means more plastic wastes.

Urbanization leads to increased plastic consumption and plastic waste generation. Around 55% of the global population now resides in cities, with that figure projected to rise to 80% by 2050 (Venditti, 2022), putting stress on waste management infrastructure in many places.

The combination of population growth, rapid urban development, and inadequate waste management infrastructure contributes to the leakage of plastic waste into nearby waterways and, ultimately, into the ocean.

13.2.1.1.3 Consumer culture and single-use plastics

The global consumer culture, characterized by convenience and disposable products, has contributed to the proliferation of single-use plastics. Items like plastic bags, food packaging, and beverage containers are used briefly and then discarded. Due to their lightweight nature, these single-use plastics are easily transported by wind and water and can make their way into rivers and oceans. Improper disposal, such as littering or inadequate waste management, further exacerbates marine plastic pollution.

13.2.1.1.4 Lack of awareness and education

Inadequate awareness and education about the environmental impact of plastic pollution can lead to improper waste disposal, littering, and a lack of demand for sustainable alternatives. These behaviors exacerbate the pressures on marine ecosystems and contribute to the accumulation of plastic pollution.

13.2.1.2 Inadequate waste management systems

It is worth noting that the dramatic increase in plastic production and consumption consequently results in a proportional rise in plastic waste generation. Around 6.9 billion metric tons of primary plastic waste have been generated globally throughout the period from 1950 to 2017 (UNEP, 2021). Consequently, the insufficient or lack of proper waste management infrastructure (waste collection, recycling, and disposal systems) is identified as the second primary catalyst for the proliferation of marine plastic pollution.

Globally, only around 9% of total plastic waste produced is recycled, 19% is burned, and 50% ends up in landfills, while the remaining 22% is either recklessly disposed of in uncontrolled dumpsites, subjected to open pit burning, or allowed to leak into the environment (Organisation for Economic Co-operation and Development, OECD, 2022). This represents a significant gap between plastic waste generation and recycling rates, highlighting the need for increased efforts to improve recycling infrastructure and practices globally. A significant portion, approximately two-thirds, of the plastic waste that plagues our environment originates from applications that have short lifespans (<5 years). These applications include packaging, which accounts for a staggering 40% of the problem, consumer products at 12%, and textiles at 11% (Organisation for Economic Co-operation and Development, OECD, 2022).

Wind, rain, and inadequate landfill sites can transport plastic waste from land to the ocean. Without effective waste management, it is projected that approximately 12.5 billion metric tons of plastic waste could potentially accumulate in the natural environment by 2050 (Geyer, Jambeck, and Lavender Law, 2017).

Accordingly, holistic solutions are needed to tackle plastic pollution at its socio-economic root causes. This includes reducing plastic production and consumption, implementing effective waste management systems, promoting recycling and proper disposal practices, fostering sustainable alternatives, and raising awareness about the impacts of plastic pollution on marine ecosystems. Governments, businesses, communities, and individuals all have a role to play in mitigating the socio-economic drivers of marine plastic pollution and working toward a more sustainable future.

13.2.2 Pressures

Pressures are the specific activities or mechanisms that release pollutants into the environment. In the context of marine plastic pollution, the above-mentioned socio-economic drivers have cumulatively created immense direct and indirect pressures on the marine environment through plastic waste entering waterways and oceans.

Generally, the sources of marine plastic waste can be categorized into land-based, sea-based, and long-range transport. More than 80% of the total plastic waste reaching the oceans originates from land (European Investment Bank, 2021).

13.2.2.1 Land-based sources

The exponential growth of plastic consumption over the past few decades, coupled with the lengthy decomposition period of plastic, has resulted in plastic waste not being properly managed in the environment (Jambeck et al., 2015).

Megacities with over 10 million people near oceans discharge large quantities of plastic waste. According to Meijer, van Emmerik, van der Ent, Schmidt, & Lebreton (2021), the top 20 countries that contribute the most to annual plastic emissions into our ocean are ranked as follows: Philippines, India, Malaysia, China, Indonesia, Myanmar, Brazil, Vietnam, Bangladesh, Thailand, Nigeria, Turkey, Cameroon, Sri Lanka, Guatemala, Haiti, Dominican Republic, Venezuela, Tanzania, and Algeria. It is clear that Asia stands as the most significant continent in terms of contributions.

The aforementioned factors have transported these waste materials to river systems, leading to their eventual presence in our seas. Meijer's global estimations revealed that 31,904 rivers contribute significantly to the influx of millions of metric tons of mismanaged plastic waste into our oceans annually (Meijer et al., 2021). This finding highlights the significant impact of these rivers on the global plastic pollution crisis.

13.2.2.2 Sea-based sources

Sea-based sources, including offshore oil production, commercial fishing, commercial shipping, and recreational activities, contribute to marine plastic pollution.

Offshore oil production involves the extraction of oil and gas from beneath the ocean floor. In this process, plastics may be used in construction, equipment, and infrastructure on offshore oil platforms. During storm events or accidents, plastic fragments can be lost from these platforms and end up in the ocean.

Commercial fishing operations generate plastic pollution through various means. Fishing gear, such as nets, lines, and traps, is often made of synthetic materials, including plastic. Accidental loss or intentional abandonment of this gear can lead to "ghost fishing," where the gear continues to trap and harm marine life. Packaging materials and other plastic waste generated during fishing operations can also contribute to marine pollution.

International trade and large-scale commercial shipping involve transporting goods across the world's oceans. Accidental spills and releases of plastic products, including packaging and other materials, can occur during loading, unloading, and transportation activities. Cargo loss overboard, such as shipping containers, can also result in the introduction of plastic waste into the marine environment. For example, according to local reports and the United Nations Environment Program, a cargo ship sank off the coast of Sri Lanka in May 2021, resulting in the deposition of 75 billion plastic pellets, known as nurdles, on the country's beaches (Subramanian, 2022).

Recreational activities such as boating, sailing, and water sports can contribute to marine plastic pollution in coastal and marine areas. Improper waste disposal by recreational boaters and cruise ships, including littering and dumping of plastic waste directly into waterways, can result in the accumulation of plastic debris in the ocean.

Addressing these pressures requires: (1) implementing proper waste management systems and recycling facilities in ports, marinas, and onboard ships to help reduce the amount of plastic waste entering the ocean; (2) encouraging responsible fishing practices and promoting the recovery and recycling of lost or abandoned fishing gear; (3) strengthening international regulations and enforcement mechanisms to prevent accidental spills; and (4) increasing public awareness among recreational boaters and cruise ship passengers about the impacts of marine pollution and the importance of responsible behavior.

13.2.2.3 Long-range transport

Long-range transport refers to moving plastic waste away from its original source over long distances—through ocean currents, wind patterns, and tidal movements—to end up in remote ocean areas, coastal regions, or even isolated islands.

Ocean currents are pivotal in redistributing plastics across vast ocean basins, carrying them over considerable distances, and ultimately causing accumulation in convergence zones, notably gyres. These gyres, characterized by large, slow-rotating currents, become hubs for collecting floating debris, forming expansive areas of plastic pollution known as "plastic patches" or "garbage patches," with the Great Pacific Garbage Patch being a prominent example in the North Pacific Ocean. As plastic waste persists in these regions, exposure to sunlight and UV radiation initiates the breakdown of larger plastic fragments into smaller particles, known as microplastics (less than 5 mm in size). Highly persistent and enduring in the environment, microplastics accumulate within gyres and other convergence zones. They can be transported atmospherically through wind and rain, depositing over long distances from their original emission sources.

The contemporary distribution of plastics in ocean waters underscores a transboundary pollution predicament. Effectively addressing challenges linked to the long-range transport of plastic pollution necessitates specific measures, including regulating shipping practices and preventing deliberate dumping. Such actions are crucial for mitigating the far-reaching environmental impact of plastic pollution.

13.2.3 State

The state represents the condition of the environment, considering the concentration and distribution of pollutants. For marine plastic pollution, the state involves assessing the quantity and location of plastic waste in the oceans.

13.2.3.1 Quantity

The annual influx of plastic waste into the world's oceans exhibits a range of estimates in various studies, spanning from 0.0061 million metric tons, as reported by Weiss et al. (2021), to a higher estimate of 8.28 million metric tons, according to UNEP (2018). It is widely acknowledged that an estimated range of 86–150 million metric tons of plastic waste has accumulated within the Earth's oceans to date (Tekman, Walther, Peter, & Gutow, 2022). In light of ongoing patterns, it is projected that by the year 2050, there is a possibility that the quantity of plastic present in the Earth's oceans may exceed the weight of fish inhabiting these aquatic ecosystems (World Economic Forum Ellen MacArthur Foundation McKinsey & Company 2016, 2016).

13.2.3.2 *Distribution*

Plastic pollution has been documented in all ocean basins and regions, from surface waters to the deep sea.

13.2.3.2.1 Surface waters and coastlines

The accumulation of floating plastic debris is most pronounced in five prominent subtropical ocean gyres: the North Pacific, South Pacific, North Atlantic, South Atlantic, and Indian Ocean. These gyres are formed by the rotational movement of ocean currents, creating vast plastic patches that are visible even from outer space. Based on aerial imagery analysis, Lebreton et al. (2018) found 1.8 trillion pieces of plastic waste within the North Pacific's Great Pacific Garbage Patch. The overall weight of these plastic pieces is ~80k metric tons (categorized as 42, ~20, ~10, and ~6.4k metric tons of megaplastics, macroplastics, mesoplastics, and microplastics, respectively).

Other regions with high levels of surface plastic pollution include the Mediterranean Sea (Li, Tse, & Fok, 2016). According to recent research conducted by Pedrotti et al. (2022), it has been estimated that approximately 650 billion plastic items, with a total weight of 660 tons, are currently adrift on the surface of the Mediterranean Sea.

13.2.3.2.2 Deep sea and ocean floor

While most research has primarily concentrated on surface waters, it is important to note that plastics have been discovered in various deep-sea ecosystems, spanning from the twilight zone to ocean trenches that reach depths exceeding 11 km (Chiba et al., 2018). Comprehensive surveys, employing remote operating vehicles and specialized sampling equipment, have retrieved plastic waste from the profound depths within the Mariana and Kermadec trenches, situated in the western Pacific region.

Plastic waste has reached every corner of the planet's oceans, and its accumulation continues to grow exponentially each year. All marine habitats, from surface waters to deep-sea trenches, are now contaminated by microplastics and larger plastic items. Its spread is global in scale, transcending geopolitical boundaries and impacting delicate ecological balances worldwide.

13.2.4 Impacts

Plastic pollution in marine environments substantially impacts the three main pillars of sustainable development (environment, society, and the economy).

13.2.4.1 Environmental impacts

13.2.4.1.1 Physical damage *via* entanglement and ingestion by marine animals

Marine animals like sea turtles, seals, sharks, dolphins, and seabirds are prone to entanglement in abandoned ghost fishing gear and other plastic litter, leading to injuries, drowning, and even death. Casale (2008) noted an alarming number of turtles, exceeding 150,000, are ensnared in ghost gear annually within the Mediterranean region. Tragically, this entrapment leads to a potential mortality rate of over 50,000 turtles.

Furthermore, it is worth noting that a wide range of marine organisms, including more than 690 species encompassing seabirds, turtles, fish, marine mammals, and invertebrates, frequently ingest plastic fragments, mistaking them for food (Carbery, O'Connor, & Palanisami, 2018). The ingestion of plastic fragments can cause physical harm, such as internal injuries, blockages in the digestive system, and reduced feeding efficiency (Puskic, Lavers, & Bond, 2020). Additionally, ingesting plastics can lead to malnutrition, creating a false feeling of satiety while providing minimal nutritional value (Puskic et al., 2020). Altogether, this can result in reduced growth rates, impaired reproductive success, and overall negative impacts on individual fitness and population dynamics.

13.2.4.1.2 Bioaccumulation

Bioaccumulation refers to the process by which microplastics progressively accumulate within the organism's tissues as time passes. Marine organisms generally lack the specific enzymes necessary to break down microplastics. As a result, when marine organisms ingest microplastics, these particles accumulate in their gut tissues and persist there for extended periods. It is worth mentioning that while microplastics can accumulate in the gut tissues, the subsequent translocation of both microplastics and/or their associated chemicals (like plasticizers, flame retardants, and other additives) from the gastrointestinal tract to other organs or tissues may take place in aquatic organisms, including fish, bivalves, and crustaceans (Lusher, Hollman, & Mendoza-Hill, 2017; Rochman, Hoh, Kurobe, & Teh, 2013), inducing toxicity (Smith, Love, Rochman, & Neff, 2018).

Moreover, microplastics can act as carriers for other pollutants in the marine environment, such as persistent organic pollutants and heavy metals. These pollutants can adsorb onto the surfaces of microplastics, leading to the potential for increased exposure and bioaccumulation of these harmful substances in marine organisms (Tanaka, Yamashita, & Takada, 2018).

13.2.4.1.3 Habitat degradation

The sheer volume of plastic waste in the oceans also leads to the physical degradation of marine habitats. Coral reefs, seagrass beds, and coastal areas can become smothered and suffocated by plastic debris, inhibiting their growth, disrupting their ecological functions, and diminishing biodiversity (Hall, Berry, Rintoul, & Hoogenboom, 2015). Plastic waste can alter light and nutrient availability, leading to shifts in species composition and reduced overall ecosystem health.

13.2.4.1.4 Transport of invasive species

Floating plastic waste serves as a substrate for the colonization of minute organisms, including algae, bacteria, and small invertebrates, facilitating their transportation across vast expanses of the ocean. Upon the arrival of plastic waste in different regions, these organisms can detach or release their propagules (such as spores, larvae, or eggs) into the surrounding environment. If these organisms are not native to the new region, they can establish populations and potentially manifest as invasive species. These species can outcompete native species, thereby disrupting food webs and altering habitats (Barnes & Milner, 2005).

13.2.4.2 Social impacts

The key social impacts of marine plastic pollution include.

13.2.4.2.1 Human health concerns

Microplastics have been found in seafood, drinking water, and salt. Once individuals consume these contaminated items, microplastics accumulate in their gastrointestinal tracts and can be absorbed, causing oxidative stress, cytotoxicity, and translocation to other tissues (Alberghini, Truant, Santonicola, Colavita, & Giaccone, 2022).

13.2.4.2.2 Cultural and recreational impacts

Marine plastic pollution can erode the cultural practices and traditions of communities with deep connections to the sea. The degradation of coastal environments can impact the cultural identity tied to the ocean.

Plastic pollution can detract from recreational activities such as swimming, diving, and boating. This affects the well-being of individuals and has implications for the social fabric of communities that rely on these activities for leisure and social interaction.

13.2.4.3 Economic impacts

Plastic pollution imposes significant economic costs, including damage to fisheries, tourism, and shipping sectors.

Coastal communities that rely on fishing as a primary source of income may experience economic losses due to the decline in fish populations caused by plastic pollution. This can lead to job losses and financial hardships for those dependent on the fishing industry.

The tourism industry, a vital source of income for many coastal communities, can suffer from the negative perception associated with plastic-laden environments.

The cost of cleaning up plastic waste from beaches and coastal regions is staggering, placing a burden on local governments and taxpayers. Additionally, industries such as shipping and aquaculture face challenges due to waterway obstructions and contamination caused by plastic debris.

13.2.5 Responses

Various policy instruments and initiatives have been implemented across scales to curb plastic pollution. International agreements, such as the United Nations' Clean Seas campaign and the Basel Convention, aim to raise awareness, promote collaboration, and strengthen regulations. Additionally, innovative technologies are being developed to remove plastic debris from oceans and rivers. However, preventing marine plastic pollution at its source is crucial.

13.3 Responses through a geospatial-participatory lens

Responses are measures taken to address environmental issues. Effective responses aim to reduce drivers, alleviate pressures, improve the state, and minimize impacts. Tackling marine plastic pollution effectively requires a holistic, multi-pronged, and comprehensive approach that combines upstream and downstream responses. Upstream responses address the root societal drivers contributing to plastic waste generation, ranging from production and consumption to waste management (such as production practices, consumption patterns, and waste management behaviors), to prevent plastic waste from entering the ocean. On the other hand, downstream responses focus on managing plastic waste after it has entered the marine environment (as clean-up efforts).

Geospatial tools, particularly GIS, offer a potent lens for comprehensively analyzing and visualizing the spatial dimensions of plastic waste generation. GIS, a technology that integrates spatial data with analytical capabilities, allows stakeholders to gain valuable insights into the distribution, patterns, and underlying factors contributing to plastic waste generation. This spatial intelligence becomes instrumental in identifying areas where interventions can be strategically targeted to maximize impact. Whether mapping plastic production facilities, identifying regions with high consumption rates, or visualizing the effectiveness of waste management systems, geospatial tools offer a strategic perspective for informed decision-making.

On the grassroots level, citizen science harnesses the power of volunteers to collect large amounts of environmental data across broad spatial and temporal scales. Citizens can play a transformative role in global plastic pollution tracking and clean-up efforts when combined with geospatial technologies.

This section explores how integrating geospatial technology and citizen science into upstream and downstream responses can tackle the problem of marine plastic pollution more effectively.

13.3.1 Upstream prevention strategies

Most plastic waste entering the oceans originates from mismanaged waste on land. Strategies are needed to curb plastic leakage at its sources.

13.3.1.1 Plastic production and consumption

Geospatial technology can be used to:

1. Track plastic production facilities by mapping their locations and monitoring their activities. This can involve using satellite imagery and drones to capture images of the factories, as well as installing sensors to monitor emissions and wastewater discharge. By analyzing this data, it is possible to identify trends in production levels and areas where there may be overproduction or waste.
2. Map the locations of retail outlets, supermarkets, and other businesses that sell plastic products to identify areas with high plastic usage and inform policy decisions to reduce plastic consumption.

Citizens can participate through:

1. Using applications like Litterati to geotag and identify litter items they encounter provides insights into which types of plastic are most commonly littered.

13.3.1.2 Waste management

Developing and enhancing waste management infrastructure is essential to prevent plastic waste from entering waterways. This includes establishing efficient collection systems, implementing recycling programs, and promoting proper waste disposal practices. Adequate waste management infrastructure helps ensure that plastic waste is properly managed and does not end up in marine environments. In this context, geospatial technology can be used to:

1. Map the connectivity and relationships between different components of the plastic supply chain to visualize the movement of materials through the supply chain, identify the most efficient paths for transporting materials between supply chain components, and identify bottlenecks or areas where the flow of materials may be impeded.
2. Pinpoint the locations of waste collection points and recycling centers in certain regions to identify gaps in the recycling infrastructure. This, in turn, enables the strategic allocation of investments toward areas with limited recycling facilities, thereby improving overall plastic waste management.
3. Map the locations of landfill sites by using satellite imagery, aerial photography, and GIS. This information can help identify areas with high landfill capacity and inform policy decisions to reduce landfill usage.
4. Using GPS-enabled waste collection trucks to track the location of waste pickups and deploying sensors to monitor landfill sites. By analyzing this data, it is possible to identify areas with high levels of plastic pollution and target clean-up efforts accordingly.

Citizen science can contribute through:

1. Data collection: Citizens can contribute by collecting data on plastic waste generation, such as by conducting waste audits, documenting litter hotspots, or reporting illegal dumping sites. This crowd-sourced data can provide valuable insights into the scale and distribution of the problem. In 2022, 397 brand audits were conducted across six continents with the help of 14,760 volunteers in 44 countries to identify the leading corporations responsible for plastic pollution (Break Free From Plastic, 2023). Analysis of 429,994 pieces of plastic waste collected revealed that Coca-Cola Company, PepsiCo, and Nestlé emerged as the three most prominent companies.
2. Behavioral observations: Citizen scientists can also monitor and report on behaviors related to plastic waste, such as consumer preferences, packaging choices, and recycling practices. This information can inform targeted interventions and behavior-change campaigns.
3. Awareness and education: Citizen science initiatives can raise awareness about plastic waste issues and educate participants about sustainable practices. This can lead to long-term behavior change and a reduction in waste generation.

By overlaying geospatial data with citizen-contributed information, it becomes possible to target interventions in high-impact areas, improve recycling infrastructure, and implement sustainable practices to effectively address and mitigate waste hotspots.

13.3.1.3 Plastic pollution sources

Advanced sensors and artificial intelligence can be used to intercept plastic in rivers to prevent it from reaching the oceans and cut off the inflow of pollution.

"The Ocean Cleanup" (https://theoceancleanup.com/), a non-profit organization founded by Dutch inventor Boyan Slat, has gained worldwide attention for its innovative approach to combating plastic pollution in the world's oceans. The organization's flagship project, the Interceptor (https://theoceancleanup.com/rivers/), is a network of autonomous systems designed to intercept plastic waste before it enters the ocean. The Interceptor system uses advanced sensors and artificial intelligence to detect and track plastic debris in rivers, which are the primary source of plastic pollution in the world's oceans. Once identified, the system deploys a series of floating barriers that capture the plastic waste before it reaches the ocean.

The Interceptor system is also equipped with drone technology, which allows for more precise targeting of plastic accumulation zones. Drones are used to survey the riverbeds and identify areas where plastic waste is concentrated, enabling the Interceptor system to focus its efforts on these areas.

13.3.2 Downstream remediation and impact mitigation

Even with improved upstream efforts, plastic debris will persist in our oceans for decades. Downstream solutions are still needed to remove what has already accumulated and minimize impacts.

13.3.2.1 Impact assessment through citizen science

Citizen science can provide critical data on the impacts of plastic pollution that enhances downstream response prioritization. Global projects that monitor the entanglement of marine animals assist in identifying the species that are most impacted.

13.3.2.2 Enhanced geospatially-enabled cleanup

Efforts to clean up the plastic that has already accumulated in the oceans have seen significant advancements, reflecting a multifaceted approach involving technology, community engagement, and global cooperation.

The "Skimmer"—the latest innovation of the Ocean Cleanup organization—is a floating device that uses a rotating drum to capture plastic waste from the water's surface more efficiently than traditional skimmers, which can become clogged with debris and require frequent maintenance.

Harnessing the power of citizen scientists on data-sharing platforms like the Marine Debris Tracker is essential for informed decision-making in clean-up efforts. By centralizing global clean-up event data and employing advanced analytics, these platforms enable the identification of priority areas, ensuring that resources are allocated where they are most needed. Over 2 million clean-ups conducted by 1143 organizations have been documented on this platform, removing over 8 million pieces of debris (https://debris-tracker.org/).

13.4 Opportunities and challenges of integration

While significant progress has been made, opportunities remain to better coordinate disparate initiatives under an integrated geospatial strategy. Standardized protocols and centralized open-access data platforms could harmonize methodologies and analyses across initiatives. Regulatory recommendations could incentivize industry compliance with mitigation needs identified through monitoring programs.

Challenges around long-term project sustainability, data accuracy, and addressing social inequities also persist. Global citizen science still suffers limitations around uneven geographic representation and potential volunteer skill variability. Ensuring meaningful participation from developing nations and indigenous communities is crucial for equitable solutions. There are also technological obstacles to mapping remote ocean regions and microdebris pathways with the necessary resolution.

With strengthened coordination and collaborative solution-setting between sectors, geospatial tools and citizen science efforts show immense promise to revolutionize downstream marine debris responses. Their continued advancement under integrated governance frameworks holds the potential to substantially curb the impacts of plastic pollution worldwide through evidence-based strategic prioritization and remediation.

13.5 Proposed method protocol for effective synergistic approach

While satellite imagery is indispensable for large-scale monitoring and initial identification of plastic waste hotspots, citizen science is crucial in providing detailed, context-specific ground information (such as types, quantities of plastic pollution, and consumer behaviors) and complementing geospatial data. The synergy between these approaches enhances the accuracy and reliability of the information collected to tackle plastic pollution comprehensively. Moreover, engaging citizens in the process fosters a sense of ownership and empowerment. It encourages collaboration among various stakeholders, including government agencies, businesses, and communities, leading to more effective solutions.

The following proposed method protocol offers a holistic approach to combating plastic pollution by harnessing the synergistic power of geospatial technology and citizen science. It covers all aspects, from project planning to implementation, monitoring, and evaluation. By combining the strengths of both approaches, the protocol aims to achieve more effective and sustainable solutions to combat plastic pollution.

Step 1: Define the project objectives and scope

Set smart targets, such as reducing plastic waste in certain regions by a certain percentage. The scope should cover the geographic area, the type of plastic (such as macroplastics or microplastics), and the target audience for citizen science.

Step 2: Geospatial tools

In the context of marine plastic pollution, the following tools should be included:

1. Recent multispectral high-resolution satellite imagery that can identify plastic waste in the designated area. Satellite imagery with 1 meter or less spatial resolution is often suitable for identifying plastic waste. Providers like DigitalGlobe, Sentinel-2, and Airbus offer multispectral, high-resolution satellite data that will be useful.
2. GIS software, such as ArcGIS, to handle the spatial data and mapping requirements.
3. Drones equipped with cameras or sensors to capture high-resolution images or data on marine plastic pollution in coastal areas.
4. Underwater robots or autonomous underwater vehicles to collect data on marine plastic pollution in deeper waters or hard-to-reach areas.

Step 3: Develop a geospatial-citizen science platform

Develop a geospatial-citizen science platform as a central hub for participant registration, training materials, and data submission, fostering a collaborative network for real-time information gathering. The platform can be either a mobile application—compatible with iOS and Android—or a web platform. It should be user-friendly, accessible, and engaging.

The platform should include the following features:

1. Simple UI: An intuitive and easy-to-use interface on both mobile and web to lower barriers to participation.
2. User profiles: Registered user accounts to track individual contributors and their observations over time.
3. Guideline resources: Information pages on plastic identification, recommended clean-up practices, project instructions identifying the information required to be collected (such as types of plastic, usage patterns, and disposal practices), etc., to guide participation and ensure consistency in data collection.

4. Location services: Allow participants to geotag observations with accurate latitude and longitude. This could be *via* GPS on a mobile app or manually entering a location on a web platform.
5. Observations categories: Pre-defined categories for common types of plastic pollution (e.g., bottles, bags, and wrappers) that can be easily tagged during reporting.
6. Image upload: allows the submission of photos along with observations to aid verification and data analysis. Especially important for plastic identification.
7. Reporting form: Allows users to submit detailed real-time information about their sighting, including location coordinates, date/time, geotagged photos or videos, and size of plastic.
8. Interactive maps on the mobile app/web platform to facilitate visual verification and real-time mapping of submitted observations. This will allow users to observe the spatial distribution of pollution, identify locations with a higher occurrence, and track changes over time.
9. Database integration: Stores all observation reports, photos, and associated location data in a backend geospatial database for mapping and analysis.
10. Verification system: Allows experts to review reports and confirm or reject them based on criteria such as accuracy, completeness, and reliability.
11. Feedback system: Allows users to receive updates on the status of their reports and any actions taken as a result of their reports.
12. Gamification elements: Incorporate game-like features such as badges, rewards, and challenges to encourage user engagement and motivation.
13. Citizen recognition: Features like profile pages and leader boards to recognize top contributors and maintain community engagement.
14. Training resources: Offer online courses, webinars, and workshops on marine plastic pollution, its impacts, prevention measures, and citizen science methods to enhance user knowledge and skills.
15. Partnerships: Collaborate with local authorities, NGOs, and other stakeholders to promote the platform's use, disseminate its findings, and implement actions based on the observation data.

Step 4: Geospatial analysis
This could involve:

1. Conducting satellite spectral signature analysis to determine the spatial distribution of marine plastic pollution over large areas.
2. Applying techniques (such as hotspot analysis, density mapping, and network analysis) to identify patterns and insights regarding plastic waste hotspots, pollution sources, and transportation routes.

3. Creating maps, charts, and visualizations to communicate the findings effectively to stakeholders and the public.

Step 5: Develop evidence-based solutions for reducing marine plastic pollution
These solutions could include:

1. Identifying priority areas for intervention based on the locations of reported sightings or hotspots identified through analysis.
2. Developing targeted clean-up campaigns in these priority areas using local resources such as volunteers, NGOs, or government agencies.
3. Implementing measures to prevent further marine plastic pollution at its source by working with local industries (such as fishing fleets, tourism operators, or waste management facilities).
4. Developing public awareness campaigns that educate local communities about the impacts of marine plastic pollution (on ecosystems and human health) and how they can reduce their use of single-use plastics.

Step 6: Monitor progress and evaluate impact
This could involve:

1. Conducting regular satellite imagery analysis to track changes in marine plastic pollution levels over time, especially in areas where clean-up campaigns have been implemented.
2. Encouraging local communities to continue reporting sightings of marine plastic pollution through the developed platform so trends can be monitored over time.
3. Conducting follow-up surveys or interviews with local communities to assess their attitudes toward marine plastic pollution prevention measures implemented.
4. Analyzing the data collected through both geospatial technology and citizen science to evaluate the impact of interventions implemented on reducing marine plastic pollution levels over time.

13.6 Conclusion

The chapter delves into the comprehensive strategies required to address marine plastic pollution, emphasizing the integration of upstream and downstream responses. Geospatial tools, particularly GIS, and citizen science are crucial components for effective environmental management. The proposed method provides a systematic and integrative approach to tackle marine plastic pollution, utilizing the strengths of geospatial tools and citizen science for evidence-based decision-making and impact evaluation. The proposed method emphasizes the importance of coordination, standardization, and inclusivity for successful environmental management. However, it is crucial to acknowledge the limitations and consider future perspectives to enhance the effectiveness of the proposed approach.

13.7 Limitations of the study

1. Incomplete coordination: While the study underscores the need for coordination, achieving seamless collaboration among disparate initiatives remains challenging. Practical barriers may impede the full integration of strategies, hindering the potential for a unified and comprehensive geospatial approach.
2. Data standardization: Achieving global consensus on methodologies and ensuring the creation of accessible platforms for sharing data require concerted efforts and may encounter resistance from various stakeholders.
3. Regulatory hurdles: Overcoming legal and regulatory barriers poses a significant challenge to successfully implementing mitigation strategies.
4. Sustainability challenges: Long-term project sustainability remains a concern, as financial, logistical, and institutional support may wane over time. Ensuring the continued effectiveness of initiatives beyond their initial phases requires ongoing commitment and resource allocation, posing a significant challenge to sustained impact.
5. Citizen science limitations: The study acknowledges the limitations of citizen science, including uneven geographic representation and variations in volunteer skill levels. Overcoming these challenges is essential for ensuring the reliability and credibility of data collected through citizen science initiatives.

13.8 Future perspectives

1. Global collaboration: Future efforts should foster global collaboration and partnerships to overcome coordination challenges. International initiatives and agreements could be pivotal in creating a unified front against marine plastic pollution.
2. Policy advocacy: Collaborative efforts with policymakers, industry stakeholders, and environmental organizations can help overcome regulatory hurdles and promote a more sustainable approach to plastic pollution mitigation.
3. Long-term funding models: Developing sustainable funding models for long-term project viability is essential. Exploring innovative financing mechanisms, public-private partnerships, and philanthropic support can contribute to the financial sustainability of initiatives to combat marine plastic pollution.
4. Capacity building: Future initiatives should prioritize capacity building in developing nations and indigenous communities, ensuring their meaningful participation in addressing marine plastic pollution. This involves providing training, resources, and support to empower these communities to contribute effectively to citizen science efforts.

References

Alberghini, L., Truant, A., Santonicola, S., Colavita, G., & Giaccone, V. (2022). Microplastics in fish and fishery products and risks for human health: A review. *International Journal of Environmental Research and Public Health, 20*(1), 789. Available from https://doi.org/10.3390/ijerph20010789.

Barnes, D. K. A., & Milner, P. (2005). Drifting plastic and its consequences for sessile organism dispersal in the Atlantic Ocean. *Marine Biology., 146*(4), 815–825. Available from https://doi.org/10.1007/s00227-004-1474-8.

Break Free From Plastic (2023). *The Brand Audit Report 2018–2022*. Available from https://brandaudit.breakfreefromplastic.org/brand-audit-2022/.

Carbery, M., O'Connor, W., & Palanisami, T. (2018). Trophic transfer of microplastics and mixed contaminants in the marine food web and implications for human health. *Environment International, 115*, 400–409. Available from https://doi.org/10.1016/j.envint.2018.03.007, http://www.elsevier.com/locate/envint.

Casale, P. (2008). *Incidental catch of marine turtles in the Mediterranean Sea: Captures, mortality, priorities*. WWF Italy. Available from: https://wwfeu.awsassets.panda.org/downloads/casale_2008_turtle_bycatch_med_wwf.pdf.

Chiba, S., Saito, H., Fletcher, R., Yogi, T., Kayo, M., Miyagi, S., ... Fujikura, K. (2018). Human footprint in the abyss: 30 year records of deep-sea plastic debris. *Marine Policy, 96*, 204–212. Available from https://doi.org/10.1016/j.marpol.2018.03.022, http://www.elsevier.com/inca/publications/store/3/0/4/5/3/.

Coleby, A. M., & Grist, E. P. M. (2019). Prioritized area mapping for multiple stakeholders through geospatial modelling: A focus on marine plastics pollution in Hong Kong. *Ocean & Coastal Management, 171*, 131–141. Available from https://doi.org/10.1016/j.ocecoaman.2018.12.021.

Emam, W. W., El-Kafrawy, S. B., & Soliman, K. M. (2021). Integrated geospatial analysis linking metal contamination among three different compartments of Lake Edku ecosystem in Egypt to human health effects. *Environmental Science and Pollution Research, 28*(16), 20140–20156. Available from https://doi.org/10.1007/s11356-020-11661-8, https://link.springer.com/journal/11356.

Emam, W. W. M., & Soliman, K. M. (2020). Applying geospatial technology in quantifying spatiotemporal shoreline dynamics along Marina El-Alamein Resort, Egypt. *Environmental Monitoring and Assessment, 192*(7). Available from https://doi.org/10.1007/s10661-020-08432-w, https://link.springer.com/journal/10661.

Emam, W. W. M., & Soliman, K. M. (2022). Geospatial analysis, source identification, contamination status, ecological and health risk assessment of heavy metals in agricultural soils from Qallin city, Egypt. *Egypt Stochastic Environmental Research and Risk Assessment, 36*(9), 2437–2459. Available from https://doi.org/10.1007/s00477-021-02097-8, https://www.springer.com/journal/477.

Emam, W. W. M., Khalil, M. T., Nassif, M. G., & Soliman, K. M. (2019). Applying geospatial technology in assessing the coastal vulnerability of zaranik protectorate to sea-level rise. *Egyptian Society for the Development of Fisheries and Human Health, Egypt Egyptian Journal of Aquatic Biology and Fisheries, 23*(4), 697–719. Available from https://doi.org/10.21608/ejabf.2019.202267. Available from:, http://ejabf.journals.ekb.eg.

European Investment Bank. (2021). *The Ocean Plastics Reduction Guide*. doi:10.2867/588812.

Fet, A. M. (2023). *Analytical frameworks, impact categories, indicators and performance evaluation. Business Transitions: A Path to Sustainability: The CapSEM Model* (pp. 77–87). Norway: Springer International Publishing, Norway Springer International

Publishing. Available from https://link.springer.com/book/10.1007/978-3-031-22245-0, 10.1007/978-3-031-22245-0_8.

Geyer, R., Jambeck, J. R., & Lavender Law, K. (2017). Production, use, and fate of all plastics ever made. *Science Advances*, *3*(7). Available from https://doi.org/10.1126/sciadv.1700782, http://advances.sciencemag.org/.

Grand View Research. (2023). Plastic market size, share & trends analysis report by product (PE, PP, PU, PVC, PET, Polystyrene, ABS, PBT, PPO, Epoxy Polymers, LCP, PC, Polyamide), by application, by end-use, by region, and segment forecasts, 2023 – 2030. Available from https://www.grandviewresearch.com/industry-analysis/global-plastics-market.

Hall, N. M., Berry, K. L. E., Rintoul, L., & Hoogenboom, M. O. (2015). Microplastic ingestion by scleractinian corals. *Marine Biology*, *162*(3), 725–732. Available from https://doi.org/10.1007/s00227-015-2619-7, http://link.springer.de/link/service/journals/00227/index.htm.

Jambeck, J. R., Geyer, R., Wilcox, C., Siegler, T. R., Perryman, M., Andrady, A., ... Law, K. L. (2015). Plastic waste inputs from land into the ocean. *Science (New York, N.Y.)*, *347*(6223), 768–771. Available from https://doi.org/10.1126/science.1260352, http://www.sciencemag.org/content/347/6223/768.full.pdf.

Kaniewski, D., Marriner, N., Cheddadi, R., Morhange, C., Vacchi, M., Rovere, A., ... Van Campo, E. (2021). Coastal submersions in the north-eastern Adriatic during the last 5200 years. *Global and Planetary Change*, *204*, 103570. Available from https://doi.org/10.1016/j.gloplacha.2021.103570.

Lebreton, L., Slat, B., Ferrari, F., Sainte-Rose, B., Aitken, J., Marthouse, R., ... Reisser, J. (2018). Evidence that the Great Pacific Garbage Patch is rapidly accumulating plastic. *Scientific Reports*, *8*(1). Available from https://doi.org/10.1038/s41598-018-22939-w, http://www.nature.com/srep/index.html.

Li, F., Bataille., & Vogt-Schilb. (2023). *Net-zero industry: Options for plastics, textiles, automobiles, and fisheries in Colombia*. IDB Publication. Available from 10.18235/0005167.

Li, W. C., Tse, H. F., & Fok, L. (2016). Plastic waste in the marine environment: A review of sources, occurrence and effects. *Science of the Total Environment*, *566–567*, 333–349. Available from https://doi.org/10.1016/j.scitotenv.2016.05.084, http://www.elsevier.com/locate/scitotenv.

Lusher, A., Hollman, P., & Mendoza-Hill, J. (2017). *Microplastics in fisheries and aquaculture: Status of knowledge on their occurrence and implications for aquatic organisms and food safety*. Food and Agriculture Organization of the United Nations.

McKinley, D. C., Miller-Rushing, A. J., Ballard, H. L., Bonney, R., Brown, H., Cook-Patton, S. C., ... Soukup, M. A. (2017). Citizen science can improve conservation science, natural resource management, and environmental protection. *Biological Conservation*, *208*, 15–28. Available from https://doi.org/10.1016/j.biocon.2016.05.015, http://www.elsevier.com/inca/publications/store/4/0/5/8/5/3.

Meijer, L. J. J., van Emmerik, T., van der Ent, R., Schmidt, C., & Lebreton, L. (2021). More than 1000 rivers account for 80% of global riverine plastic emissions into the ocean. *Science Advances*, *7*(18). Available from https://doi.org/10.1126/sciadv.aaz5803, https://advances.sciencemag.org/content/7/18/eaaz5803.

Mohamed, H. M., Khalil, M. T., El-Zeiny, A. M., Khalifa, N., Kafrawy, S. B. E., & Emam, W. W. M. (2023). Trophic state and potential productivity assessment for Qaroun Lake using spatial techniques. *Environmental Monitoring and Assessment*, *195*(8). Available from https://doi.org/10.1007/s10661-023-11504-2, https://www.springer.com/journal/10661.

Mohamed, H. M., Khalil, M. T., El-Kafrawy, S. B., El-Zeiny, A. M., Khalifa, N., & Mahmoud Emam, W. W. (2022). Can statistical remote sensing aid in predicting the potential productivity of inland lakes? Case study: Lake Qaroun, Egypt. *Stochastic Environmental Research and Risk Assessment, 36*(10), 3221–3238. Available from https://doi.org/10.1007/s00477-022-02189-z.

Organisation for Economic Co-operation and Development (OECD) (2022). *Economic drivers, environmental impacts and policy options global plastics outlook.* OECD Publishing, Paris, doi:10.1787/de747aef-en.

Pedrotti, M. L., Lombard, F., Baudena, A., Galgani, F., Elineau, A., Petit, S., ... Gorsky, G. (2022). An integrative assessment of the plastic debris load in the Mediterranean Sea. *Science of The Total Environment, 838*, 155958. Available from https://doi.org/10.1016/j.scitotenv.2022.155958.

Plastics Europe. (2023). *Plastics – The fast facts 2023.* Available from https://plasticseurope.org/knowledge-hub/plastics-the-fast-facts-2023/.

Puskic, P. S., Lavers, J. L., & Bond, A. L. (2020). A critical review of harm associated with plastic ingestion on vertebrates. *Science of the Total Environment.*, 743. Available from https://doi.org/10.1016/j.scitotenv.2020.140666, http://www.elsevier.com/locate/scitotenv.

Rochman, C. M., Hoh, E., Kurobe, T., & Teh, S. J. (2013). Ingested plastic transfers hazardous chemicals to fish and induces hepatic stress. *Scientific Reports*, 3. Available from https://doi.org/10.1038/srep03263.

Smith, M., Love, D. C., Rochman, C. M., & Neff, R. A. (2018). Microplastics in seafood and the implications for human health. *Current Environmental Health Reports, 5*(3), 375–386. Available from https://doi.org/10.1007/s40572-018-0206-z, http://link.springer.com/journal/40572.

Subramanian, M. (2022). Plastics tsunami: Can a landmark treaty stop waste from choking the oceans? *Nature Research, undefined Nature, 611*(7937), 650–653. Available from https://doi.org/10.1038/d41586-022-03793-3, https://www.nature.com/nature/.

Tanaka, K., Yamashita, R., & Takada, H. (2018). Transfer of hazardous chemicals from ingested plastics to higher-trophic-level organisms. *Handbook of Environmental Chemistry, 78*. Available from https://doi.org/10.1007/698_2018_255, http://www.springer.com/series/698.

Tekman, B. A., Walther, C., Peter, C., Gutow, L., Bergmann, M. (2022). Impacts of plastic pollution in the oceans on marine species, biodiversity and ecosystems, pp. 1–221, WWF Germany. Available from https://doi.org/10.5281/zenodo.5898684.

Troian, A., Gomes, M., Tiecher, T., Berbel, J., & Gutiérrez-Martín, C. (2021). The drivers-pressures-state-impact-response model to structure cause – Effect relationships between agriculture and aquatic ecosystems. *Sustainability, 13*(16), 9365. Available from https://doi.org/10.3390/su13169365.

Tscherning, K., Helming, K., Krippner, B., Sieber, S., & Paloma, S. Gy (2012). Does research applying the DPSIR framework support decision making? *Land use Policy, 29*(1), 102–110. Available from https://doi.org/10.1016/j.landusepol.2011.05.009.

UNEP. (United Nations Environment Programme) (2018). *Mapping of global plastics value chain and plastics losses to the environment (with a particular focus on marine environment).* Nairobi, Kenya. Available from https://wedocs.unep.org/handle/20.500.11822/26745.

UNEP (United Nations Environment Programme). (2021). *From pollution to solution: A global assessment of marine litter and plastic pollution.* Nairobi. Available from https://www.unep.org/interactives/pollution-to-solution/.

UNEP (United Nations Environment Programme). (2023). *Turning off the Tap. How the world can end plastic pollution and create a circular economy.* Nairobi. Available from https://www.unep.org/resources/turning-off-tap-end-plastic-pollution-create-circular-economy.

World Economic Forum; Ellen MacArthur Foundation McKinsey & Company 2016. (2016). *The new plastics economy: Rethinking the future of plastics*. Available from https://www.weforum.org/publications/the-new-plastics-economy-rethinking-the-future-of-plastics/.

Venditti, B. (2022). *This chart shows the impact rising urbanization will have on the world*. World Economic Forum.

Weiss, L., Ludwig, W., Heussner, S., Canals, M., Ghiglione, J. F., Estournel, C., ... Kerhervé, P. (2021). The missing ocean plastic sink: Gone with the rivers. *Science (New York, N.Y.)*, *373*(6550), 107–111. Available from https://doi.org/10.1126/science.abe0290, https://science.sciencemag.org/content/373/6550/107.full.

Williams, A. T., & Rangel-Buitrago, N. (2022). The past, present, and future of plastic pollution. *Marine Pollution Bulletin*, *176*, 113429. Available from https://doi.org/10.1016/j.marpolbul.2022.113429.

Yu, R.-S., & Singh, S. (2023). Microplastic pollution: Threats and impacts on global marine ecosystems. *Sustainability*, *15*(17), 13252. Available from https://doi.org/10.3390/su151713252.

Chapter 14

Application of ground egg shell as an efficient photocatalyst for wastewater purification

Rohini Singh[1] **and Suman Dutta**[2]

[1]*Department of Chemical Engineering, R.N.G. Patel Institute of Technology, Bardoli, Gujarat, India,* [2]*Department of Chemical Engineering, Indian Institute of Technology (Indian School of Mines), Dhanbad, India*

14.1 Introduction

Huge amounts of pollutants, such as heavy metals, organic compounds, and several hazardous microorganisms, are being released into the water sources due to increased urbanization and industrial development. The scarcity of clean water is being faced in society despite 70% of the earth's surface is water. The effluents of textile industry are considered a major source of toxic dyes. Wastewater treatment became a global issue to be solved for environmental and societal benefits. Based on the research, the demand for drinking water will increase by 35% by 2050 (Baburaj et al., 2023). Several physical, chemical, and biological methods such as ozonation (Jiang et al., 2023), membrane separation (Ma, Cao, Lu, Xiong, & Huang, 2022), desalination (Zhang, Luo, et al., 2023), electrodialysis (Hu, Yan, et al., 2023), distillation (Meng, Wang, & Wang, 2023), reverse osmosis (Hu, Guan, et al., 2023), and biodegradation (Cui, Fu, Hao, & Zhou, 2023) are being investigated for wastewater treatment and proved to be costly and not practical for large-scale applications. Removal of toxic contaminants requires efficient adsorbents and photocatalysts (Perera, Madhushani, Kumar, & Gupta, 2023). He et al. proposed that wastewater purification techniques, including net zero greenhouse gases, that is, carbon dioxide (CO_2), nitrous oxide (N_2O), and methane (CH_4) emissions, are highly essential (He et al., 2023).

Photocatalytic degradation is a versatile and cost-effective approach toward wastewater purification. Due to efficient organic contamination removal efficiency, semiconductor photocatalysis has attracted considerable attention. Heterogeneous photocatalysis, as a green technology, possesses the capability of

Water Resources Monitoring, Management, and Sustainability. DOI: https://doi.org/10.1016/B978-0-443-23665-5.00014-4

textile wastewater treatment. The technology of photocatalysis is one of the numerous advanced oxidation processes that have proven themselves to be cost-effective and environmentally friendly for domestic and industrial wastewater purification against various highly hazardous organic pollutants. In addition, heterogeneous photocatalysis utilizes non-polluting oxidants. Numerous studies have been carried out worldwide regarding the synthesis, characterization, and utilization of visible light-active photocatalysts. As a major portion of the solar spectrum consists of visible light, the researchers must find the economic methodology to synthesize photocatalysts with visible light responses. Eggs are a globally consumed food and generate a significant amount of eggshell waste. Eggshells consist of calcium carbonate that is transformed into CaO. Grounded eggshell waste includes numerous properties, i.e., enhanced adsorption capability and thermal stability. It is therefore efficiently utilized in wastewater treatment as a heterogeneous catalyst (Arami, Limaee, & Mahmoodi, 1999; Chen et al., 2010, 2012; Li, Zhou, Fan, Ye, & Tang, 2017; Sinha & Ahmaruzzaman, 2015; Ummartyotin & Manuspiya, 2018).

In the future, the supply of sufficient clean water for society's survival will be a challenging task. Therefore, it is essential to conduct research on eco-friendly and economical techniques to purify wastewater. Based on significant research worldwide, photocatalysis exhibits the potential to replace conventional techniques of wastewater treatment (Pan, 2016; Yahya et al., 2018). Textile industry effluents contaminate water with hazardous dyes. Textile wastewater contains high BOD and COD. The accumulation of toxins in the food chain threatens both human health and the aquatic ecosystem. Saratale et al. (2021) utilized eggshell membrane as a clean and efficient adsorbent for the removal of RR120 from the contaminated water. Mohamed et al. (2021) successfully synthesized a novel and eco-friendly photocatalyst for dye degradation from coffee and eggshell waste. Based on the research performed worldwide, artificial intelligence possesses the capability to remove existing and emerging organic and inorganic pollutants (Zhang, Jin, et al., 2023). CaO/$SrTiO_3$ heterostructures were prepared by Junior et al. along with their photocatalytic activity analysis toward the degradation of Rhodamine B dye. In addition to the heterostructures, several researchers studied pure CaO as photocatalysts. During the analysis, it was observed that other phases, that is, hydroxides and carbonates of calcium, also exhibit photoactivity (Junior et al., 2019). Various analyses revealed that $CaCO_3$ gets decomposed into CaO after sintering (Bathla, Singla, & Pal, 2019). TiO_2-loaded eggshells were prepared by the sol-gel method and degraded Acid Red B under solar light irradiation (Chen et al., 2010). In addition, anatase TiO_2 supported on eggshell membranes was utilized as an efficient photocatalyst of organic dyes (Li et al., 2017). Porous TiO_2 using eggshell membrane has been used for the degradation of phenol (Chen et al., 2012). Eggshells with membranes possess unique properties due to their large surface area and enhanced porosity (Baláž, 2014).

Song et al. (2009) synthesized a highly efficient, visible light-active NiO-CaO photocatalyst *via* the precipitation-calcination method for the degradation of organic pollutants. In addition, Song et al. also prepared a photoactive NiO-CaO photocatalyst *via* the mechanical mixing of NiO and CaO powder. Peralta et al. (2018) successfully prepared CaO and treated it at different temperatures. Thereafter, the synthesized CaO was tested *via* the degradation of rhodamine 6 G. In addition, the effects of catalyst dosage, calcination temperature, and pH were also studied. Ahmed et al. (2020) synthesized a novel photocatalyst for the degradation of methylene blue (MB), that is, eggshell-based activated carbon.

In the present work, the preparation, characterization, and application of ground eggshell as an efficient photocatalyst have been discussed. This paper reviews the recent advances in the area of nanotechnology and low-cost solid waste used as effective photocatalysts for water purification.

Applications and advantages of eggshells are mentioned in Figs. 14.1 and 14.2.

14.2 Synthesis and modification of eggshell

In our previous study, nano-sized grounded eggshell powder was successfully synthesized in a ball mill and modified *via* calcination (Singh, Kumari, Chavan, Datta, & Dutta, 2017). Milling enhances the adsorption capability of eggshell

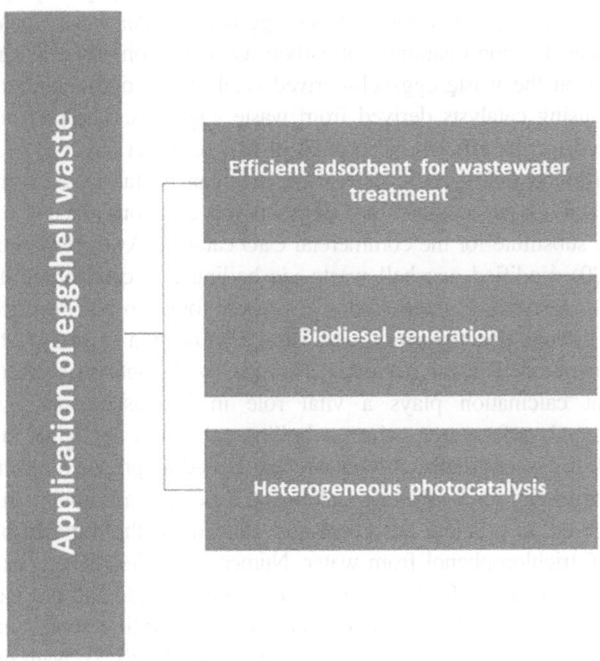

FIGURE 14.1 Applications of eggshell waste.

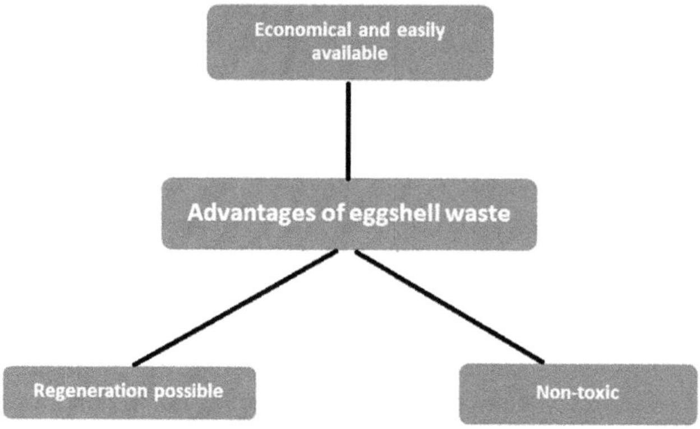

FIGURE 14.2 Advantages of eggshell waste.

waste (Baláž, 2014). Pressure, temperature, concentration, and contact time influence the activity of eggshell (Abdel-Khalek, Abdel Rahman, & Francis, 2017; Ahmad, Kumar, & Haseeb, 2012; Ahmed, Suso, Maqbool, & Hincke, 2019; Asgari & Dayari, 2018). In addition, the eggshell possesses the ability to be utilized as an effective adsorbent and provides excellent support for photocatalysts. Chicken eggshell powder and iron oxide hydroxide doped on the same were synthesized by Praipipat et al. (2023) by washing thoroughly, drying overnight, grinding, and sieving. In addition, doped eggshell powder was prepared using an orbital shaker. Carbon quantum dot-carbon nanotube composite adsorbents were synthesized on the waste eggshell-derived catalysts by one-step chemical vapor deposition using catalysts derived from waste eggshells. The prepared catalysts were utilized for the efficient adsorption of MB (Chen et al., 2021). Caldas et al. (2023) illustrated that specific heterostructures can be fabricated from eggshells for photocatalytic dye degradation. Eggshell waste is considered a cost-free and sustainable substitute for the commercial CaO catalyst. Awogbemi, Inambao, and Onuh (2020) modified eggshell waste *via* boiling and calcination at high temperatures for a specific time period. $CaCO_3$ was found to be converted into CaO due to the thermal decomposition with similar absorption capability after calcination. The mentioned research work performed by Awogbemi et al. (2022) concluded that calcination plays a vital role in increased pore size, thermal deposition, and surface area, whereas boiling exhibits a negligible effect on the catalytic ability of eggshells. Adebowale and Egbedina prepared biocarbon material from eggshells for wastewater treatment. After the adsorption study, carbon was considered an efficient adsorbent and utilized for the removal of bisphenol A and 2,4,6-trichlorophenol from water. Numerous studies have considered eggshells as a raw material for the synthesis of calcium oxide and the source of calcium for the preparation of antimicrobial materials. Antimicrobial materials can further be utilized for water disinfection (Shang et al., 2022). Nath et al. (2023) studied eggshell powder as an efficient heterogeneous catalyst for the degradation of crystal violet dye-contaminated wastewater at room temperature in the presence of H_2O_2. Nanocomposites of eggshell/starch/Fe_3O_4 were utilized for heavy metals decontamination, i.e., Cd2 + and Pb2 + from industrial wastewater. Prepared eggshell nanocomposites act as an efficient adsorbent. Due to their recyclability, eggshells can be utilized as an efficient and economical catalyst for wastewater treatment (Hosseini, Hamadi, Foroutan, Peighambardoust, & Ramavandi, 2022). Sudhakaran et al. (2023) synthesized eggshell composites for phosphate adsorption from wastewater. Novel iron-pillared eggshell and bentonite clay biocomposites were tested for the decontamination of phosphate from the aqueous medium. Electrostatic interaction, diffusion, and ion exchange play a vital role in the adsorption process. The prepared composite possesses excellent recyclability and is considered a green adsorbent. Makin, Adam, Xu, Wan, and Zhang (2022) successfully prepared a Pd–Ru nanocatalyst supported on CaO for the electro-oxidation of ethanol. CaO was derived from waste eggshells. Morphology and particle size could be the factors responsible for the efficient catalytic performance of the prepared nanoparticle. Eggshell waste could be utilized effectively due to the porous structure and the synthesized catalyst exhibiting a large surface area and well-defined morphology. In addition, agglomeration on metal particles was observed on the surface of CaO, probably due to the rapid reduction process. Seid and Gonfa (2022) prepared eggshell-based zinc oxide–cobalt oxide nanocomposites utilizing zinc nitrate hexahydrate and cobalt nitrate hexahydrate, respectively. The prepared nanocomposite was utilized for the removal of Cr(VI) from the contaminated aqueous medium. In addition, the effect of various parameters such as adsorbent dosage, adsorption time, initial pollutant concentration, and pH of the solution was investigated against the removal of Cr(VI). The Cr(VI) percentage of adsorption increases with adsorption time up to a specific time along with increased adsorption dosage. The optimum pH for the removal of the target pollutant was found to be 2, and a reduction in Cr(VI) adsorption was observed at further increased pH. Phenol, along with its derivatives, is dangerous for human health and the environment. Therefore, cost-effective adsorbents are required for the abatement of such toxic substances from the aqueous medium. Tamang and Paul (2022) prepared an adsorbent from chitosan and calcined waste eggshell for effective wastewater treatment by simple precipitation technique. The preparation of adsorbent from waste depends on numerous factors, such as the ratio of chemical impregnation, duration, and temperature. So, optimization of the mentioned parameters is essential to attain maximum adsorption efficiency. Dye-contaminated wastewater is considered dangerous for aquatic life and human health throughout the food chain. Therefore, it is necessary to remove the toxic dyes from the aqueous effluents generated due to enhanced textile sector industrialization. It is required to treat the dye wastewater before releasing it into the open water resources. Praipipat, Ngamsurach, Saekrathok, and Phomtai (2022) synthesized several adsorbents, such as chicken and duck eggshell beads, chicken and duck eggshell beads mixed with iron oxide hydroxide, and chicken or duck eggshell beads mixed with zinc oxide, for the effective removal of dyes. Solvothermal-derived TiO_2 nanoparticles were supported on the powdered nanoeggshell waste in our previous work (Singh et al., 2017). Discarded eggshells were collected, washed, and dried. Furthermore, the membranes of eggshells were separated manually before crushing in the mortar, grinding in the planetary ball mill, and calcination. The finally prepared nanopowder was utilized as an efficient support for the solvothermal-derived TiO_2 nanoparticles. In continuation of our mentioned research, the grounded eggshell was prepared in the presence of an eggshell membrane following a similar methodology. The eggshell support in our previous work was considered a photocatalyst without any additional catalyst. The eggshell membranes possess the porous property and fibrous structure that enhance the photocatalytic activity of the eggshell powder. It is considered agricultural waste. Thus, utilizing the eggshell with its membrane can be an effective remedy for solid waste management, as the eggshell waste disposal issue could be solved in addition to the attainment of enhanced photoactivity for wastewater purification. Fig. 14.3 represents the crushed eggshell, and Fig. 14.4 illustrates the eggshell membrane.

14.3 Characterization of eggshell

Awogbemi et al. (2020), as discussed in the earlier section, revealed that surface area is enhanced as raw eggshell < boiled eggshell < calcined eggshell. The increased surface area indicated the enhanced catalytic capability. Singh et al. (2017) described the characterization of nano-sized grounded eggshells. The utilization of various nanoparticles in the wastewater purification process has enhanced the possibilities for the effective removal of harmful contaminants such as dyes, microbial matter, heavy metals, and several other organic contaminants. Nanotechnology is a rapidly emerging technique due to its excellent physical, chemical, and biological capabilities as compared to bulk particles. In addition, nanoparticle-based technology is considered to be cost-effective and possesses greater efficiency as compared to the other existing techniques for wastewater treatment. Several nanomaterials, such as nano-sized membranes, catalysts, photocatalysts, and adsorbents, are being

FIGURE 14.3 Crushed eggshell.

FIGURE 14.4 Eggshell membrane.

investigated worldwide for the treatment of wastewater (Mondal et al., 2023). Nanotechnology-based water purification techniques are being researched globally as they provide clean, accessible, and affordable water for societal requirements. Nanoparticles possess enhanced sustainability and an affinity to absorb the pollutants. Contaminated water affects human health and aquatic life. Various human activities and industrial effluents are responsible for wastewater generation. Materials with a size range between 1 and 100 nm are considered nanoparticles, which could be confirmed by the field emission scanning electron microscopy (FESEM) image illustrated in our previous work (Singh et al., 2017). The characteristics of nanoparticles exhibit improved optical, physical, and electromagnetic properties. Features of nanoparticles include highly stable and accelerated adsorption, catalysis, and reactivity. Therefore, the implication of nanotechnology, such as nanocatalysts, nanosorbents, and nanomembranes, for the treatment of wastewater is necessary. However, nanoparticles have certain limitations, such as their enhanced agglomeration tendency and the separation of nano-sized catalysts after the reaction. The porosity of eggshell waste plays an important role in the enhancement of its adsorption capability. FE-SEM images of calcined and non-calcined grounded eggshell powder are shown in Fig. 14.5.The figure illustrates that the agglomeration tendency of the ground eggshell nanoparticles decreases after calcination due to the removal of moisture from the calcined sample. Nanoparticles agglomerate rapidly due to their small particle size. It causes the loss of adsorption sites. Due to the removal of moisture, the particles are separated properly. Therefore, the adsorbent or the reaction site increases, exhibiting enhanced adsorption and degradation. FESEM images of grounded nanoeggshells before and after calcination without eggshell membrane were illustrated in our previous research (Singh

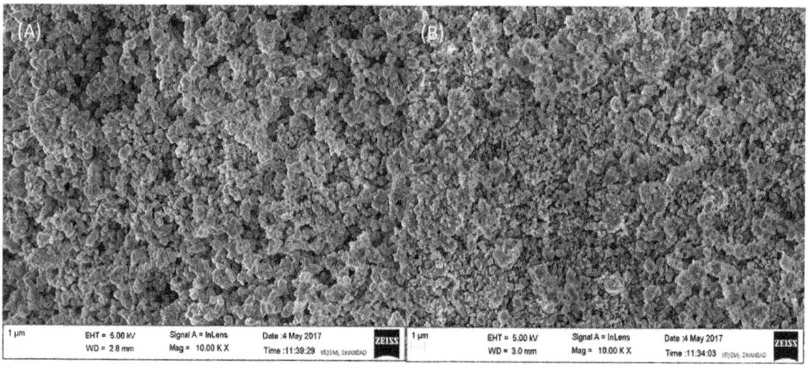

FIGURE 14.5 FE-SEM images of (A) calcined and (B) non-calcined grounded eggshell powder. *FE-SEM*, field emission scanning electron microscopy.

et al., 2017). The FESEM image with membrane showed a greater agglomeration tendency as compared to that without membrane, probably due to the additional moisture content present in the eggshell membranes. In addition, the energy dispersive X-ray (EDX) image of calcined eggshell was also included in our research (Singh et al., 2017), where it is utilized as support.

Numerous characterization techniques have been adopted by researchers worldwide, including UV-Visible (UV-Vis) absorption spectroscopy, FESEM, X-ray diffraction (XRD), and energy dispersive spectrometry (Ayodeji, Ojewumi, Rasheed, & Ayodele, 2018; De Angelis, Medeghini, Conte, & Mignardi, 2017; Eletta, Ajayi, Ogunleye, & Akpan, 2016; Elkady, Ibrahim, & El-Latif, 2011; Flores-Cano et al., 2013).

14.3.1 X-ray diffraction

XRD investigates crystallinity and distinguishes between the crystalline and amorphous natures of the material. Similar to fingerprints, each material possesses a unique XRD pattern. Diffraction patterns assist in the identification of a particular material. XRD is used for structural material characterization in solid-state chemistry for phase detection and dimension calculations of crystalline materials. In addition, peak analysis of diffraction patterns determines different phases through peak ratio calculation and crystalline size through peak broadening.

14.3.2 Ultraviolet-visible spectroscopy

UV-Vis spectroscopy is the optical material characterization analysis within the ultra-violet and visible light ranges. The absorbance curve collected from the analysis confirms the photoactivity of the tested material. The photocatalyst gets excited after the incidence of light at a particular wavelength. It is mentioned in various research studies that a visible light-responsive photocatalyst is required to harvest the solar spectrum effectively. However, visible-light active photocatalysts possess a narrow band gap. The band gap is calculated *via* the mentioned optical spectroscopy technique.

14.3.3 Field emission scanning electron microscopy

FESEM is a material characterization analytical technique that confirms the surface morphology, particle size, and agglomeration tendency. In FESEM, the materials are scanned by electrons in a zig-zag pattern. Different morphologies, such as spherical, rod, wire, fibers, pyramids, and many more, have been observed in various research works. FESEM provides clearer images with high resolution as compared to conventional scanning electron microscopy (SEM). In FESEM, the surface of the samples is scanned, whereas inner morphology, size distribution, and agglomeration tendency are confirmed in transmission electron microscopy (TEM) as electrons transmit through the sample. Therefore, FESEM, SEM, and TEM are frequently utilized in material characterization.

14.3.4 Energy dispersive X-ray analysis

EDX is the material characterization technique that confirms the composition of prepared unknown material. It is frequently done for the compositional analysis of the sample. EDX identifies the components along with their percentage present in the sample. EDX systems are attached to SEM or TEM instruments so that the surface and compositional analysis of the selected sample can be done simultaneously. The EDX could be considered the initial analysis that a researcher performs after the synthesis of the desired material. It works on the fundamental principle that each element possesses a unique atomic structure.

14.3.5 Fourier transform infrared spectroscopy

The Fourier transform infrared spectroscopy (FTIR) analytical technique confirms the nature of the functional group present in the prepared sample. In addition, it detects the residual organic compounds present in the synthesized material. The collected information from FTIR spectra provides a better understanding of the mechanisms involved in the synthesis of materials by various methods. The prepared materials are pressed into thin pallets to obtain an infrared absorption or emission spectrum. FTIR is an infrared spectrometer that collects data over a wide scan range.

14.3.6 Brunauer-Emmett-Teller/Barrett-Joyner-Halenda analysis

Brunauer-Emmett-Teller (BET) and Barrett-Joyner-Halenda (BJH) are significant material characterization techniques utilized for the measurement of the specific surface area and porosity of the synthesized sample, respectively. BET analysis assists in the investigation of the material's specific surface area. BJH analysis determines the pore volume and area *via* the adsorption and desorption processes. This material characterization technique determines the pore size distribution of the sample. The oxidation and reduction reactions executed in the photocatalytic water splitting occur on the catalyst surface. Therefore, large surface area and porosity play a vital role in the enhancement of the adsorption process, that is, a surface phenomenon that results in increased dye degradation for water purification.

14.4 Photocatalytic activity of eggshell

Application of eggshell catalysts includes the synthesis of hydrogen, biodiesel, and dimethyl carbonate, along with wastewater purification. Eggshell is a hydrophobic compound. Therefore, organic and aprotic solvents are required to remove the impurities from the same. The research revealed that calcination-hydration-dehydration is an effective method to enhance the catalytic activity of the eggshells. Numerous advantages of eggshell membranes include non-toxicity, abundance, cost-effectiveness, and high adsorption capability due to functional groups present on the surface (Mosaddegh, 2013; Niju, Meera, Begum, & Anantharaman, 2014; Park et al., 2016). Eggshell is considered a novel source of bioactive compounds such as proteins, glucosamine, collagen, hyaluronic acid, and chondroitin surface. The composition of the eggshell shell and the eggshell membrane plays an important role in improved photocatalytic activity (Ruff, Endres, Clewell, Szabo, & Schauss, 2012; Sah & Rath, 2016; Tizo et al., 2018; Torres, Troncoso, Piaggio, & Hijar, 2010). Biosorption includes the dead cell and the functional groups responsible for the adsorption of dyes. Eggshells have been considered an excellent bioadsorbent for wastewater purification. However, various low-cost adsorbents are utilized for water treatment, such as agricultural waste, fruit remains and peels, tea waste, and many more. Eggshells can be utilized to synthesize different materials that exhibit efficient sorption properties. Pyrolysis of eggshells results in the formation of CaO, which is used to synthesize calcium zeolite. A better understanding of the anatomy of eggs and eggshells is essential before initiating the research on their effective utilization in wastewater purification. Eggs consist of four basic constituents, that is, yolk, egg white, a protective layer (eggshell), and a membrane attached to the eggshell. The nutrient value of an egg lies in the yolk and egg white. Eggshells and their membrane are solid waste that needs to be utilized for the benefit of the environment and society in an environmentally friendly manner. The eggshell consists of three layers: (1) the cuticle, (2) the testa, and (3) the mammillary layer from the outside to the inside. The cuticle is the outermost layer, a dried mucus that regulates moisture evaporation. The testa is also known as the calcium carbonate layer that assists in the growth and coloration of eggs. The mammillary layer remains in direct contact with the eggshell membrane. Matrixes composed of protein fibers are formed together by the cuticle and mammillary layer that form a bond with calcium carbonate. These 2 layers consist of, that is, approximately 7000 to 17000 pores for the transpiration of gaseous exchanges through the eggshell. The thickness of the eggshell is due to the absorption of calcium during its growth, whereas the size depends on the yolk size. Its color depends on the pigment deposition during the formation of eggs in the oviduct. The major contents of the eggshell are calcium and magnesium carbonates, sulfates, and phosphates.

Eggshell and eggshell membranes are nontoxic and possess greater potential for the removal of hazardous contaminants. This technique addresses the issue of solid waste management along with wastewater purification. According to the global research work, eggshell and eggshell membranes exhibited excellent capability in the successful removal of various pollutants, such as phenol, oxalic acid, heavy metals, and harmful dyes, released in industrial effluents. Toxic acid dyes are widely used in the textile industry during the dyeing process due to their high solubility in water. They are unable to be removed by conventional, physical, or chemical methods. According to the reports, eggshells possess the ability for the same due to their unique adsorption capabilities (Guru & Dash, 2014; Laca, Laca, & Díaz, 2017; Mittal, Teotia, Soni, & Mittal, 2016). The photocatalytic activity of prepared grounded eggshells was evaluated by monitoring the photodegradation of the MB and Rhodamine 6 G mixture in an aqueous solution. The solution was stirred continuously at 500 rpm by a magnetic stirrer in the dark for 60 min for the powder to absorb the dye in the solution and reach absorption-desorption equilibrium. Fig. 14.6 illustrates the degradation

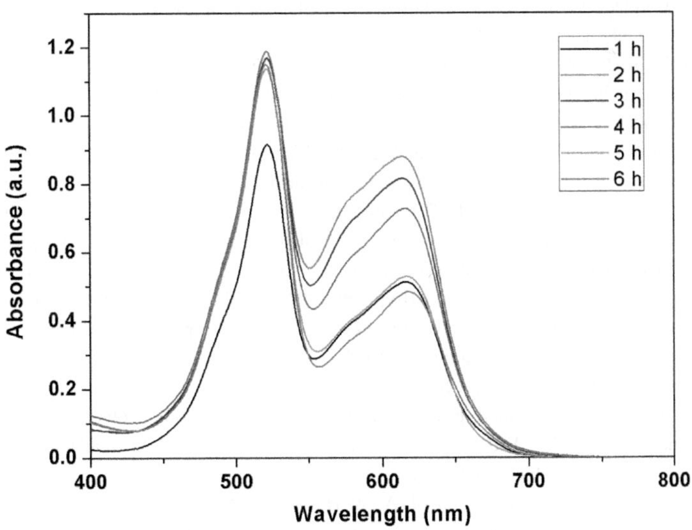

FIGURE 14.6 Illustration of degradation during the photocatalytic reaction.

during the photocatalytic reaction. During 6 h of the experiment, the absorbance was reduced continuously, which indicates the reduction in the pollutant dye concentration with time.

14.5 Conclusion

Clean water is essential for the existence of all living species. The purification of wastewater has become a serious global issue, and researchers are continuously experimenting with improved, economical, and scalable techniques regarding the same. Nanotechnology is considered to play a vital role in the efficient purification of water. The grounded eggshell can be considered an efficient photocatalyst due to its enhanced porosity and surface area. Nano-sized eggshell powder exhibits enhanced photoactivity. Powdered eggshells can be utilized as a photocatalyst for organic pollutant degradation. In addition, solid waste management can also be performed simultaneously by utilizing low-cost grounded eggshell waste. Recovery of nanoparticles is being implemented *via* membrane filtration, magnetic separation, thermal decomposition, desorption, microbial adsorbent regeneration, and advanced oxidation techniques. Recovered nanoparticles can then be recycled, which exhibits the economical aspect of nanosorbents.

References

Abdel-Khalek, M. A., Abdel Rahman, M. K., & Francis, A. A. (2017). Exploring the adsorption behavior of cationic and anionic dyes on industrial waste shells of egg. *Journal of Environmental Chemical Engineering*, *5*(1), 319–327. Available from https://doi.org/10.1016/j.jece.2016.11.043, http://www.journals.elsevier.com/journal-of-environmental-chemical-engineering/.

Adebowale, K. O., & Egbedina, A. O. (2022). Facile green synthesis of bio-carbon material from eggshells and its application for the removal of Bisphenol A and 2,4,6-trichlorophenol from water. *Environmental Nanotechnology, Monitoring and Management*, *17*. Available from https://doi.org/10.1016/j.enmm.2021.100622, http://www.journals.elsevier.com/environmental-nanotechnology-monitoring-and-management/.

Ahmad, A., Jini, D., Aravind, M., Parvathiraja, C., Ali, R., Kiyani, M. Z., & Alothman, A. (2020). A novel study on synthesis of egg shell based activated carbon for degradation of methylene blue via photocatalysis. *Arabian Journal of Chemistry*, *13*(12), 8717–8722. Available from https://doi.org/10.1016/j.arabjc.2020.10.002, http://colleges.ksu.edu.sa/Arabic%20Colleges/CollegeOfScience/ChemicalDept/AJC/default.aspx, ScienceDirect, http://www.sciencedirect.com/science/journal/18785352.

Ahmad, R., Kumar, R., & Haseeb, S. (2012). Adsorption of Cu 2+ from aqueous solution onto iron oxide coated eggshell powder: Evaluation of equilibrium, isotherms, kinetics, and regeneration capacity. *Arabian Journal of Chemistry*, *5*(3), 353–359. Available from https://doi.org/10.1016/j.arabjc.2010.09.003.

Ahmed, T. A. E., Suso, H. P., Maqbool, A., & Hincke, M. T. (2019). Processed eggshell membrane powder: Bioinspiration for an innovative wound healing product. *Materials Science and Engineering C*, *95*, 192–203. Available from https://doi.org/10.1016/j.msec.2018.10.054.

Arami, M., Limaee, N. Y., & Mahmoodi, N. M. (1999). Investigation on the adsorption capability of egg shell membrane towards model textile dyes. *Chemosphere*, *65*(11).

Asgari, G., & Dayari, A. (2018). Experimental dataset on acid treated eggshell for removing cyanide ions from synthetic and industrial wastewaters. *Data in Brief.*, *16*, 442–452. Available from https://doi.org/10.1016/j.dib.2017.11.048, https://www.journals.elsevier.com/data-in-brief.

Awogbemi, O., Inambao, F., & Onuh, E. I. (2020). Modification and characterization of chicken eggshell for possible catalytic applications. *Heliyon*, *6*(10). Available from https://doi.org/10.1016/j.heliyon.2020.e05283, http://www.journals.elsevier.com/heliyon/.

Ayodeji, A. A., Ojewumi, M. E., Rasheed, B., & Ayodele, J. M. (2018). Data on CaO and eggshell catalysts used for biodiesel production. *Data in Brief.*, *19*, 1466–1473. Available from https://doi.org/10.1016/j.dib.2018.06.028, https://www.journals.elsevier.com/data-in-brief.

Baburaj, M. S., Veeran, M. G., Painuly, D., Sreelekshmi, S., Rajkumar, R. J., & Aprem, A. S. (2023). Fabrication and characterisation of polycaprolactone/gelatin/chitosan (PCL/GEL/CHI) electrospun nano-membranes for wastewater purification. *Desalination*, *563*. Available from https://doi.org/10.1016/j.desal.2023.116709.

Baláž, M. (2014). Eggshell membrane biomaterial as a platform for applications in materials science. *Acta Biomaterialia*, *10*(9), 3827–3843. Available from https://doi.org/10.1016/j.actbio.2014.03.020, http://www.journals.elsevier.com/acta-biomaterialia.

Bathla, A., Singla, D., & Pal, B. (2019). Highly efficient CaCO 3 -CaO extracted from tap water distillation for effective adsorption and photocatalytic degradation of malachite green dye. *Materials Research Bulletin*, *116*, 1–7. Available from https://doi.org/10.1016/j.materresbull.2019.04.010, http://www.sciencedirect.com/science/journal/00255408.

Caldas, A. M., Dos Santos, D. F., Castro, M. A. M., Teodoro, M. D., Motta, F. V., & Bomio, M. R. D. (2023). Fabrication of CN-HAp heterostructures from eggshells with improved photocatalytic performance in degrading of mixing dyes under sunlight. *Materials Science in Semiconductor Processing*, *165*, 107660. Available from https://doi.org/10.1016/j.mssp.2023.107660.

Chen, Q., Wang, H., Tang, X., Ba, Z., Zhao, X., Wang, Y., & Deng, H. (2021). One-step synthesis of carbon quantum dot-carbon nanotube composites on waste eggshell-derived catalysts for enhanced adsorption of methylene blue. *Journal of Environmental Chemical Engineering*, *9*(5), 106222. Available from https://doi.org/10.1016/j.jece.2021.106222.

Chen, X., Li, C., Wang, J., Li, J., Luan, X., Li, Y., ... Wang, B. (2010). Investigation on solar photocatalytic activity of TiO2 loaded composite: TiO2/Eggshell, TiO2/Clamshell and TiO2/CaCO3. *Materials Letters*, *64*(13), 1437–1440. Available from https://doi.org/10.1016/j.matlet.2010.03.048.

Chen, Z., Liu, C., Chen, F., Cao, Y., Fan, J., & Zhou, J. (2012). Preparation of porous TiO2 using eggshell membrane as template and its photocatalytic activity. *Procedia Engineering*, *27*, 512–518. Available from https://doi.org/10.1016/j.proeng.2011.12.480, http://www.sciencedirect.com/science/journal/18777058.

Cui, B., Fu, S., Hao, X., & Zhou, D. (2023). Synergistic effects of simultaneous coupling ozonation and biodegradation for coking wastewater treatment: Advances in COD removal, toxic elimination, and microbial regulation. *Chemosphere*, *318*, 137956. Available from https://doi.org/10.1016/j.chemosphere.2023.137956.

De Angelis, G., Medeghini, L., Conte, A. M., & Mignardi, S. (2017). Recycling of eggshell waste into low-cost adsorbent for Ni removal from wastewater. *Journal of Cleaner Production*, *164*, 1497–1506. Available from https://doi.org/10.1016/j.jclepro.2017.07.085.

de Lourdes Ruiz Peralta, M., Sánchez-Cantú, M., Puente-López, E., Rubio-Rosas, E., & Tzompantzi, F. (2018). Evaluation of calcium oxide in Rhodamine 6G photodegradation. *Catalysis Today*, *305*, 75–81. Available from https://doi.org/10.1016/j.cattod.2017.09.057, http://www.sciencedirect.com/science/journal/09205861.

Eletta, O. A. A., Ajayi, O. A., Ogunleye, O. O., & Akpan, I. C. (2016). Adsorption of cyanide from aqueous solution using calcinated eggshells: Equilibrium and optimisation studies. *Journal of Environmental Chemical Engineering*, *4*(1), 1367–1375. Available from https://doi.org/10.1016/j.jece.2016.01.020.

Elkady, M. F., Ibrahim, A. M., & El-Latif, M. M. A. (2011). Assessment of the adsorption kinetics, equilibrium and thermodynamic for the potential removal of reactive red dye using eggshell biocomposite beads. *Desalination*, *278*(1–3), 412–423. Available from https://doi.org/10.1016/j.desal.2011.05.063.

Flores-Cano, J. V., Leyva-Ramos, R., Mendoza-Barron, J., Guerrero-Coronado, R. M., Aragón-Piña, A., & Labrada-Delgado, G. J. (2013). Sorption mechanism of Cd(II) from water solution onto chicken eggshell. *Applied Surface Science*, *276*, 682–690. Available from https://doi.org/10.1016/j.apsusc.2013.03.153, http://www.journals.elsevier.com/applied-surface-science/.

Guru, P. S., & Dash, S. (2014). Sorption on eggshell waste - A review on ultrastructure, biomineralization and other applications. *Advances in Colloid and Interface Science*, *209*, 49–67. Available from https://doi.org/10.1016/j.cis.2013.12.013.

He, Y., Li, Y., Li, X., Liu, Y., Wang, Y., Guo, H., ... Liu, Y. (2023). Net-zero greenhouse gas emission from wastewater treatment: Mechanisms, opportunities and perspectives. *Renewable and Sustainable Energy Reviews*, *184*, 113547. Available from https://doi.org/10.1016/j.rser.2023.113547.

Hosseini, S. S., Hamadi, A., Foroutan, R., Peighambardoust, S. J., & Ramavandi, B. (2022). Decontamination of Cd2 + and Pb2 + from aqueous solution using a magnetic nanocomposite of eggshell/starch/Fe3O4. *Journal of Water Process Engineering*, *48*, 102911. Available from https://doi.org/10.1016/j.jwpe.2022.102911.

Hu, L., Yan, G., Chauhan, B. S., Elbadawy, I., Abouelela, M., Marefati, M., & Salah, B. (2023). Development and evaluation of an electro-Fenton-based integrated hydrogen production and wastewater treatment plant coupled with the solar and electrodialysis units. *Institution of Chemical Engineers, China Process Safety and Environmental Protection*, *177*, 568–580. Available from https://doi.org/10.1016/j.psep.2023.06.072, http://www.elsevier.com/wps/find/journaldescription.cws_home/713889/description#description.

Hu, Z., Guan, D., Sun, Z., Zhang, Z., Shan, Y., Wu, Y., Ren, X. (2023). Osmotic cleaning of typical inorganic and organic foulants on reverse osmosis membrane for textile printing and dyeing wastewater treatment. Chemosphere.177, 139162.

Jiang, Z., Abdoulaye, M. G., Wei, T., Xiao, X., Liang, Y., Xiong, J., ... Chen, Y. (2023). Treatment of Eucalyptus chemical-mechanical pulp wastewater by coupling system of advanced catalytic oxidation and biodegradation: Synergistic effect of ozonation photocatalytic-microbial. *Journal of Water Process Engineering*, *55*, 104071. Available from https://doi.org/10.1016/j.jwpe.2023.104071.

Junior, U. C., Amoresi, R. A., Pereira, C. A., Simões, A. Z., Zaghete, M. A., & Elias Filho, S. M. (2019). Influence of defects on photoluminescent and photocatalytic behavior of CaO/SrTiO3 heterojunctions. *Ceramics International*.

Laca, A., Laca, A., & Díaz, M. (2017). Eggshell waste as catalyst: A review. *Journal of Environmental Management*, *197*, 351–359. Available from https://doi.org/10.1016/j.jenvman.2017.03.088, http://www.elsevier.com/inca/publications/store/6/2/2/8/7/1/index.htt.

Li, Y., Zhou, J., Fan, Y., Ye, Y., & Tang, B. (2017). Preparation of environment-friendly 3D eggshell membrane-supported anatase TiO2 as a reusable photocatalyst for degradation of organic dyes. *Chemical Physics Letters*, *689*, 142–147. Available from https://doi.org/10.1016/j.cplett.2017.10.019, http://www.elsevier.com/locate/cplonline.

Ma, W., Cao, W., Lu, T., Xiong, R., & Huang, C. (2022). Multifunctional nanofibrous membrane fabrication by a sacrifice template strategy for efficient emulsion oily wastewater separation and water purification. *Journal of Environmental Chemical Engineering*, *10*(6), 108908. Available from https://doi.org/10.1016/j.jece.2022.108908.

Makin, A. M., Adam, M. I., Xu, Z., Wan, J. K., & Zhang, G. (2022). Pd-Ru anchored on CaO derived from waste-eggshells for ethanol oxidation electrocatalysis. *Molecular Catalysis*, *528*. Available from https://doi.org/10.1016/j.mcat.2022.112437, http://www.elsevier.com/locate/issn/24688231.

Meng, L., Wang, L., & Wang, Z. (2023). Membrane distillation with electrospun omniphobic membrane for treatment of hypersaline chemical industry wastewater. *Desalination*, *564*, 116782. Available from https://doi.org/10.1016/j.desal.2023.116782.

Mittal, A., Teotia, M., Soni, R. K., & Mittal, J. (2016). Applications of egg shell and egg shell membrane as adsorbents: A review. *Journal of Molecular Liquids*, *223*, 376–387. Available from https://doi.org/10.1016/j.molliq.2016.08.065.

Mohamed, F., Shaban, M., Aljohani, G., & Ahmed, A. M. (2021). Synthesis of novel eco-friendly CaO/C photocatalyst from coffee and eggshell wastes for dye degradation. *Journal of Materials Research and Technology*, *14*, 3140–3149. Available from https://doi.org/10.1016/j.jmrt.2021.08.055, http://www.elsevier.com/journals/journal-of-materials-research-and-technology/2238-7854.

Mondal, P., Nandan, A., Ajithkumar, S., Siddiqui, N. A., Raja, S., Kola, A. K., & Balakrishnan, D. (2023). Sustainable application of nanoparticles in wastewater treatment: Fate, current trend & paradigm shift. *Environmental Research*, *232*. Available from https://doi.org/10.1016/j.envres.2023.116071, https://www.sciencedirect.com/journal/environmental-research.

Mosaddegh, E. (2013). Ultrasonic-assisted preparation of nano eggshell powder: A novel catalyst in green and high efficient synthesis of 2-aminochromenes. *Ultrasonics Sonochemistry*, *20*(6), 1436–1441. Available from https://doi.org/10.1016/j.ultsonch.2013.04.008, http://www.elsevier.com/inca/publications/store/5/2/5/4/5/1.

Nath, A., Biswas, S., & Pal, A. (2023). Eggshell powder as an efficient recyclable catalyst generates H2O2 prompted radicals for selective oxidative mineralization of crystal violet dye at room temperature. *Materials Chemistry and Physics*, *303*, 127785. Available from https://doi.org/10.1016/j.matchemphys.2023.127785.

Niju, S., Meera, K. M., Begum, S., & Anantharaman, N. (2014). Modification of egg shell and its application in biodiesel production. *Journal of Saudi Chemical Society*, *18*(5), 702–706. Available from https://doi.org/10.1016/j.jscs.2014.02.010.

Pan, H. (2016). Principles on design and fabrication of nanomaterials as photocatalysts for water-splitting. *Renewable and Sustainable Energy Reviews*, *57*, 584–601. Available from https://doi.org/10.1016/j.rser.2015.12.117, https://www.journals.elsevier.com/renewable-and-sustainable-energy-reviews.

Park, S., Choi, K. S., Lee, D., Kim, D., Lim, K. T., Lee, K. H., … Kim, J. (2016). Eggshell membrane: Review and impact on engineering. *Biosystems Engineering*, *151*, 446–463. Available from https://doi.org/10.1016/j.biosystemseng.2016.10.014, http://www.elsevier.com/inca/publications/store/6/2/2/7/9/5/index.htt.

Perera, A. A. P. R., Madhushani, K. A. U., Kumar, A., & Gupta, R. K. (2023). Metal-organic frameworks for wastewater treatment: Recent developments, challenges, and future prospects. *Chemosphere*, *339*, 139713. Available from https://doi.org/10.1016/j.chemosphere.2023.139713.

Praipipat, P., Ngamsurach, P., & Pratumkaew, K. (2023). The synthesis, characterizations, and lead adsorption studies of chicken eggshell powder and chicken eggshell powder-doped iron (III) oxide-hydroxide. *Arabian Journal of Chemistry*, *16*(5), 104640. Available from https://doi.org/10.1016/j.arabjc.2023.104640.

Praipipat, P., Ngamsurach, P., Saekrathok, C., & Phomtai, S. (2022). Chicken and duck eggshell beads modified with iron (III) oxide-hydroxide and zinc oxide for reactive blue 4 dye removal. *Arabian Journal of Chemistry*, *15*(11), 104291. Available from https://doi.org/10.1016/j.arabjc.2022.104291.

Ruff, K. J., Endres, J. R., Clewell, A. E., Szabo, J. R., & Schauss, A. G. (2012). Safety evaluation of a natural eggshell membrane-derived product. *Food and Chemical Toxicology*, *50* (3–4), 604–611. Available from https://doi.org/10.1016/j.fct.2011.12.036.

Sah, M. K., & Rath, S. N. (2016). Soluble eggshell membrane: A natural protein to improve the properties of biomaterials used for tissue engineering applications. *Materials Science and Engineering C*, *67*, 807–821. Available from https://doi.org/10.1016/j.msec.2016.05.005.

Saratale, R. G., Sun, Q., Munagapati, V. S., Saratale, G. D., Park, J., & Kim, D. S. (2021). The use of eggshell membrane for the treatment of dye-containing wastewater: Batch, kinetics and reusability studies. *Chemosphere*, *281*. Available from https://doi.org/10.1016/j.chemosphere.2021.130777, http://www.elsevier.com/locate/chemosphere.

Seid, S. M., & Gonfa, G. (2022). Adsorption of Cr (V) from aqueous solution using eggshell-based cobalt oxide-zinc oxide nano-composite. *Environmental Challenges.*, *8*.

Shang, B., Wang, S., Lu, L., Ma, H., Liu, A., Zupanic, A., … Yu, Y. (2022). Poultry eggshell-derived antimicrobial materials: Current status and future perspectives. *Journal of Environmental Management*, *314*, 115096. Available from https://doi.org/10.1016/j.jenvman.2022.115096.

Singh, R., Kumari, P., Chavan, P. D., Datta, S., & Dutta, S. (2017). Synthesis of solvothermal derived TiO2 nanocrystals supported on ground nano egg shell waste and its utilization for the photocatalytic dye degradation. *Optical Materials*, *73*, 377–383. Available from https://doi.org/10.1016/j.optmat.2017.08.040.

Sinha, T., & Ahmaruzzaman, M. (2015). High-value utilization of egg shell to synthesize Silver and Gold-Silver core shell nanoparticles and their application for the degradation of hazardous dyes from aqueous phase-A green approach. *Journal of Colloid and Interface Science*, *453*, 115–131. Available from https://doi.org/10.1016/j.jcis.2015.04.053, http://www.elsevier.com/inca/publications/store/6/2/2/8/6/1/index.htt.

Song, L., Zhang, S., Chen, B., & Sun, D. (2009). Highly active NiO-CaO photocatalyst for degrading organic contaminants under visible-light irradiation. *Catalysis Communications*, *10*(5), 421–423. Available from https://doi.org/10.1016/j.catcom.2008.08.032.

Sudhakaran, S., Mahadevan, H., Fathima S, L., & Krishnan, K. A. (2023). Performance of novel pillared eggshell-bentonite clay bio-composite for enhanced phosphate adsorption from aqueous media. *Groundwater for Sustainable Development*, *22*. Available from https://doi.org/10.1016/j.gsd.2023.100960, http://www.journals.elsevier.com/groundwater-for-sustainable-development/.

Tamang, M., & Paul, K. K. (2022). Adsorptive treatment of phenol from aqueous solution using chitosan/calcined eggshell adsorbent: Optimization of preparation process using Taguchi statistical analysis. *Journal of the Indian Chemical Society*, *99*(1), 100251. Available from https://doi.org/10.1016/j.jics.2021.100251.

Tizo, M. S., Blanco, L. A. V., Cagas, A. C. Q., Dela Cruz, B. R. B., Encoy, J. C., Gunting, J. V., ... Mabayo, V. I. F. (2018). Efficiency of calcium carbonate from eggshells as an adsorbent for cadmium removal in aqueous solution. *Chinese Institute of Environmental Engineering, Philippines Sustainable Environment Research*, *28*(6), 326–332. Available from https://doi.org/10.1016/j.serj.2018.09.002, https://link.springer.com/journal/42834.

Torres, F. G., Troncoso, O. P., Piaggio, F., & Hijar, A. (2010). Structure-property relationships of a biopolymer network: The eggshell membrane. *Acta Biomaterialia*, *6*(9), 3687–3693. Available from https://doi.org/10.1016/j.actbio.2010.03.014, http://www.journals.elsevier.com/acta-biomaterialia.

Ummartyotin, S., & Manuspiya, H. (2018). Eggshell waste: An effective source of hydroxyapatite for photocatalyst. *Journal of Metals, Materials and Minerals.*, *28*(1).

Yahya, N., Aziz, F., Jamaludin, N. A., Mutalib, M. A., Ismail, A. F., Salleh, W. N., ... Ludin, N. A. (2018). A review of integrated photocatalyst adsorbents for wastewater treatment. *Journal of Environmental Chemical Engineering*, *6*(6), 7411–7425. Available from https://doi.org/10.1016/j.jece.2018.06.051, http://www.journals.elsevier.com/journal-of-environmental-chemical-engineering/.

Zhang, H., Luo, W., Du, Y., Shi, H., Zeng, G., Yan, X., & Li, X. (2023). The g-C3N4 decorated carbon aerogel with integrated solar steam generation and photocatalysis for effective desalination and water purification. *Desalination*, *564*, 116821. Available from https://doi.org/10.1016/j.desal.2023.116821.

Zhang, S., Jin, Y., Chen, W., Wang, J., Wang, Y., & Ren, H. (2023). Artificial intelligence in wastewater treatment: A data-driven analysis of status and trends. *Chemosphere*, *336*, 139163. Available from https://doi.org/10.1016/j.chemosphere.2023.139163.

Chapter 15

Global solutions for integrated water resources management in cities throughout the world

Sukhendu Dey[1], Megha Santra[1], Apurba Ratan Ghosh[1] and Palas Samanta[2]

[1]*Department of Environmental Science, The University of Burdwan, Burdwan, West Bengal, India,* [2]*Department of Environmental Science, Sukanta Mahavidyalaya, University of North Bengal, Dhupguri, West Bengal, India*

15.1 Introduction

Water is very much essential in our lives and plays an important role, especially in human activities such as domestic uses, agriculture, industry, sanitation, transportation, energy production, and recreation. Approximately 13×10^{13} million liters of water is present on our planet, in between of this water, 97% is seawater, and from the remaining 3%, 0.4% is accessible to human beings (Aguiar, Martins, Almeida, Ribeiro, & Marto, 2022). The use of water depends on factors such as quality, quantity, and availability. Due to the increasing population in cities, urbanization, industrialization, water demand, energy and food production, and the effects of climate change, water scarcity has become a pressing issue, putting pressure on water resources throughout the world (Setegn & Donoso, 2015).

In particular, as a consequence of population increase and urbanization, cities worldwide are growing quickly, at remarkable speeds of as high as 8% per decade, ("World City Populations 2021," 2021). According to UNDESA's 2018 report, by the year 2050, there will be 9.8 billion people on the planet, 68% of whom will reside in cities. Water availability is put at risk because of these demographic trends, which are compounded by human-caused climate change. Metropolitan thirst for water is expected to rise by 80% by 2050 (Flörke, Schneider, & McDonald, 2018). A shortage of water is a significant issue that has an impact on both the human condition and the world's economy ("The United Nations World Water Development Report 2021," 2021). Urban areas are especially vulnerable

Water Resources Monitoring, Management, and Sustainability. DOI: https://doi.org/10.1016/B978-0-443-23665-5.00015-6

to the threats of water shortages because of overuse and depletion of resources, water contamination, flooding nowadays, and a lack of funding for the upkeep of crumbling infrastructure (Koop & van Leeuwen, 2017). Based on this present scenario, integrated water resource management (IWRM) could be a viable solution to ensuring global water problems to obtain a sustainable use and management of water resources in urban development. IWRM is basically a comprehensive and holistic approach to the sustainable development, allocation, and management of water resources. It considers the entire water cycle, from the source (such as rivers, lakes, and aquifers) to the various uses of water, including drinking water, irrigation, industrial processes, and environmental conservation during the fulfillment of urban requirements/development. However, the use of IWRM for meeting the global crisis of water resources in urban development is underrepresented worldwide because of several concerns. Accordingly, the present chapter will address the challenges and a comprehensive guideline about the global solutions and strategies for IWRM in cities throughout the world. This chapter will also address the probable future prospects of IWRM in urban development.

15.2 Methodology

The article is an amalgamation of data gathered till 2022 through resources belonging to academic, governmental, and international organizations using review-related search terms by adopting Scopus, Web of Science, and Google. The data gathered was carefully examined and verified at multiple points in order to create an unbiased assessment of the subject. In order to assure scientific rigor and eliminate papers that were redundant or unrelated to the subject, the results underwent a rigorous manual screening process.

15.3 What is integrated water resources management?

IWRM is a process that encourages the coordinated use and management of water, land, and related resources in order to optimize economic and social welfare in a fair way without jeopardizing the sustainability of essential ecosystems. IWRM, to put it simply, is a strategy that seeks to ensure the equitable and sustainable management of water resources by taking into account all facets of the water cycle, including its diverse uses, the preservation of ecosystems, and the inclusion of all pertinent stakeholders. It acknowledges that in order to address the complicated issues related to water resources, water management must be integrated and comprehensive.

According to an international organization the Global Water Partnership (GWP), IWRM is defined as "a process that promotes the coordinated development and management of water, land and related resources in order to maximize the resultant economic and social welfare in an equitable manner without compromising the sustainability of vital ecosystems" (Agarwal et al., 2000).

On the other hand, American Water Resources Association (AWRA) defines the term as "The coordinated planning, development, protection, and management of water, land, and related resources in a manner that fosters sustainable economic activity, improves or sustains environmental quality, ensures public health and safety, and provides for the sustainability of communities and ecosystems" (Grigg, 2016).

Generally, IWRM is a term that refers to the holistic approach to managing water resources so that they are used in a sustainable, equitable, and efficient manner. This involves looking at all aspects of water resource use, from the catchment to the consumer, and taking into account physical, chemical, economic, and social factors (Al-Jawad, Alsaffar, Bertram, & Kalin, 2019; Dungumaro & Madulu, 2003). IWRM seeks to optimize the benefits of water resources across all users, including the environment, by considering the entire water cycle. This can involve the development of management plans, watershed management strategies, and the implementation of various policies and regulations (Savenije & Van der Zaag, 2008). Additionally, IWRM requires the development of cross-sectoral collaboration and cooperation between various stakeholders, such as government, industry, and civil society (Suhardiman, Clement, & Bharati, 2015). IWRM is essential for achieving resilient, sustainable water resource management and for adapting to the impacts of climate change (Shams, 2022). The process of implementing IWRM requires an understanding of the water cycle, the user groups, and the regulatory frameworks in place. By taking a holistic approach to water management, IWRM ensures that water resources are managed in a sustainable, equitable, and efficient manner that benefits all stakeholders (Mbaigoto, Alsharif, & Landry, 2023). The steps involved in IWRM process are illustrated in Table 15.1.

IWRM has four pillars: system analysis, science and technology of water, biological aspects of water, health, and nutrition, and planning and policy of water. System analysis covers optimization approaches, system evaluation, simulation modeling, risk assessment, decision analysis, statistical analysis, multicriteria analysis, and the development of indicators, and so on. Science and technology of water include hydrology, water chemistry, quality, conservation, water resources engineering, and environmental contaminants. Biological aspects of water, health, and nutrition involve ecology, epidemiology of water-borne diseases, animal-to-human communication through water, eco-hydrology, and environmental impacts on food and nutrition. Planning and policy of water involve water scarcity at the place of households or regional, national, and international levels and dealing with water security, legal and institutional frameworks, environmental and water resource economy, and so on (Kirshen, Vogel, & Rogers, 2004). The three "E"s or key

TABLE 15.1 Steps involved in integrated water resources management strategy.

Process of developing an IWRM strategy	
Steps	**Tasks**
Establish status and overall goals	Progressive framework toward IWRM along with recent international developments and goals related to water resource issues
Build commitment to reform process analyze gaps	Political awareness, multistakeholder dialog, and water resource management functions required
Prepare strategy and action plan	Links to national policies and institutional roles to enabling environment through instrumental management
Build commitment to actions	Identify financing development, political adoptation, and stakeholder acceptance
Implement frameworks	IWRM framework which is related to building capacity and water infrastructure development
Monitor and evaluate progress	Progress indicators toward IWRM and develop the water infrastructure framework

IWRM, Integrated water resources management.

policy goals of IWRM are Equity, Ecological Integrity, and Efficiency (Kashem & Mondal, 2022):

- **Equity** refers that water is a basic need and right for everyone to access water based on quality and quantity.
- **Ecological integrity** refers to enough quantity and quality of water to persist in the environment but the process of using water should be proper way so that the future generation can be able to use water just like the present generation.
- **Efficiency** refers that water must be used based on maximum efficiency, finite, and vulnerable nature. Cost recovery of water service should be included and depending on economic value, water should be priced.

Practically Global Water Partnership (GWP) created a toolbox to support the implementation of IWRM principles Fig. 15.1 with three categories: enabling environment, institutional roles, and management instruments (Agarwal et al., 2000):

- **Enabling environment** refers to securing the rights and amount of all stakeholders, and protecting public assets, and establishing a general framework of national policies, legislation, and regulation.

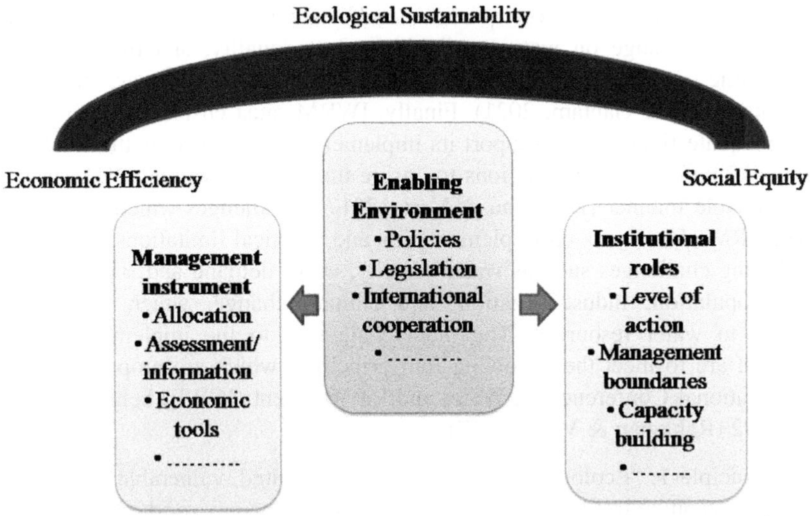

FIGURE 15.1 General framework. Modified from *Agarwal, A., Angeles, Bhatia, R., Chéret, I., Davila-Poblete, S., Falkenmark, M.,... Wright, M. (2000). Stockholm: Global water partnership.* Integrated Water Resources Management, *1–67.*

- **Institutional role** refers to the consideration of a whole range of formal rules, regulations, customs, practices, ideas, information, interests, and community group networks which together provide decision-makers and an institutional framework.
- **Management instruments** involve operational instruments for effective regulation, monitoring, and supporting decision-makers.

15.4 Challenges of integrated water resources management

The challenges of IWRM are becoming increasingly more complex and demanding (Fulazzaky, 2014). IWRM seeks to balance the needs of both human activities and the environment and ensure that water resources are used sustainably. This requires the integration of a multitude of stakeholders, including government agencies, civil society groups, industry, and water users, at all levels of governance (Mishra, 2023). Furthermore, IWRM must also take into account the full life cycle of water resources, from the source of the water to its eventual disposal or reuse. This often involves the optimization of existing infrastructure, the development of new infrastructure, and the management of surface and groundwater resources (Saritas & Proskuryakova, 2017). At the same time, this must be done in a manner that is sensitive to environmental and social concerns. This requires an understanding of the complex interconnections between different parts of a water system, as well as how different land uses affect the availability and quality of water resources. It also requires an understanding of the potential impacts of climate change on water availability, water quality, and the associated economic, social, and environmental costs (Mishra, Kumar, Saraswat, Chakraborty, & Gautam, 2021). Finally, IWRM must ensure the availability of adequate financing to support its implementation, as well as the establishment of policies and regulations to ensure that water resources are used in a sustainable manner (Rahaman & Varis, 2017). Challenges which are related to IWRM often focus on implementation and practical limitations. The world is facing challenges such as water scarcity, water demand and supply due to overpopulation, industrialization, and climate change. which are posing threat to water resources. The main challenges to the implementation of IWRM are to meet the following four principles which are proposed by the International Conference on Water and Environment (ICWE) held in Dublin, in 1992 (Rahaman & Varis, 2017):

- Principle-1: "Ecological"—fresh water is a limited, vulnerable and important resource to live life, development and for the environment.
- Principle-2: "Institutional"—The development and management of water resources must be participatory, involving all levels those who plan, use, and decide.
- Principle-3: "Gender"—Women play a crucial role in water supply, management, and safeguarding.
- Principle-4: "Economic"—Water has an economic value from all aspects, from all its competing uses, and should be recognized as an economic and social good.

The organization also specifies some different areas to pay attention, such as protection of water resources, water quality, and aquatic ecosystems, integrated water resources development and management, water resources assessment, drinking water supply and sanitation, water for sustainable food production, urban and rural development (Makanda, Nzama, & Kanyerere, 2022). Other challenges are identified to the implementation of the IWRM which are as follows (Fulazzaky, 2014; García, 2008; McDonnell, 2008; Rahaman & Varis, 2017):

- Sustainable and optimum distribution and uses of surface and groundwater resources without quantitative-qualitative overuse, considering ecological functions of water resources.
- Plan-making to manage water resources for the distribution and multiple uses of water and their multiple objectives including social, economic, and environmental aspects.
- Defining integrated political actions adapted to regional conditions and cultures involving coordination with other areas at government levels.
- Increasing water use efficiencies across sectors by dramatically reducing waste and increasing water availability and incorporate irrigation

management into water management in its three dimensions, such as scarcity, excess, and quality.
- Balancing productive development with the human right of access to water and preservation of the ecosystem.
- Establishing tools and legal framework to face climate change, and
- Establishing qualitative and quantitative information systems.

15.5 Integrated water resources management and city development for sustainability

IWRM is crucial in the development of cities for several reasons:

Sustainable Water Supply: To meet the demands of its citizens, cities depend on a reliable and hygienic water supply. IWRM ensures that water resources are managed sustainably to meet present and future demands. Planning, conservation, and effective water usage are all included, which is crucial for urban development.

Water quality: It's crucial for everyone's health to keep water sources at a high standard. IWRM involves steps to stop pollution and contaminating water sources, ensuring that urban inhabitants have a safe and dependable water supply.

Urban Development: As cities expand, the demand for water supplies rises. By preparing for an adequate water supply, wastewater treatment, and stormwater management, IWRM aids in controlling this increase. It makes sure that the city's growth won't deplete water supplies or worsen the environment.

Climate Resilience: IWRM considers how climate change may affect the occurrence of extreme weather events and increased precipitation variability. Cities can better plan for and respond to difficulties related to climate change by integrating climate resilience into water management.

Environmental protection: Major natural ecosystems, such as wetlands, lakes, and rivers, are frequently found close to cities. IWRM takes into account how water use affects the environment with the goal of preserving these ecosystems and the biodiversity they support.

Infrastructure Efficiency: Putting IWRM concepts into practice helps make water infrastructure more effective, lowering water losses from leaks and ineffective distribution systems. This can help the city save money and resources.

Conflict Prevention: Water resources can be a source of conflict in many areas, particularly in urban areas where competition for few resources is fierce. IWRM promotes collaboration and negotiation among many parties to avoid disputes over the usage of water.

Inclusivity: IWRM encourages fair access to water resources through inclusivity. This is crucial in urban areas where unequal access to water

can result in social and economic inequality. No matter their socioeconomic standing, all inhabitants should have access to clean and safe water, which is made possible by proper management.
Economic growth: Adequate and dependable water supplies are essential for urban economic growth. Water is essential for businesses, industries, and agriculture, and IWRM works to assure a consistent supply, which in turn promotes economic expansion and job development.
Resilience to Water Scarcity: Water scarcity can be a serious problem in some areas. Resilience to it. IWRM encourages the effective use and conservation of water resources, assisting in making cities more resilient to water shortages.

In conclusion, IWRM is crucial for the resilient and sustainable growth of cities, covering a range of issues including socioeconomic fairness, environmental preservation, water availability, and quality. To meet the varied requirements of urban populations while preserving the environment and maintaining long-term viability, it encourages the prudent use of water resources.

15.6 Water-efficient policies and technologies toward global solution of water crisis

Water conservation has been a growing concern and priority for many countries around the world due to the increasing demand for freshwater resources. As such, the implementation of water-efficient policies and technologies has become a necessity in order to address the impending crisis of water scarcity (Ingrao, Strippoli, Lagioia, & Huisingh, 2023). One of the most important tools for water efficiency is the use of efficient irrigation systems. For example, drip irrigation systems provide a much more efficient means of delivering water to crops compared to traditional surface irrigation methods. This technology not only allows for water to be delivered to the roots of the plants more directly but also eliminates the potential for water loss through runoff, evaporation, and wind drift. Additionally, this method allows for more precise control over the amount of water delivered to each plant, resulting in an optimization of water usage and crop yield (Yang et al., 2023). Other water-efficient technologies include the use of rainwater harvesting systems, greywater systems, and wastewater treatment and reuse systems (Maniam et al., 2022). Rainwater harvesting systems are designed to capture, filter, and store rainwater in order to reduce the reliance on traditional water sources (Zabidi, Goh, Chang, Chan, & Zakaria, 2020). Greywater systems are systems that collect, filter, and reuse the wastewater from showers, baths, and washing machines (Rakesh, Ramesh, Murugaragavan, Avudainayagam, & Karthikeyan, 2020). Finally, wastewater treatment and reuse systems are designed to capture, treat, and reuse wastewater from industrial and commercial operations in order to reduce water consumption. The implementation of these water-efficiency policies and technologies is essential to reducing water usage and ensuring that our water resources are used in the most efficient and sustainable way possible (Shahangian, Tabesh, Yazdanpanah, Zobeidi, & Raoof, 2022). In order to promote these initiatives, governments and organizations must invest in research and development, provide incentives to those who adopt these technologies, and create regulations that require companies and individuals to use these water-efficient technologies. By doing so, we can ensure that our water resources are used wisely and sustainably (Sahoo & Goswami, 2024). IWRM is a holistic approach to manage water resources involving economic, financial, technical, and institutional prospects (Loucks & van Beek, 2017b). IWRM has management instruments which are "tools and methods that enable and help decision-makers to make rational and informed choices between alternative actions." A wide area of qualitative and quantitative approaches is available and combines disciplines such as hydrology, hydraulics, economics, environmental science, and sociology (Agarwal et al., 2000). Water scarcity is rising year by year due to the increasing demand for water resources throughout the world. Serious attention is needed to manage the water resources though there is no specific model or tools to prevent the water shortage but some advanced methods or water-efficient technologies can be carried out including rainwater harvesting, drip irrigation, and use of treated wastewater for nonpotable purposes.

Rainwater harvesting: Water is a vital resource for our planet, and it is essential for sustainable development. To ensure that this resource is managed effectively, IWRM has proven to be an important concept in recent decades. IWRM is a process that encourages the coordinated development and management of water, land, and related resources in order to maximize the economic and social welfare of a given region (Galvez & Rojas, 2019). One of the strategies used in IWRM is rainwater harvesting, that is, collecting and storing rainwater to meet future needs. Rainwater harvesting is a very important element of IWRM as it has several benefits. First, it helps to offset the use of freshwater from nonsustainable sources, such as groundwater. Secondly, it helps to reduce the potential for flooding due to excessive runoff. Thirdly, it helps to reduce water pollution by reducing the amount of contaminants that enter water bodies during heavy rain events. Finally, it helps to ensure that water is available during periods of drought or other extreme weather conditions (Shaari, 2020). Rainwater harvesting systems can be set up in a variety of ways, depending on the needs of a particular area. For example, a simple system may involve simply collecting rainwater on a roof and channeling it into an underground tank or other collection container (Campisano et al., 2017). A more complex system may involve a network of tanks and pipes that collect and store rainwater and then pump it into an irrigation system for agricultural use. No matter the specific type of rainwater harvesting system, it is important to ensure that it is well-maintained in order to maximize its effectiveness. This can include regularly cleaning the collection surface, ensuring that the water is stored in a suitable environment, and regularly monitoring the water quality. Proper maintenance is key to making sure that a rainwater harvesting system is able to meet its full potential (García-Ávila, Guanoquiza-Suárez, Guzmán-Galarza, Cabello-Torres, & Valdiviezo-Gonzales, 2023). Nowadays rainwater harvesting is seeking extra attention as an alternative source in IWRM. Rainwater harvesting means collecting the rainwater, and then transporting and storing it. Collecting and storing rainwater in urban areas mainly those cities that suffer water scarcity can be a supplemental water resource. Collection of rainwater is an old technique and the complexity and scope of this technology may vary from garden irrigation to large-scale systems where rainwater harvesting is used as a primary source (Rahman et al., 2014). For rainwater harvesting, rainwater is collected from different rainfall-runoff sources such as rooftops, rocks, ground surfaces, and other surfaces and used for potable and nonpotable purposes (Li, Boyle, & Reynolds, 2010). Rainwater harvesting has five controllable components such as catchment surface, conveyance system, filter, storage tank, and water pumps. The catchment surface means how much amount of rainwater can be collected. Depend on amount rainwater storage tank required. Rainwater is conveyed from the roof to the storage tanks. Each inch of rainwater produces 623 gallons of water per 1000 square feet of roof surface. Basically, the filters are used to prevent pollutants such as airborne particles, tree leaves, animal dung, and pumps are used to discharge the rainwater to destinations (Gibney, O'Toole, Sinclair, & Leder, 2014; Gleick, 1993; Grabow, Taylor, & De Villiers, 2001; Jofre, Lucena, Blanch, & Muniesa, 2016; Laws, Dickenson, Johnson, Snyder, & Drewes, 2011; Lazarova et al., 2001).

Drip irrigation: IWRM is an approach that can be applied to the use of water in agricultural production (Candido, Coêlho, De Moraes, & Florêncio, 2022). In the case of drip irrigation, IWRM encompasses many aspects, including the conservation of water resources, the efficient and effective management of water in different areas of the agricultural sector, and the use of water-saving techniques such as drip irrigation. When it comes to the conservation of water resources, drip irrigation has been shown to be one of the most efficient and sustainable methods of water resource management (Alcon, Navarro, De-Miguel, & Balbo, 2019). It utilizes the natural landscape to conserve water and provide a more consistent supply of water for crop production. Through the use of geo-referenced soil moisture sensor, water is only applied when it is needed, reducing the risk of overirrigating. This helps to ensure the long-term health of the water resource and reduce the amount of water needed for crop production (Yang et al., 2023). In addition to water conservation, IWRM through the use of drip irrigation can help improve soil fertility and reduce the risk of water contamination. By delivering precise amounts of water at the root level, drip irrigation can help to reduce the risk of leaching nutrients from the soil. Also, because the water is delivered directly to the roots of the plants, there is less opportunity for accidental contamination by runoff or overirrigation (Yang et al., 2023). Finally, drip irrigation can help to reduce the labor required for crop production. By delivering water at the root level, the need for manual irrigation is significantly reduced, saving time and labor costs for farmers (Dawit, Dinka, & Leta, 2020). Overall, IWRM through drip irrigation can help to provide a more sustainable and efficient system for agricultural production. By utilizing modern technology to conserve resources, reduce the risk of water contamination, and reduce labor costs, IWRM through drip irrigation can help to create a more resilient and sustainable agricultural sector. Drip irrigation is a modern technique for method of irrigation where small quantity of water can be spread out near the root of plant drop by drop and maintain the soil moisture (Manda, Addanki, & Srivastava, 2021). In drip irrigation water is distributed as a daily requirement of plant in small quantities and concentrated in root zone which can help to produce abundant products with high quality. Drip irrigation is dependent on some factors as a condition such as the irrigation water quality, soil quality, climatic conditions, and the species of crops (Liu et al., 2017). Some components required for drip irrigation include a pumping station, fertilizer equipments, filtration unit, valves, pressure gauge, main and subline, lateral line, and emitters or drippers (Shareef, Ma, & Zhao, 2019). The most valuable advantages of drip irrigation are as follows: limited patch of soil can irrigate; during irrigation, small quantity and low pressure of water is required; fertilizers are distributed with water; there are no wet plant leaves and low amount of evaporated water; after irrigation, improved quality and quantity of production can be found; there are opportunities for double cropping; reduced need for manual labors; and it is an automatic technique (Mačkić et al., 2023). Moreover, drip irrigation is the greatest water-conserving and water-efficient technique along with no wind-blown, decreased weed growth, and slight evaporation of water. Drip irrigation is the most energy-sufficient and can save up to 50% of water compared to other irrigation techniques (Çetin & Akalp, 2019).

Use of treated wastewater for nonpotable purposes: The reuse of treated wastewater for nonpotable purposes has become increasingly popular in recent years as a way of conserving water and protecting the environment (Maniam et al., 2022). There are a number of different ways to treat wastewater, depending on the type of contaminants present. Once treated, water can be used in a variety of ways, including irrigation, groundwater recharge, aquaculture, industrial cooling, and more (Mohsin et al., 2023). The process of wastewater treatment begins with a physical separation of solid materials from the liquid. This is followed by chemical bonding that captures dissolved contaminants and then a biological process involving bacteria or other microorganisms that break down organic material. After settling, the water is disinfected and filtered to remove any remaining impurities, then stored and tested for safety (Crini, Lichtfouse, Wilson, & Morin-Crini, 2019). The safe use of treated wastewater for nonpotable purposes is critically important, and it is essential that the treatment process is done correctly. It is also important to ensure that any residual contaminants are removed so that the water is safe for its intended purpose. As such, treated wastewater should be regularly monitored for safety, to ensure it is not posing a risk to public health or the environment. The use of treated water and reuse of wastewater is an alternative resource throughout the world to reduce water scarcity and to manage water resources which are renamed as recycled water. This water reuse is also a part of IWRM including wastewater management. Treated water can be used for a nonpotable purpose such as irrigation and industry, as well as potable purposes. This is known as water recycling (Rasul, Neupane, Hussain, & Pasakhala, 2021). Water reuse naturally is the process of using treated wastewater which is commonly known as reclaimed water and used in various areas. In this 20th century, the water is treated by some modern wastewater treatment processes and these processes effectively remove biodegradable materials, pathogens, and nutrients (Sun, Chen, & Chen, 2011). Additionally, some innovative and modern water purification ideas and water treatment technologies can improve the water quality. The innovative ideas are ultraviolet (UV) disinfection, advanced oxidation processes, nanotechnology-based filtration, and membrane technologies which can also help to remove the contaminants. Moreover, desalination is the process of removing impurities and salt from seawater and brackish water to produce freshwater as a resource. Advancements in desalination technologies, such as forward and reverse osmosis and membrane distillation, can help to make the process more energy-efficient and cost-effective and expand access to freshwater in wide arid regions (Nallakukkala, Kassim, Othman, & Lal, 2020).

Spatial planning: IWRM is an important tool for improving the sustainability of water resources in a given region. It involves the coordination of different stakeholders to ensure effective and efficient management of water resources, with an emphasis on sustainability and equitable access (Ben-Daoud et al., 2023). In Spatial Planning, IWRM is a key factor in ensuring that water resources are managed in a way that best meets the needs of all stakeholders (Kidd & Shaw, 2007). IWRM in Spatial Planning involves the development of strategies and policies to ensure that water resources are managed in a sustainable and equitable way. This includes the identification of areas where the water resources are at risk, and the implementation of strategies and policies to reduce these risks. It also involves the development of plans for the efficient use of water, for example, through water harvesting, wastewater management, desalination, and other water-saving measures (Cameron & Katzschner, 2017). In addition to strategies and policies, IWRM in Spatial Planning also requires the consideration of social and economic impacts associated with water resources. This includes the assessment of the potential impacts on livelihoods, public health, and local economies. Furthermore, it requires the integration of water resources into land-use planning processes, to ensure that water resources are managed in a way that is socially and economically beneficial for all stakeholders (Seigerman et al., 2023). Overall, IWRM is an integral part of Spatial Planning and ensures that water resources are managed in a way that meets the needs of all stakeholders. It is essential for sustainable and equitable water resource management and is key to ensuring that the potential economic, social, and environmental impacts associated with water resources are taken into account (Kidd & Shaw, 2007).

Urban development: Urban development has long relied upon IWRM to ensure that water is managed in a sustainable, efficient, and equitable manner (Batalhão, Bouloumytis, Zuffo, & Pimentel da Silva, 2023). IWRM is a holistic approach to water management that takes into account all aspects of water use such as water supply, wastewater management, agriculture, industry, and recreation (Fulazzaky, 2014). It requires coordination between different stakeholders to ensure that water resources are managed in a way that benefits the entire community. In urban settings, IWRM seeks to balance the needs of citizens, businesses, and industry while striving for environmental sustainability. This is done by addressing issues such as water security, pollution prevention, and protection of quality of life. IWRM encourages stakeholders to work together to develop solutions to manage water resources in an effective, efficient, and equitable manner (Ngene et al., 2021). For example, cities may work together to develop water supply systems that provide citizens with access to reliable and safe sources of water. Wastewater management strategies can be implemented to reduce pollution and ensure safe disposal of wastewater (Khan, Ponce, Yu, Golpîra, & Mathew, 2022). Similarly, IWRM strategies can be used to ensure that agricultural and industrial activities are conducted in ways that minimize negative impacts on water resources (Ferreyra, de Loë, & Kreutzwiser, 2008). Ultimately, IWRM strategies seek to ensure that water resources are managed in ways that benefit the entire community while protecting the environment. By implementing these strategies, cities can better meet the needs of their citizens and ensure that water is managed in a sustainable and equitable manner (du Plessis, 2023).

Disaster risk reduction: IWRM is a holistic approach to managing water resources sustainably. It seeks to reduce the vulnerability of people and communities to natural disasters by improving the management of water resources. IWRM considers all sources of water, including surface water, groundwater, and precipitation, along with water-dependent ecosystems and social and economic activities (Lenton & Muller, 2012). Key components of IWRM include developing policies to manage and protect water resources, engaging stakeholders, and strengthening local capacity for water management (Katusiime & Schütt, 2020). A key focus of IWRM is Disaster Risk

Reduction (DRR), which is an approach to reducing the risk of disasters through preparedness, planning, and response measures. The United Nations Office for Disaster Risk Reduction (UNISDR) defines DRR as the "systematic effort to reduce disaster risks and increase resilience through the rational and coordinated management of natural, technological and social hazards" (Rubert & Beetlestone, 2014). IWRM takes a comprehensive approach to DRR, focusing on preventive measures and early warning systems, as well as the implementation of strategies to reduce risks and improve resilience (Widiyono, Renaud, & Sudmeier-Rieux, 2022). Efforts to incorporate IWRM into DRR include creating early warning systems, assessing water-related disasters, and developing strategies to reduce the impact of disasters on water resources (Grobicki, MacLeod, & Pischke, 2015). IWRM also seeks to reduce the risk of potential disasters through preventive measures, such as restoring ecosystems, developing improved land use and land management practices, and fostering better water governance. These efforts can lead to improved water quality, increased water availability, and decreased water-related risks. Ultimately, IWRM is an important element of DRR, as it provides a comprehensive framework for managing water resources in a sustainable way (Nath, Vyas, Deogade, & Chandra, 2023). IWRM has identified some successful strategies such as capturing society's opinion, recognizing water quality and quantity linkage, rearranging the planning process, coordinating land and water resources management, concerted use of surface and groundwater resources, restoring and protecting the natural system, and including consideration of climate change and disaster risk reduction.

Green infrastructure: The importance of IWRM in Green Infrastructure cannot be overstated. Green Infrastructure can provide communities with access to water, reduce the risk of flooding and water pollution, promote biodiversity, and provide natural beauty to urban environments (Pitman, Daniels, & Ely, 2015). IWRM is a holistic approach to managing water resources that draws upon existing knowledge and resources while taking into account the multiple objectives of stakeholders (Saravanan, Mcdonald, & Mollinga, 2009). It involves balancing the different interests of people, industry, agriculture, and the environment. The key to effective IWRM is a comprehensive and coordinated approach to water management. This requires multistakeholder collaboration and empowerment, with the aim of finding solutions that are mutually beneficial to all stakeholders (Sigalla, Tumbo, & Joseph, 2021). In order to facilitate this collaboration, IWRM should include a wide range of stakeholders, including government, industry, environmental organizations, academia, and local communities (Lankford, Bakker, Zeitoun, & Conway, 2013). At the project level, IWRM should involve an iterative cycle of analysis, planning, implementation, monitoring, and evaluation (Trimble et al., 2020). This process should involve identifying and assessing current and future water needs, developing strategies to improve water management, and monitoring and evaluating the effectiveness of strategies (Ben-Daoud et al., 2023). Green Infrastructure is an invaluable tool for enhancing the effectiveness of IWRM. Green infrastructure can provide a range of benefits to water resources, including improving water quality, reducing stormwater runoff and flooding, providing wildlife habitat, and providing aesthetic value to urban environments (Cascone, 2019). IWRM in Green Infrastructure is essential for sustaining healthy water resources. By making use of Green Infrastructure, we can ensure that our water resources are managed sustainably and equitably (Liu & Jensen, 2018). Green infrastructure (GI) plays an important role in water management strategy. GI captured and stored the rainwater to reuse the rainwater as water resources instead of waste by using natural substances such as soil and plants. Reduced flooding, increased water supply, smog and heat mitigation, health, cost, and quality of life improvements all are the benefits of GI.

Wetlands as reservoirs of water: The wetlands were an important source of freshwater and a critical habitat for many species of wildlife. After decades of unregulated exploitation, the Wetlands were in dire need of better management practices (Lemly, Kingsford, & Thompson, 2000). The IWRM model was adopted as a means to rehabilitate the wetlands and ensure their sustainable development (Batalhão et al., 2023). The IWRM model aimed to balance the development of water resources with conservation and human use, taking into account the natural complexities of the Wetlands as an environmental system (Overton et al., 2014). It incorporated various strategies such as public awareness campaigns, improved irrigation systems, stricter enforcement of regulations, and the rehabilitation of damaged areas. One of the key components of the IWRM model was the involvement of local communities in the management of the wetlands (Katusiime & Schütt, 2020). By ensuring that the inhabitants of the Wetlands had a stake in its protection, it was hoped that the conservation goals of the IWRM model would be more successful. To this end, a network of local stakeholders was created to provide information and advice on proposed initiatives. The local stakeholders also had the opportunity to contribute their own recommendations and solutions in order to make the IWRM model more effective (Watson, Shrubsole, & Mitchell, 2019). In addition, the IWRM model included a process of monitoring and evaluation to ensure that the desired goals were met. This included regular surveys of the wetlands to assess the ecological conditions, population surveys of flora and fauna, and water quality and quantity measurements (Overton et al., 2014). By gathering data in this way, it was possible to adjust the strategies employed in the IWRM model in order to ensure that the wetlands would be managed sustainably (Petit, 2016). By taking into account both the environment and the needs of the people, the IWRM model succeeded in rehabilitating the Wetlands, making them a healthy and vibrant source of freshwater (Volenzo & Odiyo, 2018). Wetlands are also another process to improve water quality and water resources management. Wetlands are the areas where water covers the soil for a certain period of time. Ramsar

Convention defines the term as "Wetlands are areas of marsh, fen, peatland or water, whether natural or artificial, permanent or temporary, with water that is static or flowing, fresh, brackish or salt, including areas of marine water the depth of which at low tide does not exceed six metres" (Wang, Hou, & Xue, 2017). Wetlands can remove various pollutants such as organics, trace elements, and nutrients to improve water quality through physical, chemical, and biological processes (Almuktar, Abed, & Scholz, 2018).

Green roofs: The concept of IWRM is an integral part of sustainable development. It seeks to ensure the efficient and sustainable use of water resources for the benefit of all stakeholders (Ben-Daoud et al., 2023). Green roofs can be a vital part of this effort, as they help to reduce water runoff from hard surfaces, reduce urban heat island effects, and promote natural sustainability (Irfeey et al., 2023). Green roofs use plants instead of traditional roofing materials to protect buildings from the elements (Besir & Cuce, 2018). The plants absorb rainfall and reduce water runoff, and they also act as a natural insulator, helping to keep buildings cooler in the summer and warmer in the winter. In addition, green roofs create an opportunity to establish urban habitats for birds, insects, and other wildlife that are often displaced due to habitat loss (Shafique, Kim, & Rafiq, 2018). The effectiveness of green roofs as a component of IWRM depends on the design and maintenance of the roof system (Zaina, Fadli, & Hosseini, 2022). The types of plants used should be carefully considered, as some species are better adapted to withstand the extreme temperature differences found on rooftops. Proper installation and regular maintenance will help ensure that the system is functioning properly and providing the desired benefits (Nagase & Dunnett, 2010). Overall, green roofs provide a valuable contribution to IWRM by reducing water runoff, increasing energy efficiency, and promoting wildlife habitats. They are a cost-effective way to make an immediate impact on an organization's environmental footprint. With thoughtful planning and ongoing maintenance, green roofs can make a significant difference in the overall sustainability of a community (Anderson & Gough, 2022). Green roofs are one of the types of GI. Green roofs are the process of installing plant or vegetation on rooftop consisting of growth medium (substrate) and waterproofing membrane. Nowadays green roof is a revitalized concept for developed countries (Getter & Rowe, 2006). Green roofs are also referred to as cool roofs, vegetated roofs, eco-roofs which have ecological benefits, and roof garden or living roofs (Badescu & Sicre, 2003; Francis & Lorimer, 2011; Getter, Bradley Rowe, & Cregg, 2009; Oberndorfer et al., 2007; Ondimu & Murase, 2007; Sailor, 2008; Voyde, Fassman, & Simcock, 2010; Xu, Sathaye, Akbari, Garg, & Tetali, 2012). Green roofs are made up of various components as layers such as waterproofing layer, insulation, drainage layer, filter or geotextile layer, growing medium, and vegetation. They have multiple benefits such as social, environmental, and economical (Cascone, 2019). Other benefits of green roofs are stormwater retention to reduce peak flow and runoff, water quality enhancement for water utilization, thermal benefits to improve the environment and reduce energy costs, air cleaning for easy comfort into urban areas, noise reduction, and so on. Green roofs have four categories such as intensive, semiintensive, single-course extensive, and multicourse extensive (Abass, Ismail, Wahab, & Elgadi, 2020).

Effective governance and institutional framework: Water resource management is a critically important part of any nation's governance framework, as it can have a dramatic impact on the lives of citizens and the overall health of the environment (Mishra et al., 2021). In this chapter, we will explore the ways in which effective governance and institutional frameworks can be put into place in order to ensure that IWRM initiatives are successful (Sukereman, Suratman, & Rahim, 2022). First, it is important to recognize that IWRM initiatives require strong political leadership and commitment in order to be successful. This can come from both the national and local levels, as different regions may have different needs when it comes to managing water resources. In addition, it is important for governments to ensure that there is an institutional framework in place that provides the necessary guidance, oversight, and coordination for all IWRM efforts (Elfithri, Mokhtar, & Zakaria, 2019). Institutional frameworks should include regulations, laws, and policies that are tailored to each region and provide the necessary guidance and oversight for IWRM initiatives. Additionally, public-private partnerships should be encouraged in order to bring the necessary institutional resources and expertise to bear on the issue of IWRM (Dirwai, Kanda, Senzanje, & Busari, 2021). Finally, it is important to recognize that IWRM initiatives require a long-term commitment in order to be successful. This includes investment in research and development, as well as education and public awareness campaigns. Additionally, it is important to ensure that IWRM initiatives are continually assessed and evaluated in order to ensure that they are successful and effective (Sukereman et al., 2022). Ultimately, effective governance and institutional frameworks are essential for successful IWRM initiatives. By investing in strong political leadership, crafting laws and regulations tailored to the region, and developing long-term strategies, governments can ensure that IWRM initiatives are successful (Adom & Simatele, 2022). Water plays an important role in human life and the environment (Chowdhary, Bharagava, Mishra, & Khan, 2019). Limited access to clean water, increased water demand, ineffective water resources management, and uncertainties are the cause of global water crisis which can alternatively called a governance crisis. According to the United Nations Conference on Environment and Development Agenda 21 of 1992, IWRM is a good approach that requires some establishment of an enabling environment, policies, strategies, legislation, stakeholders' participation, institutional framework, and governance and management instruments (Agarwal et al., 2000). The success of IWRM depends on the quality of the governance foundation which has been recognized as "IWRM is not applied in a vacuum and the broader picture, as described by governance, provides the context in which the IWRM approach can be applied" (Allan & Rieu-Clarke, 2010). The UNDP Water Governance Facility also noticed that more information is needed about "the type of governance that is favorable to improved water resources management and water services provisions," and "those governance components which are more critical to address than others to improve water supply and sanitation coverage as well as the sustainable use of water resources." The World Bank Institute defines the term "governance" as "the traditions and institutions by which authority in a country is exercised for the common good" and highlights three key elements such as process, capacity, and respect (Allan & Rieu-Clarke, 2010). Law, organizations, networks, social compacts, and culture are the five types of institutional evidence for water governance (Lautze, De Silva, Giordano, & Sanford, 2011). According to UN Water Development Report, some effective criteria for water governance such as participation, transparency, equity, efficiency, effectiveness, accountability, integration, rule of law, coherency, responsiveness, and ethical considerations (Keser, 2023). Based on stakeholder engagement, twelve water governance principles (Fig. 15.2) serve as a catalyst to mention good

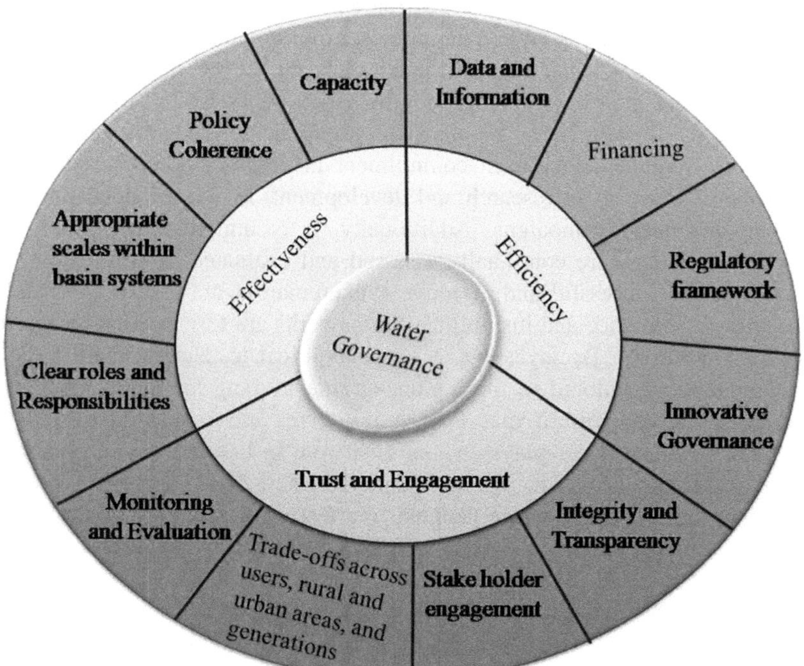

FIGURE 15.2 Water governance. Modified from *Akhmouch, A., & Correia, F. N. (2016). The 12 OECD principles on water governance – When science meets policy.* Utilities Policy, 43, *14–20. http://www.elsevier.com/inca/publications/store/3/0/4/7/8/index.htt. doi:10.1016/j.jup.2016.06.004.*

practices and initiate reform processes toward good water management at various levels of government or country (Akhmouch & Correia, 2016).

Of the three pillars, The institution also plays an important role in implementing the IWRM approach. IWRM is involved in a virtuous circle of knowledge by participatory approaches so the institution can play their decision-making role most transparently. The role of institutions is to create an organizational framework for maintaining forms and functions and institutional building capacity to develop human resources (Agarwal et al., 2000). In IWRM, institutional arrangement is very important to succeeding social equity, ecological sustainability, and economic efficiency. They refer to water laws, regulations, standards, and corresponding linkage mechanisms that inform the institutional framework between agencies or structures, including the allocation of tasks and responsibilities at various levels of government. Other necessary arrangements are cross-sectional coordination, planning, infrastructure development, building capacity, planning, pricing, monitoring, and evaluation (Araral & Wu, 2016). The key goals for the institutional framework are:

- To separate the water resources management functions from service delivery functions such as irrigation, water supply, sewerage, and hydropower generation and to establish the government as the owner of the water resources, rather than the provider of services.
- To promote stakeholders' involvement and public participation in planning and management decisions, as well as to manage surface water resources within the boundaries of a catchment, rather than within administrative boundaries. Additionally, the framework aims to decentralize regulatory functions to the lowest appropriate level.

To implement IWRM approach, institutional arrangement is needed to improve organizational structure at basin and subbasin levels. This will improve decision-making at the lowest level and enable functioning stakeholder involvements in decision-making with the representation of all sections of society and gender balance. This approach is based on some hydrological boundaries (Pellegrini, Bortolini, & Defrancesco, 2019).

Involvement of stakeholders: The stakeholders in IWRM are those individuals, groups, or organizations that have an interest in a particular water resource, and are affected by, or have the potential to affect, the water resource in question. These stakeholders can include government agencies, local communities, businesses, interest groups, and even individuals (Katusiime & Schütt, 2020). The involvement of stakeholders is an essential part of IWRM, as it allows for a greater understanding of the different perspectives and needs of the various involved parties. This understanding can then be used to create an effective strategy for managing the water resource that meets the interests of all stakeholders (Conallin, Dickens, Hearne, & Allan, 2017). For example, a stakeholder engagement plan might involve the use of meetings, workshops, or other forms of communication to solicit input from all stakeholders. This input can then be used to create a plan that takes into consideration all of the different needs, interests, and perspectives of the parties involved. The plan should also provide a clear outline of goals, objectives, and strategies for managing the water resource. In addition, it is important to ensure that all stakeholders have a voice in the decision-making process so that their opinions and perspectives are taken into account. A good stakeholder engagement plan should also allow for ample opportunity for public input and the consideration of local knowledge (Ben-Daoud et al., 2023). Finally, it is important to maintain open communications between all stakeholders and to provide regular updates on progress and outcomes. This will help to ensure that all stakeholders stay informed and feel included in the process. Stakeholder participation is one of the most important principles for IWRM which is based on some principles such as inclusiveness, equity, accountability, transparency, legitimacy, conflict minimization, social learning, effectiveness, and efficiency. Some levels of stakeholder participation in IWRM are manipulation, therapy, informing, consultation, placation, partnership, delegated power, and citizen control (Arnstein, 1969). This is necessary to share and transfer the information regarding IWRM, to enhance inclusivity, confidence, transparency, and accountability for IWRM to take decision-making processes, to gain knowledge from the starting level related to IWRM, to minimize conflict regarding IWRM and to protect land and ecosystems in the catchment area and also water resources. Stakeholder participation follows some steps to implement the IWRM approach (Table 15.2), as shown below:

1. Identification of key stakeholders,
2. Assess stakeholder interests and the potential impact of the project on these interests
3. Assessing stakeholder influence and importance
4. Outline a participation strategy

The benefits of stakeholder participation are as follows (Galvez & Rojas, 2019):

- Stakeholder participation can make long-term collaborative relationships between the government, civilians and the private sector, as most of the actions are monitored by different stakeholders.
- Due to a lack of water resources or poor management of water resources, stakeholder participation is affected more but leads to informed decision-making which may often exhibit wealth of information.

Stakeholder participation indeed involvement of all water user groups and decision-making bodies which may improve the water quality and decisions by some stages such as mobilization of stakeholder support in the IWRM process, participation of selected committees, establishing trust, and

TABLE 15.2 Levels of stakeholder participation.

Levels of participation	Characteristics
1. Manipulation	Stakeholder participation is simply a pretense to pretend the economic or political interests of more powerful role players where community members are gathered to make plans and decisions on IWRM.
2. Therapy	Stakeholders are gathered to confirm what more powerful stakeholders think about IWRM and their opinions and ability to make decisions are not entertained.
3. Informing	Stakeholders are ready to give information and oblized with already-made decisions by powerful stakeholders on IWRM processes where communication is one way.
4. Consultation	To acquire information, stakeholders gathered at a time when they are not actively involved in decision-making processes.
5. Placation	From communities of high-level decision-making bodies token represented by individuals where their representative voices are ignored.
6. Partnership	Community stakeholders with equal representation and powerful key stakeholders provide opportunities for interaction, joint planning, negotiation, and decision-making processes in IWRM.
7. Delegated Power	In local IWRM processes, community representatives become the dominant decision-making authority which indeed decentralized decision-making processes.
8. Citizen control	Make channelization for self-management where communities have total control and accountability to take part in the decision-making process on IWRM.

IWRM, Integrated water resources management.
Source: Modified from Arnstein, S. R. (1969). A ladder of citizen participation. *Journal of the American Planning Association, 35*(4), 216–224, https://doi.org/10.1080/01944366908977225.

building capacity which means skill gap identification, recognizing traditional authority structure, providing necessary training and resonance, problem identification means profiling the area, conduct a water resource audit, development of an IWRM action plan which may identify possible solutions and trying to implement those plan, implementation of the action plan which indeed stakeholder commitment, participatory learning with action and also monitoring and evaluation (Ben-Daoud et al., 2023).

Establishment of regulations: The water resource management practices of any society must be guided by a series of regulations to ensure that water is used efficiently and equitably. The IWRM approach requires the consideration of social, economic, and environmental objectives when constructing regulations to ensure that a sustainable balance is achieved between water quantity and quality (Candido et al., 2022). To establish regulations for water resource management, a comprehensive legal framework must be established. This framework should make provisions for water allocation, water usage, and water conservation. Additionally, it should provide for the implementation of community-based programs, the establishment of water-related fees, and the development of public education and awareness campaigns (Fox, 2019). Furthermore, regulations must define the responsibilities and rights of all stakeholders in the water resource management process. This includes farmers, industry, municipalities, government agencies, and other affected parties. All regulations should be designed to account for the particular geographic, hydrological, and social characteristics of an area (Loucks & van Beek, 2017a). The ultimate goal of IWRM regulations is to help protect and sustainably manage water resources. By setting guidelines and rules for water consumption, pollution, and reuse, regulations ensure that water resources will be available to meet the needs of current and future generations (Silva, 2023).

Development of mechanisms for monitoring and evaluation: IWRM is a systematic approach to the sustainable development and utilization of water resources within a basin. It is based on an integrated approach to planning, development, and management of water resources to meet multiple objectives and demands. The main aim is to ensure the optimal use of water resources and the equitable distribution of these resources among all stakeholders (Apostolaki, Koundouri, & Pittis, 2019). The success of IWRM depends on establishing effective mechanisms for monitoring and evaluation, which are crucial for decision-making as well as for ensuring that the goals of IWRM are achieved. The key elements of such mechanisms include the identification of indicators that measure the performance of different aspects of IWRM, the design of processes for data collection, analysis, and reporting, and the selection of appropriate evaluation methods (Katusiime & Schütt, 2020). The development of effective monitoring and evaluation mechanisms requires the involvement of all stakeholders, including government, industry, civil society, and other users (Aziz & Habib, 2023). Such mechanisms should be tailored to the specific context and should be regularly updated to ensure that they remain relevant and effective. The use of standardized tools, such as the Water Evaluation and Planning System (WEAP), can facilitate this process (Zegait et al., 2023). Finally, the monitoring and evaluation of IWRM must include an assessment of its effectiveness in addressing water-related issues and achieving the desired outcomes. This assessment should be conducted in an objective and transparent manner, with the participation of stakeholders (Ben-Daoud et al., 2023). It should also take into account the social, economic, and environmental impacts of the IWRM and any changes in the water resources over time (Bilalova et al., 2023). IWRM is the allocation and monitoring, as well as the sustainable development of water resources management, which is used in the background of the social, economic, and environmental objectives. By the influences of these three spheres such as social, economic, and environmental, IWRM should perform in a balanced way. It is a cross-sectoral approach that may differ from the traditional approach and be adopted by different developed and developing countries (Seigerman et al., 2023). After that, it has become broader to actualize participatory decision-making of all stakeholders, making a paradigm shift. It differs from traditional approaches in three ways:

- IWRM has multiple goals and objectives that are cross-cutting to implement, making it different from the traditional sectoral approach.
- The spatial focus is on the river basin instead of on single water resources.
- This means breaking away from narrow professional and political boundaries and perspectives, and expanding them to include participatory decision-making among all stakeholders.

To implement the policies for IWRM, some components and policies are also required to maintain the balance of water use between basic and overlapping needs in social, economic, and environmental conditions (Fig. 15.3). This is to manage water use at different sector such as irrigation, agriculture, sanitation, domestic purpose, hydropower, and industrial consumption in a holistic way, to improve the development of surface and groundwater

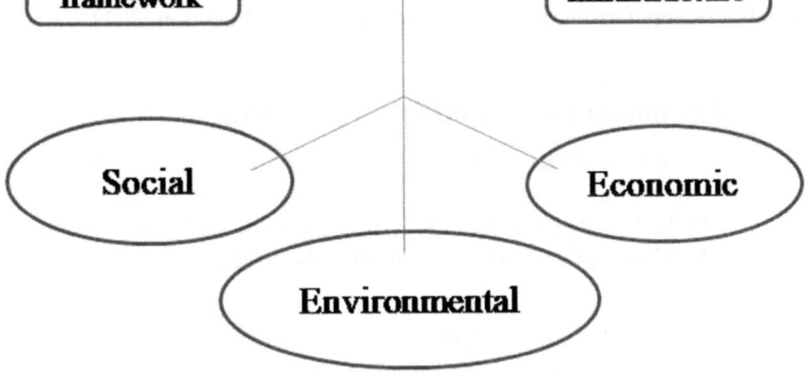

FIGURE 15.3 Basic needs of water resources management. *IWRM*, Integrated water resources management.

resources, and to apply and develop the modern instruments, most appropriate technical infrastructure, institutional framework, and stakeholders participation for water resources management (Cacal & Taboada, 2022). The policies depend on two principles: water should be treated as a social, economic, and environmental aspect, and options and policies should guide how water resources are managed and analyzed through a comprehensive framework.

Public awareness and participation: The need for public awareness and participation in the IWRM is essential (Savenije & Van der Zaag, 2008). It is the key to ensure the sustainability of the resources. Without public engagement, the resources will not be used efficiently and will be misused or wasted (Vyas & Nath, 2021). At the heart of IWRM is the process of public awareness and participation. The stakeholders must be educated and motivated to take part in the process of IWRM. They need to be aware of the types of resources available, the existing legal frameworks, and the need for sustainable management of the resources (Mersha, 2021). The government plays an important role in providing the necessary information and motivating the public to participate in the process (Uhlendahl, Salian, Casarotto, & Doetsch, 2011). It can also support organizations in organizing workshops and seminars to create awareness about the process and its benefits. Financial resources can also be made available to institutions and organizations to assist in their activities. Public participation in the IWRM process should not be seen as a one-off activity but rather as a continuous process. This means that the participation of stakeholders should be ongoing and not just limited to the initial stages of the process. The public should be encouraged to remain engaged in the process and be motivated to actively participate in decision-making and implementation (Swatuk & Qader, 2023). In conclusion, public awareness and participation are essential elements of successful IWRM. Government and nongovernmental organizations need to work together to ensure that people are educated and motivated to take part in the process and to ensure that the resources are used efficiently and sustainably (Koop, Grison, Eisenreich, Hofman, & van Leeuwen, 2022).

15.7 Recommendation and future challenges

In order to handle water scarcity, promote sustainable development, and safeguard the welfare of urban residents, integrated water resources management, or IWRM, is essential in cities. Here are some global suggestions for putting IWRM into practice in towns all over the world:

- Encourage the use of integrated planning strategies that take water into account as a cross-cutting aspect of urban development.
- Create robust governance structures that involve a range of stakeholders, such as corporations, nongovernmental organizations, local governments, and communities.

- Involve local communities in water management decision-making to make sure their needs and concerns are taken into account.
- Organize educational initiatives to increase public knowledge of sustainable water usage, pollution control, and water conservation.
- Make investments in stormwater management, sewage treatment facilities, and resilient and sustainable water infrastructure.
- Create plans for adjusting to the ways that climate change may affect water supplies, such as altered precipitation patterns, increasing sea levels, and more intense weather.
- Plan and construct urban water infrastructure with climate resilience in mind.
- Encourage the implementation of water recycling and reuse systems in order to minimize wastewater discharge and lessen the pressure on freshwater resources.
- Encourage businesses and local governments to embrace water-efficient technologies by putting policies and incentives in place.
- Create creative funding channels, such as public-private partnerships, to support water infrastructure initiatives and guarantee long-term viability.
- Examine the implementation of water pricing schemes that incentivize responsible use and represent the real worth of water.
- Incorporate the considerations for water management into more general urban policies, such as those pertaining to transportation, land use, and disaster risk reduction.
- Water policy should be in line with other sustainable development objectives to prevent trade-offs and promote synergies.
- Encourage cooperation and information exchange amongst cities throughout the world so that best practices and successful case studies can be shared.
- Cooperate across borders to address shared difficulties, particularly for cities that are close to shared water resources.
- To improve their ability to adopt and manage integrated water solutions, local governments, water utilities, and other stakeholders should invest in capacity-building initiatives.
- In order to continuously improve water management techniques and technologies, support research and development projects.

It will need a mix of political will, financial support, community involvement, and technical innovation to address these issues. The effective execution of integrated water resources management in urban settings demands a thorough, interdisciplinary approach that takes into account the particular circumstances and difficulties of each metropolitan region while taking inspiration from international best practices. International cooperation and collaboration can be very helpful in resolving common issues and promoting an increasingly resilient and sustainable urban water management future. Cities can give their citizens a resilient and water-secure future by managing their water resources in an integrated and sustainable manner.

15.8 Conclusion

IWRM in urban development is an essential approach for addressing the complex challenges associated with water resources in cities. In addition, IWRM in urban development is a multifaceted and inclusive approach that recognizes the centrality of water in urban life. It strives for the balanced and sustainable use of water resources to support the well-being and prosperity of urban populations, while also preserving the environment for future generations. It is a dynamic and evolving framework that adapts to the ever-changing urban landscape and its water-related challenges. Finally, the key is to implement a flexible, adaptive, and inclusive approach that promotes sustainability and resilience in urban water management.

References

Abass, F., Ismail, L. H., Wahab, I. A., & Elgadi, A. A. (2020). A review of green roof: Definition, history, evolution and functions. *IOP Conference Series: Materials Science and Engineering.*, *713*(1). Available from https://doi.org/10.1088/1757-899x/713/1/012048.

Adom, R. K., & Simatele, M. D. (2022). The role of stakeholder engagement in sustainable water resource management in South Africa. *Natural Resources Forum.*, *46*(4), 410–427. Available from https://doi.org/10.1111/1477-8947.12264, http://onlinelibrary.wiley.com/journal/10.1111/(ISSN)1477-8947.

Agarwal, A., Angeles., Bhatia, R., Chéret, I., Davila-Poblete, S., Falkenmark, M., ... Wright, M. (2000). Stockholm: Global water partnership. *Integrated Water Resources Management*, 1–67.

Aguiar, J. B., Martins, A. M., Almeida, C., Ribeiro, H. M., & Marto, J. (2022). Water sustainability: A waterless life cycle for cosmetic products. *Sustainable Production and Consumption*, *32*, 35–51. Available from https://doi.org/10.1016/j.spc.2022.04.008, http://www.journals.elsevier.com/sustainable-production-and-consumption/.

Akhmouch, A., & Correia, F. N. (2016). The 12 OECD principles on water governance – When science meets policy. *Utilities Policy.*, *43*, 14–20. Available from https://doi.org/10.1016/j.jup.2016.06.004, http://www.elsevier.com/inca/publications/store/3/0/4/7/8/index.htt.

Alcon, F., Navarro, N., De-Miguel, M. D., & Balbo, A. L. (2019). *Drip irrigation technology: Analysis of adoption and diffusion processes. Sustainable Solutions for Food Security: Combating Climate Change by Adaptation* (pp. 269–285). Spain: Springer International Publishing. Available from http://www.springer.com/in/book/9783319778778, 10.1007/978-3-319-77878-5_14.

Al-Jawad, J. Y., Alsaffar, H. M., Bertram, D., & Kalin, R. M. (2019). A comprehensive optimum integrated water resources management approach for multidisciplinary water resources management problems. *Journal of Environmental Management.*, *239*, 211–224. Available from https://doi.org/10.1016/j.jenvman.2019.03.045, https://www.sciencedirect.com/journal/journal-of-environmental-management.

Allan, A., & Rieu-Clarke, A. (2010). Good governance and IWRM-a legal perspective. *Irrigation and Drainage Systems*, *24*(3-4), 239–248. Available from https://doi.org/10.1007/s10795-010-9096-4.

Almuktar, S. A. A. A. N., Abed, S. N., & Scholz, M. (2018). Wetlands for wastewater treatment and subsequent recycling of treated effluent: A review. *Environmental Science and*

Pollution Research., *25*(24), 23595–23623. Available from https://doi.org/10.1007/s11356-018-2629-3, http://www.springerlink.com/content/0944-1344.

Anderson, V., & Gough, W. A. (2022). A typology of nature-based solutions for sustainable development: An analysis of form, function, nomenclature, and associated applications. *Land.*, *11*(7), 1072. Available from https://doi.org/10.3390/land11071072.

Apostolaki, S., Koundouri, P., & Pittis, N. (2019). Using a systemic approach to address the requirement for integrated water resource management within the water framework directive. *Science of The Total Environment*, *679*, 70–79. Available from https://doi.org/10.1016/j.scitotenv.2019.05.077.

Araral, E., & Wu, X. (2016). Comparing water resources management in China and India: Policy design, institutional structure and governance. *Water Policy.*, *18*(S1), 1–13. Available from https://doi.org/10.2166/wp.2016.001.

Arnstein, S. R. (1969). A ladder of citizen participation. *Journal of the American Planning Association.*, *35*(4), 216–224. Available from https://doi.org/10.1080/01944366908977225.

Aziz, S. A., & Habib, M. (2023). Policymaking process and methods: An institutional analysis. *Pakistan Journal of International Affairs.*, *6*(2). Available from https://doi.org/10.52337/pjia.v6i2.791.

Badescu, V., & Sicre, B. (2003). Renewable energy for passive house heating: II. Model. *Energy and Buildings*, *35*(11), 1085–1096. Available from https://doi.org/10.1016/j.enbuild.2003.09.004.

Batalhão, A. C. S., Bouloumytis, V., Zuffo, A. C., & Pimentel da Silva, L. (2023). *Integrated water resources management and urban sustainability. The route towards global sustainability: Challenges and management practices* (pp. 289–312). Portugal: Springer International Publishing. Available from https://link.springer.com/book/10.1007/978-3-031-10437-4, 10.1007/978-3-031-10437-4_15.

Ben-Daoud, M., El Mahrad, B., Moroêanu, G. A., Ben-Daoud, A., Swe, W. M., Elhassnaoui, I., ... Eljaafari, S. (2023). Water resources planning and management: from stakeholders' local actions to the global perspective. *Sustainable Water Resources Management*, *9*(5). Available from https://doi.org/10.1007/s40899-023-00919-x, https://www.springer.com/journal/40899.

Besir, A. B., & Cuce, E. (2018). Green roofs and facades: A comprehensive review. *Renewable and Sustainable Energy Reviews.*, *82*, 915–939. Available from https://doi.org/10.1016/j.rser.2017.09.106, https://www.journals.elsevier.com/renewable-and-sustainable-energy-reviews.

Bilalova, S., Newig, J., Tremblay-Lévesque, L.-C., Roux, J., Herron, C., & Crane, S. (2023). Pathways to water sustainability? A global study assessing the benefits of integrated water resources management. *Journal of Environmental Management.*, *343*. Available from https://doi.org/10.1016/j.jenvman.2023.118179.

Cacal, J. C., & Taboada, E. B. (2022). Assessment and evaluation of IWRM implementation in Palawan, Philippines. *Civil Engineering Journal (Iran)*, *8*(2), 290–307. Available from https://doi.org/10.28991/CEJ-2022-08-02-08, http://www.civilejournal.org.

Cameron, R., & Katzschner, T. (2017). Every last drop: The role of spatial planning in enhancing integrated urban water management in the City of Cape Town. *South African Geographical Journal.*, *99*(2), 196–216. Available from https://doi.org/10.1080/03736245.2016.1231622, http://www.tandfonline.com/loi/rsag20.

Campisano, A., Butler, D., Ward, S., Burns, M. J., Friedler, E., DeBusk, K., ... Han, M. (2017). Urban rainwater harvesting systems: Research, implementation and future perspectives. *Water Research.*, *115*, 195–209. Available from https://doi.org/10.1016/j.watres.2017.02.056, http://www.elsevier.com/locate/watres.

Candido, L. A., Coêlho, G. A. G., De Moraes, M. M. G. A., & Florêncio, L. (2022). Review of decision support systems and allocation models for integrated water resources management focusing on joint water quantity-quality. *Journal of Water Resources Planning and Management, 148*(2). Available from https://doi.org/10.1061/(ASCE)WR.1943-5452.0001496, https://ascelibrary.org/journal/jwrmd5.

Cascone, S. (2019). Green roof design: State of the art on technology and materials. *Sustainability., 11*(11). Available from https://doi.org/10.3390/su11113020.

Çetin, Ö., & Akalp, E. (2019). Efficient use of water and fertilizers in irrigated agriculture: Drip irrigation and fertigation. *Acta Horticulturae et Regiotecturae., 22*(2), 97–102. Available from https://doi.org/10.2478/ahr-2019-0019.

Chowdhary, P., Bharagava, R. N., Mishra, S., & Khan, N. (2019). *Role of industries in water scarcity and its adverse effects on environment and human health* (pp. 235–256). Springer Science and Business Media LLC. Available from 10.1007/978-981-13-5889-0_12.

Conallin, J. C., Dickens, C., Hearne, D., & Allan, C. (2017). *Stakeholder engagement in environmental water management. Water for the environment: From policy and science to implementation and management* (pp. 129–150). Netherlands: Elsevier Inc.. Available from http://www.sciencedirect.com/science/book/9780128039076, 10.1016/B978-0-12-803907-6.00007-3.

Crini, G., Lichtfouse, E., Wilson, L. D., & Morin-Crini, N. (2019). Conventional and non-conventional adsorbents for wastewater treatment. *Environmental Chemistry Letters, 17*(1), 195–213. Available from https://doi.org/10.1007/s10311-018-0786-8, http://springerlink.metapress.com/app/home/journal.asp?wasp = d86tgdwvtg0yvw9gvkwp &referrer = parent&backto = browsepublicationsresults,140,541.

Dawit, M., Dinka, M. O., & Leta, O. T. (2020). Implications of adopting drip irrigation system on crop yield and gender-sensitive issues: The case of Haramaya district, Ethiopia. *Journal of Open Innovation: Technology, Market, and Complexity, 6*(4), 1–17. Available from https://doi.org/10.3390/joitmc6040096, https://www.mdpi.com/2199-8531/6/4/96/pdf.

Dirwai, T. L., Kanda, E. K., Senzanje, A., & Busari, T. I. (2021). Water resource management: Iwrm strategies for improved water management. A systematic review of case studies of east, west and southern Africa. *PLoS ONE, 16*(5). Available from https://doi.org/10.1371/journal.pone.0236903, https://journals.plos.org/plosone/article/file?id = 10.1371/journal.pone.0236903&type = printable.

Dungumaro, E. W., & Madulu, N. F. (2003). Public participation in integrated water resources management: The case of Tanzania. *Physics and Chemistry of the Earth, 28*(20-27), 1009–1014. Available from https://doi.org/10.1016/j.pce.2003.08.042, http://www.journals.elsevier.com/physics-and-chemistry-of-the-earth/.

du Plessis, A. (2023). *Integrated water resource management, water service provision and delivery* (pp. 27–40). Springer Science and Business Media LLC. Available from 10.1007/978-3-031-24019-5_2.

Elfithri, R., Mokhtar, M. B., & Zakaria, S. (2019). The need for awareness raising, advocacy, and capacity building in integrated water resources management toward sustainable development: A case study in Malaysia. *World Water Policy, 5*(1), 43–54. Available from https://doi.org/10.1002/wwp2.12002, onlinelibrary.wiley.com/journal/2639541x.

Ferreyra, C., de Loë, R. C., & Kreutzwiser, R. D. (2008). Imagined communities, contested watersheds: Challenges to integrated water resources management in agricultural areas. *Journal of Rural Studies, 24*(3), 304–321. Available from https://doi.org/10.1016/j.jrurstud.2007.11.001.

Flörke, M., Schneider, C., & McDonald, R. I. (2018). Water competition between cities and agriculture driven by climate change and urban growth. *Nature Sustainability, 1*(1), 51–58. Available from https://doi.org/10.1038/s41893-017-0006-8.

Fox, I. (2019). *Institutions for water management in a changing world. Water in a developing world: The management of a critical resource* (p. 9) Canada: Taylor and Francis. Available from http://www.tandfebooks.com/doi/book/10.4324/9780429267277.

Francis, R. A., & Lorimer, J. (2011). Urban reconciliation ecology: The potential of living roofs and walls. *Journal of Environmental Management, 92*(6), 1429–1437. Available from https://doi.org/10.1016/j.jenvman.2011.01.012, https://www.sciencedirect.com/science/journal/03014797.

Fulazzaky, M. (2014). Challenges of integrated water resources management in Indonesia. *Water, 6*(7), 2000–2020. Available from https://doi.org/10.3390/w6072000.

Galvez, V., & Rojas, R. (2019). Collaboration and integrated water resources management: A literature review. *World Water Policy, 5*(2), 179–191. Available from https://doi.org/10.1002/wwp2.12013.

García, L. E. (2008). Integrated water resources management: A 'small' step for conceptualists, a giant step for practitioners. *International Journal of Water Resources Development., 24*(1), 23–36. Available from https://doi.org/10.1080/07900620701723141.

García-Ávila, F., Guanoquiza-Suárez, M., Guzmán-Galarza, J., Cabello-Torres, R., & Valdiviezo-Gonzales, L. (2023). Rainwater harvesting and storage systems for domestic supply: An overview of research for water scarcity management in rural areas. *Results in Engineering., 18*, 101153. Available from https://doi.org/10.1016/j.rineng.2023.101153.

Getter, K. L., Bradley Rowe, D., & Cregg, B. M. (2009). Solar radiation intensity influences extensive green roof plant communities. *Urban Forestry and Urban Greening, 8*(4), 269–281. Available from https://doi.org/10.1016/j.ufug.2009.06.005.

Getter, K. L., & Rowe, D. B. (2006). The role of extensive green roofs in sustainable development. *HortScience., 41*(5), 1276–1285. Available from https://doi.org/10.21273/hortsci.41.5.1276, https://journals.ashs.org/.

Gibney, K. B., O'Toole, J., Sinclair, M., & Leder, K. (2014). Disease burden of selected gastrointestinal pathogens in Australia, 2010. *International Journal of Infectious Diseases., 28*, 176–185. Available from https://doi.org/10.1016/j.ijid.2014.08.006.

Gleick, P. H. (1993). *Water in crisis—A guide to the world's fresh water resources*. Oxford University Press. Available from 10.2307/2623756.

Grabow, W. O. K., Taylor, M. B., & De Villiers, J. C. (2001). New methods for the detection of viruses: Call for review of drinking water quality guidelines. *Water Science and Technology, 43*(12), 1–8. Available from https://doi.org/10.2166/wst.2001.0703, https://iwaponline.com/wst.

Grigg, N. S. (2016). Water as a connector among societal needs. *Nature*, 1–18. Available from https://doi.org/10.1057/978-1-137-57615-6_1.

Grobicki, A., MacLeod, F., & Pischke, F. (2015). Integrated policies and practices for flood and drought risk management. *Water Policy., 17*(S1), 180–194. Available from https://doi.org/10.2166/wp.2015.009.

Ingrao, C., Strippoli, R., Lagioia, G., & Huisingh, D. (2023). Water scarcity in agriculture: An overview of causes, impacts and approaches for reducing the risks. *Heliyon., 9*(8). Available from https://doi.org/10.1016/j.heliyon.2023.e18507.

Irfeey, A. M. M., Chau, H. W., Sumaiya, M. M. F., Wai, C. Y., Muttil, N., & Jamei, E. (2023). Sustainable mitigation strategies for urban heat island effects in urban areas. *Sustainability (Switzerland), 15*(14). Available from https://doi.org/10.3390/su151410767, http://www.mdpi.com/journal/sustainability/.

Jofre, J., Lucena, F., Blanch, A., & Muniesa, M. (2016). Coliphages as model organisms in the characterization and management of water resources. *Water, 8*(5). Available from https://doi.org/10.3390/w8050199.

Kashem, S., & Mondal, M. S. (2022). Development of a water-pricing model for domestic water uses in Dhaka city using an IWRM framework. *Water., 14*(9), 1328. Available from https://doi.org/10.3390/w14091328.

Katusiime, J., & Schütt, B. (2020). Integrated water resources management approaches to improve water resources governance. *Water*, *12*(12), 3424. Available from https://doi.org/10.3390/w12123424.

Keser, I. (2023). Interdependence and complementarity of a multi-dimensional concept of sustainable development and the integrated approach to urban governance—Case study city of Zagreb. *Sustainability*, *15*(12), 9213. Available from https://doi.org/10.3390/su15129213.

Khan, S. A. R., Ponce, P., Yu, Z., Golpîra, H., & Mathew, M. (2022). Environmental technology and wastewater treatment: Strategies to achieve environmental sustainability. *Chemosphere*, *286*. Available from https://doi.org/10.1016/j.chemosphere.2021.131532, http://www.elsevier.com/locate/chemosphere.

Kidd, S., & Shaw, D. (2007). Integrated water resource management and institutional integration: Realising the potential of spatial planning in England. *Geographical Journal.*, *173*(4), 312–329. Available from https://doi.org/10.1111/j.1475-4959.2007.00260.x.

Kirshen, P. H., Vogel, R. M., & Rogers, B. L. (2004). Challenges in graduate education in integrated water resources management. *Journal of Water Resources Planning and Management.*, *130*(3), 185–186. Available from https://doi.org/10.1061/(asce)0733-9496(2004)130:3(185).

Koop, S. H. A., Grison, C., Eisenreich, S. J., Hofman, J., & van Leeuwen, K. (2022). Integrated water resources management in cities in the world: Global solutions. *Sustainable Cities and Society*, *86*. Available from https://doi.org/10.1016/j.scs.2022.104137.

Koop, S. H. A., & van Leeuwen, C. J. (2017). The challenges of water, waste and climate change in cities. *Environment, Development and Sustainability.*, *19*(2), 385–418. Available from https://doi.org/10.1007/s10668-016-9760-4.

Lankford, B., Bakker, K., Zeitoun, M., & Conway, D. (2013). *Water security: Principles, perspectives and practices* (pp. 1–357). United Kingdom: Taylor and Francis. Available from http://www.tandfebooks.com/doi/book/10.4324/9780203113202, 10.4324/9780203113202.

Lautze, J., De Silva, S., Giordano, M., & Sanford, L. (2011). Putting the cart before the horse: Water governance and IWRM. *Natural Resources Forum*, *35*(1), 1–8. Available from https://doi.org/10.1111/j.1477-8947.2010.01339.x.

Laws, B. V., Dickenson, E. R. V., Johnson, T. A., Snyder, S. A., & Drewes, J. E. (2011). Attenuation of contaminants of emerging concern during surface-spreading aquifer recharge. *Science of the Total Environment.*, *409*(6), 1087–1094. Available from https://doi.org/10.1016/j.scitotenv.2010.11.021.

Lazarova, V., Levine, B., Sack, J., Cirelli, G., Jeffrey, P., Muntau, H., ... Brissaud, F. (2001). Role of water reuse for enhancing integrated water management in Europe and Mediterranean countries. *Water Science and Technology*, *43*(10), 25–33. Available from https://doi.org/10.2166/wst.2001.0571, https://iwaponline.com/wst.

Lemly, A. D., Kingsford, R. T., & Thompson, J. R. (2000). Irrigated agriculture and wildlife conservation: Conflict on a global scale. *Environmental Management.*, *25*(5), 485–512. Available from https://doi.org/10.1007/s002679910039.

R. Lenton, M. Muller, Integrated water resources management in practice: Better water management for development. (2012).

Li, Z., Boyle, F., & Reynolds, A. (2010). Rainwater harvesting and greywater treatment systems for domestic application in Ireland. *Desalination*, *260*(1-3), 1–8. Available from https://doi.org/10.1016/j.desal.2010.05.035.

Liu, H., Wang, X., Zhang, X., Zhang, L., Li, Y., & Huang, G. (2017). Evaluation on the responses of maize (Zea mays L.) growth, yield and water use efficiency to drip irrigation water under mulch condition in the Hetao irrigation District of China. *Agricultural Water Management, 179*, 144–157. Available from https://doi.org/10.1016/j.agwat.2016.05.031.

Liu, L., & Jensen, M. B. (2018). Green infrastructure for sustainable urban water management: Practices of five forerunner cities. *Cities, 74*, 126–133. Available from https://doi.org/10.1016/j.cities.2017.11.013, https://www.sciencedirect.com/journal/cities/issues.

Loucks, D. P., & van Beek, E. (2017a). *Water resources planning and management: An overview* (pp. 1–49). Springer Science and Business Media LLC. Available from 10.1007/978-3-319-44234-1_1.

Loucks, D. P., & van Beek, E. (2017b). *Water resource systems planning and management: An introduction to methods, models, and applications* (pp. 1–624). United States: Springer International Publishing. Available from https://www.springer.com/in/book/9783319442327, 10.1007/978-3-319-44234-1.

Mačkić, K., Bajić, I., Pejić, B., Vlajić, S., Adamović, B., Popov, O., & Simić, D. (2023). Yield and water use efficiency of drip irrigation of pepper. *Water., 15*(16), 2891. Available from https://doi.org/10.3390/w15162891.

Makanda, K., Nzama, S., & Kanyerere, T. (2022). Assessing the role of water resources protection practice for sustainable water resources management: A review. *Water., 14*(19), 3153. Available from https://doi.org/10.3390/w14193153.

Manda, R. R., Addanki, V. A., & Srivastava, S. (2021). Role of drip irrigation in plant health management, its importance and maintenance. *Plant Archives, 21*(Suppliment-1), 1294–1302. Available from https://doi.org/10.51470/plantarchives.2021.v21.s1.204.

Maniam, G., Zakaria, N. A., Leo, C. P., Vassilev, V., Blay, K. B., Behzadian, K., & Poh, P. E. (2022). An assessment of technological development and applications of decentralized water reuse: A critical review and conceptual framework. *Wiley Interdisciplinary Reviews: Water, 9*(3). Available from https://doi.org/10.1002/wat2.1588, http://wires.wiley.com/WileyCDA/WiresJournal/wisId-WAT2.html.

Mbaigoto, N., Alsharif, K., & Landry, S. (2023). An assessment of the degree of the implementation of the integrated water resources management principles in the Lake Chad Basin: The case of the Republic of Chad. *Sustainable Water Resources Management, 9*(4). Available from https://doi.org/10.1007/s40899-023-00894-3.

McDonnell, R. A. (2008). Challenges for integrated water resources management: How do we provide the knowledge to support truly integrated thinking? *International Journal of Water Resources Development., 24*(1), 131–143. Available from https://doi.org/10.1080/07900620701723240.

A.N. Mersha, Integrated water resources management: A systems perspective of water governance and hydrological conditions. (2021).

Mishra, B., Kumar, P., Saraswat, C., Chakraborty, S., & Gautam, A. (2021). Water security in a changing environment: Concept, challenges and solutions. *Water, 13*(4), 490. Available from https://doi.org/10.3390/w13040490.

Mishra, M. K. (2023). Water governance: Effective needs for sustainable development. *SSRN Electronic Journal.*, 1556–5068. Available from https://doi.org/10.2139/ssrn.4363897.

Mohsin, M., Bhatti, I. A., Zeshan, M., Ashar, A., Younis, S., Amjed, N., & Alam, M. (2023). *Wastewater treatment and reuse in future cities. Industrial wastewater reuse: Applications, prospects and challenges* (pp. 339–369). Pakistan: Springer Nature. Available from https://link.springer.com/book/10.1007/978-981-99-2489-9, 10.1007/978-981-99-2489-9_16.

Nagase, A., & Dunnett, N. (2010). Drought tolerance in different vegetation types for extensive green roofs: Effects of watering and diversity. *Landscape and Urban Planning, 97*(4), 318–327. Available from https://doi.org/10.1016/j.landurbplan.2010.07.005.

S. Nallakukkala, Z. Kassim, N.A. Othman, B. Lal, Advancement in gas hydrate water based produced water desalination: An overview. Third International Conference on Separation Technology 2020. Atlantis Press, Atlantis Press, (2020), 190–197, doi:org/10.2991/aer.k.201229.027.

Nath, S., Vyas, J. N., Deogade, R. B., & Chandra, P. (2023). *Integrated water resources management in developing nation: Status and challenges toward water sustainability. The route towards global sustainability: Challenges and management practices* (pp. 367–378). India: Springer International Publishing. Available from https://link.springer.com/book/10.1007/978-3-031-10437-4, 10.1007/978-3-031-10437-4_18.

Ngene, B. U., Nwafor, C. O., Bamigboye, G. O., Ogbiye, A. S., Ogundare, J. O., & Akpan, V. E. (2021). Assessment of water resources development and exploitation in Nigeria: A review of integrated water resources management approach. *Heliyon*, *7*(1). Available from https://doi.org/10.1016/j.heliyon.2021.e05955, http://www.journals.elsevier.com/heliyon/.

Oberndorfer, E., Lundholm, J., Bass, B., Coffman, R. R., Doshi, H., Dunnett, N., Gaffin, S., Köhler, M., Liu, K. K. Y., & Rowe, B. (2007). Green roofs as urban ecosystems: Ecological structures, functions, and services. *BioScience.*, *57*(10), 823–833. Available from https://doi.org/10.1641/B571005.

Ondimu, S. N., & Murase, H. (2007). Combining Galerkin methods and neural network analysis to inversely determine thermal conductivity of living green roof materials. *Biosystems Engineering.*, *96*(4), 541–550. Available from https://doi.org/10.1016/j.biosystemseng.2006.12.007.

Overton, I. C., Smith, D. M., Dalton, J., Barchiesi, S., Acreman, M. C., Stromberg, J. C., & Kirby, J. M. (2014). Implementing environmental flows in integrated water resources management and the ecosystem approach. *Hydrological Sciences Journal*, *59*(3-4), 860–877. Available from https://doi.org/10.1080/02626667.2014.897408.

Pellegrini, E., Bortolini, L., & Defrancesco, E. (2019). Coordination and participation boards under the European water framework directive: Different approaches used in some EU countries. *Water*, *11*(4), 833. Available from https://doi.org/10.3390/w11040833.

Petit, O. (2016). Paradise lost? The difficulties in defining and monitoring integrated water resources management indicators. *Current Opinion in Environmental Sustainability.*, *21*, 58–64. Available from https://doi.org/10.1016/j.cosust.2016.11.006, http://www.elsevier.com/wps/find/journaldescription.cws_home/718675/description#description.

Pitman, S. D., Daniels, C. B., & Ely, M. E. (2015). Green infrastructure as life support: Urban nature and climate change. *Transactions of the Royal Society of South Australia.*, *139*(1), 97–112. Available from https://doi.org/10.1080/03721426.2015.1035219, http://www.tandfonline.com/loi/TRSS20#.V2I-mk1f2DZ.

Rahaman, M. M., & Varis, O. (2017). Integrated water resources management: Evolution, prospects and future challenges. *Sustainability: Science, Practice and Policy*, *1*(1), 15–21. Available from https://doi.org/10.1080/15487733.2005.11907961.

Rahman, S., Khan, M. T. R., Akib, S., Din, N. B. C., Biswas, S. K., & Shirazi, S. M. (2014). Sustainability of rainwater harvesting system in terms of water quality. *The Scientific World Journal*. Available from https://doi.org/10.1155/2014/721357, 2014.

Rakesh, S. S., Ramesh, P. T., Murugaragavan, R., Avudainayagam, S., & Karthikeyan, S. (2020). Characterization and treatment of grey water: A review. *International Journal of Chemical Studies.*, *8*(1), 34–40. Available from https://doi.org/10.22271/chemi.2020.v8.i1a.8316.

Rasul, G., Neupane, N., Hussain, A., & Pasakhala, B. (2021). Beyond hydropower: Towards an integrated solution for water, energy and food security in South Asia. *International Journal of Water Resources Development.*, *37*(3), 466–490. Available from https://doi.org/10.1080/07900627.2019.1579705.

Rubert, A., & Beetlestone, P. (2014). Tools to improve the management of transboundary river basins for disaster risk reduction. *Water Science and Technology: Water Supply*, *14*(4), 698–707. Available from https://doi.org/10.2166/ws.2014.023, http://www.iwaponline.com/ws/01404/0698/014040698.pdf.

Sahoo, S. K., & Goswami, S. S. (2024). Theoretical framework for assessing the economic and environmental impact of water pollution: A detailed study on sustainable development of India. *Journal of Future Sustainability*, *4*(1), 23–34. Available from https://doi.org/10.5267/j.jfs.2024.1.003.

Sailor, D. J. (2008). A green roof model for building energy simulation programs. *Energy and Buildings*, *40*(8), 1466–1478. Available from https://doi.org/10.1016/j.enbuild.2008.02.001.

Saravanan, V. S., Mcdonald, G. T., & Mollinga, P. P. (2009). Critical review of integrated water resources management: Moving beyond polarised discourse. *Natural Resources Forum.*, *33*(1), 76–86. Available from https://doi.org/10.1111/j.1477-8947.2009.01210.x.

Saritas, O., & Proskuryakova, L. N. (2017). Water resources – An analysis of trends, weak signals and wild cards with implications for Russia. *Federation Foresight*, *19*(2), 152–173. Available from https://doi.org/10.1108/FS-07-2016-0033, http://www.emeraldinsight.com/info/journals/fs/fs.jsp.

Savenije, H. H. G., & Van der Zaag, P. (2008). Integrated water resources management: Concepts and issues. *Physics and Chemistry of the Earth, Parts A/B/C*, *33*(5), 290–297. Available from https://doi.org/10.1016/j.pce.2008.02.003.

Seigerman, C. K., McKay, S. K., Basilio, R., Biesel, S. A., Hallemeier, J., Mansur, A. V., Piercy, C., Rowan, S., Ubiali, B., Yeates, E., & Nelson, D. R. (2023). Operationalizing equity for integrated water resources management. *Journal of the American Water Resources Association*, *59*(2), 281–298. Available from https://doi.org/10.1111/1752-1688.13086, http://onlinelibrary.wiley.com/journal/10.1111/(ISSN)1752-1688.

Setegn, S. G., & Donoso, M. C. (2015). *Sustainability of integrated water resources management: Water governance, climate and ecohydrology* (pp. 1–606). United States: Springer International Publishing. Available from http://www.springer.com/in/book/9783319121932, 10.1007/978-3-319-12194-9.

Shaari, S. M. (2020). An overview of rainwater harvesting for sustainable future in Malaysia. *Environment-Behaviour Proceedings Journal.*, *5*(14), 255–261. Available from https://doi.org/10.21834/ebpj.v5i14.2232.

Shafique, M., Kim, R., & Rafiq, M. (2018). Green roof benefits, opportunities and challenges – A review. *Renewable and Sustainable Energy Reviews.*, *90*, 757–773. Available from https://doi.org/10.1016/j.rser.2018.04.006.

Shahangian, S. A., Tabesh, M., Yazdanpanah, M., Zobeidi, T., & Raoof, M. A. (2022). Promoting the adoption of residential water conservation behaviors as a preventive policy to sustainable urban water management. *Journal of Environmental Management*, *313*. Available from https://doi.org/10.1016/j.jenvman.2022.115005, https://www.sciencedirect.com/journal/journal-of-environmental-management.

Shams S. 2022 Impact of climate change on water resources in the context of Brunei Darussalam: IWRM perspectives. SRICOENV 2022: Proceedings of the 3rd Sriwijaya International Conference on Environmental Issues, SRICOENV 2022.

Shareef, T. M. E., Ma, Z., & Zhao, B. (2019). Essentials of drip irrigation system for saving water and nutrients to plant roots: As a guide for growers. *Journal of Water Resource and Protection.*, *11*(09), 1129–1145. Available from https://doi.org/10.4236/jwarp.2019.119066.

Sigalla, O. Z., Tumbo, M., & Joseph, J. (2021). Multi-stakeholder platform in water resources management: A critical analysis of stakeholders' participation for sustainable water resources. *Sustainability.*, *13*(16), 9260. Available from https://doi.org/10.3390/su13169260.

Silva, J. A. (2023). Wastewater treatment and reuse for sustainable water resources management: A systematic literature review. *Sustainability.*, *15*(14), 10940. Available from https://doi.org/10.3390/su151410940.

Suhardiman, D., Clement, F., & Bharati, L. (2015). Integrated water resources management in Nepal: Key stakeholders' perceptions and lessons learned. *International Journal of Water Resources Development.*, *31*(2), 284–300. Available from https://doi.org/10.1080/07900627.2015.1020999.

Sukereman, A. S., Suratman, R., & Rahim, N. A. (2022). Applying good governance practice in the implementation of integrated water resource management: Analysing the relationship. *Journal of Sustainability Science and Management*, *17*(3), 243–265. Available from https://doi.org/10.46754/jssm.2022.03.019, http://jssm.umt.edu.my/.

Sun, F., Chen, M., & Chen, J. (2011). *Integrated management of source water quantity and quality for human health in a changing world* (pp. 254–265). Elsevier BV. Available from 10.1016/b978-0-444-52272-6.00286-5.

L. Swatuk, A.I.A. Qader, IWRM: Ideology or methodology. Oxford research encyclopedia of environmental science. (2023), doi:10.1093/acrefore/9780199389414.013.620.

The United Nations world water development report 2021. UNESCO. (2021).

Trimble, M., Jacobi, P. R., Olivier, T., Pascual, M., Zurbriggen, C., Garrido, L., & Mazzeo, N. (2020). *Reconfiguring water governance for resilient social-ecological systems in South America* (pp. 113–135). Springer Science and Business Media LLC. Available from 10.1007/978-3-030-48110-0_6.

Uhlendahl, T., Salian, P., Casarotto, C., & Doetsch, J. (2011). Good water governance and IWRM in Zambia: Challenges and chances. *Water Policy*, *13*(6), 845–862. Available from https://doi.org/10.2166/wp.2011.155, http://www.iwaponline.com/wp/01306/0845/013060845.pdf.

Volenzo, T., & Odiyo, J. (2018). Ecological public health and participatory planning and assessment dilemmas: The case of water resources management. *International Journal of Environmental Research and Public Health.*, *15*(8). Available from https://doi.org/10.3390/ijerph15081635.

Voyde, E., Fassman, E., & Simcock, R. (2010). Hydrology of an extensive living roof under sub-tropical climate conditions in Auckland, New Zealand. *Journal of Hydrology*, *394*(3-4), 384–395. Available from https://doi.org/10.1016/j.jhydrol.2010.09.013.

Vyas, J. N., & Nath, S. (2021). *The role of government and the public in water resource management in India. Pollutants and Water Management: Resources, Strategies and Scarcity* (pp. 401–415). India: Wiley. Available from https://onlinelibrary.wiley.com/doi/book/10.1002/9781119693635, 10.1002/9781119693635.ch16.

Wang, C., Hou, Y., & Xue, Y. (2017). Water resources carrying capacity of wetlands in Beijing: Analysis of policy optimization for urban wetland water resources management. *Journal of Cleaner Production.*, *161*, 1180–1191. Available from https://doi.org/10.1016/j.jclepro.2017.03.204.

Watson, N., Shrubsole, D., & Mitchell, B. (2019). Governance arrangements for integrated water resources management in Ontario, Canada, and Oregon, USA: Evolution and lessons. *Water*, *11*(4), 663. Available from https://doi.org/10.3390/w11040663.

Widiyono, W., Renaud, F., & Sudmeier-Rieux, K. (2022). Mainstreaming of eco-drr to implement integrated water resources management in Indonesia. *Ecodevelopment*, *3*(1). Available from https://doi.org/10.24198/ecodev.v3i1.39116.

World City Populations 2021. World Population Review. (2021).

Xu, T., Sathaye, J., Akbari, H., Garg, V., & Tetali, S. (2012). Quantifying the direct benefits of cool roofs in an urban setting: Reduced cooling energy use and lowered greenhouse gas emissions. *Building and Environment.*, *48*(1), 1–6. Available from https://doi.org/10.1016/j.buildenv.2011.08.011.

Yang, P., Wu, L., Cheng, M., Fan, J., Li, S., Wang, H., & Qian, L. (2023). Review on drip irrigation: Impact on crop yield, quality, and water productivity in China. *Water*, *15*(9), 1733. Available from https://doi.org/10.3390/w15091733.

Zabidi, H. A., Goh, H. W., Chang, C. K., Chan, N. W., & Zakaria, N. A. (2020). A review of roof and pond rainwater harvesting systems for water security: The design, performance and way forward. *Water (Switzerland).*, *12*(11), 1–22. Available from https://doi.org/10.3390/w12113163, http://www.mdpi.com/journal/water.

Zaina, S. M., Fadli, F., & Hosseini, S. M. (2022). Evaluation of smart irrigation systems in hot-arid climates for green roofs and walls: Case of Doha, Qatar. *Smart and Sustainable Built Environment.*, *11*(2), 346–367. Available from https://doi.org/10.1108/SASBE-11-2021-0201, http://www.emeraldinsight.com/products/journals/journals.htm?id = sasbe.

Zegait, R., Bouznad, I. E., Remini, B., Bengusmia, D., Ajia, F., Guastaldi, E., Lopane, N., & Petrone, D. (2023). Comprehensive model for sustainable water resource management in Southern Algeria: integrating remote sensing and WEAP model. *Modeling Earth Systems and Environment.* Available from https://doi.org/10.1007/s40808-023-01826-y, https://www.springer.com/journal/40808.

Chapter 16

Assessment of spatial pattern of groundwater recharge potential zones for watershed management: a study on Kuya River Basin, Eastern India

Niladri Das[1], Ranajit Ghosh[2], Subhasish Sutradhar[3], Sajal Ghosh[1] and Prolay Mondal[4]

[1]*Department of Geography, Hiralal Bhakat College, Nalhati, Birbhum, West Bengal, India,* [2]*Department of Geography, Suri Vidyasagar College, Suri, Birbhum, West Bengal, India,* [3]*NATMO, Rashtriya Atlas Bhavan, Kolkata, West Bengal, India,* [4]*Department of Geography, Raiganj University, Raiganj, India*

16.1 Introduction

Groundwater, a crucial resource essential for supporting life, faces significant threats due to factors such as population growth, rapid industrialization, and the expansion of agricultural irrigation systems (Abijith et al., 2020; Chezgi, Pourghasemi, Naghibi, Moradi, & Kheirkhah Zarkesh, 2016; Venkateswaran & Ayyandurai, 2015). In the contemporary era, the depletion of groundwater and resulting water scarcity present substantial challenges for global inhabitants. The escalating demand for groundwater across diverse economic sectors over the past few decades has led to a protracted crisis, rendering it increasingly challenging to meet the needs of the growing population (Das & Mukhopadhyay, 2020). Developing countries, like India, have experienced rapid growth in agriculture, industry, and urbanization, intensifying the demand for water resources (Venkateswaran & Ayyandurai, 2015). In India, approximately 90% of the rural population and 30% of the urban population depend on groundwater for daily necessities, including drinking and domestic activities (Bhunia, Samanta, Pal, & Pa, 2012; Reddy, Kosgi, & Zhou, 1995). Hence, it is crucial to reevaluate the management of this essential hydrological resource and formulate sustainable development plans to address these urgent issues.

Water Resources Monitoring, Management, and Sustainability. DOI: https://doi.org/10.1016/B978-0-443-23665-5.00016-8

The escalating demand for groundwater is leading to a severe crisis, affecting all states in our nation (Kumar, 2021). The irrigation sector, particularly in India, plays a significant role in the depletion of groundwater (Arulbalaji, Padmalal, & Sreelash, 2019). To address this crisis, it is essential to identify and delineate potential groundwater recharge zones, enabling the sustainable use of this vital resource. Researchers worldwide are actively engaged in this endeavor, employing remote sensing and Geographic Information System (GIS) technology to identify potential groundwater recharge zones not only for India but also for regions across the globe (Ally, Yan, Bennett, Lyimo, & Mayunga, 2024; Pande, Khadri, Moharir, & Patode, 2018; Waikar & Nilawar, 2014).

Remote sensing and GIS technologies have significantly eased the work of geoscientists (Jha, Chowdhury, Chowdary, & Peiffer, 2007; Krishnamurthy, Venkatesa Kumar, Jayaraman, & Manivel, 1996; Manap et al., 2014; Panigrahi, Nayak, & Sharma, 1995; Rahmati, Nazari Samani, Mahdavi, Pourghasemi, & Zeinivand, 2015). GIS technology, in particular, serves as an advanced map-making tool capable of handling vast datasets and generating reliable outputs (Magesh, Chandrasekar, & Soundranayagam, 2012). Concurrently, remote sensing technology offers a dependable multi-spectral and multispectral database, instrumental in identifying optimal groundwater recharge zones (Das & Mukhopadhyay, 2020). With the advancement of remote sensing and GIS techniques, this process has become both cost-effective and time-efficient (Adiat, Nawawi, & Abdullah, 2012; Chaudhary & Kumar, 2018; Thapa, Gupta, Guin, & Kaur, 2017; Verma & Singh, 2013). In the current era of advanced science and technology, accessing global spatial, spectral, and temporal data has become remarkably swift (Gupta & Srivastava, 2010; Panahi, Mousavi, & Rahimzadegan, 2017; Sener, Davraz, & Ozcelik, 2005). GIS techniques serve as valuable tools for capturing, manipulating, storing, analyzing, managing, and representing various spatial data types (Chowdhury, Jha, Chowdary, & Mal, 2008; Gupta & Srivastava, 2010). Through the integration of multiple layers, such as geology, soil type, rainfall, lineament density, aquifer thickness, drainage density, land use and land cover (LULC), slope, and Normalized differences vegetation index (NDVI), groundwater potential zones can be effectively demarcated, enabling comprehensive analysis (Abijith et al., 2020; Chaudhary & Kumar, 2018; Das & Mukhopadhyay, 2020; Krishnamurthy et al., 1996; Rahmati et al., 2015; Venkateswaran & Ayyandurai, 2015).

In addition to these, various statistical methods are employed to identify potential groundwater resource zones for sustainable development. These methods encompass a wide range of statistical tools utilized to detect artificial recharge zones across different regions worldwide (Prasad, Mondal, Banerjee, Nandakumar, & Singh, 2008).

There are a number of quantitative methods that researchers have used to find groundwater recharge potential zones. These include the Analytic Hierarchy

Process (AHP) and the Technique for Order Preference by Similarity to the Ideal Solution (Agarwal, Agarwal, Garg, & Garg, 2013; Alrawi, Chen, & Othman, 2022; Doke, Zolekar, Patel, & Das, 2021; Ghosh, Sutradhar, Das, Mondal, & Sana, 2023; Guru, Seshan, & Bera, 2017; Hasanuzzaman, Mandal, Hasnine, & Shit, 2022; Kudamnya, Edet, & Ekwere, 2019; Maity, Mallick, Das, & Rudra, 2022; Melese & Belay, 2022; Muavhi, Thamaga, & Mutoti, 2022; Rajesh, Pande, Kadam, Gorantiwar, & Shinde, 2021; Sajil Kumar, Elango, & Schneider, 2022). The AHP method was used to give parameters and their feature weights so that groundwater recharge potential zones could be found in this study, which builds on earlier research. The AHP is very important for evaluating these zones because it organizes and ranks different factors that affect them in a way that makes it easier to make decisions (Zghibi et al., 2020). AHP utilizes mathematical computations to assign weighted scores to each criterion, facilitating the identification of areas with the highest potential for groundwater recharge (Dar, Rai, & Bhat, 2021). This method takes into account multiple factors and their relative importance, ensuring a systematic and rigorous evaluation process (Selvachandran, Quek, Smarandache, & Broumi, 2018).

According to Ghosh et al. (2023), the Kuya River basin in Rarh, Bengal, plays a crucial role in agriculture. Given the crucial role that groundwater plays in supporting agriculture in this area, it is urgent to address the threat that excessive exploitation poses. Consequently, it is essential to identify zones with high groundwater recharge potential to ensure sustainable water resource management in the area. In this case, the main goal of the study is to create a unique groundwater recharge potential map for the study area by combining the best parts of different models to get a complete and accurate picture.

16.2 Database and methodology

16.2.1 Location of the study area

The Kuya River basin has great significance within the Rarh Bengal region. The Khajuri River in Jharkhand is widely recognized as the originating point of the river, but the Babla River in Murshidabad is acknowledged as the point of confluence. The river has an estimated length of 176 km, while the basin it encompasses spans an area of roughly 1533 km^2. The basin exhibits longitudinal extension ranging from 87°13′00″ East to 88°09′30″ East and latitudinal extension ranging from 23°26′18″ North to 23°56′30″ North. Fig. 16.1 depicts the geographical location of the Kuya River basin.

16.2.2 Data source

During the data collection phase, all obtained information was documented and converted into the Universal Transverse Mercator Projection, specifically employing the northern zone 45 with the utilization of the WGS84 datum. The preparation of drainage density and drainage frequency maps was

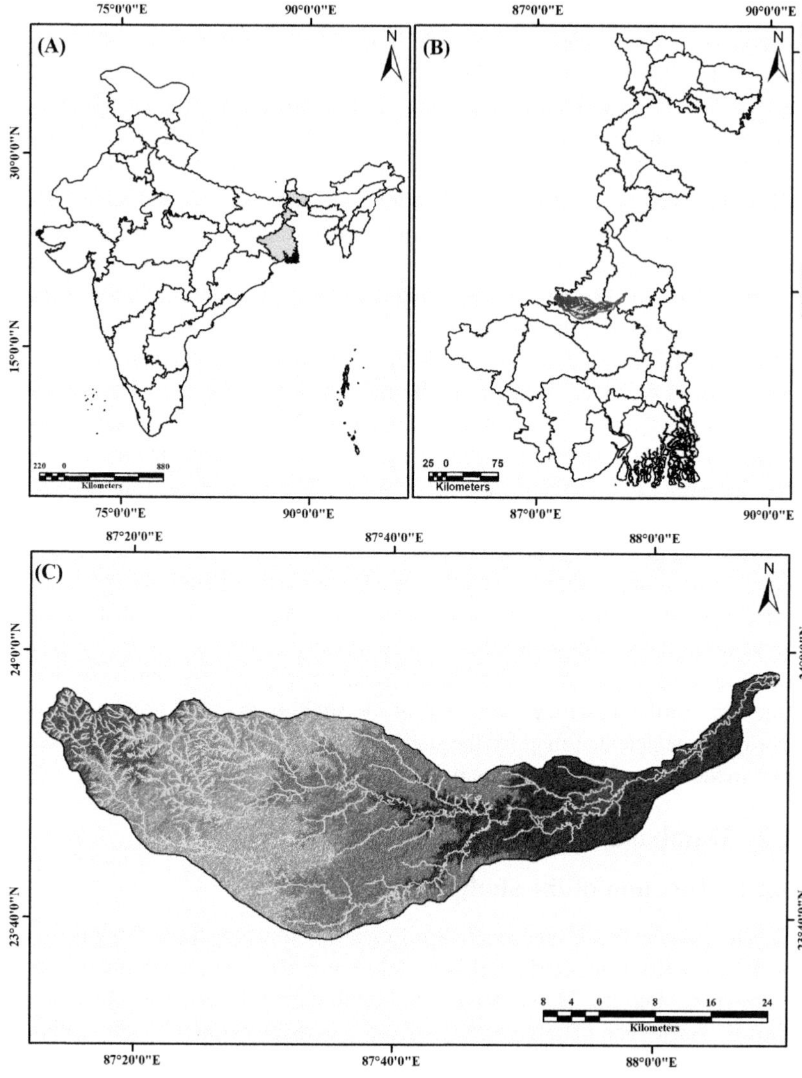

FIGURE 16.1 Location of the study area: (A) India, (B) West Bengal, and (C) Kuya River basin. Map lines delineate study areas and do not necessarily depict accepted national boundaries.

conducted using ArcGIS software and the grid index approach, utilizing the existing drainage map as a basis. The rainfall distribution map was generated by utilizing data obtained from the official website of NASA Giovanni (https://www.earthdata.nasa.gov/technology/giovanni). The soil texture map was acquired from the official website of the Water Resource Information System, accessible at https://www.earthdata.nasa.gov/technology/giovanni. The Shuttle Radar Topography Mission Digital Elevation Model (SRTM

DEM V3, 1-arc sec) was utilized within the ArcGIS software to conduct spatial analysis and generate a slope and drainage map. The creation of the LULC map involved the utilization of Landsat-8 satellite images obtained through the USGS Earth Explorer platform (https://earthexplorer.usgs.gov/). The maximum likelihood classification technique was employed within the ArcGIS environment for this purpose. The aquifer map was acquired *via* the WRIS WebGIS Portal, which is managed by the National Remote Sensing Center under the Government of India (http://india-wris.nrsc.gov.in). Subsequently, it underwent the process of digitization utilizing ArcGIS software. The lineament map, obtained from the Geo-Portal Bhuvan, which is administered by the Indian Space Research Organization and can be accessed at http://bhuvan.nrsc.gov.in, was utilized to produce a lineament density map using the spatial analysis capabilities provided by ArcGIS. The initial cartographic representation of ponds was created by utilizing a high-resolution QuickBird picture obtained from Google Earth. Subsequently, a pond frequency map was developed through the application of spatial analytic tools in ArcGIS. The geological map was obtained through the utilization of data provided by the Geological Survey of India.

16.2.3 Methodology

16.2.3.1 Application of analytical hierarchy process

The study dealt with the identification and preparation of the groundwater potential areas by applying the AHP for the Kuye River basin. The dominant fourteen factors were selected mainly based on professional experts judgment, analysis of available information, field reviews, and determined factors such as rainfall, lineament density, geology, drainage density, soil, LULC, slope, pond frequency, aquifer, surface runoff, elevation, NDVI, and Normalized Difference Water Index (NDWI). AHP is a semiquantitative approach where decisions are taken through hierarchical analysis of associated variables by assigning numerical scores through a pairwise comparison matrix, which denotes the individual factor's importance (Saaty, 1987). The AHP method includes both comparison and evaluation. The pairwise comparison method is also called the Saaty method. The mentioned method is used for multicriteria decision-making (Stofkova, Krejnus, Stofkova, Malega, & Binasova, 2022). AHP is a widely accepted model used to assign a normalized weight for each thematic layer of groundwater prospecting factor (Melese & Belay, 2022). The complicated decision-making procedure between the criteria is simplified to a single level with this pairwise comparison matrix, and the comparative relevance values of the criteria are derived (Ifediegwu, 2022). The reliability of the judgment is dependent on the consistency ratio (CR), and its value must be less than or equal to 0.1 (Sajil Kumar et al., 2022).

16.2.3.2 Normalization of weight for thematic layers by analytical hierarchy process model

The application of the AHP model involves the assignment of normalized weights to each groundwater prospecting factor or theme layer (Singh et al., 2024). The metrics associated with selected elements that impact groundwater potentiality were assigned numerical rankings ranging from 1 to 9, and their respective weight influences were subsequently determined by calculation. According to the study conducted by Ngwijabagabo et al. (2023), the presence of subjectivity in determining the assigned weight to various thematic maps, together with their respective characteristics, hinders the standardization process. According to Kessar, Benkesmia, Blissag, & Kébir (2021) and Kumari (2018), the concept was initially proposed by Saaty in the year 1980. According to Saaty (1987), the AHP model is a widely recognized multiple-criteria decision-making technique that is commonly employed in the domain of resource and environmental management for decision-making. The AHP encompasses a series of sequential steps, which are as follows (Saaty, 1987):

Step I: To construct the AHP model, elements or parameters (thematic layers) should be well defined.

Step II: To generate a pairwise comparison matrix by comprising all thematic layers based on Saaty's 1–9 scale (Table 16.1), where score 1 represents equal importance and score 9 indicates extreme importance of one theme compared to the other one (Saaty, 1987).

Step III: The final weight of the thematic layers has been obtained from the normalized value of the eigenvalue or the priority vectors, which are

TABLE 16.1 Saaty's 1–9 scale of relative importance.

Scale	Importance
1	Equal importance
2	Weak
3	Moderate importance
4	Moderate plus
5	Strong importance
6	Strong plus
7	Very strong importance
8	Very very strong importance
9	Extreme importance

associated with the maximum eigenvalue of the ratio matrix. To calculate CR, the following formula has been used:

$$CR = CI/RI$$

where RI indicates the random consistency index, whose value depends on the order of the matrix (Table 16.2), and CI indicates the consistency index, which can be obtained from the following formula:

$$CI = \lambda_{\max} - n/n - 1$$

where λ is the principal eigenvalue of the matrix and n is the number of groundwater prospecting factors. The value of CR must be <0.1 (Saaty, 1987).

16.2.3.2.1 Normalization of the weight of different features of thematic layers by ranking method

Each thematic layer map is categorized into distinct groups based on their range values, and a ranking method is employed to assign normalized weights to each feature within a thematic layer. The ranking is determined according to the groundwater potential of each criterion, with two commonly used methods: straight ranking (where the most important is assigned a rank of 1, the second most important is ranked 2, and so on) or inverse ranking (where the least important is assigned a rank of 1, the next least important is ranked 2, and so forth). In this study, the inverse ranking method is utilized, employing a rating scale from 1 to 5, where 1, 2, 3, 4, and 5 correspond to very low, low, medium, high, and very high groundwater storage potential (Jhariya et al., 2021; Kumar, Herath, Avtar, & Takeuchi, 2016).

16.2.3.3 Computation of groundwater potential index

The groundwater potential index (GWPI) is a dimensionless quantitative method that helps to delineate the groundwater potential zone in the study area (Razandi, Pourghasemi, Neisani, & Rahmati, 2015). To get groundwater potential zones, a weighted linear combination technique is employed to demarcate groundwater potential zones. The following formula is used for this (Raj, Rawat, Singh, & Mishra, 2022; Razandi et al., 2015; Shekhar & Pandey, 2015):

$$\begin{aligned} \mathrm{GWPI} = {} & \mathrm{Rf_w Rf_{wi}} + \mathrm{A_w A_{wi}} + \mathrm{NDWI_w NDWI_{wi}} + \mathrm{Sm_w Sm_{wi}} + \mathrm{St_w St_{wi}} + \mathrm{LULC_w LULC_{wi}} + \mathrm{Ge_w Ge_{wi}} \\ & + \mathrm{Dd_w Dd_{wi}} + \mathrm{Pf_w Pf_{wi}} + \mathrm{El_w El_{wi}} + \mathrm{Sr_w Sr_{wi}} + \mathrm{Ld_w Ld_{wi}} + \mathrm{NDVI_w NDVI_{wi}} + \mathrm{Sl_w Sl_{wi}} \end{aligned}$$

where W = normalized weight of each thematic layer, Wi = normalized weight of the features, Rf = rainfall, NDWI = Normalized Difference Water Index, S = Soil moisture, LULC = land use/land cover, Ge = Geology, Dd = drainage density, Pf = pond frequency, El = elevation, Sr = surface runoff, Ld = lineaent density, NDVI = Normalized Difference Vegetation Index, and Sl = slope.

TABLE 16.2 Saaty's ratio index for different values of *n*.

n	1	2	3	4	5	6	7	8	9	10	11	12	13	14
RI	0	0	0.58	0.89	1.12	1.24	1.32	1.41	1.45	1.49	1.51	1.48	1.56	1.57

16.2.3.4 Deriving weight of theme by analytic hierarchy process model

Based on the AHP model, the pairwise comparison matrix of the groundwater prospecting thematic layer has been derived, and the normalized weight of each thematic layer has been determined by calculating the eigenvector or priority vector.

Table reveals that rainfall occupies 13.50% of the weightage and is the most important parameter for storing groundwater. Aquifer (11.40%) stands as the second most important parameter for storage groundwater potential. In this way, NDWI (8.60%), soil moisture (9.80%), soil type (8.30%), LULC (7.30%), geology (6.70%), drainage density (5.70%), pond frequency (6.10%), elevation (3.30%), surface runoff (4.50%), lineament density (4.10%), NDVI (5.20%), and slope (5.50%), respectively, occupy their weightage and position regarding the influence of groundwater potential.

To check the consistency of the decision regarding the influence of groundwater prospecting factors or the selected thematic layers, the formula of Consistency Index (CI) has been implied.

The entire methodology has been shown in Fig. 16.2.

16.2.3.5 Generation of geospatial database

To get the final output, all 14 geospatial layers were overlaid on one another. The influence of each factor on groundwater occurrence has been discussed in this section.

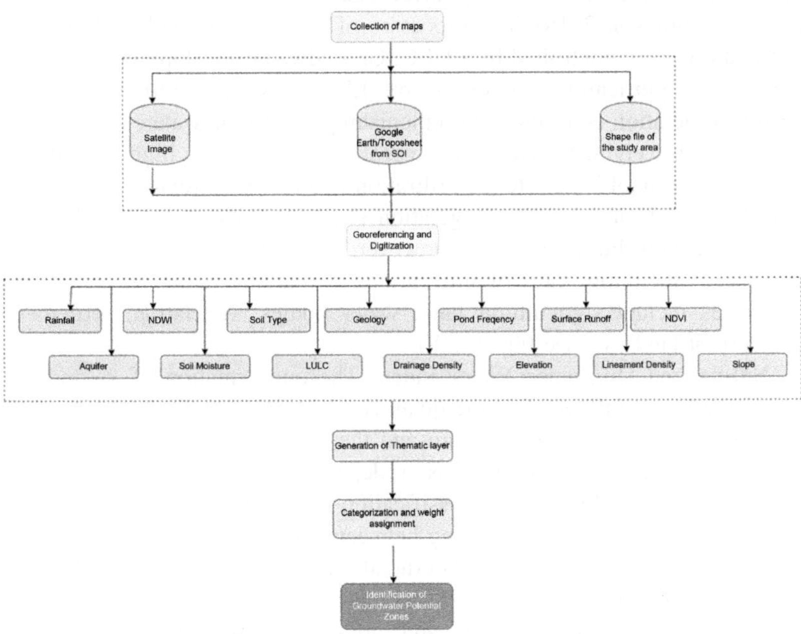

FIGURE 16.2 Methodological framework of the study.

Rainfall: Rainfall serves as a primary source for groundwater recharge and plays a crucial role in the hydrological cycle. In assessing groundwater potential, areas with lower rainfall are assigned lower ranks compared to those with higher rainfall, acknowledging its significance in influencing aquifer recharge rates (Rana, Hossain, & Nasher, 2022). Rainfall is a major contributor to the hydrological cycle and holds a dominant influence on groundwater in a given area (Arulbalaji et al., 2019). It is a critical factor that significantly affects the recharge rate into aquifers, with the percolation of water into the groundwater system being largely dependent on the amount of rainfall. Rainfall is a key component for evaluating potential groundwater zones, and its correlation with groundwater table fluctuation is well established, emphasizing its role in determining groundwater storage and fluctuation. In the study area, which experiences 80% of its rainfall from the southwest monsoon wind from June to September, the average annual rainfall is categorized into three classes: low (<1200 mm), medium (1200–1400 mm), and high rainfall (>1400 mm) (Fig. 16.3I).

Land use land cover: Use of land since LULC specifically regulates hydrogeological processes, mapping land cover plays a critical role in defining the groundwater potential zone (Chatterjee, Behera, Patra, & Goswami, 2023). Because land use and cover have an impact on evapotranspiration, runoff, and groundwater system recharge, they are important factors in the groundwater recharge process (Serele, Pérez-Hoyos, & Kayitakire, 2020). The regulation of several hydrogeological processes involved in the completion of a water cycle is greatly influenced by land use (Kamila, Bal, Sen, Parvin, & Sultana, 2018). The processes of permeability and infiltration, or percolation, were controlled by LULC. Groundwater development and occurrence are significantly influenced by LULC. An agricultural area with healthy vegetation contributes to replenishing the groundwater, which is why it has a high weight. Because more vegetation promotes a higher infiltration rate and less runoff, the rate of infiltration is directly correlated with vegetation density. As a result, the vegetation covered is also given good weight. Wasteland and built-up areas, however, have been given low weights because of their low rates of infiltration. In this study, the primary LULC categories include agricultural land, water bodies, vegetation, settlements, and current land use (see Fig. 16.4F).

Normalized Difference Water Index: The augmentation of groundwater has a significant influence on the quantity of moisture present in the soil. To evaluate the levels of soil moisture and the occurrence of water bodies, the NDWI is frequently employed as a dependable metric (Sresto, Siddika, Haque, & Saroar, 2021). The classification of NDWI pictures is performed with the objective of distinguishing between surface water and land, as well as other features. This technique is crucial for achieving accurate cartography of surface water. Consequently, NDWI images can be classified into two distinct categories, namely, water and non-water. According to Ashtekar,

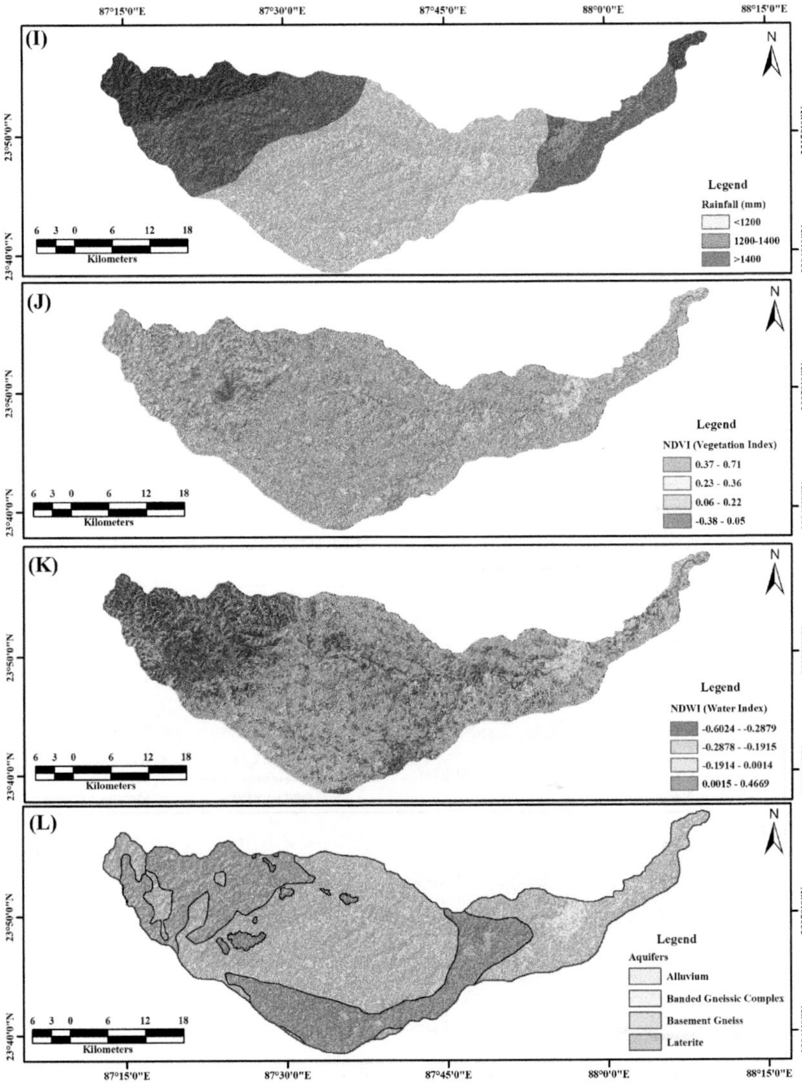

FIGURE 16.3 (I) Rainfall, (J) NDVI, (K) NDWI, and (L) aquifer types. *NDVI*, Normalized differences vegetation index; *NDWI*, Normalized Difference Water Index. Map lines delineate study areas and do not necessarily depict accepted national boundaries.

Mohammed-Aslam, & Moosvi (2019), it is apparent that non-water pixels belong to categories that are unrelated to water in the given context. The NDWI has been classified into four unique categories within the specified study region. The delineation of these groups is based on distinct intervals of NDWI values, namely, −0.6024 to −0.2879, −0.2878 to −0.1915, −0.1914 to 0.0014, and 0.0015 to 0.4669, as depicted in Fig. 16.3K.

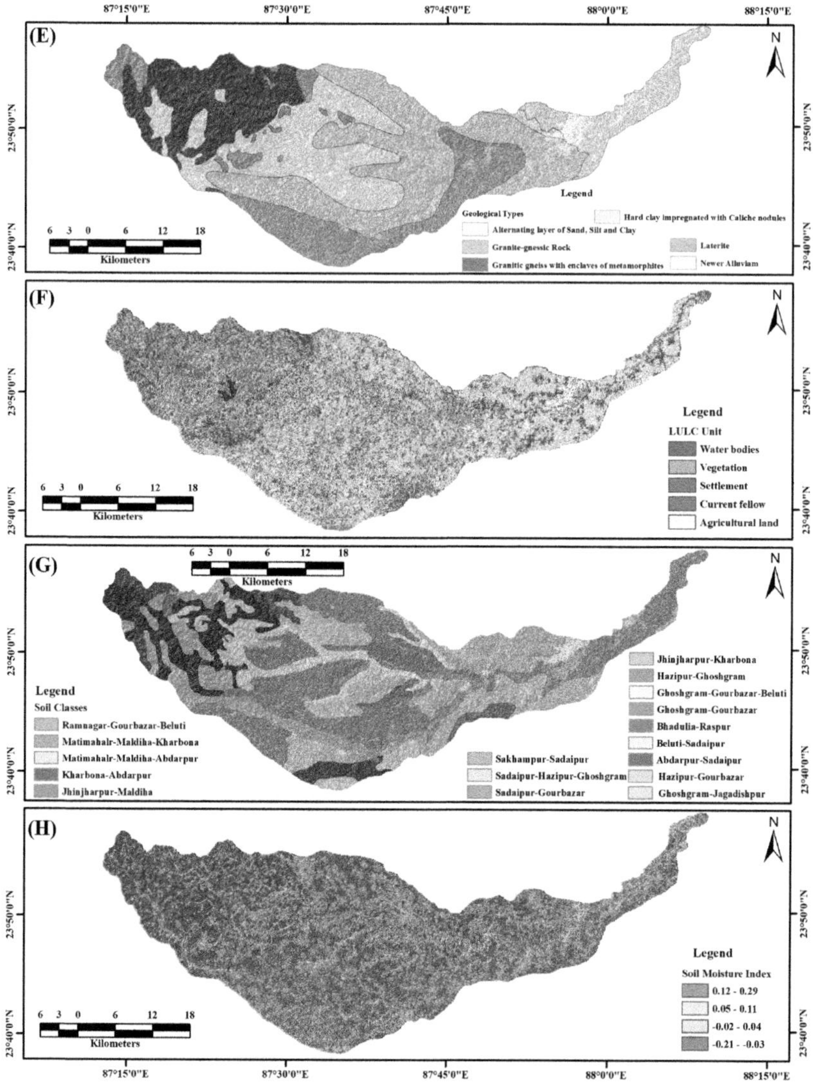

FIGURE 16.4 (E) Geology, (F) land use and land cover, (G) soil types, and (H) soil moisture index. Map lines delineate study areas and do not necessarily depict accepted national boundaries.

Geology: The study of geology assumes a critical role in the assessment of groundwater potential by examining the processes of infiltration and percolation (Ifediegwu, 2022). The utilization of improved aeromagnetic data facilitates the production of precise maps of foundation rocks and enables the distinct demarcation of sedimentary rocks, owing to the variations in magnetic properties across several geological environments (Owolabi, Madi,

Kalumba, & Orimoloye, 2020). The geological divisions within the research region include several primary formations. These include an archaeing layer composed of sand, silt, and clay; granite-gneiss rock; granite gneiss with enclaves of metamorphites; hard clay infused with caliche nodules; laterite; and younger alluvium (as seen in Fig. 16.4E).

Drainage density: The metric known as drainage density has a significant impact on the assessment of groundwater potential distribution within a specific area (Suliman, Samiullah, & Ali, 2022). The hydrological characteristics of a landscape, which are influenced by both subsurface and surface geological formations, play a crucial role in determining the patterns of water runoff and infiltration into the groundwater (Sarwar et al., 2021). The variable in question has a strong correlation with surface runoff and permeability, so it serves as an indirect indicator of the groundwater potential within a given region. Consequently, it is regarded as a key factor in evaluating the presence of groundwater (Nag & Chowdhury, 2019). The drainage density, which is characterized as the ratio of stream length to the area of the watershed, is subject to various influences, including the geological composition of the rocks, the qualities of the soil, and the steepness of the slope. The inverse function of permeability is of significant importance in the delineation of prospective groundwater zones. There exists a positive correlation between higher drainage density and increased surface runoff, resulting in a decrease in groundwater recharge, and conversely, a negative correlation between lower drainage density and decreased surface runoff, leading to an increase in groundwater recharge. The study region has been categorized into three distinct categories based on the measurement of drainage density: low (0–0.54 km^2), medium (0.55–1.33 km^2), and high drainage density (1.34–4.57 km^2) (Fig. 16.5D).

Lineament density: Lineament studies offer a wide range of applications within different fields of geoscience. These applications include the identification of tectonic features, the detection of folds and faults, as well as the exploration of mineral deposits, petroleum possibilities, and groundwater resources (Waikar & Nilawar, 2014). The thickness of lineaments within a given area might serve as an indicator of the groundwater potential, as the occurrence of lineaments frequently suggests the presence of a permeable zone. According to Kumari & Singh (2021), areas characterized by a significant thickness of lineaments are regarded as conducive for the identification of groundwater potential zones. The presence of groundwater in hard rock terrains is determined by both the composition of the rocks and the level of tectonic activity (Nag & Ghosh, 2013). Consequently, an increase in lineament density corresponds to an increased significance in assessing groundwater potential, whereas a decrease in lineament density corresponds to a decreased significance in assessing groundwater potential. Within the designated study region, the lineament density is classified into two primary categories: high, with a value of 128.799, and low, with a value of 0 (as depicted in Fig. 16.6M).

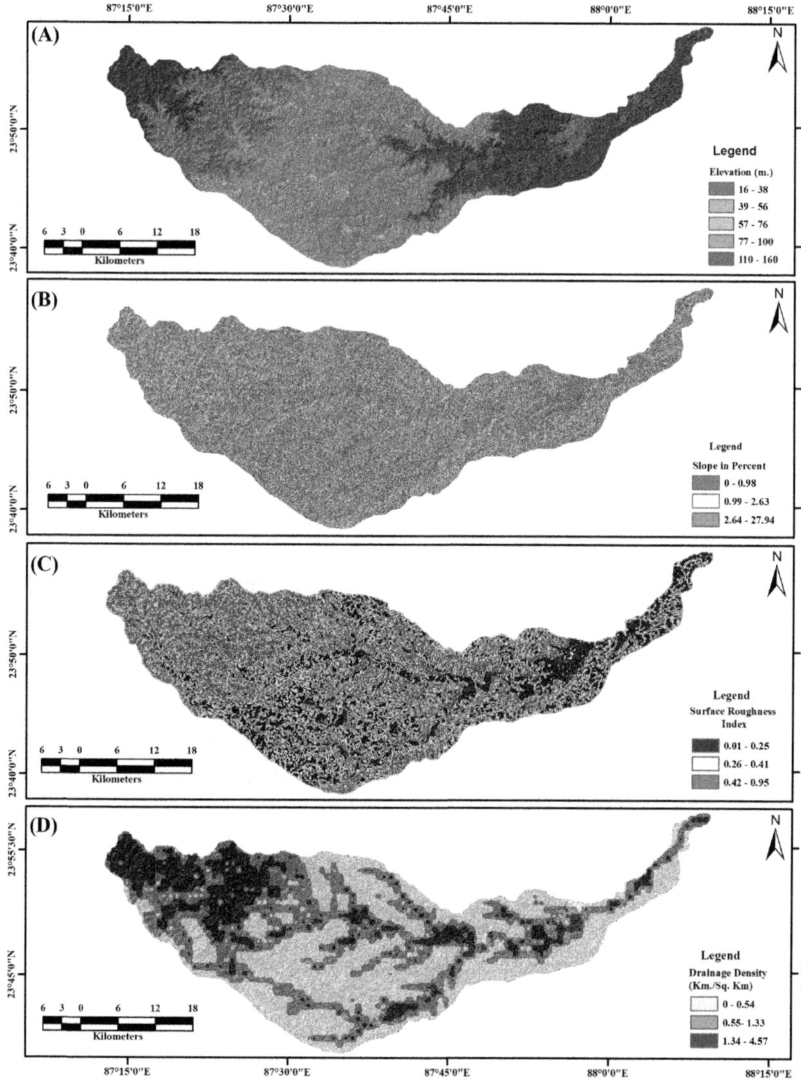

FIGURE 16.5 (A) Elevation, (B) slope, (C) surface roughness, and (D) drainage density. Map lines delineate study areas and do not necessarily depict accepted national boundaries.

Elevation: The elevation has a significant impact on the availability of groundwater in a particular area (Sohail, Singh, & Kanga, 2019). The elevation of the research region is a crucial factor in determining groundwater potential. Regions characterized by lower elevations, namely those exhibiting more level topography, have a propensity to exhibit prolonged water retention, hence facilitating enhanced infiltration and replenishment of

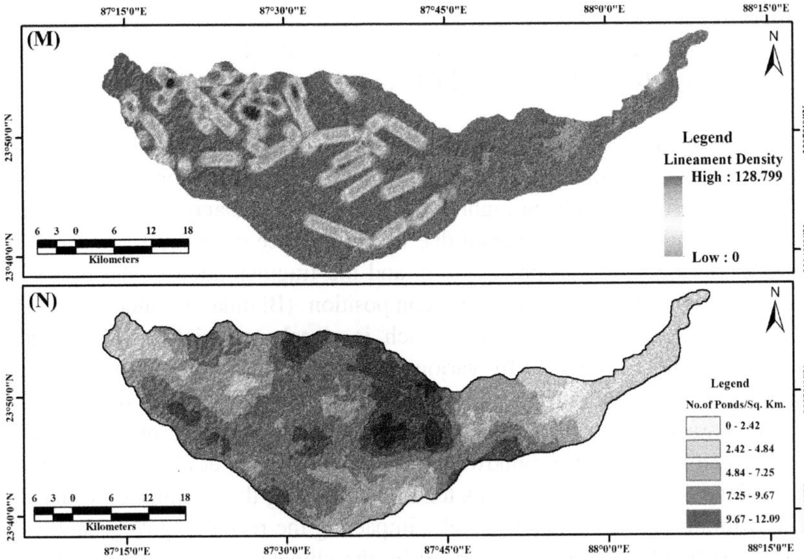

FIGURE 16.6 (M) lineament density and (N) pond frequency. Map lines delineate study areas and do not necessarily depict accepted national boundaries.

groundwater resources. On the other hand, it has been observed that regions characterized by more pronounced elevations tend to exhibit an increased volume of runoff and a decreased level of infiltration (Thapa et al., 2017). According to elevation, the study region has been divided into five subclasses, which are as follows: 16–38, 39–56, 57–76, 77–100, and 110–160 m (Fig. 16.5A).

Slope: The measurement of slope length serves as a valuable tool for assessing the physical attributes of a slope, including its extent and magnitude. This information is crucial for comprehending the dynamics of slope processes and identifying areas with potential for groundwater retention (Maqsoom et al., 2022). The steepness of the slope gradient plays a significant role in influencing the infiltration of rainfall, hence serving as a vital indicator for evaluating the potential of groundwater resources. Regions characterized by more pronounced inclines encounter reduced recharge rates as a result of increased surface runoff, which limits the duration available for water to permeate the surface and replenish the saturated zone (Murmu, Kumar, Lal, Sonker, & Singh, 2019). According to Thapa et al. (2017), the infiltration of surface water is notably affected by the slope gradient. Steeper slopes result in a swift downhill flow of water, allowing limited time for infiltration. Conversely, flatter locations offer a longer duration for rainwater infiltration. The gradient of a given area plays a pivotal role in the process of groundwater recharge, exerting influence on both the surface and subsurface movement of precipitation and its contribution to the replenishment of underground water storage (Teja & Singh, 2019). Within the designated study region, the slopes present can be classified into three primary categories, namely, 0–0.98%, 0.99–2.63%, and 2.64–27.94%, as depicted in Fig. 16.5B.

Soil type: The role of soil in delineating regions with groundwater potential and in assessing groundwater storage capacity and infiltration rates based on soil characteristics has been highlighted in a study by Gaurav & Singh (2022). The examination of soil is crucial due to its influence on the restriction of penetration during monsoon precipitation and the fragmentation of the impact of surface runoff based on material composition (Bhunia, Chatterjee, & Pal, 2014). The permeability of soil, which is closely associated with effective porosity, is greatly affected by various parameters, including grain form, size, absorbed water, void ratio, degree of saturation, and the presence of impurities. These characteristics collectively contribute to the significance of soil in identifying suitable zones for groundwater delineation (Thapa et al., 2017). The texture of the soil is a significant factor in determining the quantity of water that can infiltrate into the ground, thereby impacting the recharging of groundwater (Das, 2017). The soil is characterized by the climate, physiography, and geology, which collectively have a significant impact on groundwater recharge and runoff (Kumar et al., 2016). The soil texture map, acquired from the official portal of Birbhum District, categorizes the study region into five primary soil textural types, namely, sandy, clay loam, sandy loam, clay, and loam. The various soil types within the study area include Ramnagar-Gourbazar-Beluti, Matimabalr-Maldiha-Kharboaa, Matimabalr-Madiha-Abdarpur, Kharbona-Abdarpur, Jhimjharpur-Maldiha, Sakhampur-Sadaipur, Sadaipur-Hazipur-Ghoshgram, Sadaipur-Gourbazar, Jhinjharpur-Kharbona, Hazipur-Ghoshgram, Ghoshgram-Gourbazar-Beluti, Ghoshgram-Gourbazar, Bhadulia-Raspur, Beluti-Sadaipur, Abdarpur-Sadaipur, Hazipur-Gourbazar, and Ghoshgram-Jagadishpur (Fig. 16.4G).

Soil moisture index (SMI): The SMI serves as a crucial determinant in identifying groundwater potential zones due to its direct correlation with the moisture content in the soil. The level of soil moisture plays a pivotal role in groundwater recharge, as areas with higher soil moisture are more likely to facilitate water infiltration and contribute to replenishing the groundwater table. By assessing the spatial distribution of SMI, researchers and water resource managers can pinpoint regions with optimal conditions for groundwater recharge, providing valuable information for sustainable water management strategies (Zhang et al., 2019). The SMI values considered in this study range from 0.12 to 0.29, 0.05 to 0.11, −0.02 to −0.04, and −0.21 to −0.03 (Fig. 16.4H).

Surface roughness: The process of infiltration is influenced by the characteristics of rainfall, namely, its intensity and duration. When rainfall is characterized by high intensity but short duration, the amount of water that infiltrates into the ground is reduced, resulting in a greater proportion of surface runoff. Conversely, when rainfall is characterized by low intensity but long duration, a larger amount of water can infiltrate into the ground, leading to a higher proportion of infiltration compared to surface runoff (Sar, Khan, Chatterjee, & Das, 2015). To ascertain the key and non-key-dependent factors, a study was conducted on five hydrological variables: river runoff, groundwater level, precipitation, evaporation, and groundwater abstraction. According to Meng et al. (2021), the analysis indicated that the dependent variables most notably impacted are river runoff and groundwater level. The comprehension of the influence of LULC on the recharging of groundwater and the runoff on the surface, specifically in relation to alterations in LULC, is of utmost importance in the pursuit of sustainable management of water resources (Owuor et al., 2016). The consideration of surface roughness and its impact on groundwater recharge and surface runoff, as well as their reaction to LULC changes, is crucial in the context of sustainable water resource management. The research area's surface roughness index is categorized into three distinct groups, namely, 0.01–0.25, 0.26–0.41, and 0.42–0.95 (see Fig. 16.5C).

Normalized Difference Vegetation Index: The NDVI is a widely employed vegetation index within the field of hydro-environmental studies. The study conducted by some scholars provides valuable insights into vegetation levels through the analysis of data obtained from satellite observations. The utilization of satellite-derived NDVI is widely recognized as an essential indicator for evaluating vegetation health and status. The status of vegetation at various stages of growth is influenced by complex interactions involving weather patterns, soil moisture levels, and the specific characteristics of the soil and crop varieties. The analysis of NDVI provides a preliminary assessment of vegetation abundance and potential groundwater availability throughout the geographical area (Sar et al., 2015). The study region has been classified into four distinct categories of NDVI: 0.37–0.71, 0.23–0.36, 0.06–0.22, and −0.38 to 0.05 (refer to Fig. 16.3J).

Pond frequency: The study examines the potential zone by taking into account the frequency of ponds in the area. Ponds are recognized as a significant parameter for groundwater recharge and are frequently used for artificial recharge. Places with higher pond frequencies have the capacity to hold a greater amount of water, leading to an increased rate of water infiltration, while the opposite is true for places with lower pond frequencies. As a result, there is a greater emphasis on higher pond frequency, whereas lower pond frequency is assigned a lower level of importance. The frequency of ponds in the eastern region of the research area is significantly higher in comparison to the western region. The study area in question exhibits five primary categories of pond frequency groups, namely, 0–2.42, 2.42–4.84, 4.84–7.25, 7.25–9.67, and 9.67–12.09 (as depicted in Fig. 16.6N).

Aquifer type: The task of modeling extensive aquifers may present itself as a formidable undertaking, yet the acquisition of crucial data pertaining to water availability hinges upon comprehending the inflows and outflows of the reservoir (Murari, 2016). The geological composition of the rock layers comprising the aquifer within the designated study area exerts a substantial influence on the potential of groundwater resources (Purnama, Rahmawati, Primacintya, & Febriarta, 2021). The research area encompasses many principal aquifer types, including alluvium, banded gneiss complex, basement gneiss, and laterite (Fig. 16.3L).

16.3 Result and discussion

The cartographic representation of groundwater potential zones has been discretized into five distinct classes: very low, low, moderate, high, and very high recharge potential zones, as illustrated in Fig. 16.7. This classification is based on the methodological framework and pixel arrangement described earlier. The categorization process was conducted using the ArcGIS 10.5 program. Each of the aforementioned zones represents a distinct portion of the overall spatial extent. For instance, the zone with very low potential encompasses 16.90% of the region, while the low potential zone covers 27.36%. The moderate potential zone constitutes 26.31% of the area,

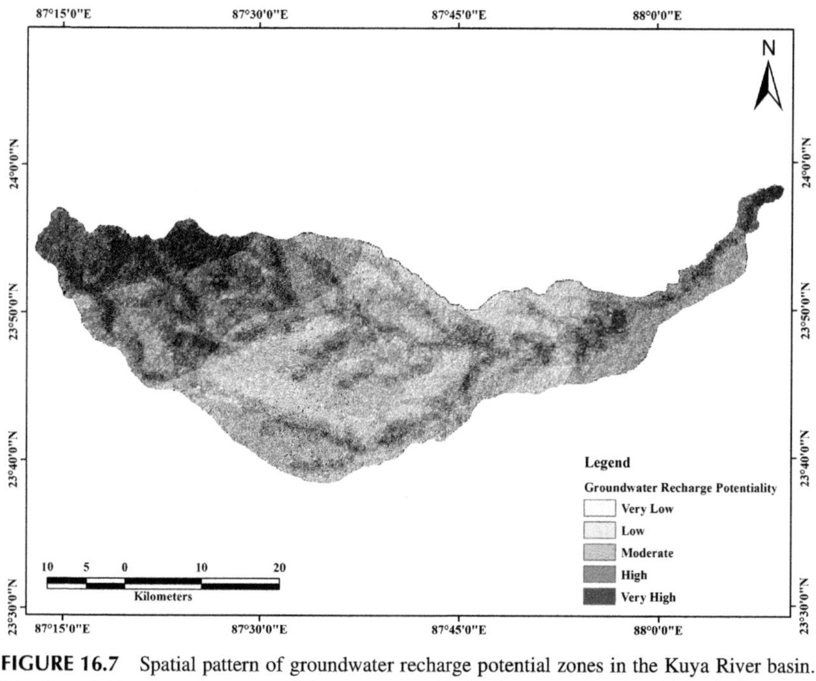

FIGURE 16.7 Spatial pattern of groundwater recharge potential zones in the Kuya River basin. Map lines delineate study areas and do not necessarily depict accepted national boundaries.

TABLE 16.3 Areal coverage of different groundwater recharge potential zone.

Groundwater Potential Zone categories	Area in km^2	Percentage of basin area
Very low	256.78	16.90
Low	415.66	27.36
Moderate	399.84	26.32
High	296.72	19.53
Very high	150.23	9.89

followed by the high potential zone, which encompasses 19.53%. Lastly, the very high recharge potential zone embodies 9.88% of the total area within the river basin (Table 16.3).

The investigation of probable recharge zones for groundwater in the Kuya River Basin reveals a comprehensive and meticulous examination, providing significant observations regarding the hydrological dynamics of the area. The upper reach of the basin has high recharge potential zones, mostly attributed to the presence of impermeable rock formations that facilitate long-term water retention. Additionally, this region is characterized by limited agricultural operations, which consequently mitigates the depletion of groundwater resources. The inclusion of a tilpara barrage also serves to augment the recharge in the central section of the basin. If we proceed over the bottom stretch of the basin, we can find that this region has primarily low recharge potential zones. The region's potential is hampered by the fact that it receives less rainfall and also by the fact that its agricultural operations cause it to draw on a considerable amount of groundwater.

16.4 Conclusion

The assessment of groundwater potential within the Kuya River Basin is critical for understanding the spatial distribution of groundwater resources. This research adopted a systematic approach, integrating various hydrogeological parameters, such as lithology, land use, slope, and drainage density, to identify potential groundwater zones. The results indicate that the upper part of the basin has a high potential for groundwater due to favorable geological and hydrogeological conditions. This is important for sustainable water resource management, especially considering the high agricultural activity in the region, which has led to decreased groundwater potential in the lower basin. Areas identified as having low groundwater potential could face challenges in meeting water demands, underscoring the need for careful planning and effective management strategies to ensure water availability for local communities and ecosystems. The study highlights the importance of considering both the quantity and quality of groundwater in sustainable development.

Effective groundwater management practices, including recharge mechanisms, conservation measures, and monitoring strategies, are essential to prevent overexploitation in high potential areas and enhance water availability in low-potential areas. Community involvement and stakeholder engagement are crucial for the success of sustainable groundwater management initiatives.

Future research should focus on using remote sensing and GIS technologies to further analyze groundwater potential in the Kuya River Basin. High-resolution satellite imagery and GIS mapping can assess land use, land cover, and geological factors affecting groundwater. Advanced spectral analysis and machine learning algorithms could enhance the accuracy of groundwater potential assessments, complemented by ground-truth data and field surveys.

Remote sensing and GIS offer comprehensive insights into the spatial distribution of groundwater resources, aiding in informed decision-making for sustainable water use. These technologies provide efficient data collection and dynamic perspectives on land cover changes affecting groundwater availability. GIS is instrumental in creating detailed maps for optimal groundwater extraction sites. The use of advanced spectral analysis and machine learning algorithms significantly increases the accuracy of groundwater potential evaluations. This study contributes to water resource planning, improving understanding of groundwater use, and building resilience against climate change impacts, thereby promoting environmental sustainability and socio-economic development in the Kuya River Basin.

References

Abijith, D., Saravanan, S., Singh, L., Jennifer, J. J., Saranya, T., & Parthasarathy, K. S. S. (2020). GIS-based multi-criteria analysis for identification of potential groundwater recharge zones - a case study from Ponnaniyaru watershed, Tamil Nadu, India. *HydroResearch*, *3*, 1–14. Available from https://doi.org/10.1016/j.hydres.2020.02.002.

Adiat, K. A. N., Nawawi, M. N. M., & Abdullah, K. (2012). Assessing the accuracy of GIS-based elementary multi criteria decision analysis as a spatial prediction tool - A case of predicting potential zones of sustainable groundwater resources. *Journal of Hydrology*, *440–441*, 75–89. Available from https://doi.org/10.1016/j.jhydrol.2012.03.028.

Agarwal, E., Agarwal, R., Garg, R. D., & Garg, P. K. (2013). Delineation of groundwater potential zone: An AHP/ANP approach. *Journal of Earth System Science*, *122*(3), 887–898. Available from https://doi.org/10.1007/s12040-013-0309-8, http://www.ias.ac.in/jess/jun2013/887.pdf.

Ally, A. M., Yan, J., Bennett, G., Lyimo, N. N., & Mayunga, S. D. (2024). Assessment of groundwater potential zones using remote sensing and GIS-based fuzzy analytical hierarchy process (F-AHP) in Mpwapwa District, Dodoma, Tanzania. *Geosystems and Geoenvironment*, *3*(1). Available from https://doi.org/10.1016/j.geogeo.2023.100232, http://www.elsevier.com/locate/issn/27728838.

Alrawi, I., Chen, J., & Othman, A. A. (2022). Groundwater potential zone mapping: Integration of multi-criteria decision analysis (MCDA) and GIS techniques for the al-qalamoun region in Syria. *ISPRS International Journal of Geo-Information*, *11*(12). Available from https://doi.org/10.3390/ijgi11120603, http://www.mdpi.com/journal/ijgi.

Arulbalaji, P., Padmalal, D., & Sreelash, K. (2019). GIS and AHP techniques based delineation of groundwater potential zones: A case study from Southern Western Ghats, India. *Scientific Reports*, *9*(1). Available from https://doi.org/10.1038/s41598-019-38567-x, http://www.nature.com/srep/index.html.

Ashtekar, A. S., Mohammed-Aslam, M. A., & Moosvi, A. R. (2019). Utility of normalized difference water index and GIS for mapping surface water dynamics in sub-upper Krishna Basin. *Journal of the Indian Society of Remote Sensing*, *47*(8), 1431−1442. Available from https://doi.org/10.1007/s12524-019-01013-6, http://www.springerlink.com/content/0255-660X.

Bhunia, G. S., Chatterjee, N., & Pal, D. K. (2014). Identification of groundwater potential zone of Nawada district, Bihar (India) - A study based on remote sensing and GIS platform. *Hydrology Research*, *45*(4−5), 631−644. Available from https://doi.org/10.2166/nh.2013.258, http://www.iwaponline.com/nh/045/0631/0450631.pdf.

Bhunia, G. S., Samanta, S., Pal, D. K., & Pa, B. (2012). Assessment of groundwater potential zone in Paschim Medinipur district, West Bengal − A Meso-scale study using GIS and remote sensing approach. *Journal of Environment and Earth Science.*, *2*(5), 41−59.

Chatterjee, T., Behera, D., Patra, P. K., & Goswami, S. (2023). Delineation of the groundwater potential zone in Kantli River Basin, Jhunjhunu district, Rajasthan: A geospatial approach. *Journal of Geosciences Research*, *8*(2), 142−152. Available from https://doi.org/10.56153/g19088-023-0144-35.

Chaudhary, B. S., & Kumar, S. (2018). Identification of groundwater potential zones using remote sensing and GIS of K-J Watershed, India. *Journal of the Geological Society of India*, *91*(6), 717−721. Available from https://doi.org/10.1007/s12594-018-0929-3, http://www.springerlink.com/content/121451/.

Chezgi, J., Pourghasemi, H. R., Naghibi, S. A., Moradi, H. R., & Kheirkhah Zarkesh, M. (2016). Assessment of a spatial multi-criteria evaluation to site selection underground dams in the Alborz Province, Iran. *Geocarto International*, *31*(6), 628−646. Available from https://doi.org/10.1080/10106049.2015.1073366, http://www.tandfonline.com/toc/tgei20/current.

Chowdhury, A., Jha, M. K., Chowdary, V. M., & Mal, B. C. (2008). Integrated remote sensing and GIS-based approach for assessing groundwater potential in West Medinipur district, West Bengal, India. *International Journal of Remote Sensing*, *30*(1), 231−250. Available from https://doi.org/10.1080/01431160802270131, https://www.tandfonline.com/loi/tres20.

Dar, T., Rai, N., & Bhat, A. (2021). Delineation of potential groundwater recharge zones using analytical hierarchy process (AHP). *Geology, Ecology, and Landscapes*, *5*(4), 292−307. Available from https://doi.org/10.1080/24749508.2020.1726562, https://www.tandfonline.com/toc/tgel20/current.

Das, N., & Mukhopadhyay, S. (2020). Application of multi-criteria decision making technique for the assessment of groundwater potential zones: A study on Birbhum district, West Bengal, India. *Environment, Development and Sustainability*, *22*(2), 931−955. Available from https://doi.org/10.1007/s10668-018-0227-7, http://www.wkap.nl/journalhome.htm/1387-585X.

Das, S. (2017). Delineation of groundwater potential zone in hard rock terrain in Gangajalghati block, Bankura district, India using remote sensing and GIS techniques. *Modeling Earth Systems and Environment*, *3*(4), 1589–1599. Available from https://doi.org/10.1007/s40808-017-0396-7, http://springer.com/journal/40808.

Doke, A. B., Zolekar, R. B., Patel, H., & Das, S. (2021). Geospatial mapping of groundwater potential zones using multi-criteria decision-making AHP approach in a hardrock basaltic terrain in India. *Ecological Indicators*, *127*. Available from https://doi.org/10.1016/j.ecolind.2021.107685, http://www.elsevier.com/locate/ecolind.

Gaurav, N., & Singh, G. (2022). Delineation of groundwater, drought and flood potential zone using Weighted Index Overlay Analysis and GIS for district Patna. *An International Quarterly Scientific Journal*, *21*(2), 813–828..

Ghosh, R., Sutradhar, S., Das, N., Mondal, P., & Sana, R. I. (2023). A comparative evaluation of GIS based flood susceptibility models: A case of Kopai river basin, Eastern India. *Arabian Journal of Geosciences*, *16*(11). Available from https://doi.org/10.1007/s12517-023-11693-7.

Gupta, M., & Srivastava, P. K. (2010). Integrating GIS and remote sensing for identification of groundwater potential zones in the hilly terrain of Pavagarh, Gujarat, India. *Water International*, *35*(2), 233–245. Available from https://doi.org/10.1080/02508061003664419.

Guru, B., Seshan, K., & Bera, S. (2017). Frequency ratio model for groundwater potential mapping and its sustainable management in cold desert, India. *Journal of King Saud University - Science*, *29*(3), 333–347. Available from https://doi.org/10.1016/j.jksus.2016.08.003, http://www.sciencedirect.com/science/journal/10183647.

Hasanuzzaman, M., Mandal, M. H., Hasnine, M., & Shit, P. K. (2022). Groundwater potential mapping using multi-criteria decision, bivariate statistic and machine learning algorithms: Evidence from Chota Nagpur Plateau, India. *Applied Water Science*, *12*(4). Available from https://doi.org/10.1007/s13201-022-01584-9, http://www.springer.com/earth + sciences + and + geography/hydrogeology/journal/13201.

Ifediegwu, S. I. (2022). Assessment of groundwater potential zones using GIS and AHP techniques: A case study of the Lafia district, Nasarawa State, Nigeria. *Applied Water Science*, *12* (1). Available from https://doi.org/10.1007/s13201-021-01556-5, http://www.springer.com/earth + sciences + and + geography/hydrogeology/journal/13201.

Jha, M. K., Chowdhury, A., Chowdary, V. M., & Peiffer, S. (2007). Groundwater management and development by integrated remote sensing and geographic information systems: Prospects and constraints. *Water Resources Management*, *21*(2), 427–467. Available from https://doi.org/10.1007/s11269-006-9024-4.

Jhariya, D. C., Mondal, K. C., Kumar, T., Indhulekha, K., Khan, R., & Singh, V. K. (2021). Assessment of groundwater potential zone using GIS-based multi-influencing factor (MIF), multi-criteria decision analysis (MCDA) and electrical resistivity survey techniques in Raipur city, Chhattisgarh, India. *Aqua Water Infrastructure, Ecosystems and Society*, *70*(3), 375–400. Available from https://doi.org/10.2166/aqua.2021.129, https://watermark.silverchair.com/jws0700375.pdf?.

Kamila,, A., Bal,, A., Sen,, T. S., Parvin,, J., & Sultana, F. (2018). An integrated GIS approach for assessment of ground water potential zone : A case study on the upper catchment area of Subarnarekha River. *Indian Journal of Geography and Environment Management*, *16*.

Kessar, C., Benkesmia, Y., Blissag, B., & Kébir, L. W. (2021). Delineation of groundwater potential zones in Wadi Saida Watershed of NW-Algeria using remote sensing, geographic information system-based AHP techniques and geostatistical analysis. *Journal of Groundwater Science and Engineering*, *9*(2), 45–64. Available from https://doi.org/10.19637/j.cnki.2305-7068.2021.01.005, http://gwse.iheg.org.cn/EN/volumn/home.shtml.

Krishnamurthy, J., Venkatesa Kumar, N., Jayaraman, V., & Manivel, M. (1996). An approach to demarcate ground water potential zones through remote sensing and a geographical information system. *International Journal of Remote Sensing*, *17*(10), 1867–1884. Available from https://doi.org/10.1080/01431169608948744.

Kudamnya, E.A., Edet, A.E., & Ekwere, A.S. (2019). Advances in sustainable and environmental hydrology, hydrogeology, hydrochemistry and water resources. In: *Proceedings of the 1st Springer Conference of the Arabian Journal of Geosciences (CAJG-1)*, 249–252, doi:10.1007/978-3-030-01572-5.

Kumar, P., Herath, S., Avtar, R., & Takeuchi, K. (2016). Mapping of groundwater potential zones in Killinochi area, Sri Lanka, using GIS and remote sensing techniques. *Sustainable Water Resources Management*, *2*(4), 419–430. Available from https://doi.org/10.1007/s40899-016-0072-5, http://springer.com/journal/40899.

Kumar, S. (2021). *Mapping the sinking groundwater resources of Northwest India.*

Kumari, A., & Singh, A. (2021). Delineation of groundwater potential zone using analytical hierarchy process. *Journal of the Geological Society of India*, *97*(8), 935–942. Available from https://doi.org/10.1007/s12594-021-1794-z, http://www.springerlink.com/content/121451/.

Kumari, L. (2018). Delineation of potential groundwater zone using RS and GIS: A review. *International Journal of Current Microbiology and Applied Sciences*, *7*(12), 196–203. Available from https://doi.org/10.20546/ijcmas.2018.712.025.

Magesh, N. S., Chandrasekar, N., & Soundranayagam, J. P. (2012). Delineation of groundwater potential zones in Theni district, Tamil Nadu, using remote sensing, GIS and MIF techniques. *Geoscience Frontiers*, *3*(2), 189–196. Available from https://doi.org/10.1016/j.gsf.2011.10.007, https://www.sciencedirect.com/journal/geoscience-frontiers.

Maity, B., Mallick, S. K., Das, P., & Rudra, S. (2022). Comparative analysis of groundwater potentiality zone using fuzzy AHP, frequency ratio and Bayesian weights of evidence methods. *Applied Water Science*, *12*(4). Available from https://doi.org/10.1007/s13201-022-01591-w, http://www.springer.com/earth + sciences + and + geography/hydrogeology/journal/13201.

Manap, M. A., Nampak, H., Pradhan, B., Lee, S., Sulaiman, W. N. A., & Ramli, M. F. (2014). Application of probabilistic-based frequency ratio model in groundwater potential mapping using remote sensing data and GIS. *Arabian Journal of Geosciences*, *7*(2), 711–724. Available from https://doi.org/10.1007/s12517-012-0795-z.

Maqsoom, A., Aslam, B., Khalid, N., Ullah, F., Anysz, H., Almaliki, A. H., ... Hussein, E. E. (2022). Delineating groundwater recharge potential through remote sensing and geographical information systems. *Water (Switzerland)*, *14*(11). Available from https://doi.org/10.3390/w14111824, https://www.mdpi.com/2073-4441/14/11/1824/pdf?version = 1654512810.

Melese, T., & Belay, T. (2022). Groundwater potential zone mapping using analytical hierarchy process and GIS in Muga Watershed, Abay Basin, Ethiopia. *Global Challenges*, *6*(1). Available from https://doi.org/10.1002/gch2.202100068, https://onlinelibrary.wiley.com/journal/20566646.

Meng, F., Xiao, C., Liang, X., Wang, G., Sun, Y., & Guo, D. (2021). Factors influencing surface water and groundwater interaction in alluvial fan. *Journal of Water and Climate Change*, *12*(3), 679–695. Available from https://doi.org/10.2166/wcc.2020.174, https://watermark.silverchair.com/jwc0120679.pdf?.

Muavhi, N., Thamaga, K. H., & Mutoti, M. I. (2022). Mapping groundwater potential zones using relative frequency ratio, analytic hierarchy process and their hybrid models: Case of Nzhelele-Makhado area in South Africa. *Geocarto International*, *37*(21), 6311–6330.

Available from https://doi.org/10.1080/10106049.2021.1936212, http://www.tandfonline.com/toc/tgei20/current.

Murari,, K. (2016). Ground water recharge potential of the Dhanbad District, Jharkhand – A case study. *International Journal of Science Technology and Engineering*, *2*(9).

Murmu, P., Kumar, M., Lal, D., Sonker, I., & Singh, S. K. (2019). Delineation of groundwater potential zones using geospatial techniques and analytical hierarchy process in Dumka district, Jharkhand, India. *Groundwater for Sustainable Development*, *9*. Available from https://doi.org/10.1016/j.gsd.2019.100239, http://www.journals.elsevier.com/groundwater-for-sustainable-development/.

Nag, S. K., & Chowdhury, P. (2019). Decipherment of potential zones for groundwater occurrence: A study in Khatra Block, Bankura District, West Bengal, using geospatial techniques. *Environmental Earth Sciences*, *78*(2). Available from https://doi.org/10.1007/s12665-018-8034-x, http://www.springerlink.com/content/121380/.

Nag, S. K., & Ghosh, P. (2013). Delineation of groundwater potential zone in Chhatna Block, Bankura District, West Bengal, India using remote sensing and GIS techniques. *Environmental Earth Sciences*, *70*(5), 2115–2127. Available from https://doi.org/10.1007/s12665-012-1713-0.

Ngwijabagabo, H., Tsinda, A., Dusengimana, A., Minani, M., Tuyishime, C. S. P., Hirwa, H., & Habimana, I. (2023). Groundwater potential mapping using geospatial and AHP techniques in eastern province of Rwanda. *Rwanda Journal of Engineering, Science, Technology and Environment*, *5*(1). Available from https://doi.org/10.4314/rjeste.v5i1.4.

Owolabi, S. T., Madi, K., Kalumba, A. M., & Orimoloye, I. R. (2020). A groundwater potential zone mapping approach for semi-arid environments using remote sensing (RS), geographic information system (GIS), and analytical hierarchical process (AHP) techniques: A case study of Buffalo catchment, Eastern Cape, South Africa. *Arabian Journal of Geosciences*, *13*(22). Available from https://doi.org/10.1007/s12517-020-06166-0, http://www.springer.com/geosciences/journal/12517?cm_mmc = AD-_-enews-_-PSE1892-_-0.

Owuor, S. O., Butterbach-Bahl, K., Guzha, A. C., Rufino, M. C., Pelster, D. E., Díaz-Pinés, E., & Breuer, L. (2016). Groundwater recharge rates and surface runoff response to land use and land cover changes in semi-arid environments. *Ecological Processes*, *5*(1). Available from https://doi.org/10.1186/s13717-016-0060-6, http://www.ecologicalprocesses.com/.

Panahi, M. R., Mousavi, S. M., & Rahimzadegan, M. (2017). Delineation of groundwater potential zones using remote sensing, GIS, and AHP technique in Tehran–Karaj plain, Iran. *Environmental Earth Sciences*, *76*(23). Available from https://doi.org/10.1007/s12665-017-7126-3, http://www.springerlink.com/content/121380/.

Pande, C. B., Khadri, S. F. R., Moharir, K. N., & Patode, R. S. (2018). Assessment of groundwater potential zonation of Mahesh River basin Akola and Buldhana districts, Maharashtra, India using remote sensing and GIS techniques. *Sustainable Water Resources Management*, *4*(4), 965–979. Available from https://doi.org/10.1007/s40899-017-0193-5, http://springer.com/journal/40899.

Panigrahi, B., Nayak, A. K., & Sharma, S. D. (1995). Application of remote sensing technology for groundwater potential evaluation. *Water Resources Management*, *9*(3), 161–173. Available from https://doi.org/10.1007/BF00872127.

Prasad, R. K., Mondal, N. C., Banerjee, P., Nandakumar, M. V., & Singh, V. S. (2008). Deciphering potential groundwater zone in hard rock through the application of GIS. *Environmental Geology*, *55*(3), 467–475. Available from https://doi.org/10.1007/s00254-007-0992-3.

Purnama, I.L. S., Rahmawati, L., Primacintya, V.A., & Febriarta, E. (2021). The influence of aquifer material to the groundwater potency in Ngawi Regency. *E3S Web of Conferences EDP Sciences Indonesia*, 325, http://www.e3s-conferences.org/. 10.1051/e3sconf/202132508011.

Rahmati, O., Nazari Samani, A., Mahdavi, M., Pourghasemi, H. R., & Zeinivand, H. (2015). Groundwater potential mapping at Kurdistan region of Iran using analytic hierarchy process and GIS. *Arabian Journal of Geosciences*, *8*(9), 7059–7071. Available from https://doi.org/10.1007/s12517-014-1668-4, http://www.springer.com/geosciences/journal/12517?cm_mmc = AD-_-enews-_-PSE1892-_-0.

Raj, S., Rawat, K. S., Singh, S. K., & Mishra, A. K. (2022). Groundwater potential zones identification and validation in Peninsular India. *Geology, Ecology, and Landscapes*. Available from https://doi.org/10.1080/24749508.2022.2097375, https://www.tandfonline.com/toc/tgel20/current.

Rajesh, J., Pande, C. B., Kadam, S. A., Gorantiwar, S. D., & Shinde, M. G. (2021). Exploration of groundwater potential zones using analytical hierarchical process (AHP) approach in the Godavari river basin of Maharashtra in India. *Applied Water Science*, *11*(12). Available from https://doi.org/10.1007/s13201-021-01518-x, http://www.springer.com/earth + sciences + and + geography/hydrogeology/journal/13201.

Rana, M. M. S. P., Hossain, M. A., & Nasher, N. M. R. (2022). Identification of groundwater potential zone using geospatial techniques of agriculture dominated area in Dinajpur district, Bangladesh. *Environmental Challenges*, *7*. Available from https://doi.org/10.1016/j.envc.2022.100475, https://www.journals.elsevier.com/environmental-challenges.

Razandi, Y., Pourghasemi, H. R., Neisani, N. S., & Rahmati, O. (2015). Application of analytical hierarchy process, frequency ratio, and certainty factor models for groundwater potential mapping using GIS. *Earth Science Informatics*, *8*(4), 867–883. Available from https://doi.org/10.1007/s12145-015-0220-8, http://www.springer.com/geosciences/computer + %26 + mathematical + applications/journal/12145.

Reddy, K. R., Kosgi, S., & Zhou, J. (1995). A review of in-situ air sparging for the remediation of VOC-contaminated saturated soils and groundwater. *Hazardous Waste and Hazardous Materials*, *12*(2), 97–118. Available from https://doi.org/10.1089/hwm.1995.12.97.

Saaty, R. W. (1987). The analytic hierarchy process-what it is and how it is used. *Mathematical Modelling*, *9*(3-5), 161–176. Available from https://doi.org/10.1016/0270-0255(87)90473-8.

Sajil Kumar, P. J., Elango, L., & Schneider, M. (2022). GIS and AHP based groundwater potential zones delineation in Chennai River Basin (CRB), India. *Sustainability (Switzerland)*, *14*(3). Available from https://doi.org/10.3390/su14031830, https://www.mdpi.com/2071-1050/14/3/1830/pdf.

Sar, N., Khan, A., Chatterjee, S., & Das, A. (2015). Hydrologic delineation of ground water potential zones using geospatial technique for Keleghai river basin, India. *Modeling Earth Systems and Environment*, *1*(3). Available from https://doi.org/10.1007/s40808-015-0024-3, http://springer.com/journal/40808.

Sarwar, A., Ahmad, S. R., Rehmani, M. I. A., Asif Javid, M., Gulzar, S., Shehzad, M. A., ... Sabagh, A. E. L. (2021). Mapping groundwater potential for irrigation, by geographical information system and remote sensing techniques: A case study of district lower dir, pakistan. *Atmosphere*, *12*(6). Available from https://doi.org/10.3390/atmos12060669, https://www.mdpi.com/2073-4433/12/6/669/pdf.

Selvachandran, G., Quek, S. G., Smarandache, F., & Broumi, S. (2018). An extended technique for order preference by similarity to an ideal solution (TOPSIS) with maximizing deviation method based on integrated weight measure for single-valued neutrosophic sets. *Symmetry*,

10(7). Available from https://doi.org/10.3390/sym10070236, http://www.mdpi.com/2073-8994/10/7/236/pdf.

Sener, E., Davraz, A., & Ozcelik, M. (2005). An integration of GIS and remote sensing in groundwater investigations: A case study in Burdur, Turkey. *Hydrogeology Journal, 13*(5-6), 826–834. Available from https://doi.org/10.1007/s10040-004-0378-5.

Serele, C., Pérez-Hoyos, A., & Kayitakire, F. (2020). Mapping of groundwater potential zones in the drought-prone areas of south Madagascar using geospatial techniques. *Geoscience Frontiers, 11*(4), 1403–1413. Available from https://doi.org/10.1016/j.gsf.2019.11.012, https://www.sciencedirect.com/journal/geoscience-frontiers.

Shekhar, S., & Pandey, A. C. (2015). Delineation of groundwater potential zone in hard rock terrain of India using remote sensing, geographical information system (GIS) and analytic hierarchy process (AHP) techniques. *Geocarto International, 30*(4), 402–421. Available from https://doi.org/10.1080/10106049.2014.894584, http://www.tandfonline.com/toc/tgei20/current.

Singh,, A., Kumar,, R., Pippal,, P. S., Sharma,, P., & Sharma,, A. (2024). Delineation of groundwater potential zone using geospatial tools and analytical hierarchy process (AHP) in the state of Uttarakhand, India. *Advances in Space Research, 73*(6), 2939–2954.

Sohail, A. T., Singh, S. K., & Kanga, S. (2019). A geospatial approach for groundwater potential assessment using multi influence factor (MIF) technique. *International Journal on Emerging Technologies, 10*(1), 183–196. Available from https://www.researchtrend.net/ijet/pdf/A%20Geospatial%20Approach%20for%20Groundwater%20Potential%20Assessment%20Using%20Multi%20Influence%20Factor%20MIF%20Technique%20SURAJ%20KUMAR%20SINGH.pdf.

Sresto, M. A., Siddika, S., Haque, M. N., & Saroar, M. (2021). Application of fuzzy analytic hierarchy process and geospatial technology to identify groundwater potential zones in north-west region of Bangladesh. *Environmental Challenges, 5*. Available from https://doi.org/10.1016/j.envc.2021.100214, https://www.journals.elsevier.com/environmental-challenges.

Stofkova, J., Krejnus, M., Stofkova, K. R., Malega, P., & Binasova, V. (2022). Use of the analytic hierarchy process and selected methods in the managerial decision-making process in the context of sustainable development. *Sustainability (Switzerland), 14*(18). Available from https://doi.org/10.3390/su141811546, http://www.mdpi.com/journal/sustainability/.

Suliman, M., Samiullah., & Ali, M. (2022). Identification of potential groundwater recharge sitein a semi-arid region of pakistan using saaty's analytical hierarchical process (Ahp). *Geomatics and Environmental Engineering, 16*(1), 53–70. Available from https://doi.org/10.7494/geom.2022.16.1.53, https://journals.bg.agh.edu.pl/GEOMATICS/2022.16.1/geom.2022.16.1.53.pdf.

Teja, K. S., & Singh, D. (2019). Identification of groundwater potential zones using remote sensing and GIS, case study: Mangalagiri mandal. *International Journal of Recent Technology and Engineering, 7*(6), 860–864. Available from https://www.ijrte.org/wp-content/uploads/papers/v7i6c2/F11590476C219.pdf.

Thapa, R., Gupta, S., Guin, S., & Kaur, H. (2017). Assessment of groundwater potential zones using multi-influencing factor (MIF) and GIS: A case study from Birbhum district, West Bengal. *Applied Water Science, 7*(7), 4117–4131. Available from https://doi.org/10.1007/s13201-017-0571-z, https://www.springer.com/journal/13201.

Venkateswaran, S., & Ayyandurai, R. (2015). Groundwater potential zoning in Upper Gadilam River Basin Tamil Nadu. *Aquatic Procedia, 4*, 1275–1282. Available from https://doi.org/10.1016/j.aqpro.2015.02.166.

Verma, A. K., & Singh, T. N. (2013). Prediction of water quality from simple field parameters. *Environmental Earth Sciences, 69*(3), 821–829. Available from https://doi.org/10.1007/s12665-012-1967-6.

Waikar,, M. L., & Nilawar,, A. P. (2014). Identification of groundwater potential zone using remote sensing and GIS technique. *International Journal of Innovative Research in Science, Engineering and Technology, 3.*

Zghibi, A., Mirchi, A., Msaddek, M. H., Merzougui, A., Zouhri, L., Taupin, J. D., ... Tarhouni, J. (2020). Using analytical hierarchy process and multi-influencing factors to map groundwater recharge zones in a semi-arid mediterranean coastal aquifer. *Water (Switzerland), 12* (9). Available from https://doi.org/10.3390/w12092525, https://res.mdpi.com/d_attachment/water/water-12-02525/article_deploy/water-12-02525-v2.pdf.

Zhang, H., Li, Y., Meng, Yl, Cao, N., Li, Ds, Zhou, Zg, ... Dou, Fg (2019). The effects of soil moisture and salinity as functions of groundwater depth on wheat growth and yield in coastal saline soils. *Journal of Integrative Agriculture, 18*(11), 2472–2482. Available from https://doi.org/10.1016/S2095-3119(19)62713-9, http://www.elsevier.com/journals/journal-of-integrative-agriculture/2095-3119.

Chapter 17

Use of DinSAR techniques to mitigate dam siltation: a case study of the Abdelkarim El Khattabi Dam in northern Morocco

Mohammed Mourjane, Hassan Tabyaoui and Fatima El Hammichi
Natural Resources and Environment Laboratory, Polydisciplinary Faculty of Taza, Sidi Mohamed Ben Abdellah University - Fez, Morocco

17.1 Introduction

The issue of sediment buildup in dams represents a significant ecological and geological challenge necessitating substantial governmental interventions (Alahiane, 2016; Hanssen & van Leijen, 2008a; Minister of Public Works and Communications, Morocco, 1976). Take, for instance, the situation in Morocco, where ongoing monitoring and interest to the Abdelkarim El khattabi Dam in the north of the country are imperative. This dam plays a pivotal role in supplying water to a large portion of the population and supporting agricultural endeavors while also mitigating flood risks (Fernandes et al., 2023). Efforts to address this issue have been ongoing since the 1980s, involving comprehensive studies and culminating in recommendations for appropriate infrastructure installations. Radar remote sensing emerges as a crucial tool for studying natural phenomena, particularly for quantifying and modeling ecological, geological, and atmospheric processes (Fernandes et al., 2023; Franceschini et al., 2023; Li, Hinnov, & Kump, 2019; Li, Wang, Chen, & Han, 2022). Significant physical parameters are derived from internal data measurements extracted from radar imagery, such as optical radiation characteristics viewed from intensity and polarimetry angles (Xiao, Liu, Ma, & Cui, 2023; Yu, Chen, & Xu, 2020). This approach was used to track sediments and waste movement in the vicinity of the Abdelkarim El Khattabi dam within the Oued Nekor watershed. The objective was to estimate siltation rates and identify areas necessitating practical intervention, utilizing Sentinel-1 data

Water Resources Monitoring, Management, and Sustainability. DOI: https://doi.org/10.1016/B978-0-443-23665-5.00017-X

and Differential SAR Image Interferometry (DINSAR) techniques (Difference in phase of Synthetic Aperture Radar images). The study primarily focused on preprocessing, processing, and ground motion measurement techniques, alongside evaluating their practical reliability (Perrone, 2023; Shi et al., 2022; Watzig et al., 2023; Hanssen & Van Leijen, 2008b; Pathier, 2014; Yang, López-Dekker, Dheenathayalan, Liao, Hanssen et al., 2022). The outcomes comprised maps depicting terrain instability between different satellite imagery acquisition dates, featuring varying degrees of positive (uplift) and negative (subsidence) values across different regions (Goldstein, Zebker, & Werner, 1988; Shi et al., 2022; Yu, Chen, & Xu, 2020; Arab-Sedze, Heggy, Bretar, Berveiller, & Jacquemoud, 2014). These finding present additional avenues for investigating dam siltation issues and mitigating the reduction of water retention capacity. Identifying zones prone to sediment movement underscores the strategic significance of selecting the Abdelkarim El Khattabi dam in northern Morocco, particularly given its geological studies (), conducted through in situ monitoring, GIS techniques, and other conventional methods. Utilizing recent Sentinel-1 radar imagery for interferometry calculations enables space-based characterization of dam perimeters, identifying areas experiencing subsidence or uplift. This analysis is performed at a millimeter scale, delineating surface change regions (Li, Hinnov, & Kump, 2019a; Massonnet & Feigl, 1998; Kaveh, 2014; Tarata and Vasile, 2017).

Nearly all reservoirs worldwide content with siltation issues, exhibiting varying sedimentation rates across region. Northern Morocco is part of a mountainous zone characterized by a volatile climate, experiences alternating dry and wet seasons. Autumn rains, particularly intense, pose significant soil erosion risks (Li et al., 2019a; Niazi, Snoussi, & Foutlane, 2005; Poupart & Royet, 2001). This study employs a novel technique utilizing radar satellite imagery, notably Sentinel, distinct from optical products due to its radiation emission and reception of reflected signals (Xiao, Liu, Ma, & Cui, 2023; Arab-Sedze et al., 2014; Small dam reservoir siltation in the Atlas Mountains of Central Morocco: analysis of factors impacting sediment yield | Springer Link, 2023; Tarata & Vasile, 2017). This approach facilitates the calculation of vertical displacement, if present, on the earth's surface. It serves as a robust method for detecting displacement, assessing soil stability, and monitoring sediment mobility in dam siltation studies (Massonnet & Feigl, 1998; Yu, Chen, & Xu, 2020; Kaveh, 2014; Pathier, 2014).

17.2 Objectives of the study

Morocco has long recognized the strategic importance of dam construction due to their multifaceted benefits, including flood mitigattion, groundwater replenishment, enhanced water storage for agriculture and livestock, and expanded domestic and recreational opportunities (Lahlou, 2014; Alahiane, 2016; Minister of Public Works and Communications, Morocco, 1976; Niazi et al., 2005). Dams sugnificantly improve community resilience, especially in rural areas prone to drought, aligning with the country's geographical reality, as two thirds of its landmass is arid or semi-arid areas (Alahiane et al., 2016; Alahiane, 2016; Minister of Public Works and Communications, Morocco, 1976; Niazi et al., 2005).

However, the transportation of sediment by dams within hydrological networks, coupled with sediments deposition in reservoirs, presents operational challenges, such as progressive reduction in dam capacity due to sediments accumulation from water or wind transport (Alahiane, Elmouden, Aitlhaj, & Boutaleb, 2016; Lahlou, 2014; Poupart & Royet, 2001). Adrerssing sedimentation issues is costly, with sediment removal often necessitating dam shutdown, a practical challenge in arid and semi-arid regions (Alahiane et al., 2016; Alahiane, 2016; Minister of Public Works and Communications, Morocco, 1976; Poupart & Royet, 2001; Xiao et al., 2023).

17.3 Study area

The Oued Nekor watershed, situated in northern Morocco, spans approximately 780 km^2, with latitudes between 35 degrees 43′ and 35 degrees 6′ north and longitudes between 3 degrees 36′ and 4 west. It is bordred to the north by the Abdelkarim El Khettabi dam, to the south by the Oued Sebou watershed, to the east by Oued Kert, and to the west by the Oued Rhis watershed (Fig. 17.1). The region's climate is predominantly Mediterranean with a semi-arid tendency, characterized by cold and humid winters and hot and dry summers. Annual rainfall averages around 340 mm in the basin, concentrated

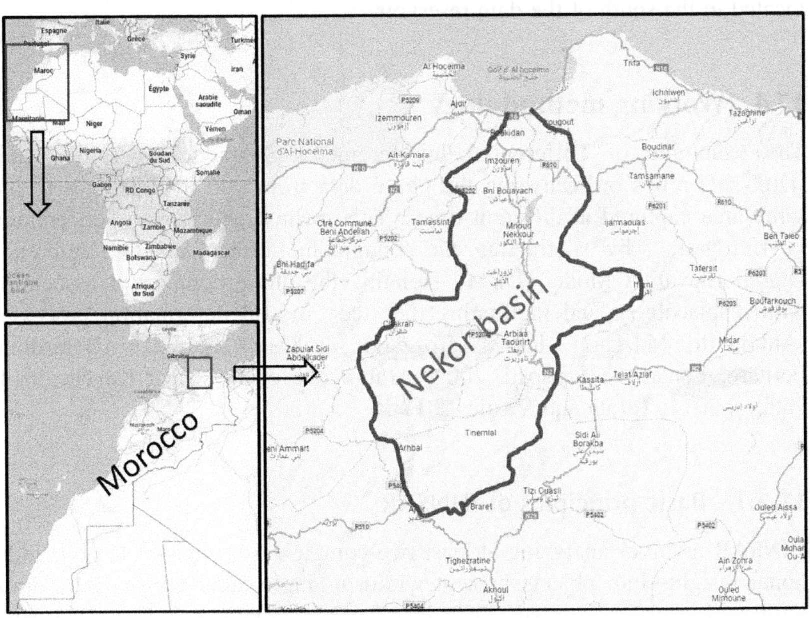

FIGURE 17.1 Geographical location of the study area. Map lines delineate study areas and do not necessarily depict accepted national boundaries.

between October and May due to the passage of cold fronts, with occasional summer showers influenced by local topography. Nekor experiences an average of 3–4 floods annually, peaking no more than 10 times per year, resulting in highly intermittent wadi flow. Monthly temperatures range from 7°C (January) to 28°C (August) across the basin, dipping to 0°C at higher elevations. Despite relatively low overall precipitation, the watershed is susceptible to significant floods due to sporadic heavy rains and a terrain conducive to runoff (Minister of Public Works and Communications, Morocco, 1976).

This region exhibits distenct characteristics specific to the Rif mountain, featuring a complex geomorphology characterized by rugged terrain and diverse lithological formations, consisting mainly of shale, limestone, and marl. (Fig. 17.2). The altitude of Nekor watershed depicts high mountains rabging from 600 to 2000 meters in the south and east, while the northern and central basin areas feature medium-altitude relief. Analysis of satellite imagery from the Google Earth platform spanning the past two decades (2004 to 2024) reveals a significant reduction in the surface area of the dam reservoir (Fig. 17.3). Specifically, there has been a 64% decrease in extent compared to October 2004. This retreat has occurred progressively from the southern to the northern regions of the reservoir, primarily due to the gradual accumulation of sediments. Over the 20-year period, this process has led to the formation of a new geomorphological landscape characterized by various features such as landslides, alluvial cones, and alluvial plains. These features have effectively transformed the previously flat-bottomed perched valley located to the south of the dam reservoir.

17.4 Working methodology

The technique of Differential Interferometry Synthetic Aperture Radar (DInSAR) relies on analyzing the phase data from two SAR images of the same area captured at different times. This method generates a "differential interferogram", by subtracting the topographic effects using a reference Digital Elevation Model (DEM), isolating the phase component associated with displacement and facilitating the study of soil deformation processes (Antonielli, Mazzanti, Rocca, Bozzano, & Dei Cas, 2019; Berardino, Fornaro, Lanari, & Sansosti, 2002; Arab-Sedze et al., 2014; Kaveh, 2014; Pathier, 2014; Tarata and Vasile, 2017).

17.4.1 Basic principles of DINSAR

DINSAR involves analyzing at least two complex radar images to gain additional insights into objects present within a single radar image (Berardino, Fornaro, Lanari, & Sansosti, 2002; Brandt, Ertel, Spore, & Stolle, 2023; Bru, 2017; Massonnet & Feigl, 1998; Shi et al., 2022; Arab-Sedze et al., 2014).

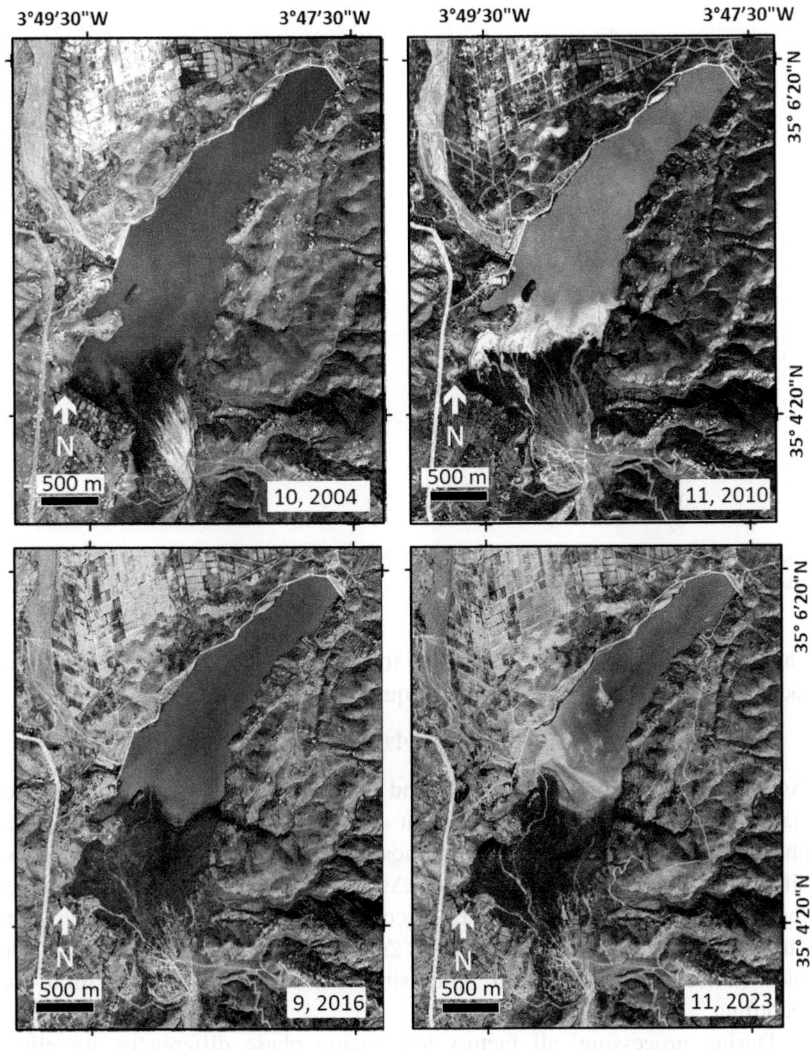

FIGURE 17.2 Successive historical evolution of Abdelkrim El Khattabi dam surface 2004-2023 (Images from Google-Earth).

The amplitude reflects the energy backscattered back to the sensor and depends on target characteristics of the target like roughness, absorption, etc.).

The Phase of the signal represents the distance (*d*) between the sensor and the target, influenced by the target's structure and dielectric properties.

Interferogram calculation involves computing the phase difference pixel by pixel from two radar images acquired under similar conditions but at different times (Antonielli et al., 2019; Berardino et al., 2002; Goldstein et al., 1988;

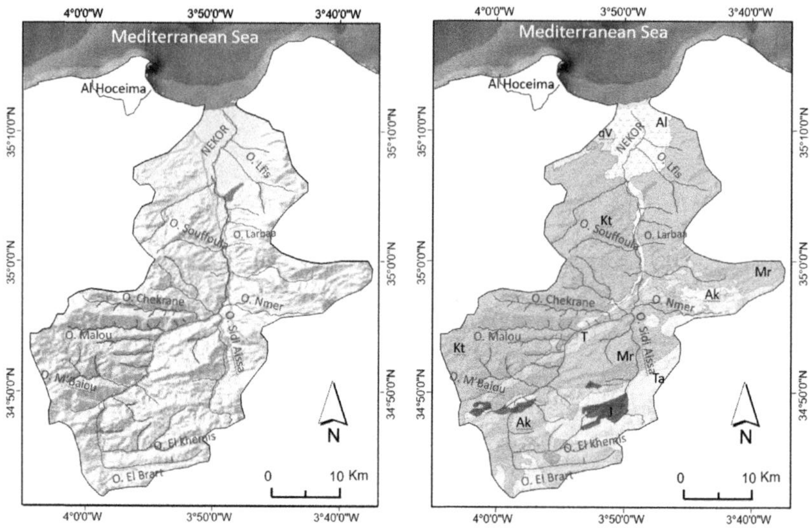

FIGURE 17.3 Nekor catchment elevation map. Map lines delineate study areas and do not necessarily depict accepted national boundaries.

Shi et al., 2022). The phase difference for a given pixel is affected by various factors as described by the following equation:

$$\Delta\Phi = \Delta\Phi\text{d} + \Delta\Phi\text{a} + \Delta\Phi\text{t} + \Delta\Phi\text{dc} + \Delta\Phi\text{r} + 2k\pi$$

$\Delta\Phi$ represents the phase difference and $\Delta\Phi d$ indicate phase difference associated to possible ground displacement along the line of sight. $\Delta\Phi a$ signifies phase difference due to change in atmospheric conditions. $\Delta\Phi t$ denote phase difference linked site topography. $\Delta\Phi dc$ accounts for phase difference caused by geometric and temporal decorrelations, arising from surface state fluctuations, and shooting variations. $\Delta\Phi r$ represents residual phase difference (noise) stemming from processing, including SAR registration and recording.

During processing, all factors influencing phase differences are eliminated, retaining only those linked to surface displacement.

17.4.2 Input data

The ground motion study utilized images acquired by Sentinel 1B mission, employing the IW acquisition mode, and SLC product type, with upward trajectory in orbit, and VV polarity.

- S1B_IW_SLC__1SDV_2018-02-10 (master image)
- S1B_IW_SLC__1SDV_2018-05-17 (slave image)

17.4.3 SNAP software

SNAP, available as free software from the European agency (ESA) website under the GNU license, is developed and supported by ESA. Its primary objective is to serve as an efficient tool for processing satellite data, encompassing both optical and radar imagery. SNAP includes:

- Sentinel-1, -2, and -3 toolboxes
- Toolboxes for SMOS, PROBA-V, and tools for a number of other sensors: RADRSAT, COSMOSYMED, etc.
- External plugins to download or activate (in particular sen2cor for Sentinel-2 for the transition to level 2, or SNAPHU for DinSAR analysis with Sentinel1).
- Classic image processing python libraries (gdal...) and a SNAP application library for python called snappy.
- The GPT library for processing graphs in XML format on the command line (very practical for massive analysis).

17.4.4 Pretreatment

The area of interest is encompassed within three fragments of the Sentinel image (iw1, iw2, and iw3), each divided into nine bursts. The initial preprocessing stage involve selecting a single segment and a limited number of burst. This approach minimizes processing time in subsequents steps and narrows the analysis to a specific area rather than the entire scene. VV polarimetry is utilized in this step.

17.4.5 Co-registration

Processing can conducted either by the top panel operators or through the graph processes available in the SNAP software. Initially, in-orbit files are applied to Sentinel-1 products to ensure accurate satellite positioning and velocity data. These products are typically distributed shortly after acquisition, with a geometric precision inferred from the pre-schedule acquisition. Subsequently, the ground segment treleases a file for each orbit (within a maximum of 20 days post-acquisition), furnishing precise satellite position and velocity information during acquisition, obtained from telemetry measurements onboard the satellite. This process is executed for both master and slave input images.

The subsequent step aims to co-register the two Sentinel-1 products, namely the master and slave images. The slave image undergoes co-registeration with the master image utilizing the orbits of both products alongside a digital elevation model (DEM). During resampling of the slave images within the master frame, initial processing involves burst synchronization and demodulation on the slave image, followed by truncated interpolation. Subsequently, re-amplification and demodulation are applied to the interpolated slave image. Additionally, an enhanced spectral diversity (EDD) operator function is incorporated post-backgeocoding. The EDD approach entails data extraction from the overlapping region of adjacent blocks, followed by range and azimuth correction for each burst.

17.4.6 Interferogram Construction

Interferometry is conducted utilizing the Repeat Pass technique, where the satellite observes the same target at two different times. An interferogram is generated between the pair of interferometric imaes, with consideration giving to estimating the coherence image of the stack of complex registered images. Processing procedures are then implemented to eliminate spacing between busts. Each sub-row image comprises a series of slaves images, whith each slave treated as an individual Single Look Complexe (SLC) image. The complex, individually focused burst images are organized in azimuth time order within a single subband image, separated by black fill boundaries. Sufficient overlap exists between adjacent blocks and sub-strips to ensure continuous ground coverage. All slaves within each subband are resampled to a uniform pixel spacing grid in both range and azimuth directions, preserving phase information and consistently utilizing VV polarization.

17.4.7 Goldstein Phase Filtering, Multilooking, and Topophase Removal

Goldstein phase filtring processing aims to mitigate potential error in interferometric coherence by enhancing the signal-to-noise ratio through phase filtering. This process enables the reconnection of interferometric fringe, where each fringe represents a complete cycle of 2π, corresponding to half the wavelength of the sensor ($\lambda = 5.66$ cm). The closer the fringes are the larger observed ground displacement. Consequently, this procedure snhances the quality of the interferogram. The Fast Fourier Transform (FFT) core select small samples of the interferogram and applies Fourier transform filtering in this new 2D domain. The alpha parameter serves as a multiplication factor within a given specified window.

Multilooking is employed to decrease the standard deviation of the noise level and promotes a more Gaussian distribution of noise in the image, albeit at the expense of reduced spatial resolution. This technique is implemented within a window (range $\times$ azimuth) and SNAP performs window averaging.

The topophase removal step involves eliminating the induced phase resulting from topographic effect in the constructed interferogram. To achieve this, the Shuttle Radar Topography Mision (SRTM) data is concorporated.

17.4.8 SNAPHU Treatment

SNAPHU, developed by Chen and Zebker (2002), is a freely available software designed for phase unwrapping, a pivotal step in interferometry processing. Phase unwinding is intricate because the interferometric phase is inherently ambiguous and estimated only modulo 2p. To establish a connection between the interferometric phase and topographic height, the phase must undergo unwrapping. By introducing an ambiguity altitude (EA), defined as the difference in altitude, the unwrapping process ensures a phase change of 2p after flattening the interferogram using the topophase removal function. Unwrapping involves integrating the phase difference between neighboring pixels, thereby converting fringe variation into altitude variation.

Following the removal of topography effect, the resulting product is exported with displacement chosen and other parameters left at default setting., consisting the SNAPHU export phase. Subsequently, SNAPHU unwrapping is executed within the same export folder, a computationally intensive task requiring substantial computing ressources and time. Finally, the file generated during export is imported back into the processing pipeline.

17.4.9 Phase to displacement

The software computes ground displacement based on the previously generated file, utilizing a Shuttle Radar Topography Mission (SRTM) 1-second Digital Elevation Model (DEM). It then, converts the digital data into a displacement map with metric scale, suitable for visual inspection or utilization with other tools such as Google Earth, ArcGIS, and similar software platforms.

17.5 Results and Discussion

Fig. 17.4 illustrates the interferogram derived from two images covering an extensive area, including the study sector. Generally, this enables precise measurement of terrain relief. However, as the topographic phase has been removed, it assesses surface deformations with vertical precision by comparing the two acquired positions. Fringes depicted on the map represent variations in the path length of optical radiation, denoted by complete cycles of differents colors, corresponding to a deformation of 2.75 cm (half-wavelength 5.5 cm, band C). The diverse coloration observed across the map and surrounding the stidy area indicates the indicative nature of interferogram.

The final calculated result is depicted in Fig. 17.5. At a millimeter scale, DINSAR interferometry processing phase yield conclusive outcomes, displaying either positive or negative nalues based on substrate movement, such

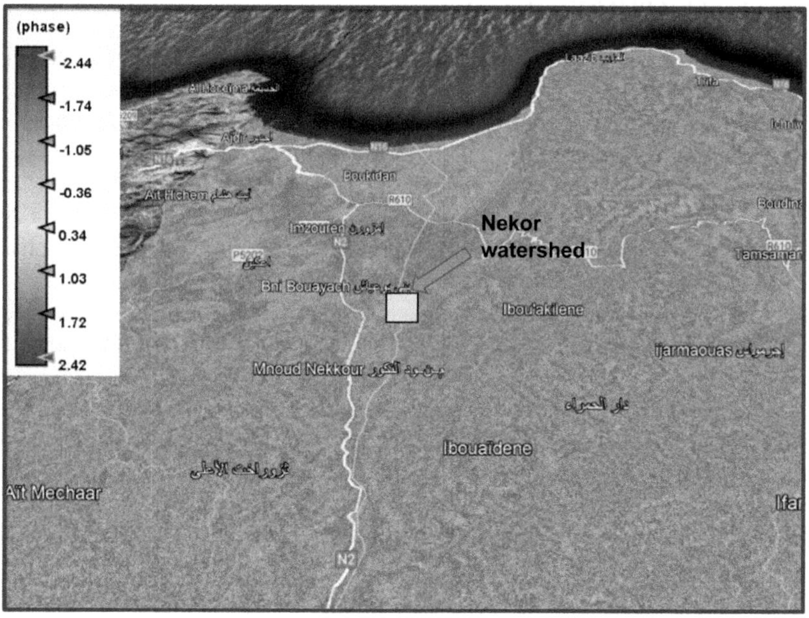

FIGURE 17.4 DINSAR Interferogram processed along the Nekor watershed in northern Moroccan. Map lines delineate study areas and do not necessarily depict accepted national boundaries.

as subsidence (negative values) or uplift (positive values). In our case, areas of subsidence are negligible compared to uplift areas, with the majority of the region remaining stable. However, considering the short detection period (between February 10, 2018 and May 17,2018), and the obtained values averaging almost 2 cm, it is evident that sediments accumulation at this rate becomes significant, potentially shortening lifespan of the dam. accumulation rates span a considerable range, from values deemed insignificant to those significantly elevated. As illustrated in Fig. 17.6, these rates are categorized into three zones: areas with substantial displacement are shaded in gray, those with moderate displacement in black, and regions with minimal displacement in white. Focusing solely on areas proximal to water retention, whiwh supply the dam with sediments and other inputs through various carriers—namely water, wind, and gravity- the most susceptible zone primarily corresponds to a sedimentation cone downstream of the wadi (Fig. 17.7). Conversely, the three displacement regions north of the dam remainrelatively unaffected (Fig. 17.8), as the movement is impeded by concrete barriers (A), vegetative obstacles (B), or topographical features (C). Notably, significant displacement are concentrated near the dam, whereas less intense displacements occur farther away, with values ranging from 0 to 4 mm. Comparing the results with the altitude map (Fig. 17.3) does not provide evidence linking sediments transport to gravity, as similar speeds are observed at

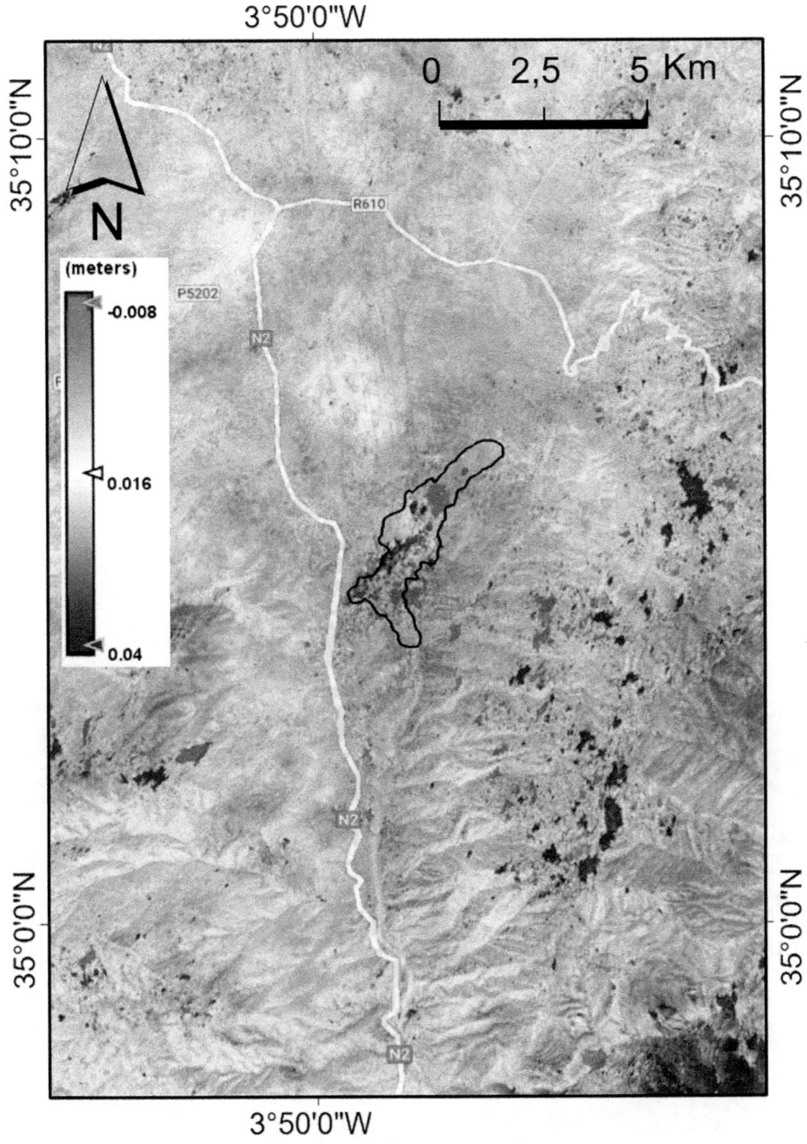

FIGURE 17.5 Ground displacement map around the Abdelkarim el Khattabi dam. Map lines delineate study areas and do not necessarily depict accepted national boundaries.

comparable altitudes. This observation may be attributed to the millimetric-scale of displacement, whereas gravity predominantly influences the transport of larger blocks over metric distances. Additionnally, the lithological composition of the area yields insights into the rocks' fragility or hardness.

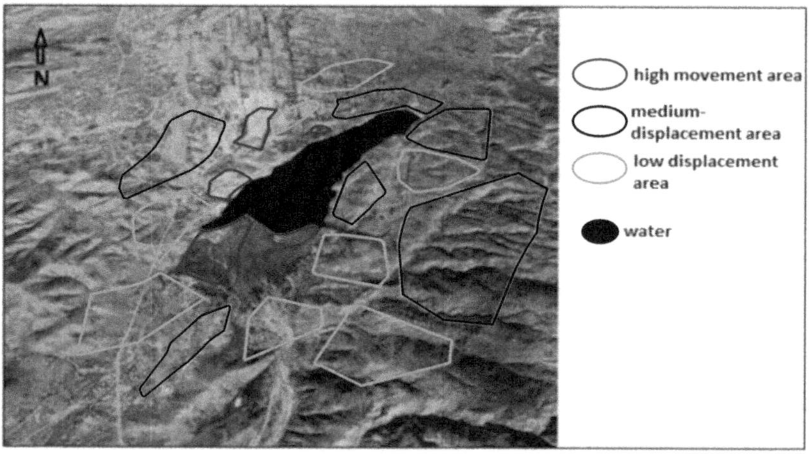

FIGURE 17.6 Distinction between the different intensities of movement.

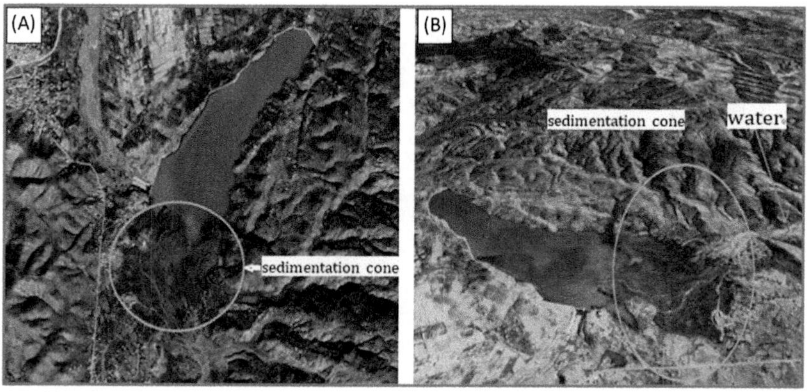

FIGURE 17.7 Sedimentation cone (A) view from space. (B) 3D view.

FIGURE 17.8 Areas of less displacement influence. (a) limitation of movements by the restrained concrete limit, (b) vegetative obstacles, (c) less weighting.

17.6 Conclusion

Soil degradation sterms from the complexe interplay of numerous factors, including rainfall patterns, rocks composition, slope inclination, and vegetation density. These variables undergo temporal fluatuation, and the Nekor watershed is currently experiencing substantial rates of degradation, positioning itself as one of the most severely affected basins by siltation issues in northen Morocco. While development endeavors aim to mitigate waste and sediment transport to water reservoir, their efficacy remain limited when addressing specific regions that emerge as new degradation hotspots. To address this, precautionary measures and vulnerablility assessments of these regions are imperative and can be informed through various studies. Remote sensing, particularly radar remote sensing, emerges as a pivotal tool in this context, aimed at safequarding the natural environment and prolonging the operational lifespan of the Abdelkrim El Khattabi dam. By providing crucial insights and generating high-resolution maps depicting debris, waste, and sediment movementat at a millimeter scale, radar remote sensing facilitates onformed decision-making processes. Integration of these findings with previous sectoral investigations, including GIS analyses, ground-based soil erosion essessments, climatological and hydrological modeling, and ecological studies, holds promise in achieving meaningful objectives to safeguard the natural environment and sustain the Abdelkrim El Khattabi dam's functionality.

Funding statement

This work was made possible thanks to the support of CNRST-Morocco (National Center for Scientific and Technical Research, Rabat) and the Hassan II Academy of Sciences and Techniques, Rabat, Morocco. The content is the sole responsibility of the authors and does not necessarily represent the official opinions of the research centers. The authors would like to thank the anonymous reviewers and editors for their assistance, comments, and suggestions.

Conflict of interest

The authors declare no conflict of interest.

Author's contributions

HT and MM planned and designed the study; MM collected and processed Sentinel 1 data and prepared other assigned data; HTMM and FE interpreted and wrote the manuscript. All authors have reviewed and approved the manuscript.

Declaration of competing interests

The authors have declared no competing interests.

References

Alahiane, N. (2016). Small dam reservoir siltation in the Atlas Mountains of Central Morocco: analysis of factors impacting sediment yield, June 16, 2016.

Alahiane, N., Elmouden, A., Aitlhaj, A., & Boutaleb, S. (2016). Small dam reservoir siltation in the Atlas Mountains of Central Morocco: Analysis of factors impacting sediment yield. *About Earth Sciences*, *75*(12), 1035. Available from https://doi.org/10.1007/s12665-016-5795-y, flight.

Antonielli., Mazzanti., Rocca., Bozzano., & Dei Cas. (2019). A-DInSAR performance for updating landslide inventory in mountain areas: An example from Lombardy region (Italy. *Geosciences*, *9*(9), 364. Available from https://doi.org/10.3390/geosciences9090364, flight.

Arab-Sedze, M., Heggy, E., Bretar, F., Berveiller, D., Jacquemoud, S. (2014). Quantification of L-band InSAR coherence over volcanic areas using LiDAR and in situ measurements, *Remote Sensing of Environment*. 152: 202–216. Available from https://doi.org/10.1016/j.rse.2014.06.011.

Berardino, P., Fornaro, G., Lanari, R., & Sansosti, E. (2002). A new algorithm for surface deformation monitoring based on small baseline differential SAR interferograms. *IEEE Transactions on Geoscience and Remote Sensing*, *40*(11), 2375–2383. Available from https://doi.org/10.1109/TGRS.2002.803792, [1] “UPE-MLV / OTIG CONTRIBUTION,” January 2009.

Brandt, J., Ertel, J., Spore, J., & Stolle, F. (2023). Wall-to-wall mapping of tree extent in the tropics with Sentinel-1 and Sentinel-2. *Remote Sensing of Environment*, *292*, 113574. Available from https://doi.org/10.1016/j.rse.2023.113574.

Bru, G., González, P. J., Mateos, R. M., Roldán, F. J., Herrera, G., Béjar-Pizarro, M., Fernández, J. (2017). A-DInSAR Monitoring of Landslide and Subsidence Activity: A Case of Urban Damage in Arcos de la Frontera, Spain. R*emote Sens*. *9*, 787. Available from https://doi.org/10.3390/rs9080787.

Fernandes, R., Brown, L., Canisius, F., Dash, J., He, L., Hong, G., Huang, L., Quynh, Le N., MacDougall, C., Meier, C., Osei Darko, P., Shah, H., Spafford, L., Sun, L. (2023). Validation of Simplified Level 2 Prototype Processor Sentinel-2 fraction of canopy cover, fraction of absorbed photosynthetically active radiation and leaf area index products over North American forests, *Remote Sensing of Environment*. 293, 113600. Available from https://doi.org/10.1016/j.rse.2023.113600.

Franceschini, S., Meier, A. C., Suan, A., Stokes, K., Roy, S., & Madin, E. M. P. (2023). A deep learning model for measuring coral reef halos globally from multispectral satellite imagery. *Remote Sensing of Environment*, *292*, 113584. Available from https://doi.org/10.1016/j.rse.2023.113584.

Goldstein, R. M., Zebker, H. A., & Werner, C. L. (1988). Satellite radar interferometry: Two-dimensional phase unwrapping. *Radio Science*, *23*(4), 713–720. Available from https://doi.org/10.1029/RS023i004p00713, flight.

Hanssen, R. F., & Van Leijen, F. J. (2008a). Monitoring deformation of water defense structures using satellite radar interferometry. May 2008a.

Hanssen, R. F., & Van Leijen, F. J. (2008b). Monitoring deformation of water defense structures using satellite radar interferometry. May 2008.

Kaveh, H. (2014). Contribution of radar interferometry (DinSAR and PSI) for the study of the effects of geotechnical drought: applications to the east of the Île-de-France region.

Lahlou, A. (2014). Siltation of the Mohamed Ben Abdelkrim Al Khattabi dam and anti-erosion control of the mountainous watershed located. *upstream*.

Li, J., Wang, W., Chen, G., & Han, Z. (2022). Spatiotemporal assessment of landslide susceptibility in Southern Sichuan, China using SA-DBN, PSO-DBN and SSA-DBN models compared with DBN model. *Advances in Space Research*, *69*(8), 3071–3087. Available from https://doi.org/10.1016/j.asr.2022.01.043.

Li, M., Hinnov, L., & Kump, L. (2019). Acycle: Time-series analysis software for paleoclimate research and education. *Computers & Geosciences*, *127*, 12–22. Available from https://doi.org/10.1016/j.cageo.2019.02.011.

Massonnet, D., & Feigl, K. L. (1998). Radar interferometry and its application to changes in the Earth's surface. *Reviews of Geophysics (Washington, D.C.: 1985)*, *36*(4), 441–500. Available from https://doi.org/10.1029/97RG03139, flight.

Minister of Public Works and Communications, Morocco. (1976). Contribution à l'étude de la crue de projet de l'oued Nekor.

Niazi, S., Snoussi, M., & Foutlane, A. (2005). Impact of climatic hazards on the water quality of a hydrosystem developed in a semi-arid zone: Case of the Nekkor watershed (Morocco). vol. 16.

Perrone, M., Di Febbraro, M., Conti, L., Divíšek, J., Chytrý, M., Keil, P., et al. (2023). The relationship between spectral and plant diversity: Disentangling the influence of metrics and habitat types at the landscape scale. *Remote Sensing of Environment*, *293*, 113591. Available from https://doi.org/10.1016/j.rse.2023.113591.

Pathier, E. Contributions of differential radar interferometry to the active tectonics of Taiwan. 2014

Poupart, M., & Royet, P. (2001). La surveillance des barrages, Colloque technique, Aix-en-Provence, CFGB, 48 p.

Shi, G., Huang, B., Leung, A. K., Ng, C. W. W., Wu, Z., & Lin, H. (2022). Millimeter slope ratcheting from multitemporal SAR interferometry with a correction of coastal tropospheric delay: A case study in Hong Kong. *Remote Sensing of Environment*, *280*, 113148. Available from https://doi.org/10.1016/j.rse.2022.113148, flight.

Small dam reservoir siltation in the Atlas Mountains of Central Morocco: analysis of factors impacting sediment yield | Springer Link. https://link.springer.com/article/10.1007/s12665-016-5795-y (accessed April 30, 2023).

Tarata, L.-I., & Vasile, G. (2017). Processing high spatial resolution interferometric SAR images for the monitoring of dams and dykes by radar remote sensing: validation of the operational chain. Available from https://doi.org/10.13140/RG.2.2.32581.04328.

Watzig, C., Schaumberger, A., Klingler, A., Dujakovic, A., Atzberger, C., & Vuolo, F. (2023). Grassland cut detection based on Sentinel-2 time series to respond to the environmental and technical challenges of the Austrian fodder production for livestock feeding. *Remote Sensing of Environment*, *292*, 113577. Available from https://doi.org/10.1016/j.rse.2023.113577.

Xiao, Y., Liu, R., Ma, Y., & Cui, T. (2023). MERRA-2 reanalysis-aided sea fog detection based on CALIOP observation over North Pacific. *Remote Sensing of Environment*, *292*, 113583. Available from https://doi.org/10.1016/j.rse.2023.113583.

Yang, M., López-Dekker, P., Dheenathayalan, P., Liao, M., Hanssen, R. F. (2019). On the value of corner reflectors and surface models in InSAR precise point positioning. ISPRS J. Photogramm. Remote Sens. 158: 113–122. Available from https://doi.org/10.1016/j.isprsjprs.2019.10.006.

Yu, B., Chen, F., & Xu, C. (2020). Landslide detection based on contour-based deep learning framework in case of national scale of Nepal in 2015. *Computers & Geosciences*, *135*, 104388. Available from https://doi.org/10.1016/j.cageo.2019.104388, flight.

Chapter 18

Examining the effects of vented dams on land use and land cover in the Shambhavi Catchment: a multitemporal sentinel imagery analysis

S. Chandana[1], Aishwarya Hegde[2], U. Pruthviraj[2] and M.C. Chandan[1]

[1]*Department of Civil Engineering, The National Institute of Engineering, Mysore, India,*
[2]*Department of Water Resource and Ocean Engineering, National Institute of Technology, Surathkal, Karnataka, India*

18.1 Introduction

As a result of human activity, the Earth's surface is changing quickly. Change detection (CD) is the process of identifying such alterations over time. Land cover change (LCC) is one of the most important indications for assessing how humans are affecting the environment (Aragaw, Goel, & Mishra, 2021). Additionally, land use and land cover (LULC) are powerful indicators of how humans are altering the ecology on Earth. With the newest cutting-edge technology, CD using Remote Sensing (RS) and Google Earth Engine (GEE) is growing considerably (Asokan & Anitha, 2019). In this way, more effective interactions and relationships between people and environmental phenomena can be created to promote improved decision-making (Campbell & Wang, 2019).

Multitemporal Sentinel images provide a valuable resource for accurately monitoring and classifying LCC over time. The combination of advanced RS technology and machine learning (ML) techniques offers a comprehensive approach to studying the Earth's evolving landscape and supports informed choices for sustainable land management and resource planning (Cavur, Duzgun, Kemec, & Demirkan, 2019). The process includes training the model, applying it to the images, and assessing its accuracy. By analyzing the results from different time periods, LCCs like urban growth,

Water Resources Monitoring, Management, and Sustainability. DOI: https://doi.org/10.1016/B978-0-443-23665-5.00018-1

deforestation, and agricultural shifts can be identified, facilitating the use of informed judgment in land management decisions (Erener, Düzgün, & Yalciner, 2012).

18.1.1 Land use land cover

The categorization and characterization of the various land use categories as well as the physical cover of the Earth's surface are referred to as "land use and land cover." It provides valuable information about the distribution, extent, and spatial patterns of various land uses, such as agriculture, urban areas, forests, water bodies, and barren land, as well as the physical cover, including vegetation, bare soil, impervious surfaces, and natural features (Asokan & Anitha, 2019).

Land use refers to human activities and the purpose for which the land is utilized. It encompasses activities such as agriculture, residential areas, industrial zones, commercial areas, transportation infrastructure, recreational spaces, and protected areas (Chowdhury & Dwarakish, 2022). The land cover (LC) reflects how humans interact with the land and its resources. LC, on the other hand, discusses the physical characteristics and materials that cover the surfaces of Earth. It describes the actual coverage of the land, including vegetation, water bodies, barren land, snow, ice, and artificial surfaces like buildings and roads. LC provides insights into the natural and built environment and helps assess environmental conditions and ecosystem health (Gashu, 2022; Geetha, Gowda Asha Nandeesha, & Nagaraj, 2021).

18.1.1.1 Definition

Land use is the umbrella term for life aspects and the different uses of land. Types of LC include natural flora, water features, rock and soil, and so on. The categories given below offer a thorough explanation of the features of a particular feature that was utilized to create the LULC feature database, as shown in Table 18.1.

The study and analysis of LULC changes are crucial for understanding the transformation of landscapes over time and its implications for various aspects, such as urbanization, agriculture, deforestation, habitat loss, climate change, and biodiversity conservation. It aids in monitoring and managing natural resources, planning land use policies, assessing environmental impacts, and supporting sustainable development (Mishra, Rai, & Rai, 2020).

For the purpose of observing and analyzing LULC patterns and changes, RS methods, like satellite imaging and aerial photography, are frequently employed. Advanced technologies like Geographic Information Systems (GIS) and ML algorithms, such as random forest (RF) classifiers, are utilized to process and classify the RS data into different LC categories. Ordinary LULC automated classification methods are based on calculating spectral

TABLE 18.1 Description of classified land cover classes.

Land cover classified classes	Description
Surface water	**Surface water** as shown in the figure is included in this category if it is flowing through streams, rivers, canals, or other water bodies rather than being impounded in ponds, lakes, and reservoirs. These can be seen in blue to dark blue or cyan on the satellite view, depending on the depth of the water
Barren land	**Barren land** is considered damaged land that can be put under natural vegetation with minimal effort and is currently underutilized, in addition to land that is deteriorating owing to a lack of sufficient soil and water management or natural factors.
Dense-forest	**Dense forests** within the designated forest limits are the locations where trees predominately coexist with other plant species to create a distinctive ecosystem.
Urban	**Urban areas** refer to the land surfaces produced by human settlement as a result of nonagricultural use. Along with water, vegetation, and vacant ground, built-up regions can also include transportation, communication, businesses, utilities, and services. It is further divided into the following groups.
Agriculture	**Agricultural land** is the locations where food, fiber production, and other industrial and agricultural crops are primarily produced. It contains plantations, irrigated and nonirrigated agriculture, and various types of land.

signatures of certain LULC classes using training data and pixel-based discriminating between various LC types when applied to RS data (Amini, Saber, Rabiei-Dastjerdi, & Homayouni, 2022). LCCs can significantly influence vented dams in these regions.

18.1.2 Vented dams

Vented dams are hydraulic structures designed to release excess water and control downstream flow while providing water storage and other benefits. They include mechanisms such as spillways or outlets to regulate water levels and prevent flooding (Asokan & Anitha, 2019). Vented dams offer advantages such as flood control, water storage, hydropower generation, and water supply management. They help manage water resources by releasing excess water during high rainfall or snowmelt periods. This controlled release prevents downstream flooding and reduces damage risks (Fonji & Taff, 2014). They also contribute to renewable energy production through hydropower generation, reducing reliance on fossil fuels (Jiang et al., 2018).

However, the construction of vented dams can impact ecosystems and communities. Changes in water quality, temperature, and sediment transport can affect aquatic habitats and fish populations. Additionally, the creation of reservoirs may lead to the displacement of flora, fauna, and human communities. To mitigate these impacts, measures such as fish passages, sediment management, and habitat restoration are implemented. Monitoring and adaptive management strategies are crucial to assess long-term environmental effects and ensure sustainable practices (Raj & Gurugnanam, 2018).

Overall, vented dams play a vital role in water management, but their development should balance benefits with environmental considerations to ensure sustainable water resource utilization and enhance water security. Alterations in land use, like urban expansion or deforestation, can disrupt the natural hydrological cycle, leading to increased runoff during rainfall events and higher water flow into the dam's reservoir (Karlsson et al., 2016). This could necessitate more frequent use of the spillway for flood management. Moreover, changes in LC can intensify sedimentation in the reservoir, potentially reducing storage capacity, impairing water quality, and affecting downstream ecosystems. The stability of the dam surroundings might also be compromised due to erosion and landslides induced by changes in land use (Koppad & Janagoudar, 2013). Climate change interactions should be considered as well, making collaboration among land use planners, dam operators, and environmental experts crucial to ensuring resilient dam infrastructure and effective management strategies amid evolving LULC patterns (Koppad & Janagoudar, 2013). However, the alteration of river systems due to the construction of vented dams can result in modifications to LULC. This analysis examines the connection between the existence of vented dams and LULC changes, focusing on the effects on vegetation, agricultural land, water bodies, human settlements, and other LC types (Shetty, Umesh, & Shetty, 2022).

18.1.3 Machine learning

Artificial intelligence in the form of ML enables computer programmers to forecast outcomes more accurately without having been expressly taught to do so. ML algorithms estimate new output values using past data as input. This can turn unstructured data into geospatial information products from which you can extract information and outcomes by combining the strengths of GIS and ML (Amini et al., 2022; Aniah, Bawakyillenuo, Codjoe, & Dzanku, 2023). The newest ML classifiers have demonstrated superior performance to the traditional maximum likelihood ones and do not rely on any presumptions about the distribution of the data (Asokan & Anitha, 2019). The most recent ML classifiers have proven to perform better than the conventional maximum likelihood ones and do not rely on any assumptions regarding the distribution of the data.

It has been identified that nonparametric ML classifiers, in particular Classification and Regression Trees (CART), RF, and Support Vector Machines (SVMs), can classify LULC with a high degree of accuracy from remotely sensed pictures. While RF is an ensemble classifier based on the combination of many CART-like trees, CART is a straightforward binary decision tree classifier that operates using a predefined threshold (Chowdhury & Dwarakish, 2022). One of the preferred classifiers for LCC, RF is quite effective and accurate and consistently outperforms other well-liked classifiers. The RF algorithm is a powerful ML technique used for both classification and regression tasks (Jiang et al., 2018). It operates by constructing "n" number of decision trees during the training phase and then combining their outputs to make predictions or classifications. The algorithm works as represented in Fig. 18.1, which consists of a collection of decision trees, including root node, entropy, decision node, and leaf node. Each sample was trained using a random selection of the training data and perhaps a subset of the feature sets that were made accessible. This randomness aids in improving the model's generalization and decreasing overfitting (Geetha et al., 2021).

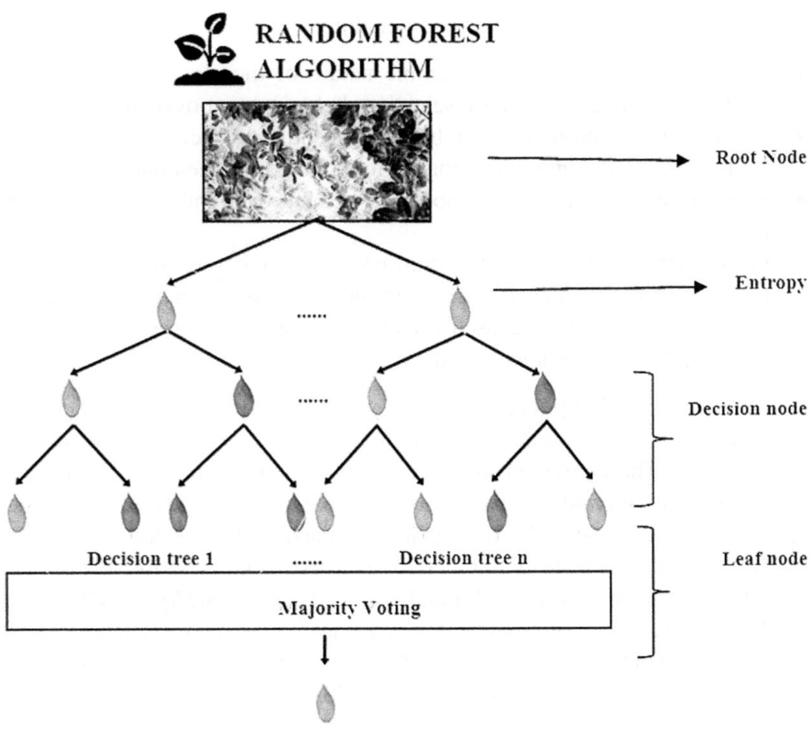

FIGURE 18.1 Random Forest algorithm.

Each decision tree is created throughout the training process using a different subset of the training data, chosen using a method called "bagging" (bootstrap aggregation). This involves randomly selecting instances from the training set with replacements. This creates diversity among the trees, enhancing the overall robustness of the model. The method will construct each decision tree and will also randomly choose a subset of features from the entire collection of features. As a result, the trees are more diverse, and the decision-making process is less dominated by some traits (Koppad & Janagoudar, 2013).

Utilizing the characteristics and the chosen data subset, each decision tree is built. The data are recursively divided into subsets depending on the chosen characteristics during the tree-building process, with each partition seeking to reduce impurity or inaccuracy. Until a stopping requirement is satisfied, such as reaching a specific depth or having a specific minimum number of instances in each leaf node, the process continues. Once the RF is trained, it can make predictions by aggregating the predictions from all the individual decision trees (Mishra *et al.* 2020). The final prediction for classification tasks is frequently regarded as the mode (most common class) of the anticipated classes. The projected values from all trees' average predictions are often utilized for regression tasks.

RF has a wide range of benefits, including:

- Reduced Overfitting: The ensemble nature of the algorithm, along with random sampling and feature selection, helps prevent overfitting.
- Good Generalization: The combination of multiple decision trees' outputs leads to improved generalization performance on unseen data.
- Robustness: RFs can handle noisy or missing data and are less sensitive to outliers.
- Feature Importance: The algorithm provides a measure of feature importance, indicating which features contribute the most to predictions.
- Versatility: They can be used for classification and regression tasks and even handle high-dimensional data.

RFs have found applications in various fields, including finance, healthcare, natural language processing, and image analysis, among others (Brown et al., 2022). The confusion matrix method, which compares the classification outputs to ground truth data, is the most used method for evaluating classification accuracy (Mohammadpour, Viegas, & Viegas, 2022). The matrix is used to create a number of analytical and descriptive statistics, including the Kappa statistic (K) and overall accuracy (OA), which may be used to analyze and summarize the accuracy level of a particular LULC map. Thus it contributes more to the GEE cloud platform for the creation of LULC maps (Shetty, Kushwaha, & Shetty, 2023).

18.1.4 Google Earth Engine

GEE is a cloud-based platform for studying and displaying earth observation data that offers geographical data, tools, and computer resources. The main uses and features of GEE are as follows:

GEE offers a vast catalog of geospatial data, including satellite imagery, climate data, LC maps, and terrain data. Users can access and explore these datasets for various analysis purposes (Gashu, 2022). It provides a powerful computational engine that enables users to process and analyze large-scale geospatial datasets, as the workflow has been represented in Fig. 18.2.

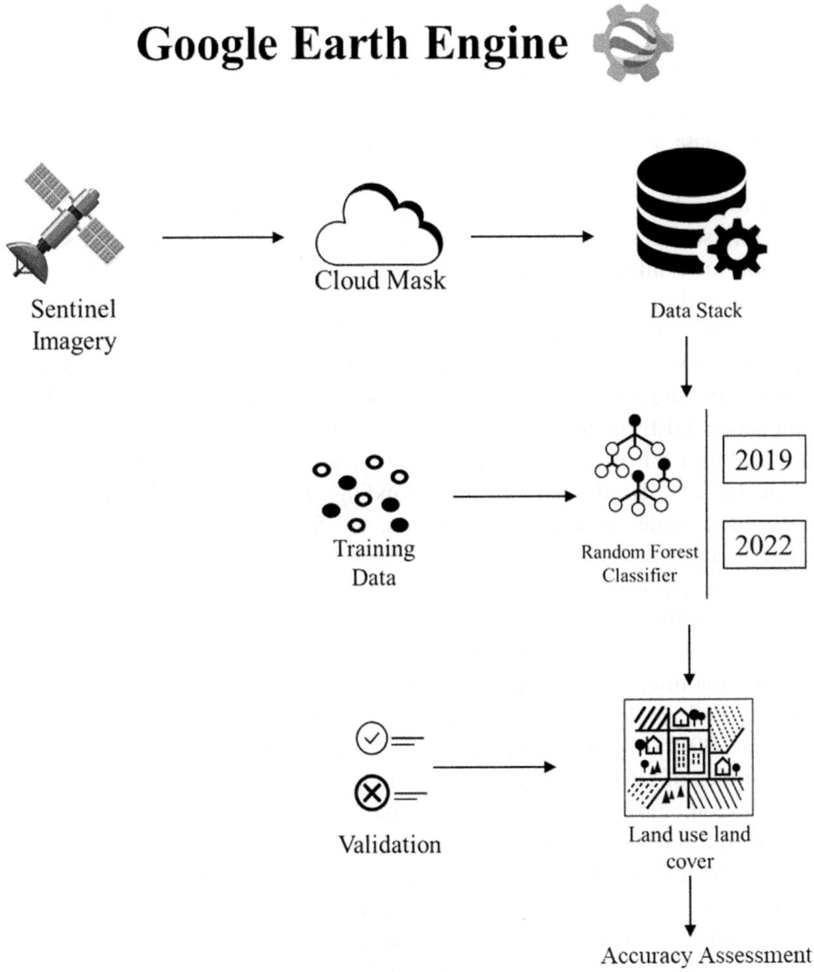

FIGURE 18.2 Workflow of Google Earth Engine.

It supports various processing tasks such as image mosaicking, filtering, reprojecting, and resampling (Geetha et al., 2021). It includes a wide range of tools and algorithms specifically designed for RS analysis. These tools allow users to extract information from satellite imagery, calculate spectral indices (e.g., NDVI, NDWI), perform radiometric and atmospheric corrections, and apply ML algorithms for classification and regression tasks. GEE supports temporal analysis by allowing users to work with time series data (Karlsson et al., 2016). It enables the visualization and analysis of changes over time, trend analysis, anomaly detection, and monitoring of vegetation dynamics, among other applications.

GEE offers interactive mapping capabilities, allowing users to visualize geospatial data, create custom maps, and generate animations to showcase temporal changes. It provides tools for overlaying and comparing different datasets, adding labels and legends, and customizing the visual appearance of maps, which facilitates collaboration among users by allowing the sharing of code, data, and analysis workflows. It provides an environment for collaborative coding and supports version control, making it easier for multiple users to work on the same project (Priya, Sasanka, & Osuri, 2023).

GEE seamlessly integrates with ML libraries and frameworks, such as TensorFlow, allowing users to apply advanced algorithms for geospatial analysis and prediction tasks. It provides a JavaScript-based API that allows users to interact with the platform programmatically (Sazib, Mladenova, & Bolten, 2018). This API enables the development of custom geospatial applications and the automation of data processing and analysis workflows. Users can access GEE for free, encouraging open data access and cooperation. It is a useful tool for researchers and others because it gives access to a variety of open datasets. To aid users in understanding and properly utilizing the platform, it provides educational resources, tutorials, and training materials. It offers documentation, code samples, and a welcoming community that promotes learning and knowledge exchange (Shetty *et al.*, 2022).

The significance of accurate and timely LULC CD cannot be overstated, as it serves as a crucial foundation for understanding the environmental, socio-economic, and ecological implications of human activities (Sharannya, Mudbhatkal, & Mahesha, 2017; Shelestov, Lavreniuk, Kussul, Novikov, & Skakun, 2017). As we navigate through this chapter, we will explore the capabilities of GEE in facilitating multitemporal analyses, leveraging spectral information, and incorporating ML techniques for robust and scalable CD.

Overall, GEE is a powerful tool for geospatial analysis, enabling users to access, process, analyze, and visualize large-scale geospatial datasets for several uses in domains like environmental monitoring, LCC mapping, agriculture, climate change analysis, and urban planning (Shetty *et al.*, 2023). By studying LCC, decision-makers, researchers, and land managers can gain valuable insights into the dynamics of land use patterns, identify areas of environmental concern, plan for sustainable land management practices, and develop strategies for conservation and land use planning to ensure the optimal use of land resources while minimizing negative impacts on ecosystems and society (Sinha & Eldho, 2018).

18.2 Literature

Using GIS tools, Ritse et al. (2020) investigated the variations in LULC in the Eastern Himalayan landscape of Nagaland, Northeast India. Remotely sensed LANDSAT pictures (TM and OLI) based on the GIS platform were employed in addition to field surveys. Four main LULC classes were identified using the supervised classification approach and the Landsat level-1 classification system. Changes in LULC were mostly brought about by the excluding of shifting agriculture, deforestation, infrastructure development, urban migration, and the rise of aquaculture cultivation. This study provides baseline data on LULC in the area of northeast India, where data is hard to locate to preserve sustainable land use planning in the face of growing urbanization. The objective of this study is to analyze the LULC changes caused by vented dams using a supervised RF classifier and quantifications of the vented dams in the study area to prepare thematic maps for illustrating various classified parameters.

This study discusses the detection of change in land usage in Davangere, Karnataka State, India, between 2016 and 2021, focusing on urbanization and shifts in traditional cropping patterns by Geetha et al. (2021). The objective of the research is to capture this change using RS and supervised ML classifiers such as Minimum Distance, Mahalanobis Distance, and Maximum Likelihood. The Maximum Likelihood classifier was found to ensure the highest accuracy of 95.2%. The study found significant changes in land use, particularly in the built-up area and area the under cultivation of paddy. Furthermore, it provides an overview of the CD of LULC using ML, which includes a study in Morocco that classified multiple crops using SVM, artificial neural network (ANN), and ML, achieving an accuracy of 89% and a Kappa coefficient of 0.85. Another study compared classifiers for LULC using RF, SVM, and maximum likelihood, finding RF to be the best with the highest accuracy. A research paper assessed classifiers for LC classification using sentinel-2A data, with ANN identified as the best classifier with 90% accuracy. A study using a Landsat-3 image found SVM to be the best classifier with 95.35% accuracy and a Kappa coefficient of 0.94.

In a study by Gohain et al. (2021), the impacts of LULC on LST were evaluated using RS and GIS techniques. LANDSAT data for Pune City from 1990 to 2019 was employed to assess the variations in LST across different LULC classifications. The results demonstrated that the classified LULC maps for 1990, 1999, 2009, and 2019 exhibited overall accuracies of 90.4%, 94.5%, 92.7%, and 93.7%, respectively. Notably, all LU categories in Pune City changed, with a considerable increasing trend observed in the built-up area while the other LU types displayed a decreasing tendency. Regarding land surface temperatures (LSTs), metropolitan areas, agricultural land, and unusable land consistently recorded higher temperatures compared to other land use classes throughout the year.

This paper presents an algorithm that uses Landsat time series data to analyze LULC change by Amini et al. (2022). The RF classifier is applied in the GEE using imagery from Landsat 5, 7, and 8 as inputs for the 1985–2019 period. The impact of different image compositions, including pan-sharpening algorithms and spectral indices, on the final classification accuracy is explored. The study shows that using pan-sharpened top-of-atmosphere Landsat products and incorporating features like LST and digital elevation models can increase the accuracy of LULC classification. The proposed algorithm outperforms the Copernicus Global LC Layers map in terms of OA, Kappa values, and F1-score. The method applies to other regions of interest for LULC mapping, and a code repository is provided for classifying Landsat 4, 5, 7, and 8 within GEE. Various Approaches for LULC Change Analysis. Researchers commonly use time series indices derived from multispectral satellite data for LULC change analysis. The role of image composition in LULC classification using RF on the GEE platform is investigated, showing that input features directly influence final classification accuracy. Gray-level cooccurrence matrix texture features are examined with standard spectral features to monitor urban LULC change using Landsat time series imagery. Features in LULC classification are categorized into spectral, textural, and geometrical features, and their selection depends on factors like dataset, area conditions, and auxiliary data availability. The relation between LST and LULC is explored, and pan-sharpening Landsat-8 (L8) imagery is found to improve LULC classification accuracy. Different pan-sharpening methods are available, with the intensity, hue, and saturation method being one of the most popular. Variable importance in RF is used to measure the impact of each variable on the overall prediction performance of the model.

This study aimed to classify LULC in the Munneru River Basin using ML algorithms on the GEE platform by Loukika et al. (2021). Three different ML algorithms, namely, SVM, RF, and CART, were used for the classification. The LULC of the study area was classified *via* supervised classification, and NDVI and NDWI indices were derived and included to improve classification accuracy. The major land use classes identified were "water bodies," "forest," "barren land," "vegetation," and "built-up." The average OA of the classifiers for Landsat-8 images was 90.88% for SVM, 94.85% for RF, and 82.88% for CART. For Sentinel-2 images, the accuracy was 93.8% for SVM, 95.8% for RF, and 86.4% for CART. RF classifiers outperformed both SVM and CART classifiers in terms of accuracy. The combination of band data and the use of high-resolution Sentinel-2 data contributed to the improved accuracy of the RF classifier. The study highlights the importance of accurate LULC classification for natural resource management and climate change research.

18.3 Study area

The Shambhavi Catchment (SC) covers 345.48 km^2 and is in the Dakshina Kannada and Udupi districts of Karnataka, India. The representation of the study area has been presented in Fig. 18.3, which consists of the river in Dakshina Kannada district, known by the name Shambhavi. It is 60 km long, located 10 km north of Surathkal, and flows through the well-known Parpappadi Falls in Nitte.

The SC is a geographic region encompassing the drainage basin of the Shambhavi River, located in a specific area or watershed. It serves as the focal point for various studies and analyses related to water resources, land use, and environmental management. Geographically, the SC is situated in a particular region, which could be described in terms of its location, topography, and land characteristics. However, the fall of its ranges is in between Udupi as well as Dakshina Kannada district, where detailed information about the location and features of the SC is not easily accessible. Nevertheless, the SC can be expected to have a diverse landscape, including river channels, tributaries, hills, plains, and potentially urban or rural settlements, depending on the surrounding land use patterns. The catchment's extent and boundaries are defined by the area that contributes runoff and water flow into the Shambhavi River. The Shambhavi River, being the central water body within the catchment, is likely to be influenced by land-use activities in its surroundings. Land use practices such as agriculture, forestry,

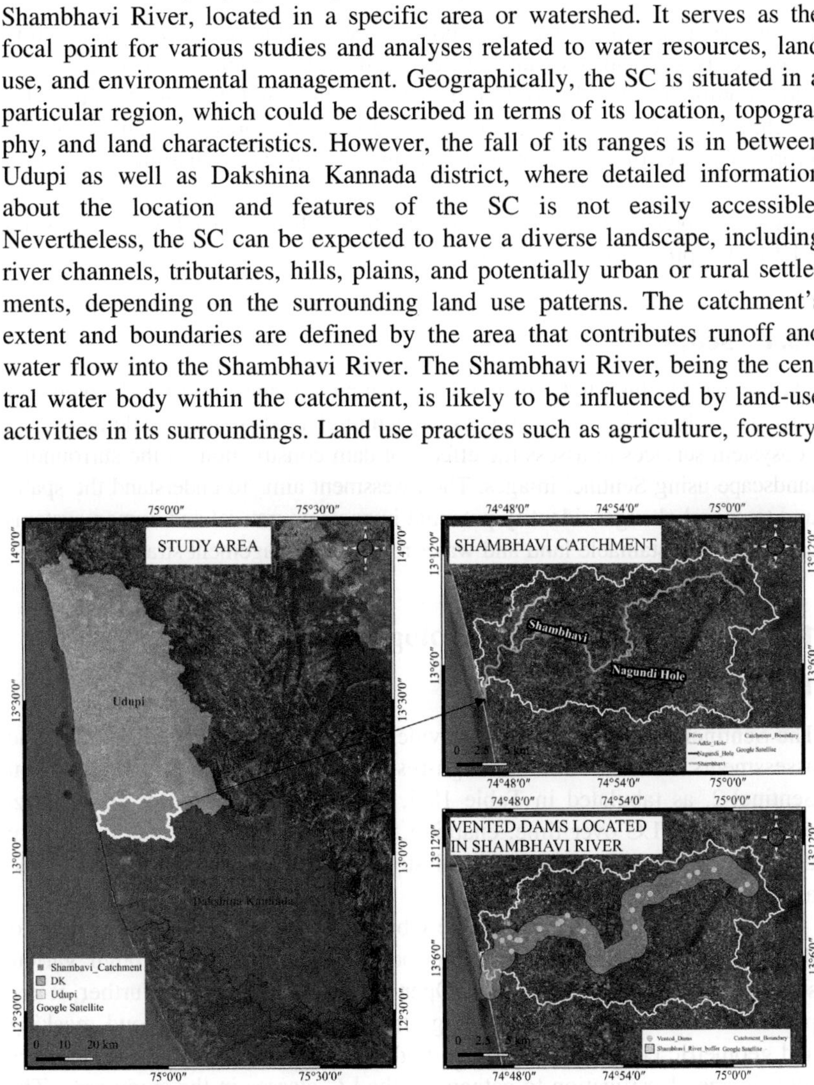

FIGURE 18.3 Study area location. Map lines delineate study areas and do not necessarily depict accepted national boundaries.

urbanization, and industrial activities within the catchment can significantly impact water quality, hydrological processes, and ecological dynamics.

It originates in the Chikkamagaluru district of Karnataka, India, at Kudremukh. Before going to the Arabian Sea, it joins the Nandini River near Mulki, Karnataka. Numerous villages close to the river depend mostly on this river for their water supply. With the use of various little dams, the river's water level is monitored. The final dam west of the river is the Palimar Dam, consisting of 68 vented dams along the water body and primary, secondary, and tertiary stream orders.

Further detailed information about the SC, including its specific location, size, LC types, and associated land use patterns, would be necessary for a more comprehensive description of the study area and its characteristics. The catchment's LC is made up of water, arid land, forests, urban areas, and crops. The LC of the catchment is composed of water, barren land, forest, urban, and cropland. The following part of this report represents the methodologies adopted and the analysis results for each task of NDVI analysis and LULC mapping of the SC in each section.

18.4 Scope

The scope of this study includes evaluating changes in LULC categories, vegetation dynamics, water bodies, agricultural land, human settlements, and ecosystem services to assess the effects of dam construction on the surrounding landscape using Sentinel images. The assessment aims to understand the spatial and temporal changes, identify potential impacts on natural and human systems, and support sustainable land and water resource management strategies.

18.5 Materials and methodology

18.5.1 Data collection

The Sentinel-2 satellite mission provides valuable data for conducting impact assessments on LULC. The high-resolution multispectral imagery from Sentinel-2, as tabulated in Table 18.2, can be used to monitor and analyze changes in LULC over time, enabling the assessment of various impacts on the landscape. Here are the general steps involved in conducting an impact assessment using Sentinel-2 data:

In terms of data acquisition, the Obtain Sentinel-2 (S2) imagery data for the desired study area and time period is freely available through various sources, including the Copernicus Open Access Hub, or GEE. Further, it preprocesses the S2 imagery, including atmospheric correction, cloud masking, and geometric correction, to ensure high-quality and consistent data. To perform image classification to categorize the LC classes in the study area. This

TABLE 18.2 Sentinel-2 image band information and its details.

Bands	Wavelength region	Central wavelength (nm)	Resolution (m)
Band-1	Coastal aerosol	443	60
Band-2	Blue	490	10
Band-3	Green	560	10
Band-4	Red	665	10
Band-5	Vegetation red edge	705	20
Band-6	Vegetation red edge	740	20
Band-7	Vegetation red edge	783	20
Band-8	NIR	842	10
Band-8A	Vegetation red edge	865	20
Band-9	Water vapor	942	60
Band-10	SWIR-Cirrus	1375	60
Band-11	SWIR	1610	20
Band-12	SWIR	2190	20

can be done using supervised or unsupervised classification algorithms, as shown in the overall methodology in Fig. 18.4. Sentinel-2 multispectral bands provide information on vegetation, water bodies, urban areas, and other LC types.

For change analysis, compare the classified LC maps from different time periods to detect and quantify changes in LULC. This analysis helps identify areas experiencing land degradation, urban expansion, deforestation, or other significant changes to assess the impacts of the observed LULC changes on various environmental and socio-economic factors. This can involve analyzing the effects on biodiversity, ecosystem services, water resources, climate change, agricultural productivity, urban planning, and other relevant indicators. It can also interpret the results of the impact assessment and communicate the findings through reports, maps, and visualizations. This information can be used to inform decision-making, land management practices, and policy development.

It is important to note that conducting a comprehensive impact assessment requires a combination of satellite data, ground truth data, and domain-specific knowledge. Additional data sources and analysis techniques may be incorporated based on the specific objectives and scope of the

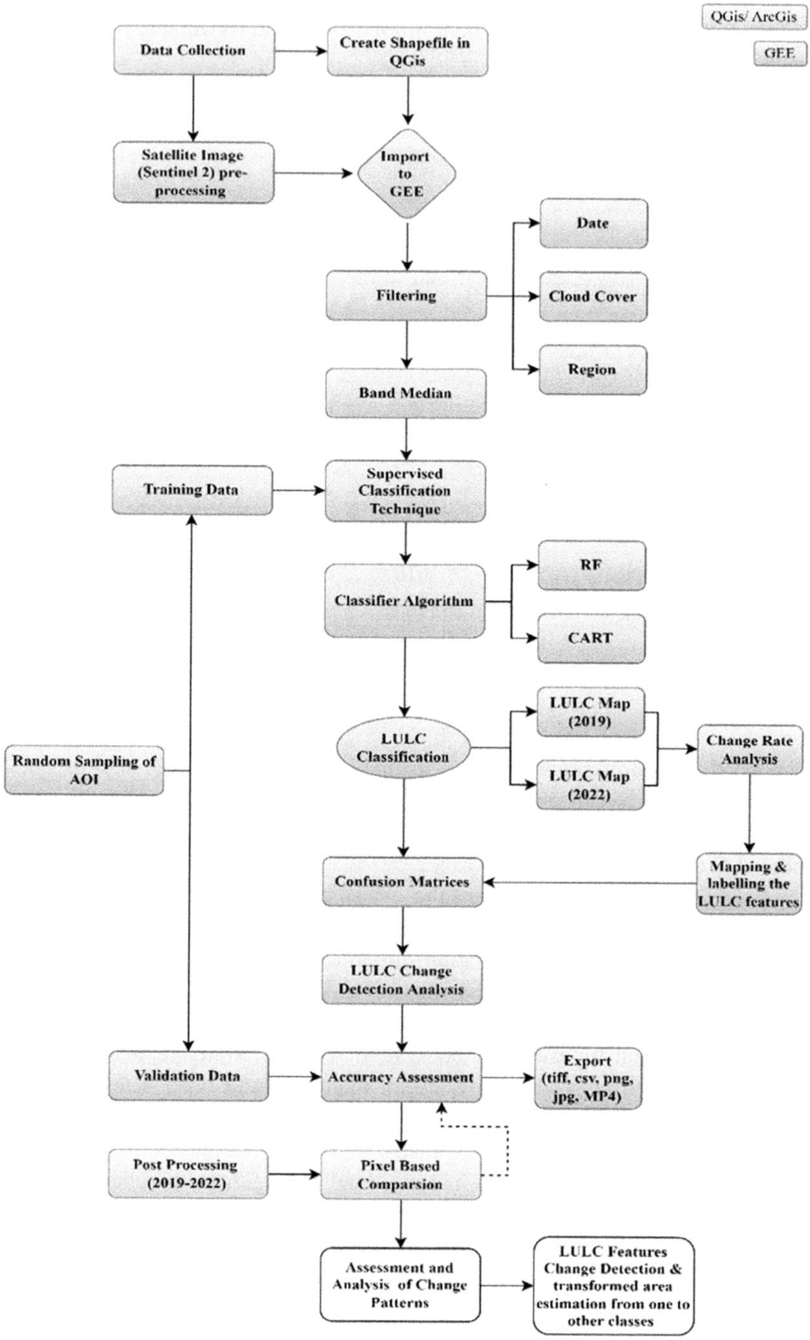

FIGURE 18.4 Flowchart of the techniques used in this study.

assessment. By leveraging the capabilities of S2 and performing a thorough analysis of LULC changes, an impact assessment can provide valuable insights into the consequences of land use practices and support sustainable land management decisions.

18.6 Methodology

This study uses maximum likelihood classification-based methodologies to quantify the change in LC for the SC. In the flow diagram, the overall technique used for the investigation is displayed in Fig. 18.4. LC patterns from multitemporal sensor sentinel imagery for 4 years (2019–22) have been used to study the process.

First, GEE was used to construct and preprocess the polygon for the required catchment. For quantitative analysis, images have been classified using supervised classification by the maximum likelihood algorithm. Using training sets or signatures that are manually provided to the application, this classification technique categorizes the data based on the types of pixels that correspond to the various types of LC. The LC change was evaluated using the postclassification comparison approach based on different dam-building phases, and it was further analyzed along the river buffer. OA means the measure of correctness measured in terms of percentage. The OA of the classified image compares how each of the pixels is classified versus the actual LC conditions obtained from their corresponding ground truth data. It can be calculated by dividing the total number of correctly classified pixels by the total number of reference pixels. Meanwhile, Eq. (18.1) represents the Kappa coefficient (K) and measures the difference between how many agreements are present (“observed” agreement) compared to how much agreement would be expected to be present by chance alone (“expected” agreement).

$$K = \frac{K1 - K2}{1 - K2} \quad (18.1)$$

where

K1 is observed agreement, which is the proportion of pixels for which the two classifiers (or raters) agree on the category.

K2 is the expected agreement, which is the proportion of agreement expected by chance.

The variations in images of the same scene taken at different periods are explained and identified by CD analysis. For the chosen study region, classified images from the past few years are utilized to assess the extent of various land uses and cover types and determine changes in each class. This analysis is useful for identifying various changes, such as an increase in land use area or a decrease in LC, that are occurring in different classifications of LULC.

18.6.1 Training and validation sample data

The study area is characterized by a complex landscape mosaic composed of five LULC classes. To test the usability and reliability of the whole procedure, the points for each class have been identified manually through the GEE interface. For gathering the training data, Google Maps high-resolution layer, composite infrared, and S2 RGB layers were used.

18.7 Results and discussion

In this study, LULC classification is performed using a supervised classification method with a maximum likelihood algorithm in the GEE. Using the GIS platform, maps were created, and picture classification was represented (Figs. 18.5 and 18.6). Image classification resulted in five LC classes, that is, urban, forest, vegetation, water, and barren land. The LC map for years is presented in Fig. 18.5, whereas Fig. 18.6 shows the class map of the area. LC mapping of the study area indicates a continuous increase in vegetation that comprises the other classes, namely, barren land. It is observed that the barren land, which was *87.1km*, was lowered to *19.2km* in 2022; similarly, the forest, agriculture, and urban areas increased by *147.2km* to *194.6km, 101.4 km* to *119.7 km*, and *6.6 km* to *8.9 km*, respectively.

The findings of this study emphasize the complex and interconnected nature of the impacts of vented dams on LCC. The changes observed have

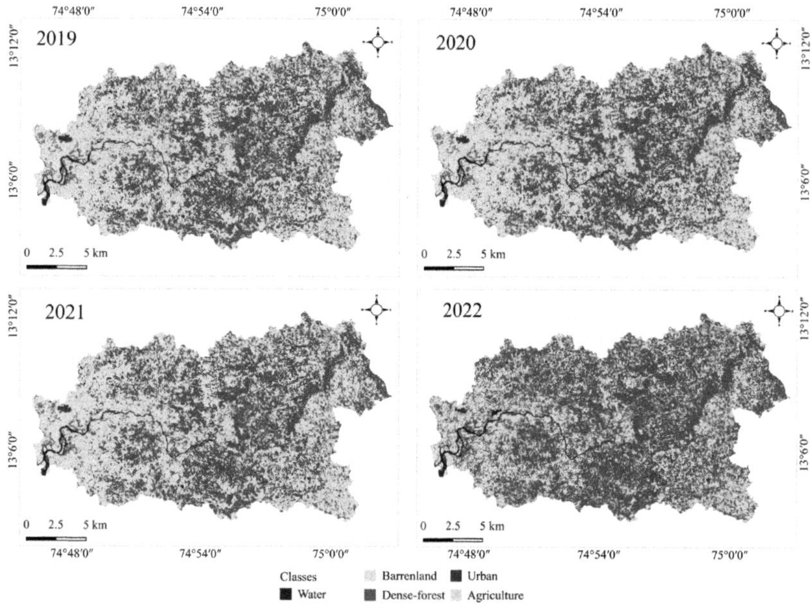

FIGURE 18.5 Classified land cover maps generated annually. Map lines delineate study areas and do not necessarily depict accepted national boundaries.

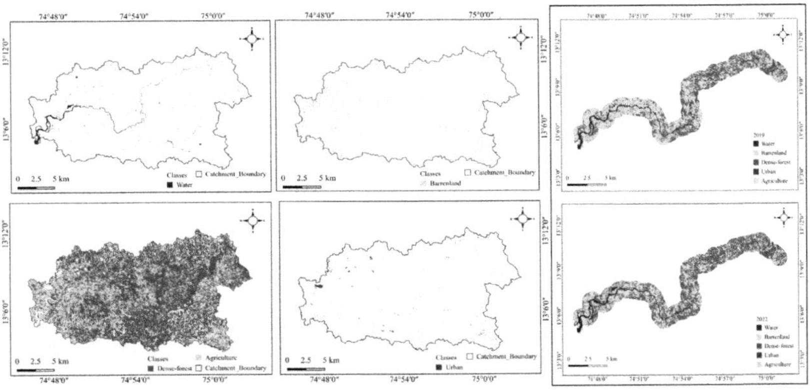

FIGURE 18.6 Change detected class maps, that is, from 2019 to 2022. Map lines delineate study areas and do not necessarily depict accepted national boundaries.

both positive and negative implications for environmental sustainability. While vented dams can provide water resources and flood control benefits, they can also disrupt natural ecosystems, affect biodiversity, and potentially displace communities. It is crucial to carefully consider these effects during the planning, design, and management of vented dams to ensure sustainable development and minimize adverse impacts.

This study's objective was to use the S2 dataset in GEE to analyze the LULC change in the study area. The high-resolution bands of S2 were chosen for agricultural water management. GEE and GIS techniques are utilized to categorize and map the time series LCC (2019–22) of the SC presented in Fig. 18.7. A classification's effectiveness was also evaluated in terms of accuracy.

The datasets of this study area (S2 for 2019 and S2 for 2022) were used by GIS. The datasets were obtained by satellite images with the latitude and longitude values of the study area. Furthermore, both datasets were trained through visual interpretation. By creating polygonal digital representations of the indicated classes, signature files were created. The following phase involved doing supervised classification in accordance with the GEE signature file to create the LCC maps.

The two datasets obtained had OA evaluation results of 96.30% and 95.47%, respectively. The attribute table was used to determine how much space various classes took up in both years. The postclassification comparison technique was used after image classification, comparing an LCC map from one data set (2019) with an LCC map from another data set (2022). The comparison presents quantifiable information on the differences between the 2 study years, as quantified in Fig. 18.8.

The image was classified into five classes and areas obtained for 2019: water (1.37%), barren land (25.10%), forest (42.42%), urban (1.89%), and agriculture (34.47%), and for 2021, water (1.36%), barren land (5.52%),

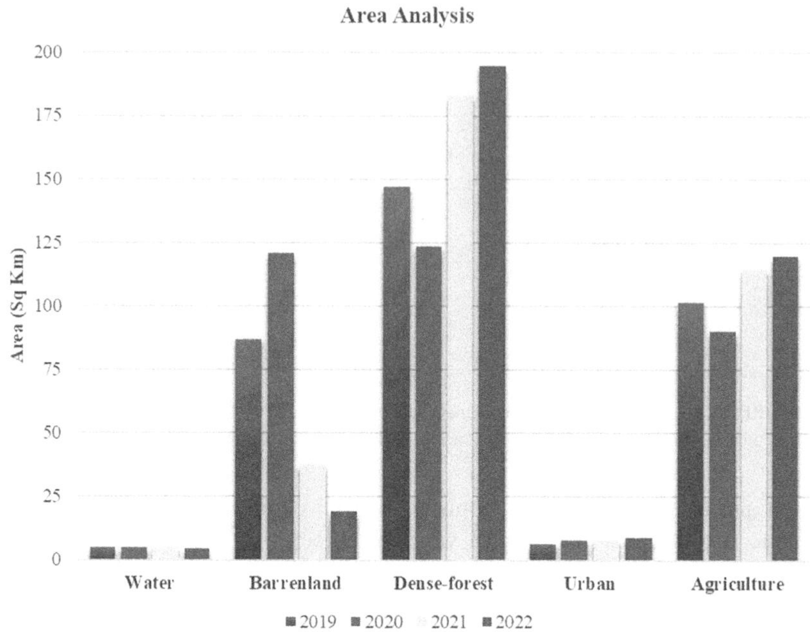

FIGURE 18.7 Area analysis from 2019 to 2022.

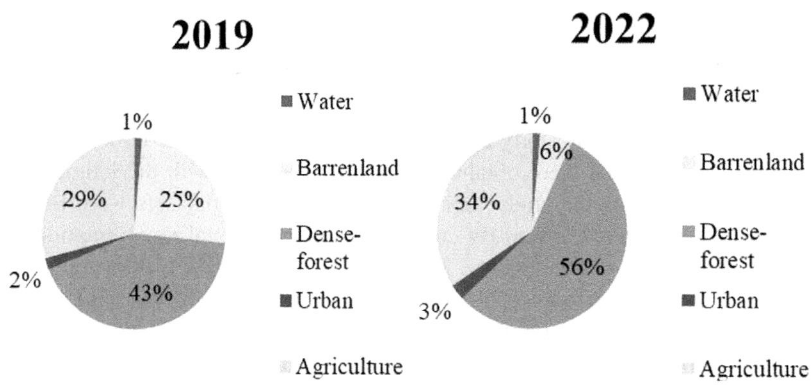

FIGURE 18.8 LULC statistical analysis of an area over a study period 2000–22. *LULC*, Land use and land cover.

forest (56.06%), urban (2.57%), and agriculture (34.47%). The supervised classification was carried out using an RF classifier. For the year 2019, the "OA" was 96.30%, and its "K" was 0.89. Similarly, 95.47% and 0.87 for 2022 are tabulated in Table 18.3, respectively. As a result, the LULC map of the watershed was well prepared using GEE. Therefore the RF classifier was good for preparing the LULC map of the study area.

TABLE 18.3 Tabulation of model accuracies estimated.

Classes	Training		Testing	
	OA	K	OA	K
	2019	**2019**	**2022**	**2022**
Water	97.84	0.95	82.73	0.74
Barren land	93.22	0.90	68.63	0.65
Dense forest	98.94	0.84	81.25	0.72
Urban	95.18	0.81	76.37	0.69
Agriculture	96.33	0.92	72.14	0.68
Overall	**96.30**	**0.89**	**76.22**	**0.70**

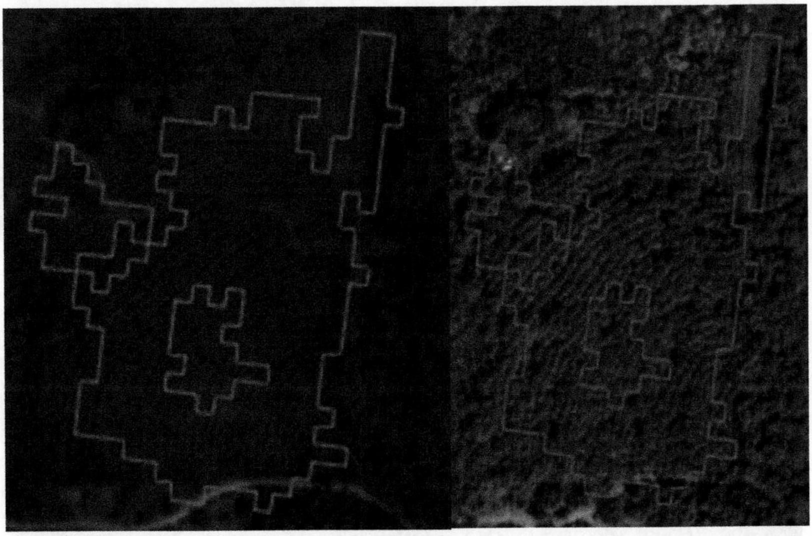

FIGURE 18.9 Agriculture to dense-forest transition.

The classified image for 2019 was produced using the Sentinel-2 dataset, in which the results obtained are as shown in Fig. 18.6 (2019), and the classified images for 2022 were produced using the same datasets as those for 2019, whereas the results are shown in Fig. 18.6 (2022) over the Shambhavi River, respectively. The few LULC classified class transitions identified have been presented in Figs. 18.9, 18.10, 18.11, and 18.12 as follows, where the classified LC classes change and are noticeable.

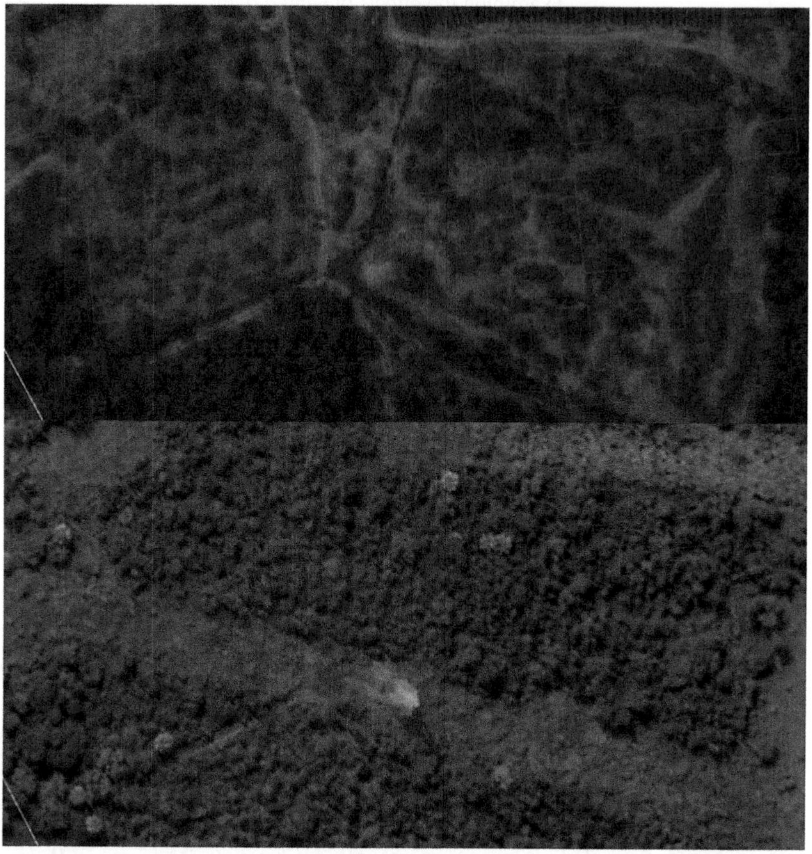

FIGURE 18.10 Barren land to dense-forest transition.

FIGURE 18.11 Barren land to urban transition.

FIGURE 18.12 Agriculture to urban transition.

The construction of vented dams significantly influences LC patterns. This report highlights the importance of understanding the effects of vented dams on vegetation, agricultural land, water bodies, and human settlements. The findings underscore the need for holistic and sustainable approaches in dam construction and management, considering the ecological, socio-economic, and environmental aspects [16]. Further research and monitoring are required to assess long-term impacts and develop appropriate mitigation measures to ensure the sustainable utilization of vented dams while minimizing adverse effects on LCC and ecosystems.

With the advancement of technology, satellite image processing for RS applications has become very prominent [27]. There are always new algorithms and applications developed so that various limitations pertaining to each of the image processing techniques can be addressed. The common challenges identified from the literature are the difficulty in image acquisition, the large image sizes, computational complexity, and the effect of noise on satellite images. The main issues with CD are the low accuracy and Kappa coefficient values that are sometimes acquired as a result of categorization errors. Though certain CD algorithms offer good accuracy and Kappa coefficients for certain image types, this is only the case for some image types. Hence, no generalized CD technique can be fully applied to different types of satellite images. In recent times, deep neural networks have gained a lot of popularity in various image processing applications. Applying deep neural networks to CD can improve CD accuracy. Also, combining different CD techniques to form a hybrid detection framework can utilize the advantages of individual CD methods and hence improve the accuracy of

CD. The LCC may not have a considerable environmental impact on the study area. The two datasets obtained had OA evaluation results of 96.30% and 95.47%, respectively. The attribute table was used to determine how much space various classes took up in both years. The postclassification comparison technique was used after image classification, comparing an LC map from one data set (2019) with an LC map from another data set (2022). The comparison presents quantifiable information on the differences between the 2 study years. Using GIS platforms, maps were created following picture classification.

18.8 Limitations

The limitations of LULC impact assessment using GEE include limited spatial and temporal resolution, potential classification errors, incomplete representation of LC dynamics, lack of ground truth data, challenges in capturing certain LC types, dependence on cloud cover and image availability, and limitations in data-intensive analyses. It is essential to acknowledge and address these limitations when conducting LULC impact assessments using GEE to ensure the accuracy and reliability of the results.

18.9 Recommendations

The work could be extended by using various machines and deep learning algorithms like CNN, R-CNN and other deep learning methods that could enhance the performance of the algorithm.

Acknowledgment

The author would sincerely like to thank one and all who have directly or indirectly supported with their guidance.

References

Amini, S., Saber, M., Rabiei-Dastjerdi, H., & Homayouni, S. (2022). Urban land use and land cover change analysis using random forest classification of landsat time series. *Remote Sensing*, *14*(11). Available from https://doi.org/10.3390/rs14112654.

Aniah, P., Bawakyillenuo, S., Codjoe, S. N. A., & Dzanku, F. M. (2023). Land use and land cover change detection and prediction based on CA-Markov chain in the savannah ecological zone of Ghana. *Environmental Challenges*, *10*. Available from https://doi.org/10.1016/j.envc.2022.100664, https://www.journals.elsevier.com/environmental-challenges.

Aragaw, H. M., Goel, M. K., & Mishra, S. K. (2021). Hydrological responses to human-induced land use/land cover changes in the Gidabo River basin, Ethiopia. *Hydrological Sciences Journal*, *66*(4), 640–655. Available from https://doi.org/10.1080/02626667.2021.1890328.

Asokan, A., & Anitha, J. (2019). Change detection techniques for remote sensing applications: A survey. *Earth Science Informatics*, *12*(2), 143–160. Available from https://doi.org/10.1007/s12145-019-00380-5.

Brown, C. F., Brumby, S. P., Guzder-Williams, B., Birch, T., Hyde, S. B., Mazzariello, J., ... Tait, A. M. (2022). Dynamic world, near real-time global 10 m land use land cover mapping. *Scientific Data, 9*(1). Available from https://doi.org/10.1038/s41597-022-01307-4, http://www.nature.com/sdata/.

Campbell, A., & Wang, Y. (2019). High spatial resolution remote sensing for salt marsh mapping and change analysis at Fire Island National Seashore. *Remote Sensing, 11*(9). Available from https://doi.org/10.3390/rs11091107.

Cavur, M., Duzgun, H. S., Kemec, S., & Demirkan, D. C. (2019). Land use and land cover classification of Sentinel 2-A: St Petersburg case study. *International Archives of the Photogrammetry, Remote Sensing and Spatial Information Sciences - ISPRS Archives, 42*(1), 13–16. Available from https://doi.org/10.5194/isprs-archives-XLII-1-W2-13-2019, http://www.isprs.org/proceedings/XXXVIII/4-W15/.

Chowdhury, A., & Dwarakish, G. S. (2022). Selection of algorithm for land use land cover classification and change detection. *International Journal of Advanced Research in Science, Communication and Technology*, 15–24. Available from https://doi.org/10.48175/ijarsct-2610.

Erener, A., Düzgün, S., & Yalciner, A. C. (2012). Evaluating land use/cover change with temporal satellite data and information systems. *Procedia Technology, 1*, 385–389. Available from https://doi.org/10.1016/j.protcy.2012.02.079.

Fonji, S. F., & Taff, G. N. (2014). Using satellite data to monitor land-use land-cover change in North-eastern Latvia. *SpringerPlus, 3*(1), 1–15. Available from https://doi.org/10.1186/2193-1801-3-61, http://www.springerplus.com/archive.

Gashu, M. W. (2022). Integration of remote sensing and Google Earth Engine for land use/ land cover change analysis in small agricultural watershed (a case of Brante watershed, Ethiopia). *Research Square, Ethiopia Research Square*. Available from https://doi.org/10.21203/rs.3.rs-2014415/v1, https://www.researchsquare.com/browse.

Geetha M., Gowda Asha Nandeesha K., Nagaraj R.V. B. 2021 Change detection of LULC using machine learning. Proceedings of the 3rd International Conference on Integrated Intelligent Computing Communication & Security (ICIIC 2021), 4(Iciic 339–347. 10.2991/ahis.k.210913.042.

Gohain, K. J., Mohammad, P., & Goswami, A. (2021). Assessing the impact of land use land cover changes on land surface temperature over Pune city, India. *Quaternary International, 575*, 259–269.

Jiang, X., Lu, D., Moran, E., Calvi, M. F., Dutra, L. V., & Li, G. (2018). Examining impacts of the Belo Monte hydroelectric dam construction on land-cover changes using multitemporal Landsat imagery. *Applied Geography, 97*, 35–47. Available from https://doi.org/10.1016/j.apgeog.2018.05.019, http://www.elsevier.com/inca/publications/store/3/0/3/9/0/index.htt.

Karlsson, I. B., Sonnenborg, T. O., Refsgaard, J. C., Trolle, D., Børgesen, C. D., Olesen, J. E., ... Jensen, K. H. (2016). Combined effects of climate models, hydrological model structures and land use scenarios on hydrological impacts of climate change. *Journal of Hydrology, 535*, 301–317. Available from https://doi.org/10.1016/j.jhydrol.2016.01.069, http://www.elsevier.com/inca/publications/store/5/0/3/3/4/3.

Koppad, A. G., & Janagoudar, B. S. (2013). Vegetation analysis and land use land cover classification of forest in Uttara Kannada district India using remote sensign and GIS techniques. *International Archives of the Photogrammetry, Remote Sensing and Spatial Information Sciences - ISPRS Archives, 42*(4), 121–125. Available from https://doi.org/10.5194/isprs-archives-XLII-4-W5-121-2017, http://www.isprs.org/proceedings/XXXVIII/4-W15/.

Loukika, K. N., Keesara, V. R., & Sridhar, V. (2021). Analysis of land use and land cover using machine learning algorithms on google earth engine for Munneru River Basin, India. *Sustainability, 13*(24), 13758.

Mishra, P. K., Rai, A., & Rai, S. C. (2020). Land use and land cover change detection using geospatial techniques in the Sikkim Himalaya, India. *Egyptian Journal of Remote Sensing and Space Science*, *23*(2), 133–143. Available from https://doi.org/10.1016/j.ejrs.2019.02.001, http://www.elsevier.com/wps/find/journaldescription.cws_home/723780/description#description.

Mohammadpour, P., Viegas, D. X., & Viegas, C. (2022). Vegetation mapping with random forest using Sentinel 2 and GLCM texture feature—A case study for Lousã Region, Portugal. *Remote Sensing*, *14*(18). Available from https://doi.org/10.3390/rs14184585, http://www.mdpi.com/journal/remotesensing.

Priya, K., Sasanka, T., & Osuri, K. K. (2023). Land use land cover representation through supervised machine learning methods: Sensitivity on simulation of urban thunderstorms in the east coast of India. *Natural Hazards*, *116*(1), 295–317. Available from https://doi.org/10.1007/s11069-022-05674-4, https://www.springer.com/journal/11069.

Raj, N., & Gurugnanam, B. (2018). Land use / land cover change detection mapping using satellite data. *Along The Thiruvallur Coastal Stretch*, *4*.

Ritse, V., Basumatary, H., Kulnu, A. S., Dutta, G., Phukan, M. M., & Hazarika, N. (2020). Monitoring land use land cover changes in the Eastern Himalayan landscape of Nagaland, Northeast India. *Environmental Monitoring and Assessment*, *192*, 1–17.

Sazib, N., Mladenova, I., & Bolten, J. (2018). Leveraging the Google Earth Engine for drought assessment using global soil moisture data. *Remote Sensing*, *10*(8). Available from https://doi.org/10.3390/rs10081265.

Sharannya T.M., Mudbhatkal, Mahesha A. 2017 Xth IGU International Thematic Conference Hydrologic impact of climate change on Gurupura river catchment.

Shetty, S. T., Kushwaha, A., & Shetty, A. (2023). Historical water body changes using spatio temporal indices - A case study of Dakshina Kannada. *Materials Today: Proceedings*, *80*, 1488–1494. Available from https://doi.org/10.1016/j.matpr.2023.01.283, https://www.sciencedirect.com/journal/materials-today-proceedings.

Shelestov, A., Lavreniuk, M., Kussul, N., Novikov, A., & Skakun, S. (2017). Exploring Google Earth Engine platform for big data processing: Classification of multi-temporal satellite imagery for crop mapping. *Frontiers in Earth Science*, *5*, 232994.

Shetty, A., Umesh, P., & Shetty, A. (2022). An exploratory analysis of urbanization effects on climatic variables: A study using Google Earth Engine. *Modeling Earth Systems and Environment*, *8*(1), 1363–1378. Available from https://doi.org/10.1007/s40808-021-01157-w.

Sinha, R. K., & Eldho, T. I. (2018). Effects of historical and projected land use/cover change on runoff and sediment yield in the Netravati river basin, Western Ghats, India. *Environmental Earth Sciences*, *77*(3). Available from https://doi.org/10.1007/s12665-018-7317-6, http://www.springerlink.com/content/121380/.

Chapter 19

Climate change vulnerability of water resource management in mature and active deltaic parts of inhabited islands of Indian Sundarbans: a geospatial analysis with special reference to Matla-Bidyadhari interfluve

Jayanta Gour
Department of Geography, Sambhu Nath College, Labpur, West Bengal, India

19.1 Introduction

Based on hydrological parameters, the Bengal Delta is divided into three (Mukherjee, 1976), viz., (1) the Moribund Delta, where land building practically ceased and drainage channels became functionally inoperative. Example: Murshidabad and Nadia districts; (2) the Mature Delta, where rivers still have well-defined channels and provide some impetus for irrigation and induce drainage disposal under normal conditions. Example: Mostly the northern part of the 24 Pgs. (S), Calcutta, roughly to the north of a line joining Kakdwip with Canning; and (3) the Active Delta, where the dynamic functions of the hydrological system are more manifested. The channels and creeks are wide, deep, and well-defined; the land building is still in its current phase, in which upland flow changes regime with marine incursions. Example: Southern Basanti, Gosaba, and so on. Matla-Bidyadhari Interfluve falls under the mature and active deltaic parts, as this interfluve provides some impetus for irrigation and induces drainage disposal, as well as the dynamic functions of delta building processes in the hydrological systems are manifested here.

Water Resources Monitoring, Management, and Sustainability. DOI: https://doi.org/10.1016/B978-0-443-23665-5.00019-3

Tides and ebbs ply along the most violent river, that is, R. Matla and R. Bidyadhari, twice regularly, which, due to the difference in water density, regularly drain in the saline water of the Bay of Bengal (BOB) inward through numerous channels and creeks and drain out the terrestrial sweet water through the same route from beneath due to the higher density. The average tide level is above 3.5 m, which exceeds during spring tides and lowers during neap tides. However, it becomes unpredictable according to the gust and wind speed of the cyclonic storms (like Aila, Amphan, and Yaas), which often originate just a few kilometers south of BOB. Recent changes in tidal ranges in the Indian Sundarbans due to the changing character of the ENSO cycle due to climate change and anthropogenic factors have been found responsible for the recent increase in vulnerability to bank erosion, which have been discussed in the following chapters below.

19.1.1 Objectives of the research work

The entire research work has been designed following the objectives below:

1. To find out the recent trends in tidal behavior over the BOB since the last decades.
2. To search for the importance and hydro-geomorphological influence of tides in configuring the Sundarban Delta.
3. To make some propositions based on the results obtained.

19.1.2 Materials and methods

For the convenience of discussion, an entire spectrum of methodology has been categorized into (1) prefield work, (2) fieldwork, and (3) postfield work.

1. **Prefield Work:** For the appraisal, the present scholar went exhaustively through library materials, including previous publications, historical records, and maps, in the prefield stage. For the measurement and analysis of fluvial landform features, the necessary SOI Topographical Sheets (Survey of India, 2020) bearing No. 79 B/10, 79 B/11, 79 B/12, 79 B/15, 79 B/16, 79 C/9, and 79 C/13 of R. F. 1: 50,000 and the Toposheet of Khula (1924) of R. F. 1: 2, 50,000 have been consulted in the research work. Administrative delineation of the study area has been done with special reference to census records and the government portal of West Bengal. Selection of objective-oriented fluvio-geomorphological environmental properties and land use themes to be stressed during field and postfield works. Determination of hydro-morphological and land use parameters for two-way impact assessment between the fluvio-geomorphological environment and land use in the research area.

Consultation of mathematical and statistical techniques to be applied. Framing of rules to conduct both fieldwork and postfield work analysis.

2. **Fieldwork:** A door-to-door survey has been done to collect the primary data for the study area. The questionnaire for the perception study has been done accordingly. Detailed field surveys over the entire area have been done to collect the present status of the research area. Careful fieldwork has been done incorporating the measurement of the height and depth of the riverbeds, embankments, locational attributes of the settlements, spatiotemporal changes in the courses of tidal rivers, channels, creeks, areal extension or shrinkage of forested and wetland's area, increase or decrease in the back swamps, marshy and swampy areas, identifying wastelands, charas, and bheris, along with the present land use pattern.
3. **Postfield Work:** Data obtained from the Port Trust of Kolkata, Govt. of India, and tidal range data obtained from the Survey of India, Govt. of India, have been correlated. To show the valley form and river patterns, large numbers of cross-sections have been done from the top to the mouth of this region. Apart from the field investigations, the land use/land cover data has been consulted from the respective offices and the published sources. Cadastral maps, Census maps, Revenue Survey maps of District 24 Parganas, Govt. of Bengal (1925–26), and NRSC satellite imageries of IRS P6 LISS-III and DEM have been used to show the vulnerable water management about the changing fluvio-geomorphological environment of the Matla-Bidyadhari interfluve. Different statistical methodologies, like mean, standard deviation, frequency distribution, and drainage frequency, have been applied. In the cartographic methodologies, different shapefiles from Sentinel Hub (2015), diagrams, and maps have been used and prepared with the help of bar diagrams, line graphs, pie diagrams, proportionate pie diagrams, choropleth maps, and so on, with the aid of Microsoft Excel, Surfer, Arc GIS, and QGIS.

19.2 Physical background

From geological perspectives, marine transgression on the Bengal Shelf occurred locally during the Late Cretaceous and widely during the Eocene. During the Miocene, only the eastern parts of the Bengal Shelf were under marine influence. Slow subsidence of the shelf area took place during the Late Cretaceous when brackish to lagoonal deposits accumulated on the Bengal shelf. In the Pliocene, with the regression of the sea, estuarine and fluvial conditions of deposition prevailed in most parts of Bengal. In the early Pleistocene, shallow marine conditions dominated only the deeper portions of the Bengal basin. Probably the sea finally receded completely from the Bengal basin area only during the very late Pleistocene. Then only erosion occurred, followed by the peneplanation of the entire Tertiary basin area of Bengal.

Finally, the older sediments were completely covered by a thick mantle of river-borne Holocene alluvium (Biswas & Agarwal, 1992). As per the seismic zone, Matla-Bidyadhari Interfluve is falling under Seismic Zone IV.

Geospatially, the study area extends between 21°57'N to 22°45'N latitudes and 88°35'E to 88°53'E longitudes. Lying along the northern triangular apex of the BOB, the Sundarbans are formed by the sediment deposits carried down and settled by the R. Ganga (part of India), R. Brahmaputra, and R. Meghna (part of Bangladesh), which is the only largest mangrove all over the world. Although there are so many controversies over the origin and formation of the Bengal Delta, the majority of geographers opine that the Sundarbans are mainly a prograding mangrove region of the Bengal Delta (India and Bangladesh), which lies to the south of northern Gangetic-Padma floodplains extending from Kanthi Coastal Plain, Damodar subbasin, and Hooghly estuary to its west and beach, sand dunes, and alluvium deposits of Chittagong of Bangladesh to its east (Paul, 2002). The geographic territory of the Sundarbans, as delineated by Dampier and Hodges (1829–30), spreads over the 24 Pgs. (N and S) districts of West Bengal (Fig. 19.1) and is located south of Kolkata, Nadia districts of West Bengal, India, and to the west of Jessore and Bakharganj districts of Khulna Subdivision of Bangladesh. It is surrounded by the Hooghly River, which separates the Sundarbans from the Hooghly, Howrah, and Purba Medinipur districts of West Bengal, India. According to the Bengal District Gazetteers of 24-Parganas by LSS O'Malley, ICS, the Sundarbans cover an area of 7617.155 km^2 out of a total area of 12,545.9 km^2 of undivided 24 Parganas (O'Malley, 1914), that is, almost 60.17% of the entire undivided 24 Parganas. However, as per the Centre for Science and Environment, the Sundarbans cover an area of 9630 km^2 in India and an area of 16,370 km^2 in Bangladesh. The study area covers an area of 728.528 km^2 and consists of 39 villages in Canning-II Block, 67 villages in Basanti Block, and 18 villages in Gosaba Block, that is, a total of 124 villages in South 24 Parganas District of West Bengal. The Herobhanga Reserved Forest covering Lot No. 124, Lot No. 125, and the southern part of Lot No. 126 is a separate island that demarcates the southern limit of the Matla-Bidyadhari interface.

The Matla-Bidyadhari Interfluve is the most unique as it bears both mature and active deltaic hydrological characters in the Indian Sundarbans. Due to its typical location in the middle of the Indian Sundarbans, it resembles all types of geomorphic features such as Oxbow lakes, meandering channels, point bars, mid-channel bars, creeks, interlinking tidal channels, and many more deltaic features. This interfluve is a part of the eastern parts of Canning-II Block to its north and north-west, the entire Basanti Block covering its entire middle, north-eastern, and southern part (up to Herobhanga Reserve Forest), and the western parts of Gosaba Block to its eastern side within the R. Matla bordering its west and the R. Bidyadhari bordering its east and south. The R. Matla flows for about 62 km from its source and thereafter joins the R. Bidyadhari at the

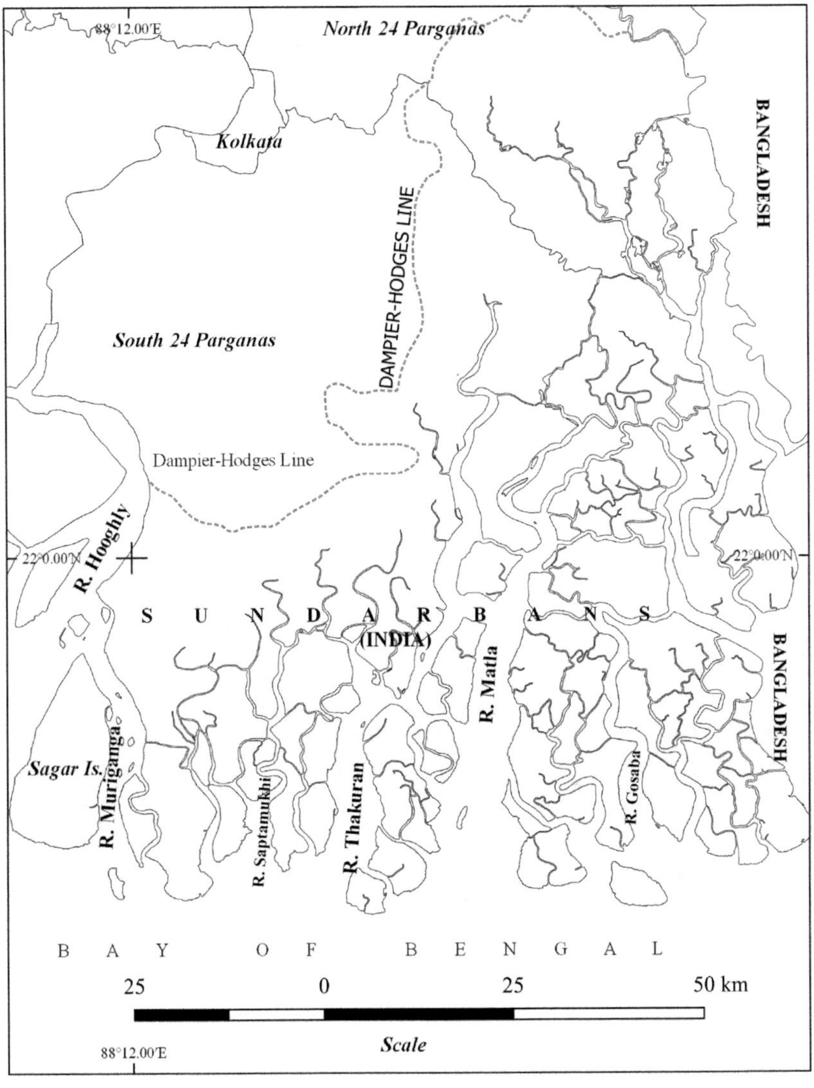

FIGURE 19.1 Location map of Indian Sundarbans within Dampier-Hodge's line. Map lines delineate study areas and do not necessarily depict accepted national boundaries.

southern margin of the Basanti mainland (south-east tip of Herobhanga Reserve Forest), ultimately flowing straight to the BOB from the north-east of Herobhanga Is. to mouth at the BOB for 45 km. Before joining the R. Bidyadhari, the R. Matla takes a sharp meander near Harbhangi village of Basanti Block (Fig. 19.2), east of Kultali Block, and slowly widens from Canning town (minimum with a channel width of only 35 m) and becomes the largest mouth (more than 21 km) from west to east, that is, from Mir's Forest

FIGURE 19.2 Location of Matla-Bidyadhari interfluves with villages in South 24 Pargana district of West Bengal, India. Map lines delineate study areas and do not necessarily depict accepted national boundaries.

of Bulcheri near Kalas Is. to Chaimari Is. near Bhangaduni Is. in Indian Sundarbans. At present, R. Matla is linked with R. Bidyadhari by the interlinking tidal channels like- Hogol Nadi-Karatal Gang across the middle of Basanti

C.D. Block, Bara Herobhanga Khal between Lot No. 126 and Lot No. 125 and by the Cattle Gang between Herobhanga Is. and Ajmalmari Is. to the south of this interfluve (Gour, 2018).

This part of the Indian Sundarbans has an elevation of 0–14 m (Fig. 19.3) above MSL, and almost 80% of the total interfluve has an average elevation of 3–5 m sloping toward east and south-east (except the southern part, which has

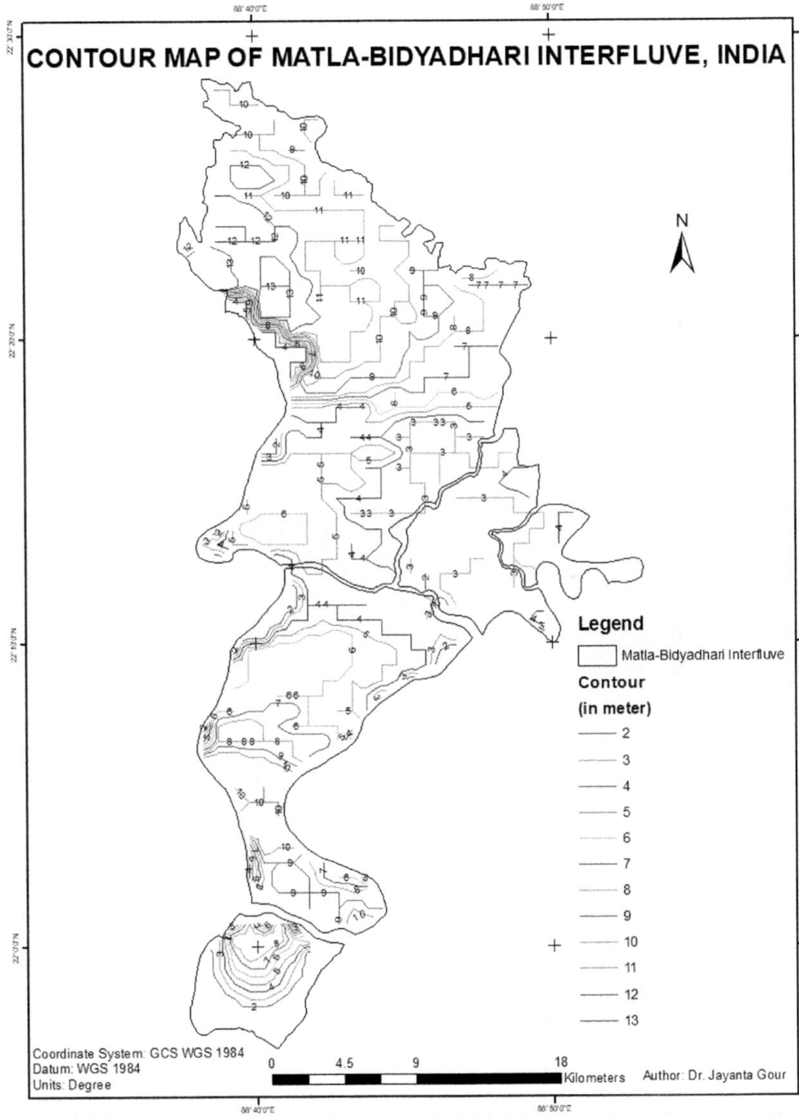

FIGURE 19.3 Contour map of MBI (based on DEM_USGS). Map lines delineate study areas and do not necessarily depict accepted national boundaries.

an average elevation of 0–2 m above MSL). Most of the embankments here have been elevated in the last few decades due to the rapid siltation of the interlinking tidal channels, which often lie at a higher elevation than surrounding settled areas. The relief and slope map (Fig. 19.4) shows that this interfluvial region is almost a flatland in its north (Fig. 19.5), but the erection of

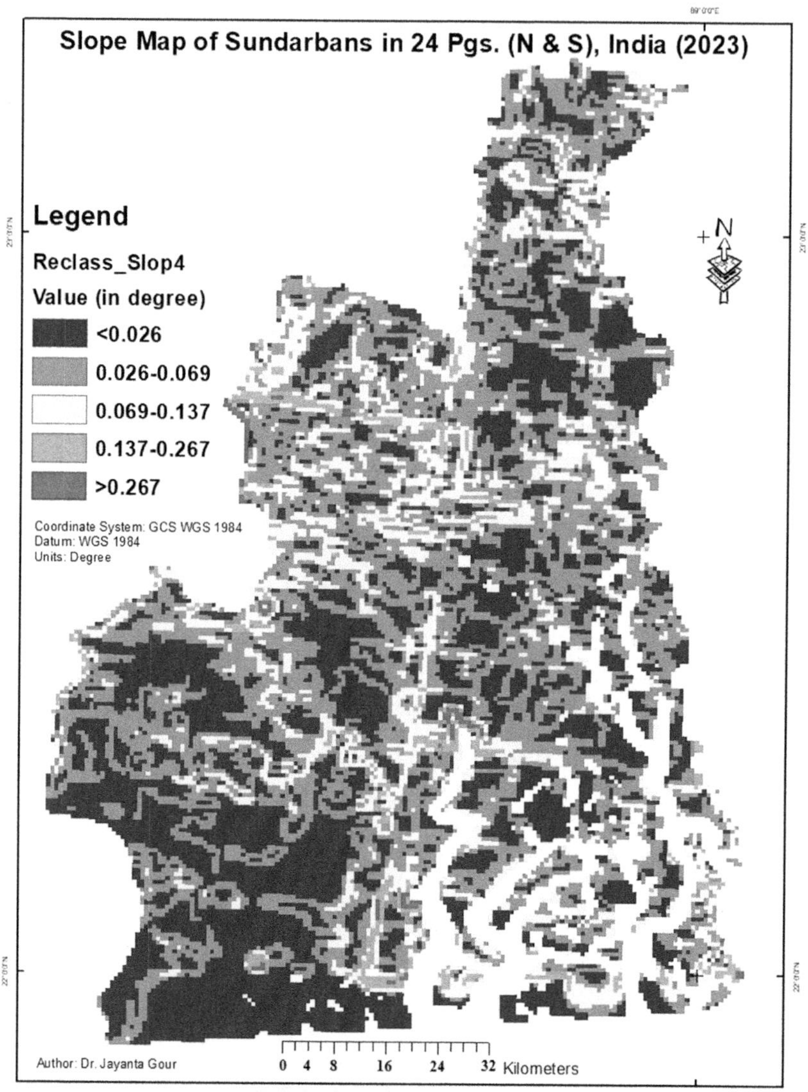

FIGURE 19.4 Slope map of Sundarbans in North and South 24 Parganas in West Bengal, India. Map lines delineate study areas and do not necessarily depict accepted national boundaries.

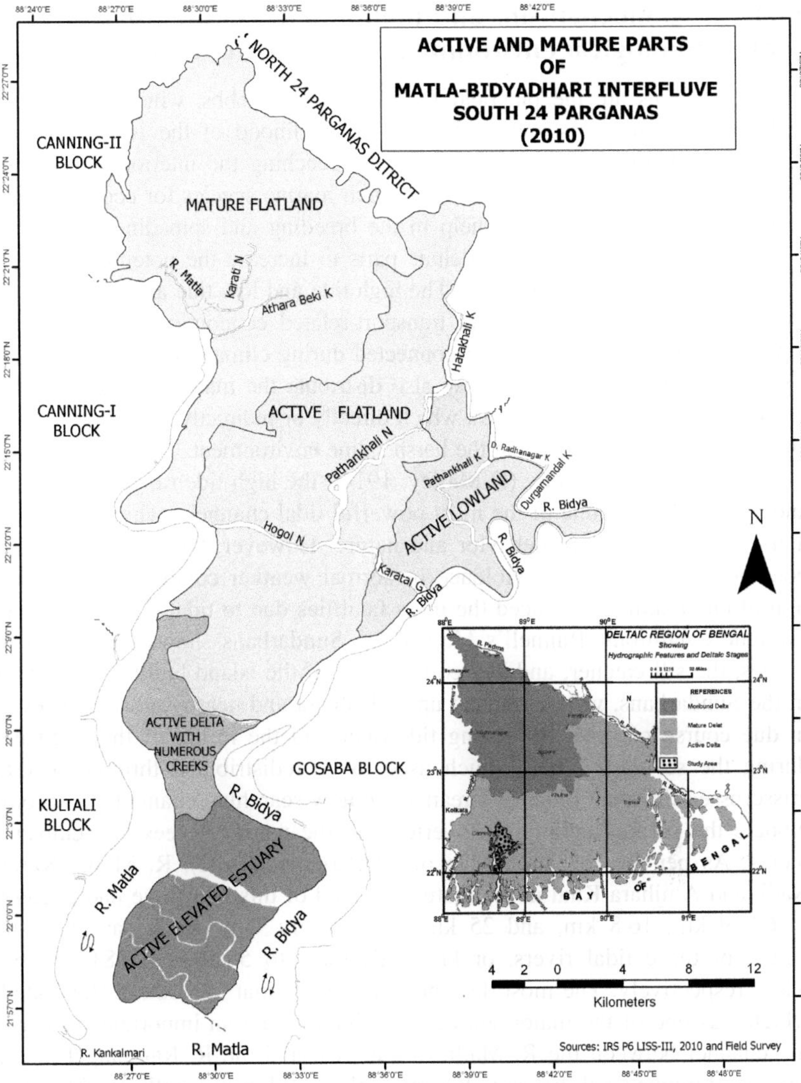

FIGURE 19.5 Major geomorphic zones of MBI in Indian Sundarbans. Map lines delineate study areas and do not necessarily depict accepted national boundaries.

embankments has disturbed the natural slope of the northern flatlands (Fig. 19.4). Due to blockage in the source of R. Matla, a vast marshland has been formed, which has been converted into a fragmented *bheri* (local saline field or pool for prawn and shrimp cultivation in the intertidal zone).

19.3 Tidal mechanisms and their impact on the fluvial-geomorphological environment in Indian Sundarbans

The Sundarbans are the playfield for the tides and ebbs, which occur twice daily and are the parts and parcels of the livelihood of the island dwellers (Gour, 2022). However, the incoming tides reaching the interior parts of the mature and moribund deltaic parts bring rich marine species for economic sustainability, and outgoing ebbs help in the breeding and spreading of the rich flora and fauna over the entire deltaic parts to increase the potentiality of the most unique and rich biodiversity. The high tide and low tide also play a major role in agriculture, industry, and transport-related economic activities in the remote islands, which become disconnected during climatic hazards and disasters. The tidal channels and creeks also distribute the mangrove seeds for the natural growth of the mangroves, which directly or indirectly feed thousands of endangered and rare species in the harsh saline environment.

According to O'Malley (O'Malley, 1914), the high tide ranged to 7.11 m, and the R. Matla is one of the most powerful tidal channels, which coined its name due to its violent behavior and nature. However, the recent tidal data does not show that much violence in normal weather conditions. The erection of embankments reduced the mass fatalities due to tidal floods in severe cyclonic conditions. Rannell's Map of the Sundarbans shows that the mangroves, the silt catcher, and key components of the island building processes in the Sundarbans, were stretched up to Kolkata and nearby areas. However, in due course of time, the spring tidewater and the influx of the tidewater during the severe cyclone, which used to be redistributed through several crisscrossed different order of streams, are now searching channels to accommodate the surplus volume by overtopping the natural levees and embankments. As per the Irrigation Map of 24 Parganas (1932), R. Matla, Karati Nadi, and Amjhara Khal in the mature flatland of this interfluve had a length of 62.09 km, 16.8 km, and 25 km, respectively. This study says that the length of these tidal rivers, or khals, declined to 59.40 km, 9.81 km, and 6 km, respectively. The most disappointing one is that the Athara Beki Khal, which was one of the major sources of irrigation and an important interlinking channel between the R. Matla to the west and the R. Raimangal to the east *via* Rampur Khal, has now become only 6.00 km in length. Besides this, in 1920, the Karati Nadi was linked with the R. Raimangal *via* the R. Bidyadhari in its upper reaches, and once the Payna Bil used to get the amount of water from this interlinking tidal channel. However, due to the indiscriminate, overharnessing, and misuse of the river water, the riverbed of this interlinking tidal channel will soon be its identity. The Kuriabhanga Nadi, which helped in both agricultural activities in the eastern part of MBI, is thriving due to poor tidal discharge. The recent study observed that some rapidly silting up tidal channel beds can be seen in Bidya Khal, Shiyalpheli

Khal, in the active lowlands (Fig. 19.5) of this interfluve, and Hatakhali, Durgamandal Khal, in both the mature and active parts of this interfluve.

19.4 Vulnerability of water resource management in Indian Sundarbans

Every type of delta has its own natural drainage system all over the world. Whenever mankind intervened in natural drainage circuits for their short-term gain, natural drainage has, on the other hand, altered its direction of flow, shifted through bank erosion and deposition, and adjusted the volume of water either by inundating the floodplains or widening its courses (Paul, 2002). The Sundarbans is not exceptional in this regard. The Sundarbans' natural tidal drainage network was repeatedly interfered with and obstructed by sluices, which confined their natural shifting by erecting embankments on both sides of tidal channel banks. The aftermath of the water resource management of the British officials can now easily be understood, as the present Sundarbans have already experienced the havoc of the destruction by tidal floods in every coastal block of South 24 Pargana in West Bengal, India. Like the Mousuni, Lohachara, Ghoramara, and Purbasha Islands, the interfluves like the Matla-Bidyadhari Interfluve are also becoming more and more vulnerable. The vulnerability is not only limited to the coastal and riverside villages, but the impact of intangible tidal flow has now been felt in the mature northern parts of the Indian Sundarbans.

19.5 Age-old water resource management in Sundarbans and issues related to recurring failure in water resource management during climatic crisis

Rivers always have a shifting tendency to redistribute their sediment load and increase the adjustability and capacity of holding seasonal variation of discharge in deltaic parts. The tidal channels in the Sundarbans always had an eastward shifting tendency (Fig. 19.6) toward the BOB, as per the cartographic records of J. Rennell. During the natural shift of river courses, bank erosion and deposition of silt and clay are a natural phenomena that were perhaps overlooked by the then rulers in the Sundarbans. Since the British colonial period in undivided Bengal, the Governor General and deputed officers have attempted to reclaim the marshes and wetlands of the Sundarbans (Mukherjee, 1976). Thousands of hectares of mangroves were cleared (Das, 2006; Milne, 2009) in different phases to create new *Abad*, or agricultural lands; *Anabadi*, or uncultivated lands, were made for settlement purposes (Hunter, 1875). The first phase of land reclamation and interference in the natural tidal network was disturbed from 1770 to 1780. In this phase, the tidal channels, which have their sources near Barasat, Bhangar, parts of Jaynagar, Sonarpur, Baruipur, Kakdwip, and Kulpi, were very much affected

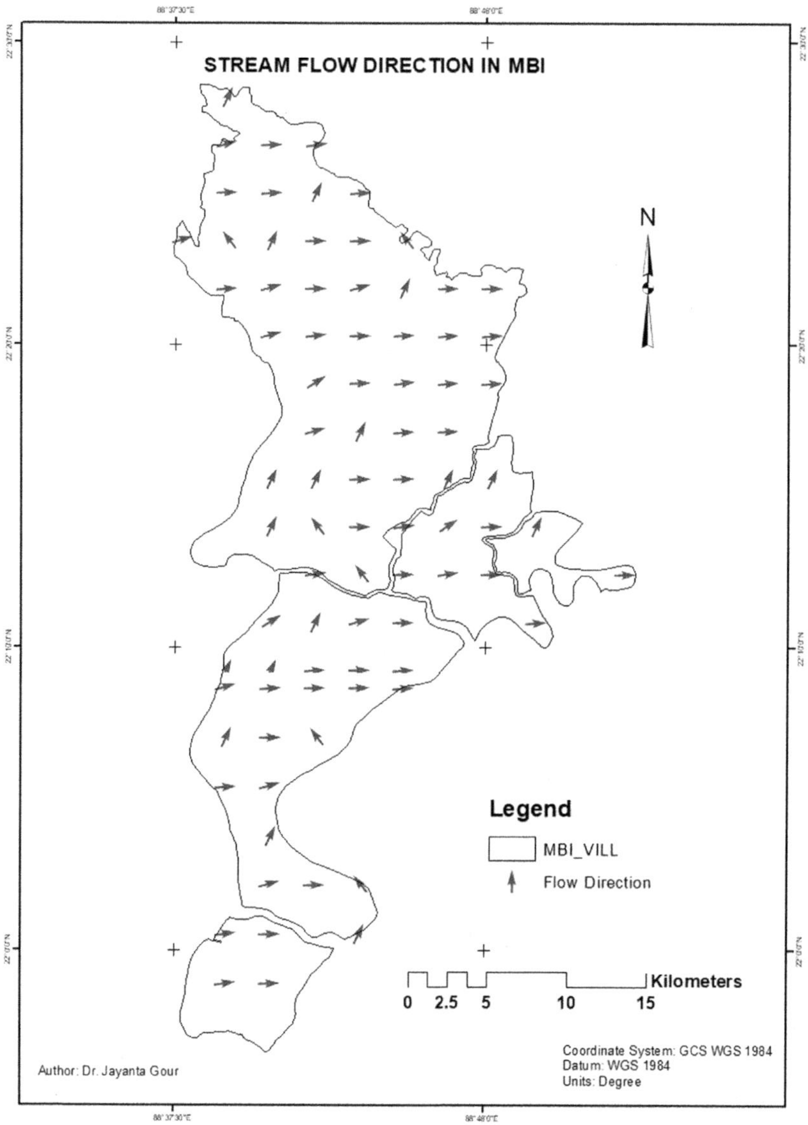

FIGURE 19.6 The stream flow direction of tidal channels and creeks in Matla-Bidyadhari interfluve. Map lines delineate study areas and do not necessarily depict accepted national boundaries.

by anthropogenic interferences (Sarkar, 2007). The mangroves of these areas, which used to collect the silt and accommodate the influx of tidal water during heavy rains in cyclonic storms and spring tides, became weak and sluggish. Later, in the second phase, the rest of Jaynagar, Mathurapur, and Sagar Is. were reclaimed. Canning, Baruipur, Kultali, and places toward the BOB, like Canning, parts of Basanti, and parts of Gosaba, were reclaimed during the third (1873–1939) and fourth (1945–51) phases, respectively (Bose, 2006). The reclamation and building of sluices and the lengthening of embankments went on even after the independence of India. Here, it is noteworthy to mention that "Every landform is unique and can only be fully understood after careful study, taking into account the general principles of the processes which have formed it, its geological structure, and the stage reached in its evolution" (Sawyer, 1978). Thus after the different stages of land reclamation, the British Government realized that planning to encroach toward the BOB was a blunder and the water resource management in reclaimed parts was becoming a failure. The stream order was also changed, which led to malfunctions in tidewater distribution in the entire Matla-Bidyadhari Interfluve (Fig. 19.7). Besides this, there was miscommunication among the local zamindars and British Government officials who were allotted to collect the revenue from the reclaimed areas of the Sundarbans and manage the irrigation as well as mitigate the flood issues in low-lying areas. The funding for embankment building and setting up of sluices was not properly and timely used, and different zamindars tried to manage their territory unscientifically and without proper management and policy. Due to the lack of a proper way of solving the saline water intrusion issues in newly set-up agricultural lands in the islands of the Sundarbans, one channel's abandonment or embankment became a matter of woe and agony to other channel systems. The drainage systems, which are supposed to follow the slope of the interfluve, have also become very intricate and complex (Fig. 19.6). Thus waterlogging problems are becoming a serious issue during tidal flood occurrences induced by severe cyclonic storms originating near the BOB (Fig. 19.6). This led to a divert or natural shifting of the tidal channels like Matla and Bidyadhari to a greater magnitude. In recent times, the drainage density of Matla-Bidyadhari has been very low, which seems to make it vulnerable during any future severe cyclonic conditions (Fig. 19.8).

In the Canning-II block, there used to be lots of tidal channels and interlinking tidal khals, which used to carry the high tidewater of R. Matla and helped to reduce the waterlogging conditions. Khals like Tambuldaha Khal, Khagra Khal, Payna Khal, Hatiamari Khal, Katharabeki Khal, Bhuban Khal, and Athara Beki Khal now carry the least amount of tidal water and become another tidewater spilling channels in the mature parts of the Matla-Bidyadhari Interfluve. The least maintenance and rapid silting up of their channel bed are unable to distribute the flood water. The Basanti block's interlinking and short tidal streams, such as Amjhara Khal, Athara Beki Khal, Kalu Khal, Baria Khal, Chunakhali Khal, Lebu Khali Khal, Piprakhali Khal, Kala Hazra Khal, Ramchandrakhali Khal, Roa Khal, Hatakhali Khal, Mokamberia Khal, Gadkhali Khal, Dundunkhali Khali Khal, Goyer Khal, Shibganj Khal, Malsoyet Khal, Kultali Khal, Bagar Khal, Shialpheli Khal, Narayantali Khal, Dhola Khal, Bidya Khal, Basanti Khal, Kalir Khal, Lotamani Khal, Jora Khal, Haldhoa Khal, Charkhali Khal, Bali Khal,

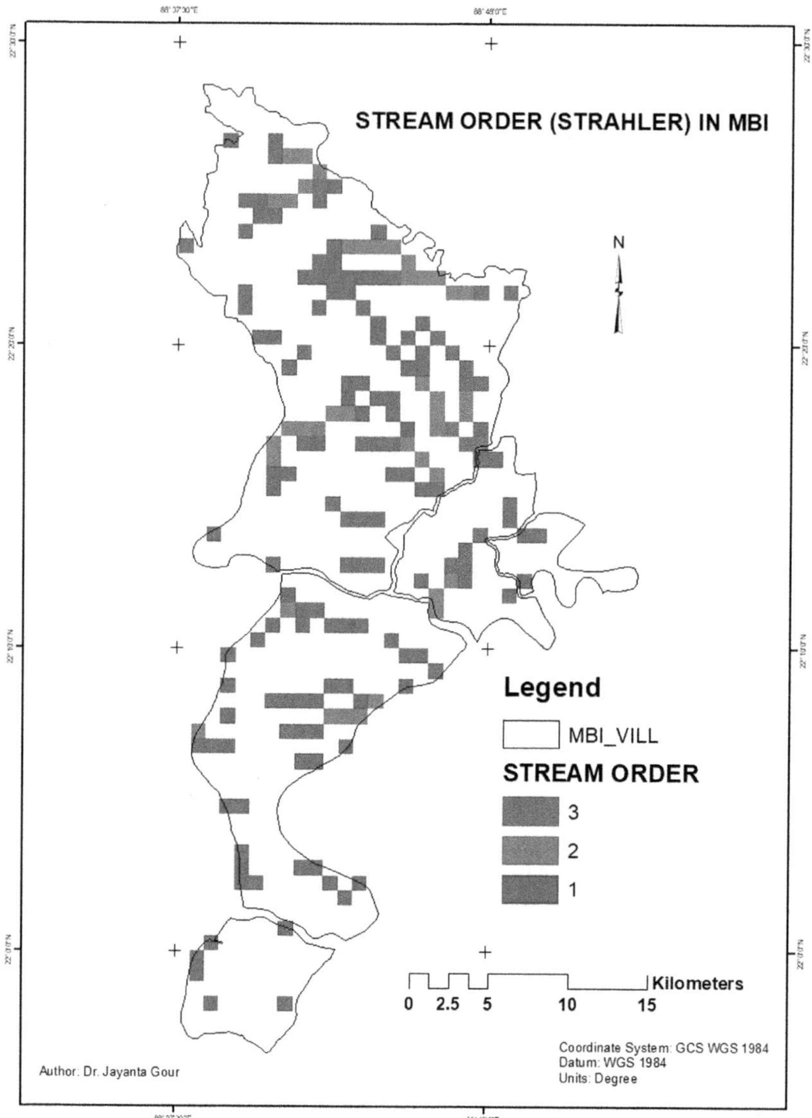

FIGURE 19.7 Stream order of tidal streams and creeks in Matla-Bidyadhari interfluve. Map lines delineate study areas and do not necessarily depict accepted national boundaries.

Benamall Khal, Bara Herobhanga Khal, Chhota Herobhanga Khal, Chhota Matla Khal, and Chhota Bidya Khal, have also been dysfunctioning due to anthropogenic interference. In Gosaba block, the channel beds of tidal streams and creeks, such as Raktabir Khal, Katyani Khal, Moukhali Khal, Jhaukhali Khal, Pathankhali Khal, Dakshin Radhanagar Khal, and Durgamandal Khal, are rising at a rapid rate, keeping the bank-side

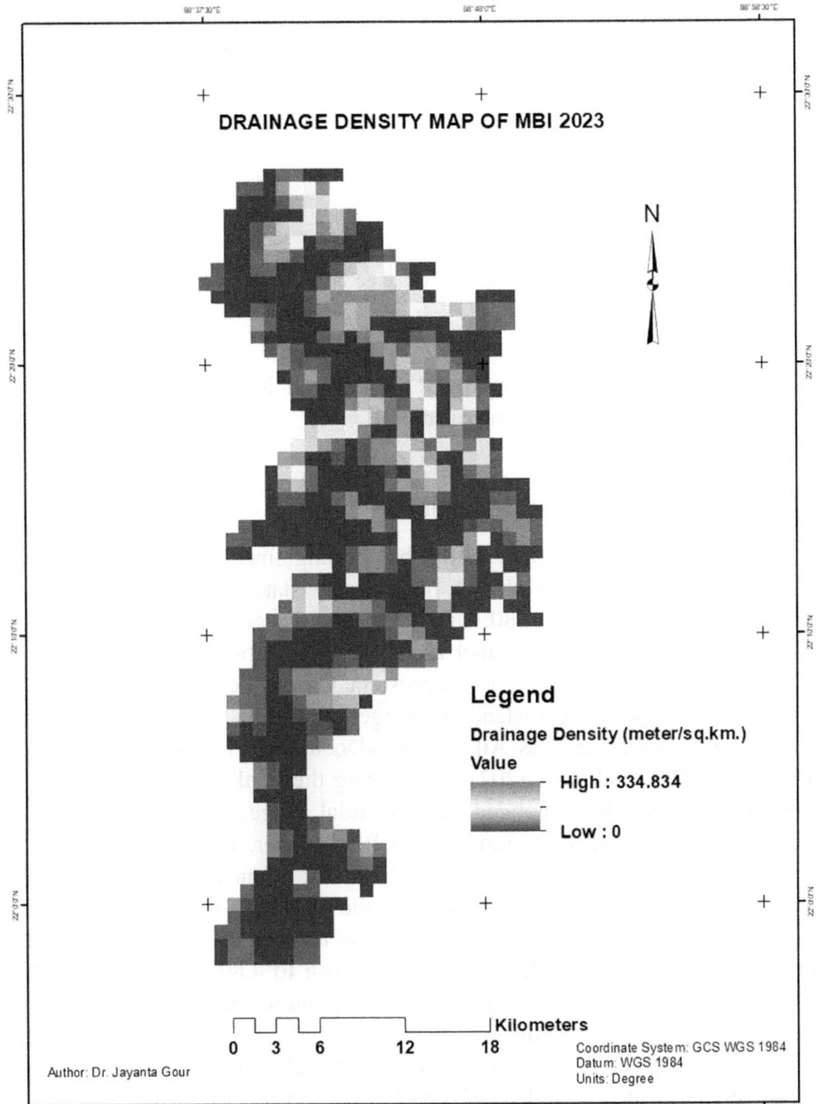

FIGURE 19.8 Drainage density map of Matla-Bidyadhari interfluve in Indian Sundarbans. Map lines delineate study areas and do not necessarily depict accepted national boundaries.

settlements below their levels. A recent study finds that these partially and almost abandoned interlinking and intertidal channels will not only be able to return the tidal water to BOB *via* R. Matla and R. Bidyadhari; instead, they will also overspill and drown the low-lying villages lying below their bed-levels. Long-term waterlogging situations are increasing day by day in the Indian Sundarbans as these tidal streams are not capable of redistributing

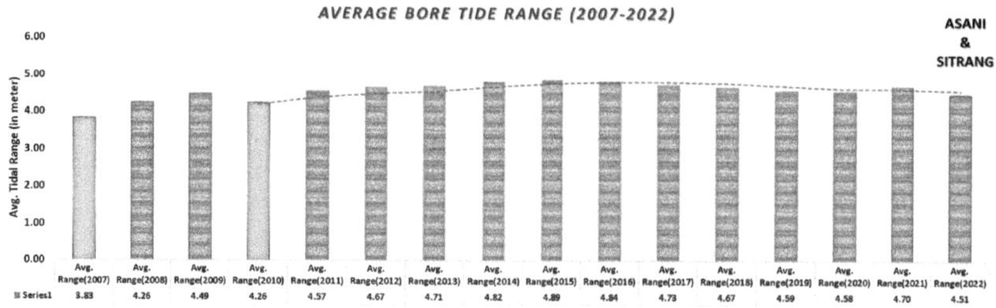

FIGURE 19.9 Trend in changing tidal range and occurrences of some vulnerable storm surges and severe cyclonic storms from 2007 to 2022.

the high tidewater brought up by R. Matla and R. Bidyadhari during any super or severe cyclonic storms over BOB.

19.6 Data analysis and results

It is a general fact that sea level has been rising since the last century at an alarming rate. The recent NOAA data clearly states that the sea level has risen to 21−24 cm since 1880. Under this circumstance, the low-lying areas and coastal regions of our planet are more susceptible to being drowned if left unchecked. Sundarbans, the largest mangrove delta, has already been surpassed through several island submergences during recent tropical severe and super cyclones such as Aila (2009) (Document, TD, and No, no date), Um-Pun (2020), and Yaas (2021). Analyzing the tidal range data of Kolkata Port Trust, it has been noticed that the tidal range, which was 3.83 m in 2007, never maintained its range thereafter. Rather, it slowly increased its level, and after the landfall of Aila in 2009 and the next year, the minimum tidal range was 4.26 m in 2010. During the last year, that is, in 2022, this tidal range was 4.51 m (Fig. 19.9). After the landfall of Hudhud and Phailin cyclonic storms over BOB, the tidal range rose to 4.89 m in the year 2015. Thus it is found that the tidal range steadily increased after Aila till 2015 (avg., 3 months mean value of El Niño was -0.3 and -0.4 from AMJ and MJJ, respectively) and again started decreasing at a very slow rate till 2020. However, it again reached a height of 2021 during Yaas (the value of El Niño was -0.1 and -0.3 during AMJ and MJJ, respectively) supercyclone over Sundarbans. Hence, within the gap of 12 years, the devastation caused by Aila was again experienced more destructively by the islanders of the Indian Sundarbans in Yaas in 2021 (Fig. 19.9).

Taking the impact of global warming and allied issues of sea level rise as per the recent NOAA data, the sea level should have risen steadily, accelerating the increase in tidal range as well. However, it did not happen as expected. Rather, the tidal range was decreasing after 2015 and again started increasing in 2021. So, it is clear that global sea level rise has obviously

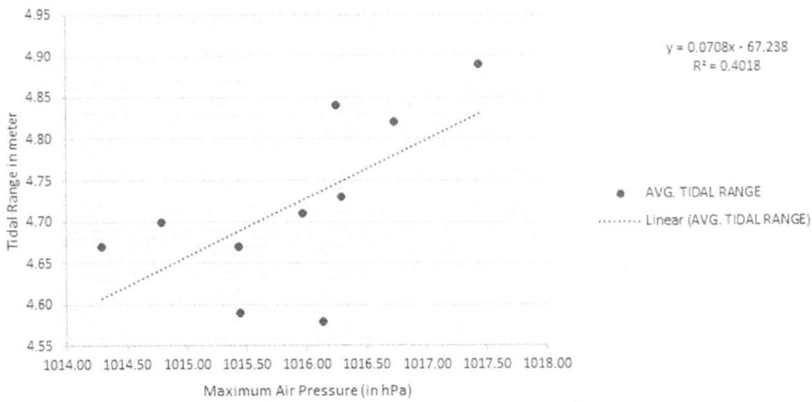

FIGURE 19.10 The correlation coefficient between maximum air pressure (in hPa) over BD08 in BOB and tidal range in Indian Sundarbans as per Kolkata Port Trust of India. *BOB*, Bay of Bengal.

increased the tidal range in the Indian Sundarbans, but ENSO factors are also controlling the rise of the tidal range at an interval of about 5 years. The INCOIS data of the heating and cooling rate of the surface water of the BOB at 1 m and 500 m also depicts the fact that the ENSO cycle changes the tidal behavior of the BOB. The correlated data (2011/2012–2021/2022 as per availability of data) obtained from the BD08 Buoys (17.82°N and 89.24°E) over BOB (distance from BD08 to the mouth of R. Matla is about 430 km away) and the tidal range data showed that the meteorological parameters often influence a lot over the tidal behavior, which are witnessed by the island dwellers of Indian Sundarbans. The following are the results depicting the correlation coefficient values:

1. The correlation between max. air temperature (@1 m) and tidal range is slightly negative, that is, R = -0.185.
2. The correlation between max. air pressure and tidal range (Fig. 19.10) is moderately strong, that is, R = +0.634.
3. The correlation between min. air pressure and tidal range is slightly positive, that is, R = +0.228.
4. The correlation between max. angle of wind direction and tidal range is moderately negative, that is, R = -0.405.
5. The correlation between max. wind speed and tidal range is slightly positive, that is, R = +0.180.
6. The correlation between max. wind gust and tidal range is slightly positive, that is, R = +0.194.
7. The correlation between max. salinity at 1 m (in PSU) and tidal range is slightly positive, that is, R = -0.192 and is +0.10 between min. salinity at 1 m (in PSU) and tidal range.

The insignificant correlation values have been aborted for better analysis of this study.

Thus the result shows that not only the sea level rise due to global warming but the changing behavior of the ENSO cycle also impacts the intangible behavior of the tidal activities in the Matla-Bidyadhari Interfluve of the Indian Sundarbans. As R. Matla has the widest opening to the incoming tides and has the maximum linear N-S stretch (about 45 km from the mouth of R. Matla to the confluence with R. Bidyadhari near Herobhanga RF) inside the Indian Sundarbans, the active deltaic parts (Fig. 19.5) and places having an average elevation of less than 5 m are most vulnerable to occasional saline water intrusion and inundation. During the formation of severe and super cyclones like Aila, Um-Pun, and Yaas over the northern BOB induced by low-pressure formation controlled by the ENSO cycle to some extent, the already decayed and rapidly decaying intertidal and intra-tidal channels and creeks in Indian Sundarbans, particularly in active deltaic parts of Namkhana, Patharpratima, Kultali, Basanti, and Gosaba blocks of South 24 Pargana district of West Bengal in India, the low-lying areas always face risks of embankment breaching, overtopping, and intrusion of tidal water.

19.7 Conclusion

The Matla-Bidyadhari Interfluve is a unique representative of the entire Sundarbans region, where one can observe almost every geomorphic feature. However, this area has been exploited extensively over the past three centuries. The sluggish downstream discharge of fresh water has further increased the bed heights, leading to the horizontal widening of the tidal channels. Besides, the rising sea level due to global warming and the unpredictable behavior of the ENSO cycle have increased the frequency and intensity of cyclones over the BOB, resulting in the failure of water management in the Sundarbans.

This recent study found that not only anthropogenic activities but also the changing character of ENSO phenomena have left direct footprints over the Sundarbans. The intensity and vulnerability of the last three major cyclones since 2009 (Aila, Amphan, and Yaas) were triggered by climate change. However, the Sundarbans region has not received much attention concerning the global changes in marine water behavior. The BOB, which is an extended part of the Indian Ocean, is highly influenced by the ENSO cycle and has witnessed an increase in the number of cyclones in its northern part over the past few decades. The NOAA data also reveals abnormal tidal ranges all over the world due to climate change.

If the rising sea level accelerates in the future, the low-lying areas and some islands of the Sundarbans may get submerged under the BOB.

19.8 Recommendations

Every research work aims to search for the major causes and consequences of the factors influencing or controlling the research problems. Likewise, this research work also went through the research methodology, and after analyzing and obtaining the results and major findings, some recommendations have been put forth, which are as follows:

1. Tidal channels and creeks are the nerves and veins of the Sundarbans and are the key players in shaping the Bengal Delta. Their behavior has been changing since the taming of the earlier creeks and khals during the British colonial period. Instead of raising the embankments along the vulnerable islands, the interlinking creeks and channels (khals) should be artificially rejuvenated either by dredging or by pumping out artificially created effluents back to the BOB. New canals, or khals, should be made or burrowed to redistribute the influx of tidal water.
2. More and more emphasis should be given to the existing slope of every island of the Sundarbans as all are interlinked with each other and play a key role in delta building processes.
3. Short-term embankments with temporary benefits may just give huge benefits for a few months, but, in the aftermath, it is proved that it increases the vulnerability of bank erosion and salinization in the Indian Sundarbans. Instead, government policies should be redesigned, orienting the natural tidal drainage circuit under the supervision of technical and environmental experts.
4. Mangroves should be planted between the natural and artificial river banks to reduce the vulnerability of tidal floods. Thalweg lines must be taken into consideration during the construction of an embankment.
5. Global warming and the changing behavior of the ENSO cycle are also responsible for changing the behavior of the tides to some extent. So, the motto of the policies must be Field-Laboratory-Field and Laboratory-Field-Laboratory to forecast the next possible river bank erosion site and situation. Tidal dynamics are very much correlated with atmospheric changes. So, more and more scientific research is welcomed, which should be interdisciplinary.
6. Encroachment toward the BOB must be prohibited, and active deltaic parts should be given more space and time to recover the proper tidal drainage system in the Sundarbans.

References

Biswas, S. K., Agarwal, A. (1992). Tectonic evolution of the Bengal Foreland Basin since the early Pliocene and its implication on the development of Bengal Fan. GSI Special Publication, 5–19.

Bose, S. (2006). Mangrove forests in Sundarbans active delta – ecological disaster and remedies.

Das, Sundarbans- environment and ecosystem. Sarat Book Distributors, (2006).

Gour, J. (2022). *Tidal mechanism and the changing fluvio-geomorphological environment of the interfluves in Indian Indian Sundarbans: A geospatial technology-based approach* (pp. 173–197). Springer Science and Business Media LLC. Available from https://doi.org/10.1007/978-3-030-77572-8_9.

Gour, J. (2018). Importance of remotely sensed data to analyze the changing fluviogeomorphological environment of the uninhabited islands in Indian Sundarbans: Special reference to Ajmalmari islands. *International Journal for Research in Engineering Application & Management (IJREAM)*, *4*(9), 278–283. Available from https://doi.org/10.18231/2454-9150.2018.1190, https://ijream.org/papers/IJREAMV04I0945067.pdf.

Hunter, W. W. (1875). Preface. Statistical account of Bengal. 1.

Milne, E. Soils, land-use & land-cover change, natural resource manage & policy, climate change (2009). Available at http://www.

Mukherjee, K. N. (1976). Harmonious solutions of the basic problems of Sundarban Reclamation. *Geographical Review of India*, *38*(3), 311–315.

Paul, A. K. (2002). Coastal geomorphology and environment.

Survey of India 2020 https://www.surveyofindia.gov.in/.

O'Malley, L. S. S. (1914). *Bengal District Gazetteers: 24-Parganas*. Logos Press.

Sarkar, K. (2007). *Itihas Kotha Koy*. Bhagabati Press.

Sentinel Hub (2015) https://sentinel2explorer.esri.com/.

Sawyer, K. E. (1978). *Studies: An introduction to geomorphology*, (Second Edition). *Landscape*, (176). London: Edward Arnold (Publications) Ltd.

Chapter 20

A scientific study on appraisal of embankment breaching and its coherent management over Indian Sundarbans

Sweta Chatterjee[1] **and Gupinath Bhandari**[2]

[1]*School of Water Resources Engineering, Faculty of Interdisciplinary Studies, Law & Management (FISLM), Jadavpur University, Kolkata, West Bengal, India,* [2]*Department of Civil Engineering, Jadavpur University, Kolkata, West Bengal, India*

20.1 Introduction

Pargiter (1934) described, in his own words, the appearance of the Sundarbans as "The scenery in the Sundarbans possesses no beauty....... few of the forest trees display a handsome bloom.....the forest closes in with breaks of sunshine streaming through the foliage." Sundarbans being declared a Sundarbans Biosphere Reserve culminates in the Tiger Reserve, Wild Life Sanctuaries, and National Park as well. Thirteen blocks of South 24 Parganas and six blocks of North 24 Parganas are administratively bound for the formation of Indian Sundarbans, geographically bordered by the Hugli River in the west, the Raimangal-Harinbhanga Rivers in the east, the Bay of Bengal in the south, and the Dampier Hodges line demarcating the Sundarbans from the mainland in the north (Fig. 20.1). Through this study, the possible mechanism behind the active embankment breaching and the case-specific study regarding the hazardscape and embankment-scape substudies of the Indian Sundarbans during the Yaas cyclone specifically mark out the possible vulnerable zones.

Brief geographical information on this vibrant ecosystem is furnished at a glance at Table 20.1.

Water Resources Monitoring, Management, and Sustainability. DOI: https://doi.org/10.1016/B978-0-443-23665-5.00020-X

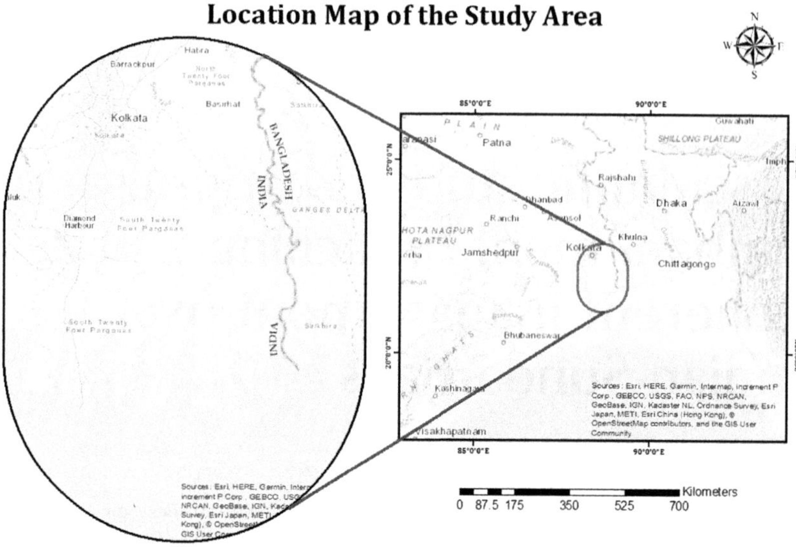

FIGURE 20.1 Location map of the study area. Map lines delineate study areas and do not necessarily depict accepted national boundaries.

TABLE 20.1 Basic geographical information on Indian Sundarbans.

Geographical extension	**21°31′to22°30′N&88°10′to 89°51′E**
Area (within West Bengal)	9630 square kilometers (60% over Bangladesh; 40% over India)
Number of Islands	54
Tidal height	+5 meters
Height above mean sea level	<3 meters
Mangrove forest area	4267 square kilometers
Reclaimed area	5363 square kilometers
Area under water	2069 square kilometers
Length of embankments	Approximated 3500 kilometers
Total population	3,311,698 (2001); 3,869,012 (2011)

Source: Modified after Rudra, K. (2011). The proposal of strengthening embankment in Sundarban: Myth and reality.

20.1.1 Importance of embankment

Embankments, being the pillars of the society of the Indian Sundarbans, act as the safeguard of life, property, and livelihood of local people against saltwater intrusion. Earthen embankments provide the road network for the villagers (Sarkar, Roy, & Siddique, 2016). According to Das (2016), the Indian Sundarbans have four basic types of embankments: *(1) Earthen Embankment (2 m Height), (2) Earthen Embankment with Brick Pitching (2.7 m Height), (3) Seafront Earthen Embankment with Brick Pitching (3.0m Height), and (4) Comparatively High Earthen Embankment with Boulder Pitching (3.67 m Height)*: however, Dhara and Paul (2016) have identified several types of embankments over Indian Sundarbans, such as earthen embankment, earthen embankment with bamboo skeleton, with or without sand fill bags, boulder gabion pitching and geo-bag gabion pitching, and brick pitching. The embankment details obtained from the literatures studied have been presented in Table 20.2.

Embankments are nothing but artificial levees, or banks raised above the surrounding island and constructed by using local alluvial soil, with or without bricks pitching to protect the people and their livelihoods against flooding. Embankments can be homogenous material-based or may be composed of a central impervious core associated with the shells of more recent materials (Das, 2016). It has also been mentioned that the textural composition of embankments is composed of fine sand with a content of more than 75%, which means less cohesion is observed and stability is easily decreased due to continuous tidal intrusion and wave attacks. It may be noted that most

TABLE 20.2 Basic details of embankment types.

Height of embankments (meters)	Embankment types	Length of embankments (km)	Position of embankments
3.67	Earthen	40	Exposed to coastal waves
3.00	Earthen	1560	Large river
2.70	Brick pitching	1200	Island margins
2.00	Block/boulder pitching	700	Small channels

Source: Ghosh, S., Mistri, B. (2020). Geo-historical appraisal of embankment breaching and its management on active tidal land of Sundarban: A case study in Gosaba Island, South 24 Parganas, West Bengal. *Space and Culture, India 7*(4), 166–180, https://doi.org/10.20896/saci.v7i40.587. Sarkar, H., Roy, A., Siddique, G. Impact of embankment breaching and rural livelihood: A case study in Ghoramara Island of the Sundarbans Delta in South 24 Parganas. *The Journal of Bengal Geographer, 5*(4) (2016), 97–117.

of the embankments of the Indian Sundarbans are earthen in nature, but these are highly susceptible to erosion and overtopping during a storm surge.

20.1.2 Process-response analysis of embankment breaching

Das (2014) also elaborated on the primary mechanism of embankment breaching with the help of a systematic approach to the process-response relationship.

- As per the river hydraulics study, the Thalweg lies in the creeks near the concave bank or sinuous river channel, where velocity distribution across the channel becomes the maximum and flows through the maximum depth of a river channel. The maximum velocity and associated gradient of the primary flow are positively related to the maximum depth, gradient, and secondary flow, as well as the off-shoaling trajectory belonging to the convex bank, which creates an eddy near the concave bank that enables the **basal scouring**, removing the base materials of the concave bank.
- In another process, **overtopping** (Fig. 20.2) occurs due to the disparity in water level height and the height of embankments due to catastrophic natural events like storm surges and spring tides. This enables the *supercritical flow* to have a *Froude Number greater than 1* that dislodges the unprotected and sensitive materials of the landward islands. Overtopping is the process when not only the swash effect of the tidal water level breaks away the unprotected slope-based materials, but also the backwash effect of returning the water level to the main river channel erodes the riverward slope of the embankment and removes the materials from it. Therefore this process was active on both sides of the embankments.
- **Tension cracks** on the steep embankment slope are another severe problem that enables the removal of lateral support that tends to be expanded.

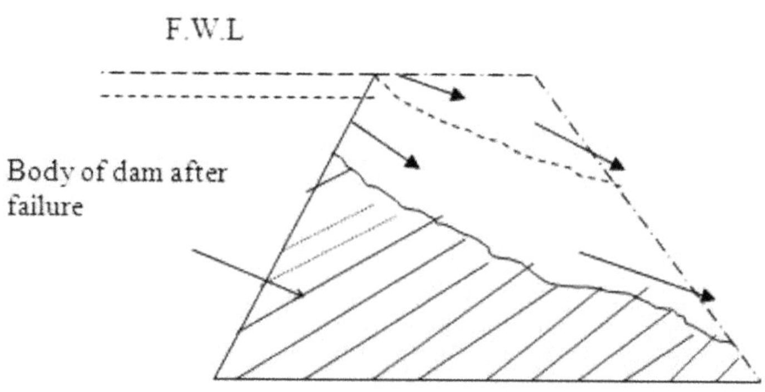

FIGURE 20.2 Embankment breaching by overtopping (Ozbek, 2017). *Modified after Ozbek, S. E. (2017). Numerical modeling of seepage in Koyunbaba Dam [Doctoral dissertation].*

- Sarkar (2018) asserted that **slump breaching** is also a common process frequently observed on the walls facing directly toward the strong tides attacked by the tidal bore, accompanied by strong wind gush. As a result, this attack creates a tensional force against the wall, almost erodes vertically, scours the basement of the wall, and crosses the threshold, ultimately falling straight to the waterside from the basement. The gap is generated between the widened walls, where saltwater intrusion is commonly observed.
- Due to *overtopping*, **hydraulic failure** of earthen embankments may be observed *via* two primary processes: *toe erosion and gulling*. This kind of failure is observed in noncohesive composition-based structures where bank undercutting, bed degradation, and basal scouring are in process. *Hydraulic toe erosion* occurs when the flow direction is at the river bend and the highest velocity is observed in the outer edge and center depth of the water. *Gulling* is the process of removing soil along the drainage level by the surface water runoff that is continued to be moved by the head-ward erosion.
- **Seepage failure** (Fig. 20.3) is another kind of process of embankment breaching that can be elaborated through the analysis of *soil permeability, shear strength, soil mechanics, slope stability, basal scouring*, and so on. *Soil permeability* is the physical property of soil that can be defined as the transmitting power of soil. Seepage failure is inevitable when the base or foundation of the embankments gets worn and torn apart and finally moves forward. "**Piping**" (Fig. 20.4) is defined as water passing

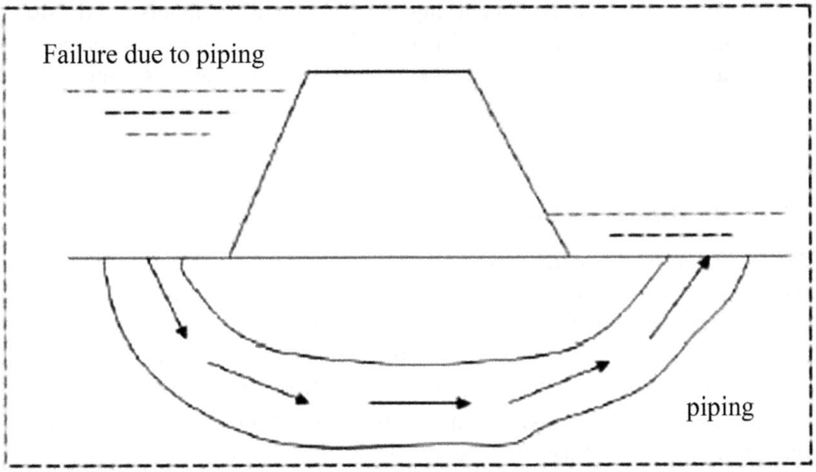

FIGURE 20.3 Formation of piping in embankment. *Reproduced with permission from Mohamed, M.A., Samuels, P.G., & Morris, M.W., (2002). Improving the accuracy of prediction of breach formation through embankment dams and flood embankments. River Flow; Ozbek, S. E. (2017). Numerical modeling of seepage in Koyunbaba Dam [Doctoral dissertation].*

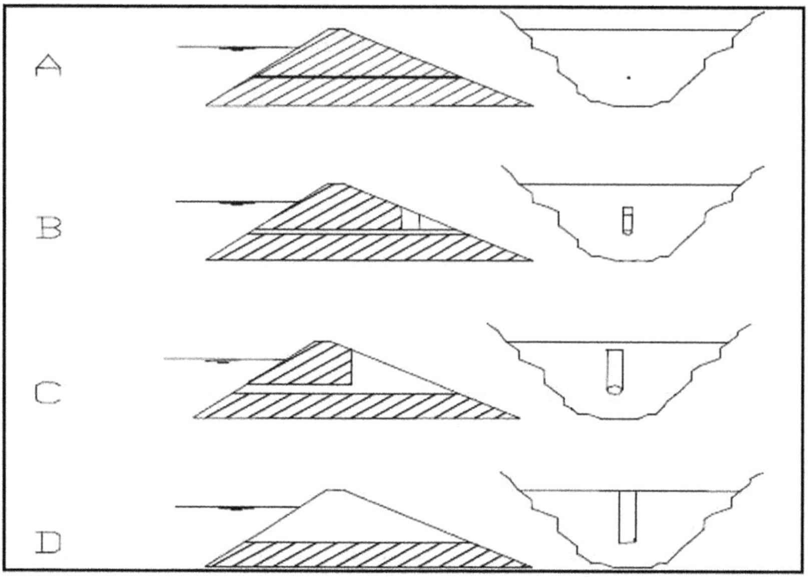

FIGURE 20.4 Formation of piping in embankment. *Data from Mohamed, M.A., Samuels, P.G., & Morris, M.W., (2002). Improving the accuracy of prediction of breach formation through embankment dams and flood embankments. River Flow; Ozbek, S. E. (2017). Numerical modeling of seepage in Koyunbaba Dam [Doctoral dissertation].*

through the narrow slice in the earthen embankments, weakening the adhesive properties of the soil and soil stability due to the saline water intrusion (Fig. 20.4). According to Mohamed, Samuels, and Morris (2002), *piping* is generally formed when the water flows beneath the embankment through the cracks, weak layers, cavities, and so on. Cumulative processes are generated after the formation of pipes beneath the embankment.

20.2 Rationale of the study

The rationale of the study lies amid the primary and secondary objectives of the present work. Assessment and analysis of the embankment breaching zones in the aftermath of the Yaas cyclone helps to locate the vulnerable breached zones and identify the distinct processes and triggering factors associated with them over the Indian Sundarbans. This study only implies the aftermath of the Yaas cyclone, to be more specific. The assessment of the aftermath of the Yaas cyclone also leads the way to decipher the possible implementable management techniques through the Public-Private Partnership (PPP) for the maintenance and conservation of the embankments in a long-term manner. The present study also makes clarify the base of the retrospective study on the land reclamation over the Indian Sundarbans, which helps to highlight the historical analysis of the same. Following the retrospective study over the study area, *hazardscape* and *embankment-scape* are also detailed in this study, where specifically possible breached zones over two blocks under the Indian Sundarbans have been incorporated to analyze the vulnerable zones of the same during the Yaas cyclone. Sustainable management techniques have been amplified in the last section through the perceptions of local people, researchers' opinions, and authors' recommendations, securing the secondary objective of the study.

20.3 Objectives

- To assess and analyze the embankment breaching zones with their associated processes and triggering factors in the Indian Sundarbans during the Yaas Cyclone.
- To identify the possible implementable management techniques in the maintenance and conservation of embankments in a long-term manner.

20.4 Materials and methods

20.4.1 Data used

After planning the study design and framework, it is important to depict the nature of the data used and make further assessments so that the study can be justified. Numerous data have been compiled for the present study, collected from secondary sources, mainly as per the requirement for analyzing several aspects of the scientific study of embankment breaching over the Indian Sundarbans. The following secondary data have been evaluated through several published and unpublished sources, for example, "*A Statistical Abstract of South Twenty Four Parganas*" *by* Hunter (1877), the *Bengal District Gazetteer, the Bengal Secretariat Book Depot, the Bengal District Gazetteers of 24-Parganas by L.S.S. O' Malley, the Irrigation and Waterways Directorate, the Government of West Bengal, and the West Bengal Disaster Management Plan.*

20.4.2 Methods

The methodological framework (Fig. 20.5), as per the nature and strategy of the data structure and its orientation, entails several distinct phases to analyze and assess the appraisal of embankment breaching and its management scenario in the Indian Sundarbans. Both qualitative and quantitative tools and techniques using software, for example, *Arc GIS 10.8, 10.3.1, QGIS Desktop 2.16.3, Google Earth Pro, and Microsoft Office*, have been used to depict the ideal factors behind the embankment breaching and destruction since the reclamation period over the Indian Sundarbans with the help of

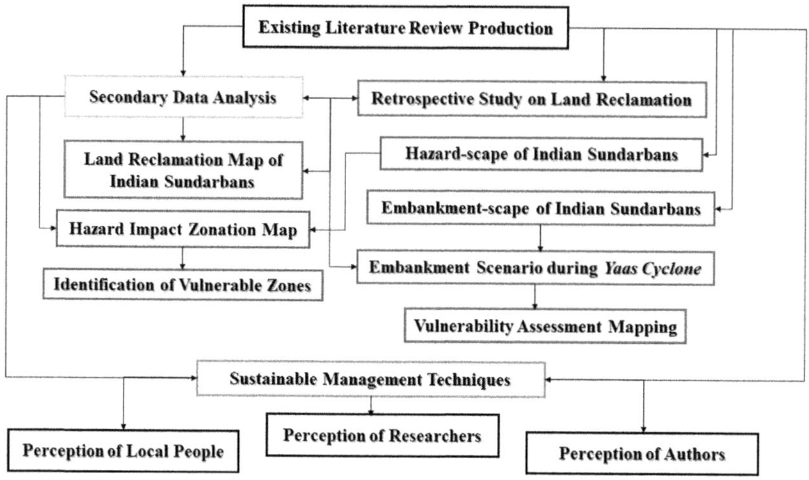

FIGURE 20.5 Methodological framework.

previous and recently published secondary data sources. The methodology of the present study has been oriented from the retrospective study of the land reclamation study over the Indian Sundarbans to the hazardscape of the study area. Particularly focusing on the aftermath of the *Yaas*, it analyzes the climatic variability and vulnerable embankment breaching zones over the Indian Sundarbans through geospatial techniques.

20.4.2.1 Retrospective study on land reclamation over Indian Sundarbans

The retrospective study of encroachment over premature land over the Indian Sundarbans as well as the historical perspective on land reclamation have sparked the recent scenario of embankment breaching over the Indian Sundarbans. This section reviews the appraisal of a retrospective study on land reclamation with distinctive phases of encroachment over the Indian Sundarbans associated with spatial distribution over the same (Fig. 20.6). Land reclamation over the premature land over the Indian Sundarbans was distinctively divided into five phases, *initially during 1770–80, subsequently during 1780–1873, 1873–1939, 1945–51, 1951–71, and so on after different extreme events of tropical cyclones, subsequent storm surges, and regular spring tides.*

According to Danda (2007), embankments were erected for the purpose of not making the reclaimed land turn into marshes; however, Bhattacharyya (1998) first approached, *"The presence of embankments seals off the possibility of these tracts ever naturally maturing into lands habitable by*

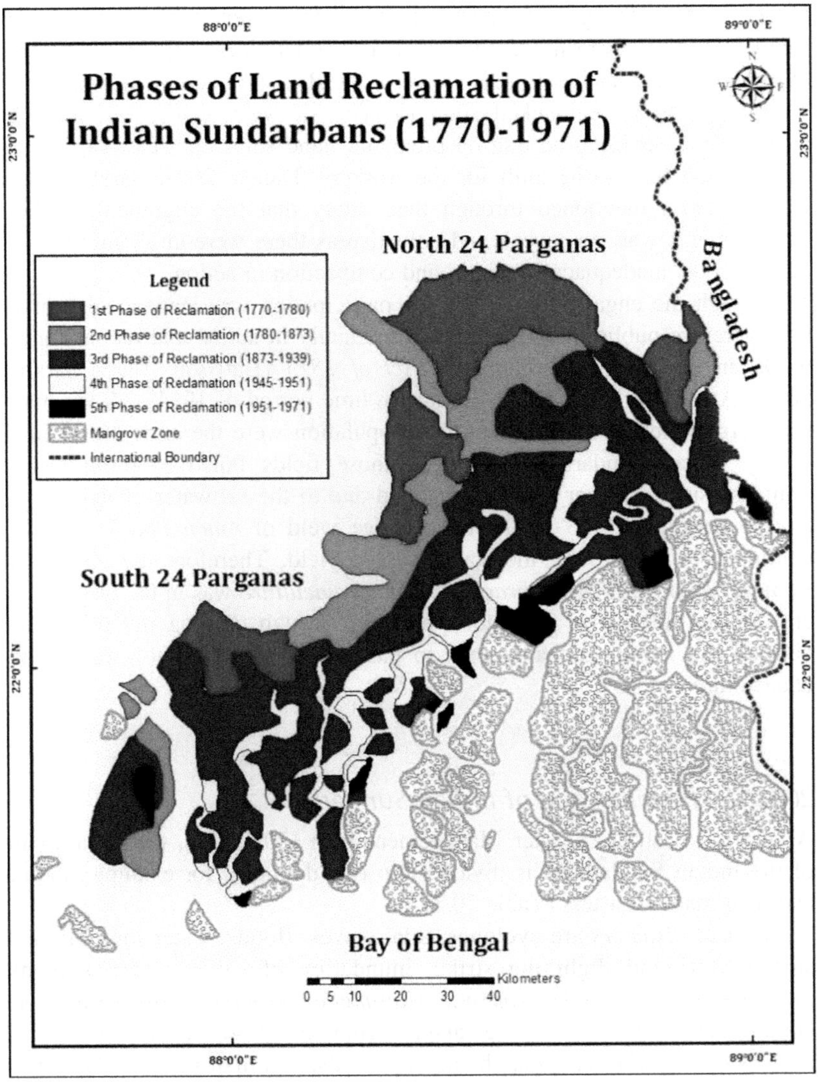

FIGURE 20.6 Land reclamation phases over Indian Sundarbans (1770–1971) (Gour, 2021). Map lines delineate study areas and do not necessarily depict accepted national boundaries. *Modified after Gour, J. Stages of land reclamation and their impact on the fluvio-geomorphological environment of Indian Sundarbans with special reference to Matla–Bidyadhari Interfluve.* Modern Cartography Series, 10 *(2021), Available from: http://www.elsevier.com/wps/find/bookdescription.cws_home/BS_MCS/description#description. http://pubs.rsc.org/10.1016/B978-0-12-823895-0.00022-1.*

humans." Actually, natural geomorphic processes over the deltaic system of southern Bengal were hindered by human habitation as well as settlements. O'Malley (1914), as the representative of the British government, wanted the virgin land to be embanked soon to keep saltwater aside, forests to be cleared soon, tanks to be dug on premature land soon for water supply and storage, and huts to be built for the workers (Danda, 2007; Sarkar, 2002). Danda (2007) mentioned through their study that the alignment of those embankments was very weak and defective as there were unscientific building methods, inadequacy in width, and compaction in action.

Though the engagement of private participation was inadequate in manner, therefore public maintenance was gradually in action and recommended works in the *Bengal Embankment Act of 1873* (Harrison, 1875; Sarkhel, 2013). As per Sarkhel (2013), during the time period of 1914–44, the major suppliers of rice to the Calcutta City population were the agricultural fields of the Indian Sundarbans; however, those fields failed to provide such immense supply as per the high demand due to the saltwater incursion, and for the period of 1993–2008, the average yield of *Aman Paddy* had been continuously lower than the state average yield. Therefore the *conversion from agricultural fields to brackish water aquaculture* was in the next hierarchy for the alternative livelihood practice through digging the tidal inlets beneath the muddy embankments to pass the tidal water flow toward the aquaculture fields.

20.4.2.2 Hazardscape of Indian Sundarbans

As per the District Disaster Management Plan (2020–21), the hazardscape of the Indian Sundarbans is divided into two domains, for example, natural and man-made disasters (Table 20.3).

Natural disasters are cyclones, tidal waves, floods, water logging, hailstorms/whirlwind, lightning strike, inundation of saline water, embankment breaching, and erosion. *Man-made disasters* are fire, food contamination, boat capsizing, arsenic contamination, accident (road and railway), industrial hazards, chemical hazards, and man-made embankment breaches.

A Flood and Cyclone *Hazard Impact Zonation Map* of Indian Sundarbans (Fig. 20.7) has also been prepared using the scales of impact (e.g., *severe, moderate, mild* and *moderate to severe*) to get a clear picture of the vulnerable zones (Table 20.4) susceptible to flood and cyclone disasters specifically over the Indian Sundarbans. Vulnerable zones of both districts are susceptible to embankment breaching and damage due to several cyclonic storms and associated flood-like conditions; specifically, blocks under the South 24 Parganas are more vulnerable than the North 24 Parganas.

TABLE 20.3 Generalized retrospective study on hazardscape of Indian Sundarbans.

Hazard type	Year of occurrence	Area affected	Scale of impact	Livestock	Remarks
Cyclonic storm	*Aila* (2009)	All blocks of South 24 Parganas and 6 blocks of North 24 Parganas (*Haroa, Sandeshkhali I, Sandeshkhali II, Hasnabad, Hingalganj, and Minakhan*)	Severe	Affected to a greater extent	Possibilities of damaging dwelling houses specifically in the coastal areas
	Fani (2019)				
	Bulbul (2019); *Yaas (2021)*				
Flood	1978	All blocks of South 24 Parganas			
	1986	*Kultali, Joynagar II, Canning I, Gosaba, Basanti, Mathurapur II, Kakdwip, Namkhana, Sagar, and Patharpratima*	Moderate to severe condition	NA	NA
Flood-like situation	2015	All blocks of South 24 Parganas	Severe	Affected to a greater extent	Possibilities of damaging dwelling houses specifically in the coastal areas

(*Continued*)

TABLE 20.3 (Continued)

Hazard type	Year of occurrence	Area affected	Scale of impact	Livestock	Remarks
Boat Capsize	2010	*Kakdwip*	Severe	NA	
	2018	*Kakdwip*	Moderate		
	2019	*Kakdwip and Mathurapur II*			
Drought	1998–99		*Kakdwip, Basanti, Canning II, Gosaba, Mathurapur I, and Patharpratima*		Possibilities of damaging dwelling houses specifically in the coastal areas
Embankment breaching and erosion	Associated with each flood-like situation	All blocks of South 24 Parganas and 6 blocks of North 24 Parganas (*Haroa, Sandeshkhali I, Sandeshkhali II, Hasnabad, Hingalganj, and Minakhan*)	Moderate to Severe	Affected to a greater extent	Possibilities of embankment damage and breaching; overtopping condition in selected parts

Source: Compiled from District Disaster Management Plan, 2020–21, South 24 Parganas, Government of West Bengal.

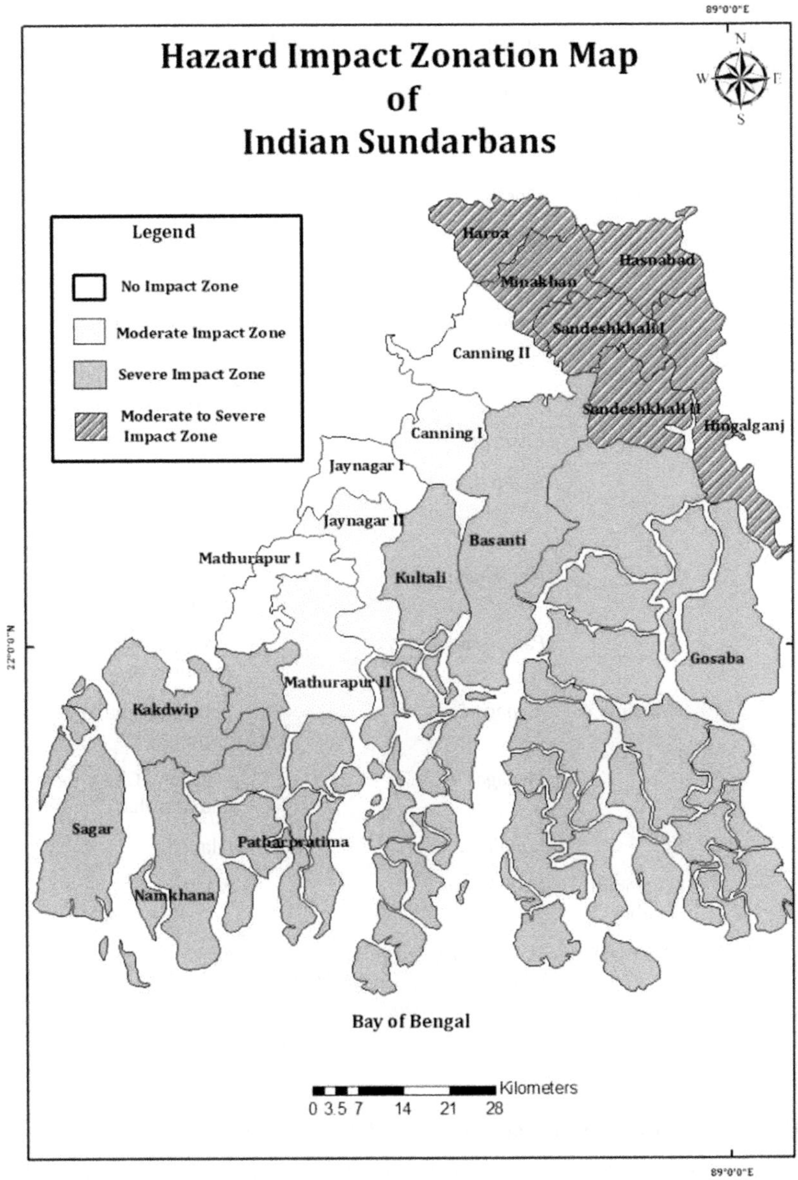

FIGURE 20.7 Hazard impact zonation map. Map lines delineate study areas and do not necessarily depict accepted national boundaries. *Modified after District Disaster Management Plan, 2020–21, South 24 Parganas, Government of West Bengal.*

TABLE 20.4 Vulnerable zones over Indian Sundarbans.

Districts	Name of sub divisions	Name of vulnerable blocks	Vulnerable area
South 24 Parganas	Canning	*Gosaba*	NA
		Basanti	Amjhara and Charvidya
		Canning I	Bansra
		Canning II	Deuli I, Deuli II GP, Atharobaki, Matherdighi, and Kalikapota
	Kakdwip	*Kakdwip*	NA
		Sagar	Muriganga II, Dhablat, Dhaspara, and Sumatinagar II
		Namkhana	Narayanpur and Mousuni
		Patharpratima	Laxmijanardanpur and Achintyanagar
	Baruipur	*JoynagarI*	Uttar Durgapur and Sripur
		Joynagar II	Nalghara, Chuprijhara, and Masirhat
		Kultali	Jalaberia
	Diamond Harbor	*Mathurapur II*	NA
North 24 Parganas	Basirhat	*Haroa, Hingalganj, Hasnabad, Sandeshkhali I, Sandeshkhali II*, and *Minakhan*	NA

20.4.2.3 Embankment-scape of Indian Sundarbans

In the present study, this section enables the embankments and sea dykes scenario of the Indian Sundarbans aftermath of Yaas cyclone that was formed on May 22, 2021, over the east-central Bay of Bengal greatly impacted coastal areas of Odisha, West Bengal, adjoining South Bihar, Jharkhand, and Chhattisgarh. It caused heavy to very heavy rainfall activity at isolated places over Gangetic West Bengal on May 25 and 26. Estimated storm surge (Table 20.5) of approximately *0.5–1.1 meters* above the

TABLE 20.5 Storm surge height information.

Mandal/Taluk	District	Nearest place of habitation	Storm surge[a] (M)	Expected inundation extent (km)
Basirhat	North 24 Parganas	Amlamethi	0.5–0.9	Up to 0.37
Diamondharbour	South 24 Parganas	Paschim Bhabanipur	0.5–1.1	Up to 0.78

[a]*Surge heights are equivalent to the astronomical tide.*
Source: Government of India, Ministry of Earth Sciences, Indian Meteorological Department (IMD), Bulletin No.: BOB/02/2021.

astronomical tide that inundated coastal tracts of West Bengal districts (East Medinipur and 24 Pargana districts) **(Government of India, Ministry of Earth Sciences, Indian Meteorological Department)**.

The aftermath of Yaas damages and breaches the embankments of both 24 Parganas, specifically hampering greatly the embankments of the North 24 Parganas districts' 6 blocks under the Indian Sundarbans. In the North 24 Parganas, a total of *76* embankments got breached, and *196* embankments got damaged due to Yaas, from which *Sandeshkhali II (30)* got the highest proportion of breaching and *Haroa (1)* got the lowest; zone wise, *Hingalganj (66)* got the highest proportion of damaged embankments, whereas *Haroa (5)* and *Minakhan (5)* got the lowest damaged embankments; there is no existence of sea dykes over the district. In the South 24 Parganas, a total of 26 embankments got breached and 351 embankments got damaged, from which *Patharpratima (16)* had the highest proportion of breaching and *Kultali* had *10* breached embankments; only *2 sea dykes* got breached at *Sagar Island* and *1* got breached at *Patharpratima*; the maximum proportion of damaged embankments were at *Gosaba* (93) and Canning II got only 1 damaged embankment; *Namkhana* got only 6 damaged embankments and *Patharpratima* got 3 damaged embankments. These statistics give a clear picture of the frequency distribution of both damaged and breached embankments as well as sea dykes, which shows that embankments were greatly impacted and damaged, and sea dykes got breached and damaged at a small scale due to Yaas in the blocks of South 24 Parganas, whereas breaching happened over the 6 blocks of North 24 Parganas (Table 20.6).

The aftermath of *Yaas* damages and breaches of several embankments are mapped through geospatial techniques and secondary data sources with respect to both districts separately. Six administrative blocks of North 24 Parganas and 12 blocks of South 24 Parganas show a clear picture of greatly

TABLE 20.6 Frequency Distribution Analysis of Breached/Damaged Embankments/Sea dykes due to *Yaas*, Indian Sundarbans.

District	Name of block	Frequency of breached embankments (zone wise)			Frequency of breached sea dykes (zone wise)	Frequency of damaged embankments (zone wise)	Frequency of damaged sea dykes (zone wise)
North 24 Parganas	Haroa	1			NA	5	NA
	Hasnabad	8				30	
	Hingalganj	21				66	
	Sandeshkhali I	9				34	
	Sandeshkhali II	30				56	
	Minakhan	7				5	
	Total		76		0	196	0
South 24 Parganas	Sagar				2	10	NA
	Kultali	10			NA	54	
	Basanti	NA		41			
	Kakdwip			46			
	Mathurapur I			1			

Mathurapur II			22			
Gosaba			93			
Canning I			12			
Canning II			1			
Joynagar II			6			
Namkhana			27	6		
Patharpratima	16			1	61	3
Total		26		3	374	9

Source: Compiled & modified after Irrigation & Waterways Directorate, Government of West Bengal.

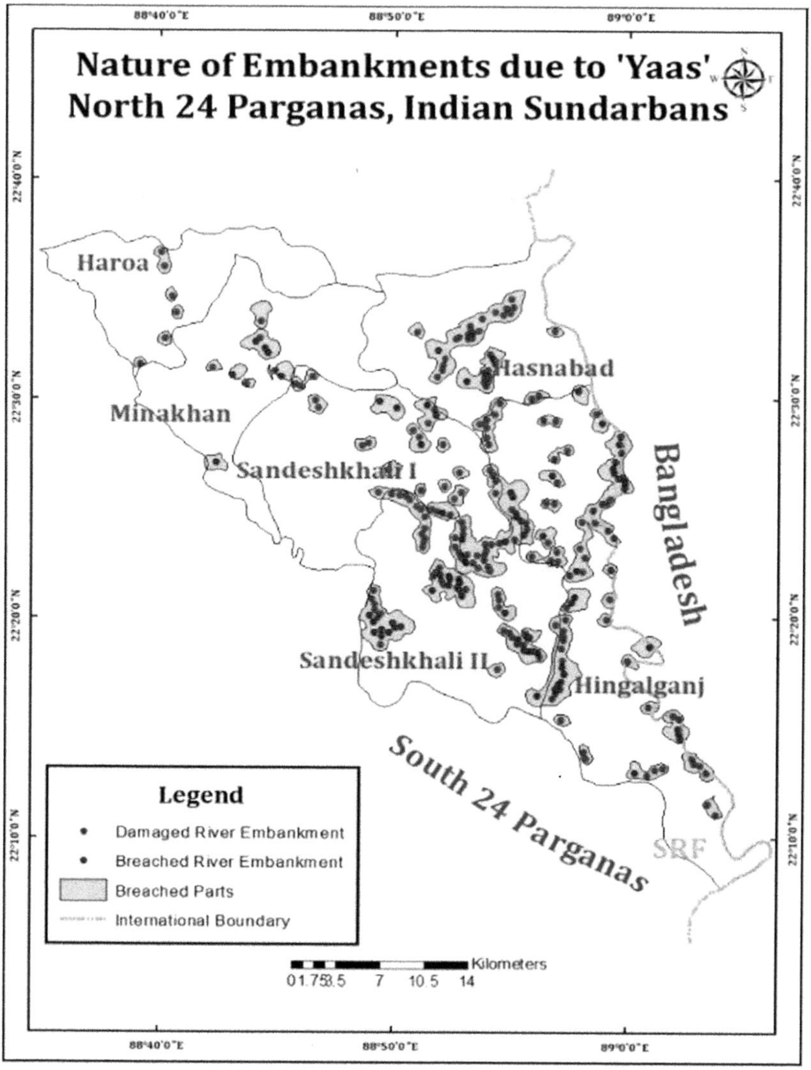

FIGURE 20.8 Embankment scenario and location of breached/damaged embankments due to *Yaas*, North 24 Parganas, Indian Sundarbans. Map lines delineate study areas and do not necessarily depict accepted national boundaries.

and partly damaged and breached embankments during Yaas. Figs. 20.8 and 20.9 assess that Yaas breached and damaged several embankments throughout the selected blocks of North 24 Parganas, where all six blocks had damaged as well as breached embankments; *Sandeshkhali II* got the highest frequency (30) of breached embankments and *Hingalganj* got the highest

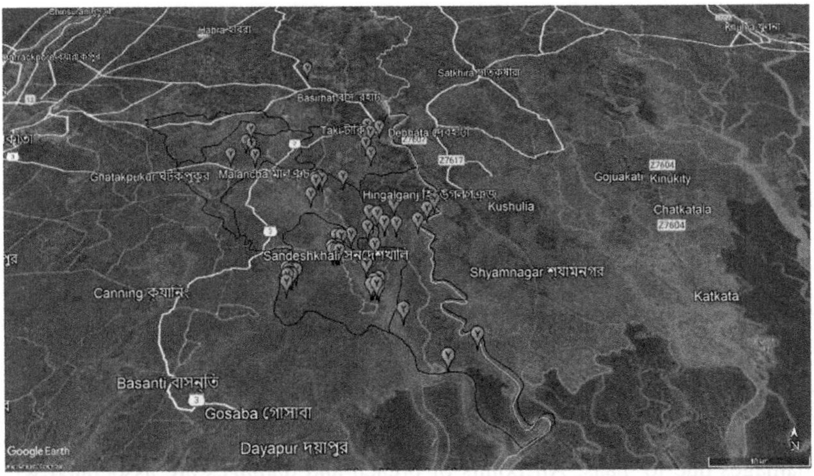

FIGURE 20.9 Embankment scenario and location of breached/damaged embankments due to *Yaas*, North 24 Parganas, Indian Sundarbans. Map lines delineate study areas and do not necessarily depict accepted national boundaries.

frequency (66) of damaged embankments. It is observed that *Sandeshkhali II* and *Hingalganj* were the most vulnerable blocks of the North 24 Parganas regarding the damage and breaching of embankments. An associated Google Earth image with validated locations of breached/damaged embankments of selected blocks under the North 24 Parganas of the Indian Sundarbans helps to imply ground-level monitoring and assessment of damages due to Yaas. All points of breach and damages are associated with valid latitudes and longitudes.

As shown in Table 20.7, all frequencies of both sectors (Breached/damaged embankments) are clubbed for the analysis of vulnerable zones over the district that is tabulated below.

Table 20.7 states that the less vulnerable zones are *Haroa* and *Minakhan*; the moderate ones are *Hasnabad* and *Sandeshkhali I;* and the most vulnerable ones are *Sandeshkhali* II and *Hingalganj*, as per the frequency distribution and analysis of breached and damaged embankments due to Yaas.

Figs. 20.10 and 20.11 assess that Yaas breached and damaged several embankments throughout the selected blocks of South 24 Parganas, where 12 blocks had damaged as well as breached embankments; *Patharpratima* got the highest frequency (16) of breached embankments and *Gosaba* got the highest frequency (93) of damaged embankments. It is observed that *Patharpratima* and *Gosaba* turn out to be the most vulnerable blocks regarding the damage to the embankment, and *Patharpratima* and *Kultali* are the most vulnerable sites for embankment breaching.

Associated Google Earth Image with validated locations of breached/damaged embankments of 12 blocks under the South 24 Parganas of the

TABLE 20.7 Hierarchy of vulnerable zones, North 24 Parganas.

Classes for breaching	Classes for damage	Block	Remarks
<8	<30	*Haroa and Minakhan*	Less vulnerable zone
8–16	30–45	*Hasnabad and Sandeshkhali I*	Moderate vulnerable zone
>16	>45	*Sandeshkhali II and Hingalganj*	Most vulnerable zone

Indian Sundarbans helps to imply ground-level monitoring and assessment of damages due to Yaas. All points of breach and damages are associated with valid latitudes and longitudes. As shown in Table 20.9, all frequencies of both sectors (Breached/damaged embankments) are clubbed for the analysis of vulnerable zones over the district that is tabulated below:

Table 20.8 states that the less vulnerable zones are *Sagar, Canning I and II, Joynagar II and Namkhana, and Mathurapur I and II*; the moderate ones are *Kultali, Basanti, and Kakdwip;* and the most vulnerable ones are *Patharpratima and Gosaba*, as per the frequency distribution and analysis of breached and damaged embankments due to Yaas.

With respect to Tables 20.7 and 20.8, a map on vulnerability assessment due to embankment failure has been drawn out to get a clear picture of the vulnerable zones of the Indian Sundarbans. Yaas impacted these zones through embankment breaching and damages of a diversified nature that delineates the overall entity in primarily four zones, such as the *less vulnerable zone, the moderate vulnerable zone, the most vulnerable zone, and the no vulnerable zone. The mangrove zone with core and buffer area* has also been demarcated Fig. 20.12.

20.5 Sustainable management techniques

Proper suitability of management techniques for the vulnerable zones over the Indian Sundarbans with respect to the embankment scenario and hazardscape during the Yaas cyclone in a sustainable manner is the final objective of this study. There are some proposals from local people as well as officials and researchers for strengthening the vulnerable embankments to get rid of the breaching and damages caused by several cyclonic disasters and associated surges that are listed in the studies (Das, 2015; Das, 2016; Ghosh & Mistri, 2020; Rudra, 2011; Sarkhel, 2013) and the authors' own perceptions (Table 20.9).

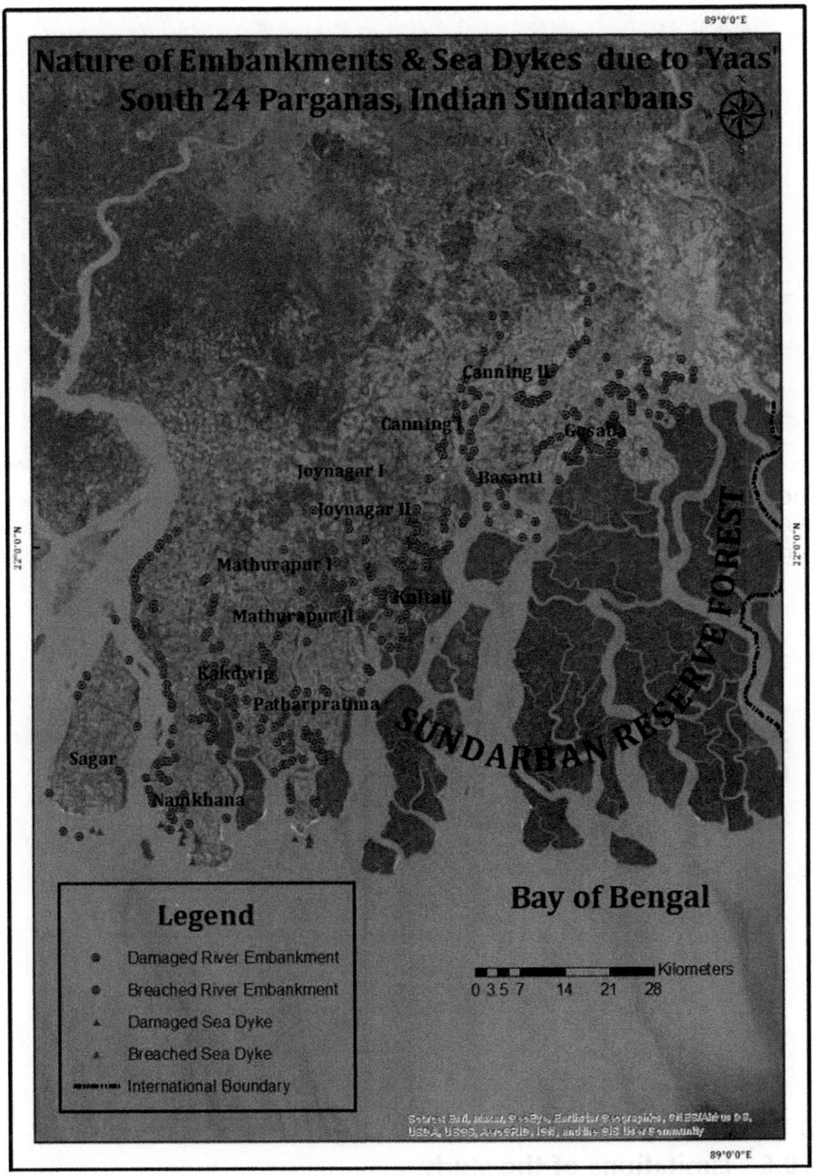

FIGURE 20.10 Embankment scenario and location of breached/damaged embankments due to *Yaas*, South 24 Parganas, Indian Sundarbans. Map lines delineate study areas and do not necessarily depict accepted national boundaries.

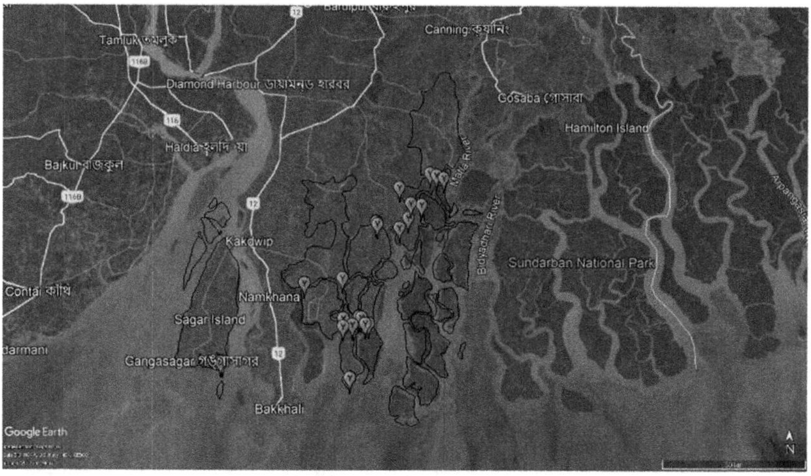

FIGURE 20.11 Embankment scenario and location of breached/damaged embankments due to *Yaas*, South 24 Parganas, Indian Sundarbans. Map lines delineate study areas and do not necessarily depict accepted national boundaries.

TABLE 20.8 Hierarchy of vulnerable zones, South 24 Parganas.

Classes for breaching	Classes for damage	Block	Remarks
NA	<30	*Sagar Island, Canning I and II, Joynagar II, Namkhana, and Mathurapur I and II*	Less vulnerable zone
	30–60	*Kultali, Basanti and Kakdwip*	Moderate vulnerable zone
	>60	*Patharpratima and Gosaba*	Most vulnerable zone

20.6 Limitations of the study

This study is only concentrated on analyzing and deciphering the appraisal of embankment breaching processes along with zones of Indian Sundarbans in the aftermath of the Yaas cyclone. Vulnerability mapping has been run on both districts in the Indian part of the Sundarbans with respect to the secondary data on breached or damaged embankments and sea dykes available from the Irrigation Department, Govt. of West Bengal. Yet other vulnerable

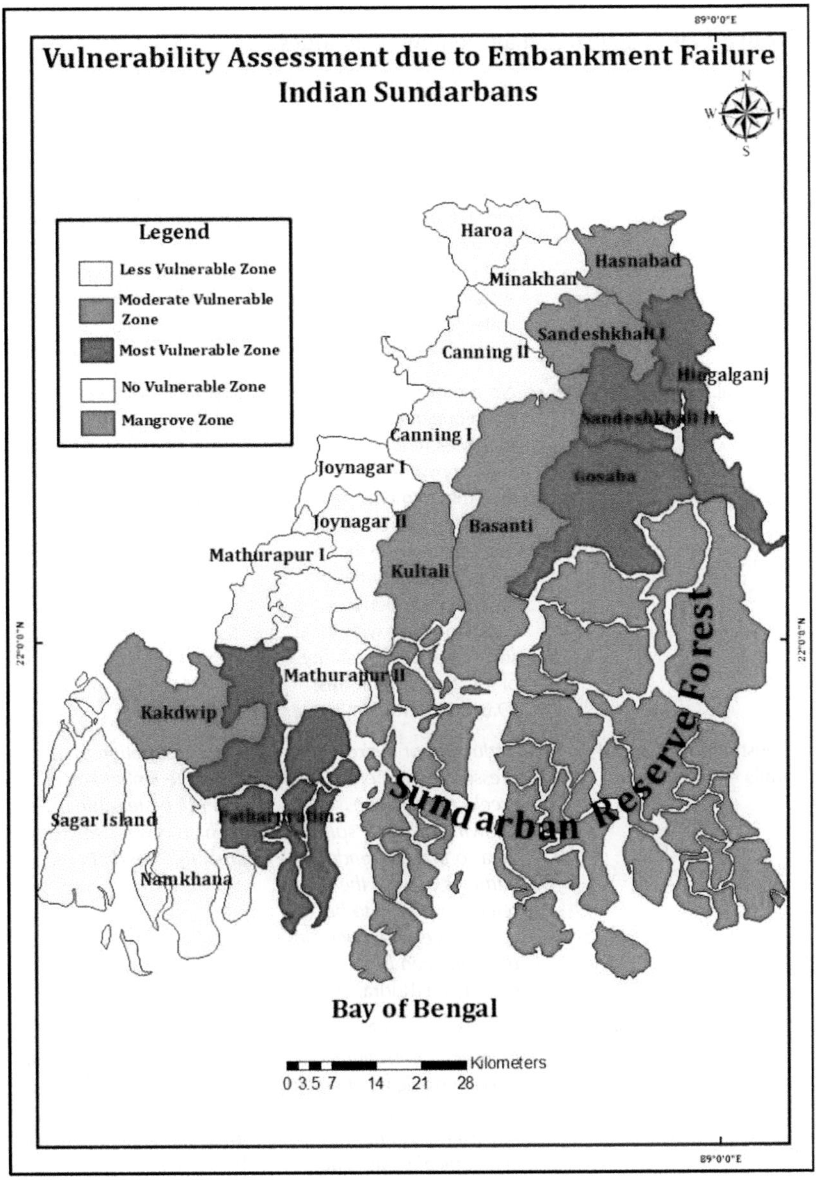

FIGURE 20.12 Vulnerability assessment mapping due to embankment failure, Indian Sundarbans. Map lines delineate study areas and do not necessarily depict accepted national boundaries.

TABLE 20.9 Sustainable management techniques.

Local people perception (Das, 2016)	Researchers' perception	Author's perception
Initiatives should be taken against the toe erosion in embankments	*The height of the embankments should be more than the highest recorded storm surges* so that embankments can sustain and resist the high water level as per the Situation (Rudra, 2011).	*Mangrove regeneration* through *natural* as well as *artificial techniques* has the potential to get rid of bank failure as well as Embankment breaches.
Construction of Porcupine mesh	*No use of nonbiodegradable materials like Polypropylene sheets* over the embankments, being the protective cover, as this cannot have the ability to anchor the basement properly; bamboo cover with local soil as per the suitability of location can hit this (Das, 2015; Rudra, 2011).	*Examination of both public-private initiatives and actions* for embankment maintenance, conservation, and management.
Construction of guard walls	*Reduction of hydrostatic pressure on the embanked creeks through the allowing of wider spill area so that suspended particles can get the retention period to be spilled over and river decaying can be obstructed* (Rudra, 2011).	*The construction of a Concrete Embankment will not be feasible in such an eco-sensitive area, like Sundarbans.*
Concrete embankment building	*Long-term-based management techniques through dredging of creeks can be an option however; it is not economically easy and geomorphologically not so viable due to tidal asymmetry conditions over the study area. Spillover of sediments from the creeks towards*	*Installation of Bio-engineering techniques as per suitability and applicability to protect the embankments against breaching.*

(Continued)

TABLE 20.9 (Continued)

Local people perception (Das, 2016)	Researchers' perception	Author's perception
	the floodplains can build the natural levee as the embankments of preferable heights that can be the best sustainable option (Rudra, 2011).	
Mangrove plantation over the vulnerable embankments as per the soil and saline suitability zonation	*Mangrove plantation can be the best option to protect the vulnerable river banks as well as embankments also* (Rudra, 2011).	*Application of bio-grouting technique.*
Protection through block/brick pitching; however, this technique is kind of costly and labor intensified and engineering and hydro-geomorphological analysis is required for the better understanding of the basement	*Usage of sandbags over the river bank by layering often reduces the probability of embankment breaching; approximately weight of 14–18 kg sandbags can be used for the stabilization; bamboo piles can also be attached to the front of the river bed for more stability* (Ghosh & Mistri, 2020).	
	Separately bamboo piles can be used for strengthening the embankments with the help of a rope or iron wire. Toe erosion can be eradicated through the bamboo piling with two parallel rows of piles and the space between the two can be filled with boulders, blocks, and pebbles (Ghosh & Mistri, 2020).	
	A protective cover of jute geotextiles can be useful for temporary embankment breaching conditions (Ghosh & Mistri, 2020).	

(Continued)

TABLE 20.9 (Continued)

Local people perception (Das, 2016)	Researchers' perception	Author's perception
	Concrete embankments with nonbiodegradable material cover is not a feasible option for Indian Sundarban (Das, 2015).	

factors due to other disasters are limited to this study, as specifications on the appraisal of embankment breaching and its management techniques due to the Yaas cyclone (2021) have only been incorporated. However, previously there were/are other devastating cyclones and their aftermath, for example, Sidr (2007), Aila (2009), Bulbul (2019), and Amphan (2020), over the study area that is not focused in this study.

20.7 Conclusion

The present study enlightens that the Sundarbans have been encroached on in a premature condition, which appears to be the primary reason for the coastal erosion of specific islands over the same area through the retrospective study on the land reclamation over the Indian Sundarbans. The hazardscape study consists of a retrospective elaboration of hazards over the Indian Sundarbans that also maps the hazard impact zonation based on flood and cyclones, where severe hazard-impacted zones are the low-lying coastal islands of the Indian Sundarbans. The embankment-scape of Indian Sundarbans broadens the embankment scenario during the Yaas cyclone as the aftermath and further impact measurement analysis. The most vulnerable zones out of the consideration of breached embankments from the South 24 Parganas are, for example, *Patharpratima and Gosaba*, and from the North 24 Parganas, for example, *Sandeshkhali II and Hingalganj*. Vulnerability assessment mapping as the final output shows that after the Yaas cyclone (2021), the most vulnerable zones as per the embankment failure are, for example, *Patharpratima, Sandeshkhali II, and Gosaba*; the moderately vulnerable zones are *Basanti, Sandeshkhali I, Kultali, Hasnabad, and Kakdwip*; the less vulnerable zones are *Sagar Island, Namkhana, Mathurapur I and II, Canning I and II, and Haroa and Minakhan*. For encroaching the island as agricultural land and human settlement formation, earthen embankments were constructed surrounding some of the premature islands for fortificating the Sundarbans. The sediment dynamics and hydrodynamics were affected due to the construction of these earthen embankments. As a result, the bed of the creeks got uplifted due to sediment deposition. Along with these sediment dynamics and hydrodynamics, several kinds of anthropogenic activities merged to enforce the degradation of the Sundarbans. Climate change has become another important parameter to create vulnerability in the Sundarbans to changes in sea surface temperature and subsequent tropical cyclone and storm surges. The tropical cyclone, with its high frequency and intensity, affected the people of the Sundarbans. To save them, there may be a plan of rehabilitation, which may appear hypothetical, even if it can be done, and there may be a professional hazard. The present study has identified the soft techniques to save the eco-sensitive Sundarbans by implementing the PPP Model with the help of indigenous materials, such as *bamboo and mud*, that can guarantee rural employment opportunities or livelihoods in a sustainable way as well. Before adopting and implementing a proper management plan, it is needed to get a clear picture of which hydrogeomorphological processes operate and how they perform a responsive system over the Indian Sundarbans. Socio-economic conditions and the betterment of the millions of Indian Sundarbans have to be in mind when implementing a long-term management plan so that their livelihood structures and opportunities cannot be hampered in near future.

Acknowledgment

The authors are grateful to the University Grant Commission (UGC) for providing the Fellowship to the First Author as NET JRF. The authors have extended their heartiest gratitude to Jadavpur University, for providing the logistics and infrastructure, along with the supervising expertise. And last but not least, would like to express their heartfelt regards to the Irrigation and Waterways Directorate (I & WD), Government of West Bengal, for providing Real-World data on the Earthen Embankments of Indian Sundarbans.

References

Bhattacharyya, S. (1998). Sunderban: Dying a slow death. *The Hindu Survey of the Environment '98*, 89–94.

Danda, A.A. (2007). Surviving in the Sundarbans: Threats and responses-An analytical description of life in an Indian Riparian commons.

Das, G. K. (2015). *Estuarine morphodynamics of the Sunderbans* Coastal Research Library(11). Springer International Publishing. Available from 10.1007/978-3-319-11343-2.

Das, K. (2016). Sundarban embankments-A study along Suryaberia River. *International Journal of Current Research.*, *8*(5), 32187–32195.

Das, T. (2014). Embankment breaching and related hazards_ A study through system approach at Indian Sundarbans.

Dhara, S., & Paul, A. K. (2016). Embankment breaching and its impact on local community in Indian Sundarban: A case study of some blocks of South West Sundarban. *IJISET.*, *3*(2), 23–32.

Ghosh, S., & Mistri, B. (2020). Geo-historical appraisal of embankment breaching and its management on active tidal land of Sundarban: A case study in Gosaba Island, South 24 Parganas, West Bengal. *Space and Culture, India*, *7*(4), 166–180. Available from https://doi.org/10.20896/saci.v7i40.587.

Gour, J. (2021). Stages of land reclamation and their impact on the fluvio-geomorphological environment of Indian Sundarbans with special reference to Matla–Bidyadhari Interfluve. *Modern Cartography Series, 10*. Available from https://doi.org/10.1016/B978-0-12-823895-0.00022-1. Available from:, http://www.elsevier.com/wps/find/bookdescription.cws_home/BS_MCS/description#description.

Harrison, H.L. (1875). The Bengal embankment manual: Containing an account of the action of the government in dealing with embankments and water-courses since the permanent settlement, a discussion of the Principles of the Act of 1873, a reprint of the act itself, with notes and appendices describing the present conditions of embankments under state control in Bengal and the chief conroversies concerning them. Bengal Secretariat Press, Bengal Secretariat Press.

Hunter, W. W. (1877). *A statistical account of Bengal, 20*.

Mohamed, M. A., Samuels, P. G., & Morris, M. W. (2002). Improving the accuracy of prediction of breach formation through embankment dams and flood embankments. *River Flow., 2002*.

O'Malley, L.S.S. (1914). Bengal District Gazetteers: 24-Parganas. Bengal Secretariat Book Depot, Bengal Secretariat Book Depot.

Ozbek, S.E. (2017). Numerical modeling of seepage in Koyunbaba Dam [Doctoral dissertation].

Pargiter, F. E. (1934). *A revenue history of the Sundarbans, from 1765 to 1870. Superintendent, Government Print*. Bengal Government Press.

Rudra, K. (2011). The proposal of strengthening embankment in Sundarban: Myth and reality.

Sarkar, H. (2018). Significance of embankments breaching in southern blocks of South 24 Parganas District, West Bengal. *Journal of Emerging Technologies and Innovative Research (JETIR).*, *5*(8), 683–696.

Sarkar, H., Roy, A., & Siddique, G. (2016). Impact of embankment breaching and rural livelihood: A case study in Ghoramara Island of the Sundarbans Delta in South 24 Parganas. *The Journal of Bengal Geographer.*, *5*(4), 97–117.

Sarkar, J. (2002). Vulnerability of Sundarban due to recurrent coastal flooding National Seminar on Vulnerability of Sundarban Mangrove Ecosystem in the Perspective of Global Climate Change. 14–15.

Sarkhel, P. (2013). Examining private participation in embankment maintenance in the Indian Sundarbans.

Further reading

Hazra, S., & Samanta, K. (2016). Temporal change detection (2001–2008): Study of Sundarban. *IDEAS*. Available from. Available from https://ideas.repec.org/p/ess/wpaper/id10526.html.

Chapter 21

Basin-scale agricultural drought across India under changing climate

Mayank Suman[1] and Rajib Maity[2]

[1]*Department of Civil and Infrastructure Engineering, Indian Institute of Technology Jodhpur, Jodhpur, Rajasthan, India,* [2]*Department of Civil Engineering, Indian Institute of Technology Kharagpur, Kharagpur, West Bengal, India*

21.1 Introduction

Agricultural water is sourced from either surface or subsurface sources. Management of this resource (such as better irrigation practices and the development of irrigation canal network), along with other technological interventions such as the introduction of improved crop breeds, synthetic fertilizers, pesticides, and others, had a positive impact on agricultural production in the recent past (Pingali, 2012). For instance, with only a 30% expansion in cultivable land, crop production has increased three times since the 1960s (Wik, Pingali, & Brocai, 2008). However, both sources of agricultural water resources are vulnerable to the impact of climate change (De Wit & Stankiewicz, 2006; Green et al., 2011; Suman & Maity, 2020; Taylor et al., 2013; Trenberth, 2011), among other factors. Additionally, the population and thus the agricultural yield to sustain the population are expected to increase in the future (Pingali, 2012; Wik *et al.*, 2008). Hence, information about agricultural water availability and its proper management will be helpful in meeting food demand in the future.

Geospatial techniques using remote sensing products have been applied to estimate and manage agricultural water on a basin or field scale (Ifediegwu, 2022; Vishwakarma, Rena, Singh, & Mukherjee, 2021). Geospatial techniques can be used for terrain analysis such as watershed delineation (Datta et al., 2022; Lehner, Verdin, & Jarvis, 2008) or demarcation of groundwater quality domains (Aravinthasamy, Karunanidhi, Subramani, & Roy, 2021). This can further be used for estimating and managing water resources on the basin level (Li, Wang, Li, & Liu, 2022;

Water Resources Monitoring, Management, and Sustainability. DOI: https://doi.org/10.1016/B978-0-443-23665-5.00021-1

Neupane & Kumar, 2015; Pumo, Arnone, Francipane, Caracciolo, & Noto, 2017; Tadese, Kumar, Koech, & Kogo, 2020). Multiple distributed hydrological models exist that can be used for simulating surface, subsurface flow, and other hydrological processes (such as infiltration and evapotranspiration) in a basin (Suliman, Jajarmizadeh, Harun, & Darus, 2015; Yang, Yang, Chen, & Zhao, 2019; Yang et al., 2020). Soil moisture at different scales (basin or field scale) has been estimated using a synthetic aperture radar remote sensing product (Pal et al., 2017; Srivastava, 2017). Similarly, remote sensing products have also been used for the identification of crops and the estimation of total crop water demand (hence, the irrigation requirement) for a large region (Calera, Campos, Osann, D'Urso, & Menenti, 2017; El-Magd & Tanton, 2005). The crop water stress is also analyzed using geospatial techniques (Shivers, Roberts, & McFadden, 2019).

Geospatial techniques can also be used for designing mitigation measures against the impact of changing climates and increasing populations on agricultural water resources. The techniques can be used to estimate future water resource availability (Suman & Maity, 2020) and irrigation requirements (Kadiyala et al., 2015). Geospatial techniques have also been used for studying the status of observed and future agricultural droughts. Agricultural drought is a long-term period of soil moisture deficit, which reduces agricultural yield. The risk of agricultural drought has increased in the recent past, as per multiple studies (Dai, 2011; Dai, Trenberth, & Qian, 2004; Vicente-Serrano et al., 2014; Wang, Chen, Pan, & Li, 2015; Xu et al., 2015). However, some studies do suggest that the risk of agricultural drought has not changed significantly in the past (Greve et al., 2014; Seneviratne et al., 2012; Sheffield, Wood, & Roderick, 2012). Despite this disagreement about the status of agricultural drought in the past, there is a general consensus about the increased risk of agricultural drought in the future. Additionally, the prediction of agricultural drought is important for ensuring food safety in the future. An increase in global temperature is expected to increase evapotranspiration, which will lead to a more drying tendency in soil moisture and surface water resources (Annamalai, Hafner, Sooraj, & Pillai, 2013; Trenberth, Dai, Rasmussen, & Parsons, 2003; Trenberth, 2011). These changes will lead to more frequent and intense droughts, with an increase in area under drought conditions (Chanda & Maity, 2017; Stocker, 2014; Suman & Maity, 2021).

In this chapter, we focus on agricultural drought and its spatial variation. A case study from Maity, Suman, and Verma (2016) and Suman and Maity (2021) is presented in this chapter to highlight the role geospatial techniques can play in accessing the availability of agricultural water resources in the future.

21.2 Case study: spatial distribution of agricultural drought on subbasin scale across Indian mainland

In the context of India, the reliable prediction of agricultural drought can have significant social-economic importance as the majority of the population is directly or indirectly related to the agricultural sector. The contribution of the agricultural sector to the national GDP is 16% (Ministry of Finance, 2018). Hence, there are multiple studies that focus on the past and future status of droughts of different kinds, such as meteorological, hydrological, or agricultural, in different regions of India (Bisht, Sridhar, Mishra, Chatterjee, & Raghuwanshi, 2019; Mallya, Mishra, Niyogi, Tripathi, & Govindaraju, 2015; Mishra, Shah, & Thrasher, 2014; Naresh Kumar, Murthy, Sesha Sai, & Roy, 2012; Ojha, Nagesh Kumar, Sharma, & Mehrotra, 2013; Pai, Sridhar, Guhathakurta, & Hatwar, 2011; Sharma & Mujumdar, 2017; Thomas, Nayak, & Ghosh, 2015; Zhang, Obringer, Wei, Chen, & Niyogi, 2017). Most of the studies were not pan-India and focused on meteorological drought (i.e., prolonged precipitation deficit in a region). However, from a food security point of view, the pan-India study of the future status of agricultural drought might have more utility. Additionally, agricultural droughts do not occur in isolation. Usually, long periods of meteorological drought may transform/propagate into an agricultural drought (Maity & Suman, 2019; Maity *et al.*, 2016). This aspect of drought propagation should also be given proper consideration while studying agricultural drought. Additionally, the use of high-resolution data for studying drought propagation may help, as drought is expected to have regional variation (Bisht et al., 2019; Mallya et al., 2015; Naresh Kumar et al., 2012). While keeping these aspects, a model, as shown in Fig. 21.1, is formulated to assess the status of agricultural drought over the Indian mainland during the future period. Details of the methodology are presented in the next section.

21.2.1 Methodology

There are five modules in the methodology: (1) bias correction of CORDEX-simulated climatic variables (precipitation, temperature, and soil moisture) and evapotranspiration estimation using bias-corrected climatic variables; (2) selection of most suitable CORDEX simulation for each basin depending upon its skill in simulating monthly precipitation on basin level; (3) characterization of drought using the bias-corrected variables from the most suitable CORDEX model; (4) modeling of temporal consequences of meteorological drought over agricultural drought; and (5) frequency and trend analysis of agricultural drought. The details about the different steps of the model and results are discussed in subsequent sections.

21.2.1.1 Bias correction of CORDEX simulation outputs

CORDEX-simulated estimates of climatic/soil variables (such as precipitation, soil moisture, and near-surface air temperature) are used to characterize drought. However, the CORDEX-simulated variables may have bias, and the bias characteristics might be different for different variables. For instance, a

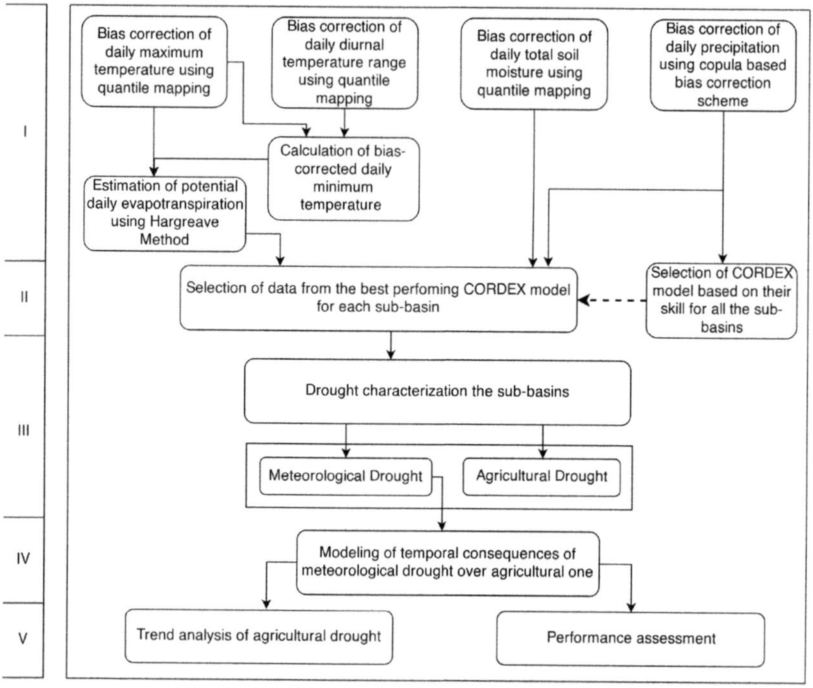

FIGURE 21.1 The methodological overview for assessing the future status of agricultural drought over the Indian mainland. *From Suman, M., & Maity, R. (2021). Assessment of basin-wise future agricultural drought status across India under changing climate.* Journal of Water and Climate Change, 12*(6), 2400–2421. https://doi.org/10.2166/wcc.2021.369.*

zero-inflated variable such as daily precipitation will have different bias characteristics compared to a bounded variable such as soil moisture. Hence, different methodologies are used for bias-correcting different variables. A copula-based bias correction scheme was used for bias-correcting daily precipitation (Maity, Suman, Laux, & Kunstmann, 2019). Quantile Mapping (QM) is used for bias-correcting soil moisture.

For bias-correcting the near-surface air temperature (both minimum and maximum air temperature), QM is applied to the near-surface maximum air temperature and near-surface diurnal temperature difference. A bias-corrected minimum air temperature can be obtained by the difference of bias-corrected values of maximum air temperature and diurnal temperature difference. The bias-corrected diurnal temperature difference is expected to always be positive. Hence, this scheme of bias correction ensures that the maximum air temperature is never smaller than the minimum air temperature for any day (Thrasher, Maurer, McKellar, & Duffy, 2012). With this bias-corrected air temperature, the potential evapotranspiration (PET) can be calculated by the Hargreaves Method (Hargreaves, 1994; Spinoni, Vogt,

Naumann, Barbosa, & Dosio, 2018). It should be noted that for estimation of PET, three widely used methods exist: Penman-Monteith (Feng, Trnka, Hayes, & Zhang, 2017; Liu, Wang, Shao, Jia, & Li, 2016), Hargreaves (Hargreaves, 1994; Spinoni et al., 2018), and Thornthwaite (Feng et al., 2017). The Penman-Monteith method is the most data-intensive and accurate among the three models; however, the Thornthwaite method requires less data but is less accurate as compared to other methods. The Hargreaves Method considers multiple factors, such as mean daily air temperature (say T_{avg}), diurnal temperature difference (D_T), and extraterrestrial radiation at a given latitude (R_a), which is a function of solar declination at the location (S_d), to estimate the daily values of PET.

$$ET_0 = 0.0023R_a(T_{avg} + 17.8)\sqrt{D_T} \tag{21.1}$$

The extraterrestrial radiation is calculated as:

$$R_a = 0.408\frac{118r_d}{\pi}[\omega\sin(l)\sin(S_d) + \cos(l)\cos(S_d)\sin(\omega)] \tag{21.2a}$$

$$\omega = arc\cos(-\tan(S_d)\tan(l)) \tag{21.2b}$$

$$S_d = M_a\sin\left(\frac{2\pi}{365}(d-80)\right) \tag{21.2c}$$

$$r_d = 1 + \frac{1}{30}\cos\left(\frac{2\pi d}{365}\right) \tag{21.2d}$$

where l, r_d, S_d, and M_a are latitude of the location, the relative distance between the Earth and the Sun for d^{th} day, the Solar declination for d^{th} day, and the mean axial tilt of the Earth (23.43673°), respectively. The estimation of PET will help in characterizing meteorological drought at a later stage. It should be noted that even after bias correction, all the CORDEX simulations being utilized may not have the same skill to simulate a climatic variable (say precipitation) on the basin level (Maraun et al., 2017). Hence, CORDEX simulations are vetted for their skill in simulating basin precipitation to select a suitable CORDEX model for a basin.

21.2.1.2 Selection of the most suitable CORDEX simulation on the basin scale

As all CORDEX models are not expected to have the same skill for capturing different climatological and hydrological phenomena at basin level, the most suitable CORDEX simulation for every basin is selected using outranking Multi-Criteria Decision-Making (MCDM) methods such as Compromise Programming (CP) (Kumar, 2010; Raju & Kumar, 2014; Srinivasa Raju, Sonali, & Nagesh Kumar, 2017) and Preference Ranking Organization Method of Enrichment Evaluation-2 (PROMETHEE2) (Brans, Vincke, &

Mareschal, 1986; Kumar, 2010; Raju & Kumar, 2014). First, the performance of different CORDEX simulations with respect to observed precipitation series is quantified using multiple performance measures (such as coefficient of determination, mean absolute error, unbiased root mean square error, and others), and then MCDM methods are used to rank them based on their relative skill in simulating basin precipitation. The details of these MCDM methods are provided in the subsequent section.

21.2.1.2.1 Compromise programming

CP ranks different CORDEX models based on the weighted p-norm between the performance measure for the CORDEX model and the best possible/ideal values of the performance measure. The CORDEX model with the lowest p-norm is selected as the best. The weighted p-norm for a model $m\left(L_p(m)\right)$ is given by:

$$L_p(m) = \left[\sum_{i=1}^{s} \left(w_i \left|f_i(m) - g_i\right|\right)^p\right]^{(1/p)} \tag{21.3}$$

where s, w_i, $f_i(m)$, and g_i are a total number of performance measures, weight for i^{th} performance measure, value of ith performance measure for model m, and best possible value of ith performance measure, respectively. In this study, L_2 norm or weighted Euclidean distance is used for ranking different CORDEX models.

The relative weight for different performance measures can be achieved using the relative measurement of the entropy of individual performance measures, which is outlined in the following steps:

1. To reduce the effect of scale, all performance measures are normalized to range [0,1] by following the expression:

$$f'_{ij} = \frac{f_{ij}}{\sum_{i=1}^{N} f_{ij}} \tag{21.4}$$

   where f_{ij} is i^{th} performance measure for j^{th} model. f'_{ij} is the normalized performance measure. N is the total number of CORDEX models being considered. For this normalization step to work effectively, the performance measure should be lower bound by zero.

2. The entropy of normalized performance measures is then calculated. The entropy for i^{th} performance measure is given by:

$$E_i = \frac{1}{\ln(N)} \sum_{j=1}^{N} f'_{ij} \tag{21.5}$$

Entropy is a direct measure of the uncertainty involved with the performance measure. The higher the entropy, the greater the uncertainty.

3. The degree of information diversification for i^{th} performance measure (D_i) is calculated as $1 - E_i$. Hence, the degree of information diversification will be low for a performance measure with high uncertainty/entropy.
4. The weight for individual performance measures is calculated as the relative degree of information diversion with respect to other performance measures. Hence, the weight for i^{th} performance measure is given by:

$$w_i = \frac{D_i}{\sum_{i=1}^{s} D_i} \tag{21.6}$$

21.2.1.2.2 PROMETHEE2

As the name suggests, preference ranking organization method of enrichment evaluation-2 (PROMETHEE2) ranks different models (say m_j and m_k) based on the preference function, which is a function of pairwise difference for a performance measure between two models (say $d_i(m_j, m_k)$ for i^{th} performance measure). In this study, a usual criteria preference function is utilized, which is given by:

$$P_i(m_j, m_k) = \mathbb{I}(d_i(m_j, m_k) > 0) \tag{21.7}$$

where $P_i(m_j, m_k)$ and $\mathbb{I}(\bullet)$ denote the usual criteria preference function for i^{th} performance measure between m_j and m_k CORDEX models, and identity function which outputs unity when the condition is true otherwise 0, respectively. The preference function for all performance measures between two CORDEX models can be weighted to estimate the multicriteria preference index for those two CORDEX models (say $I(m_j, m_k)$).

$$I(m_j, m_k) = \frac{\sum_{i=1}^{s} w_i P_i(m_j, m_k)}{\sum_{i=1}^{s} w_i} \tag{21.8}$$

where s and w_i are the total number of performance measures, and weight assigned to i^{th} performance measure, respectively. Like CP, the weights for performance measures are estimated using their entropy.

Multicriteria preference index ($I(m_j, m_k)$) is a measure of the relative strength of the model m_j over m_k for all performance measures, as it was calculated using $d_i(m_j, m_k)$. Hence, if $I(m_j, m_k) > I(m_k, m_j)$, then the model m_j can be considered better than the model m_k. Hence, the ranking of one model with respect to another can be obtained by ranking the mean difference of multicriteria preference indices for all the pairs of models involving the selected model. This quantity is called the outranking index for the selected model ($\phi(m_j)$).

$$\phi(m_j) = \frac{\sum_{k=1,k\neq j}^{N} \left(I(m_j, m_k) - I(m_k, m_j)\right)}{N-1} \text{ for } j \in \{1, 2, ., N\} \quad (21.9)$$

The CORDEX model with the highest outranking index can be selected as a CORDEX model demonstrating good skill in simulating basin precipitation.

21.2.1.3 Drought characterization

After the selection of the most suitable CORDEX model at the basin level, the variables from the selected CORDEX model are used for the characterization of drought in different basins. Additionally, being a long-lasting dynamic phenomenon, drought can be characterized by different time periods. As this study focuses on assessing agricultural drought status across India, a time period of 3 months is used for characterizing drought, which is the average period for cropping in India. It should be noted that a number of drought indices for characterizing different types of droughts exist; however, none of them can be considered universal, and the selection of a drought index depends upon the use case (Keyantash & Dracup, 2002). In this study, as a temporal consequence or interrelation of different droughts will be used for assessing future agricultural drought status, the drought indices used to characterize meteorological and agricultural droughts should be mathematically consistent.

The standardized Precipitation Evapotranspiration Index (SPEI) is used for meteorological drought characterization. SPEI is calculated from the monthly water deficit or climatic water balance, which is calculated as a difference between monthly precipitation and PET (Beguería, Vicente-Serrano, Reig, & Latorre, 2014; Vicente-Serrano, Beguería, & López-Moreno, 2010). Hence, in this study, bias-corrected precipitation from the most suitable CORDEX model and PET calculated using the Hargreaves method are used to calculate the monthly climatic water balance. Mathematically, SPEI is similar to the Standardized Precipitation Index (McKee, Doesken, & Kleist, 1993); however, SPEI is more suitable for studying meteorological drought under a warming climate (Beguería et al., 2014; Liu et al., 2016; Vicente-Serrano *et al.*, 2010). SPEI is calculated using the following steps:

1. The monthly climatic water balance time series is moving averaged for 3-month time period.
2. A three-parameter log-logistic PDF is fitted to the moving averaged time series. The CDF is then calculated after fitting the PDF.

3. The nonexceeding cumulative probability or reduced variate for every month is calculated from the fitted CDF.
4. SPEI is obtained by transforming the reduced variate into the standard normal variate. Hence, SPEI shows the times of the standard deviation that the monthly climatic water balance is away from the mean condition.

Hence, a positive value of SPEI represents a wetter condition compared to the average, and a negative value represents a drier condition. Meteorological drought conditions can be defined as conditions of SPEI being less than some selected negative threshold. Different thresholds can be selected to identify droughts of different intensities. The agricultural drought is characterized using the Standardized Soil Moisture Index (SSMI), which is calculated in similar manner as compared to SPEI. For calculating SSMI, bias-corrected soil moisture from the most suitable CORDEX model is utilized.

21.2.1.4 Modeling of drought interrelationship

This interrelationship between the drought indices is then studied using their wavelet components using the methodology developed by Maity et al. (2016). Three models, as listed in Table 21.1, are formulated for modeling the temporal consequences of meteorological drought on agricultural drought.

It should be noted that the interrelationship or temporal consequence analysis at the wavelet component level produces better results compared to the study of parent drought indices without wavelet decomposition (Maity et al. 2016). All three models presented in Table 21.1 differ in their assumptions. For instance, model 1 assumes that the agricultural drought index at the current time should be related to meteorological drought component values at past time steps. Similarly, model 2 additionally incorporates the memory of components of the agricultural drought index while estimating SSMI. The third model estimates the SSMI components and then calculates the SSMI series based on wavelet reconstruction.

The temporal consequences of drought may vary with time and can be affected by intermittent events like a short spell (usually a day) of rainfall. Hence, the modeling of temporal consequences is done using a moving window scheme. In the scheme, a calibration period length is selected, and the model is trained on that calibration period. The trained model is then used for estimating the SSMI at the next (only one) time step. Next, the calibration window is moved by one (by keeping the calibration period length the same), and the model is re-calibrated. The re-calibrated model is used for one prediction again. This cycle is continued until we have an estimate for the entire length of SSMI. This scheme will respond better to modeling dynamic relations, which may be affected by intermittent events like 1-day rainfall. Additionally, this modeling scheme will capture short-moving changes in time series and is hence desirable in a changing climate.

TABLE 21.1 Modeling of the temporal consequence of meteorological drought on agricultural drought.

Model No.	Model formulation
1	$SSMI\left(t\right) = f\begin{pmatrix} SPEI_{a_2}(t-T_1), SPEI_{d_2}(t-T_1), SPEI_{d_1}(t-T_1), \\ SPEI_{a_2}(t-T_2), SPEI_{d_2}(t-T_2), SPEI_{d_1}(t-T_2) \end{pmatrix}$
2	$SSMI\left(t\right) = f\begin{pmatrix} SPEI_{a_2}(t-T_1), SPEI_{d_2}(t-T_1), SPEI_{d_1}(t-T_1), \\ SPEI_{a_2}(t-T_2), SPEI_{d_2}(t-T_2), SPEI_{d_1}(t-T_2), \\ SSMI_{a_2}(t-T_1), SSMI_{d_2}(t-T_1), SSMI_{d_1}(t-T_1) \end{pmatrix}$
3	$SSMI_C(t) = f\begin{pmatrix} SPEI_{a_2}(t-T_1), SPEI_{d_2}(t-T_1), SPEI_{d_1}(t-T_1), \\ SPEI_{a_2}(t-T_2), SPEI_{d_2}(t-T_2), SPEI_{d_1}(t-T_2), \\ SSMI_{a_2}(t-T_1), SSMI_{d_2}(t-T_1), SSMI_{d_1}(t-T_1) \end{pmatrix} \text{ for } C \in \{a_2, d_2, d_1\}$ $SSMI(t) = g\left(SSMI_{d_1}, SSMI_{d_2}, SSMI_{a_2}\right)$

Notes: The function *f* stands for multiple linear regression. T_1 is the minimum time lag to ensure that past data is not utilized given the multi-level wavelet decomposition. $T_1 = 2^\wedge 1$ and $T_2 = T_1 + 1$, where l is the level of decomposition. For this study, $l = 2$.

The performance of these models (as presented in Table 21.1) is evaluated by the coefficient of determination (R^2), degree of agreement (D_r), Nash-Sutcliffe efficiency (*NSE*), and unbiased root mean square error (*uRMSE*). The details about these performance measures can be found in the literature (Krause, Boyle, & Bäse, 2005; Maity *et al.* 2016; Willmott, Robeson, & Matsuura, 2012). In brief, R^2 measures the percentage of variability in the original time series explained by the predicted time series. Similarly, D_r compares the deviation between observed and predicted time series with respect to the deviation of the observed series from its mean.

$$D_r = \begin{cases} 1 - D_{rf} \text{for} D_{rf} \leq 1 \\ \dfrac{1}{D_{rf}} - 1 \text{for} D_{rf} > 1 \end{cases} D_{rf} = \frac{\sum_{t=1}^{n} \left|\hat{Y}(t) - Y(t)\right|}{2\sum_{t=1}^{n} \left|Y(t) - \overline{Y}\right|} \tag{21.10}$$

where $Y(t)$ and $\hat{Y}(t)$ are observed and predicted values at the time step t, respectively. The $\overline{Y}$ represents the mean of the observed time series. *NSE* measures the predictive skill of the hydrological model by comparing the square of the deviation of the observed and predicted time series and the square of the deviation of the observed time series from its mean.

$$NSE = 1 - \frac{\sum_{t=1}^{n} \left(\hat{Y}(t) - Y(t)\right)^2}{\sum_{t=1}^{n} \left(Y(t) - \overline{Y}\right)^2} \tag{21.11}$$

uRMSE is *RMSE* between "deviation from mean" series of observed and predicted time series.

21.2.2 Study area and data used

The Indian mainland is divided into 226 basins, as shown in Fig. 21.2. Six CORDEX simulations (listed in Table 21.2) of climatic variables, such as precipitation, total soil moisture content, and air temperature (minimum and maximum) at daily scale, are re-gridded to 0.5° latitude × 0.5° longitude during 1961–2100 (1961–2005 is termed the historical period, and future simulations for RCP8.5 and RCP4.5 are utilized for 2006–2100).

For bias-correcting these CORDEX simulations, observed precipitation records from IMD (Pai, Rajeevan, Sreejith, Mukhopadhyay, & Satbha, 2014) and reanalysis estimates of soil moisture and air temperature from ERA5 (Hersbach, 2016) are used. ERA5 data is available after 1979; hence, ERA5 data from 1979–95 is used for bias correction, and the rest of the data from the historical period (1996–2005) is used for assessing the bias correction performance.

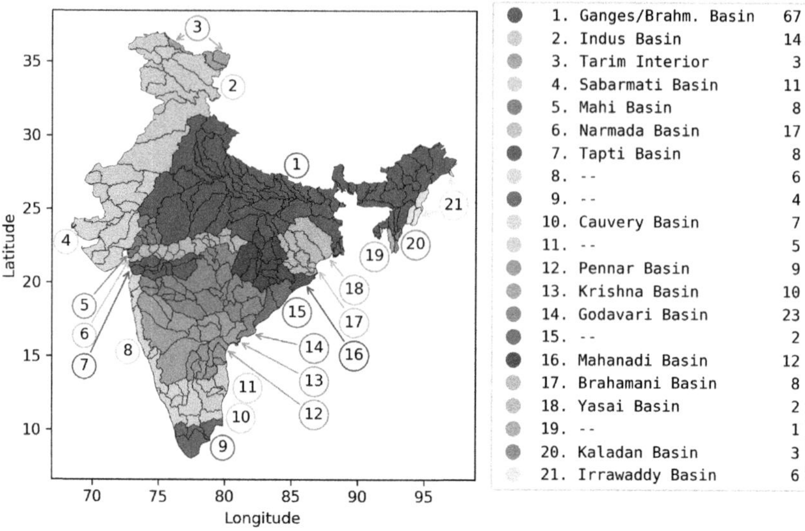

FIGURE 21.2 226 subbasins and their 21 major groups across the Indian mainland. If the major group is a larger basin, then the name is provided in legend, otherwise, the group name is marked as names such as group no. 8, 9, and so on. Map lines delineate study areas and do not necessarily depict accepted national boundaries. *From Suman, M., & Maity, R. (2021). Assessment of basin-wise future agricultural drought status across India under changing climate.* Journal of Water and Climate Change, 12*(6), 2400–2421. https://doi.org/10.2166/wcc.2021.369.*

TABLE 21.2 Details of CORDEX simulations with their general circulation model (GCM) forcing utilized in the study.

S. No.	RCM used	GCM forcing	Spatial resolution	Temporal resolution	Institute
1.	REMO2009	MPI-ESM-LR	0.5° lat by 0.5° lon	Daily	MPI-CSC
2.	RegCM4	GFDL-ESM2M	50 km by 50 km	Daily	IITM
3.	RegCM4	MPI-ESM-MR	50 km by 50 km	Daily	IITM
4.	RegCM4	CNRM-CM5	50 km by 50 km	Daily	IITM
5.	RegCM4	CSIRO-Mk3.6.0	50 km by 50 km	Daily	IITM
6.	RegCM4	IPSL-CM5A-LR	50 km by 50 km	Daily	IITM

Source: Adapted from Suman, M., & Maity, R. (2021). Assessment of basin-wise future agricultural drought status across India under changing climate. *Journal of Water and Climate Change, 12*(6), 2400–2421. https://doi.org/10.2166/wcc.2021.369.

However, while modeling the temporal consequence of meteorological drought (characterized by SPEI) on agricultural drought, the calibration period length is selected as 21 years (1979–99). The remaining historical period (i.e., 2000–05) is considered a validation period. If modeling the temporal consequences of meteorological drought yields similar performance during the calibration and validation periods, then the model is used for predicting and assessing the future status of agricultural drought. This is done as the uncertainty involved with precipitation prediction in CORDEX is comparatively less compared to the uncertainty involved in soil moisture estimation; hence, foreseeing the status of agricultural drought based on the information of meteorological drought will be more useful compared to assessing agricultural drought based on CORDEX soil moisture simulations directly.

21.2.3 Results and discussion

21.2.3.1 Bias correction of CORDEX simulation and selection of the most suitable CORDEX model for basin

Among all the variables being bias-corrected, daily precipitation is a zero-inflated variable. Hence, daily precipitation was bias-corrected using a stochastic bivariate copula-based approach as proposed by Maity et al. (2019). The bias-corrected approach works better in correcting bias for different quantiles of precipitation and retaining the seasonal variation across the country (Suman, Maity, & Kunstmann, 2022). The bias-corrected precipitation dataset can be obtained from Suman, Maity, and Kunstmann (2019).

Other variables (soil moisture content, maximum air temperature, and diurnal temperature range) are bias-corrected using QM methodology. The bias-corrected maximum air temperature is found to match the spatio-temporal variation of the observed maximum air temperature better when compared to CORDEX simulations (Fig. 21.3). Similar results are obtained for the diurnal temperature range and soil moisture content. Hence, by virtue of modeling the spatio-temporal variability of observed variables much better, the bias-corrected values of these variables are desirable for hydrological modeling when compared to CORDEX simulations. A bias-corrected minimum air temperature is obtained by subtracting bias-corrected values of the diurnal temperature range from the bias-corrected maximum air temperature. This scheme of bias-correcting the minimum air temperature ensures that the maximum air temperature is never smaller than the minimum air temperature. The PET is estimated for every grid point using the Hargreaves method.

The bias-corrected variables at the monthly scale are obtained by accumulating daily values over the month for each grid point. These values are then spatially averaged over each subbasin by using the Thiessen polygon

FIGURE 21.3 Spatio-temporal variation of bias-corrected maximum temperature throughout the Indian mainland. The results are shown for the RegCM4 RCM model forced by MPI-ESM-Mr GCM. Map lines delineate study areas and do not necessarily depict accepted national boundaries. *From Suman, M., & Maity, R. (2021). Assessment of basin-wise future agricultural drought status across India under changing climate.* Journal of Water and Climate Change, 12 *(6), 2400–2421. https://doi.org/10.2166/wcc.2021.369.*

method to get a time series of bias-corrected variables for each subbasin at a monthly scale. The skill of the CORDEX model is then evaluated on monthly precipitation using two MCDM approaches: (1) CP, and (2) PROMETHEE2. The performance measures used to characterize the skill of CORDEX models are R^2, Taylor Skill Score (*TSS*) (Taylor, 2001), Mean Absolute Error (*MAE*), and *uRMSE*. The spatial distribution of the most suitable CORDEX simulation is shown in Fig. 21.4.

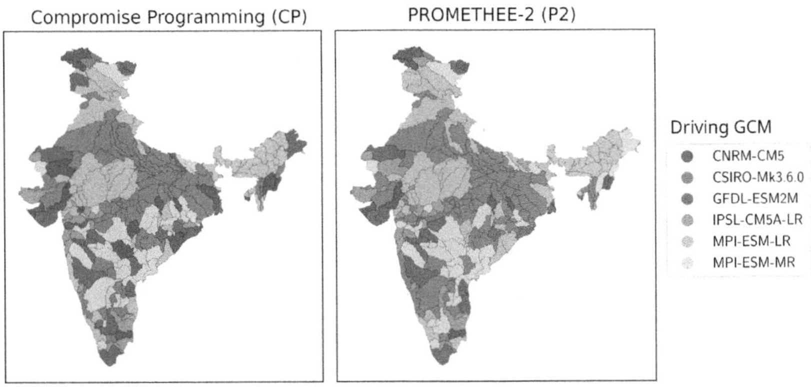

FIGURE 21.4 Most suitable CORDEX models (along with driving GCM in legend) for basins across the Indian mainland. Map lines delineate study areas and do not necessarily depict accepted national boundaries. *GCM*, General circulation model.

The most suitable CORDEX simulation might not be the same for a subbasin for both MCDM approaches, as they use different criteria for ranking CORDEX simulations. To reiterate, CP ranks CORDEX simulations based on the deviation of their performance from the best possible performance. However, PROMETHEE2 intercompares CORDEX simulations based on differences in their performance and the criteria function selected. Despite this difference in ranking criteria, RegCM4 RCM, driven by MPI-ESM-LR and CSIRO-Mk3.6.0 General Circulation Models, is simulating the monthly precipitation, which is the best in most of the subbasins across northern and northeastern India. However, the most suitable CORDEX simulation shows more variability in the case of southern India. The bias-corrected data from the most suitable CORDEX simulation for different subbasins is then used for modeling the temporal consequence of meteorological drought on agricultural drought. It should be noted that the skill of CORDEX simulation is tested for only one variable (i.e., monthly precipitation) for each basin. If the skill of the CORDEX simulation is tested for another variable, it may result in a situation where the selected data for further steps might come from different CORDEX simulations. Hence, evaluation of the CORDEX skill in predicting precipitation only ensures that all the selected variables are from the same CORDEX models. The temporal consequence of meteorological drought is then modeled using the bias-corrected values from selected CORDEX models.

21.2.3.2 Modeling of the temporal consequence of drought

Using the bias-corrected monthly precipitation and PET, SPEI is calculated at a 3-month averaging period for characterizing meteorological drought. Similarly, bias-corrected values of soil moisture are utilized to calculate SSMI at a 3-month averaging period. The temporal consequence modeling

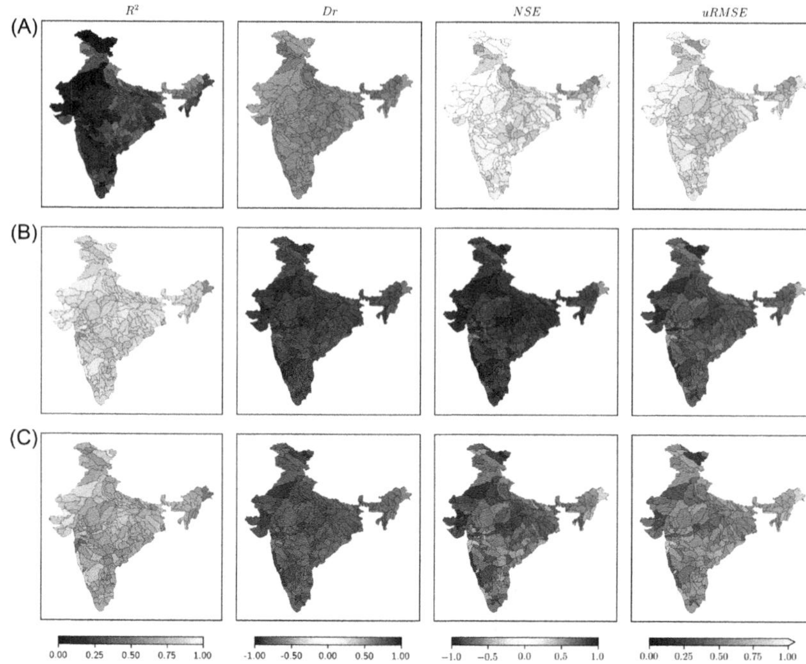

FIGURE 21.5 Development period performance of Model 1–3 (A–C) for predicting the status of agricultural drought from meteorological drought (characterized by SPEI) across the Indian mainland using data from the CORDEX model selected using PROMETHEE2. Map lines delineate study areas and do not necessarily depict accepted national boundaries.

of the effect of meteorological drought over agricultural drought is then modeled using the models listed in Table 21.1. The calibration period for the model is selected as 21 years, and the performance of the models is validated over 2000–05 by using the moving window approach. The performance of all models during the development period for the CORDEX model selected using PROMETHEE2 is shown in Fig. 21.5.

Despite varying climatology across the Indian mainland, model 2 outperformed other models, as evidenced by high values of R^2, D_r, NSE, and a low value of $uRMSE$ for subbasins across the Indian mainland. It should be noted that model 2 included the past values of SSMI while predicting future values of SSMI. Hence, the better performance of model 2 reveals that SSMI, or soil moisture, has a high innate memory. This conclusion is found to be true across subbasins in India. The performance of model 2 during the validation period for the most suitable CORDEX models selected from both MCDM methods is also shown in Fig. 21.6.

Comparing the Figs. 21.5 and 21.6, the performance of model 2 is comparable during the development and validation periods. Hence, model 2 is neither under- nor overfitting. Furthermore, among all the model options,

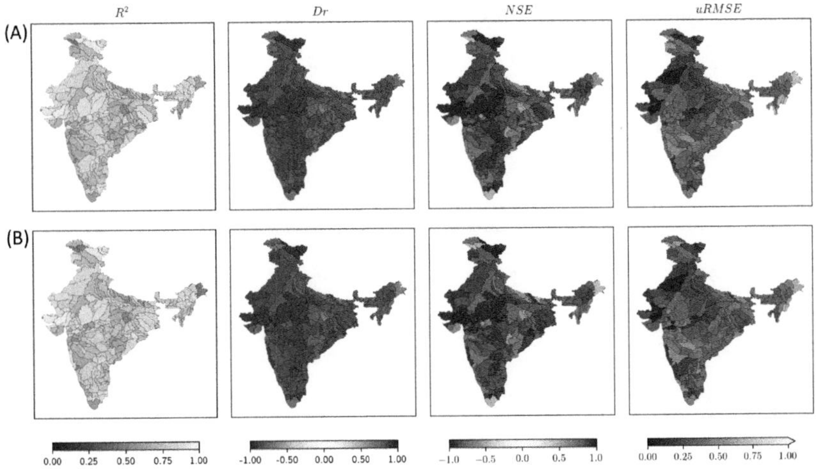

FIGURE 21.6 Performance of model 2 during the validation period for most suitable CORDEX models selected from (A) Compromise Programming and (B) PROMETHEE2 across the Indian mainland. Meteorological drought is characterized by the SPEI. Map lines delineate study areas and do not necessarily depict accepted national boundaries.

model 2 performs the best and, thus, is utilized for the assessment of future agricultural droughts.

21.2.3.3 Assessment of future agricultural drought

The assessment of future agricultural drought in basins across India is carried out using the estimated SSMI series from model 2. Four realizations of the SSMI series are obtained for two MCDM models (CP and PROMETHEE2) and two RCP scenarios (RCP 4.5 and RCP 8.5). The statistical significance of the trend in the SSMI series is worked out using Mann-Kendall's test at a 5% significance level. It should be noted that an increasing (or positive) trend in SSMI shows wetting conditions whereas a decreasing (or negative) trend in SSMI shows drying conditions. The trend analysis of the future SSMI series (Fig. 21.7) shows a geographically contrasting change in the distribution of agricultural drought. For instance, more wet conditions are expected to prevail in most of the subbasins of southern India. However, a drying trend is expected for most northern, central, and northeastern Indian basins. These geographically contrasting changes in agricultural drought might be related to the higher change in mean precipitation expected in southern India compared to north and central India (Suman & Maity, 2020). To study the temporal change in the trend of agricultural drought, the entire future period (2006–2100) is divided into three epochs: E1 (2006–35), E2 (2036–70), and E3 (2071–2100). Epochwise analysis suggests either a drying or insignificant SSMI trend in South Indian basins by the end of the century.

FIGURE 21.7 Trend of SSMI series for different future periods using model 2. Columns indicate different time period in the future. The row name shows the MCDM method (CP for Compromise Programming and P2 for PROMETHEE2) and RCP scenario. The significance of the trend is assessed using Mann-Kendall Test at a 5% level of significance. Subbasins showing no significant trend are shown in white patches. Map lines delineate study areas and do not necessarily depict accepted national boundaries. *From Suman, M., & Maity, R. (2021). Assessment of basin-wise future agricultural drought status across India under changing climate.* Journal of Water and Climate Change, 12*(6), 2400–2421. https://doi.org/10.2166/wcc.2021.369.*

To assess the changes in the intensity of agricultural drought, two thresholds are selected: a moderate drought condition when SSMI ≤ -1, and an extreme drought condition when SSMI ≤ -2. The percentage area under either drought conditions (moderate [SSMI ≤ -1] or extreme conditions [SSMI ≤ -2]) on the monthly scale is expected to increase continuously in the future (Fig. 21.8). For instance, moderate and extreme drought conditions

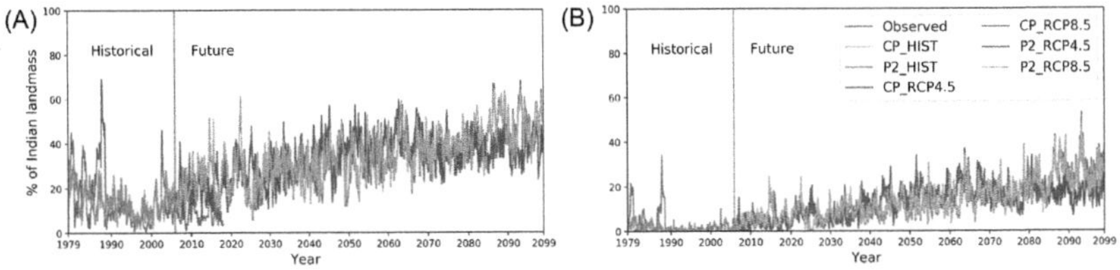

FIGURE 21.8 Monthly series of percentage area of Indian mainland under (A) moderate (SSMI ≤ − 1), and (B) extreme (SSMI ≤ − 2) agricultural drought conditions. "Observed" indicates the SSMI calculated using ERA5 reanalysis data for 1979–2018. *From Suman, M., & Maity, R. (2021). Assessment of basin-wise future agricultural drought status across India under changing climate.* Journal of Water and Climate Change, 12*(6), 2400–2421. https://doi.org/10.2166/wcc.2021.369.*

are expected to prevail in at least 50% and 20% of the Indian mainland, respectively, for the RCP 8.5 scenario by the century end. Also, a similar trend (but in a lesser area) is visible for RCP 4.5. Additionally, the result is consistent for the CORDEX model selected on the basis of different MCDM methods.

Furthermore, the percentage time a subbasin spends in extreme drought conditions was calculated, and its spatial variation is shown in Fig. 21.9. As expected, most of the subbasins of north, central, and north-east India are showing a higher percentage of time under extreme drought conditions. However, subbasins in south India and some large basins in western India are expected to spend less time under extreme drought conditions. Additionally, it should be noted that many subbasins with high vulnerability to agricultural drought fall in important agricultural regions of the country. Hence, an adverse impact on the food security of India is expected in the future due to these geographically distinct changes in agricultural drought.

21.3 Conclusion

In this chapter, a review is presented about the use of geospatial techniques for managing agricultural water resources, followed by a case study dealing with the basin-wise assessment of agricultural drought with a changing climate over the Indian mainland. Major conclusions from the study are:

1. The risk of agricultural drought is increasing throughout different subbasins across India, except the subbasins situated in the southern part of the country.
2. Agricultural drought conditions will prevail on the larger part of the Indian mainland. However, drought vulnerability may change from region to region. For instance, agricultural drought is expected to become more frequent in the northern, central, and northeastern parts of India. However, the same is not expected for the subbasins of southern India.

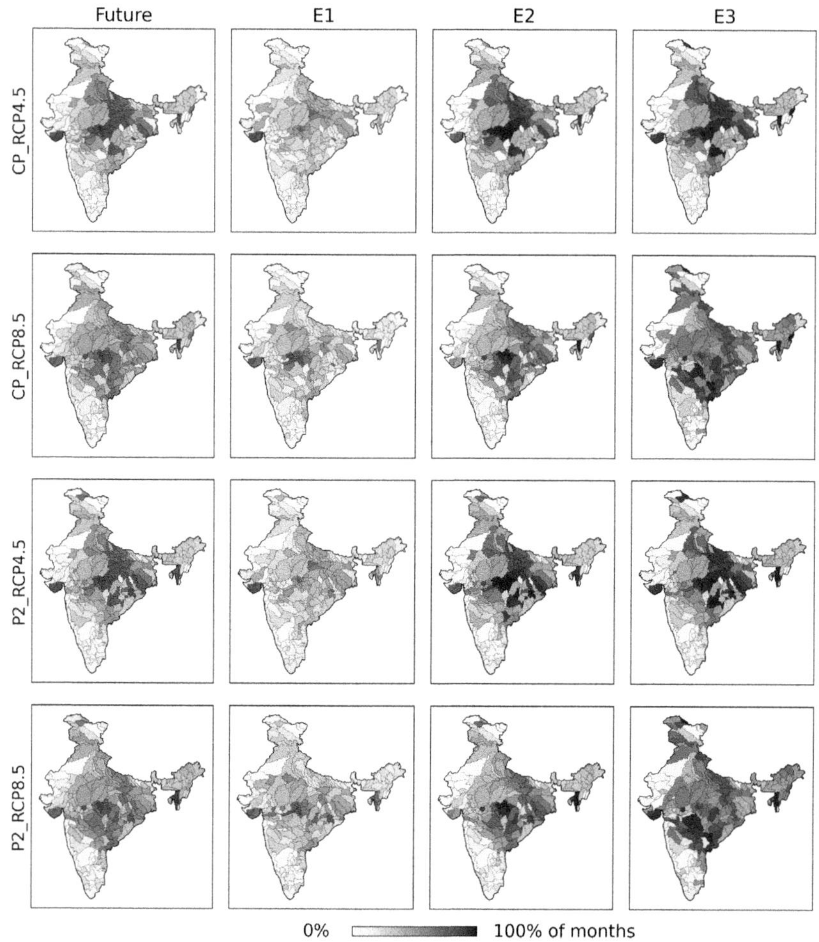

FIGURE 21.9 Percentage time a subbasin remains under extreme drought conditions for different future time periods. Map lines delineate study areas and do not necessarily depict accepted national boundaries. *From Suman, M., & Maity, R. (2021). Assessment of basin-wise future agricultural drought status across India under changing climate.* Journal of Water and Climate Change, 12*(6), 2400–2421. https://doi.org/10.2166/wcc.2021.369.*

3. About 20% and 50% of the Indian mainland are expected to suffer from moderate and extreme drought conditions, respectively, by the end of this century. Most of the subbasins in northern, central, and northeastern India are expected to spend more time under drought conditions as compared to subbasins in southern India.

In this study, irrigation frequency and irrigation pattern due to canals are not considered, which may affect the status of agricultural drought on the field level. Uneven distribution of precipitation (Suman & Maity, 2020) and geographically contrasting distribution of agricultural drought over Indian landmass may result in overexploitation of soil water (Goldin, 2016). This groundwater overexploitation may result in the reduction of the water table, which will adversely affect agricultural yield. This aspect can be investigated in future studies. Additionally, soil moisture has an important feedback relationship with multiple other climatic/hydrological variables. Hence, the effect on other basin phenomena can be studied, given the expected decrease in soil moisture.

References

Annamalai, H., Hafner, J., Sooraj, K. P., & Pillai, P. (2013). Global warming shifts the monsoon circulation, drying South Asia. *Journal of Climate*, *26*(9), 2701–2718. Available from https://doi.org/10.1175/JCLI-D-12-00208.1UnitedStates, http://journals.ametsoc.org/doi/pdf/10.1175/JCLI-D-12-00208.1.

Aravinthasamy, P., Karunanidhi, D., Subramani, T., & Roy, P. D. (2021). Demarcation of groundwater quality domains using GIS for best agricultural practices in the drought-prone Shanmuganadhi River basin of South India. *Environmental Science and Pollution Research*, *28*(15), 18423–18435. Available from https://doi.org/10.1007/s11356-020-08518-5, https://link.springer.com/journal/11356.

Beguería, S., Vicente-Serrano, S. M., Reig, F., & Latorre, B. (2014). Standardized precipitation evapotranspiration index (SPEI) revisited: Parameter fitting, evapotranspiration models, tools, datasets and drought monitoring. *International Journal of Climatology*, *34*(10), 3001–3023. Available from https://doi.org/10.1002/joc.3887, http://onlinelibrary.wiley.com/journal/10.1002/(ISSN)1097-0088.

Bisht, D. S., Sridhar, V., Mishra, A., Chatterjee, C., & Raghuwanshi, N. S. (2019). Drought characterization over India under projected climate scenario. *International Journal of Climatology*, *39*(4), 1889–1911. Available from https://doi.org/10.1002/joc.5922, http://onlinelibrary.wiley.com/journal/10.1002/(ISSN)1097-0088.

Brans, J. P., Vincke, P., & Mareschal, B. (1986). How to select and how to rank projects: The Promethee method. *European Journal of Operational Research*, *24*(2), 228–238. Available from https://doi.org/10.1016/0377-2217(86)90044-5.

Calera, A., Campos, I., Osann, A., D'Urso, G., & Menenti, M. (2017). Remote sensing for crop water management: From ET modelling to services for the end users. *Sensors*, *17*(5), 1104. Available from https://doi.org/10.3390/s17051104.

Chanda, K., & Maity, R. (2017). Assessment of trend in global drought propensity in the twenty-first century using drought management index. *Water Resources Management*, *31*(4), 1209–1225. Available from https://doi.org/10.1007/s11269-017-1571-3, http://www.wkap.nl/journalhome.htm/0920-4741.

Dai, A. (2011). Drought under global warming: A review. *Wiley Interdisciplinary Reviews: Climate Change*, *2*(1), 45–65. Available from https://doi.org/10.1002/wcc.81, http://onlinelibrary.wiley.com/journal/10.1002/(ISSN)1757-7799.

Dai, A., Trenberth, K. E., & Qian, T. (2004). A global dataset of Palmer Drought Severity Index for 1870-2002: Relationship with soil moisture and effects of surface warming. *Journal of Hydrometeorology*, *5*(6), 1117–1130. Available from https://doi.org/10.1175/JHM-386.1.

Datta, S., Karmakar, S., Mezbahuddin, S., Hossain, M. M., Chaudhary, B. S., Hoque, M. E., ... Baul, T. K. (2022). The limits of watershed delineation: Implications of different DEMs,

DEM resolutions, and area threshold values. *Hydrology Research*, *53*(8), 1047–1062. Available from https://doi.org/10.2166/nh.20220.126, https://watermark.silverchair.com/nh0531047.pdf?.

De Wit, M., & Stankiewicz, J. (2006). Changes in surface water supply across Africa with predicted climate change. *Science (New York, N.Y.)*, *311*(5769), 1917–1921. Available from https://doi.org/10.1126/science.1119929.

El-Magd, I. A., & Tanton, T. (2005). Remote sensing and GIS for estimation of irrigation crop water demand. *International Journal of Remote Sensing*, *26*(11), 2359–2370. Available from https://doi.org/10.1080/0143116042000298261, https://www.tandfonline.com/loi/tres20.

Feng, S., Trnka, M., Hayes, M., & Zhang, Y. (2017). Why do different drought indices show distinct future drought risk outcomes in the U.S. Great Plains? *Journal of Climate*, *30*(1), 265–278. Available from https://doi.org/10.1175/JCLI-D-15-0590.1, http://journals.ametsoc.org/doi/pdf/10.1175/JCLI-D-15-0590.1.

Goldin, T. (2016). India's drought below ground. *Nature Geoscience*, *9*(2), 98. Available from https://doi.org/10.1038/ngeo2648, 1752-0894.

Green, T. R., Taniguchi, M., Kooi, H., Gurdak, J. J., Allen, D. M., Hiscock, K. M., … Aureli, A. (2011). Beneath the surface of global change: Impacts of climate change on groundwater. *Journal of Hydrology*, *405*(3-4), 532–560. Available from https://doi.org/10.1016/j.jhydrol.2011.050.002.

Greve, P., Orlowsky, B., Mueller, B., Sheffield, J., Reichstein, M., & Seneviratne, S. I. (2014). Global assessment of trends in wetting and drying over land. *Nature Geoscience*, *7*(10), 716–721. Available from https://doi.org/10.1038/NGEO2247, http://www.nature.com/ngeo/index.html.

Hargreaves, G. H. (1994). Defining and using reference evapotranspiration. *Journal of Irrigation and Drainage Engineering*, *120*(6), 1132–1139. Available from https://doi.org/10.1061/(ASCE)0733-9437(1994)120:6(1132).

Hersbach, H. (2016). The ERA5 Atmospheric Reanalysis. AGU fall meeting abstracts. 2016.

Ifediegwu, S. I. (2022). Assessment of groundwater potential zones using GIS and AHP techniques: A case study of the Lafia district, Nasarawa State, Nigeria. *Applied Water Science*, *12*(1). Available from https://doi.org/10.1007/s13201-021-01556-5, http://www.springer.com/earth + sciences + and + geography/hydrogeology/journal/13201.

Kadiyala, M. D. M., Nedumaran, S., Piara, S., Chukka, S., Irshad, M. A., & Bantilan, M. C. S. (2015). An integrated crop model and GIS decision support system for assisting agronomic decision making under climate change. *Science of The Total Environment*, *521-522*, 123–134. Available from https://doi.org/10.1016/j.scitotenv.2015.03.097, 00489697.

Keyantash, J., & Dracup, J. A. (2002). The quantification of drought: An evaluation of drought indices. *Bulletin of the American Meteorological Society*, *83*(8), 1167–1180. Available from http://www.ams.allenpress.com, https://doi.org/10.1175/1520-0477(2002)083 < 1191:TQODAE > 2.3.CO;2.

Krause, P., Boyle, D. P., & Bäse, F. (2005). Comparison of different efficiency criteria for hydrological model assessment. *Advances in Geosciences*, *5*, 89–97. Available from https://doi.org/10.5194/adgeo-5-89-2005, http://www.adv-geosci.net/volumes.html.

Kumar, D. N. (2010). *Multicriterion analysis in engineering and management*. PHI Learning Pvt. Ltd.

Lehner, B., Verdin, K., & Jarvis, A. (2008). New global hydrography derived from spaceborne elevation sata. *Eos, Transactions American Geophysical Union*, *89*(10), 93–94. Available from https://doi.org/10.1029/2008eo100001.

Li, P., Wang, D., Li, W., & Liu, L. (2022). Sustainable water resources development and management in large river basins: An introduction. *Environmental Earth Sciences*, *81*(6). Available from https://doi.org/10.1007/s12665-022-10298-9, https://link.springer.com/journal/12665.

Liu, Z., Wang, Y., Shao, M., Jia, X., & Li, X. (2016). Spatiotemporal analysis of multiscalar drought characteristics across the Loess Plateau of China. *Journal of Hydrology*, *534*, 281–299. Available from https://doi.org/10.1016/j.jhydrol.2016.010.003, http://www.elsevier.com/inca/publications/store/5/0/3/3/4/3.

Maity, R., Suman, M., (2019). Predictability of hydrological systems using the wavelet transformation: Application to drought prediction springer water. Springer Nature, India, 109–137. Available from: http://www.springer.com/series/13419. 10.1007/978-3-030-02197-9_5.

Maity, R., Suman, M., & Verma, N. K. (2016). Drought prediction using a wavelet based approach to model the temporal consequences of different types of droughts. *India Journal of Hydrology*, *539*, 417–428. Available from https://doi.org/10.1016/j.jhydrol.2016.05.042, http://www.elsevier.com/inca/publications/store/5/0/3/3/4/3.

Maity, R., Suman, M., Laux, P., & Kunstmann, H. (2019). Bias correction of zero-inflated RCM precipitation fields: A copula-based scheme for both mean and extreme conditions. *Journal of Hydrometeorology*, *20*(4), 595–611. Available from https://doi.org/10.1175/JHM-D-18-0126.1, https://journals.ametsoc.org/doi/pdf/10.1175/JHM-D-18-0126.1.

Mallya, G., Mishra, V., Niyogi, D., Tripathi, S., & Govindaraju, R. S. (2015). Trends and variability of droughts over the Indian monsoon region. *Weather and Climate Extremes*, *12*, 43–68. Available from https://doi.org/10.1016/j.wace.2016.010.002, http://www.journals.elsevier.com/weather-and-climate-extremes/.

Maraun D. Shepherd T.G. Widmann M. Zappa G. Walton D. Gutiérrez J.M., ... Mearns L.O. 2017, Towards process-informed bias correction of climate change simulations, Nature Climate Change, Nature Publishing Group Austria. 11, 764–773 http://www.nature.com/nclimate/index.html, Available from https://doi.org/10.1038/nclimate3418.

McKee T.B. Doesken N.J. Kleist J. 1993 The relationship of drought frequency and duration to time scales, Proceedings of the 8th Conference on Applied Climatology, 179–183, 17.

Ministry of Finance, India. (2018). Economic Survey 2017-18. http://mofapp.nic.in:8080/economicsurvey/.

Mishra, V., Shah, R., & Thrasher, B. (2014). Soil moisture droughts under the retrospective and projected climate in India. *Journal of Hydrometeorology*, *15*(6), 2267–2292. Available from https://doi.org/10.1175/JHM-D-13-0177.1, http://journals.ametsoc.org/doi/pdf/10.1175/JHM-D-13-0177.1.

Naresh Kumar, M., Murthy, C. S., Sesha Sai, M. V. R., & Roy, P. S. (2012). Spatiotemporal analysis of meteorological drought variability in the Indian region using standardized precipitation index. *Meteorological Applications*, *19*(2), 256–264. Available from https://doi.org/10.1002/met0.277, http://onlinelibrary.wiley.com/journal/10.1002/(ISSN)1469-8080.

Neupane, R. P., & Kumar, S. (2015). Estimating the effects of potential climate and land use changes on hydrologic processes of a large agriculture dominated watershed. *Journal of Hydrology*, *529*(1), 418–429. Available from https://doi.org/10.1016/j.jhydrol.2015.070.050, http://www.elsevier.com/inca/publications/store/5/0/3/3/4/3.

Ojha, R., Nagesh Kumar, D., Sharma, A., & Mehrotra, R. (2013). Assessing severe drought and wet events over india in a future climate using a nested bias-correction approach. *Journal of Hydrologic Engineering*, *18*(7), 760–772. Available from https://doi.org/10.1061/(ASCE)HE.1943-5584.0000585, 10840699.

Pai, D. S., Sridhar, L., Guhathakurta, P., & Hatwar, H. R. (2011). District-wide drought climatology of the southwest monsoon season over India based on standardized precipitation index (SPI. *Natural Hazards*, *59*(3), 1797–1813. Available from https://doi.org/10.1007/s11069-011-9867-8, 0921030X.

Pai, D. S., Rajeevan, M., Sreejith, O. P., Mukhopadhyay, B., & Satbha, N. S. (2014). Development of a new high spatial resolution (0.25° × 0.25°) long period (1901-2010) daily gridded rainfall data set over India and its comparison with existing data sets over the region. *MAUSAM*, *65*(1), 1–18. Available from https://doi.org/10.54302/mausam.v65i10.851.

Pal, M., Maity, R., Suman, M., Das, S. K., Patel, P., & Srivastava, H. S. (2017). Satellite-based probabilistic assessment of soil moisture using C-band quad-polarized RISAT1 data. *IEEE Transactions on Geoscience and Remote Sensing*, *55*(3), 1351–1362. Available from https://doi.org/10.1109/TGRS.2016.2623378.

Pingali, P. L. (2012). Green revolution: Impacts, limits, and the path ahead. *Proceedings of the National Academy of Sciences*, *109*(31), 12302–12308. Available from https://doi.org/10.1073/pnas.0912953109, 0027-8424.

Pumo, D., Arnone, E., Francipane, A., Caracciolo, D., & Noto, L. V. (2017). Potential implications of climate change and urbanization on watershed hydrology. *Journal of Hydrology*, *554*, 80–99. Available from https://doi.org/10.1016/j.jhydrol.2017.090.002, http://www.elsevier.com/inca/publications/store/5/0/3/3/4/3.

Raju, K. S., & Kumar, D. N. (2014). Ranking of global climate models for India using multicriterion analysis. *Inter-Research, India Climate Research*, *60*(2), 103–117. Available from https://doi.org/10.3354/cr01222, http://www.int-res.com/articles/cr2014/60/c060p103.pdf.

Seneviratne, S. I., Nicholls, N., Easterling, D., Goodess, C. M., Kanae, S., Kossin, J., ... Zwiers, W. (2012). Changes in climate extremes and their impacts on the natural physical environment. *Managing the Risks of Extreme Events and Disasters to Advance Climate Change Adaptation: Special Report of the Intergovernmental Panel on Climate Change*. Available from https://doi.org/10.1017/CBO97811391772450.006, http://doi.org/10.1017/CBO9781139177245.

Sharma, S., & Mujumdar, P. (2017). Increasing frequency and spatial extent of concurrent meteorological droughts and heatwaves in India. *Nature Publishing Group, India Scientific Reports*, *7*(1). Available from https://doi.org/10.1038/s41598-017-15896-3, http://www.nature.com/srep/index.html.

Sheffield, J., Wood, E. F., & Roderick, M. L. (2012). Little change in global drought over the past 60 years. *Nature*, *491*(7424), 435–438. Available from https://doi.org/10.1038/nature11575.

Shivers, S. W., Roberts, D. A., & McFadden, J. P. (2019). *Using paired thermal and hyperspectral aerial imagery to quantify land surface temperature variability and assess crop stress within California orchards*, *Remote Sensing of Environment* (222 pp. 215–231). Available from https://doi.org/10.1016/j.rse.2018.12.030, http://www.elsevier.com/inca/publications/store/5/0/5/7/3/3.

Spinoni, J., Vogt, J. V., Naumann, G., Barbosa, P., & Dosio, A. (2018). Will drought events become more frequent and severe in Europe? *International Journal of Climatology*, *38*(4), 1718–1736. Available from https://doi.org/10.1002/joc.5291, http://onlinelibrary.wiley.com/journal/10.1002/(ISSN)1097-0088.

Srinivasa Raju, K., Sonali, P., & Nagesh Kumar, D. (2017). Ranking of CMIP5-based global climate models for India using compromise programming. *Theoretical and Applied*

Climatology, *128*(3-4), 563–574. Available from https://doi.org/10.1007/s00704-015-1721-6, http://link.springer.de/link/service/journals/00704/.

Srivastava, P. K. (2017). Satellite soil moisture: Review of theory and applications in water resources. *Water Resources Management*, *31*(10), 3161–3176. Available from https://doi.org/10.1007/s11269-017-1722-6, http://www.wkap.nl/journalhome.htm/0920-4741.

Stocker, T. Climate change 2013: The physical science basis: Working Group I contribution to the Fifth assessment report of the Intergovernmental Panel on Climate Change. Cambridge University Press (2014).

Suliman, A. H. A., Jajarmizadeh, M., Harun, S., & Darus, I. Z. M. (2015). Comparison of semi-distributed, GIS-based hydrological models for the prediction of streamflow in a large catchment. *Water Resources Management*, *29*(9), 3095–3110. Available from https://doi.org/10.1007/s11269-015-0984-0, http://www.wkap.nl/journalhome.htm/0920-4741.

Suman, M., & Maity, R. (2020). Southward shift of precipitation extremes over south Asia: Evidences from CORDEX data. *Nature Research, India Scientific Reports*, *10*(1). Available from https://doi.org/10.1038/s41598-020-63571-x, http://www.nature.com/srep/index.html.

Suman, M., & Maity, R. (2021). Assessment of basin-wise future agricultural drought status across india under changing climate. *Journal of Water and Climate Change*, *12*(6), 2400–2421. Available from https://doi.org/10.2166/wcc.20210.369, https://watermark.silverchair.com/jwc0122400.pdf?.

Suman, M., Maity, R., & Kunstmann, H. (2019). Bias corrected daily most expected and extreme CORDEX precipitation data set for India. *Figshare*. Available from https://doi.org/10.6084/m9.figshare.9686648.v2.

Suman, M., Maity, R., & Kunstmann, H. (2022). Precipitation of mainland India: Copula-based bias-corrected daily CORDEX climate data for both mean and extreme values. *Geoscience Data Journal*, *9*(1), 58–73. Available from https://doi.org/10.1002/gdj30.118, http://onlinelibrary.wiley.com/journal/10.1002/(ISSN)2049-6060.

Tadese, M., Kumar, L., Koech, R., & Kogo, B. K. (2020). Mapping of land-use/land-cover changes and its dynamics in Awash River Basin using remote sensing and GIS. *Remote Sensing Applications: Society and Environment*, *19*, 100352. Available from https://doi.org/10.1016/j.rsase.2020.100352.

Taylor, K. E. (2001). Summarizing multiple aspects of model performance in a single diagram. *Journal of Geophysical Research Atmospheres*, *106*(7), 7183–7192. Available from https://doi.org/10.1029/2000JD900719, http://onlinelibrary.wiley.com/journal/10.1002/(ISSN)2169-8996.

Taylor, R. G., Scanlon, B., Döll, P., Rodell, M., Van Beek, R., Wada, Y., ... Treidel, H. (2013). Ground water and climate change. *Nature Climate Change*, *3*(4), 322–329. Available from https://doi.org/10.1038/nclimate1744.

Thomas, T., Nayak, P. C., & Ghosh, N. C. (2015). Spatiotemporal analysis of drought characteristics in the bundelkhand region of central india using the standardized precipitation index. *Journal of Hydrologic Engineering*, *20*(11). Available from https://doi.org/10.1061/(ASCE)HE.1943-5584.0001189, https://ascelibrary.org/journal/jhyeff.

Thrasher, B., Maurer, E. P., McKellar, C., & Duffy, P. B. (2012). Technical note: Bias correcting climate model simulated daily temperature extremes with quantile mapping. *Hydrology and Earth System Sciences*, *16*(9), 3309–3314. Available from https://doi.org/10.5194/hess-16-3309-2012.

Trenberth, K. E. (2011). Changes in precipitation with climate change. *Climate Research*, *47*(1-2), 123–138. Available from https://doi.org/10.3354/cr00953UnitedStates, http://www.int-res.com/articles/cr_oa/c047p123.pdf.

Trenberth, K. E., Dai, A., Rasmussen, R. M., & Parsons, D. B. (2003). The changing character of precipitation. *Bulletin of the American Meteorological Society, 84*(9), 1205–1218. Available from https://doi.org/10.1175/BAMS-84-9-1205.

Vicente-Serrano, S. M., Beguería, S., & López-Moreno, J. I. (2010). A multiscalar drought index sensitive to global warming: The standardized precipitation evapotranspiration index. *Journal of Climate, 23*(7), 1696–1718. Available from https://doi.org/10.1175/2009JCLI2909.1Spain, http://journals.ametsoc.org/doi/pdf/10.1175/2009JCLI2909.1.

Vicente-Serrano, S. M., Lopez-Moreno, J. I., Beguería, S., Lorenzo-Lacruz, J., Sanchez-Lorenzo, A., García-Ruiz, J. M., ... Espejo, F. (2014). Evidence of increasing drought severity caused by temperature rise in southern Europe. *Environmental Research Letters, 9*(4). Available from https://doi.org/10.1088/1748-9326/9/4/044001, http://iopscience.iop.org/1748-9326/9/4/044001/pdf/1748-9326_9_4_044001.pdf.

Vishwakarma, C.A., Rena, V., Singh, D., Mukherjee, S., Exploration of water resources using remote sensing and geographic information system Groundwater Geochemistry: Pollution and Remediation Methods. Wiley, India (2021), 364–378. Available from: https://www.wiley.com/en-us/Groundwater + Geochemistry%3A + Pollution + and + Remediation + Methods-p-9781119709701. Available from: https://doi.org/10.1002/9781119709732.ch18.

Wang, H., Chen, Y., Pan, Y., & Li, W. (2015). Spatial and temporal variability of drought in the arid region of China and its relationships to teleconnection indices. *Journal of Hydrology, 523*, 283–296. Available from https://doi.org/10.1016/j.jhydrol.2015.010.055, http://www.elsevier.com/inca/publications/store/5/0/3/3/4/3.

Wik, M., Pingali, P., Brocai, P., Global agricultural performance: past trends and future prospects. (2008).

Willmott, C. J., Robeson, S. M., & Matsuura, K. (2012). A refined index of model performance. *International Journal of Climatology, 32*(13), 2088–2094. Available from https://doi.org/10.1002/joc.2419.

Xu, K., Yang, D., Yang, H., Li, Z., Qin, Y., & Shen, Y. (2015). Spatio-temporal variation of drought in China during 1961-2012: A climatic perspective. *Elsevier, China Journal of Hydrology, 526*, 253–264. Available from https://doi.org/10.1016/j.jhydrol.2014.090.047, http://www.elsevier.com/inca/publications/store/5/0/3/3/4/3.

Yang, S., Yang, D., Chen, J., Zhao, B., Real-time reservoir operation using recurrent neural networks and inflow forecast from a distributed hydrological model. Journal of Hydrology. 00221694. 579 (2019), doi:10.1016/j.jhydrol.2019.124229, 124229.

Yang, S., Yang, D., Chen, J., Santisirisomboon, J., Lu, W., & Zhao, B. (2020). A physical process and machine learning combined hydrological model for daily streamflow simulations of large watersheds with limited observation data. *Journal of Hydrology, 579*, 124229. Available from https://doi.org/10.1016/j.jhydrol.2020.125206.

Zhang, X., Obringer, R., Wei, C., Chen, N., & Niyogi, D. (2017). *Droughts in India from 1981 to 2013 and implications to wheat production* 20452322*China Scientific Reports* (p. 7). Available from http://www.nature.com/srep/index.html, 10.1038/srep44552.

Chapter 22

Roles and challenges of drainage basin management institutions in promoting the integrated Fish Rice-Vegetable farming technology model in the Lower Ankobra Basin

Dickson Okoree Mireku
Department of Academic Planning, Directorate of Academic Planning and Quality Assurance, University of Cape Coast, Cape Coast, Ghana

22.1 Introduction

Integrated farming systems (IFSs) have garnered significant attention in recent years as sustainable and efficient approaches to agricultural production (Green, Webb, Johnson, Wynn, & Bogen, 2023). One notable example is the Integrated Fish, Rice, and Vegetable (IFRV) farming technology model, emerging as a climate-smart solution for enhancing productivity, improving resource utilization, and promoting environmental sustainability (Nguyen et al., 2019). IFRV farming model entail the simultaneous cultivation of fish, rice, and vegetables within a single farming unit, harnessing natural interactions and complementarities among these components (Gill, Singh, & Gangwa, 2009). This multitiered farming system mimics ecological interactions found in nature, striving to optimize resource utilization, maximize production, and minimize environmental impact (Sarkar et al., 2020). IFRV model often employ sustainable practices such as organic farming techniques, water management strategies, and the integration of aquaculture, agriculture, and livestock production (Shahbaz, Rashid, Saleem, Mackey, & McKay, 2023).

Several countries in Europe, North America, Australia, and New Zealand have successfully implemented IFRV farming technology models, yielding

Water Resources Monitoring, Management, and Sustainability. DOI: https://doi.org/10.1016/B978-0-443-23665-5.00022-3

positive results (Halbrendt et al., 2014). Notable achievements of this practice include the reduction in synthetic fertilizer usage in the IFS model in China and the optimization of land (Van Hong, Thao Nguyen, Tran Ngoc Huy, Thu Thuy, & Thanh Huong, 2021) and water resources in Vietnam. Similar successes are observed in rice–shrimp agroecosystem farmers in Brazil (Fierro-Sañudo, Rodríguez-Montes de Oca, & Páez-Osuna, 2020) and North American countries, including the United States, where fish pond sediments and by-products contribute to rice and vegetable production (Van Hong et al., 2021). The smallholders' practices of integrated agriculture aquaculture systems in sub-Saharan African food production have also shown significant success (Kinkela, Dochain, Rollin, Mafwila, & Bindelle, 2019). For example, empirical and documentary evidence from the Forum for Agriculture Research in Africa's (2010) Report indicates increased food production and rural incomes after the implementation of IFS to support indigenous communities (Diaz-Chavez, Mutimba, Watson, Rodriguez-Sanchez, & Nguer, 2010).

Besides providing a symbiotic relationship between the crops and fish farm products, the adoption of the IFS has been observed to increase crop yields, reduce greenhouse gas emissions, and improve food security in many developing economies, including African countries (Ghosh et al., 2020). An example of IFRV farming in Africa is in Nigeria, where the Feed the Future Innovation Lab for Fish has established six on-farm adaptive research plots to increase food production for surrounding local communities. In Tanzania, the World Fish Center has been working with farmers to promote integrated fish-rice farming systems, which have been shown to increase yields and improve food security. Finally, the Integrated Fish Farming Project has been implemented along many river courses to promote the businesses of smallholder farmers toward increasing their incomes as a poverty reduction measure. In Ghana, there are few cases of the implementation of the IFS with documented evidence of some positive results when vegetables and rice cultivations were integrated into aquaculture farming systems (Callo-Concha, Gaiser, & Ewert, 2012). These studies highlight ecological sustainability, increased yields (Shoko, Getabu, & Mwayuli, 2011), and an increase in income levels for local farmers.

Recognizing the numerous benefits associated with IFS, the Ghanaian government has shown a strong commitment to promoting sustainable IFS in every region. The government's efforts became evident through the implementation of strategic agricultural innovation programs and policies, such as the Rapid Appraisal of Agricultural Innovation policy (Schut et al., 2015). These initiatives aimed to revamp key IFS projects at notable farm sites within river catchments to facilitate the adoption of IFS by local farmers to increase productivity, enhance livelihoods, and promote the sustainability of water resources in catchments like river basins.

In the Western Region of Ghana, the lower portions of the Ankobra Basin were identified as having several resources conducive to supporting the IFRV farming system for increased food production. These resources include extensive rice fields, arable farmlands suitable for vegetable cultivation, aquatic resources for fish farming, traditional knowledge and experiences in mixed farming practices, favorable climatic conditions, a significant agricultural workforce offering cheap farm labor, ready market access, and support from agro-processing industries and institutions. It is crucial to consider climate change impacts and adapt farming practices accordingly. The availability of skilled labor, combined with supportive institutions and market access, enhances the potential success of implementing the IFRV farming system in the Lower Ankobra Basin.

22.2 Literature review

22.2.1 Institutional framework for resource utilization in the Lower Ankobra River Basin

This research is underpinned by the Integrated Drainage Basin Governance (IDBG) Framework developed by Mireku, Acheampong, Antwi, and Hlovor (2022) to evaluate the institutional framework supporting the implementation of the IFRV farming system. Specifically known as the IDBG framework. It is tailored to assess the institutional dynamics required for overseeing water resource utilization in the Lower Ankobra River Basin. The subsequent discussions provide details on the structures or components of the IDBG Framework. To properly assess the principle of accountability in the management of the Lower Ankobra River Basin, two routes of accountability, namely, the short route and the long route, have been included in the IDBG framework. The inclusion of these routes was necessary because they can serve as common pathways through which local people may follow to demand accountability from social service providers. As such, the United Nations Development Program and Sustainable Human Development (1997) opined that avenues opened to resource users can be used to determine the approach that they may use to demand accountability. Consequently, these two avenues (the short and long routes) have been included for the users of the Lower Ankobra River Basin to select the one that is ideal to demand accountability from management institutions.

22.3 Results and discussion

22.3.1 The roles of the top drainage basin resource management institutions

The literature review identifies three key drainage basin management institutions: the Water Resources Commission (WRC), the National Development Planning Commission (NDPC), and the Ministry of Works and Housing

(MWH), with the latter two being subsidiaries of the WRC. The WRC, established in 1996, was mandated to plan and manage projects related to water resource use in catchment areas, such as drainage basins in Ghana. It was also empowered to coordinate with the NDPC and MWH for national development plans and policy formulation.

The WRC, in addition to policy and planning, was entrusted with soliciting funds from the Ministry of Finance, according to Section 12 of the Water Resource Management Act 522 of 1996. This legal mandate allowed the WRC to support its collaborative management institutions, including the Forestry Commission (FC) and the Environmental Protection Agency (EPA). The legal framework was further strengthened by additional laws and regulations, such as Supreme Military Council Decree 85 of 1977, empowering the Irrigation Development Authority, and National Liberation Council Decree 293 of 1969, empowering the EPA. The Ghana Water Company Limited (GWCL) and FC were also assigned specific roles. While the legal framework established a strong foundation for the management institutions, the study raises questions about their logistical and human resource capacity to effectively implement government-formulated policies. This leads to an assessment of the institutions' capacity to manage resources in the Lower Ankobra River Basin.

22.3.2 The roles of local-level drainage basin management institutions

The successful realization of the IFRV farming system project within the Lower Ankobra Basin hinges on the strategic and collaborative roles of local drainage basin-level management institutions. The following details the roles of the local drainage basin management institutions. Following the documentary literature review approach, the following roles of the local basin management institutions were identified: The GWCL's role within the project involves the extension of water supply and distribution infrastructure to previously underserved rural areas within the Lower Ankobra Basin. This expansion aims to support the essential irrigation needs of rice and vegetable farming (Gross, Senadza, Hodey, & Abdalla, 2015). The EPA plays a pivotal role in regulating and monitoring farming activities within the project to ensure strict compliance with environmental standards designed to safeguard water quality and protect the local ecosystems within the Lower Ankobra Basin. The MoFA is instrumental in this context, as it should actively engage in developing agricultural policies and programs that align with and encourage the IFS. This includes promoting sustainable agricultural practices, technologies, and research that are directly applicable to fish, rice, and vegetable production.

22.3.3 Challenges of local water resources management institutions

Evidence from the review of the literature shows that Ghana has a Water Use Regulations and Legislative Instrument LI, 1692 of 2001, which supports local water resources management institutions to promulgate by-laws to address unapproved uses of water resources that are likely to affect projects in river basins. Section 10 of the Legislative Instrument stipulates that a person commits an offense when he or she:

- Fails to follow the right procedure to obtain a use permit before extracting surface or underground water from any drainage basin;
- Refuses to produce a permit upon demand by any agent authorized by the government to regulate the general development of drainage basin resources;
- Fails or refuses to pay the prescribed penalty for illegal use of drainage basin resources;
- Indiscriminately discharges solid or liquid waste in drainage basins; and
- Refuses to pay the Water Use Permit and Drilling Licensing fee before using resources in any drainage basin.

Additionally, Section 2 of the Drilling License and Groundwater Development Regulations, 1827, goes on to prescribe penalties for any person who contravenes the above-stated by-laws governing the use of resources in the Lower Ankobra River Basin. It states that:

> "Any person who contravenes any of the provisions of the water resource management by-laws commits an offense and shall be liable on summary conviction, to a fine of 100 penalty units (GH¢ 200.00: equivalent to US$150 at the time) or a term of imprisonment not exceeding six months or to both. In the case of a continuing offense, a further fine of one penalty unit (GH¢ 2.00 equivalent to US$1.5 at the time) shall be imposed after a notice in writing of the offense has been served on the offender."

The implication of the above is that there are sufficient deterrent laws and regulations to control the unlawful extraction of resources from drainage basins in Ghana. The question now is: to what extent have these laws been enforced to achieve results such as supporting projects such as the IFRV farming system in the Lower Ankobra River Basin? Besides, the state laws to protect projects in drainage basins are local traditional laws, and they argue that the extent to which traditional laws can be enforced largely depends on traditional rules. Therefore it was deemed important to assess the level of support received through the use of local customary laws and regulations to support the management of the IFRV project.

An in-depth interview with the head of the Environmental Assessment Department of the EPA revealed that the local institutions faced challenges when they tried to enforce the management laws to support the IFRV farming system. The following submission, given by this head, illustrates the sentiments of the heads in the sister institutions who tried to enforce the basin management by-laws:

> Indeed, there is nothing more deterrent than paying GH¢ 200 for just touching a resource in the basin without permission. When some of the local people are caught fishing or farming at unauthorized places, they call politicians who always plead for them to be set free. Indeed, we have no option but to set them free. If this negative practice does not cease, we shall continue to face challenges in enforcing the laws that govern the use of the basin's resources. The problem has always been with the enforcement and arrest of offenders!

The inability of institutions to implement by-laws is highlighted in the IWM framework (WRC, 2000) as the strongest factor that undermines the strength of management systems. Consequently, the inability of local institutions to enforce the aforementioned laws when political party representatives intervened partly gave way to the indiscriminate extraction and utilization of resources in the basin. The activities of the miners destroyed the valuable resources reserved for developing the IFRV farming system targeted by the farmers.

This finding on the interference from the local party representatives is underpinned by the IDBG Framework developed by Mireku et al. (2022), which identifies clientelism as one of the most influential factors that allow politicians to influence management decisions to favor political party members. When the government realized the numerous legal challenges in the management of the Lower Ankobra River Basin to support the IFRV farming system, the WRC was mandated to establish a supervisory body officially called the Ankobra River Basin Secretariat. The main duty of the Secretariat was to enforce all local by-laws and promote community participation using noncoercive strategies toward supporting the implementation of the IFRV farming activities. The Secretariat was to persuade and educate local users of the basin to accept their responsibilities to support local management institutions. Despite that special arrangement, the reports given by the head of the Ankobra River Basin Secretariat were that the noncoercive approach became discriminatory and finally failed, even when the police and community tribunals in the district interfered with the management system. The head's disappointments from the poor work of the police and the community tribunals (to him) explained the lapses in the by-laws.

It is clear from the above that, even though the local management institutions established close links with the law-enforcement agencies to deal with offenses in the basin to pave the way for the IFRV farmers, they were unsuccessful. The hindrance faced by local management institutions in enforcing basin use by-laws primarily stemmed from the "clientelistic" practices of political party leaders. These activities had created an environment conducive to persistent illegal gold and sand mining activities, leading to the degradation of precious resources earmarked for the IFRV farming initiatives outlined by the government. The influence exerted by political figures had, in turn, undermined the effective implementation of regulations designed to protect the integrity of the basin. When IFRV farmers confronted the challenges posed by these illicit mining activities, they tended to channel their grievances through the lengthy accountability route. This route, while comprehensive, was marred by numerous delays and bureaucratic complexities. In doing so, the direct and shorter accountability route, which could have offered a more expeditious resolution to their concerns, often got overlooked.

22.3.4 Recommendation

Based on the findings of the study, the following recommendations are worth considering:

There are sufficient laws backing the WRC's work. Therefore the WRC should act by implementing stringent measures to curb the "clientelistic" practices of political party leaders that compromise the integrity of basin management. The WRC should take a leading role in coordinating and implementing policies that explicitly address water allocation and quality standards for the IFRV farming system within the Lower Ankobra Basin.

The GWCL should take up the commitment to providing reliable and potable water supplies, crucial for irrigation purposes within the IFRV project areas. Collaborative efforts with local communities and project stakeholders are essential to ensuring the proper maintenance and sustainability of the water infrastructure.

Regular environmental impact assessments should be conducted by the EPA to evaluate the effects of the IFRV farming projects on the environment. Collaborating with the IFRV farmers, the EPA should strive to mitigate any adverse environmental effects and actively promote sustainable water and land use for the growth of the IFRV farms.

The Ankobra Basin Secretariat must work closely with the project stakeholders to develop comprehensive basin-level plans that intricately consider the water needs of all users. Through effective coordination and management of water resources within the Lower Ankobra Basin, these authorities should ensure equitable distribution among different users.

Collaborating with the project, MMDAs should plan and execute water infrastructure development initiatives tailored to the IFS's specific requirements within their respective jurisdictions.

The Ministry of Food and Agriculture is instrumental in this context, as it should actively engage in developing agricultural policies and programs that align with and encourage the IFS. This includes promoting sustainable agricultural practices, technologies, and research that are directly applicable to fish, rice, and vegetable production. The MoFA should also provide the necessary technical support and extension services to farmers participating in the project, thereby enhancing their productivity. Collaborative efforts with the MSWR are critical to ensuring that water resource management aligns seamlessly with the agricultural objectives outlined in the Lower Ankobra Basin.

22.4 Conclusion

The study investigates the roles and challenges faced by local water resource management institutions in the Lower Ankobra River Basin, Ghana. Established in 1996, the WRC was tasked with planning the use of resources in catchment areas, empowering the NDPC and MWH to coordinate national development plans. Supported by legal frameworks, the WRC obtained funds from the Ministry of Finance to distribute to its subordinate agencies. In the Lower Ankobra River Basin, the WRC played a crucial role in managing projects and soliciting funds. However, the study raised concerns among some basin resource users about the capacity of these institutions to effectively implement government-formulated policies. The roles of local drainage basin management institutions are outlined, emphasizing collaborative efforts for IFRV farming. The study identifies challenges faced by local water resource management institutions, citing Water Use Regulations and Legislative Instrument LI, 1692 of 2001. While legal frameworks exist, enforcement is hindered by political interference and an inability to implement by-laws effectively. The establishment of the Ankobra River Basin Secretariat aimed to address these challenges, but noncoercive strategies proved discriminatory and unsuccessful. The study suggests a need for effective enforcement mechanisms and community involvement to ensure sustainable water resource management to support the development of the IFRV farms in the basin.

References

Callo-Concha, D., Gaiser, T., & Ewert, F. (2012). Farming and cropping systems in the West African Sudanian savanna. WASCAL research area: northern Ghana, southwest Burkina Faso and northern Benin. *ZEF working paper series.*

Diaz-Chavez., R., Mutimba, S,. Watson, H., Rodriguez-Sanchez, S., & Nguer, M. (2010). Mapping Food and Bioenergy in Africa. A report prepared on behalf \tof FARA. Forum for Agricultural Research in Africa, Ghana: ERA-ARD.

Fierro-Sañudo, J. F., Rodríguez-Montes de Oca, G. A., & Páez-Osuna, F. (2020). Co-culture of shrimp with commercially important plants: A review. *Reviews in Aquaculture*, *12*(4), 2411–2428. Available from https://doi.org/10.1111/raq.12441, http://onlinelibrary.wiley.com/journal/10.1111/(ISSN)1753-5131.

Ghosh, A., Misra, S., Bhattacharyya, R., Sarkar, A., Singh, A. K., Tyagi, V. C., ... Meena, V. S. (2020). Agriculture, dairy and fishery farming practices and greenhouse gas emission footprint: A strategic appraisal for mitigation. *Environmental Science and Pollution Research*, *27*(10), 10160–10184. Available from https://doi.org/10.1007/s11356-020-07949-4, https://link.springer.com/journal/11356.

Gill, M. S., Singh, J. P., & Gangwa, K. S. (2009). Integrated farming system and agriculture sustainability. *Indian Journal of Agronomy*, *54*(2), 128–139. Available from https://doi.org/10.59797/ija.v54i2.4790.

Green, L., Webb, E., Johnson, E., Wynn, S., & Bogen, C. (2023). Cost-effective approach to explore key impacts on the environment from agricultural tools to inform sustainability improvements: Inversion tillage as a case study. *Environmental Sciences Europe*, *35*(1). Available from https://doi.org/10.1186/s12302-023-00784-7.

Gross, E., Senadza, B., Hodey, L. S., & Abdalla, S. (2015). Drinking water infrastructure projects in Ghana and Sudan. University of Bayreuth, Germany.

Halbrendt, J., Gray, S. A., Crow, S., Radovich, T., Kimura, A. H., & Tamang, B. B. (2014). Differences in farmer and expert beliefs and the perceived impacts of conservation agriculture. *Global Environmental Change*, *28*(1), 50–62. Available from https://doi.org/10.1016/j.gloenvcha.2014.05.001, http://www.elsevier.com/inca/publications/store/3/0/4/2/5.

Kinkela, M., D. Dochain, X. Rollin, J. Mafwila, J. Bindelle, Smallholders' practices of integrated agriculture aquaculture system in peri-urban and rural areas in sub Saharan Africa. (2019).

Mireku, D. O., Acheampong, P. K., Antwi, K. B., & Hlovor, I. K. (2022). Evaluating the effectiveness of principles used for sharing resources of the Lower Ankobra Basin in the Western Region, Ghana. *Natural Resources Forum*, *46*(4), 428–449. Available from https://doi.org/10.1111/1477-8947.12265, http://onlinelibrary.wiley.com/journal/10.1111/(ISSN)1477-8947.

Nguyen, T. T., Ngo, H. H., Guo, W., Wang, X. C., Ren, N., Li, G., ... Liang, H. (2019). Implementation of a specific urban water management - Sponge City. *Science of the Total Environment*, *652*, 147–162. Available from https://doi.org/10.1016/j.scitotenv.2018.10.168, http://www.elsevier.com/locate/scitotenv.

Sarkar, S., Skalicky, M., Hossain, A., Brestic, M., Saha, S., Garai, S., ... Brahmachari, K. (2020). Management of crop residues for improving input use efficiency and agricultural sustainability. *Sustainability*, *12*(23). Available from https://doi.org/10.3390/su12239808.

Schut, M., Klerkx, L., Rodenburg, J., Kayeke, J., Hinnou, L. C., Raboanarielina, C. M., ... Bastiaans, L. (2015). RAAIS: Rapid appraisal of agricultural innovation systems (Part I). A diagnostic tool for integrated analysis of complex problems and innovation capacity. *Agricultural Systems*, *132*, 1–11. Available from https://doi.org/10.1016/j.agsy.2014.08.009, http://www.elsevier.com/inca/publications/store/4/0/5/8/5/1.

Shahbaz, M., Rashid, N., Saleem, J., Mackey, H., & McKay, G. (2023). Tareq Al-Ansari, A review of waste management approaches to maximise sustainable value of waste from the oil and gas industry and potential for the State of Qatar. *Fuel*, *332*. Available from https://doi.org/10.1016/j.fuel.2022.126220.

Shoko, A. P., Getabu, A., & Mwayuli, G. (2011). Growth performance, yields and economic benefits of Nile tilapia Oreochromis niloticus and Kales Brassica oleracea cultured under vegetable-fish culture integration. *Tanzania Journal of Science*.

Van Hong, P., Thao Nguyen, N., Tran Ngoc Huy, D., Thu Thuy, N., & Thanh Huong, L. T. (2021). Evaluating several models of quality management and impacts on lychee price applying for Vietnam agriculture products value chain sustainable development. *Alinteri Journal of Agricultural Sciences*, *36*(1), 122–130. Available from https://doi.org/10.47059/alinteri/v36i1/ajas21018.

United Nations Development Programme & Sustainable Human Development (1997). *Water governance for poverty reduction* (MDGR Working Paper No. 5). New York, NY: The United Nations Development Programme and the Sustainable Human Development.

Chapter 23

Altitudinal gradient and Himalayan vegetation in changing climate: A short overview

Divyata Yadav[1] and Manish Pandey[2,3]

[1]*School of Environmental Sciences, Jawaharlal Nehru University, New Delhi, India,* [2]*University Center for Research and Development (UCRD), Chandigarh University, Mohali, Punjab, India,* [3]*Department of Civil Engineering, Chandigarh University, Mohali, Punjab, India*

23.1 Introduction

23.1.1 The Himalayas: a biodiversity hotspot

Significance of the Himalayan Mountain Range: The Himalayan Mountain Range holds immense significance as a biodiversity hotspot due to its exceptional richness in plant and animal life. Its unique topography, spanning a wide range of elevations from tropical forests to alpine meadows, has fostered the evolution of a multitude of species, including numerous endemics and endangered wildlife. Beyond their ecological value, the Himalayas are of great cultural importance, influencing the traditions and spirituality of the region's people. Moreover, these mountains are a vital source of freshwater for millions, regulate climate patterns, serve as a hub for scientific exploration, and contribute to tourism, making their conservation a matter of global concern, ensuring both the well-being of local communities and the preservation of our planet's biodiversity.

23.1.1.1 Altitudinal gradients and vegetation diversity

The altitudinal gradient along the Himalayan Mountain Ranges plays a significant role in shaping the vegetation diversity in the region (Rawat et al., 2021; Sekar et al., 2023). This diversity is influenced by a range of environmental factors (Bhandari, Mehta, & Tiwari, 2000; Mandal & Joshi, 2014; Singh & Singh, 1987) that change with elevation, including temperature, precipitation, soil types,

Water Resources Monitoring, Management, and Sustainability. DOI: https://doi.org/10.1016/B978-0-443-23665-5.00023-5

and topography. Here is a breakdown of how altitudinal gradients contribute to vegetation diversity in the Himalayas (Bhat et al., 2020):

- Tropical and subtropical zones (low elevations): at lower elevations, the Himalayas feature tropical and subtropical zones. Here, you will find lush, broad-leaved evergreen forests, dominated by species like sal (Shorea robusta), teak, and various species of bamboo (Bhattarai & Vetaas, 2003). The climate is warm and humid, supporting a variety of plant life.
- Temperate zones (midelevations): as you ascend the mountains, you will encounter temperate zones characterized by deciduous forests and coniferous forests. Oak, rhododendron, and various species of pine and fir are common in this zone. The climate becomes cooler, and the vegetation reflects the changes in temperature and precipitation.
- Subalpine and alpine zones (higher elevations): at even higher elevations, the subalpine and alpine zones dominate the landscape. These areas have short growing seasons, harsh winters, and lower temperatures. Plant life includes shrubs, alpine meadows, and hardy herbaceous plants adapted to extreme conditions. Notable species include juniper, dwarf rhododendron, and various alpine herbs.
- Nival zone (high elevations): at the highest elevations, the nival zone represents areas covered in snow and ice. Here, vegetation is scarce, consisting mainly of mosses, lichens, and some cold-adapted plants.

23.1.1.2 Importance of understanding vegetation patterns

Vegetation patterns provide valuable insights into the health and dynamics of ecosystems (Ahmad et al., 2020; Guirado, Blanco-Sacristán, Rigol-Sánchez, Alcaraz-Segura, & Cabello, 2019), as well as the impacts of environmental changes and human activities (Mishra & Mainali, 2017). Studying the distribution, composition, and changes in vegetation helps scientists monitor and predict climate change, habitat loss, and species distribution shifts (Afuye, Kalumba, & Orimoloye, 2021; Zhou & Wang, 2000). Moreover, it is essential for effective land management and conservation strategies, as it allows for the identification of areas with unique biodiversity as well as those at risk. Knowledge of vegetation patterns also informs agriculture, forestry, and urban planning, enabling sustainable resource management and minimizing environmental degradation. In essence, understanding vegetation patterns is a crucial foundation for the sustainable coexistence of human societies and the natural world.

23.1.2 What determines vegetation characteristics

Vegetation characteristics refer to the various features and attributes of plant communities in a specific area. These characteristics can vary significantly depending on the type of ecosystem, climate, and location. Table 23.1 lists some common vegetation characteristics.

TABLE 23.1 List of common vegetation characteristics.

S. no	Characteristics	Description
1.	Species diversity	The number of different plant species present in a given area can range from low diversity (monoculture) to high diversity (diverse ecosystems).
2.	Species composition	The specific plant species present in a given location can provide valuable information about the type of ecosystem and its health.
3.	Plant density	The number of individual plants per unit area can indicate how crowded or sparse the vegetation is.
4.	Vegetation structure	The physical arrangement of plants in an area includes aspects like canopy height, layering (e.g., shrub layer and tree layer), and overall architecture.
5.	Life forms	The different life forms of plants, such as trees, shrubs, herbs, grasses, and vines impact the overall structure of the vegetation.
6.	Canopy cover	The amount of ground covered by the crowns of plants affects light availability and understory vegetation.
7.	Plant growth forms	Characteristics like growth habit (e.g., woody and herbaceous), leaf type (e.g., deciduous, evergreen), and stem characteristics (e.g., succulent and woody).
8.	Leaf characteristics	Features of leaves, including size, shape, color, and texture, are important for plant identification.
9	Flowering and fruit production	The timing of flowering and fruiting is critical for reproductive success and interactions with pollinators and seed dispersers.
10.	Plant adaptations	Specialized features that enable plants to survive in their environment, such as drought tolerance, cold resistance, and fire adaptation.
11.	Elevation and altitude range	The specific altitude or elevation range at which a plant species or community is found can be a crucial characteristic in mountainous regions like the Himalayas.
12.	Endemism	The presence of plant species that are unique to a specific geographic area makes them of high conservation value.
13.	Invasive species	The presence of nonnative or invasive plant species that can outcompete and disrupt native vegetation.

(*Continued*)

TABLE 23.1 (Continued)

S. no	Characteristics	Description
14.	Successional stage	The stage of ecological succession that a plant community is in, such as early successional (pioneer species) or late successional (climax) vegetation.
15.	Habitat and microclimate characteristics	The type of habitat in which vegetation occurs (e.g., wetland, forest and desert) and microclimate conditions (e.g., temperature and humidity) that influence plant growth.
16.	Plant associations	The groupings or assemblages of plant species often co-occur due to similar ecological requirements.
17.	Ecosystem services	The functions that vegetation provides to the ecosystem include carbon sequestration, water purification, and soil stability.
18.	Phenology	The seasonal timing of key plant events, such as leaf emergence, flowering, fruiting, and leaf fall.
19.	Health and condition	The overall health and condition of the vegetation can be assessed through factors like disease, pest infestations, and stress.
20.	Conservation status	The conservation status of plant species or communities can range from critically endangered to least concern.

23.1.3 Environmental gradients shaping Himalayan vegetation

Some environmental variables and how they shape vegetation characteristics along environmental gradients are described in brief in the following lines:

1. Temperature and its influence on vegetation zones: The Himalayas are warming at a faster rate than the global average, with a mean temperature increase of 1.7°C in the past century. This has a significant impact on high-altitude ecosystems, particularly on the vegetation structure and biodiversity (Aryal, Brunton, & Raubenheimer, 2014; Bhutiyani, Kale, & Pawar, 2007).
2. Precipitation patterns and water availability: Factors such as temperature, soil texture, soil depth, snow cover, and snowmelt timing complicate the availability of water in high mountains. This makes it difficult to accurately measure water availability in dry mountains, especially in remote areas with limited climate data. Therefore it is important to identify simple and reliable indicators of water availability to understand how species are distributed and

how ecosystems function along elevation gradients in dry mountain regions (Ale et al., 2018; Williams, Mcnamara, & Chandler, 2009).

3. Soil conditions and nutrient availability: mountain ecosystems exhibit pronounced spatial variations in species richness and composition, a phenomenon attributed to the dynamic interplay of eco-physiological factors along altitudinal gradients. Several studies found that the amount of nutrients in the soil was highest at the treeline and was lower at lower altitudes (Kewlani, Negi, Bhatt, Rawal, & Nandi, 2021). The ratios of nitrogen (N) to phosphorus (P) and carbon (C) to phosphorus in the soil decrease as altitude increases. This decrease is especially noticeable in the transition from Krummholz, a stunted tree formation, to the Alpine tundra, a treeless zone. In contrast, the ratio of carbon to nitrogen in the soil increases significantly with altitude, particularly from closed forests to Krummholz and the Alpine tundra. These changes suggest that nutrient availability, especially phosphorus, is becoming increasingly scarce in the alpine tundra. This scarcity is likely due to the acidic nature of soils at high altitudes (Müller, Oelmann, Schickhoff, Böhner, & Scholten, 2017).

23.1.4 Ecological principles governing vegetation patterns

1. Species adaptation to specific altitudinal zones: We must study species and their functional diversity in high-altitude areas to identify indicator communities that preserve ecosystem stability and functionality (Thakur & Chawla, 2019). The high-altitude plant communities harbor a smaller proportion of nonnative plant species compared to low-altitude regions due to the challenging climatic conditions at higher elevations (Pauchard et al., 2009). Based on the observed patterns of species niche width across different elevations, it can be deduced that higher elevations tend to have narrower niches. Additionally, it is more probable that the rare species found at the highest altitudes and on either of the aspects will be negatively impacted by future changes in climate change (Thakur & Chawla, 2019).
2. Competition and resource partitioning among plants: The degree of niche overlap between species undoubtedly varies with species densities and the level of competition, whether inter- or intra-specific, that occurs (Morris, 1999). Native species have a higher niche overlap due to invading species using more resources. The dominant species exhaust all nutrients, including those used by their associates, leading to a shortage and niche overlap. *Capillipedium parviflorum* shows more overlap across all sites in the Garhwal region of the Himalayas (Anthwal, Bhatt, Nautiyal, & Anthwal, 2008).
3. Disturbance regimes and their impact on vegetation: The occurrence of fire has a significant impact on the type and number of species that exist within a forest ecosystem (Danthu et al., 2003). In the Bhagirathi

Catchment Area of Garhwal Himalaya, disturbance regimes have a clear and strong relationship with the regeneration of tree species as well as the diversity and composition patterns across altitudes (Om Prakash Tiwari, 2019). Disturbance intensities had a significant impact on Quercus leucotrichophora and Quercus semecarpifolia forests, resulting in a decline in tree density, dominance, and natural recruitment. These forests were highly sensitive to disturbances, and increasing levels of disruption harmed their overall health (Rawal, Gairola, & Dhar, 2012).

4. Biodiversity and endemism along altitudinal gradients: The pattern of plant richness exhibits a distinct hump-shaped curve, with noticeable shifts in the mid-elevation peaks attributed to variations in spatial extent. The decline in tree height and richness along the elevation gradient is a direct consequence of woody plants being more susceptible to low temperatures and limited liquid water availability. At higher elevations, herb richness prevails, demonstrating their enhanced adaptability to climatic stressors compared to trees (Sharma, Behera, Das, & Panda, 2019). It has been well-established that peak endemic species richness is found at higher elevations in the Himalayas, specifically in flowering plants (Vetaas & Grytnes, 2002), liverworts (Grau, Grytnes, & Birks, 2007), and terrestrial orchids (Acharya, Vetaas, & Birks, 2011). One of the main drivers of endemism in the region is the complex topography shaped by morphotectonic processes and climatic events (Pandit, Manish, & Koh, 2014).

23.1.5 Theories explaining altitudinal vegetation gradient

1. Massenerhebung Effect and local climate modification: the Massenerhebung Effect is the phenomenon of increased precipitation at higher elevations due to the lifting of moist air masses as they are forced to rise over mountains. This effect is most pronounced on the windward side of mountains, where the air masses are forced to rise more rapidly. As the air rises, it cools and condenses, forming clouds and releasing precipitation. The Massenerhebung Effect is responsible for the lush vegetation found in many mountain regions (Grubb, 1971). The positions of tree lines and snow lines in the interiors of mountainous regions are displaced to higher elevations compared to their respective locations on the outer fringes. Mountains exert a significant influence on the local climate through various mechanisms. They effectively obstruct the movement of cold air masses, preventing them from reaching lower altitudes. As a result, frost-free zones develop in the valleys situated below the mountains, enabling the cultivation of crops that would otherwise not be viable. Furthermore, mountains create rain shadow regions on the leeward side of mountains that receive less rainfall than the surrounding areas. This is because the air masses have already relinquished most of their

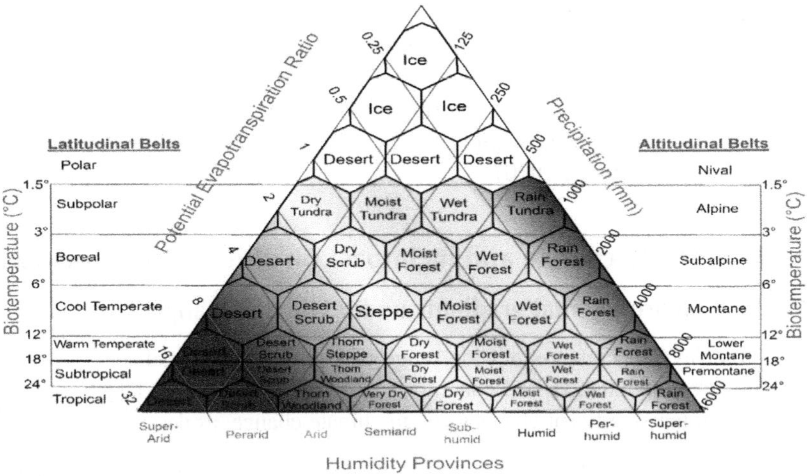

FIGURE 23.1 Holdridge life zone (Kadri Tekin, Tatli, & Koç, 2021).

moisture as they ascended over the windward side of the mountains (Pepin et al., 2022).

2. Life zone theory and distinct vegetation zones: the Holdridge life zone (Fig. 23.1) system firmly asserts that climate is the main determining factor of plant life. It confidently employs a combination of temperature and precipitation to define life zones while also taking into account potential evapotranspiration, which is a measure of the maximum amount of water that can be lost to the atmosphere through evaporation and transpiration (Macmahon & Wieboldt, 1978).
3. Altitudinal relay floristics and species turnover: altitudinal relay floristics is a phenomenon observed in mountainous regions where closely related plant species replace each other along an altitudinal gradient. This concept suggests that distinct vegetation zones, each characterized by unique plant assemblages, are not entirely isolated but rather exhibit overlapping distributions. Species turnover, also known as beta diversity, is a measure of the change in species composition between different sites or habitats (Matthews, Hill, Winkler, Owen, & Vater, 2018). For instance, in the Himalayas, *Rhododendron arboreum*, a broadleaf evergreen shrub, dominates lower elevations, while *Rhododendron campanulatum*, a closely related species, replaces it at higher altitudes.
4. Climatic vicariance and speciation patterns: climatic vicariance is an indisputable mechanism of speciation that occurs when populations of a species become geographically separated due to climate changes. These changes inevitably result in the formation of barriers to gene flow, leading to the independent evolution of the isolated populations over time.

5. Bioclimatic law and temperature-vegetation relationships: the bioclimatic law is an ecological principle that explains the impact of elevation, latitude, and longitude on the timing of seasonal occurrences in the natural environment. This law was formulated by Andrew Delmar Hopkins, an entomologist from the United States, during the late 19th and early 20th centuries (Hopkins, 1920).

23.2 Climate change and its impact on altitudinal vegetation

23.2.1 Environmental gradients under climate change

The Himalayas encompass a wide range of environmental gradients, including temperature, precipitation, soil conditions, and sunlight exposure. These gradients are undergoing significant changes due to climate change, as described below:

1. Temperature: rising temperatures are causing a shift in temperature zones, with warmer conditions encroaching on higher altitudes. This can lead to the upward migration of plant species and the potential loss of alpine habitats.
2. Precipitation: precipitation patterns are becoming more erratic, with some areas experiencing increased rainfall and others facing droughts. This can disrupt water availability for plants and alter the distribution of vegetation zones.
3. Soil conditions: changing temperatures and precipitation patterns can affect soil moisture, nutrient cycling, and erosion rates. These changes can impact plant growth and species composition.
4. Sunlight exposure: changes in cloud cover and atmospheric conditions can affect the intensity and duration of sunlight reaching different altitudes. This can influence plant photosynthesis, growth patterns, and species distribution.

23.2.2 Ecological principles in a changing climate

Ecological principles that govern vegetation patterns are being challenged by climate change and are described below:

1. Species adaptation: plants may not be able to adapt quickly enough to the pace of climate change, leading to range shifts, reduced fitness, and potential extinctions.
2. Competition and resource partitioning: as environmental conditions change, competition for resources like water and nutrients may intensify, leading to shifts in species composition and the potential loss of biodiversity.
3. Disturbance regimes: climate change can increase the frequency and intensity of natural disturbances like wildfires, landslides, and insect outbreaks, further disrupting vegetation patterns.

4. Biodiversity and endemism: climate change poses a significant threat to biodiversity, especially for endemic species adapted to specific altitudinal niches. Range shifts and habitat loss can lead to species decline and extinction.

23.2.3 Vegetation characteristics under climate change

Vegetation characteristics under climate change are a subject of critical concern in the fields of ecology and environmental science. As global temperatures rise and weather patterns shift, plant communities are undergoing significant alterations in their distribution, growth patterns, and flowering times. These changes can have profound impacts on ecosystems as well as on agriculture and forestry. Researchers are closely monitoring these shifts to understand the adaptive responses of plant species and to assess the potential consequences for biodiversity, food security, and the overall health of the planet. Such studies play a vital role in shaping strategies for climate adaptation and conservation, ensuring that ecosystems can adjust to the changing climate and continue to provide essential services to humanity and the natural world. To sum up, the overall climate change impact on vegetation characteristics can be summed up in three points:

1. Lower elevations: expansion of tropical and subtropical species.
2. Middle elevations: shifts in species composition due to precipitation changes.
3. Higher elevations: vulnerability of alpine habitats and upward migration.

23.2.4 Theories in the face of climate change

The salient theories explaining variability in vegetation characteristics along the altitudinal gradients under changing climatic conditions are briefly assessed under the following points:

1. Reassessing the Massenerhebung Effect under climate change: Panigrahy, Anitha, Kimothi, and Singh (2010) employed satellite imagery spanning nearly two decades to map treeline changes. Their findings reveal a striking increase in vegetation cover and a substantial reduction in snow cover in the glacier areas of Nanda Devi Biosphere Reserve, central Indian Himalaya, between 1986 and 2004. Alpine plant species exhibited an upward shift in distribution, with the rate of change varying among species and their climate sensitivity. This study underscores the profound transformations that have reshaped Himalayan ecosystems since 1960. The timberlines in the Himalayan region rank among the highest globally due to the pronounced Massenerhebung Effect induced by the Tibetan Plateau. Despite projections of timberline advancement under global

warming scenarios, observations suggest that increasing tree recruitment and growth are more prevalent than a distinct upward shift in timberline (Peili, Ning, & Rawat, 2020).

2. Shifting boundaries of life zones and vegetation reorganization: The seminal study by Telwala Yasmeen and Brook (2013) on climate-induced species distribution changes in alpine Sikkim provides compelling evidence of warming-driven elevational range shifts in 87% of the 124 endemic plant species examined over the past 150 years. The study documented substantial upward shifts in species' upper elevation limits, ranging from 23 to 998 meters. These findings underscore the profound transformation of plant assemblages and community structures in the region, with present-day communities deviating significantly from those of the nineteenth century. The upslope expansion of woody species from dwarf-scrub communities into alpine meadows is attributed to the indisputable influence of climate warming. The retreat of glaciers has initiated vegetation succession processes on glacier forelands, a phenomenon that has been largely neglected in the existing body of literature (Schickhoff, Singh, & Mal, 2016).
3. Potential for further speciation or lineage loss due to climate change: contrary to expectations, the speciation of Roscoea in the Himalayas occurred primarily from highland to lowland, linked to the rapid uplift of the Himalayas and subsequent climate changes. This finding challenges traditional views of speciation in mountains and highlights the importance of considering environmental factors in understanding the evolution of species. The study's findings on Roscoea speciation suggest that vicariance driven by the rapid uplift of the Himalayas may be a key factor in the region's extraordinary biodiversity. This highlights the importance of vicariance in speciation in mountainous regions, particularly in areas with a history of rapid uplift (Zhao, Paudel, Yu, Zhang, & Li, 2021).
4. Complexities in bioclimatic law under changing temperature gradients: it is important to note that the Maxent model does not directly predict the actual distribution of Rhododendrons but rather identifies potential climate spaces or "bioclimate envelopes" that are suitable for their growth and survival. The study employed model extrapolation to assess the potential distribution of Rhododendrons under climate change. This approach allowed researchers to predict suitable environmental conditions beyond the range of available observational data, providing preliminary insights into the potential impact of climate change on Rhododendron distribution. However, it is essential to acknowledge the uncertainties associated with model extrapolation, and further research is needed to refine these predictions and fully understand the interplay between climate change and Rhododendron distribution dynamics (Kumar, 2012).

23.3 Conclusion

23.3.1 Summary of key findings

23.3.1.1 Impacts of climate change on altitudinal vegetation gradients

High-altitude plant species have evolved to thrive under harsh conditions, but as these conditions abate due to warming, generalist species are becoming more abundant, while specialized species are declining. This loss of unique biodiversity is a major concern for the long-term health of high-altitude ecosystems (Thakur & Chawla, 2019). While increased functional diversity in high-altitude plant communities is often associated with enhanced stability and productivity, the loss of rare and endemic species triggered by warming poses a significant threat to these ecosystems. Preserving both functional diversity and specialized species is crucial for ensuring the long-term health of these fragile environments (Thakur & Chawla, 2019).

Alpine plant diversity declines with altitude, peaking at lower elevations (3200–3500 meters). This pattern holds true for herbs, shrubs, and trees. Beta diversity, a measure of species turnover, increases with altitude, indicating a higher proportion of unique species adapted to harsh conditions. Herbs and shrubs exhibit higher beta diversity than trees, suggesting greater tree population homogeneity at high altitudes (Sekar et al., 2023).

Hamid, Khuroo, Malik, Ahmad, and Singh (2020) studied the temporal changes in the flora of Kashmir (the north–western extreme of the Himalayas) from 2014 to 2018. They observed certain specific dynamism in the floral ecological traits in the wake of climate change, which is provided below in the following paragraphs:

1. A rise in the number of different species was observed on mountain summits over four years. This increase was especially noticeable on lower peaks. The study suggests that warming temperatures are causing species to migrate mid-elevations from lower elevations, leading to the observed biodiversity boost. They conclude that, if warming continues at its current pace, we can expect an even faster increase in species richness on mountain summits.
2. Most common shrubs, grasses, and wildflowers have significantly expanded their coverage. This likely happened because the study area's warmer and drier weather extended the growing season, giving these plants more time to colonize new areas and grow bigger.
3. High-altitude plant life is changing. Summits, especially the lower ones, are seeing a rise in plants that prefer warmer climates. This likely reflects the overall warming trend.
4. Species like *Juniperus squamata*, *Rhododendron anthopogon*, *Polygonum affine*, *Sibbaldia cuneata*, and *Swertia petiolataare* are commonly

widespread and generalist, meaning they can thrive in a variety of environmental conditions. This adaptability makes them relatively resilient to climate change.

Similarly, a study by Naithani, Nainwal, Sati, and Prasad (2001) found that rising temperatures (up to 2°C) with less snowfall and earlier snowmelt in the Tungnath alpine region (Garhwal Himalayas) led to the introduction and establishment of new plant species migrating from lower altitudes. These changing conditions created a longer growing season and earlier plant growth, which benefited plants requiring extended growth cycles (Nautiyal, Nautiyal, & Prakash, 2004).

The research by Satti, Naveed, Shafeeque, and Li (2024) investigates trends in vegetation growth across Gilgit-Baltistan (GB) using two types of data: satellite-based vegetation indexes (NDVI) and ground-based weather observations from 1991 to 2020. By considering different scales (regional, zone, altitude, and individual pixels) and employing various statistical models, the study analyzes how vegetation responds to different climatic factors at various levels. Overall, GB's vegetation showed a gradual greening trend, with the NDVI increasing by an average of 0.19% per year over the studied period. The vegetation's response to weather varied according to season and elevation. In summer, it was more sensitive to changes in precipitation, while in spring, temperature had a stronger influence, especially at lower altitudes. Additionally, winter precipitation played a more significant role in vegetation growth at lower altitudes compared to higher elevations.

Interestingly, the researchers observed a hump-shaped pattern in species richness in the Dhauladhar Mountains. It peaked at around 2800 meters and then declined at higher and lower altitudes. This finding aligns with observations from other Himalayan regions like Bhabha Valley (Chawla et al., 2008), Kashmir (Dar & Sundarapandian, 2016; Hamid, Khuroo, Malik, Ahmad, Singh, et al., 2020), Arunachal Pradesh (Saikia et al., 2017), and Sikkim (Sharma et al., 2019). Evenness followed a different path than species richness. While diversity peaked at mid-elevations, the number of individuals within each species was more balanced at lower and higher altitudes. Analysis of beta diversity showed lower rates of species replacement within mid-altitude communities, suggesting greater environmental homogeneity and ecological overlap compared to lower and higher elevations (Dar & Sundarapandian, 2016).

The work of Li et al. (2019) studied how alpine grasslands on the Qinghai-Tibet Plateau (QTP) respond to recent climate changes, adding a valuable vertical layer to global studies of sensitive ecosystems. Between 2000 and 2016, the study found grasslands on the southern QTP to be the most susceptible to climatic swings. Interestingly, the main culprits varied across the region: temperature, precipitation, and sunlight each played a dominant role in different areas. Notably, sensitivity climbed with altitude, implying that higher grasslands were particularly vulnerable, especially to rising temperatures. In short, this work sheds light on the diverse and delicate dance between climate and high-altitude grasslands on the QTP.

23.3.1.2 Challenges and complexities in predicting and mitigating climate change effects

While climate change research in the northwestern Himalayas has been substantial, studies in the eastern Himalayas remain limited. Existing literature suggests rising temperatures, increasing monsoon precipitation, and decreasing winter precipitation, potentially leading to increased monsoon flooding. Policy assessments should extend beyond mitigation and adaptation to consider the trade-offs of various measures. However, the complex topography of the Himalayas introduces uncertainty in model-based climate projections, necessitating further research (Negi, Tiwari, Singh, Thakur, & Bhatt, 2022).

23.3.2 Conservation implications

Importance of understanding climate change impacts on Himalayan vegetation.

Investigations into Himalayan fauna have uncovered range shifts, phenological shifts, and community shifts as hallmarks of climate change. Species distribution models are prevalent but fail to consider species adaptability and dispersal capabilities. An integrated framework is essential for accurate range-shift assessments. Research on community composition and phenological change is still in its infancy. Metagenomics and genomics integration hold immense promise in advancing this field (Dahal, Lamichhaney, & Kumar, 2021).

23.3.2.1 Need for conservation efforts to protect biodiversity and ecosystem services

While few studies incorporate socio-ecological and ground-based data into climate assessments, these are crucial for adaptation and management planning. Perception-based studies provide valuable insights into the impacts of climate change on natural resources, socio-ecological systems, and livelihoods in rural and remote areas, particularly in the absence of weather data. Promoting access and benefit-sharing of natural resources under the CBD Act, establishing Biodiversity Management Committees, and preparing People's Biodiversity Registers, along with promoting restoration of degraded lands and ecosystems, will enhance the role of local communities in biodiversity management and conservation, contributing to climate change adaptation and mitigation (Negi et al., 2022).

23.3.3 Future research directions

23.3.3.1 Monitoring vegetation changes along altitudinal gradients

The escalating plant species richness on Himalayan mountain summits attributed to climate change underscores the necessity of elucidating plant traits that influence their colonization or disappearance under these shifting conditions. Our study unveils the variation in plant traits along elevation gradients, shedding light on two critical questions: (1) identifying traits that enhance plant adaptation and survival in challenging alpine environments under projected climate change scenarios; and (2) predicting future alterations in alpine vegetation under climate warming, given the utilization of air and soil temperature differentials along alpine elevations as a proxy for climate warming (Hamid et al., 2020). Technological and methodological advancements are propelling emerging research topics, with a notable shift from conventional ecological monitoring to remote sensing, particularly on the Tibetan Plateau. This shift is broadening the research landscape and attracting a wider range of researchers to the field. However, this diversification may lead to further fragmentation of the literature (Verrall & Pickering, 2020).

23.3.3.2 Investigating species-specific responses to climate change

Recent studies indicate that climate warming in alpine regions has triggered species range shifts, including plant colonization at higher elevations. The successful colonization of higher elevations hinges on plant traits and local ecological conditions, which act as filters, selecting species with traits suited to the site's local conditions. This study, by exploring spatial trends of functional traits, species composition, diversity, and soil temperature along elevation, can serve as a surrogate to understand the potential future impacts of climate change on alpine plant diversity in this ecologically vulnerable Himalayan region. Additionally, reassessing plant traits of newly appearing or disappearing species in the future will aid in comprehending the pace of warming-induced changes in alpine plant assemblages (Hamid et al., 2020).

23.3.3.3 Developing predictive models for vegetation shifts under future climate scenarios

The term "Hindu Kush Himalaya" (HKH) encompasses a broader region than the Himalayas, including Tibet, Karakoram, and other areas. This expanded scope introduces significant heterogeneity and complexity, altering the overall landscape perspective. While forests predominate in the Himalayas, grasslands become the dominant system when referring to the HKH region. This distinction has far-reaching implications for understanding climate change impacts and developing effective adaptation strategies. It necessitates a more nuanced and regionally specific approach to climate change research, ensuring that resource allocation and adaptation measures align with the diverse needs of different areas within the HKH. Recent studies on climate change in the HKH region have revealed that our understanding of climate change and its impacts is more comprehensive in certain areas, including the eastern Tibetan Plateau, the Yunnan-Guizhou Plateau, and the low-lying plains of northern India (Ren & Shrestha, 2017). However, a general lack of data on climate change impacts on ecosystems and biodiversity, as well as a scarcity of systematic studies and empirical observations on species-level impacts of climate change in the Himalayas, remains a significant challenge (Gautam Govinda, Timilsina, & Acharya, 2013). All this leads to a larger bias in regional projection models of CC impact toward the biome/landform richly represented in regional/global databases (e.g., Tibetan Plateau). The limited availability of data from forested landscapes, such as the Himalayas, hinders the applicability of global climate change projections and scenarios to these regions. To address this issue, it is crucial to include a more balanced representation of data from diverse landforms and biomes across the HKH region. This will enhance the accuracy and relevance of climate change projections, enabling more effective adaptation and mitigation strategies (Goel, Ravindra, & Chattopadhyay, 2019).

References

Acharya, K. P., Vetaas, O. R., & Birks, H. J. B. (2011). Orchid species richness along Himalayan elevational gradients. *Journal of Biogeography*, *38*(9), 1821–1833. Available from https://doi.org/10.1111/j.1365-2699.2011.02511.x.

Afuye, G. A., Kalumba, A. M., & Orimoloye, I. R. (2021). Characterisation of vegetation response to climate change: A review. *Sustainability*, *13*, 7265. Available from https://doi.org/10.3390/su13137265.

Ahmad, M., Uniyal, S. K., Batish, D. R., Singh, H. P., Jaryan, V., Rathee, S., ... Kohli, R. K. (2020). Patterns of plant communities along vertical gradient in Dhauladhar Mountains in Lesser Himalayas in North-Western India. *Science of The Total Environment*, *716*, 136919. Available from https://doi.org/10.1016/j.scitotenv.2020.136919.

Ale, R., Zhang, L., Li, X., Raskoti, B. B., Pugnaire, F. I., & Luo, T. (2018). Leaf δ13C as an indicator of water availability along elevation gradients in the dry Himalayas. *Ecological Indicators*, *94*, 266–273. Available from https://doi.org/10.1016/j.ecolind.2018.07.002.

Anthwal, S., Bhatt, A. B., Nautiyal, B. P., & Anthwal, A. (2008). Vegetation structure, niche width, niche overlap and types of competition in temperate grazingland of Garhwal Himalaya, India. *The Environmentalist*, *28*(3), 261–273. Available from https://doi.org/10.1007/s10669-007-9137-1.

Aryal, A., Brunton, D., & Raubenheimer, D. (2014). Impact of climate change on human-wildlife-ecosystem interactions in the Trans-Himalaya region of Nepal. *Theoretical and Applied Climatology*, *115*(3–4), 517–529. Available from https://doi.org/10.1007/s00704-013-0902-4.

Bhandari, B. S., Mehta, J. P., & Tiwari, S. C. (2000). Dominance and diversity relations of woody vegetation structure along an altitudinal gradient in a Montane forest of Garhwal Himalaya. *Journal of Tropical Forest Science*, *12*, 49–61.

Bhat, J. A., Kumar, M., Negi, A. K., Todaria, N. P., Malik, Z. A., Pala, N. A., ... Shukla, G. (2020). Species diversity of woody vegetation along altitudinal gradient of the Western Himalayas. *Global Ecology and Conservation, 24*, e01302. Available from https://doi.org/10.1016/j.gecco.2020.e01302.

Bhattarai, K. R., & Vetaas, O. R. (2003). Variation in plant species richness of different life forms along a subtropical elevation gradient in the Himalayas, East Nepal. *Global Ecology and Biogeography, 12*, 327–340. Available from https://doi.org/10.1046/j.1466-822X.2003.00044.x.

Bhutiyani, M. R., Kale, V. S., & Pawar, N. J. (2007). Long-term trends in maximum, minimum and mean annual air temperatures across the Northwestern Himalaya during the twentieth century. *Climatic Change, 85*(1–2), 159–177. Available from https://doi.org/10.1007/s10584-006-9196-1.

Chawla, A., Rajkumar, S., Singh, K. N., Lal, B., Singh, R. D., & Thukral, A. K. (2008). Plant species diversity along an altitudinal gradient of Bhabha Valley in western Himalaya. *Journal of Mountain Science, 5*(2), 157–177. Available from https://doi.org/10.1007/s11629-008-0079-y.

Dahal, N., Lamichhaney, S., & Kumar, S. (2021). Climate change impacts on Himalayan biodiversity: Evidence-based perception and current approaches to evaluate threats under climate change. *Journal of the Indian Institute of Science, 101*(Issue 2), 195–210. Available from https://doi.org/10.1007/s41745-021-00237-1, Springer.

Danthu, P., Ndongo, M., Diaou, M., Thiam, O., Sarr, A., Dedhiou, B., ... Vall, M. (2003). Impact of bush ®re on germination of some West African acacias. *Forest and Ecology Management.*

Dar, J. A., & Sundarapandian, S. (2016). Patterns of plant diversity in seven temperate forest types of Western Himalaya, India. *Journal of Asia-Pacific Biodiversity, 9*(3), 280–292. Available from https://doi.org/10.1016/j.japb.2016.03.018.

Gautam Govinda, M.R., Timilsina, R., & Acharya, K. (2013). *Climate change in the Himalayas current state of knowledge*. http://econ.worldbank.org.

Goel, P. S., Ravindra, R., & Chattopadhyay, S. (2019). *Climate change and the white world. Climate Change and the White World.* Springer International Publishing. Available from https://doi.org/10.1007/978-3-030-21679-5.

Grau, O., Grytnes, J. A., & Birks, H. J. B. (2007). A comparison of altitudinal species richness patterns of bryophytes with other plant groups in Nepal, Central Himalaya. *Journal of Biogeography, 34* (11), 1907–1915. Available from https://doi.org/10.1111/j.1365-2699.2007.01745.x.

Grubb, P. J. (1971). Interpretation of the 'Massenerhebung' Effect on tropical mountains. *Nature, 229*(5279), 44–45. Available from https://doi.org/10.1038/229044a0.

Guirado, E., Blanco-Sacristán, J., Rigol-Sánchez, J. P., Alcaraz-Segura, D., & Cabello, J. (2019). A multi-temporal object-based image analysis to detect long-lived shrub cover changes in Drylands. *Remote Sensing, 11*, 2649. Available from https://doi.org/10.3390/rs11222649.

Hamid, M., Khuroo, A. A., Malik, A. H., Ahmad, R., & Singh, C. P. (2020). Assessment of alpine summit flora in Kashmir Himalaya and its implications for long-term monitoring of climate change impacts. *Journal of Mountain Science, 17*(8), 1974–1988. Available from https://doi.org/10.1007/s11629-019-5924-7.

Hopkins, A. D. (1920). The bioclimatic law. *Source: Journal of the Washington Academy of Sciences, 10*(Issue 2).

Kadri Tekin, M., Tatli, H., & Koç, T. (2021). Climate classification in Turkey: A case study evaluating Holdridge life zones. *Theoretical and Applied Climatology*. Available from https://doi.org/10.1007/s00704-021-03565-5/Published.

Kewlani, P., Negi, V. S., Bhatt, I. D., Rawal, R. S., & Nandi, S. K. (2021). Soil nutrients concentration along altitudinal gradients in Indian Western Himalaya. *Scandinavian Journal of Forest Research*, *36*(2–3), 98–104. Available from https://doi.org/10.1080/02827581.2020.1871065.

Kumar, P. (2012). Assessment of impact of climate change on Rhododendrons in Sikkim Himalayas using Maxent modelling: Limitations and challenges. *Biodiversity and Conservation*, *21*(5), 1251–1266. Available from https://doi.org/10.1007/s10531-012-0279-1.

Li, L., Zhang, Y., Wu, J., Li, S., Zhang, B., Zu, J., ... Paudel, B. (2019). Increasing sensitivity of alpine grasslands to climate variability along an elevational gradient on the Qinghai-Tibet Plateau. *Science of The Total Environment*, *678*, 21–29. Available from https://doi.org/10.1016/j.scitotenv.2019.04.399.

Macmahon, J.A., & Wieboldt, T.F. (1978). Applying biogeographic principles to resource management: a case study evaluating holdridge's life zone modeL. In Source: Great Basin Naturalist Memoirs (Issue 2).

Mandal, G., & Joshi, S. P. (2014). Analysis of vegetation dynamics and phytodiversity from three dry deciduous forests of Doon Valley, Western Himalaya, India. *Journal of Asia-Pacific Biodiversity*, *7*, 292–304. Available from https://doi.org/10.1016/j.japb.2014.07.006.

Matthews, J. A., Hill, J. L., Winkler, S., Owen, G., & Vater, A. E. (2018). Autosuccession in alpine vegetation: Testing the concept on an altitudinal bioclimatic gradient, Jotunheimen, southern Norway. *CATENA*, *170*, 169–182. Available from https://doi.org/10.1016/j.catena.2018.06.012.

Mishra, N. B., & Mainali, K. P. (2017). Greening and browning of the Himalaya: Spatial patterns and the role of climatic change and human drivers. *Science of The Total Environment*, *587–588*, 326–339. Available from https://doi.org/10.1016/j.scitotenv.2017.02.156.

Morris, D. W. (1999). Has the ghost of competition passed? *Evolutionary Ecology Research*, *1*.

Müller, M., Oelmann, Y., Schickhoff, U., Böhner, J., & Scholten, T. (2017). Himalayan treeline soil and foliar C:N:P stoichiometry indicate nutrient shortage with elevation. *Geoderma*, *291*, 21–32. Available from https://doi.org/10.1016/j.geoderma.2016.12.015.

Naithani, A. K., Nainwal, H. C., Sati, K. K., & Prasad, C. (2001). Geomorphological evidences of retreat of the Gangotri glacier and its characteristics. *Current Science*, *80*(1), 87–94. Available from http://www.jstor.org/stable/24105561.

Nautiyal, M. C., Nautiyal, B. P., & Prakash, V. (2004). Effect of grazing and climatic changes on alpine vegetation of Tungnath, Garhwal Himalaya, India. *The Environmentalist*, *24*.

Negi, V. S., Tiwari, D. C., Singh, L., Thakur, S., & Bhatt, I. D. (2022). Review and synthesis of climate change studies in the Himalayan region. *Environment, Development and Sustainability*, *24*(Issue 9), 10471–10502. Available from https://doi.org/10.1007/s10668-021-01880-5, Springer Science and Business Media BV.

Om Prakash Tiwari, C. M. S. Y. S. R. R. K. (2019). Disturbance, diversity, regeneration and composition in temperate forests of Western Himalaya, India. *Journal of Forest and Environmental Science*.

Pandit, M. K., Manish, K., & Koh, L. P. (2014). Dancing on the roof of the world: Ecological transformation of the himalayan landscape. *BioScience*, *64*(Issue 11), 980–992. Available from https://doi.org/10.1093/biosci/biu152, Oxford University Press.

Panigrahy, S., Anitha, D., Kimothi, M. M., & Singh, S. P. (2010). Timberline change detection using topographic map and satellite imagery. *Tropical Ecology*, *51*(1), 87–91. Available from http://www.tropecol.com.

Pauchard, A., Kueffer, C., Dietz, H., Daehler, C. C., Alexander, J., Edwards, P. J., ... Seipel, T. (2009). Ain't no mountain high enough: Plant invasions reaching new elevations. *Frontiers in Ecology and the Environment*, *7*(Issue 9), 479–486. Available from https://doi.org/10.1890/080072.

Peili, S., Ning, W., & Rawat, G. S. (2020). The distribution patterns of timberline and its response to climate change in the Himalayas. *Journal of Resources and Ecology*, *11*(4), 342. Available from https://doi.org/10.5814/j.issn.1674-764x.2020.04.002.

Pepin, N. C., Arnone, E., Gobiet, A., Haslinger, K., Kotlarski, S., Notarnicola, C., … Adler, C. (2022). Climate changes and their elevational patterns in the mountains of the World. *Reviews of Geophysics*, *60*(Issue 1). Available from https://doi.org/10.1029/2020RG000730, John Wiley and Sons Inc.

Rawal, R. S., Gairola, S., & Dhar, U. (2012). Effects of disturbance intensities on vegetation patterns in oak forests of Kumaun, west Himalaya. *Journal of Mountain Science*, *9*(2), 157–165. Available from https://doi.org/10.1007/s11629-012-2029-y.

Rawat, D. S., Bagri, A. S., Parveen, M., Nautiyal, M., Tiwari, P., & Tiwari, J. K. (2021). Pattern of species richness and floristic spectrum along the elevation gradient: A case study from Western Himalaya, India. *Acta Ecologica Sinica*, *41*, 545–551. Available from https://doi.org/10.1016/j.chnaes.2021.03.012.

Ren, G. Y., & Shrestha, A. B. (2017). Climate change in the Hindu Kush Himalaya. *Advances in Climate Change Research*, *8*(Issue 3), 137–140. Available from https://doi.org/10.1016/j.accre.2017.09.001, National Climate Center.

Saikia, P., Deka, J., Bharali, S., Kumar, A., Tripathi, O. P., Singha, L. B., … Khan, M. L. (2017). Plant diversity patterns and conservation status of eastern Himalayan forests in Arunachal Pradesh, Northeast India. *Forest Ecosystems*, *4*(1), 28. Available from https://doi.org/10.1186/s40663-017-0117-8.

Satti, Z., Naveed, M., Shafeeque, M., & Li, L. (2024). Investigating the impact of climate change on trend shifts of vegetation growth in Gilgit Baltistan. *Global and Planetary Change*, *232*, 104341. Available from https://doi.org/10.1016/j.gloplacha.2023.104341.

Schickhoff, U., Singh, R. B., & Mal, S. (2016). *Climate change and dynamics of glaciers and vegetation in the Himalaya: An overview. Climate Change, Glacier Response, and Vegetation Dynamics in the Himalaya: Contributions Toward Future Earth Initiatives* (pp. 1–26). Springer International Publishing. Available from https://doi.org/10.1007/978-3-319-28977-9_1.

Sekar, K. C., Thapliyal, N., Pandey, A., Joshi, B., Mukherjee, S., Bhojak, P., … Bahukhandi, A. (2023). Plant species diversity and density patterns along altitude gradient covering high-altitude alpine regions of west Himalaya, India. *Geology, Ecology, and Landscapes*, 1–15. Available from https://doi.org/10.1080/24749508.2022.2163606.

Sharma, N., Behera, M. D., Das, A. P., & Panda, R. M. (2019). Plant richness pattern in an elevation gradient in the Eastern Himalaya. *Biodiversity and Conservation*, *28*(8–9), 2085–2104. Available from https://doi.org/10.1007/s10531-019-01699-7.

Singh, J. S., & Singh, S. P. (1987). Forest vegetation of the Himalaya. *The Botanical Review*, *53*, 80–192. Available from https://doi.org/10.1007/BF02858183.

Telwala Yasmeen and Brook, B. W. A. N. D. M. K. A. N. D. P. M. K. (2013). Climate-induced elevational range shifts and increase in plant species richness in a Himalayan biodiversity epicentre. *PLoS One*, *8*(2), 1–8. Available from https://doi.org/10.1371/journal.pone.0057103.

Thakur, D., & Chawla, A. (2019). Functional diversity along elevational gradients in the high altitude vegetation of the western Himalaya. *Biodiversity and Conservation*, *28*(8–9), 1977–1996. Available from https://doi.org/10.1007/s10531-019-01728-5.

Verrall, B., & Pickering, C. M. (2020). *Alpine vegetation in the context of climate change: A global review of past research and future directions, . Science of the total environment* (Vol. 748). Elsevier B.V. Available from https://doi.org/10.1016/j.scitotenv.2020.141344.

Vetaas, O. R., & Grytnes, J.-A. (2002). Distribution of vascular plant species richness and endemic richness along the Himalayan elevation gradient in Nepal. *Global Ecology & Biogeography, 11*. Available from http://www.blackwell-science.com/geb.

Williams, C. J., Mcnamara, J. P., & Chandler, D. G. (2009). Hydrology and Earth System Sciences Controls on the temporal and spatial variability of soil moisture in a mountainous landscape: The signature of snow and complex terrain. *Hydrology and Earth System Sciences, 13*. Available from http://www.hydrol-earth-syst-sci.net/13/1325/2009/.

Zhao, J. L., Paudel, B. R., Yu, X. Q., Zhang, J., & Li, Q. J. (2021). Speciation along the elevation gradient: Divergence of Roscoea species within the south slope of the Himalayas. *Molecular Phylogenetics and Evolution, 164*. Available from https://doi.org/10.1016/j.ympev.2021.107292.

Zhou, G., & Wang, Y. (2000). Global change and climate-vegetation classification. *Chinese Science Bulletin, 45*, 577–585. Available from https://doi.org/10.1007/BF02886031.

Chapter 24

Assessment of drinking water quality using GIS technique and water quality index in West Bengal, India

Bappa Sarkar
Department of Geography, Dinhata College, Koch Bihar, India

24.1 Introduction

Groundwater, a plentiful and invaluable natural resource in India, is primarily used for domestic and irrigation needs in rural and urban settings (Xie et al., 2023). Groundwater is the main source of sources of drinking water globally (Kreitler, 2023). Safe drinking water is fundamental for human health and the environment (Marobhe, Renman, & Jacks, 2007). India initiated a decade-long effort in 1981 to ensure safe drinking water delivery through piped systems to urban and rural regions. However, despite increased water availability, the safety of water sources remains uncertain (Elala, Labhasetwar, & Tyrrel, 2011). Many developing countries rely on unreliable, costly, low-quality water supplies (WHO, 2013). More particularly, urban populations face higher health risks related to drinking water hazards due to inadequate water testing and mitigation (WHO, 2013). Cities often use untreated or treated municipal sewage, leading to waterborne diseases (Mudiam, Pathak, Gopal, & Murthy, 2012). Contaminated water has caused significant harm, with millions affected and millions of deaths in emerging countries (Dzwairo, Hoko, Love, & Guzha, 2006). Waterborne diseases contribute substantially to global illnesses, with an estimated 2.2 million fatalities and over 72 million cases of diarrhea and gastrointestinal disorders (WHO, 2011). Bacterial contamination is often higher in commercial water samples than in residential and industrial ones (Mudiam et al., 2012). Access to safe drinking water is closely linked to socioeconomic development (WHO, 2011a) *Guidelines for Drinking-water Quality*, no date; "(WHO (2011b) Guidelines for drinking-water quality," no date). While financially advanced individuals may opt for filtered water, disadvantaged sections

Water Resources Monitoring, Management, and Sustainability. DOI: https://doi.org/10.1016/B978-0-443-23665-5.00024-7

often consume water directly from sources. Extensive research worldwide has investigated water quality for drinking and distribution (Abd El-Aziz, 2017; Adimalla, Li, & Venkatayogi, 2018; He, Wu, & He, 2019; Khan & Jhariya, 2017; Rahman et al., 2023; Sao et al., 2023). Most of the drinking water is extracted from groundwater. Excessive extraction of groundwater is causing rapid declination in groundwater levels. The rapid decline in groundwater appears to be an emerging problem in recent decades. However, owing to the primary concern of drinking water quality, it is indispensable to investigate water quality, especially in highly populated States like West Bengal.

The Water Quality Index (WQI) is a crucial tool for converting vast water characterization data into a single numerical representation of water quality (Şener, Şener, & Davraz, 2017). Various WQIs, such as the U.S. National Sanitation Foundation Water Quality Index (NSFWQI) and Canadian Council of Ministers of the Environment Water Quality Index (CCMEWQI), have been developed globally (Brown, McClelland, Deininger, & Tozer, 1970; Debels, Figueroa, Urrutia, Barra, & Niell, 2005; Khan, Paterson, & Khan, 2003). Examining water quality is essential in hydrological investigations to understand chemical composition. Drinking water quality is a pressing concern, with some areas relying on purchased water due to inadequate supply. In many parts of the town, iron, manganese, and total hardness levels exceed permissible limits, emphasizing the need for continuous water quality monitoring and remedial measures to protect human health and life.

24.2 Study area

West Bengal is located in the eastern part of India (Fig. 50.1). This State is bordered by Bangladesh to the east, Nepal to the north, and the Bay of Bengal to the south. The latitudes of the State extend from 21°25° N to 27°13°N and longitude 89°25°E to 89°28° E. The State has 91 million inhabitants within 88,752 km^2, 22 districts. The northern part of the State has a relatively high altitude, having hilly tracks of the sub-Himalayan region, while the southern part has a very low altitude. The Ganga River cut across the State through the middle portion. The southwestern part is a part of the plateau region. The population density is greater in the southern part, having Kolkata metropolitan (Fig. 24.1).

24.3 Methodology

24.3.1 Water quality parameters maps by using geospatial techniques

846 water samples were gathered to estimate the WQI to show the geospatial distribution of water quality across the State. The water sample data was

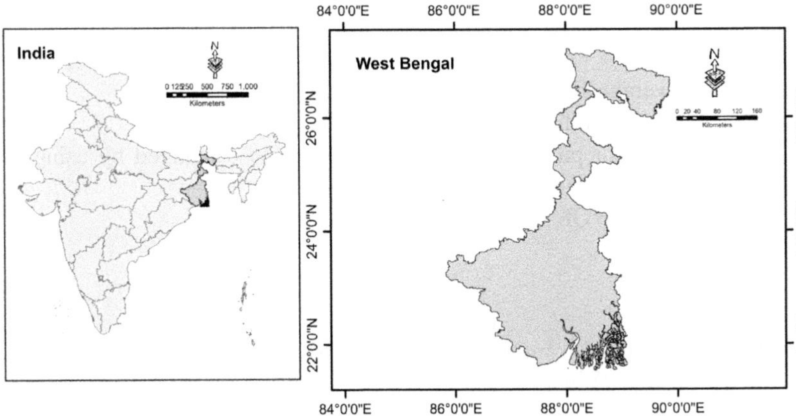

FIGURE 24.1 Location map of the Study Area, West Bengal. Map lines delineate study areas and do not necessarily depict accepted national boundaries.

retrieved from the Central Ground Water Board for 2018. The samples have been tested according to specific procedures.

24.3.2 WQI computation equations

WQI is defined as a tool that provides the composite influence of the overall quality of water. In every water sample, twelve parameters were chosen for the calculation of the WQI.

24.3.3 Calculation of subindex of quality rating (q_n)

The water quality parameter and the quality rating for the nth parameter is a number reflecting the relative value of this parameter. The value is calculated using the following equation.

$$Q_n = 100\left[(V_n - V_{\mathrm{i}}0)/(S_n - V_i0)\right] \tag{24.1}$$

Where,

Q_n, quality rating for the nth water quality parameter.

V_n, observed value of the nth parameter at a given sampling station.

S_n, standard value of the nth parameter.

V_i0, ideal value of the nth parameter. All the ideal values (V_i0) will be zero (0) for drinking water except P.H., where it is 7.0 and dissolved oxygen 16.6 mg/dm^3 (Tripathy & Sahu, 2005).

Calculation of Unit Weight (W_n)

Unit weight calculation for water quality parameters is conversely proportional to the specified standard S_n for the selected parameters.

$$W_n = K/S_n \tag{24.2}$$

Where,

W_n, Unit weight for the nth parameter

S_n, Standard value for the nth parameter

K = Constant for proportionality and can also be calculated by using the following equation.

Calculation of WQI

$$K = \frac{1}{\sum\left(\frac{1}{Sn}\right)}$$

WQI is calculated from the following expression

$\mathrm{WQI} = \sum W_n q_n / \sum wn$

24.3.4 Assessment of water quality

WQI has been classified into five categories (Table 24.1) shows the five water quality classes based on the arithmetic WQI method. The WQI values are classified into five categories: excellent, good, poor, very poor, and unsuitable for drinking (Table 24.2).

24.4 Result and discussion

The distribution and spatial pattern indicate groundwater chemical distribution arrangement across various parts of West Bengal. All the figures revealed the spatial pattern of groundwater chemicals. The interpolation method was employed to depict spatial variations based on random data. The interpolation was conducted using the IDW tool in ArcGIS 10.3 software, as detailed in the methodology section. This section analyses the distribution of all chemical components and examines the factors influencing their formations.

The Fig. 24.2 represents the spatial distribution of pH, E.C., T.H., and TDS in West Bengal Fig. 24.2. The pH of the water varies between 6.08 and 8.66, indicating that the water is somewhat acidic to slightly alkaline. The average pH of the West Bengal is 7.73. Looking at the pH map of West Bengal, we can see that the southern Districts have higher pH levels, and the northern districts have lower pH levels. Hills and foothills such as Jalpaiguri, Alipurduar, and Koch Bihar District have pH between 6 and 7, and most 24 Parganas Murshidabad Uttar Dinajpur have pH above 8. While the pH value usually doesn't directly impact users, it remains crucial in determining water quality for work-related purposes ("WHO (2011b) Guidelines for drinking-water quality," no date). The permissible limit of pH is 6.5–8.5 (WHO (2011a) *Guidelines for Drinking-water Quality*, no date). The permissible limit of E.C. is 300 (WHO (2011a) *Guidelines for Drinking-water Quality*, no date). The E.C. ranges between 25.3 at Purulia to 9036 at 24 Pargana.

TABLE 24.1 Permissible limit of drinking water quality, BIS, 2012.

Sl. no.	Parameter	Permissible limit (BIS, 2012)
1	pH	8.50
2	EC	300.00
3	TH	200.00
4	TDS	500.00
5	Ca	75.00
6	Mg	30.00
7	Bicarbonate	600.00
8	Cl	250.00
9	SO_4	200.00
10	NO_3	45.00
11	F	1.00
12	SiO_2	200.00
13	Na	200.00
14	K	55.00
15	Total alkalinity	200.00

Source: Permissible Limit of Drinking Water Quality, BIS, 2018.

TABLE 24.2 Water quality index range, status and possible usage of water sample.

WQI range	Status	Grading	Possible usage
0–25	Excellent	A	Drinking, irrigation, industrial
26–50	Good	B	Drinking, irrigation, industrial
51–75	Poor	C	Irrigation and industrial
76–100	Very poor	D	Irrigation
Above 100	Unsuitable for drinking	E	Proper treatment is required before use

Based on Khatri, N., Tyagi, S., Rawtani, D., & Tharmavaram, M. (2020). Assessment of river water quality through application of indices: A case study River Sabarmati, Gujarat, India. Sustainable Water Resources Management, 6(6), 101.

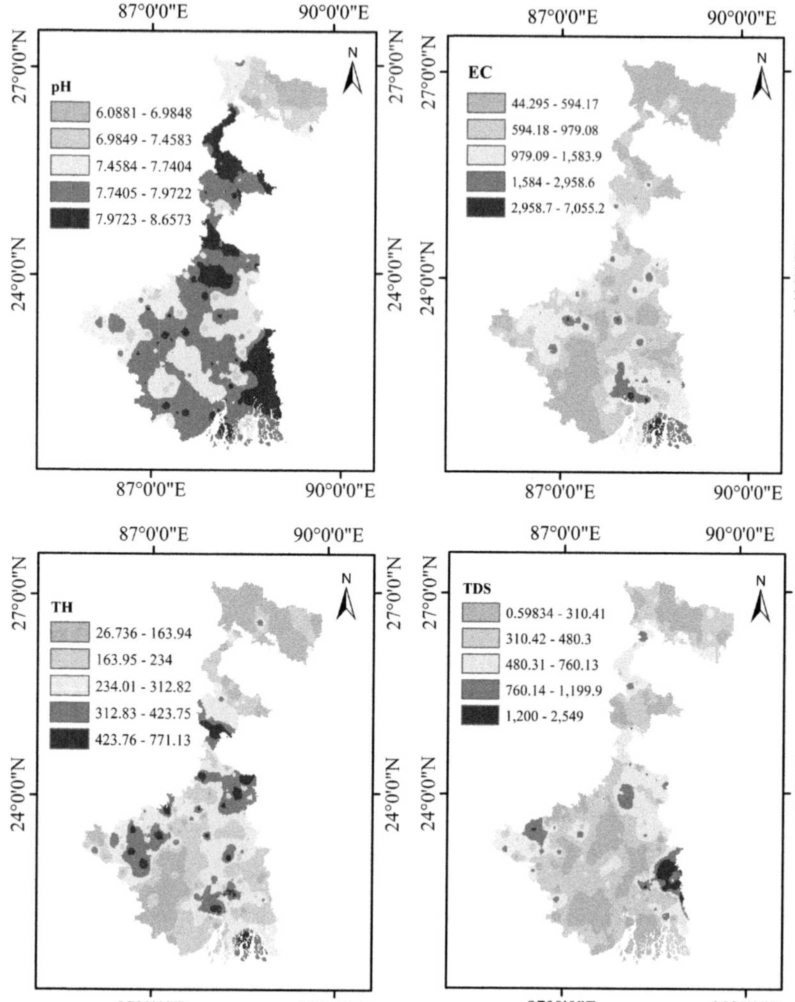

FIGURE 24.2 Spatial distribution of pH, E.C., T.H. and TDS of West Bengal. Map lines delineate study areas and do not necessarily depict accepted national boundaries.

The average E.C. of West Bengal is 851.86. The concentration of E.C. is found in most of the Southern District of West Bengal, such as 24 Pargana, Hawra, Bakura, Birbhum etc and lower E.C. is found in Koch Bihar, Jalpaiguri, Alipuduar, Darjeeling, Kalimpong. Urban areas may contribute to increased E.C. due to road salts, wastewater discharges, industrialization, and other anthropogenic influences. Total hardness (T.H.) defines the concentration of divalent cations, including magnesium (Mg), calcium (Ca), ferrous iron (Fe^{2+}), manganese (Mn), and strontium (Sr), in water. It is expressed as mg/L of $CaCO_3$. The primary sources of total hardness are Magnesium and

Calcium (WHO (2011a) *Guidelines for Drinking-water Quality*, no date). Elevated water hardness levels contribute to the formation of scales in water systems due to the precipitation of calcium-magnesium salts (Antony et al., 2011). From Darjeeling Kalimpong in the hilly districts of North Bengal to Jalpaiguri, Koch Bihar, North Dinajpur, and South Dinajpur, T.H. values are within permissible limits. In comparison, a higher concentration of E.C. was recorded at Purulia, South 24 Pargana, Hawra, Bakura, Murshidabad, and Malda. The rest of the districts have sparsely distributed T.H. values greater than the permissible limit Table 24.3.

Elevated Total Dissolved Solids (TDS) in water have multiple adverse effects on water quality. High concentrations of mineralized salts, indicative of elevated TDS levels, can lead to an unpleasant taste in water (Swamy, 1991). Water sources with a Total Dissolved Solids (TDS) concentration exceeding 500 mg/L are deemed unsuitable for drinking water supplies according to the World Health Organization (WHO) guidelines from 2011. However, a TDS value of up to 1000 mg/L in developing nations is acceptable (Farid, Baloch, & Ahmad, 2012). About 36.52% of the sample exceeded the permissible TDS limit in West Bengal. The highest TDS was recorded at 6359 at Bardhaman, and the lowest TDS in Jalpaiguri, that is, 25. Murshidabad, the eastern part of South 24 Parganas, West Bengal and some parts of Purulia have a TDS above 3000 (Fig. 24.2). In the districts of North Bengal, the amount of TDS is within permissible limits. Only in some places of North Dinajpur the amount of TDS is more than 500.

Fig. 24.3 represents the spatial distribution of K, Na, and Total Alkali in West Bengal. Getting enough calcium is crucial to prevent osteoporosis and achieve optimal bone health (WHO, 2009). About 53.83% of the sample exceeded the permissible limit of Ca in West Bengal. The permissible limit of Ca is 35 (BIS, 2012). The average Ca concentration of West Bengal is 48.66. A higher concentration of Ca was recorded at Purulia, Some patches of Malda, Murshidabad, and Purba Mednipur. And except for Malda, all the district of North Bengal has within the permissible limit of Ca. Mg, Cl, K and Na are higher in the Southern part of West Bengal and Lower in the State's Northern District. At the same time, the Total alkali is highly concentrated in most of the Eastern districts of West Bengal, such as Malda, Mushidabad, Nadia, and South 24 Pargona. Except for Dakshin Dinajpur (2.9), all the district has lower fluoride concentration (Fig. 24.4). The permissible limit of fluoride is 1 (BIS, 2012).

The correlation matrix Table 24.4 unveils statistical associations among diverse water quality parameters. Electrical Conductivity (E.C.), exhibiting strong positive correlations with Total Hardness ($r = 0.78$), Chloride ($r = 0.87$), Bicarbonate ($r = 0.57$), Sodium ($r = 0.81$), and a perfect correlation with Silica ($r = 1.00$). This implies that elevated E.C. values indicate increased concentrations in this parameter. Total Dissolved Solids (TDS) positively correlates with E.C. ($r = 0.02$), affirming that heightened electrical

TABLE 24.3 District-wise sample site distribution in percentage.

Name of the district	Excellent (0–25)	Good (26–50)	Poor (51–75)	Very poor (76–100)	Unsuitable for drinking (>100)
Bakura	47.92	46.88	4.17	1.04	0.00
Bardhaman	39.77	34.09	3.41	22.73	0.00
Birbhum	12.50	28.57	21.43	26.79	10.71
Dakshin Dinjapur	0.00	23.81	9.52	14.29	52.38
Darjeeling	61.29	35.48	3.23	0.00	0.00
East Midnapur	37.50	62.50	0.00	0.00	0.00
Haora	10.53	63.16	21.05	5.26	0.00
Hugli	37.93	55.17	0.00	6.90	0.00
Jalpaigui	76.74	18.60	4.65	0.00	0.00
Koch Bihar	52.38	42.86	4.76	0.00	0.00
Malda	0.00	35.71	57.14	7.14	0.00
Murshidabad	32.76	53.45	12.07	1.72	0.00
Nadia	58.00	36.00	4.00	0.00	2.00
South 24 Pargona	3.92	62.75	15.69	15.69	1.96
Uttar Dinajpur	0.00	18.75	12.50	50.00	18.75
West Midnapur	68.75	29.69	0.00	1.56	0.00
Kolkata	11.54	46.15	19.23	23.08	0.00
Purulia	55.56	31.11	8.89	2.22	2.22

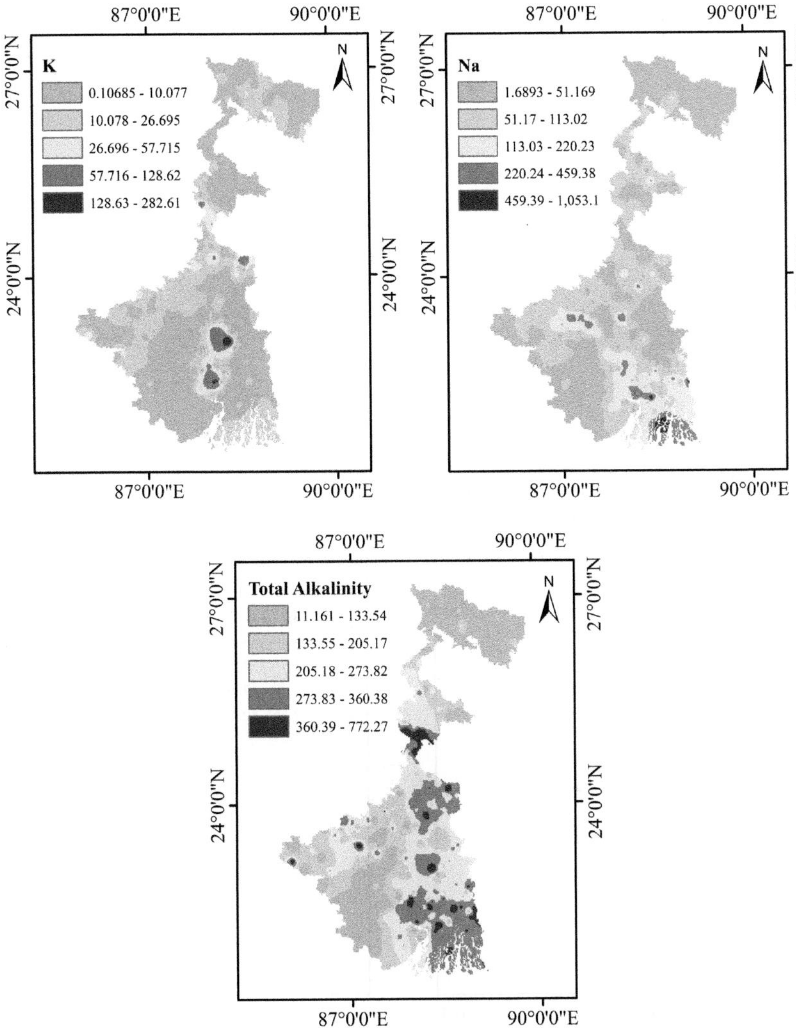

FIGURE 24.3 Spatial distribution of K, Na, total alkali of West Bengal. Map lines delineate study areas and do not necessarily depict accepted national boundaries.

conductivity corresponds to a rise in dissolved solids. Silica positive correlation with E.C. ($r = 1.00$) underscores a consistent relationship between electrical conductivity and silica levels. Fluoride (F) demonstrates positive correlations with Sodium ($r = 0.81$) and negative correlations with Total Alkalinity ($r = -0.44$), suggesting that higher fluoride concentrations align with elevated sodium levels and reduced alkalinity. In contrast, pH exhibits weak correlations with most parameters, indicating a relatively independent behavior, with exceptions noted in its negative correlation with Silica ($r = -0.08$) and a positive correlation with Sodium ($r = 0.04$).

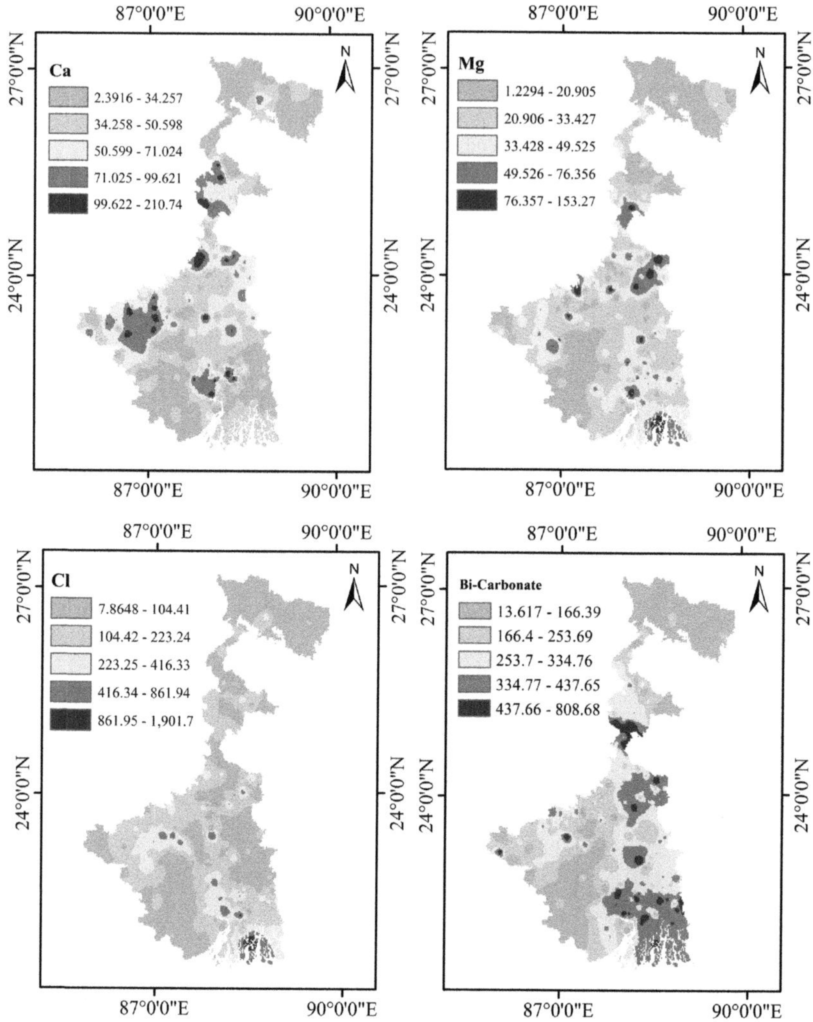

FIGURE 24.4 Spatial distribution of sulfate, nitraite, fluride and SiO_2 of West Bengal. Map lines delineate study areas and do not necessarily depict accepted national boundaries.

24.5 Spatial distribution of water quality index across the state

South Dinajpur district of West Bengal has the highest water quality index, reaching 244.23. In other places of South Dinajpur district, water quality is more than 200, which is unfit for drinking according to WQI class. Also, the WQI value is more than 150 in various places of Nadia Birbhum South 24 Parganas, which is considered unfit for drinking (Fig. 24.5). Places where the WQI is greater than 100 are respectively: Nadia 210.58, Dubrajpur in

TABLE 24.4 Correlation matrix among the hydrochemical parameter of groundwater.

	pH	E.C.	TH	TDS	Ca	Mg	Cl	HCO_3	SO_2	NO_3	F	SiO_2	k	Na	TA
pH	1.00														
EC	−0.07	1.00													
TH	−0.13	0.78	1.00												
TDS	0.09	0.02	0.04	1.00											
Ca	−0.16	0.52	0.77	0.04	1.00										
Mg	−0.05	0.70	0.81	0.03	0.26	1.00									
Cl	−0.06	0.87	0.67	−0.01	0.46	0.59	1.00								
HCO_3	0.07	0.57	0.58	0.10	0.35	0.55	0.33	1.00							
SO_2	−0.15	0.28	0.35	−0.01	0.25	0.30	0.24	0.07	1.00						
NO_3	−0.09	0.17	0.30	0.01	0.31	0.18	0.15	0.00	0.61	1.00					
F	0.19	0.05	0.08	0.06	0.00	0.12	0.04	0.15	−0.02	−0.10	1.00				
SiO_2	−0.08	1.00	0.78	0.02	0.52	0.70	0.87	0.56	0.28	0.18	0.05	1.00			
k	−0.05	0.35	0.27	−0.06	0.20	0.23	0.32	0.33	0.17	0.13	0.05	0.36	1.00		
Na	0.04	0.81	0.46	0.04	0.23	0.48	0.90	0.45	0.22	0.09	0.06	0.81	0.25	1.00	
TA	0.09	0.54	0.55	0.11	0.34	0.52	0.31	0.99	0.06	0.00	0.17	0.54	0.32	0.44	1

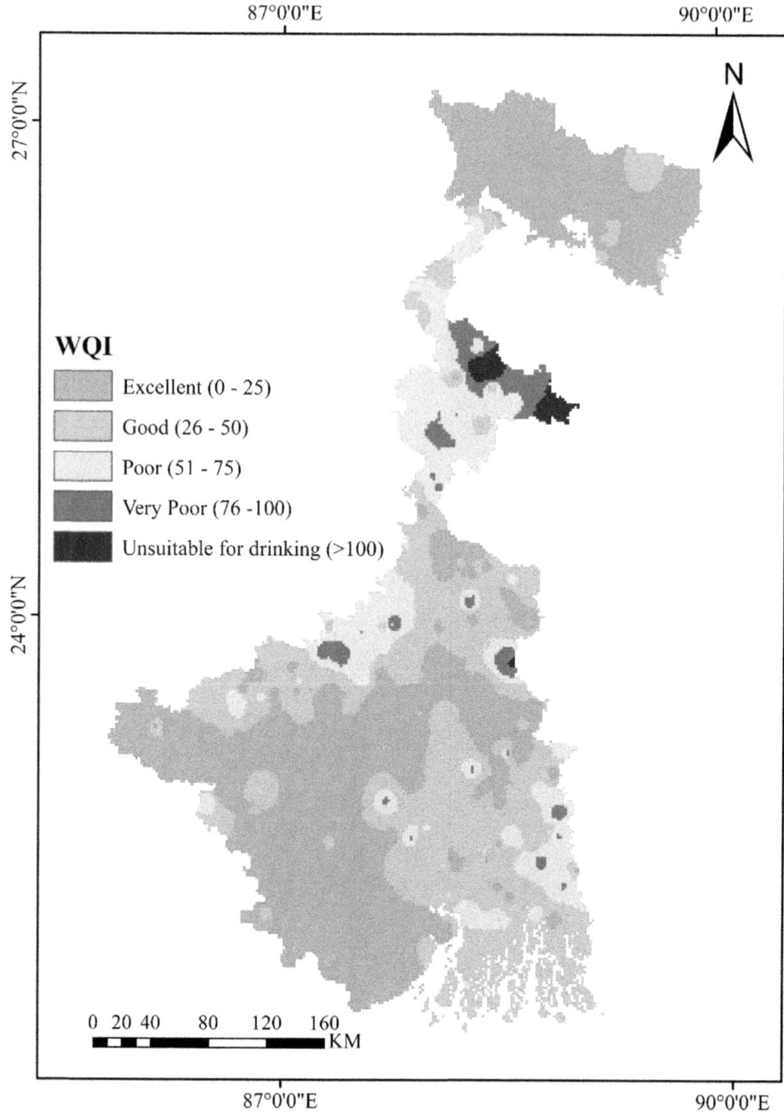

FIGURE 24.5 Spatial distribution of WQI, West Bengal, 2018. Map lines delineate study areas and do not necessarily depict accepted national boundaries.

Birbhum 194.17, Chopra in North Dinajpur 166.31 Purulia Para 163.83 Mald English Bazar 164.40, Kaliaganj in North Dinajpur 154.97 South 24 Parganas Mathurapur 126.51. North districts of West Bengal, especially Koch Bihar Jalpaiguri Alipurdua Darjeeling Kalimpong all these places have WQI values between zero and 25, which is very good for drinking and moderate water quality according to the West Bengal WQI some parts of

Murshidabad, Bankura, Burdwan, Purulia, the western part of Jhargram, and most of the Southern part of Nadia Midnipur have Excellent water quality. The map also shows good water locations in South 24 Parganas, Hooghly, Howrah, Malda, North Dinajpur and North Purulia, where values range from 51 to 75, indicating poor water quality. Table 24.3 revealed the district-wise water sample quality distribution, which indicates that Dakshin Dinjapur, Malda, has good to poor water quality and excellent water quality found in the northern part of the State, West Medinapur, Nadia, Purulia.

24.6 Conclusion

The water quality assessment in West Bengal showed that about 18.2% (16,151.8 km^2) of areas were considered excellent water sources, and 21.2% (18,815.6 km^2) of waterfalls in the Good category. In comparison, 22.4% (19,879.90 km^2) were recorded as the poor category, 19.68% (17,464.90 km^2) fell under the very poor category, and 18.52% (16,440.80 km^2) were classified as unsuitable for drinking. The hilly districts of West Bengal and Jalpaiguri Alipurduar Koch Bihar located at the foothills of the mountains, have relatively better water quality in these districts than the other districts. In contrast, the water quality of the western part and southern districts of South Bengal is good. Some parts of Pargana Howrah and Birbhum are unsuitable for drinking, so the Government and the policymakers should be kept in mind for installing Drinking water sources and the filtration process.

References

Abd El-Aziz, S. H. (2017). Evaluation of groundwater quality for drinking and irrigation purposes in the north-western area of Libya (Aligeelat). *Environmental Earth Sciences, 76* (4). Available from https://doi.org/10.1007/s12665-017-6421-3, http://www.springerlink.com/content/121380/.

Adimalla, N., Li, P., & Venkatayogi, S. (2018). Hydrogeochemical evaluation of groundwater quality for drinking and irrigation purposes and integrated interpretation with water quality index studies. *Environmental Processes*, *5*(2), 363–383. Available from https://doi.org/10.1007/s40710-018-0297-4.

Antony, A., Low, J. H., Gray, S., Childress, A. E., Le-Clech, P., & Leslie, G. (2011). Scale formation and control in high pressure membrane water treatment systems: A review. *Journal of Membrane Science*, *383*(1–2), 1–16. Available from https://doi.org/10.1016/j.memsci.2011.08.054.

BIS, I. (2012). 10500: 2012. Indian standard drinking water-specification (second revision). New Delhi: Bureau of Indian Standards.

Brown, R. M., McClelland, N. I., Deininger, R. A., & Tozer, R. G. (1970). A water quality index – Do we dare. *Water, Water*.

Debels, P., Figueroa, R., Urrutia, R., Barra, R., & Niell, X. (2005). Evaluation of water quality in the Chillán River (Central Chile) using physicochemical parameters and a modified water quality index. *Environmental Monitoring and Assessment, 110*(1–3), 301–322. Available from https://doi.org/10.1007/s10661-005-8064-1.

Dzwairo, B., Hoko, Z., Love, D., & Guzha, E. (2006). Assessment of the impacts of pit latrines on groundwater quality in rural areas: A case study from Marondera district, Zimbabwe. *Physics and Chemistry of the Earth, Parts A/B/C, 31*(15–16), 779–788. Available from https://doi.org/10.1016/j.pce.2006.08.031.

Elala, D., Labhasetwar, P., & Tyrrel, S. F. (2011). Deterioration in water quality from supply chain to household and appropriate storage in the context of intermittent water supplies. *Water Science and Technology: Water Supply, 11*(4), 400–408. Available from https://doi.org/10.2166/ws.2011.064, http://www.iwaponline.com/ws/01104/0400/011040400.pdf.

Farid, S., Baloch, M. K., & Ahmad, S. A. (2012). Water pollution: Major issue in urban areas. *International Journal of Water Resources and Environmental Engineering, 4*(3), 55–65.

He, X., Wu, J., & He, S. (2019). Hydrochemical characteristics and quality evaluation of groundwater in terms of health risks in Luohe aquifer in Wuqi County of the Chinese Loess Plateau, northwest China. *Human and Ecological Risk Assessment, 25*(1–2), 32–51. Available from https://doi.org/10.1080/10807039.2018.1531693. Available from, http://www.tandf.co.uk/journals/titles/10807039.asp.

Khan, A. A., Paterson, R., & Khan, H. (2003). Modification and application of the CCMEWQI for the communication of drinking water quality data in Newfoundland and Labrador. *Central Symposium on Water QualityResearch, 38.*

Khan, R., & Jhariya, D. C. (2017). Groundwater quality assessment for drinking purpose in Raipur city, Chhattisgarh using water quality index and geographic information system. *Journal of the Geological Society of India, 90*(1), 69–76. Available from https://doi.org/10.1007/s12594-017-0665-0, http://www.springerlink.com/content/121451/.

Kreitler, C. W. (2023). *Geochemical techniques for identifying sources of ground-water salinization*. Routledge. Available from http://doi.org/10.1201/9780203753668.

Marobhe, N. J., Renman, G., & Jacks, G. (2007). The study of water supply and traditional water purification knowledge in selected rural villages in Tanzania. Indigenous knowledge systems and sustainable development: Relevance for Africa. *Tribes and Tribals, 1.*

Mudiam, M. K. R., Pathak, S. P., Gopal, K., & Murthy, R. C. (2012). Studies on urban drinking water quality in a tropical zone. *Environmental Monitoring and Assessment, 184*(1), 461–469. Available from https://doi.org/10.1007/s10661-011-1980-3.

Rahman, M. M., Haque, T., Mahmud, A., Al Amin, M., Hossain, M. S., Hasan, M. Y., ... Bai, L. (2023). Drinking water quality assessment based on index values incorporating WHO guidelines and Bangladesh standards. *Physics and Chemistry of the Earth, 129.* Available from https://doi.org/10.1016/j.pce.2022.103353, http://www.journals.elsevier.com/physics-and-chemistry-of-the-earth/.

Sao, V., Chey, C. O., Sriv, T., The, N., Phin, S., Hoeng, S., & Phan, K. (2023). Assessing drinking water quality and health risks of contaminants in the coastal areas of Cambodia. *Journal of Water and Health, 21*(2), 224–234. Available from https://doi.org/10.2166/wh.2023.215, https://watermark.silverchair.com/jwh0210224.pdf?.

Swamy, K. (1991). An approach towards assessment of dug-well water quality by physico-chemical characteristics—A case study. *Pollution Research, 10*(1), 13–20.

Tripathy, J. K., & Sahu, K. C. (2005). Seasonal hydrochemistry of ground water in the barrier spit system of the Chilika lagoon, India. *Journal of Environmental Hydrology, 13.* Available from http://www.hydroweb.com/protect/pubs/jeh/jeh2005/tripathy.pdf.

WHO. (2011). Valuing water, valuing livelihoods.

WHO. (2011a). Guidelines for drinking-water quality.

WHO. (2011b). Guidelines for drinking-water quality. World Health.

World Health Organization. (2009). Bromide in drinking-water: Background document for development of WHO Guidelines for Drinking-water Quality *(No. WHO/HSE/WSH/09.01/6)*. World Health Organization.

WHO. (2013). Proposal for consolidated drinking water, sanitation and hygiene targets, indicators and definitions.

Xie, X., Jianbo Shi, K., Pi, Y., Deng, B., Yan, L., Tong, L., … Jiang, G. (2023). Groundwater quality and public health. *Annual Review of Environment and Resources*, *48*(1), 395–418. Available from https://doi.org/10.1146/annurev-environ-112321-114701.

Şener, Ş., Şener, E., & Davraz, A. (2017). Evaluation of water quality using water quality index (WQI) method and GIS in Aksu River (SW-Turkey). *Science of the Total Environment*, *584–585*, 131–144. Available from https://doi.org/10.1016/j.scitotenv.2017.01.102.

Chapter 25

Water resource mapping, monitoring, and modeling using geospatial approaches

Suyog Gupta[1,2] and Sunil Kumar Gupta[1]
[1]Department of Environmental Science and Engineering, Indian Institute of Technology (Indian School of Mines), Dhanbad, Jharkhand, India, [2]Department of Civil Engineering, Harcourt Butler Technical University, Kanpur, Uttar Pradesh, India

25.1 Introduction

Water is an indispensable resource for sustaining life, supporting ecosystems, and fueling economic and social development. As the water demand continues to escalate due to populace growth, urbanization, and industrialization, the need for efficient management and utilization of water resources becomes increasingly critical (Bijekar et al., 2022; Gupta and Gupta, 2021; Mehla et al., 2023; Ram and Irfan, 2021). Water resource mapping, monitoring, and modeling using geospatial approaches have appeared as influential tools to comprehensively analyze and manage these valuable resources (Feizizadeh et al., 2021; Mashala et al., 2023; Santos et al., 2023). Fortunately, advancements in geospatial technologies have paved the way for a paradigm shift in understanding and managing water resources through mapping, monitoring, and modeling. By harnessing geographic information systems (GIS), remote sensing (RS) technologies, and advanced modeling techniques, we can gain a holistic understanding of water resources, that is, their spatial distribution, quality, availability, and behavior over time (Gupta and Gupta, 2021; Santos et al., 2023). This integrated approach empowers stakeholders, including government bodies, environmental organizations, researchers, and policymakers, to make informed decisions, develop sustainable water management strategies, and mitigate challenges associated with water scarcity, pollution, and climate change.

In the wake of environmental challenges and a growing consciousness of the need for responsible resource management, the integration of geospatial techniques into water resource management has become pivotal. It offers a transformative opportunity to holistically understand the dynamics of water

Water Resources Monitoring, Management, and Sustainability. DOI: https://doi.org/10.1016/B978-0-443-23665-5.00025-9

resources, manage them efficiently, and ensure their sustainability for future generations. Water resource mapping, monitoring, and modeling using geospatial approaches are critical components of water resource management (Feizizadeh et al., 2021; Murmu et al., 2019; Elhag, 2016; Singh et al., 2020). These techniques leverage geospatial technologies such as GIS, RS, and global positioning systems (GPS) to collect, analyze, and visualize spatial data related to water resources (Mashala et al., 2023; Matlhodi et al., 2021; Fu et al., 2020; (Zhu et al., 2021)). The process begins with data acquisition, which involves gathering various types of spatial data such as satellite imagery, digital elevation models (DEMs), land use/land cover data, hydrological data (e.g., streamflow and precipitation), and other relevant datasets. These data sources provide valuable information about the spatial distribution, availability, and quality of water resources. Once the data is acquired, it undergoes preprocessing to ensure its quality and compatibility for analysis. This may include tasks such as geometric correction, radiometric correction, mosaicking, and data integration to create a coherent dataset for further analysis. Satellite imagery can be classified into different land cover/use classes using classification techniques, allowing for the identification of water bodies, urban areas, agricultural lands, and other features relevant to water resource mapping. Additionally, features like rivers, lakes, and reservoirs can be extracted from the imagery using image processing algorithms, providing valuable spatial information for water resource management (Zeng et al., 2022; Zhu et al., 2021).

Hydrological modeling is a key application of geospatial approaches in water resource management. Hydrological models simulate the movement of water through the hydrological cycle, including processes such as precipitation, infiltration, runoff generation, and streamflow routing (Zeng et al., 2022; Aderemi et al., 2021). These models use geospatial data to represent the physical characteristics of the watershed and predict the impact of various factors on water availability and quality (Zhu et al., 2021). In addition to quantity, geospatial techniques are also used for water quality assessment. This involves analyzing satellite imagery and field measurements to monitor changes in water quality parameters such as turbidity, chlorophyll-a concentration, and water temperature. By integrating spatial and temporal data on water quality, stakeholders can identify trends, assess the impact of land use changes, and make informed decisions regarding water quality management. GIS software is used to analyze and visualize the geospatial data related to water resources, allowing for tasks such as overlaying different layers to identify spatial relationships, creating thematic maps to visualize the distribution of water resources and related factors, and conducting spatial analysis to derive meaningful insights from the data. GIS-based analysis and visualization tools are essential for communicating complex spatial information to stakeholders and decision-makers. Fasipe et al. (2021) explain how the RS and GIS approaches are paired with the curve number method for the identification of small hydropower potential sites in view of different catchment vulnerabilities. Moreover, Zhu et al., 2021 have provided a review of the application of GIS to water resources. They determined that GIS can be utilized efficiently in water resource management due to the spatial nature of the essential data.

The application of RS and GIS techniques in water resources management has increased significantly during the last few years with the advancement of high-performance computing devices (Fu et al., 2020; Zhu et al., 2021; Fasipe et al., 2021; Mashala et al., 2023). Several studies have confirmed the efficacy of RS and GIS techniques in water quality monitoring (Dandge and Patil, 2022; Gupta and Gupta, 2021; 2023; Yan et al., 2022), surface water mapping (Ashtekar et al., 2019; Al-Ghobari et al., 2020), groundwater exploration and management (Al-Djazouli et al., 2021; Khan et al., 2022; Oyedele, 2019), flood risk assessment and management (Tomar et al., 2021; Subraelu et al., 2023), drought monitoring (Adisa et al., 2020; De et al., 2021), water use efficiency in agriculture (Bonfante et al., 2019; Abdelhaleem et al., 2021), hydrological modeling (Fasipe et al., 2021; Rautela et al., 2022), water infrastructure planning (Subraelu et al., 2023; Yadav et al., 2022), wetland monitoring and conservation (Orimoloye et al., 2019; Singh et al., 2020), and water resource conflict resolution (Gorgoglione et al., 2019; Haseeb et al., 2023) due to their capability to capture information in both temporal and spatial scales (Abbas, 2013).

This exploration embarks on a comprehensive journey into water resource mapping, monitoring, and modeling, aiming to elucidate their profound significance and multifaceted applications. We will delve into the methodologies that underpin these approaches, unveiling the intricacies of data collection, analysis, and interpretation. The discussion will further highlight the critical role of geospatial technologies in assessing water quality, quantifying groundwater resources, predicting hydrological patterns, and supporting policy formulation. This chapter explores the conceptual background, state-of-the-art techniques, and tools comprising geospatial approaches to water resource mapping, monitoring, and modeling for sustainable water resource management. Moreover, the challenges involved in geospatial approaches and their future directions are also discussed for the continual state of transition in computing technology.

25.2 Materials and methods

Water resource mapping and assessment using geospatial approaches involve leveraging GIS, RS, and other geospatial technologies to analyze and manage water-related data. This approach allows for a comprehensive understanding of water resources, their availability, quality, and distribution, which is crucial for sustainable water management and decision-making. Fig. 25.1 shows the flow diagram, including different elements involved in water resource mapping, monitoring, and modeling using geospatial approaches.

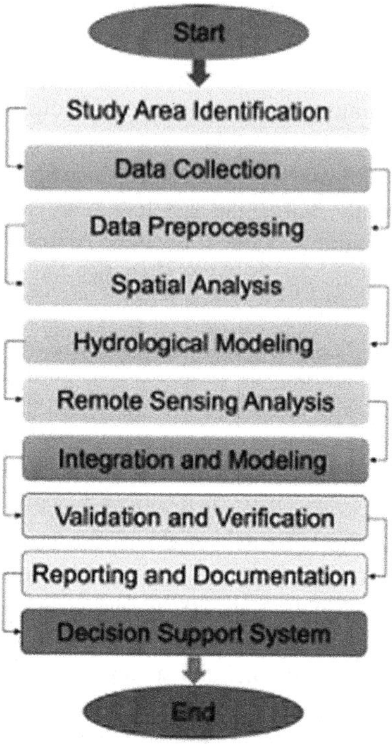

FIGURE 25.1 Flow diagram of materials and methods.

25.2.1 Data collection and preparation

Satellite imagery and aerial photography are used to obtain information on water bodies, land use, vegetation cover, and other relevant features using RS techniques. Ground truthing is required to validate and augment RS data with ground-based observations and measurements, ensuring accuracy and reliability (Mehla et al., 2023). An overview of the processes and technologies involved in collecting these images is described in various studies (Haseeb et al., 2023; Zhu, Zhao, Liang, & Qin, 2021; Fasipe et al., 2021; Ashtekar et al., 2019).

25.2.1.1 Satellite imagery collection

Various Earth-observing satellites equipped with different sensors are used to capture Earth's surface imagery. These sensors include optical, multispectral, hyperspectral, thermal, radar, and LIDAR instruments, among others. These satellites orbit the Earth at different altitudes, providing global or regional coverage. Depending on the mission and sensor, they can capture images at various spatial and temporal resolutions. To collect satellite imagery, organizations and researchers plan acquisition schedules based on their specific needs. This may involve specifying the geographic area, date, and time of interest. Various space agencies, such as NASA, ESA, and commercial satellite companies like DigitalGlobe, operate satellites and provide access to satellite imagery. Researchers can access these images through dedicated portals or data providers. Raw satellite imagery is often preprocessed to correct for distortions, radiometric calibration, and georeferencing. This ensures the data's accuracy and usability. Users can access satellite imagery data through online platforms, data subscriptions, or data purchases. Some platforms (e.g., Bhuvan and Earth Explorer) offer free access to publicly available data, while others require paid subscriptions.

25.2.1.2 *Aerial photography collection*

Aerial photography is carried out using aircraft, unmanned aerial vehicles, or helicopters. The choice of platform depends on the application and the desired level of detail. The cameras used in aerial photography are specially designed to capture high-resolution images from above. They can be film-based or digital, with features like multiple lenses for stereoscopic imaging. The aerial missions involve careful planning to determine the flight path, altitude, and camera settings. Factors such as weather conditions, the area of interest, and imaging goals are considered in this phase. These photographs are georeferenced to accurately represent the Earth's surface. Ground control points, GPS data, and inertial navigation systems are used to achieve georeferencing. The aerial platform captures a series of overlapping images that cover the entire area of interest. Overlap is essential for creating mosaics and 3D models. Once collected, aerial photos undergo postprocessing to enhance image quality, remove distortion, and ensure consistency across the dataset. Users can access aerial photography data through various channels, including government agencies, private companies, and aerial survey providers. Data may be provided in digital formats or printed maps.

Both satellite imagery and aerial photography are essential tools for various applications, that is, environmental assessment, disaster management, agronomy, land use planning, forestry, and infrastructure development. These high-resolution images provide valuable information for decision-makers and researchers and support a wide range of geospatial analyses and modeling efforts.

25.2.2 Data integration and processing

Data integration and processing are crucial components of data management, particularly in the context of GIS, RS, and other spatial data applications. These processes involve combining and transforming data from various sources into a consistent format, making it suitable for analysis, visualization, and decision-making. The GIS software, that is, ArcGIS, QGIS, GRASS GIS, and ILWIS, helps to integrate various spatial data sources, including RS data, topographic maps, hydrological data, and existing GIS datasets, to create a comprehensive database. Data can come from different sources, that is, satellite imagery, aerial photography, field surveys, sensor networks, public databases, and proprietary datasets. Therefore, different data types and formats (e.g., raster, vector, and tabular) need to be harmonized and transformed for consistency. This may involve converting data from one format to another, ensuring a common coordinate system, and handling missing or incomplete data. Data integration often requires cleaning and preprocessing to remove duplicates, correct errors, and address inconsistencies in attribute values. Geospatial data may need to be resampled, reprojected, and rescaled to match a common reference frame and resolution. Transformation methods vary based on the specific data types involved. Integrating data often requires establishing links between related datasets, such as associating land cover data with soil data to support agricultural modeling. Accurate and standardized metadata are essential to keeping track of the source, quality, and characteristics of integrated data. Metadata provides important context for understanding and using the data. Data processing involves the manipulation and analysis of integrated data to extract meaningful information and insights.

Data processing includes a wide range of analytical techniques, from simple queries and filtering to complex geospatial analyses. This can involve operations like buffer creation, spatial joins, and network analysis. Geoprocessing tools are used to perform operations on geographic data. GIS software provides a variety of geoprocessing tools for spatial analysis, data conversion, and modeling. In RS, image processing techniques are applied to satellite or aerial imagery to extract information about land cover, vegetation, terrain, and more. This may involve classification, feature extraction, and image enhancement. Data processing can involve the development and execution of spatial models to simulate or predict real-world phenomena. Models can range from simple regression models to complex hydrological, ecological, or transportation models. Once data is processed, it is often visualized using maps, charts, and graphs to convey information and insights effectively. Data visualization is crucial for conveying complex geospatial information to a broad audience. In many applications, processed data and results are used to generate reports, make decisions, and communicate findings to stakeholders.

Data integration and processing play a vital role in harnessing the value of spatial data for a wide range of applications. Properly managed and processed data allows organizations and researchers to derive actionable insights and make decisions based on precise and up-to-date information.

25.2.3 Water resource mapping

Water resource mapping is the process of creating visual representations or maps that depict the distribution, quality, and spatial characteristics of water resources in a given area. It plays a critical role in understanding, managing, and conserving water sources, whether for drinking water, agriculture, industrial use, or environmental preservation. Water resource mapping involves a combination of data collection, analysis, and visualization techniques. Land use and land cover (LULC) classification is the key aspect of water resource mapping (Aghsaei et al., 2020; Daba & You, 2022; Darem, Alhashmi, Almadani, Alanazi, & Sutantra, 2023; Ngondo et al., 2021; Zhu, Zhao, Liang, & Qin, 2021) (Woldesenbet, 2022). The RS data, GIS techniques, and machine learning (ML) classification algorithms are extensively used to classify and map different LULC categories, including water bodies, urban areas, and forests. This methodology involves the use of satellite or aerial imagery to categorize different land cover types and land uses within a particular region. ML algorithms are then applied to automatically classify and interpret the imagery. A step-by-step description for performing LULC classification for water resource mapping using RS and GIS-based ML techniques is shown in Fig. 25.2.

A high-resolution satellite or aerial imagery is required to cover the entire study area. The choice of the satellite sensor depends on the spatial and spectral resolution requirements of the study. This data is preprocessed to remove noise, correct atmospheric distortions, and enhance the overall quality of the data. Common preprocessing steps include radiometric calibration,

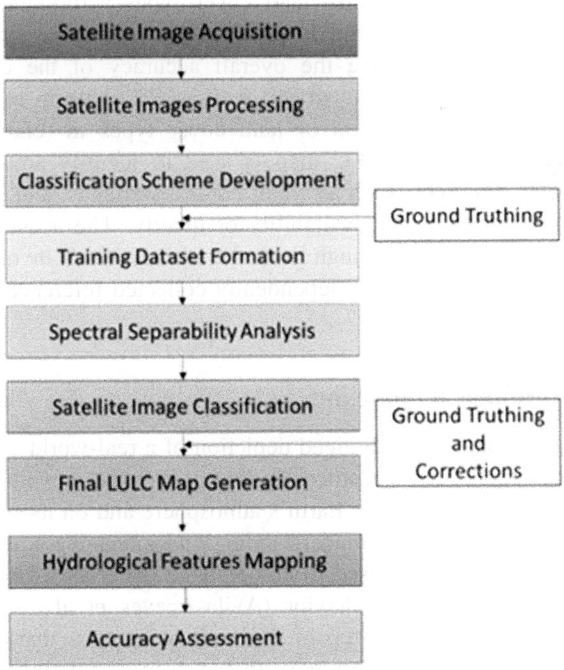

FIGURE 25.2 Flow diagram of water resource mapping.

geometric correction, and image normalization. If using multispectral imagery, stack the bands to create a composite image. This can improve the classification accuracy by combining information from different spectral bands (Darem et al., 2023; Ashtekar et al., 2019; Fasipe et al., 2021). Representative samples are collected from the region for each land cover class. These samples are used for the ML model's training. Ground-truth data collected through field surveys can aid in this process. The relevant spectral, textural, and spatial features are extracted from the satellite imagery, which serve as input parameters for these models. This data is fragmented into training and validation datasets, where the training dataset is used to train the model while the validation dataset supports assessing the model's performance (Darem, Alhashmi, Almadani, Alanazi, & Sutantra, 2023; Gupta & Gupta, 2021). A suitable ML algorithm is selected for classification. Commonly used algorithms for LULC classification include random forest, decision trees, support vector machines, and artificial neural networks (CNN). The selected ML model is trained using the training dataset and associated ground truth labels. The model learns to classify pixels based on the features extracted from the satellite imagery. The model is validated using the validation dataset, and the parameters are fine-tuned to optimize classification accuracy. This step involves adjusting hyperparameters and evaluating the model's performance. The trained ML model is applied to the entire satellite image to classify land cover types across the study area, and postclassification processing is conducted to eliminate small, isolated misclassifications and improve the overall accuracy of the classification. The classified LULC map is integrated with a GIS platform to analyze and visualize the spatial distribution of land cover types in relation to water resources. The generated LULC map is used to identify areas of interest related to water resources, such as water bodies, wetlands, and land-use patterns that may impact water availability or quality. The accuracy of water resource maps is validated through ground truthing, which involves comparing the classified results with independently collected reference data (Power & Sharda, 2023).

25.2.4 Hydrological modeling

A hydrological model is an abridged depiction of a real-world system, which mathematically represents the processes that govern the movement, distribution, and quality of water in the Earth's atmosphere and on its surface. These models are essential tools for understanding and managing water resources. These models are primarily used to comprehend different hydrological processes and forecast system behavior (Avila-Aceves et al., 2023) (Horton et al., 2022). These are made up of different parameters that specify their properties. A collection of formulas that aid in estimating runoff as a function of different parameters used to describe the watershed features is known as a runoff model. The drainage area and rainfall data are two crucial inputs used by all models. In addition to these, features of the watershed, such as the soil's moisture content, vegetation cover, the geography of the watershed, and characteristics of the groundwater aquifer, are also considered. These days, hydrological models are regarded as a crucial and essential instrument for managing water and environmental resources. These models can be classified based on various criteria, such as their complexity, spatial scale, and purpose (Table 25.1).

TABLE 25.1 Characteristics of hydrological models.

Model type properties	Physically-based model	Conceptual model	Empirical model
Description	These models explicitly represent the underlying physical processes of the hydrological cycle, such as precipitation, evaporation, and runoff, using governing equations.	These models represent the hydrological system using simplified conceptualizations of physical processes. They often involve lumped parameters.	These models are based on observed associations between input and output variables without explicit consideration of the underlying physical processes.
Characteristic	Black box/data-based/metric model.	Gray box/ parametric model.	Black box/data-based/metric model.
Basic principle	Parameters are derived from the physical characteristics of the catchment based on spatial distribution. It requires morphological data.	Parameters are derived from field data using semi-empirical mathematical equations.	Parameters are derived from time series using mathematical equations.
Advantage and limitation	Complex models, which require human expertise and suffer from scaling.	Simple models, which require huge meteorological and hydrological data.	High predictive power with low explanation.
Example			

(Continued)

TABLE 25.1 (Continued)

Model type properties	Physically-based model	Conceptual model	Empirical model
	MIKE SHE, SWAT, HEC-HMS, MODFLOW, PRMS, and MIKE 11.	PRMS, GR4J, HBV, VSA, TOPMODEL, SCS, Snyder's Synthetic Unit Hydrograph Model, and Clark's Unit Hydrograph Model.	Rating Curve Models, Unit Hydrograph Models, Flood Frequency Models,
Application	Best suited for regions with well-defined physical characteristics. Used in research and for understanding complex interactions in the hydrological system.	Suitable for regions with limited data and where a balance between simplicity and accuracy is required. Commonly used for watershed-scale modeling.	Often used for quick assessments and predictions in data-rich environments where the relationships between variables are well-established.

25.2.5 Water quality monitoring and modeling

Water quality monitoring involves the application of GIS, RS, and spatial analysis to assess, manage, and visualize water quality data in the context of its geographical location (Fig. 25.3). Geospatial techniques provide valuable insights into the distribution, changes, and sources of water pollution, enabling more effective water quality management. Geospatial techniques allow for the integration of various datasets, including water quality parameters (e.g., temperature, pH, dissolved oxygen, and turbidity), land use, hydrological features, and topography. GIS software provides spatial visualization tools that allow users to create maps and visual representations of water quality data. These maps help identify trends, hotspots of pollution, and spatial patterns in water quality (Balla, Zichar, Kiss, Szabó, & Mester, 2022; Bojago, Tyagi, Ahamad, & Chandniha, 2023; Korchenko et al., 2019). Moreover, temporal analysis tools in GIS enable the tracking of water quality changes over time. This is critical for assessing seasonal variations, long-term trends, and the impact of climate change on water quality (Korchenko et al., 2019). These techniques can also help in identifying potential sources of water pollution, such as industrial discharges, agricultural runoff, or urban

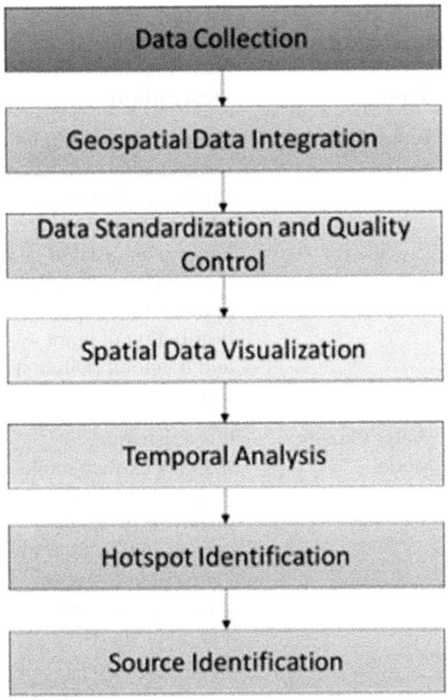

FIGURE 25.3 Flow diagram of water quality monitoring.

development. By analyzing spatial relationships, it becomes easier to pinpoint pollution origins.

RS and GIS play crucial roles in water quality modeling by providing spatially explicit information and facilitating the integration of various environmental factors. When utilizing RS and GIS in water quality modeling, it is essential to integrate accurate and up-to-date spatial information, validate models with observed data, and continuously improve model accuracy through calibration processes. Additionally, collaboration between hydrologists, GIS specialists, and RS experts is crucial for successful model implementation. A brief description of water quality models that incorporate RS and GIS is described in Table 25.2 (Bai, Zhao, Zhang, & Tian, 2022; Shabani et al., 2021).

25.2.6 Groundwater assessment and potential zone mapping

Groundwater assessment using geospatial techniques involves the use of GIS, RS, and spatial analysis to evaluate and manage groundwater resources. Geospatial techniques are valuable tools for understanding the distribution, quality, and sustainability of groundwater, as well as for making informed decisions regarding its management and protection (Berhanu & Hatiye,

TABLE 25.2 Description of water quality models.

Model name	Type	Description	Application
CE-QUAL-W2	Water Quality Model	It is a two-dimensional water quality simulation model for lakes and reservoirs. GIS is often used to input bathymetric data and define the spatial distribution of point and nonpoint pollution sources.	Lake and reservoir water quality modeling, prediction of nutrient cycling, and impact assessment.
WASP (Water Quality Analysis Simulation Program)	Water Quality Model	It is a dynamic, spatially explicit model for simulating water quality in rivers and streams. GIS is used to input data on stream geometry, land use, and point source discharges.	Assessing the impact of point source pollution, understanding nutrient dynamics, and designing water quality management strategies.
HEC-RAS (Hydrologic Engineering Center-River Analysis System)	River Hydraulic and Water Quality Model	It is commonly used for river hydraulics, but it also has water quality modeling capabilities. GIS is used for inputting river geometry and representing point sources.	Simulating river hydraulics and predicting water quality changes due to point source discharges.
AQUATOX	Ecological and Water Quality Model	It simulates the fate of pollutants in aquatic ecosystems and their impact on aquatic organisms. GIS is used for inputting spatial data on land use and hydrology.	Assessing the impact of contaminants on aquatic ecosystems and designing water quality management strategies.

2020; Murmu, Kumar, Lal, Sonker, & Singh, 2019; Achu et al., 2020). Groundwater potential zones (GWPZ) mapping using RS and GIS-based ML techniques is an innovative approach that combines satellite imagery, GIS, and ML algorithms to assess and predict the likelihood of finding groundwater in a given area (Fig. 25.4).

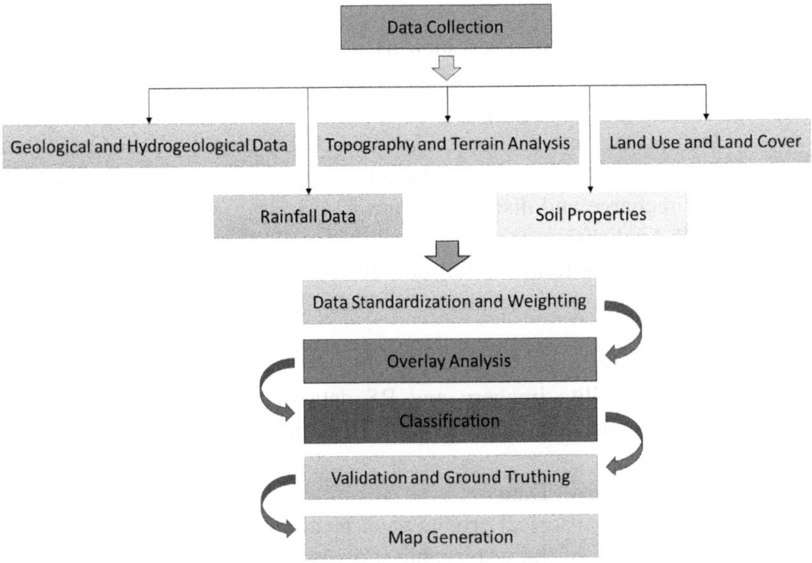

FIGURE 25.4 Flow diagram of groundwater potential zone mapping.

This methodology is valuable for water resource management, particularly in regions where access to groundwater is crucial for agricultural, domestic, and industrial purposes. Generating GWPZ maps involves the integration of geological, hydrological, and environmental data using GIS and spatial analysis techniques. The following key components and steps are involved in producing GWPZ maps:

1. Data Collection

   Collect relevant data, including geological maps, topographic maps, hydrogeological maps, LULC maps, soil maps, rainfall data, and existing groundwater well data.
2. Geological and Geomorphological Mapping

   Analyze geological and geomorphological features to identify formations that are conducive to groundwater storage and movement, such as alluvial deposits, fractured rock, and permeable layers.
3. Topographic Analysis

   Use DEMs to analyze terrain characteristics. Identify areas with gentle slopes and low relief, as these may be indicative of potential groundwater recharge zones.
4. Land Use and Land Cover Analysis

   Assess LULC patterns to understand the impact of human activities on groundwater recharge. Urbanized or heavily cultivated areas may have reduced recharge potential.

5. Soil Characteristics

   Analyze soil properties such as texture, porosity, and permeability to identify areas with soils conducive to groundwater recharge.

6. Hydrogeological Mapping

   Incorporate information on aquifer characteristics, including lithology, aquifer thickness, and hydraulic conductivity. Identify potential aquifer recharge and discharge zones.

7. Rainfall Analysis

   Analyze long-term rainfall data to identify areas with high and consistent rainfall, as these areas are likely to contribute to groundwater recharge.

8. RS Data Analysis

   Utilize satellite imagery and RS data to identify surface features, vegetation, and land use patterns that may influence groundwater recharge.

9. Groundwater Level Data

   Incorporate historical groundwater level data to understand the depth and fluctuations of the water table. Low water tables may indicate areas with potential groundwater extraction.

10. Spatial Analysis

    Use GIS tools to integrate and analyze spatial data layers. Overlay different thematic maps to identify areas where favorable conditions overlap.

11. Weighting and Ranking

    Assign weights to each factor based on its importance in influencing groundwater potential. Rank the factors based on their significance in contributing to groundwater availability.

12. Modeling and Indexing

    Develop a groundwater potential index by combining the weighted and ranked factors. This index represents the overall groundwater potential of different areas within the study area.

13. Map Validation

    Validate the groundwater potential map using independent datasets or field measurements. Compare the model predictions with actual groundwater conditions to assess the map's accuracy.

14. Potential Zonation

    Classify the study area into different GWPZs based on the index values. Common classifications include high, moderate, and low potential zones.

15. Map Generation and Interpretation

    Generate GWPZ maps using GIS software. Interpret the maps and provide information on the characteristics and potential uses of each zone.

16. Report and Documentation

Document the methodology, data sources, and results in a comprehensive report. Include recommendations for sustainable groundwater management practices in each potential zone.

17. Community Consultation

Engage with local communities and stakeholders to gather additional insights and localknzowledge and validate the findings. This step enhances the social acceptance of the GWPZ maps.

Generating GWPZ maps is a multidisciplinary process that involves the integration of geological, hydrogeological, climatic, and anthropogenic factors. The accuracy of the maps depends on the quality of the data, the appropriateness of methods, and the validation process (Saranya and Saravanan, 2020; Allafta et al., 2020).

25.2.7 Continuous monitoring and updating

Continuous monitoring and updating are essential aspects of water resource management, environmental monitoring, and various other fields where data accuracy and timely information are critical. This refers to the ongoing and real-time process of collecting and refreshing data to ensure its accuracy and relevance. This practice is critical in various fields, including environmental monitoring, infrastructure management, public safety, and data-driven decision-making. Continuous monitoring and updating help organizations and individuals stay informed, adapt to changing conditions, and make timely decisions. The key aspects of continuous monitoring and updating can be described below.

1. Real-Time Data Collection

Real-time data collection is instrumental in various water resource applications, providing organizations and individuals with the ability to respond promptly to varying conditions and make verdicts based on up-to-date information. This can be achieved through various technologies, including sensors, telemetry systems, RS, and IoT devices.

2. Data Sources

Data sources may vary depending on the application. They can include weather stations, water level sensors, air quality monitors, traffic cameras, social media feeds, and more. These sources provide a wide range of information.

3. Data Transmission

Collected data is often transmitted to a central database or control center, where it can be analyzed and accessed by relevant parties. Data transmission can occur through wired or wireless communication networks.

4. Quality Control

   Continuous monitoring systems typically include quality control mechanisms to confirm the accuracy and consistency of the collected data. Data validation and error correction processes are common.

5. Data Storage

   The collected data is stored in databases or on cloud-based platforms, making it easily accessible for analysis and decision-making. Data storage infrastructure should be robust and secure.

6. Data Analysis

   Continuous monitoring data is subject to real-time or periodic analysis to identify trends, anomalies, and potential issues. Analytical tools and algorithms are often used for this purpose.

7. Early Warning Systems

   In applications like disaster management and public safety, continuous monitoring is used to create early warning systems. These systems provide alerts and notifications in the event of emerging threats or abnormal conditions.

8. Dashboard and Visualization

   Data is often presented in real-time dashboards and visualizations to help users interpret and understand the information quickly. These interfaces may include maps, charts, and graphs.

9. Automation and Alerts

   Automated processes and alerts can be set up to trigger specific actions when certain conditions or thresholds are met. For example, automated flood warnings can trigger evacuation procedures.

10. Predictive Modeling

    In some cases, continuous monitoring data is integrated into predictive models that forecast future conditions or events. This is common in weather forecasting, energy management, and financial markets.

11. Maintenance and Calibration

    Continuous monitoring systems require regular maintenance and calibration to ensure the accuracy of the data. Sensors and equipment may drift over time and need adjustment.

12. Feedback Loop

    Continuous monitoring and updating often involve a feedback loop where the collected data is used to adapt strategies and make informed decisions. This helps organizations respond proactively to changing conditions.

13. Regulatory Compliance

    In certain industries and applications, continuous monitoring is required to meet regulatory and compliance standards, such as environmental regulations and safety standards.

Continuous monitoring and updating are invaluable for keeping systems, processes, and decision-making current and responsive to changes. They help mitigate risks, improve efficiency, and enhance safety and the quality of services in various domains.

25.2.8 Decision support systems

Decision Support Systems (DSS) are computer-based tools and programs that process and analyze data and information to help people and organizations make decisions (Power & Sharda, 2023; Siciliani, Taccardi, Basile, Di Ciano, & Lops, 2023). These systems integrate data from various sources, apply analytical techniques, and present the results in a format that helps decision-makers evaluate options and choose the most appropriate course of action. DSS is widely used in various fields, including business, healthcare, environmental management, and public policy (Gupta & Panigrahi, 2022; Rezaei & Vahidnia, 2023). The key components of DSS are model integration (integrating hydrological and water quality models into a DSS to simulate scenarios and assess potential impacts on water resources) and policy planning support (utilizing the DSS to provide decision-makers with valuable insights for sustainable water resource management, policy formulation, and future planning). By following this approach, stakeholders can effectively map, monitor, model, and manage water resources using geospatial techniques.

DSS plays a vital role in water resources management by providing tools and capabilities for collecting, analyzing, and visualizing data to support informed decision-making (Candido, Coêlho, De Moraes, & Florêncio, 2022; Reljić et al., 2023). Water resource management is complex, often involving the allocation of limited resources to meet various needs, such as drinking water supply, irrigation, industrial use, and environmental conservation. The key elements and functions of DSS for water resources management can be described below.

1. Data Integration and Management

   DSS collects and integrates data from various sources, including hydrological measurements, weather data, water quality data, and geographical information. This data is stored, managed, and updated to provide a comprehensive view of the water resource system.
2. Real-time Monitoring

   DSS can incorporate real-time data from sensors, telemetry systems, and RS technologies to monitor water levels, flow rates, and water quality. Real-time data allows for an immediate response to changing conditions.
3. Hydrological Modeling

   DSS often includes hydrological models that simulate water flow, infiltration, and groundwater recharge. These models help predict the impact of various management decisions on water resources.

4. Water Allocation and Optimization
   DSS assists in the allocation of water resources among different users and sectors. It can optimize water allocation to balance competing demands, taking into account legal and regulatory constraints.
5. Scenario Analysis
   DSS enables scenario analysis, allowing decision-makers to evaluate the potential outcomes of different management strategies, such as dam operations, irrigation scheduling, or drought response.
6. Risk Assessment
   DSS assesses and quantifies risks related to water resources, including floods, droughts, pollution incidents, and climate change impacts. Risk assessments help in planning for resilience and mitigation.
7. Decision Support Models
   DSS often incorporates decision-support models that consider various criteria and objectives to recommend the best management options. Multi-criteria decision analysis is a common approach.
8. Early Warning Systems
   DSS can be configured to create early warning systems for floods, water quality degradation, or other potential crises. These systems trigger alerts and recommend actions when predefined conditions are met.
9. Water Quality Management
   DSS assists in water quality management by monitoring water quality parameters, identifying pollution sources, and recommending remediation actions.
10. Geographic Information Systems
    Integration with GIS allows spatial visualization and analysis of water resource data. This is particularly valuable for understanding the geographical distribution of water resources and their associated features.
11. Stakeholder Engagement
    DSS often provides tools for engaging and communicating with stakeholders, including water authorities, local communities, and environmental organizations. Public participation is essential in water management decisions.
12. Regulatory Compliance
    DSS supports compliance with water-related regulations and standards. It helps organizations meet reporting requirements and demonstrate adherence to environmental laws.
13. Decision Support Dashboards
    User-friendly dashboards and visualization tools make it easier for decision-makers to access and interpret data. These tools often include maps, charts, and interactive features.

DSS for water resources management is essential for addressing the growing challenges related to water availability, quality, and sustainability. They enable more informed, efficient, and sustainable management of water resources, supporting the needs of society, the economy, and the environment.

25.3 Challenges and future directions

25.3.1 Challenges

Water resource mapping, monitoring, and modeling using geospatial approaches offer numerous benefits but are not without their challenges. These challenges can vary depending on the region, the specific application, and the data available. Some of the key challenges are described below.

1. Quality Data Availability
   Geospatial approaches heavily rely on data, and the availability and quality of data can be a major challenge. In some regions, the data may be sparse or outdated. Ensuring data accuracy and consistency is crucial.
2. Remote Location Assessment
   Mapping and monitoring water resources in remote or inaccessible areas can be challenging. Limited infrastructure and logistical difficulties can hinder data collection efforts.
3. Costing and Resource Allocation
   Geospatial technology, equipment, and skilled personnel can be expensive. Small organizations or developing countries may struggle to allocate the necessary resources for geospatial approaches.
4. Data Integration
   Integrating data from various sources, such as satellite imagery, ground-based measurements, and hydrological models, can be complex. Different data formats and coordinate systems may require standardization.
5. Spatial and Temporal Data Resolution
   Balancing the need for high spatial and temporal resolution with cost constraints is a common challenge. High-resolution data can be costly and may not be needed for all applications.
6. Sensor Calibration and Maintenance
   RS sensors and monitoring equipment require regular calibration and maintenance to ensure data accuracy. This can be challenging in remote or harsh environments.
7. Model calibration and Validation
   Hydrological models used in geospatial approaches must be validated with ground-truth data. Obtaining reliable validation data can be difficult, and model performance can vary.

8. Data Privacy and Security

   Geospatial data often has privacy and security implications, especially in areas related to water infrastructure or sensitive ecosystems. Managing and protecting such data can be a challenge.
9. Interagency Coordination

   Effective water resource management often involves multiple agencies and organizations. Coordinating efforts and data sharing among these entities can be complex.
10. Climate Change Uncertainty

    Climate change introduces uncertainty into water resource modeling. Accurate projections of future climate conditions are challenging, and models must account for this uncertainty.
11. Environmental and Human Impacts

    Geospatial approaches need to consider the environmental and social impacts of data collection and modeling efforts. Environmental degradation and community displacement can result from some monitoring activities.
12. Data Ownership and Access Rights

    Data ownership and access rights can be contentious, particularly when working with indigenous or marginalized communities. Respecting these rights while still collecting the necessary data can be a challenge.
13. Public Education and Understanding

    The public's understanding and acceptance of geospatial methods and their results can vary. Educating the public and building trust is essential, especially when making policy decisions based on geospatial data.
14. Regulatory Policies

    Compliance with regulatory policies related to water resource management and geospatial data can be complex, and navigating these legal frameworks can pose challenges.

Overcoming these challenges in water resource mapping, monitoring, and modeling using geospatial approaches requires collaboration among various stakeholders, robust data management practices, investment in technology and training, and a commitment to ethical and responsible data collection and use. Addressing these challenges is essential for effective water resource management and ensuring the sustainability of this critical natural resource.

25.3.2 Future directions

The field of water resource mapping, monitoring, and modeling is evolving rapidly, driven by technological advances, increased awareness of water-related challenges, and the need for sustainable management of this vital resource. Advances in satellite and RS technology provide more frequent and higher-resolution data for mapping and monitoring water resources. Future developments include more accurate and real-time data, enhanced spectral and spatial resolution, and improved data fusion techniques. AI and ML techniques are increasingly used to process and analyze water-related data. These technologies can help improve predictive models, data interpretation, and early warning systems for droughts, floods, and water quality issues. Creating integrated data platforms that bring together information from various sources (e.g., satellite data, ground-based sensors, social media, and citizen science) will be a crucial trend. These platforms will facilitate better decision-making and resource management.

Water resource systems are becoming increasingly complex. Digital twin technology, which creates a virtual replica of a physical system, will help model and simulate water infrastructure, optimizing its performance, resilience, and sustainability. With growing concerns about water pollution, there is a need for advanced water quality monitoring systems. Miniaturized and affordable sensors, as well as autonomous monitoring devices, will become more widespread. Continuous real-time monitoring will help detect and respond to pollution events promptly. Hydrological models will continue to advance, incorporating climate change projections and improving accuracy. Models will be used for predicting and managing water resources in the face of changing climate patterns. Smart water grids and networks, equipped with sensors and control systems, will optimize water distribution, reduce leaks, and improve overall water resource management in urban areas. Understanding the interdependencies between water and energy resources will be critical. Modeling and monitoring systems will need to address the implications of water use in energy production and vice versa. Public engagement and citizen science initiatives will play a more significant role in data collection and monitoring. Communities will be involved in reporting water-related issues and collaborating in conservation efforts. The integration of water resource data into policy and governance will be crucial.

Decision-makers will rely on more data-driven approaches to develop and implement water management strategies. There will be a stronger focus on building resilience in water resource systems. Models will need to account for extreme weather events and long-term climate change impacts to support adaptive management. International cooperation for managing shared water resources will continue to be important. Advanced monitoring and modeling tools can help in equitable resource allocation and conflict resolution. There will be an increasing emphasis on sustainable water use, with models and monitoring systems helping to optimize water allocation, reduce waste, and promote conservation. Monitoring and modeling efforts will incorporate a broader perspective on water resource management, considering aquatic health and the biodiversity they support. Blockchain technology may be used to enhance transparency and trust in water transactions and the management of water rights.

The future of water resource mapping, monitoring, and modeling will be characterized by a combination of advanced technologies, data integration, and a growing awareness of the importance of sustainable water management in the face of climate change and population growth. These developments will be essential in ensuring the availability and quality of water resources for future generations.

25.4 Limitation

Geospatial approaches have significantly advanced water resource mapping, monitoring, and modeling but also have limitations. Geospatial analysis relies heavily on the availability and quality of input data, such as satellite imagery, DEMs, and hydrological data. In some regions, especially in developing countries or remote areas, data may be scarce or of poor quality, which can limit the accuracy and reliability of the analyses. Geospatial data often comes at different scales and resolutions, and selecting the appropriate scale for analysis can be challenging. For example, satellite imagery may have a high spatial resolution but limited temporal resolution, while hydrological models may require data at a finer spatial scale than what is available. Hydrological processes are complex and can be influenced by numerous factors, such as topography, land cover, soil properties, and climate. Geospatial models may oversimplify these processes, leading to water resource assessment and prediction inaccuracies. All models, including geospatial models, have inherent uncertainties due to simplifications, assumptions, and input data errors. These uncertainties can propagate through the modeling process, leading to uncertainty in the results. Water resource management requires interdisciplinary hydrology, ecology, engineering, and policy knowledge. Geospatial approaches may only sometimes integrate these disciplines effectively, leading to gaps in understanding or decision-making. Geospatial analysis and modeling require specialized technical expertise in GIS, RS, and hydrological modeling. Access to this expertise may be limited in some regions or organizations. Implementing geospatial approaches for water resource mapping, monitoring, and modeling can be costly, requiring equipment, software, data acquisition, and personnel training. This can be a barrier for some organizations or countries with limited resources. Geospatial data often raises ethical and privacy concerns, especially concerning sensitive information such as water sources, infrastructure, or vulnerable populations. Ensuring data security and privacy while using geospatial approaches is essential but challenging. Addressing these limitations requires careful consideration of data sources, model assumptions, validation techniques, and the integration of interdisciplinary knowledge. Collaboration between researchers, practitioners, and policymakers is also crucial for advancing geospatial approaches in water resource management.

25.5 Summary and conclusion

Water resource mapping, monitoring, and modeling using geospatial approaches have become crucial tools in managing water resources sustainably. By integrating satellite imagery, GIS data, and RS techniques, these approaches offer a comprehensive understanding of water availability, quality, and usage patterns. The process begins with the collection of various geospatial data, such as satellite images, DEMs, land use/land cover maps, hydrological parameters, and meteorological data. These datasets are then processed and analyzed using GIS software to create detailed maps depicting water resources, including surface water bodies, groundwater aquifers, watersheds, and water quality parameters. Monitoring is a continuous process that involves updating the spatial databases with new information to track changes in water resources over time. This is essential for identifying trends, assessing the impact of human activities and climate change, and making informed decisions for water resource management. Modeling plays a crucial role in simulating different scenarios and predicting future trends based on various factors such as population growth, land use changes, and climate variability. These models help in forecasting water availability, identifying vulnerable areas, and developing strategies for sustainable water management. In conclusion, geospatial approaches offer powerful tools for water resource mapping, monitoring, and modeling. They provide valuable insights into the complex dynamics of water resources, enabling informed decision-making for sustainable development and environmental conservation. However, challenges such as data availability, accuracy, and integration across different platforms need to be addressed to maximize the potential of geospatial technologies in water resource management.

Acknowledgments

The authors acknowledge the Indian Institute of Technology (Indian School of Mines), Dhanbad, India and Harcourt Butler Technical University, Kanpur, India for providing support in carrying out the study.

References

Abbas, J. (2013). Assessment of water quality in Tigris River-Iraq by using GIS mapping. *Natural Resources*, 2013. Available from https://doi.org/10.4236/nr.2013.46054.

Abdelhaleem, F. S., Basiouny, M., Ashour, E., & Mahmoud, A. (2021). Application of remote sensing and geographic information systems in irrigation water management under water scarcity conditions in Fayoum, Egypt. *Journal of Environmental Management*, *299*, 113683. Available from https://doi.org/10.1016/j.jenvman.2021.113683.

Achu, A. L., Thomas, J., & Reghunath, R. (2020). Multi-criteria decision analysis for delineation of groundwater potential zones in a tropical river basin using remote sensing, GIS and analytical hierarchy process (AHP. *Groundwater for Sustainable Development, 10*, 100365. Available from https://doi.org/10.1016/j.gsd.2020.100365.

Aderemi, B. A., Olwal, T. O., Ndambuki, J. M., & Rwanga, S. S. (2021). A review of groundwater management models with a focus on IoT-based systems. *Sustainability, 14*(1), 148. Available from https://doi.org/10.3390/su14010148.

Adisa, O. M., Masinde, M., Botai, J. O., & Botai, C. M. (2020). Bibliometric analysis of methods and tools for drought monitoring and prediction in Africa. *Sustainability, 12*(16), 6516. Available from https://doi.org/10.3390/su12166516.

Aghsaei, H., Mobarghaee Dinan, N., Moridi, A., Asadolahi, Z., Delavar, M., Fohrer, N., & Wagner, P. D. (2020). Effects of dynamic land use/land cover change on water resources and sediment yield in the Anzali wetland catchment, Gilan, Iran. *Science of the Total Environment, 712*. Available from https://doi.org/10.1016/j.scitotenv.2019.136449. Available from, http://www.elsevier.com/locate/scitotenv.

Al-Djazouli, M. O., Elmorabiti, K., Rahimi, A., Amellah, O., & Fadil, O. A. M. (2021). Delineating of GWPZs based on remote sensing, GIS and analytical hierarchical process: a case of Waddai, eastern Chad. *GeoJournal, 86*, 1881–1894. Available from https://doi.org/10.1007/s10708-020-10160-0.

Al-Ghobari, H., Dewidar, A., & Alataway, A. (2020). Estimation of surface water runoff for a semi-arid area using RS and GIS-based SCS-CN method. *Water, 12*(7), 1924. Available from https://doi.org/10.3390/w12071924.

Allafta, H., Opp, C., & Patra, S. (2020). Identification of groundwater potential zones using remote sensing and GIS techniques: a case study of the Shatt Al-Arab Basin. *Remote Sensing, 13*(1), 112. Available from https://doi.org/10.3390/rs13010112.

Ashtekar, A. S., Mohammed-Aslam, M. A., & Moosvi, A. R. (2019). Utility of normalized difference water index and GIS for mapping surface water dynamics in sub-upper Krishna Basin. *Journal of the Indian Society of Remote Sensing, 47*, 1431–1442. Available from https://doi.org/10.1007/s12524-019-01013-6.

Avila-Aceves, E., Plata-Rocha, W., Monjardin-Armenta, S. A., & Rangel-Peraza, J. G. (2023). Geospatial modelling of floods: a literature review. *Stochastic Environmental Research and Risk Assessment, 37*(11), 4109–4128. Available from https://doi.org/10.1007/s00477-023-02505-1. Available from: https://www.springer.com/journal/477.

Bai, J., Zhao, J., Zhang, Z., & Tian, Z. (2022). Assessment and a review of research on surface water quality modeling. *Ecological Modelling, 466*. Available from https://doi.org/10.1016/j.ecolmodel.2022.109888.

Balla, D., Zichar, M., Kiss, E., Szabó, G., & Mester, T. (2022). Possibilities for Assessment and Geovisualization of Spatial and Temporal Water Quality Data Using a WebGIS Application. *ISPRS International Journal of Geo-Information, 11*(2). Available from https://doi.org/10.3390/ijgi11020108.

Berhanu, K. G., & Hatiye, S. D. (2020). Identification of Groundwater Potential Zones Using Proxy Data: Case study of Megech Watershed, Ethiopia. *Journal of Hydrology: Regional Studies, 28*. Available from https://doi.org/10.1016/j.ejrh.2020.100676. Available from: https://www.journals.elsevier.com/journal-of-hydrology-regional-studies.

Bijekar, S., Padariya, H. D., Yadav, V. K., Gacem, A., Hasan, M. A., Awwad, N. S., ... Jeon, B. H. (2022). The state of the art and emerging trends in the wastewater treatment in developing nations. *Water, 14*(16), 2537. Available from https://doi.org/10.3390/w14162537.

Bojago, E., Tyagi, I., Ahamad, F., & Chandniha, S. K. (2023). GIS based spatial-temporal distribution of water quality parameters and heavy metals in drinking water: Ecological and health modelling. *Physics and Chemistry of the Earth, Parts A/B/C, 130*. Available from https://doi.org/10.1016/j.pce.2023.103399.

Bonfante, A., Monaco, E., Manna, P., De Mascellis, R., Basile, A., Buonanno, M., ... Brook, A. (2019). LCIS DSS—An irrigation supporting system for water use efficiency improvement in precision agriculture: A maize case study. *Agricultural Systems, 176*, 102646. Available from https://doi.org/10.1016/j.agsy.2019.102646.

Candido, L. A., Coêlho, G. A. G., De Moraes, M. M. G. A., & Florêncio, L. (2022). Review of Decision Support Systems and Allocation Models for Integrated Water Resources Management Focusing on Joint Water Quantity-Quality. *Journal of Water Resources Planning and Management, 148*(2). Available from https://doi.org/10.1061/(ASCE)WR.1943-5452.0001496. Available from: https://ascelibrary.org/journal/jwrmd5.

Daba, M. H., & You, S. (2022). Quantitatively Assessing the Future Land-Use/Land-Cover Changes and Their Driving Factors in the Upper Stream of the Awash River Based on the CA–Markov Model and Their Implications for Water Resources Management. *Sustainability, 14*(3). Available from https://doi.org/10.3390/su14031538.

Dandge, K. P., & Patil, S. S. (2022). Spatial distribution of ground water quality index using remote sensing and GIS techniques. *Applied Water Science, 12*, 1–18. Available from https://doi.org/10.1007/s13201-021-01546-7.

Darem, A. A., Alhashmi, A. A., Almadani, A. M., Alanazi, A. K., & Sutantra, G. A. (2023). Development of a map for land use and land cover classification of the Northern Border Region using remote sensing and GIS. *Egyptian Journal of Remote Sensing and Space Science, 26*(2), 341–350. Available from https://doi.org/10.1016/j.ejrs.2023.04.005. Available from: http://www.elsevier.com/wps/find/journaldescription.cws_home/723780/description#description.

De, A., Upadhyaya, D. B., Thiyaku, S., & Tomer, S. K. (2021). *Use of multi-sensor satellite remote sensing data for flood and drought monitoring and mapping in India. Civil Engineering for Disaster Risk Reduction* (pp. 27–41). Singapore: Springer Nature Singapore. Available from https://doi.org/10.1007/978-981-16-5312-4_3.

Elhag, M. (2016). Evaluation of different soil salinity mapping using remote sensing techniques in arid ecosystems, Saudi Arabia. *Journal of Sensors, 2016*. Available from https://doi.org/10.1155/2016/7596175.

Fasipe, O. A., Izinyon, O. C., & Ehiorobo, J. O. (2021). Hydropower potential assessment using spatial technology and hydrological modelling in Nigeria river basin. *Renewable Energy, 178*, 960–976. Available from https://doi.org/10.1016/j.renene.2021.06.133.

Feizizadeh, B., Omarzadeh, D., Ronagh, Z., Sharifi, A., Blaschke, T., & Lakes, T. (2021). A scenario-based approach for urban water management in the context of the COVID-19 pandemic and a case study for the Tabriz metropolitan area, Iran. *Science of the Total Environment, 790*, 148272. Available from https://doi.org/10.1016/j.scitotenv.2021.148272.

Fu, B., Horsburgh, J. S., Jakeman, A. J., Gualtieri, C., Arnold, T., Marshall, L., ... Rashleigh, B. (2020). Modeling water quality in watersheds: From here to the next generation. *Water Resources Research, 56*(11). Available from https://doi.org/10.1029/2020WR027721, e2020WR027721.

Gorgoglione, A., Crisci, M., Kayser, R. H., Chreties, C., & Collischonn, W. (2019). A new scenario-based framework for conflict resolution in water allocation in transboundary watersheds. *Water, 11*(6), 1174. Available from https://doi.org/10.3390/w11061174.

Gupta, B. B., & Panigrahi, P. K. (2022). Analysis of the Role of Global Information Management in Advanced Decision Support Systems (DSS) for Sustainable Development. *Journal of Global Information Management (JGIM), 31*(2), 1–13.

Gupta, S., & Gupta, S. K. (2021). A critical review on water quality index tool: Genesis, evolution and future directions. *Ecological Informatics, 63*. Available from https://doi.org/10.1016/j.ecoinf.2021.101299. Available from: http://www.elsevier.com/wps/find/journaldescription.cws_home/705192/description#description.

Gupta, S., & Gupta, S. K. (2023). Application of Monte Carlo simulation for carcinogenic and non-carcinogenic risks assessment through multi-exposure pathways of heavy metals of river water and sediment, India. *Environmental Geochemistry and Health, 45*(6), 3465–3486. Available from https://doi.org/10.1007/s10653-022-01421-7.

Haseeb, M., Farid, H. U., Khan, Z. M., Anjum, M. N., Ahmad, A., & Mubeen, M. (2023). Quantifying irrigation water demand and supply gap using remote sensing and GIS in Multan, Pakistan. *Environmental Monitoring and Assessment, 195*(8). Available from https://doi.org/10.1007/s10661-023-11546-6. Available from: https://www.springer.com/journal/10661.

Horton, P., Schaefli, B., & Kauzlaric, M. (2022). Why do we have so many different hydrological models? A review based on the case of Switzerland. *Wiley Interdisciplinary Reviews: Water, 9*(1), e1574. Available from https://doi.org/10.1002/wat2.1574.

Khan, M. Y. A., ElKashouty, M., Subyani, A. M., Tian, F., & Gusti, W. (2022). GIS and RS intelligence in delineating the groundwater potential zones in Arid Regions: a case study of southern Aseer, southwestern Saudi Arabia. *Applied Water Science, 12*, 1–23. Available from https://doi.org/10.1007/s13201-021-01535-w.

Korchenko O., Pohrebennyk V., Kreta D., Klymenko V., Anpilova Y. 2019. International Multidisciplinary Scientific Geoconference Poland GIS and remote sensing as important tools for assessment of environmental pollution. International Multidisciplinary Scientific GeoConference Surveying Geology and Mining Ecology Management, SGEM. 19. 2 297–304 http://www.sgem.org/.

Mashala, M. J., Dube, T., Mudereri, B. T., Ayisi, K. K., & Ramudzuli, M. R. (2023). A Systematic Review on Advancements in Remote Sensing for Assessing and Monitoring Land Use and Land Cover Changes Impacts on Surface Water Resources in Semi-Arid Tropical Environments. *Remote Sensing, 15*(16), 3926. Available from https://doi.org/10.3390/rs15163926.

Matlhodi, B., Kenabatho, P. K., Parida, B. P., & Maphanyane, J. G. (2021). Analysis of the future land use land cover changes in the gaborone dam catchment using ca-markov model: Implications on water resources. *Remote Sensing, 13*(13), 2427. Available from https://doi.org/10.3390/rs13132427.

Mehla, M. K., Kothari, M., Singh, P. K., Bhakar, S. R., & Yadav, K. K. (2023). Water footprint assessment and its importance in Indian context: a meta-review. *Water Supply, 23*(8), 3113–3127. Available from https://doi.org/10.2166/ws.2023.174. Available from: https://iwaponline.com/ws/article/23/8/3113/96288/Water-footprint-assessment-and-its-importance-in.

Murmu, P., Kumar, M., Lal, D., Sonker, I., & Singh, S. K. (2019). Delineation of groundwater potential zones using geospatial techniques and analytical hierarchy process in Dumka district, Jharkhand, India. *Groundwater for Sustainable Development, 9*, 100239. Available from https://doi.org/10.1016/j.gsd.2019.100239.

Ngondo, J., Mango, J., Liu, R., Nobert, J., Dubi, A., & Cheng, H. (2021). Land-use and land-cover (LULC) change detection and the implications for coastal water resource management in the Wami–Ruvu Basin, Tanzania. *Sustainability, 13*(8), 4092. Available from https://doi.org/10.3390/su13084092.

Orimoloye, I. R., Mazinyo, S. P., Kalumba, A. M., Nel, W., Adigun, A. I., & Ololade, O. O. (2019). Wetland shift monitoring using remote sensing and GIS techniques: landscape dynamics and its implications on Isimangaliso Wetland Park, South Africa. *Earth Science Informatics*, *12*, 553–563. Available from https://doi.org/10.1007/s12145-019-00400-4.

Oyedele, A. A. (2019). Use of remote sensing and GIS techniques for groundwater exploration in the basement complex terrain of Ado-Ekiti, SW Nigeria. *Applied Water Science*, *9*, 1–13. Available from https://doi.org/10.1007/s13201-019-0917-9.

Power, D. J., & Sharda, R. (2023). Decision Support and Analytics. *Springer Handbooks*, *674*. Available from https://doi.org/10.1007/978-3-030-96729-1_66. Available from: https://www.springer.com/series/15558.

Power, D. J., & Sharda, R. (2023). Decision Support and Analytics. *Springer Handbook of Automation* (pp. 1401–1410). Cham: Springer International Publishing. Available from https://doi.org/10.1007/978-3-030-96729-1_66.

Ram, S. A., & Irfan, Z. B. (2021). Application of System Thinking Causal Loop Modelling in understanding water Crisis in India: A case for sustainable Integrated Water resources management across sectors. *HydroResearch*, *4*, 1–10. Available from https://doi.org/10.1016/j.hydres.2021.02.001.

Rautela, K. S., Kumar, M., Khajuria, V., & Alam, M. A. (2022). Comparative geomorphometric approach to understand the hydrological behaviour and identification of the Erosion prone areas of a coastal watershed using RS and GIS tools. *Discover Water*, *2*(1), 1. Available from https://doi.org/10.1007/s43832-021-00009-z.

Reljić, M., Romić, M., Romić, D., Gilja, G., Mornar, V., Ondrasek, G., ... Zovko, M. (2023). Advanced Continuous Monitoring System—Tools for Water Resource Management and Decision Support System in Salt Affected Delta. *Agriculture*, *13*(2). Available from https://doi.org/10.3390/agriculture13020369.

Rezaei, Z., & Vahidnia, M. H. (2023). Effective medical center finding during COVID-19 pandemic using a spatial DSS centered on ontology engineering. *GeoJournal*, *88*(3), 2721–2735. Available from https://doi.org/10.1007/s10708-022-10777-3. Available from: https://www.springer.com/journal/10708.

Santos, E., Carvalho, M., & Martins, S. (2023). Sustainable Water Management: Understanding the Socioeconomic and Cultural Dimensions. *Sustainability*, *15*(17), 13074. Available from https://doi.org/10.3390/su151713074.

Saranya, T., & Saravanan, S. (2020). Groundwater potential zone mapping using analytical hierarchy process (AHP) and GIS for Kancheepuram District, Tamilnadu, India. *Modeling Earth Systems and Environment*, *6*(2), 1105–1122. Available from https://doi.org/10.1007/s40808-020-00744-7. Available from: springer.com/journal/40808.

Shabani, A., Woznicki, S. A., Mehaffey, M., Butcher, J., Wool, T. A., & Whung, P. Y. (2021). A coupled hydrodynamic (HEC-RAS 2D) and water quality model (WASP) for simulating flood-induced soil, sediment, and contaminant transport. *Journal of Flood Risk Management*, *14*(4). Available from https://doi.org/10.1111/jfr3.12747. Available from: http://onlinelibrary.wiley.com/journal/10.1111/(ISSN)1753-318X.

Siciliani, L., Taccardi, V., Basile, P., Di Ciano, M., & Lops, P. (2023). AI-based decision support system for public procurement. *Information Systems*, *119*. Available from https://doi.org/10.1016/j.is.2023.102284.

Singh, S., Bhardwaj, A., & Verma, V. K. (2020). Remote sensing and GIS based analysis of temporal land use/land cover and water quality changes in Harike wetland ecosystem, Punjab, India. *Journal of Environmental Management*, *262*, 110355. Available from https://doi.org/10.1016/j.jenvman.2020.110355.

Subraelu, P., Ahmed, A., Ebraheem, A. A., Sherif, M., Mirza, S. B., Ridouane, F. L., & Sefelnasr, A. (2023). Risk Assessment and Mapping of Flash Flood Vulnerable Zones in

Arid Region, Fujairah City, UAE-Using Remote Sensing and GIS-Based Analysis. *Water, 15*(15), 2802. Available from https://doi.org/10.3390/w15152802.

Tomar, P., Singh, S. K., Kanga, S., Meraj, G., Kranjčić, N., Đurin, B., & Pattanaik, A. (2021). GIS-Based Urban Flood Risk Assessment and Management—A Case Study of Delhi National Capital Territory (NCT), India. *Sustainability, 13*(22), 12850. Available from https://doi.org/10.3390/su132212850.

Wai, K. P., Chia, M. Y., Koo, C. H., Huang, Y. F., & Chong, W. C. (2022). Applications of deep learning in water quality management: A state-of-the-art review. *Journal of Hydrology, 613*. Available from https://doi.org/10.1016/j.jhydrol.2022.128332. Available from, http://www.elsevier.com/inca/publications/store/5/0/3/3/4/3.

Woldesenbet, T. A. (2022). Impact of land use and land cover dynamics on ecologically-relevant flows and blue-green water resources. *Ecohydrology and Hydrobiology, 22*(3), 420–434. Available from https://doi.org/10.1016/j.ecohyd.2022.03.002.

Yadav, S., Yadav, K. K., Perez, G. A. C., Muñoz-Arriola, F., & Yadava, R. N. (2022). In V. P. Singh (Ed.), *Application of Remote Sensing and GIS in Natural Resources and Built Infrastructure Management*. Springer. Available from https://doi.org/10.1007/978-3-031-14096-9.

Yan, T., Shen, S. L., & Zhou, A. (2022). Indices and models of surface water quality assessment: Review and perspectives. *Environmental Pollution, 308*, 119611. Available from https://doi.org/10.1016/j.envpol.2022.119611.

Zeng, C., Zhang, F., & Luo, M. (2022). A deep neural network-based decision support system for intelligent geospatial data analysis in intelligent agriculture system. *Soft Computing, 26* (20), 10813–10826. Available from https://doi.org/10.1007/s00500-022-07018-7.

Zhu, A. X., Zhao, F. H., Liang, P., & Qin, C. Z. (2021). Next generation of GIS: must be easy. *Annals of GIS, 27*(1), 71–86. Available from https://doi.org/10.1080/19475683.2020.1766563. Available from: http://www.tandfonline.com/toc/tagi20/current.

Chapter 26

An overview of geothermal energy as technology and its use in the current and modern context

Ana Carolina Borges Monteiro[1], Reinaldo Padilha França[1], Rodrigo Bonacin[1] and Rangel Arthur[2]

[1]Renato Archer Information Technology Center (CTI), Computing Methodologies Division-DIMEC, Campinas, Brazil, [2]School of Technology (FT), University of Campinas (UNICAMP), Campinas, Brazil

26.1 Introduction

Geothermal energy is specified by the heat coming from the planet Earth. It is the heat energy produced less than 64 km from the surface in a layer of melted rocks (magma), which reaches up to 6000°C and is a type of renewable energy. Magma results from the tremendous pressures below the surface and the heat generated by the decomposition of radioactive substances such as uranium and thorium. Geo signifies earth, and thermal signifies heat. In this sense, geothermal is the heat energy from the planet. Finding cracks in the earth's crust, magma explodes in volcanic eruptions, or the gases released by cooling it cool underground waters that emerge in the form of geysers or hot water mines. The process of harnessing this energy is done through large perforations in the soil since the heat of our planet exists in a part below the Earth's surface (Dickson & Fanelli, 2013; Geothermal Energy and Society, 2019).

In such a way, this type of energy resource is obtained by employing hot rocks (dry and wet) and hot steam coming from the interior of the Earth. Geothermal energy began to be explored in the early 20th century, although it was already used by ancient peoples for bathing and cooking food in the so-called hot springs. Electric energy can be obtained by drilling the soil in places where there is a large amount of hot water and even steam, which are drained to the surface through specific pipes. The steam

Water Resources Monitoring, Management, and Sustainability. DOI: https://doi.org/10.1016/B978-0-443-23665-5.00026-0

is then transported to a geothermal power plant, where it rotates the blades of a turbine. Finally, this mechanical energy (turbines) is transformed into electrical energy through the generator. It is currently produced in geothermal plants, where it is transformed into electrical energy (Geothermal Energy and Society, 2019).

In some systems of production of electric energy through the heat of the Earth, water is injected into the heated subsoil so that it turns into heat and returns in the form of steam, which, in the same way as in the previous case, drives the turbines that activate the generator. We are still evaluating whether geothermal energy, in some cases, is also used to heat water in residential areas or even in entire cities during the winter when hot water is used. It is currently produced at geothermal plants, where it is transformed into electrical energy. Plants, or geothermal plants, are the places of production of this type of energy, being implanted close to places where there is a great deal of steam and hot water. Thus geothermal energy can be used as an important source of electricity production, in addition to allowing domestic and industrial use of heated water (Ismail, 2019).

The positive aspects of this type of energy are that the emission of polluting gases (SO_2 and even CO^2) is approximately zero, not intensifying the greenhouse effect; on the contrary, fossil fuels. Thus the area necessary for the installation of the plant is small, allowing it to supply isolated communities. On the other hand, the negative aspects are that energy is very expensive and not very profitable, as it requires high structural investments and its efficiency is low. This can cause the depletion of the geothermal field since the lost heat increases the temperature of the environment, causing the emission of hydrogen sulfide (H_2S), which is corrosive and harmful to human health (Mohammadzadeh Bina, Jalilinasrabady, Fujii, & Pambudi, 2018).

In summary, this high temperature is taken from water resources, which can come from reservoirs, geothermal plants, or employing geothermal heat pumps. In these places, wells are used that allow access to steam from the temperature of the Earth's interior. Thus this steam is converted, utilizing plants, pumps, and other resources, into electrical energy, which can be used for various functions. Among the functions, it is possible to mention the heating of buildings, electricity for industries, heating of water resources, and other purposes of the type, as mentioned and among others. It is still worth mentioning that water resources are used only for the capture of steam, but, after the conversion process, they return to the reservoirs. Thus it is possible to guarantee the preservation of this resource, since this does not make geothermal energy fully sustainable (Dickson & Fanelli, 2013).

Therefore this chapter aims to provide a panorama and overview of geothermal energy, showing the importance and relevance of this disruptive technology and demonstrating a landscape view of key concerns, advantages, and challenges with a concise bibliographic background.

26.2 Geothermal energy concept

As it is known, the planet Earth is covered by the Earth's crust, corresponding to a thin layer of rocks that are located above the Earth's mantle, which is a layer with high depth and composed basically of magma. Magma is relative to rocks that are in a fluid or pasty state due to the high temperatures inside the planet. In this sense, geothermal energy is a category of renewable energy on the planet Earth, resulting from the procedure of using this type of energy carried out through large drilling in the soil, whereas the heat of the planet Earth exists in a part below the Earth's surface (Brown, 2000).

All this internal heat eventually manifests itself in some areas of the Earth's surface; in this sense, geothermal energy is related to the use of all this internal heat for the production of electricity, generally related to volcanic eruptions, geological cracks, or areas of internal heating, causing the emergence of geysers of steam and hot springs. From a historical point of view, this type of geothermal energy began to be explored in the early 20th century, obtained through hot rocks (dry and wet) and hot steam from the interior of the Earth, although it was already used by the people for bathing and cooking food in the so-called hot springs. Currently, geothermal energy is produced in geothermal plants, where it is transformed into electrical energy (Glassley, 2014).

As it is a renewable natural resource, it has a low environmental impact and is currently used for cooking, heating buildings, and swimming pools, as well as for the production of greenhouses for vegetables. Geothermal energy is seen as one of the options for the employment of fossil fuels to generate electricity. It can be used for heating water in residential areas or even in entire cities during the winter and also considering the use of internal (geothermal) heating for heat production and use in heaters or thermal appliances. Considering that this can also be used for the production of heat and use in heaters or thermal devices in greenhouses, fishing fields, or leisure areas, among others (Stefansson, 2000).

Still existing for leisure criteria, the well-known ones that use thermal sources (natural hot waters) for tourism correspond to places with the emergence of waters heated by the geothermal process, which are examples of geothermal water heated inside the earth by contact with very hot rocks. In addition to the high temperatures, these waters have large amounts of minerals that are good for the skin and the whole organism, such as potassium, selenium, calcium, zinc, chlorides, and magnesium (Bahadori, Zendehboudi, & Zahedi, 2013).

With respect to geothermal plants, they correspond to the production sites of this type of geothermal energy, which is responsible for converting the Earth's internal heat into electrical energy, which is implanted close to places where there is a large amount of steam and hot water. In this way, the geothermal reservoirs provide the necessary energy to power the turbine generators. Geothermal plants, therefore, produce energy by capturing hot water or steam in the interior of the Earth through specially designed tubes that supports directing this vapor that is released under strong pressure. From there, this released steam moves the turbines that rotate mechanically, considering that these turbines drive the generator that produces electricity and also considering that in some cases the steam is stored and reused when it is converted into water (Johnston, Narsilio, & Colls, 2011).

Or even considering that in some systems of production of electric energy through the heat of the Earth (geothermal), water is injected in the heated subsoil so that this water turns into heat and returns in the form of steam, that is, in the same way, that capturing hot water or already in the form of steam inside the Earth through tubes. In this case, it is also possible to start the turbines that activate the electric energy generator ("Geothermal Energy Systems: Exploration, Development, & Utilization," 2011).

Among the advantages of geothermal energy for the production of electricity, it is possible to mention that this does not operate through the burning of fuels, as there is also no need to import and purchase raw materials, which consequently reduces production costs. Since oil or nuclear plants have a high cost of purchasing primary products, the advantage is that it does not harm the soil, even considering that despite the internal perforations, the geothermal energy does not attack the relief, wear out the soil, flood large areas, or contaminate the water table, as occurs with other energy sources. Just as the released steam does not severely harm the atmosphere, it does not emit carbon dioxide (CO_2) or even methane gas (CH_4) (Harlow & Pracht, 1972).

However, it is important to note that although geothermal energy is less polluting, it can cause damage to the planet if it is removed irregularly, thus changing its geology. Even with regard to drilling the soil, which is carried out by very noisy machines, it also generates noise pollution generated by the geothermal plants, even considering that the water found below the Earth's surface must be treated before it is released into rivers and seas since many ores found are harmful to health, in addition to several dissolved gases, for example, hydrogen sulfide (H_2S) (Awerbuch, Lindemuth, May, & Rogers, 1976).

Geothermal energy is also a benefit for remote areas, considering that in areas where there is not ample access to the electricity grid, geothermal plants can meet the needs of the population quickly. In the same way, geothermal energy has flexible production since the production of electricity in these geothermal plants can vary according to demand, not depending on water reservoirs or the availability of raw materials, among other aspects and characteristics. Thus it is possible to point out that geothermal energy is not vulnerable to the climate, such as wind and solar energy, since even in the presence of climatic variations, this does not interfere with the functioning of geothermal plants (Ungemach, 1985).

However, with regard to the disadvantages of geothermal energy, it is possible to mention the possible sinking of the terrain, despite not wearing down the soil. In some cases, it is necessary to inject water or another component to fill the endogenous compositions since the geothermal plants can wear out the internal areas of the earth's crust, causing possible shocks on the surface. Also considering that with the water vapor, it is common to release sulfur dioxide (SO_2), which may not attack the atmosphere, causing environmental damage; however, it is harmful to human health in addition to being highly corrosive and having an unpleasant odor (Zeiler & Boxem, 2009).

Noise pollution and high local heating caused by geothermal plants, in addition to generating a lot of noise, add to the high local heating, making it impossible to install houses and communities in the vicinity of the implantation of this type of plant. In addition, even considering that thermal fluids can release mineral compositions, which possibly tend to contaminate rivers and lakes and can affect watercourses in areas close to the plants. Although the maintenance of geothermal plants is small, their construction and installation are expensive, that is, high investment costs due to expensive technology. It also considers the infeasibility of installation and use in most places of this type of geothermal plant since it can only be operated in suitable areas, that is, with high internal (terrestrial) heating and where access to the thermal areas is easy and less expensive. However, even though geothermal power plants do not require large spaces for energy production, this is very expensive compared to other power generation plants, has low efficiency, and is thus not very profitable (Gehlin & Andersson, 2016).

However, despite geothermal energy presenting these disadvantages, it is being used more and more because it is a good alternative to obtaining electric energy since it is a renewable resource and, therefore, less polluting. Still considering the comparison with energy from fossil fuels, which contributes to the increase of the greenhouse effect, the emission of gases into the atmosphere by geothermal energy is practically nil, although if the heat removed from the earth is lost, without a doubt, there will be an increase in the temperature of the Earth's surface (Berman, 1975).

26.3 Introductions to energy harvesting systems

Geothermal energy results from the heat retained under the earth's crust, where the earth's heat appears close to surfaces and uses that heat directly, which is brought to the earth's surface as steam or hot water; this is created when water flows through the groundwater near heated rocks, considering that it is possible to use this type of energy directly to heat houses and buildings or even convert it into electricity (Kazmierski & Beeby, 2014).

The existence of a large temperature gradient is not the only criterion for judging the usefulness of a geothermal area. The presence of sources of hot water or gases is also required, and the association of water with a high gradient depends on geothermal conditions. Additionally, the existence of fields dominated by water (semithermal fields containing water and no steam) or by gases (hyperthermic fields vary in the proportion of water or gas) has an important role in determining the choice of the factory to be used (Priya & Inman, 2009).

However, temperature gradients vary widely over the earth's surface, resulting from local melting due to friction and pressure, and even the movements of the neighboring tectonic plates against each other, or even with respect to the localization of neighboring plates, also matches regions where volcanic activities are verified. Still considering that the heat measured close to the surface arising from the magma and other factors affecting the heat flow as well as the thermal gradient. Even in some scenarios, convection from a natural water source disrupts the heat flow pattern, or the release of hot, deep stone gases can increase the flow (Moser, 2007).

In areas where there are volcanoes, water is pumped from the surface to the depths of the subsoil where there are magmatic chambers since the temperature is very high, and for this reason, the water is transformed into steam, returning to the surface by pressure through pipes and driving turbines in geothermal plants. In regions where there are geysers, that is, where water vapor under pressure from deep layers of the Earth's crust, through cracks in it, periodically explodes on the Earth's surface, this water vapor is used for domestic heating. In this sense, as can be seen, in relation to nonrenewable natural resources, geothermal energy must also be employed rationally (Sanyal, 2010).

Another fundamental mechanism is the production of heat from radioactive isotopes of elements (thorium, uranium, and even potassium); however, this is not yet fully scientifically understood, but specified areas of the crust undergo sequential melting, and even related recrystallization over time has led to the accumulation of these elements at determining levels of the Earth's crust, since, to a lesser extent, exothermic chemical reactions contribute to local warming. Areas (regions) classified as hyperthermic exhibit very high gradients (generally as large as nonthermal areas) and are commonly close to neighboring tectonic plates, and even semithermal areas can have anomalies in the thickness of the crust in otherwise stable regions or due to effects related to radioactivity (Glassley, 2014).

Wet hyperthermic fields contain pressurized water, and when depressurization occurs on the surface, a mixture of gases and boiling water occurs, unlike dry hyperthermic fields, which contain saturated or overheated gas. Whether the hyperthermic field is dry or wet depends on the properties and shapes of the aquifers. Or even considering that if a geothermal hole or a natural fissure connecting the hyperthermic field with the surface and filled with water is in the form of a long tube with insulating walls, heated by boiling water at the bottom, it is possible that sometimes they flow spontaneously with the emission of vapors (Kuo, 2012).

Or even relating that if in the lowest part of the tube the water is boiling, a little steam will rise and find a colder water column level, considering that by condensation the latent heat produced will increase the water temperature at that level, but because the hydrostatic pressure (occurring inside the liquids, being exerted by the weight of the liquid itself) of the water at that level is slightly lower, the water will boil at a lower temperature. Or even, if the steam at the top of the liquid column can only be ejected through a narrow surface, the steam must expel into the atmosphere with some force (Kuo, 2012).

Just as considering the convection (heat transmission process) that occurs by the internal movement of a fluid, through the internal movement of fluids (water or atmospheric air) due to atmospheric buoyancy, it would occur in the column, there would be heat transfer on the walls, and the heating at the bottom of the column would not be maintained. Dry geothermal fields generally flow without stimulation because of the vapor pressure added to the water pressure, indicating that the extraction rate of a geothermal field influences its useful life. One method that, in principle, ensures that water in the field will not run out quickly is the re-injection of water into the aquifer. Although expensive, it can lead to a decrease in temperature within the aquifer, considering the advantage of minimizing the environmental effect (Moser, 2007).

However, despite geological information, heat flow measurements, and the use of geophysical tools as important geochemical estimates in the exploration of areas with geothermal potential, holes must often be drilled to confirm evidence and start production effectively. Considering that the depth of drilling of the hole depends on the nature of the geothermal field, it is pointed out that in hyperthermic fields, it is generally necessary to drill at a depth close to 2000 m, and in semithermal fields, it is necessary to drill just beyond this estimate. Still relating to the fact that the holes narrow with depth, in this way the surface diameter depends on the depth of the field, and the cost of drilling is one of the determining factors in the construction of geothermal plants (Awerbuch et al., 1976).

The space between the holes in the surface depends on many factors, as small separations reduce the number of connections, and even on the other hand, if the holes are too close, there may be interference between them. So, an estimate of the exact separation ratio depends on the cracking system of the aquifer. Still depending on some factors like good temperature and pressure as a function of the heart rate; fluid enthalpy (it is a physical quantity determined in the scope of classical thermodynamics, the amount of energy in a given reaction, measuring the maximum energy of a thermodynamic system); and chemical composition (Dickson & Fanelli, 2013).

26.3.1 Comparison with renewables

Renewable energy sources, including biomass, biofuels, hydroelectric, geothermal, solar, and wind energy, are the cleanest, as they do not require combustion and therefore do not have direct emissions of greenhouse gases or air pollutants compared to the burning of biomass and biofuels, which emits greenhouse gases and particles that pollute the air and cause climate change (Bandyopadhyay, 2019).

Energy resources are those that provide energy and are classified as renewable natural resources (clean energy) with less environmental impact. It is considered to regenerate spontaneously or through human intervention, that is, clean energy, since the residues left in nature are zero, as they are not exhausted in nature because they renew in a certain time, highlighting the geothermal, solar, wind, hydroelectric energy, among others, and not renewable (dirty energy). In turn, nonrenewable natural resources, which cannot be renewed in nature, that is, if that source is exploited indiscriminately, may disappear over time, with nature unable to renew them. This type of energy resource causes great environmental impact and unbalances the Earth's ecosystems, relating fossil fuels (oil, natural gas, and even coal) to nuclear energy (Beiter, Elchinger, & Tian, 2017).

It is important to note that the term "energy" is closely related to work, action, or movement, so anything that is working is producing energy. There are several types of energy, such as kinetic energy related to movement, mechanical energy related to force, thermal energy related to temperature, electric energy related to an electrical potential, hydroelectric energy related to water, chemical energy related to chemical reactions, wind energy related to wind, solar energy related to the sun, and even nuclear or atomic energy related to the disintegration of the nucleus of matter (Márquez, García, & Karyotakis, 2018).

Some examples of renewable energy, such as hydroelectricity, originated from the force of the water in the rivers, produced by the use of hydraulic potential, that is, the strength of the water in the rivers, mediated by the construction of hydroelectric plants, those that will supply electric energy to the population. Considering that hydroelectric energy is generated in hydroelectric plants with respect to the process of transformation of potential, mechanical and kinetic energy that is supplied by hydroelectric plants that use hydraulic energy for the production of electric energy since the country has a large number of large rivers being its hydraulic potential. Despite being considered renewable, clean energy (which does not emit pollutants into the atmosphere), this requires the construction of hydroelectric plants to be produced, taking into account certain locations that are far from urban centers but close to many communities that live on the land, hunting, and fishing. This causes some environmental impacts, such as flooding of areas, destruction of ecosystems (fauna, flora, and soil), species extinction, and destruction of riverside, quilombola, and indigenous communities (Oliveira Gomes, 2021).

Solar energy is obtained by heat and sunlight; it is obtained through solar panels, consisting of renewable energy obtained by sunlight, used for heating water (thermal energy, through the solar heating method), or used as a source of electrical energy. In addition, this type of energy can be obtained from solar plants composed of numerous panels that capture the sun's energy. This can also be used to produce electrical energy directly, through photovoltaic solar panels, or indirectly, through plants that use heliothermic energy (Kabir, Kumar, Kumar, Adelodun, & Kim, 2018).

Solar heating comes from a solar collector for solar water heating in residences and is used to heat water in homes, pools, or industries. It consists of a boiler above the plates. Solar collectors are basically systems with a dark surface that absorb sunlight and transmit heat to water, which in turn is stored in thermal reservoirs. Photovoltaic panels or plates use a direct method of producing electric energy, considering that sunlight is absorbed in solar cells (photoelectric or even photovoltaic) that are made of semiconductor material, generally silicon crystals. In sequence, the particles of sunlight (photons) come into contact with the silicon atoms, causing the electrons to shift, thus producing an electric current, which is employed to charge a battery (Kabir et al., 2018).

However, considering the disadvantages related to the high cost of the plates, they are still little used. If there are several days without the sun, there will be no energy. They also require a large quantity of raw materials for manufacturing, such as silicon, for the manufacture of equipment, demanding the exploitation of resources (Lakatos, Hevessy, & Kovács, 2011).

In this sense, when comparing solar or geothermal energy, it is necessary to consider the need for use; thus it is worth noting that looking for a source of energy only for certain times of the year, such as summer or winter, it is possible to choose geothermal energy, as it applies to the heating of rooms, water, or the use of electricity for air conditioning. However, solar energy can be an ally for the generation of electric energy by taking advantage of sunlight throughout the property, generating electricity every day, and helping to save on the electricity bill, contributing to a more complete solution. The idea is to count on the installation of a photovoltaic system in residences or establishments (Tester, 2006).

Wind energy is associated with ancient times in systems (pumping water, moving boats, and even grinding grains), considering the process of using the wind (a natural resource) as an energy generator, transforming it into kinetic energy, and from there into electricity with the use of specific equipment, classified as a Clean Development Mechanism by the United Nations (Razmjooy, Estrela, Padilha, & Monteiro, 2020).

Wind energy is generated by moving large turbines, known as wind turbines, installed in regions where there are prevailing winds in the shape of a vane or windmill. Producing electricity by transforming kinetic energy into mechanical energy or electrical energy. There are no limitations on supplying homes and commercial units with wind energy for all needs, such as water heating, lighting, and autonomy for electrical and electronic equipment (Blaabjerg & Ma, 2017).

Still considering that the wind is the result of conversion currents in the Earth's atmosphere that are driven by the thermal energy generated by the Sun, that is, movement of the air appears on the Earth's surface where there is soil and heat. In this sense, the wind is a type of solar energy, noting that the heat of the sun warms the soil faster than water and the heated air is lighter, so it rises, and when there is no sun, like at night, the air over the water is warmer and is replaced by fresher soil air. Thus this temperature change causes the movement of the wind, which generates kinetic energy and is transformed into electrical energy from the movement of the turbines (Al-Dousari et al., 2019).

The advantages of wind energy consist of an inexhaustible energy source that does not emit gases and does not generate waste, whereas generators can be installed in areas without the need for displacement by the community, and the permanence of activities such as agriculture and livestock is compatible, reducing dependence on fossil fuels. Even with regard to the disadvantages associated with the intermittency of the wind and integration for the constant generation of energy, which produces noise pollution and impacts bird migration (Blaabjerg & Ma, 2017).

In this sense, when comparing wind or geothermal energy, it is necessary to consider, in addition to the need for use, the variations in the intensity of the wind throughout the year and the constant voltage in the system. Thus wind turbines are considered a variable source of energy, not a safe supplier. However, geothermal energy is also a renewable resource that also has a low environmental impact, being used to heat houses and buildings, among others, and also in the production of greenhouses for vegetables (Bundschuh et al., 2017).

However, the capital cost of wind development, despite having decreased over the years, is still expensive and still associated with the cost of other components for wind turbines and even their manufacture. In contrast, under other perspectives, geothermal energy has better energy efficiency in relation to investment in development and even concerning the cost of components, indicating that geothermal energy is really competitive (Stehly & Beiter, 2018).

Also, geothermal energy has several advantages in comparative aspects to solar and wind energy, including weatherproof, initial investment, basic load, greater stability, time to capital return and construction time, difficulty in modularizing, and even thermal efficiency. However, the total installed capacity of electricity from geothermal energy is much less than wind energy and even solar energy (Geothermal Energy and Society, 2019).

26.4 Hot dry rock

Generally, in developed countries, the most economically important sources of electricity generation have already been developed. In these cases, because of the high temperature of the flow rate and the proximity of the surface (low drilling cost), the energy generated should cost little, considering that the economy of the exploration of a particular field depends on several factors, including exploration cost, purity and water temperature, reservoir development, flow rate, the type of plant, and even the degree of environmental protection required (Brown, 2009).

In this respect, conventional geothermal production considers the existence of permeable rocks in which there are sources of water that can be extracted since the sources of geothermal energy could grow a lot if the heat of such rocks could be used; however, many parts of the Earth's surface are covered by impermeable rocks. In this context, electricity could be generated directly from 'heat mines', meaning that deep holes could be drilled, given that magmatic formations that exhibit high thermal gradients could be explored, and energy power could be produced from these deep rocks (Johnston et al. 2011).

However, it is worth noting that the main technological barrier is to establish reproductive methods of fracking (a method that allows the extraction of liquid and gaseous fuels from the subsoil) from rocks, providing paths for the circulation of liquids. Still considering that fracking is used to perform deep drilling of the soil, where a pipe is inserted that crosses the water table to the rock. The difference between this technique and conventional drilling is that it allows access to sedimentary rocks in the subsoil and, consequently, explores reservoirs that were previously unattainable. During this process, the water used, the solvents used, and the extraction residues return to the surface, or even any leakage of the polluting gases and contributing to the greenhouse effect is an additional risk. They may bring risks of contamination of groundwater, a risk that is already common in the traditional drilling methodology ("Geothermal Energy Systems: Exploration, Development, & Utilization," 2011).

In this context, HDR technology is the latest development in mineral science that helps in the production of geothermal energy using very high temperatures and is found in rocks that inhabit a few kilometers below the earth's surface. In this way, the process is carried out in the following way: water is pumped into hot crystalline rocks (granite hot stones) underground with high pressure through an injection well in the heat of the rock. HDR geothermal energy is also a renewable energy source, as it is possible to extract large amounts of heat stored in the form of thermal energy from hard rock in the earth's crust, providing a significant amount of energy for the future with minimal environmental results (Guiling et al., 2016).

This water then passes through the fracture holes, and the heat recovery of the rock is pushed out of a second hole, converting it into very hot water. When this heated water flows through the open joints in the hot rock tank through wells, it is converted into electrical energy using either a steam turbine or a central binary system. Thus HDR technology uses natural geothermal systems and is used in places where there are cracks or pores being filled with hot enough water to create electricity, even considering that on the surface, the necessary heat is extracted from conventional mechanisms and this same water from refrigeration is recycled to the mine with more thermal energy (Guiling et al., 2016).

The main feature of a successful heat mine is forming a geothermal reservoir (HDR) projected on a hot body (impermeable rock), relating that the surface of a hot rock body is specified by the depth inside the well-hole since the water is pumped into the hot rock and converted into electrical energy without carbon emissions through the employment of traditional steam turbine technology and even by choosing a localization on the surface where the well is drilled (Aydemir, Bilim, Kosaroglu, & Buyuksarac, 2019).

In addition, as opposed to traditional fossil fuel plants that burn oil, natural gas, or coal, any fuel would be needed. Thus on the contrary, with wind and photovoltaic systems, a geothermal plant operates 24 hours, providing a continuous source of electricity. Relating to the need for deep knowledge and understanding of local geology to implement the creation of an HDR energy reserve. In this sense, drilling a well in hot rock is usually to connect it to a fractured rock region to allow water to flow through it, creating a heat exchanger (exchanger) for the electrical generators on the surface (Aydemir et al., 2019).

Or tanks and the addition of additional wells capable of increasing the power directly proportional to the total amount of water in the rock (at the time of its formation), as well as considering factors such as shape, orientation, and the internal structure of the tank, are the natural results of local geological conditions. Since, during the formation of a reservoir, HDR blocks are moved uniformly by the pumped water, the movements of these rocks generate low-frequency waves of stress that are similar in nature but on a smaller and minimal scale to those caused by earthquakes (Santamarina, Carlos, & Rached, 2021).

In this approach, it is possible to highlight that the HDR geothermal energy system is environmentally safe and viable because of the lower emission rates and because the absorbed and removed energy is positioned entirely underground and the surface equipment necessary for the conversion of electricity is relatively compact. Still reflecting on microseismic technology to recognize signals and identify their points of origin, collecting this data generates a sense of the size, shape, and orientation of the tank, allowing the execution of the formation of energy production wells to make efficient use of the heated water that was pumped into the rock (Zhang, Wan, & Elsworth, 2020).

Thus HDR geothermal energy is a very productive source with enormous potential, consisting of a preferable solution as a substitute energy source for fossil fuels related to the fact that these traditional energy sources become expensive to import oil and gas from foreign sources, or even with respect to environmental concerns regarding fossil fuels and other pollutants into the atmosphere (Lu, Lin, Gan, Liu, & Wang, 2017).

HDR geothermal electricity is different from traditional geothermal energy since it is installed in thermal regions, whereas traditional geothermal energy is produced from naturally occurring steam and hot water in rocks close to volcanic regions. However, traditional geothermal power plants are commonly limited in size and linked to emissions of toxic elements and even volcanic gases. Still, the same environmental effects caused by the development of these HDR factories prevent the emission of these toxic gases (Santamarina et al., 2021).

However, despite HDR techniques offering optimal possibilities, many problems still need to be considered and solved, as it has yet to be proven that a suitable surface can be formed and reproduced in depth, or drilling and instrumentation techniques at great depths in hot rock have yet to be developed. The nature of the fracking system must vary from area to area since in some formations there is water loss, and due to this, it decreases with the increase in depth. This is still related to environmental problems such as the visual rejection of waste heat, limiting the use of HDR geothermal energy (Aydemir et al., 2019).

26.5 Discussion

Even though the production of geothermal energy is dependent on water, there are two ways: the first is the existence of sheets of hot water in the subsoil since the water contained in the underground reservoirs can heat up or even boil when it contacts the hot rock; and the second is the injection of water, which, in contact with the high temperatures found inside the planet, evaporates and returns to the surface through high-pressure ducts. From then on, the process of generating electricity takes place in the usual way, considering that the steam is taken to the power station where the turbine that drives a generator moves, and this same steam is again condensed as water is reintroduced in the pipelines to generate more energy. In the face of all this, geothermal energy is the only source for the production of electricity in many countries, stating that to obtain this, it is not necessary to burn fossil fuels or even that the exploitation of geothermal energy can only be done in a limited number of places that are normally identified by the presence of volcanoes or geysers (hot water sprays).

In a broad sense, the concept of geothermal energy has principles related to exploration at the technological level in the generating of electricity and cogeneration (production process and combined use of heat and electricity) by geothermal plants from the associated geothermal energy. Cogeneration is the heat produced in an electrical generator that is used in the production process in the form of steam; that is, it is a process that allows, from a single fuel, the simultaneous production of heat and electrical energy. The advantage of this solution is that the consumer saves the fuel he would need to produce the heat from the process. Energy efficiency is thus much higher, as it makes fuel energy useful.

In this context, geothermal energy is the use of these natural conditions to take advantage of this thermal energy, which is in the form of heat or electricity. In any case, it is a renewable energy source since the thermal resources of the Earth's interior are practically inexhaustible. Of the five forms of geothermal energy, only two are used commercially, considering hydrothermal reservoirs consisting of large pools of steam or hot water (by turning a turbine of an electric generator to generate electricity) trapped in porous rocks and earth energy (the technology of direct use) relating to the places where the heat of the earth appears close to the surfaces and it is possible to use this heat directly to heat houses and buildings. The other forms are pressurized seas, heated rocks, and magma.

Moreover, the thermal gradient, which depends on thermal conductivity and heat flow, varies with respect to the type of rock or soil, observing a different thermal gradient (temperature variation depending on the depth). This gradient can vary from the surface part of the Earth's crust or even from the seabed in the vicinity of the oceanic mountain ranges, considering that the geothermal gradient is influenced by the circulation of internal terrestrial fluids.

Still considering that the temperature measurement of the thermal gradient is complex and is disturbed by the heat of the drilling and even by the presence of liquids (water, moisture, and air) that may be present in the soil, it is carried out with the suitable instruments that are discarded in the wells geothermal, and the accuracy of temperature measurement is a function of the machinery used. The instruments used to measure temperature in geothermal wells are generally thermometers employed in geothermal investigations and have two essential characteristics. They must have low thermal inertia due to their speed to reach thermal equilibrium quickly, and even reading should be possible when outside the well, being commonly thermocouples, resistance thermometers, and even oscillators.

Thus reflecting on the pressure factor optimizes the power of the hole, which is not necessarily the most economical pressure that can be used, since for higher pressures, more expensive fittings such as valves and thicker insulation tend to be required on the turbine plant leaf. However, the production of maximum power will reduce the number of cavities required; other factors that are also important, or even in relation to low pressures, tend to require bulky tubes for the transport of steam. In this sense, the useful energy of a reservoir depends on the flow rate, temperature, and specific latent heat of the fluid.

Even considering the geothermal energy originated from the heat coming from the interior of the Earth through the exploration of the geothermal potential of aquifers, the design of a power system considering engineering details such as pipe connections, silencers, separators, and other typical parts of the plant's geothermal systems depends on many factors. Certainly, the pressure and temperature of the aquifer are the most important, considering the performance of an aquifer to provide the maximum electrical power, implying the use of the pressure of the water source where the maximum heat can be obtained. Since electricity can be obtained from this, however, it leads to different optimization pressures, reflecting that the increase in pressure reduces the amount of useful steam but increases the useful heat in a given steam mass.

Even pondering an enhanced geothermal system (EGS) generating geothermal electricity without the need for natural convective hydrothermal resources, since until then, geothermal energy systems exploited only resources where there was permeability to natural heat, water, and rocks as a sufficient source to allow energy extraction. However, most of the geothermal energy available for conventional techniques is only in dry, impermeable rock; that is, it is not only seismically active hotspots that are the only ones with geothermal energy.

There is a constant supply of milder heat, that is, there is a large amount of thermal energy available, which is also useful for immediate heating purposes considering depths of tens to a few hundred kilometers below the surface containing enough heat to control the heat of the climate in a house or even buildings in a city. Since using the emerging EGS technology, it is possible to capture this heat for electricity generation on a larger scale than traditional technologies currently allow, representing a huge source of electricity production capacity.

26.6 Conclusions

Geothermal energy is a way of extracting energy from the temperature of the interior of the planet, consisting of a practically inexhaustible natural resource since this heat can be converted into electricity, and it is shown as one of the options for the employment of fossil fuels for production. However, this energy resource (which does not need finite resources to be extracted) is not yet popular. In summary, this is what geothermal energy consists of; the capture of heat and the withdrawal of water resources, which can come from reservoirs and geothermal plants, which will be converted into mechanical energy and later transformed into electricity.

In consideration of geothermal plants, in these places, wells are used that allow access to steam from the temperature of the Earth's interior. This steam is then converted by means of plants, pumps, and other resources into electrical energy. Still, the use of water resources is only for the capture of steam; after the conversion process, they return to the reservoirs, guaranteeing the preservation of this resource; however, this does not make geothermal energy totally sustainable. Even so, it is possible to consider that geothermal energy represents a more lasting energy source, in addition to not depleting resources. In terms of efficiency, geothermal energy does not lose anything and can be used for several functions, including domestic electrical supply, electricity for industries, heating water resources, and other purposes.

Geothermal energy is also "cleaner," as long as it does not emit pollutants during its production. Considering that it is only heated without the use of chemical substances, it is not necessary to carry out burning as occurs with oil, which is the most used resource currently, and thus fewer polluting substances are emitted. Thus resources occur naturally in the forms of steam, hot water (aquifers), and hot stones, and the development phase is dictated by the natural availability and the cost of extraction.

In this sense, it is also worth mentioning that in recent years, geophysics has advanced rapidly, and together, the scientific knowledge of the structure and composition of the planet has increased. In particular, the theory of plate tectonics allows an understanding of why certain regions have greater volcanic (heat) and seismic activity than others. In this context, indicating soil water sources, deep hot water systems, steam fields, or even a geyser, it has expanded the production of electricity through geothermal means.

It is also worth noting that geothermal energy cannot be transported, so it can only be used in the production sites that surround the plant and in its surroundings, because to be captured and produced, it requires a specific area. As well as considering that these areas are impacted since they are very modified and, consequently, lose their natural state. In addition, it changes the temperature of the environment that surrounds their extraction site. Or even another factor that makes this type of energy less used is its very high cost, as it requires specific infrastructure. It is also worth mentioning that these are not the only disadvantages of using this source of geothermal energy, although they are the most relevant.

References

Al-Dousari, A., Al-Nassar, W., Al-Hemoud, A., Alsaleh, A., Ramadan, A., Al-Dousari, N., & Ahmed, M. (2019). Solar and wind energy: Challenges and solutions in desert regions. *Energy, 176*, 184–194. Available from https://doi.org/10.1016/j.energy.2019.03.180.

Awerbuch, L., Lindemuth, T. E., May, S. C., & Rogers, A. N. (1976). Geothermal energy recovery process. *Desalination, 19*(1–3), 325–336. Available from https://doi.org/10.1016/s0011-9164(00)88041-8.

Aydemir, A., Bilim, F., Kosaroglu, S., & Buyuksarac, A. (2019). Thermal structure of the Cappadocia region, Turkey: A review with geophysical methods. *Mediterranean Geoscience Reviews, 1*(2), 243–254. Available from https://doi.org/10.1007/s42990-019-00011-7.

Bahadori, A., Zendehboudi, S., & Zahedi, G. (2013). RETRACTED: A review of geothermal energy resources in Australia: Current status and prospects. *Renewable and Sustainable Energy Reviews, 21*, 29–34. Available from https://doi.org/10.1016/j.rser.2012.12.020.

Bandyopadhyay, S. (2019). The first step towards energy revolution. *Clean Technologies and Environmental Policy, 21*(2), 227–228. Available from https://doi.org/10.1007/s10098-019-01674-4.

Beiter P., Elchinger M., Tian T., 2016 Renewable energy data book. National Renewable Energy Lab. (NREL). (2017).

Berman E.R., Geothermal energy. Park Ridge. (1975).

Blaabjerg, F., & Ma, K. (2017). Wind energy systems, Institute of Electrical and Electronics Engineers Inc., Denmark*Proceedings of the IEEE*, *105*(11), 2116–2131. Available from https://doi.org/10.1109/JPROC.2017.2695485, http://ieeexplore.ieee.org/xpl/RecentIssue.jsp?punumber = 5.

Brown D.W. 2000 A hot dry rock geothermal energy concept utilizing supercritical CO2 instead of water. Proceedings of the Twenty-Fifth Workshop on Geothermal Reservoir Engineering.

Brown D.W. 2009 Hot dry rock geothermal energy: important lessons from Fenton Hill. Thirty-Fourth Workshop on Geothermal Reservoir Engineering.

Bundschuh J., Chen G., Tomaszewska B., Ghaffour N., Mushtaq S., Hamawand I., ... Antille D. L., Solar, wind and geothermal energy applications in agriculture: Back to the future? Geothermal, wind and solar energy applications in agriculture and aquaculture. Taylor and Francis, Australia (2017), 1–32, https://www.routledge.com/products/9781138029705. doi:10.1201/9781315158969.

Dickson M.H., Fanelli M., Geothermal energy: Utilization and technology. Taylor and Francis, Italy. (2013), http://www.taylorandfrancis.com/books/details/9781315065786/. doi:10.4324/9781315065786.

Gehlin S., Andersson O., Geothermal energy use, country update for Sweden. European Geothermal Congress. (2016).

Geothermal Energy and Society. Springer International Publishing, (2019).

Geothermal energy systems: Exploration, Development, and utilization. John Wiley & Sons, (2011).

Glassley, W. E. (2014). *Geothermal energy: Renewable energy and the environment*. CRC Press.

Guiling, W., Wenjing, L., Wei, Z., Chuan, L., Feng, M., & Haonan, G. (2016). Research on formation mechanisms of hot dry rock resources in China. *Acta Geologica Sinica - English Edition*, *90*(4), 1418–1433. Available from https://doi.org/10.1111/1755-6724.12776.

Harlow, F. H., & Pracht, W. E. (1972). A theoretical study of geothermal energy extraction. *Journal of Geophysical Research*, *77*(35), 7038–7048. Available from https://doi.org/10.1029/jb077i035p07038.

Ismail B., Renewable geothermal energy explorations. BoD–Books on Demand. (2019).

Johnston, I. W., Narsilio, G. A., & Colls, S. (2011). Emerging geothermal energy technologies. *KSCE Journal of Civil Engineering*, *15*(4), 643–653. Available from https://doi.org/10.1007/s12205-011-0005-7.

Kabir, E., Kumar, P., Kumar, S., Adelodun, A. A., & Kim, K.-H. (2018). Solar energy: Potential and future prospects. *Renewable and Sustainable Energy Reviews*, *82*, 894–900. Available from https://doi.org/10.1016/j.rser.2017.09.094.

Kazmierski, T. J., & Beeby, S. (2014). *Energy harvesting systems*. Springer.

Kuo, G. (2012). Geothermal energy. *World Futures Review*, *4*(1), 5–7. Available from https://doi.org/10.1177/194675671200400102.

Lakatos, L., Hevessy, G., & Kovács, J. (2011). Advantages and disadvantages of solar energy and wind-power utilization. *World Futures*, *67*(6), 395–408. Available from https://doi.org/10.1080/02604020903021776.

Lu, C., Lin, W., Gan, H., Liu, F., & Wang, G. (2017). Occurrence types and genesis models of hot dry rock resources in China. *Environmental Earth Sciences*, *76*(19). Available from https://doi.org/10.1007/s12665-017-6947-4, http://www.springerlink.com/content/121380/.

Márquez, F. P., García, A., & Karyotakis. (2018). *Renewable energies: Business outlook 2050*. Springer.

Mohammadzadeh Bina, S., Jalilinasrabady, S., Fujii, H., & Pambudi, N. A. (2018). Classification of geothermal resources in Indonesia by applying exergy concept. *Renewable and Sustainable Energy Reviews, 93*, 499–506. Available from https://doi.org/10.1016/j.rser.2018.05.018.

Moser, C. (2007). *Design, automation & test in Europe conference & exhibition*. IEEE.

Oliveira Gomes 2021 A study on investigations carried out in dams from the perspective of risk analysis. Proceedings of the 5th Brazilian Technology Symposium, Springer.

Priya S., Inman D.J., Energy harvesting technologies. Springer, United States (2009), 1–517, http://springerlink.com/openurl.asp?genre = book&isbn = 978-0-387-76463-4. doi: 10.1007/978-0-387-76464-1.

Razmjooy N., Estrela V.V., Padilha R., Monteiro A.C.B., World cup optimization algorithm: Application for optimal control of pitch angle in hybrid renewable PV/wind energy system. Springer Science and Business Media LLC, 696, (2020), doi:10.1007/978-3-030-56689-0_3.

Santamarina, J., Carlos, R., & Rached. (2021). *International conference of the international association for computer methods and advances in geomechanics*. Springer.

Sanyal S.K., Future of geothermal energy. Proceedings. (2010).

Stefansson V., The renewability of geothermal energy. Proceedings of the World Geothermal Congress. (2000).

Stehly T.J., Beiter P.C., Cost of wind energy review. No. NREL/TP-5000-74598. National Renewable Energy Lab. (NREL). (2018).

Tester, J. W. (2006). The future of geothermal energy. *Massachusetts Institute of Technology, 358*.

Ungemach, P. (1985). An overview of the state of geothermal energy. *International Journal of Energy Research, 9*(3), 223–228. Available from https://doi.org/10.1002/er.4440090304.

Zeiler W., Boxem G., Geothermal active building concept. Springer Science and Business Media LLC, (2009), 305–314, doi: 10.1007/978-3-642-03454-1_32.

Zhang, H., Wan, Z., & Elsworth, D. (2020). Failure behavior of hot-dry-rock (HDR) in enhanced geothermal systems: Macro to micro scale effects. *Geofluids, 2020*, 1–13. Available from https://doi.org/10.1155/2020/8878179.

Chapter 27

GIS-based modeling for water resource monitoring and management: a critical review

Manju Pandey[1], Saroj Senapati[2] and Gouri Sankar Bhunia[3]
[1]*Department of Geography, Government College Barpali, Korba, Chhattisgarh, India,*
[2]*Department of Geography, Atal Bihari Vajpayee Viswavidyalaya, Chhattisgarh, India,*
[3]*Independent Researcher, Paschim Medinipur, West Bengal, India*

27.1 Introduction

A key element of the socioeconomic-environmental framework is water resources, which are increasingly under threat from problems arising from the intricacy of hydrological cycles, socioeconomic factors, divergent stakeholder opinions, and concerns about the use of water for multiple uses (Mishra, Kumar, Saraswat, Chakraborty, & Gautam, 2021). Because of its incredibly irregular spatial and cyclical distribution, water supplies are under increasing pressure on a worldwide scale. However, people may not be aware of the locations and amounts of local water resources, especially when it comes to deep-confined groundwater, remote alpine glaciers, and snow. Understanding the constantly shifting interactions and feedback mechanisms among the interconnected hydrological, social, economic, and ecological components is crucial for efficiently regulating and planning the use of water resources. (Sun & Yang, 2019). As a result, it is sometimes difficult to develop comprehensive water management plans since the underlying causes of the issues are not completely recognized (Gohari, Mirchi, & Madani, 2017). Using cutting-edge techniques like remote sensing (RS), geographic information systems (GIS), geostatistics, and hydrologic models is essential to effectively managing resource availability.

A method called system dynamics modeling helps to identify the many connections between the different yet connected subsystems that generate the dynamic activity of the system as a whole. RS and GIS are being used in the assessment, development, and planning of accessible water resources. It is essential to use water resources wisely in order to meet future demand.

Water Resources Monitoring, Management, and Sustainability. DOI: https://doi.org/10.1016/B978-0-443-23665-5.00027-2

Satellite RS systems can collect current, continuous data with global coverage depending on their orbital features; nevertheless, to create algorithms and achieve calibration, they need measurements from the ground (Dubovik et al., 2021; Tang, Gao, Lu, & Lettenmaier, 2009). For instance, since February 2000, the Moderate Resolution Imaging Spectroradiometer (MODIS) onboard the Terra and Aqua satellites has given daily global snow cover estimates (Fu, Ma, Chen, & Chen, 2019). They have found widespread application in diverse fields such as hydrology, agriculture, and climate research (Wang, Wang, Wu, Xu, & Gao, 2017). With relatively high-resolution data from the Landsat series, mountain glaciers could be restocked and their conditions assessed globally, enabling updates to the second Chinese Glacier Inventory, Glacier Area Mapping for Discharge from the Asian Mountains, and other global glacier inventories to be updated with great accuracy and confidence. The Soil Moisture Active and Passive Mission, the Advanced Microwave Scanning Radiometer for NASA's Earth Observing System, and the Soil Moisture Ocean Salinity Mission of the European Space Agency may all offer worldwide soil moisture assessment (Burgin et al., 2017). Although groundwater is the hardest material for satellite sensors to find, GRACE has been able to track both the Three Gorges Reservoir's refilling and groundwater depletion rather well. With ground-penetrating RADAR, it is possible to acquire even more exact assessments of groundwater, soil moisture, and glacier thickness.

Surface water mapping studies at the local scale are often carried out, and the proposed methodologies work well in the respective specific research areas. Even after decades of study, the flexibility and dependability of traditional methods such as machine learning and spectral indices remain limited. To map surface water using satellite images, a deep convolutional neural network (ConvNet), which has led to advancements in computer vision (Lecun, Bengio, & Hinton, 2015), has just been introduced. Since deep learning algorithms require large amounts of labeled data and computational resources to train models and learn features, respectively, they have not been extensively used. Many studies tracked and analyzed geographic changes in surface water characteristics using nongeostatistical approaches. Splines and Kriging are the two primary interpolation techniques used to map spatial variability (Ordinary Kriging) (Boumpoulis, Michalopoulou, & Depountis, 2023). These techniques will be used in the current study to extrapolate the region's whole surface from known locations (samples/monitoring stations). Utilizing RS and GIS data to their fullest, hydrologic models may conduct several physical investigations and scenario assessments (Biernacik, Kazimierski, & Włodarczyk-Sielicka, 2023). They can provide a broad variety of modeling skills for both projected future occurrences and past events. The availability of model simulations over an extended period of time can also provide a meaningful estimation of low-probability instances that escaped ground observation.

There is currently a significant gap in the literature, though, as no reviews have comprehensively assessed the potential of GIS in the management of water resources, particularly with regard to geographic distribution, model subsystems, model calibration and validation techniques, policy assessment, and optimization of different strategies for decision-making regarding water-related issues under various conditions. The purpose of this study is to perform a comprehensive, quantitative assessment of GIS-based modeling usage in the area of water resource management. More precisely, study objectives, distribution across regions, temporal and geographical scales, modeling subsystems, model validation, and calibration techniques were synthesized to identify limits in the use of the GIS modeling methodology in water resource management.

27.2 Methods

A comprehensive empirical literature evaluation was conducted to locate and categorize the relevant material and critically assess the use of GIS models that are currently in use for the management of water resources. Scholarly electronic databases (Scopus, Web of Sciences, and Google Scholar) were utilized to identify original research and academic publications that were published in English-language journals until August 2023. This assessment did not take into account books, book chapters, review papers, conference papers, or theoretical publications without any GIS applications in the actual world. The following keywords were combined to search for pertinent material in these databases: "GIS modeling*," "geo-intelligence approach," "GIS mapping and monitoring model," "Geospatial model," "Machine learning" or "AI," and the change-related search terms "Water*," "Water resource," or "Water resource management." For the keywords "GIS model" and "Water," the wildcard "*" was used to determine every phrase around these two keywords.

The articles' keywords, abstract, and title were all included in these searches. Papers were also identified using the reference lists of research publications found in the initial database search. We double-checked the title and abstract of every article that sprang up from these searches to make sure that the use of GIS modeling in water resource management was the primary topic. The key research articles used to create the database that described how GIS modeling was used to manage water resources were then entered into the database using the following fields: (1) citation information (authors, journal, and year of publication); (2) geographic distribution (continent and country); (3) research objectives (water quality, water supply, and water demand); (4) modeling subsystems (hydrological, social, economic, and environmental); (5) decision-making context (i.e., scenario-based, optimization-based, or predictive only); (6) model calibration and validation techniques (e.g., structure tests or behavior-pattern tests); and (7) model output.

27.3 Results

Because of the rapid urbanization and population growth in recent years, there has been an increased demand for water supplies. The average yearly availability of water per person was predicted to be 1816 cubic meters in 2001 and 1545 cubic meters in 2011. These numbers may drop even lower in 2021 and 2031, to 1486 cubic meters and 1367 cubic meters, respectively. India aims to guarantee universal access to affordable and safe drinking water by 2030, in compliance with the Sustainable Development Goal for sanitation and clean water. Furthermore, in 2025, 72.48% of the water used for irrigation will be accounted for by research conducted by the National Commission for Integrated Water Resources Development in an environment of high demand. According to Abdulkareem, Pradhan, Sulaiman, and Jamil (2018), water models are crucial parts and tools for addressing groundwater concerns and environmental planning strategies. Hydrologists, engineers, and other academics are particularly interested in accurate hydrological model categorization since it will aid in comprehending the distinctive aspects of the models before making a judgment about employing them (Gupta, Sharma Jethoo, Tyagi, & Gupta, 2015; Mishra et al., 2021).

Within the WoS database over the past 10 years, there has been a noticeable and rapid growth in the quantity of research articles published in this area (Fig. 27.1).

There were just five 1980s papers found in the database, which accounts for 0.16% of total publications on this topic. The number of publications published in the 1990s increased by 10–43, or 1.38% of all publications. In the 2000s, this number more than quadrupled to 78 publications, or 2.78% of all publications. The decade of the 2010s had the highest increase in publications, with 110 publications (3.53% of the total) throughout that decade (the year 2023 is included in this series). Stakeholders' growing interest in developing

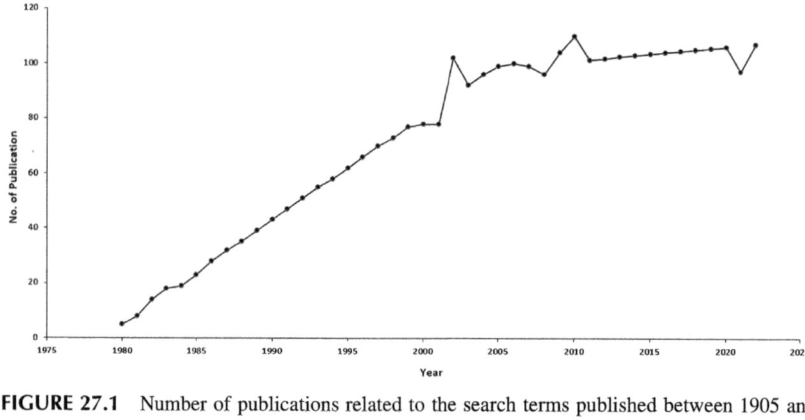

FIGURE 27.1 Number of publications related to the search terms published between 1905 and 2022—data from the Web of Sciences database.

tools for monitoring the water quality in inland water bodies (such as lakes and reservoirs) is the reason for the remarkable increase in papers between 2010 and 2020. EO-derived data is becoming increasingly important for understanding the environment because of this information gap. A growth in publications using EO data to monitor inland waters has coincided with technological advancements, especially the development of new orbital sensors, in addition to sociopolitical concern (Greb & Dekker, 2018).

The bulk of the studies were published in five journals: Water Resources Management (WRM), Journal of Water Resources Planning and Management, Journal of Water Resource Engineering and Management, Water Resources Research, and Water Resources and Economics. There were not more than five manuscripts from each of the other journals. The use of spatial models in water resource management has attracted scholarly interest from all around the world. Most research has been conducted in Asia and North America. Conversely, there has not been much research done in South America. The proportion of foreign studies that are taken into consideration is less than 30%.

27.3.1 Spectral indices and water resource monitoring

In the evaluation and continuous surveillance of surface water resources, satellite RS plays a significant role. When using satellite data, data pertaining to surface water resources is gathered using a variety of sensors and resolutions (Chawla, Karthikeyan, & Mishra, 2020; Huang, Chen, Zhang, & Wu, 2018; Khalid, Khalil, & Qureshi, 2021). Landsat images are typically used to identify surface water features using a variety of techniques (Acharya, Subedi, & Lee, 2018; Jiang et al., 2018). A number of RS methods have been used in a variety of applications involving water resources, including water quality evaluation (He et al., 2012), pre-post-flood impact observation (Ji, Zhang, & Wylie, 2009), and modifications in surface WRM (Acharya, Subedi, Huang, & Lee, 2019). Water bodies are identified and extracted from RS data in a variety of ways. Water indices' techniques (Acharya et al., 2018) such as Normalized Difference Water Index (NDWI) (Xu, 2006), Automated Water Extraction Index (AWEI) (Feyisa, Meilby, Fensholt, & Proud, 2014), Modified NDWI (MNDWI) (Xu, 2006), AWEI (Feyisa et al., 2014), Normalized Difference Moisture Index (Gang & Dong-sheng, 2012), New Water Index (Ding, Cuo, Zhang, & Zhu, 2018), and Water Ratio Index (Shen & Li, 2010) are applied extensively. To collect data regarding water bodies, classification approaches involving decision trees (Mueller et al., 2016; Tulbure, Broich, Stehman, & Kommareddy, 2016), maximum likelihoods (Yang & Du, 2017), and statistical pattern recognition approaches (Acharya et al., 2018) are also utilized. Recent developments in automation have led to the use of a variety of machine learning algorithms for obtaining water bodies from RS images, including neural networks (Rokni, Ahmad, Solaimani, & Hazini, 2015), artificial neural networks (Skakun, 2010), support vector machines (Sun, Zhao, Gong, Ma, & Dai, 2014), naive Bayes, random forests, gradient-boosted machines, recursive partitioning and regression trees, and constraint energy minimizations (CEMs) (Acharya et al., 2019).

Although surface water can be identified using both optical and microwave (passive and radar) sensors, cloud pollution hinders the use of optical sensors, which makes it difficult to utilize them effectively for small-scale temporal-scale flood detection and monitoring. Examples of spectral band differences used in optical techniques include the NDWI (McFeeters, 1996). Depending on the significant absorption and low radiation of water bodies in these spectral areas, the NDWI uses the green and NIR bands, despite the fact that it is prone to incorrect identification of built-up (urban) locations. Satellite RS of relevant factors for water management and hydrological hazard monitoring has come a long way, starting with the few early perpetrated missions that focused on waterbody extent and height and continuing to the current routine monitoring of almost all the elements that make up the water balance. This is true even if there are not enough particular missions for a lot of crucial aspects, and missions need to coordinate with one another to benefit from complementary data. Here, we provide a thorough assessment of the state-of-the-art for relevant variables, active satellite missions, and goods. Table 27.1 lists a few current initiatives and goods that are helpful for WRM.

27.3.2 Water resources and spatial modeling

Over the years, many geographical hydrological models have been developed. These models have been the subject of several published evaluations. Applications for watershed size and groundwater systems are common with these models. The first dynamic hydrological model that was realistically usable across large territories (such as a whole continent) and showed promise for applications involving water resources was the variable infiltration capacity model, which was developed in the early 1990s. For investigating, forecasting, and regionally estimating the occurrence of water, RS and GIS approaches may be used as rapid and inexpensive tools by utilizing different prediction models (Abdelkareem & Abdalla, 2021; Avand et al., 2021). Groundwater potentiality mapping is a common use of methodologies that are data- and knowledge-driven. Data-driven strategies encompass the frequency ratio, machine learning models (Lee et al., 2017), logistic regression, weight-of-evidence method (Chen et al., 2018), evidential belief function (Mogaji et al., 2016), and artificial neural network model. Knowledge-driven techniques consist of overlay analysis, the analytical hierarchy process (AHP) (Kumar et al., 2020), Boolean logic, index overlays, and fuzzy approaches.

When evaluating water management difficulties, academics frequently choose to use spatial models such as regression models, GIS, RS, and satellite pictures. Care was also taken while using APEX, AHP, Wet-Spass, CN-SCS,

TABLE 27.1 Summary of selected current satellite missions and satellite-based products that are useful for water resource monitoring.

Parameters	Satellite	Period	Scale	Repetition	Objectives	Source
Surface water height	Jason-2/3	2008/2016	Lakes >100 km^2	10 days	There are numerous near real-time (7-day latency) and high-quality research items available.	https://ipad.fas.usda.gov/cropexplorer/global_reservoir/
	Sentinel-3	2016	350 m along the track	27 days	Ensures the European remote sensing (ERS), Envisat, and SPOT missions have access to continuous data	https://sentinel.esa.int/web/sentinel/sentinel-data-access
Surface water extent	Landsat	2013 (Landsat-8), 2022 (Landsat-9)	30 (multispectral) to 100 m (thermal)	16 days	Indicators like NDWI can be used to determine the area of surface water	https://landsat.usgs.gov
	MODIS	2000 to present	500 m	Daily	The extent of surface water from NDWI. As an illustration, consider the NASA MODIS Near Real Time Global Flood Mapping Project	https://floodmap.gsfc.nasa.gov
	Sentinel-2	2015/2017	10–60 m	5/10 days	Water surface area as determined by NDWI. From two satellites, a 5-day repetition	https://sentinel.esa.int/web/sentinel/sentinel-data-access

TOPMODEL, Multi-Criteria Evaluation, FREEWAT, and the Soil and Water Assessment Tool model; primarily, these models were combined with one another or with other models or methodologies. Furthermore, these models have become a useful tool for forecasting, simulating, and evaluating the availability and scarcity of water resources throughout the last few decades. Additionally, using these models is a critical and significant step in developing a thorough water resource management plan.

Hydrological phenomena are dynamic and changing on both spatial and time scales. Conventional techniques like in situ or point evaluations were employed in the past to study hydrological phenomena. The development of RS and GIS technologies has created new opportunities for research into the hydrological cycle, including the measurement of water balance components and the geographical and temporal distribution of meteorological indicators, among other things (Kushwaha, Bhardwaj, & Verma, 2016).

27.3.3 Water quality monitoring and geospatial technology

Surface water quality, a crucial component of the ecosystem, is now quite concerning. To put it simply, variations in water quality are caused by a combination of natural and man-made variables. In general, human activities constitute a continuous source of pollution; nevertheless, the climate has an impact on the water catchment basin's seasonal surface runoff (Audu and Usman). RS has demonstrated potent capacities to track and assess the caliber of inland waters for more than 40 years. To establish robust relationships between water column reflection (and sometimes emission) and physical and biogeochemical elements, such as transparency, chlorophyll concentration (phytoplankton), and the organic matter and minerals suspended in sediments in different water bodies, many scientists frequently employ the visible and near-infrared bands of the solar spectrum (mainly from the blue to near-infrared region).

Much scholarly work has recently shown an increasing interest in the use of multivariate statistical techniques and geostatistical approaches to achieve sustainable water resource use (Bordalo, Nilsumranchit, & Chalermwat, 2001; Kumar & Pal, 2012). The combined use of multivariate statistics and geostatistical techniques, which help detect probable sources that might influence water environmental systems, is an essential tool for the proper management of water resources and the prompt resolution of contamination issues. Numerous studies have combined water quality indicators (WQIs) and GIS to create maps (Ketata, Gueddari, & Bouhlila, 2012), evaluate the quality of groundwater, evaluate drinking water facilities (Avery & Smith, 2018), and explain the temporal and spatial differences in water quality (Singh, Malik, Mohan, & Sinha, 2004). WQIs are another word for data from water quality monitoring that condenses a lot of variables into a single or double-digit figure. Such measures help examine local variations in water quality (Akter et al., 2016), highlight particular environmental issues (Othman, Alaa Eldin, & Mohamed, 2012), and assist decision-makers in assessing the outcomes of various water quality intervention measures (Damo & Icka, 2013), which makes them helpful in monitoring and managing water quality.

27.3.4 Hydrological models

Hydrological modeling did not link to the economic, environmental, or social subsystems; instead, it concentrated primarily on the hydrological processes and performance of a watershed. Water-related issues such as reservoir operation (Keith, Ford, & Horton, 2017; Sidle, 2021), flooding prevention (Ahmad & Simonovic, 2015; Banda, Dzwairo, Singh, & Kanyerere, 2022), water supply management (Rusuli, Li, Li, & Eziz, 2016), groundwater management (Bates, Beruvides, & Fedler, 2019), and water quality management (Hallas, Mackowiak, & Wilkie, 2020) have all been addressed using hydrological modeling. Hydrological-environmental geographic modeling was developed to understand the links between ecosystems and hydrological cycles at different spatiotemporal scales. The relationship between environmental flows and the management of water quality in the water resource system was the focus of much of the environmental hydrology research that was examined. To comprehend the effects of reservoir management on fish habitats, sediment flushing, and landslip stability, Jiang et al. (2018) established a geographical model. To evaluate the interaction between water supply, flood detention, and a healthy aquatic ecosystem in man-made lakes in Taiwan, Chen, Hsu, & Chen, 2020; Chu, Chang, Lin, Wang, & Chen (2010) used spatial modeling strategy.

27.4 Discussion/summary

Numerous studies on the scheduling and management of water resources have been carried out on several continents using RS and GIS (Kushwaha, Elbeltagi, Mehan, Malik, & Yousuf, 2022). In this regard, Senthilkumar, Gnanasundar, and Arumugam (2019) used RS and GIS approaches to evaluate the prospective groundwater recharge region in the Amaravathi aquifer system in southern India. Yousaf et al. (2021) made an additional effort to track the crop water requirements in real time for effective crop water allocation. They conducted their investigation using open-source data from Landsat-8 and the MODIS. To preserve water resources sustainably, Kushwaha et al. (2016) used RS and GIS to analyze the hydrological behavior of the Takarla-Ballowal watershed in the Shivalik Hills. The results indicated that the watershed has medium to high relief and moderately permeable soil, which leads to an average runoff volume that exacerbates the problem of soil erosion in the watershed. In the Sundarban region of India, Sahana and Sajjad (2019) used RS and GIS to map and recognize the storm surge flood-prone zones. They discovered that low-lying places near rivers, such as communities, were particularly vulnerable to surge floods.

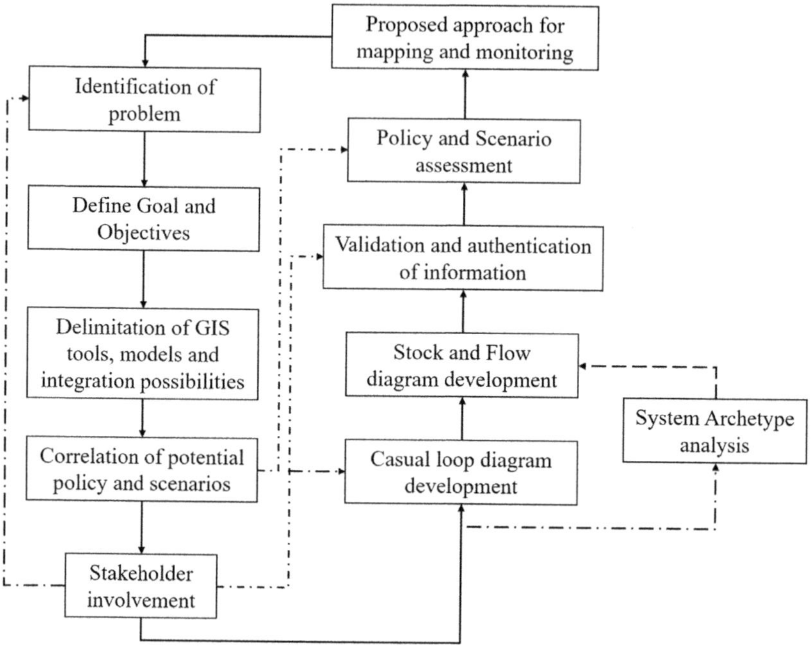

FIGURE 27.2 Framework proposed for the management of complex and uncertain water resource systems.

The growing use of GIS has opened up new opportunities and challenges, but it has also resulted in a number of new issues. It is getting more and more crucial to administer a growing variety of geographic datasets and software applications. These datasets and software are very expensive to develop but very easy to distribute, so there is a strong incentive to avoid repeating production efforts. It seems that managing these elements within a spatial library is in vogue right now. The spatial library will include information on map automation standards, data storage, manipulation, and updating procedures, and a catalog of spatial data holdings.

Moreover, a unified geographic modeling approach with various phases is put forward to represent complicated, uncertain, and spatial water resource systems, as illustrated in Fig. 27.2, which is further detailed below. This framework is intended to manage complicated and uncertain water resource systems.

- Outline the difficulties in managing water resource systems by gathering and assessing data and information to comprehend system issues and find stakeholders to participate in the modeling procedure.
- Determine the model's nature (such as optimization algorithms for management measure refinement or simulation models for managing metric assessments of efficacy under various scenarios).

- Consider integrating geospatial data with other modeling methods to make up for the shortcomings of water resource implementation (such as uncertain and geographical characteristics) in water resource systems by conceiving and articulating linkages among aspects of the system as a whole.
- With the appropriate stakeholders, propose possible regulations and strategies for the systems of water resources under study.
- By identifying crucial elements that advance comprehension of convoluted water resource systems (such as problematic borders and model layout) and increase the precision of simulation models, one can create spatial models, establish system archetypes, and more.
- To comprehend the dynamic behavior of the framework across time, create stock and flow diagrams by including critical variables from spatio-temporal systems and outcomes from past information analysis.
- To improve the reliability and credibility of the geospatial solutions and to identify the variables or model elements that have the biggest effects on results, calibrate the models, evaluate them, and do a sensitivity analysis.
- To cope with ambiguities in simulation and policy evaluation and help decision-makers better comprehend changing behavior and management alternatives for difficult water resource challenges, consider various circumstances and policies for a variety of values.
- Models and management mechanisms should be updated to reflect changing conditions and management objectives, and the efficacy of the implemented management measures should be regularly assessed.

27.5 Conclusion and future prospects

An extensive analysis was conducted on the possible applications of RS and GIS in many water resource disciplines, including planning, developing, implementing, assessing, and managing water resources. The results of this study have shown how important it is to assess the water resources in tropical and subtropical areas to better understand and track global hydrological processes. Only a few percent of hydrological models use RS data at this time. The reason behind this might be that most operational hydrological modeling techniques are not made to work with data that is dispersed geographically, which is a need for making efficient use of remotely sensed data. Based on an analysis of current approaches for evaluating water resources, a plethora of programs have been created to assess plans for their development and management on a larger scale. It is well known that the study of water resources and the goal of sustainable development require conceptual hydraulic and hydrologic models. Furthermore, it was shown that precipitation efficiency plays a major role in determining the accuracy of a hydrological model simulation, and a variety of models are now available to researchers. Most studies have discovered that water allocation and scarcity may be resolved at the basin level by using both intra- and interbasin strategies. These techniques have many advantages for allocating the water strategy within and between basins, determining the competition and relationship between hillslopes and channels as well as between the nearest basins, resolving conflicts and difficulties associated with water scarcity, and developing plans for water policy intervention and management. Researchers need to create more generic tools and approaches to hydrological modeling that are more appropriate for everyday usage with RS data. Model creation and verification are the two main pillars of policymaking and resource appraisal, and RS has the chance to advance the field of water resource modeling. For RS to be implemented successfully, high-resolution spatial and temporal data is required for image processing, which depends on the sensor calibration to supply the required information about the object under observation.

References

Abdelkareem, M., & Abdalla, F. (2021). Revealing potential areas for water resources using integrated remote-sensing data and GIS-based Analytical Hierarchy Process. *Geocarto International*. Available from https://doi.org/10.1080/10106049.2021.2005155.

Abdulkareem, J. H., Pradhan, B., Sulaiman, W. N. A., & Jamil, N. R. (2018). Review of studies on hydrological modelling in Malaysia. *Modeling Earth Systems and Environment*, *4*(4), 1577–1605. Available from https://doi.org/10.1007/s40808-018-0509-y, http://springer.com/journal/40808.

Acharya, T. D., Subedi, A., Huang, H., & Lee, D. H. (2019). Application of water indices in surface water change detection using Landsat imagery in Nepal. *Sensors and Materials*, *31*(5), 1429–1447. Available from https://doi.org/10.18494/SAM.2019.2264, https://myukk.org/SM2017/sm_pdf/SM1869.pdf.

Acharya, T., Subedi, A., & Lee, D. (2018). Evaluation of water indices for surface water extraction in a Landsat 8 scene of Nepal. *Sensors*, *18*(8), 2580. Available from https://doi.org/10.3390/s18082580.

Ahmad, S. S., & Simonovic, S. P. (2015). System dynamics and hydrodynamic modelling approaches for spatial and temporal analysis of flood risk. *International Journal of River Basin Management*, *13*(4), 443–461. Available from https://doi.org/10.1080/15715124.2015.1016954, http://www.tandfonline.com/loi/trbm20.

Akter, T., Jhohura, F. T., Akter, F., Chowdhury, T. R., Mistry, S. K., Dey, D., ... Rahman, M. (2016). Water quality index for measuring drinking water quality in rural Bangladesh: A crosssectional study. *Journal of Health, Population, and Nutrition*, *35*(1). Available from https://doi.org/10.1186/s41043-016-0041-5, https://jhpn.biomedcentral.com/.

Avand, M, et al. (2021). Using machine learning models, remote sensing, and GIS to investigate the effects of changing climates and land uses on flood probability. Journal of Hydrology, 595. Available from https://doi.org/10.1016/j.jhydrol.2020.125663.

Avery, D. C., & Smith, C. D. (2018). Access to public drinking water fountains in Berkeley, California: A geospatial analysis. *BioMed Central, United States BMC Public Health*, *18*(1). Available from https://doi.org/10.1186/s12889-018-5087-4, http://www.biomedcentral.com/bmcpublichealth.

Banda, V. D., Dzwairo, R. B., Singh, S. K., & Kanyerere, T. (2022). Hydrological modelling and climate adaptation under changing climate: A review with a focus in sub-Saharan Africa. *Water, 14*(24), 4031. Available from https://doi.org/10.3390/w14244031.

Bates, G., Beruvides, M., & Fedler, C. B. (2019). System dynamics approach to groundwater storage modeling for basin-scale planning. *Water (Switzerland), 11*(9). Available from https://doi.org/10.3390/w11091907, https://res.mdpi.com/d_attachment/water/water-11-01907/article_deploy/water-11-01907.pdf.

Biernacik, P., Kazimierski, W., & Włodarczyk-Sielicka, M. (2023). Comparative analysis of selected geostatistical methods for bottom surface modeling. *Sensors, 23*(8), 3941. Available from https://doi.org/10.3390/s23083941.

Bordalo, A. A., Nilsumranchit, W., & Chalermwat, K. (2001). Water quality and uses of the Bangpakong River (Eastern Thailand). *Water Research, 35*(15), 3635–3642. Available from https://doi.org/10.1016/S0043-1354(01)00079-3, http://www.elsevier.com/locate/watres.

Boumpoulis, V., Michalopoulou, M., & Depountis, N. (2023). Comparison between different spatial interpolation methods for the development of sediment distribution maps in coastal areas. *Earth Science Informatics, 16*(3), 2069–2087. Available from https://doi.org/10.1007/s12145-023-01017-4, https://www.springer.com/journal/12145.

Burgin, M. S., Colliander, A., Njoku, E. G., Chan, S. K., Cabot, F., Kerr, Y. H., ... Yueh, S. H. (2017). A comparative study of the SMAP passive soil moisture product with existing satellite-based soil moisture products. *IEEE Transactions on Geoscience and Remote Sensing, 55*(5), 2959–2971. Available from https://doi.org/10.1109/TGRS.2017.2656859.

Chawla, I., Karthikeyan, L., & Mishra, A. K. (2020). A review of remote sensing applications for water security: Quantity, quality, and extremes. *Journal of Hydrology, 585*, 124826. Available from https://doi.org/10.1016/j.jhydrol.2020.124826.

Chen, T.-T., Hsu, W.-L., & Chen, W.-K. (2020). An assessment of water resources in the Taiwan Strait Island using the water poverty index. *Sustainability, 12*(6), 2351. Available from https://doi.org/10.3390/su12062351.

Chen, W., Li, H., Hou, E., Wang, S., Wang, G., Panahi, M., et al. (2018). GIS-based groundwater potential analysis using novel ensemble weights-of-evidence with logistic regression and functional tree models. *Science of The Total Environment, 1*(634), 853–867.

Chu, H. J., Chang, L. C., Lin, Y. P., Wang, Y. C., & Chen, Y. W. (2010). Application of system dynamics on shallow multipurpose artificial lakes: A case study of detention pond at Tainan, Taiwan. *Environmental Modeling and Assessment, 15*(3), 211–221. Available from https://doi.org/10.1007/s10666-009-9196-4, http://www.kluweronline.com/issn/1420-2026.

Damo, R., & Icka, P. (2013). Evaluation of water quality index for drinking water. *Polish Journal of Environmental Studies, 22*(4), 1045–1051.

Ding, J., Cuo, L., Zhang, Y., & Zhu, F. (2018). Monthly and annual temperature extremes and their changes on the Tibetan Plateau and its surroundings during 1963–2015. *Scientific Reports, 8*(1). Available from https://doi.org/10.1038/s41598-018-30320-0, http://www.nature.com/srep/index.html.

Dubovik, O., Schuster, G. L., Xu, F., Hu, Y., Bösch, H., Landgraf, J., & Li, Z. (2021). Grand challenges in satellite remote sensing. *Frontiers in Remote Sensing, 2*. Available from https://doi.org/10.3389/frsen.2021.619818.

Feyisa, G. L., Meilby, H., Fensholt, R., & Proud, S. R. (2014). Automated water extraction index: A new technique for surface water mapping using Landsat imagery. *Remote Sensing of Environment, 140*, 23–35. Available from https://doi.org/10.1016/j.rse.2013.080.029.

Fu, W., Ma, J., Chen, P., & Chen, F. (2019). *Remote sensing satellites for digital earth* (pp. 55–123). Springer Science and Business Media LLC. Available from 10.1007/978-981-32-9915-3_3.

Gang, W., & Dong-sheng, G. (2012). Effects of vegetation cover and normalized difference moisture index on thermal landscape pattern: A case study of Guangzhou South China. *Yingyong Shengtai Xuebao.*

Gohari, A., Mirchi, A., & Madani, K. (2017). System dynamics evaluation of climate change adaptation strategies for water resources management in Central Iran. *Water Resources Management, 31*(5), 1413–1434. Available from https://doi.org/10.1007/s11269-017-1575-z, http://www.wkap.nl/journalhome.htm/0920-4741.

Greb S. Dekker A. 2018 International Ocean Colour Coordinating Group. Earth observations in support of global water quality monitoring.

Gupta S.K. Sharma Jethoo A.S. Tyagi J. Gupta N.K. 2015 Hydro 2015 International IIT. 20th International Conference on Hydraulics. A critical review of hydrological models.

Hallas, J. F., Mackowiak, C. L., & Wilkie, A. C. (2020). Mitigating rural WWTP impacts: System dynamics modeling of downstream nutrient outputs. *Science of the Total Environment, 744*. Available from https://doi.org/10.1016/j.scitotenv.2020.140809, http://www.elsevier.com/locate/scitotenv.

He, B., Oki, K., Wang, Y., Oki, T., Yamashiki, Y., Takara, K., ... Kawasaki, N. (2012). Analysis of stream water quality and estimation of nutrient load with the aid of Quick Bird remote sensing imagery. *Hydrological Sciences Journal, 57*(5), 850–860. Available from https://doi.org/10.1080/02626667.2012.683792.

Huang, C., Chen, Y., Zhang, S., & Wu, J. (2018). Detecting, extracting, and monitoring surface water from space using optical sensors: A review. *Reviews of Geophysics, 56*(2), 333–360. Available from https://doi.org/10.1029/2018RG000598, http://agupubs.onlinelibrary.wiley.com/agu/journal/10.1002/(ISSN)1944-9208/.

Ji, L., Zhang, L., & Wylie, B. (2009). Analysis of dynamic thresholds for the normalized difference water index. *Photogrammetric Engineering and Remote Sensing, 75*(11), 1307–1317. Available from https://doi.org/10.14358/PERS.75.11.1307, http://www.asprs.org/Photogrammetric-Engineering-and-Remote-Sensing/PE-RS-Journals.html.

Jiang, W., He, G., Long, T., Ni, Y., Liu, H., Peng, Y., ... Wang, G. (2018). Multilayer perceptron neural network for surface water extraction in Landsat 8 OLI satellite images. *Remote Sensing, 10*(5), 755. Available from https://doi.org/10.3390/rs10050755.

Keith, B., Ford, D. N., & Horton, R. (2017). Considerations in managing the fill rate of the Grand Ethiopian renaissance dam reservoir using a system dynamics approach. *Journal of Defense Modeling and Simulation, 14*(1), 33–43. Available from https://doi.org/10.1177/1548512916680780, http://dms.sagepub.com/.

Ketata, M., Gueddari, M., & Bouhlila, R. (2012). Use of geographical information system and water quality index to assess groundwater quality in el khairat deep aquifer (enfidha, central east tunisia). *Journal of Geosciences, 5*(6), 1379–1390. Available from https://doi.org/10.1007/s12517-011-0292-9, http://www.springer.com/geosciences/journal/12517?cm_mmc = AD-_-enews-_-PSE1892-_-0.

Khalid, H. W., Khalil, R. M. Z., & Qureshi, M. A. (2021). Evaluating spectral indices for water bodies extraction in western Tibetan Plateau. *Egyptian Journal of Remote Sensing and Space Science, 24*(3), 619–634. Available from https://doi.org/10.1016/j.ejrs.2021.090.003, http://www.elsevier.com/wps/find/journaldescription.cws_home/723780/description#description.

Kumar, D., Roshni, T., Singh, A., et al. (2020). Predicting groundwater depth fluctuations using deep learning, extreme learning machine and gaussian process: a comparative study. *Earth Science Informatics, 13*, 1–14. Available from https://doi.org/10.1007/s12145-020-00508-y.

Kumar, J., & Pal. (2012). Water quality monitoring of Ken River of Banda District. *Elixir Pollution.*, *42*, 6360–6364.

Kushwaha, N. L., Bhardwaj, A., & Verma, V. K. (2016). Hydrologic response of Takarla-Ballowal watershed in Shivalik foot-hills based on morphometric analysis using remote sensing and GIS. *Journal of Indian Water Resources Society*, *36*(1), 17–25.

Kushwaha, N. L., Elbeltagi, A., Mehan, S., Malik, A., & Yousuf, A. (2022). Comparative study on morphometric analysis and RUSLE-based approaches for micro-watershed prioritization using remote sensing and GIS. *Arabian Journal of Geosciences*, *15*(7). Available from https://doi.org/10.1007/s12517-022-09837-2.

Lecun, Y., Bengio, Y., & Hinton, G. (2015). Deep learning. *Nature*, *521*(7553), 436–444. Available from https://doi.org/10.1038/nature14539, http://www.nature.com/nature/index.html.

Lee, S., Hong, S. M., & Jung, H. S. (2017). GIS-based groundwater potential mapping using artificial neural network and support vector machine models: the case of Boryeong city in Korea. *Geocarto International*, *33*(8), 847–861.

McFeeters, S. K. (1996). The use of the normalized difference water index (NDWI) in the delineation of open water features. *International Journal of Remote Sensing*, *17*(7), 1425–1432. Available from https://doi.org/10.1080/01431169608948714.

Mishra, B., Kumar, P., Saraswat, C., Chakraborty, S., & Gautam, A. (2021). Water security in a changing environment: Concept, challenges and solutions. *Water*, *13*(4), 490. Available from https://doi.org/10.3390/w13040490.

Mogaji, K. A., Omosuyi, G. O., Adelusi, A. O., & Lim, H. S. (2016). Application of GIS-based evidential belief function model to regional groundwater recharge potential zones mapping in hardrock geologic terrain. *Environmental Processes*, *3*(1), 93–123.

Mueller, N., Lewis, A., Roberts, D., Ring, S., Melrose, R., Sixsmith, J., … Ip, A. (2016). Water observations from space: Mapping surface water from 25years of Landsat imagery across Australia. *Remote Sensing of Environment*, *174*, 341–352. Available from https://doi.org/10.1016/j.rse.2015.110.003, http://www.elsevier.com/inca/publications/store/5/0/5/7/3/3.

Othman, F., Alaa Eldin, M. E., & Mohamed, I. (2012). Trend analysis of a tropical urban river water quality in Malaysia. *Journal of Environmental Monitoring*, *14*(12), 3164–3173. Available from https://doi.org/10.1039/c2em30676j.

Rokni, K., Ahmad, A., Solaimani, K., & Hazini, S. (2015). A new approach for surface water change detection: Integration of pixel level image fusion and image classification techniques. *International Journal of Applied Earth Observation and Geoinformation*, *34*(1), 226–234. Available from https://doi.org/10.1016/j.jag.2014.080.014, http://www.elsevier.com/locate/jag.

Rusuli, Y., Li, L., Li, F., & Eziz, M. (2016). Water-level regulation for freshwater management of Bosten Lake in Xinjiang, China. *Water Supply*, *16*(3), 828–836. Available from https://doi.org/10.2166/ws.20160.002.

Sahana, M., & Sajjad, H. (2019). Vulnerability to storm surge flood using remote sensing and GIS techniques: A study on Sundarban Biosphere Reserve, India. *Remote Sensing Applications: Society and Environment*, *13*, 106–120. Available from https://doi.org/10.1016/j.rsase.2018.100.008, http://www.journals.elsevier.com/remote-sensing-applications-society-and-environment/.

Senthilkumar, M., Gnanasundar, D., & Arumugam, R. (2019). Identifying groundwater recharge zones using remote sensing & GIS techniques in Amaravathi aquifer system, Tamil Nadu, South India. *Sustainable Environment Research*, *1*(1). Available from https://doi.org/10.1186/s42834-019-0014-7, https://link.springer.com/journal/42834.

Shen L. Li C. 2010 18th International Conference on Geoinformatics, Geoinformatics 2010, 10.1109/GEOINFORMATICS.2010.5567762, China. Water body extraction from Landsat ETM + imagery using adaboost algorithm.

Sidle, R. C. (2021). Strategies for smarter catchment hydrology models: Incorporating scaling and better process representation. *Tajikistan Geoscience Letters*, *8*(1). Available from https://doi.org/10.1186/s40562-021-00193-9, http://geoscienceletters.springeropen.com/.

Singh, K. P., Malik, A., Mohan, D., & Sinha, S. (2004). Multivariate statistical techniques for the evaluation of spatial and temporal variations in water quality of Gomti River (India) - A case study. *Water Research*, *38*(18), 3980–3992. Available from https://doi.org/10.1016/j.watres.2004.060.011, http://www.elsevier.com/locate/watres.

Skakun, S. (2010). A neural network approach to flood mapping using satellite imagery. Slovak Academy of Sciences. *Ukraine Computing and Informatics*, *29*(6), 1013–1024. Available from http://www.cai.sk/ojs/index.php/cai/issue/archive.

Sun, B., & Yang, X. (2019). Simulation of water resources carrying capacity in Xiong'an New Area based on system dynamics model. *Water (Switzerland)*, *11*(5). Available from https://doi.org/10.3390/w11051085, https://res.mdpi.com/water/water-11-01085/article_deploy/water-11-01085.pdf?filename = &attachment = 1.

Sun, F., Zhao, Y., Gong, P., Ma, R., & Dai, Y. (2014). Monitoring dynamic changes of global land cover types: Fluctuations of major lakes in China every 8 days during 2000–2010. *Chinese Science Bulletin*, *59*(2), 171–189. Available from https://doi.org/10.1007/s11434-013-0045-0, http://link.springer.com/journal/11434, http://www.springerlink.com/content/1001-6538.

Tang, Q., Gao, H., Lu, H., & Lettenmaier, D. P. (2009). Remote sensing: Hydrology. *Progress in Physical Geography*, *33*(4), 490–509. Available from https://doi.org/10.1177/0309133309346650, https://journals.sagepub.com/home/PPG.

Tulbure, M. G., Broich, M., Stehman, S. V., & Kommareddy, A. (2016). Surface water extent dynamics from three decades of seasonally continuous Landsat time series at subcontinental scale in a semi-arid region. *Remote Sensing of Environment*, *178*, 142–157. Available from https://doi.org/10.1016/j.rse.2016.020.034, http://www.elsevier.com/inca/publications/store/5/0/5/7/3/3.

Wang, X., Wang, G., Wu, Y., Xu, Y., & Gao, H. (2017). Comprehensive assessment of regional water usage efficiency control based on game theory weight and a matter-element model. *Water*, *9*(2), 113. Available from https://doi.org/10.3390/w9020113.

Xu, H. (2006). Modification of normalised difference water index (NDWI) to enhance open water features in remotely sensed imagery. *International Journal of Remote Sensing*, *27*(14), 3025–3033. Available from https://doi.org/10.1080/01431160600589179, https://www.tandfonline.com/loi/tres20.

Yang, J., & Du, X. (2017). An enhanced water index in extracting water bodies from Landsat TM imagery. *Annals of GIS*, *23*(3), 141–148. Available from https://doi.org/10.1080/19475683.2017.1340339, http://www.tandfonline.com/toc/tagi20/current.

Yousaf, W., Awan, W. K., Kamran, M., Ahmad, S. R., Bodla, H. U., Riaz, M., ... Chohan, K. (2021). A paradigm of GIS and remote sensing for crop water deficit assessment in near real time to improve irrigation distribution plan. *Agricultural Water Management*, *243*. Available from https://doi.org/10.1016/j.agwat.2020.106443, http://www.journals.elsevier.com/agricultural-water-management/.

Chapter 28

Soil bacteria are more conducive to cowpea (*Vigna unguiculata* L. Walp) growth than the physico-chemical factors in agricultural fields of India

Priya Chatterjee[1], Paramita Mandal[1], Ayan Mondal[2], Soumendranath Chatterjee[1] and Sudipto Mandal[2]

[1]Department of Zoology, The University of Burdwan, Purba Bardhaman, India, [2]Ecology and Environmental Modelling Laboratory, Department of Environmental Science, The University of Burdwan, Purba Bardhaman, West Bengal, India

28.1 Introduction

Cowpea (*Vigna unguiculata* L. Walp) is one of the legumes that serve as fodder and a green manure crop. Various parts of cowpea are consumed by humans and livestock (Maia *et al.*, 2000; Tarawali, Peters, & Blade, 1997), as these are rich in protein, ranging from 22% to 30% (Nielsen, Ohler, & Mitchell, 1997; Tarawali, Peters, & Blade, 1997). Cowpea is known to have the potential to fix nitrogen in the soil and requires little or no nitrogen anthropogenically. Concurrently, the ecological role of soil bacteria toward cowpea growth is indubitable.

With the continuous cropping of the land, there is a gradual depletion of soil nutrients. Thus it requires additional amendments of soil nutrients during its growth for maximum production. Tiwari *et al.* (2008) pointed out that extensive use of chemical fertilizers decreased organic matter and soil fertility and increased soil erosion and environmental degradation. There are several studies in relation to chemical fertilizers on soil and human health, like chronic and acute poisoning, carcinogenicity, mutagenicity, and reproductive defects (Agbulu & Elaigwu, 2013), poor cropping systems (Abbasi & Khizar, 2012; Ogbodo, 2013; Savci, 2012), and groundwater pollution

Water Resources Monitoring, Management, and Sustainability. DOI: https://doi.org/10.1016/B978-0-443-23665-5.00028-4

(Azraf-ul-Haq, Ahmad, & Naeem, 2007). As these fertilizers are considered to release nutrients relatively quickly and enhance the growth of plants and production, their negative effects are often ignored (Azuaga, Wombo, & Agbulu, 2017).

The quality of the soil in terms of chemical (electrical conductivity, soil pH [SpH], and soil organic carbon), physical (bulk density and aggregate stability), and biological activity is greatly enhanced by the use of organic manures in the form of compost (Arif *et al.*, 2014; Fereidooni, Raiesi, & Fallah, 2013; Shrestha, Lal, & Rimal, 2013). This in turn improves the soil fertility and carbon pool of the soil system (Olal, 2015). In addition, cowpea productivity and soil fertility were greatly improved by the combined doses of chemical fertilizer (NPK) and organic manure (Azuaga, Wombo, & Agbulu, 2017; Liu *et al.*, 2017).

Besides the NPK availability, pH and environmental conditions also influence the legume growth (Brady, Edwards, & Asher, 1994; Kunert *et al.*, 2016). The overwhelming effect of the bacterial communities on the growth of cowpea plants was demonstrated in many research works (Deepa, Dastager, & Pandey, 2010; Dimkpa *et al.*, 2008; Yadav *et al.*, 2015; Yang, Kloepper, & Ryu, 2009). Leite *et al.* (2017) put forward that the nonrhizobia bacterial communities are shaped up by the specific soil type.

Gago et al. (2010) advocated the use of artificial neural network (*ANN*) as a tool in modeling intricate and nonlinear relationships contingent on data that are not observed superficially. There are several studies on the application of *ANN* in the field of plant science (Diamantopoulou, 2005; Ferreira & Galo, 2013; Goltsev *et al.*, 2012; Li & He, 2008). Earlier, the *ANN*-based studies were focused on plant stress, the amount of water in plant leaves, and fluorescence kinetics (Goltsev *et al.*, 2012). Barroso *et al.* (2016) used an *ANN* modeling approach to identify the cowpea genotypes with high adaptability and phenotypic stability. They identified cowpea genotypes that were specific to the environment and produced the highest yield.

Hitherto, the studies were based on the application of different types of fertilizers (organic or chemical), the role of leguminous symbiotic and endophytic bacteria, and the application of plant growth-promoting rhizobacteria on cowpea growth and production. However, the importance of physico-chemical factors (Pcf) and the role of nonsymbiotic rhizospheric bacteria that influence cowpea growth at different phenophases are not clearly understood. Here, a study was conducted to understand the relative importance of abiotic (physico-chemical) and biotic (nonsymbiotic rhizospheric bacteria) factors that shape the plant at different growth phases. The main objectives are as follows: (1) to study the diversity and distribution of various groups of nonsymbiotic rhizospheric bacteria under control, chemical, and organic treatment at various phases of cowpea growth; (2) to understand the relationship between Pcf and soil bacteria; and (3) to compare the relative importance of Pcf and nonsymbiotic soil bacterial groups that influence the vegetative, reproductive, and mature phases of cowpea growth.

28.2 Materials and methods

28.2.1 Study site

The research work was carried out in the Crop Research and Seed Multiplication Farm (CRSMF) on the campus of the University of Burdwan, West Bengal, located in the eastern part of India (latitude of 87°50′37.35″ E and longitude of 23°15′7.29″N, with an average altitude of 30 m above mean sea level) during March–June for the 2 consecutive years 2016 and 2017 with only one variety of cowpea (var. *DEB 113, Kashi Kanchan*).

28.2.2 Plot design and treatment

Three replications were done for the treatment plots of control, chemical, and organic, and they were arranged in a randomized block design. The individual plot sizes were 4 × 3 m^2. The seeds were collected from CRSMF, Burdwan, and soaked in distilled water for 1 day. Seeds were sown in the control, chemical, and organic plots on March 10, 2016 and March 15, 2017. The crops were harvested as per the attainment of pod maturity in the organic plots (June 5, 2016 and June 10, 2017), chemical plots (June 25, 2016 and June 27, 2017), and control plots (June 30, 2016 and June 30, 2017). Two seeds were sown per hill at a depth of 0.03–0.04 m with row spacing of 0.3 m and plant-to-plant spacing of 0.2 m. To ensure the uninterrupted flow of irrigation for each individual plot, irrigation channels of 0.45 m wide were made in between the replications.

Since cowpea is a leguminous crop, it needs a small quantity of nitrogen (*N*) for early growth. In all the chemical plots, the recommended doses of *N* in the form of urea at the rate of 20 kg/ha, phosphorus (*P*) in the form of single superphosphate at the rate of 40 kg/ha, and potassium (*K*) in the form of muriate of potash at the rate of 20 kg/ha were applied as a starter dose. The fertilizers were applied by drilling at the time of sowing in such a way that they were placed around 0.06 m below the seed.

In the organic plots, vermicompost and cow dung mixtures in the ratio of 1:1 were applied at a rate of 600 g/m^2, as this combination produced better growth of cowpeas. Control plots were left without any treatment. The same treatments were followed for 2 consecutive years. Hand weeding was carried out twice, at 15–20 DAS (days after sowing) and 45–50 DAS in all the field plots. The first irrigation was made after the seeds were sown. Later, the frequency of irrigation was maintained at a regular interval of 15 days. Meanwhile, the crop did not exhibit any sign of insect/pest attack or disease incidence; hence, no crop protection measures were adopted.

28.2.3 Soil sample collection

Five soil samples were collected from each plot at a depth of 0.05–0.1 m from the rhizosphere zone of cowpea plants. The soil samples were collected using sterilized polythene bags and mixed thoroughly for the determination of microbial diversity. The soil samples were brought to the Parasitology and Microbiology Research Laboratory at the University of Burdwan for microbial analysis.

28.2.4 Physico-chemical factors

Pcf or environmental factors like root temperature (*RT*), air temperature (*AT*), surface solar irradiance, ambient CO_2 (CO_2), relative humidity (*H*), *SpH*, rainfall (*PPT*), and soil water content (*SWC*) were measured using soil and *AT* meters (LUTRON and PTM-816), lux meters (LUTRON and LX-101), CO_2 meters (LUTRON and GC-2028), *SpH* (Eutech, pH, and Test-10), rain gauges, and volumetric method, respectively. The data were collected at 15-day intervals throughout the study period. The average values were used for vegetative, reproductive, and mature phases. In situ measurements of *RT* and *SpH* were done at the rhizosphere zone of the soil. These Pcf influencing cowpea growth were selected after consultation with agricultural scientists, local farmers, and a thorough literature survey.

Soil available *N*, *P*, *K*, and organic carbon (*OrgC*) of baseline and three different phases of control, chemical, and organic plots were analyzed in the laboratory following the alkaline permanganate method (Subbiah & Asija, 1956), Bray II method (Bray & Kurtz, 1945), flame photometry (Rich, 1965), and modified wet digestion method (Walkley & Black, 1934), respectively. The soil texture of the field plots was dominated by silt, followed by clay and sand.

28.2.5 Nonsymbiotic rhizospheric bacteria (Sb) analysis

The soil samples were analyzed for different bacterial groups like heterotrophic bacteria (*Hb*), nitrate-reducing bacteria (*Nr*), phosphate-solubilizing bacteria (*Ps*), nitrifying bacteria (*Ni*), asymbiotic nitrogen-fixing bacteria (*ANf*), fat hydrolyzing bacteria (*Fh*), and starch hydrolyzing bacteria (*Sh*). The soil samples were kept in sterilized polythene bags and stored at 4°C for a brief period, where the microbiological analysis was not performed immediately. The *Hb* group was determined by diluting the soil samples up to 10^{-3}, and a 20 μL soil suspension (10^{-3}) was mixed with 100 mL nutrient agar (peptone 5 g/L, beef extract 3 g/L, agar 2 g/L, pH 7) and incubated at 30°C $\pm$ 1°C in the BOD incubator. The *Sh* group was enumerated by incubating the soil on starch agar media for 24 h. Those bacterial colonies were counted that produced a halo zone after fusing with Gram's iodine. The *Ni* group was determined after 5–30 days (5-day intervals) from the date of incubation. The bacterial population was enumerated on Winogradsky's medium containing $(NH_4)_2SO_4$ (1.0 g/L), and the colonies were identified (pink color) by labeling the plates with sulfanilic acid reagent. The *Ps* group was determined by observing the halo zone formation around the colonies on the insoluble phosphate $[Ca_3(PO_4)_2]$-containing medium. The *ANf* group was enumerated in a nitrogen-free medium. The *Nr* and *Fh* were cultured in nitrate agar medium and spirit blue agar medium with TWEEN 20, respectively. These groups of bacterial populations were enumerated after 1–3 days of incubation (Azmi & Chatterjee, 2016; Chatterjee, Syed, & Mukhopadhyay, 2014; Lacey, 1997; Ogbodo, 2013).

28.2.6 Plant growth indicators

The phenological events of the cowpea plant were divided into three phases. The vegetative phase (7 DAS to 45 DAS) comprises six stages: VE: seedling emergence; VC: 2 cotyledons and primary leaves at nodes 1 and 2, V1 to V4: 1st and 4th trifoliate unfolded at nodes 2nd and 6th plus branching. The reproductive phase (46 DAS to 75 DAS) comprises six stages, R1 to R6: From one open flower on the plant (early bloom) to a 50% pod with fully developed seeds. The mature phase (76 DAS to 110 DAS) comprises two stages: R7 and RH: From one pod at mature color (physiological maturity) to 80% of pods at mature color (harvest mature).

Plant morpho-physiological attributes like plant height (*PH*), leaf area index (*LAI*), number of flowers per plant (*NOF*), and number of pods per plant (*POP*) were measured directly from the field.

28.2.7 Statistical analysis

To understand the association among the bacterial population groups in the control, organic, and chemically treated plots of the experimental field, hierarchical cluster analysis was done using the Unweighted Pair Group Method with Arithmetic Mean method with Morisita index and 10,000 bootstraps. The model used different sets of data for adjusting the weights, called training sets, and some parts of the randomly assigned data were used to test the performance of the model, called test data sets.

The *ANN* model was separately applied for vegetative, reproductive, and mature phases using the hyperbolic tangent function for hidden layer activation and the identity function for output layer activation on the SPSS platform. In the model, *Pcf* and soil bacteria (*Sb*) were considered as inputs or covariates, and plant growth indicators were considered output or dependent variables. Here, a multilayered perceptron neural network was performed using the aforesaid covariates. The following equations (Eqs. 28.1–28.3) were used to construct the *ANN* model:

$$pcf \times w = \sum_{n=0}^{i} pcf_n \times w_n \tag{28.1}$$

or,

$$Sb \times w = \sum_{n=0}^{i} Sb_n \times w_n \tag{28.2}$$

where *Pcf* are the physico-chemical factors ($Pcf_n = Pcf_0.Pcf_i$), *w* is the synaptic weight assigned for each *Pcf* ($w_n = w_0.w_i$) or *Sb* ($w_n = w_0.w_i$).

$$PGI = \varphi(PcforSb \times w) \tag{28.3}$$

where φ is the hidden function.

The diversity ordering technique developed by Patil and Taillie (1977) was applied to bacterial groups. Here, multiple diversity measures were implemented at once, and communities or samples were displayed as curves. These diversity curves are called diversity profiles and allow a visual and quantitative comparison of multiple aspects of diversity. The exponential Renyi diversity profiles for the different treatment types were done using PAST software. H-alpha (Hα) is the Renyi diversity index. Alpha (α) is a scale parameter ranging from zero to infinity. Diversity profile values (H-alpha) were calculated from the frequencies of each treatment plot and a scale parameter (α) ranging from zero to infinity (Tóthmérész, 1995). The four diversity measures are related to the respective values of the scale parameter α: H_0 = species richness, H_1 = Shannon Diversity, H_2 = Simpson Diversity, and H_∞ = Berger-Parker Index (Legendre & Legendre, 1998). The principal component analysis (*PCA*) was done using STATISTICA software.

28.3 Results

28.3.1 Distribution of soil bacteria across vegetative, reproductive, and mature phases of cowpea

The result of the hierarchical cluster analysis (Fig. 28.1) indicated the distribution pattern of seven bacterial groups across vegetative, reproductive, and mature phases in three distinct clusters. The first cluster was formed by the *Hb*, *Nr*, and *Ps* groups of bacteria. The *Ps* group at the mature phase stood apart from the other two clusters and formed a separate cluster alone. The third cluster was formed by the assemblage of *Ni*, *ANf*, *Fh*, and *Sh* groups of bacteria. The clusters correspond to the trend of abundance of bacterial groups across vegetative, reproductive, and mature phases. Within the first cluster, the nitrate-reducing group of bacteria at vegetative (*NrV)* and reproductive (*NrR*) phases were quite alike, forming a common clade with a

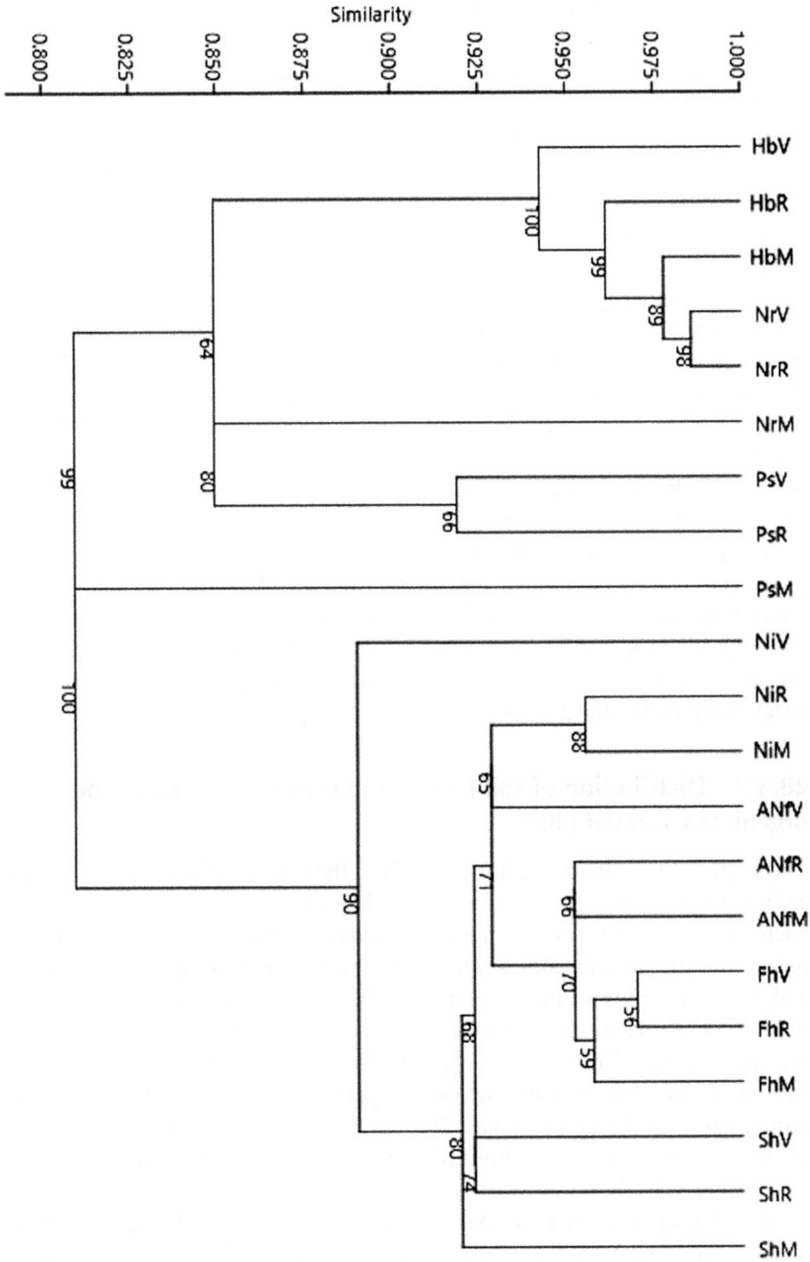

FIGURE 28.1 Dendrogram representation of hierarchical cluster analysis using the UPGMA method and Morisita index.

similarity of 98%. However, at the mature (*NrM*) phase, the population was distantly related to this clade. The heterotrophic bacterial groups at vegetative (*HbV*), reproductive (*HbR*), and mature phases (*HbM*) were similar with a little difference. The phosphate-solubilizing (*Ps*) group at the vegetative phase (*PsV*) and reproductive phase (*PsR*) showed 66% similarity, forming a separate subcluster, but this subcluster was distantly related to the population at the mature phase (*PsM*).

The nitrifying bacterial group (*Ni*) of the soil showed a different picture. Here, at the reproductive (*NiR*) and mature (*NiM*) phases, the populations showed a similarity of 88%, forming a separate subcluster. The subcluster was far from the population at the vegetative phase (*NiV*). A similar trend was shown by the asymbiotic nitrogen-fixing bacteria (*ANf*) in soil. The subcluster was formed by the bacteria at reproductive phase (*ANfR*) and mature phase (*ANfM*) with 66% similarity, while at vegetative phase (*ANfV*), the same bacterial group stood far away from the subcluster formed by *ANfR* and *ANfM*. In the fat hydrolyzing group (*Fh*) at vegetative and reproductive phases of plant growth, *FhV* and *FhR* showed 56% similarity, while in the mature phase (*FhM*), the bacterial population showed 59% similarity to the population at vegetative and reproductive phases. A common subcluster was formed by the starch hydrolyzing bacterial group (*Sh*) at vegetative (*ShV*) and reproductive (*ShR*) phases with 74% similarity. The same bacterial group, at the mature phase (*ShM*), showed 80% similarity with the other bacterial group of the third cluster.

28.3.2 Distribution of soil bacteria in control, chemical, and organically treated plots

The bacterial populations (cfu/g $\times$ 10^5) in the control, chemical, and organically treated plots showed some interesting variations. At control plots, the bacterial groups, *Ni*, *ANf*, *Fs*, and *Sh*, showed a significant increase in number across the growth phases. However, the bacterial population of *Hb*, *Nr*, and *Ps* decreased from the vegetative to the mature phase (Fig. 28.2A).

At chemically treated plots, only *Fs* showed a rise in population. An initial rise was noted for *Nr*, *Ps*, *ANf*, and *Sh* from the vegetative to the reproductive phase, but a decrease in the population was observed from the reproductive to the mature phase. The decrease was insignificant in the case of *Ni*. The *Hb* population showed a gradual decrease throughout the study period (Fig. 28.2B).

The bacterial groups of *Nr*, *Ni*, *ANf*, *Fs*, and *Sh* showed a gradual increase in population throughout the growth phases. However, the *Ps* group showed an initial decrease in population from the vegetative to reproductive phase, but a significant rise in population (37%) was noted from the reproductive to mature phase. A little rise in population was observed in the case of the *Hb* group (Fig. 28.2C). The organically treated plots showed an

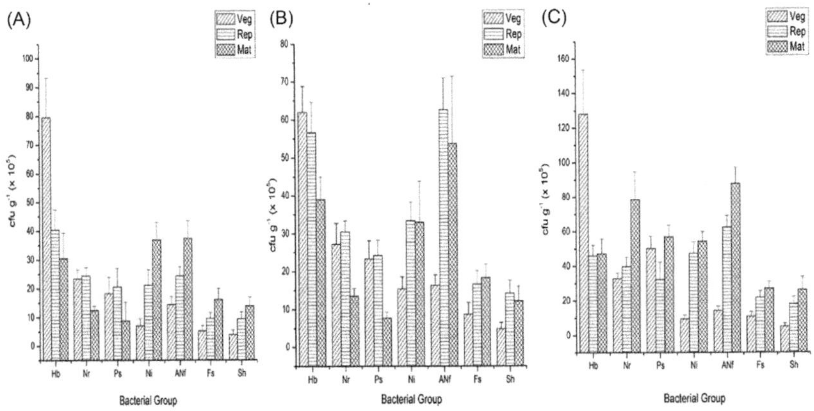

FIGURE 28.2 (A) Comparison of relative abundance of soil bacterial groups at vegetative (Veg), reproductive (Rep) and mature (Mat) phases in the (A) control, (B) chemical, and (C) organic field plots. The X-axis indicates the bacterial groups and the Y-axis indicates the bacterial populations (cfu/g $\times$ 10^5).

overall increase in bacterial population in all three phases when compared to the chemical and control plots.

28.3.3 Diversity plots of bacterial populations in control, chemical, and organically treated plots

The Renyi diversity profiles serve as an indicator of richness, heterogeneity, dominance, and evenness. The diversity profile was calculated for control, chemical, and organically treated plots. At $\alpha = 0$, the profiles were not distinctly separated, indicating that the richness of the bacterial groups (H_0) is the same for the three treatments. At $\alpha = 1$, the separation of the profiles was negligible, indicating low Shannon's diversity (H_1). The dominance indices (H_2 and larger) showed that the control values were largely different from chemical and organically treated plots (Fig. 28.3).

28.3.4 Principal component analysis

The *PCA* result indicated that the first and second components, or descriptors, explained 37.18% and 25.19% of the total variance, with eigenvalues of 7.06 and 4.78, respectively. The first descriptor was strongly correlated with 14 of the original variables. The first descriptor was positively correlated with *Ni* (the nitrifying bacteria group) with a value of 0.896, followed by *ANf* (0.874), *SWC* (0.836), *Sh* (0.637), *Ps* (0.590), *IR* (0.557), and *H* (0.555). Negative correlations were observed with *ST* (0.779), *OrgC* (0.750), *Nr* (0.739), *P* (0.672), *N* (0.620), *AT* (0.596), and *SpH* (0.557).

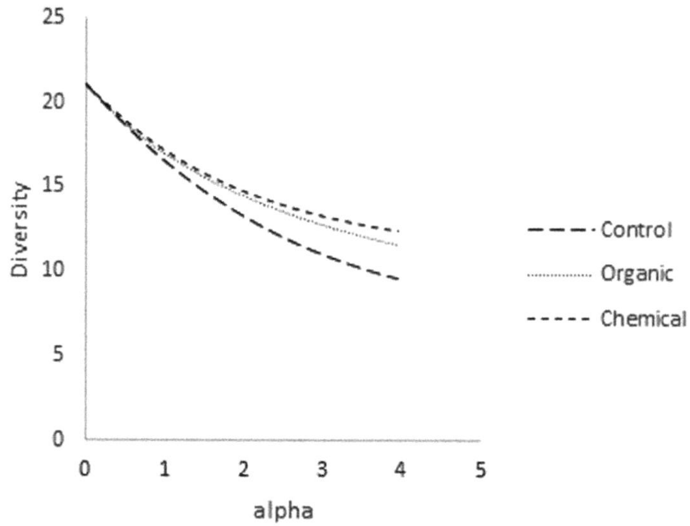

FIGURE 28.3 Diversity profiles of soil bacterial groups at control, organic, and chemical plots using the Renyi index. The Y-axis represents diversity and the X-axis represents "alpha." When the "alpha" = 0, the function shows the richness of nonsymbiotic rhizospheric soil bacterial groups; when the "alpha" = 1, the function shows the Shannon index for bacterial groups; when the "alpha" = 2, the function shows the evenness index among the bacterial groups.

The second descriptor was positively correlated with *IR*, *Ps*, and *SpH* with values of 0.754, 0.558, and 0.551, respectively. Negative correlations were observed with *PPT* (0.906), CO_2 (0.763), *Hb* (0.798), *Fh* (0.756), and *K* (0.525).

The close association was indicated by the four clusters, as shown in Fig. 28.4. Environmental factors like *OrgC*, *RT*, *AT*, and *SpH* were correlated. Similarly, *Ps* and *Ni* were correlated with *IR*. *Hb* and *Fh* were closely related to *K*, CO_2, and *PPT*. The *Sh* and *ANf* groups were correlated with *H* and *SWC*. The *Nr* group stood apart from other groups and was weakly correlated with *N* and *P*.

28.3.5 The artificial neural network model

To understand the influence of Pcf and nonsymbiotic rhizospheric bacteria on cowpea growth distinctly, separate *ANN* models were constructed for the abiotic factors and soil bacterial groups as inputs or covariates and the growth indicators like *PH*, *LAI*, *POP*, and *NOF* as outputs or dependent variables.

The sensitivity analysis of the *ANN* models of vegetative, reproductive, and mature phases of soil bacteria is shown in Fig. 28.5 (A, B, and C), respectively. The network structures of the *ANN* models are shown in Fig. S1a, S1b, and S1c. In the vegetative phase, two output variables were considered: *PH* and *LAI*. In addition to these dependent variables, *NOF* and

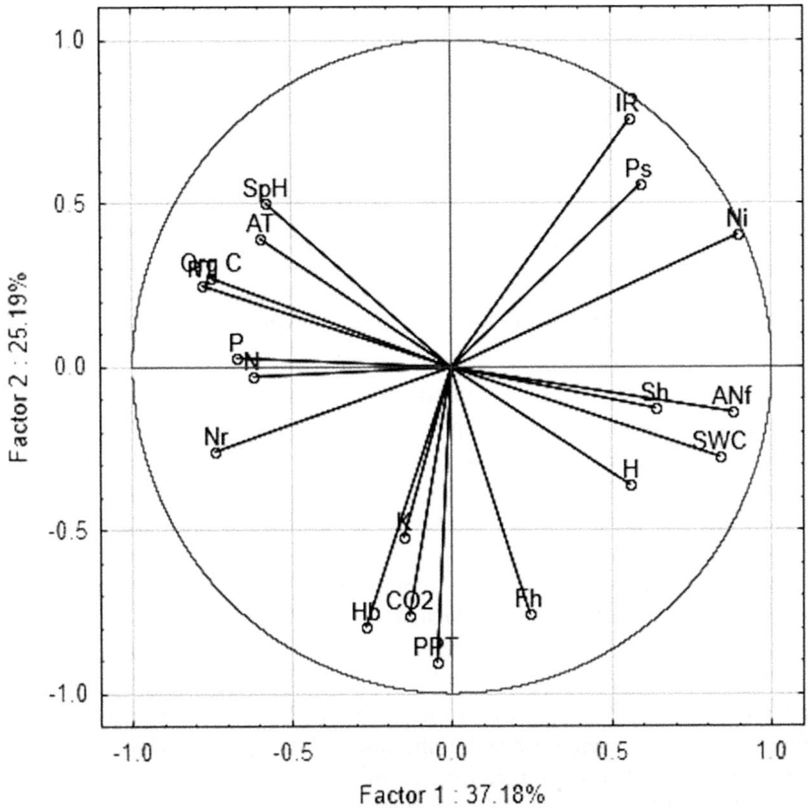

FIGURE 28.4 PCA biplot of abiotic factors and nonsymbiotic rhizospheric soil bacterial groups influencing cowpea growth. *PCA*, Principal component analysis.

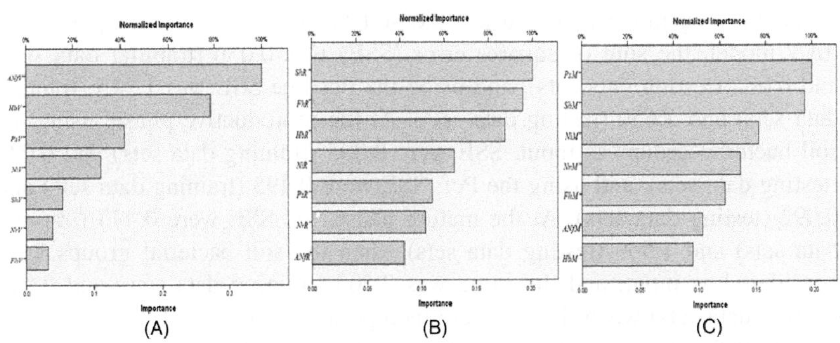

FIGURE 28.5 The sensitivity analysis of nonsymbiotic rhizospheric soil bacterial groups at (A) vegetative phase, where the *ANf* group of bacteria is the most sensitive factor; (B) reproductive phase, where the *Sh* group of bacteria is the most sensitive factor; and (C) mature phase, where the *Ps* group of bacteria is the most sensitive factor.

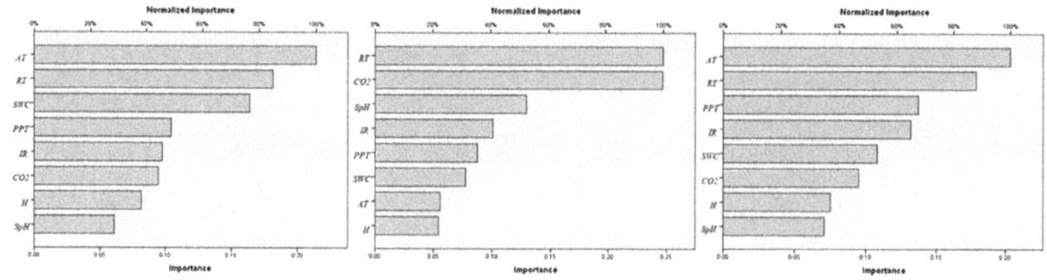

(A) (B) (C)

FIGURE 28.6 The sensitivity analysis of physico-chemical factors at (A) the vegetative phase, where the *AT* is the most sensitive factor; (B) the reproductive phase, where the *RT* is the most sensitive factor; and (C) the mature phase, where the *AT* is the most sensitive factor.

POP were considered at the reproductive and mature phases, respectively. During the computation of synaptic weights between covariates and dependent variables, four hidden layers were noted for vegetative and mature phases, while five hidden layers were observed in the case of the reproductive phase. Results indicated that the *ANf* group was the most sensitive bacteria during the vegetative phase. The *Sh* and *Ps* groups of bacteria were the sensitive bacterial groups during reproductive and mature growth phases, respectively. Their R^2 values indicated a good fit between the observed and predicted values of the output variables for all the cases ($P < .05$).

The sensitivity analysis of the *ANN* models of vegetative, reproductive, and mature phases of Pcf is shown in Fig. 28.6 (A, B, and C), respectively. Here, four hidden layers were observed in the case of the vegetative phase, while five layers were detected in the reproductive and mature phases. Sensitivity analysis showed that *AT* was the governing factor for the vegetative and mature growth of plants, and *RT* was the most sensitive factor during reproductive growth. The observed and predicted results of the output variables were in good agreement ($P < .05$). The sensitive factors determined through the *ANN* model were the same for control, chemical, and organically treated field plots.

At the vegetative phase, using the soil bacterial groups as input in the *ANN* model, the sum of squares error (SSE) was 0.036 (training data sets) and 0.002 (testing data sets), and using the Pcf, the SSE was 1.945 (training data sets) and 1.650 (testing data sets). At the reproductive phase, using the soil bacterial groups as input, SSE were 0.005 (training data sets) and 0.035 (testing data sets), and using the Pcf, SSE were 0.495 (training data sets) and 0.092 (testing data sets). At the mature phase, the SSE were 0.473 (training data sets) and 1.569 (testing data sets) when the soil bacterial groups were considered as input, and the same was 2.505 (training data sets) and 2.013 (testing data sets) when the Pcf were incorporated as input.

The Pcf and the morpho-physiological attributes of cowpea plants at vegetative, reproductive, and mature phases are detailed in Table 28.1. The average values of Pcf associated with the control, chemical, and organically treated plots are shown in Table 28.2. The Pcf like *AT*, CO_2, *IR*, *H*, and *PPT*

TABLE 28.1 Average values of physico-chemical factors and growth indicators during vegetative, reproductive, and mature phases of cowpea.

Physico-chemical factors (unit)	Description	Vegetative phase mean ± SE	Reproductive phase mean ± SE	Mature phase mean ± SE
RT (°C)	Root temperature	36.18 ± 0.29	29.87 ± 0.35	28.4 ± 0.38
AT (°C)	Air temperature	33.72 ± 0.24	31.88 ± 0.82	30.1 ± 0.53
CO_2 (ppm)	Ambient carbon dioxide	408.60 ± 9.62	396.2 ± 2.81	439.6 ± 11.09
IR (lux)	Surface solar irradiance	58,020 ± 8909.43	76,480 ± 17,970.74	62,860 ± 11,143.44
SWC (%)	Soil water content	26.30 ± 0.19	56 ± 3.90	64 ± 3.53
H (%)	Relative humidity	51.97 ± 3.61	63.32 ± 0.87	77.02 ± 2.48
SpH	Soil pH	6.34 ± 0.03	5.76 ± 0.08	5.79 ± 0.08
PPT (mm)	Rainfall	1.64 ± 0.91	1.92 ± 0.08	9.68 ± 2.75
Growth indicators				
PH (cm)	Plant height	7.72 ± 0.41	97.58 ± 14.41	119.01 ± 9.92
NOF	Number of flowers	–	5.54 ± 0.64	–
POP	Number of pods per plant	–	–	35.73 ± 2.29
LAI	Leaf area index	109.12 ± 25.10	513.10 ± 206.39	808.76 ± 127.3

TABLE 28.2 Average values of the physico-chemical factors and growth indicators at control, organic, and chemical field plots.

Physico-chemical factors (unit)	Description	Control mean ± S.E.	Organic mean ± S.E.	Chemical mean ± S.E.
RT (°C)	Root temperature	31.09 ± 3.73	32.49 ± 3.21	31.25 ± 2.18
AT (°C)	Air temperature	31.9 ± 3.50	31.9 ± 3.50	31.9 ± 3.50
CO_2 (ppm)	Ambient carbon dioxide	414.8 ± 37.26	414.8 ± 37.26	414.8 ± 37.26
IR (lux)	Surface solar irradiance	$65,786.67 \pm 5262.93$	$65,786.67 \pm 5262.93$	$65,786.67 \pm 5262.93$
SWC (%)	Soil water content	55.56 ± 2.77	52.33 ± 1.56	38.73 ± 2.71
H (%)	Relative humidity	64.10 ± 5.76	64.10 ± 5.76	64.10 ± 5.76
SpH	Soil pH	5.50 ± 0.20	6.38 ± 0.40	6.06 ± 0.46
PPT (mm)	Rainfall	4.41 ± 0.35	4.41 ± 0.35	4.41 ± 0.35
N (kg/ha)	Available nitrogen	149 ± 17.88	164 ± 18.04	133 ± 21.28
P (kg/ha)	Available phosphorus	253 ± 27.83	254 ± 20.32	260 ± 33.8
K (kg/ha)	Available potassium	601.5 ± 90.22	502.5 ± 35.14	466.5 ± 55.92
OrgC (%)	Organic carbon	0.92 ± 0.04	0.93 ± 0.09	0.62 ± 0.07
Growth indicators				
PH (cm)	Plant height	44.53 ± 2.67	96.2 ± 6.73	87.84 ± 7.02
NOF	Number of flowers	3.6 ± 0.10	7.4 ± 0.37	6 ± 0.12
POP	Number of pods per plant	26.4 ± 2.63	43.8 ± 0.87	37 ± 4.44
LAI	Leaf area index	191.22 ± 21.01	690.58 ± 62.1	563.72 ± 45.09

were measured on the same date at control, chemical, and organically treated plots, so the average values were found to be the same in these plots.

28.4 Discussion

28.4.1 Soil bacterial distribution and diversity

The soil bacteria harbored by legume rhizobia and nonrhizobia have been studied by many authors (Girvan et al., 2003; Marschner et al., 2001). The isolation of bacteria from legume nodules has been studied since 1888 (Beijerinck, 1890; Frank, 1889). Previous research revealed that there exist more than a hundred described species of bacteria (Peix et al., 2015). De Meyer et al. (2015) found that there exists a higher diversity of nonsymbionts than the symbionts in the rhizosphere of leguminous plants, which are responsible for growth promotion. Moreover, Leite et al. (2017) added that these nonsymbiont bacterial communities are reformed by soil types. Zerihun and Haile (2017) worked on various combinations of fertilizer and manure in soybean production at different ranges of rainfall and concluded that soybean production is affected by the application of low organic manure and high chemical fertilizer doses during the short rainy season and that high organic manure input is needed in the long rainy season. In the present study, the combination of vermicompost and cow dung at a 1:1 ratio in the organically treated plots showed a significant rise in the overall bacterial population. The above combination showed a high yield of maize in the earlier study (Chatterjee *et al.*, 2019). The results indicated the variation in abundance of the bacterial groups under chemical and organic treatments. Thus the treatment of chemical fertilizer and organic manure on cowpea field plots was assumed to alter the nonsymbiotic bacterial communities in the cowpea rhizosphere.

The cluster analysis revealed the close association of soil bacterial groups at different phases of cowpea growth. The *Ni* and *ANf* groups of bacteria were found to show similar trends of abundance at the reproductive and mature phases of cowpea growth, while the other groups of bacteria showed similarities during the vegetative and reproductive phases. The *Ps* group of bacteria at the mature phase stood separate from all other clusters formed by the groups of soil bacteria.

The Renyi diversity profiles of the study area revealed that the three curves of chemical, organic, and control fields were nonoverlapping. The overall diversity of chemical and organically treated plots was always higher than the control plots. Though the curve of chemically treated plots was closer to that of organically treated plots, it appears that chemically treated plots were less diverse than organic and more diverse than the control. Since all the plots have an equal number of bacterial groups, the plots did not differ in terms of richness. All the profile curves were curvilinear and concave, although the degree of concavity varied among the three types of treatment. The slope of the curve of chemical and organically treated plots was less steep in comparison to control plots, indicating more evenness (Fig. 28.3).

28.4.2 Relationship between physico-chemical factors and soil bacteria

The *PCA* study revealed a close association of Pcf like *SWC* and *H* with the bacterial groups *ANf* and *Sh*. Studies indicated that high *SWC* and copious supply of plant residue promote the growth of the *ANf* group of bacteria. The physiological state of microorganisms and plants is affected by *SWC* (Walker *et al.*, 2003). Moreover, the *SWC* is also essential for the active state of enzymes present in the soil (Jiang & Zhang, 2002). Borowik and Wyszkowska (2016) showed that the optimum level of *SWC* for the growth of nonsymbiotic bacteria is 40%. In the present study, the average *SWC* was higher in the organic than the chemically treated plots throughout the study period. Consequently, the alpha-amylase activity by the *Sh* group of bacteria was distinct in the organic plots when compared to the chemical plots. The solar irradiance influences the *RT* and indirectly regulates the activity of the *Ps* and *Ni* groups of bacteria. Sahrawat (2008) studied the rate of the nitrification process at different levels of *RT* in agricultural fields. The study showed the appearance of a peak in the nitrification rate around 35°C. The increase in *RT* influences the microbial activity of soil over the typical ambient temperature range (Gilliam, Galloway, & Sarmiento, 2015). *RT* determines both the fate of available nitrogen and the quantity of nitrogen content present in the soil for plant use. The optimum temperature for the nitrification activity of ammonia oxidizers ranges from 20°C to 37°C. The present study was conducted between March and June, which represents the summer season, and the *RT* for all the treatments ranged between 31°C and 33°C. This shows that the activity of the *Ps* and *Ni* groups was in favorable condition during the study period. Phosphate solubilization and nitrification may be considered important soil processes regulating cowpea growth. Rainfall, or *PPT*, releases labile carbon and nitrogen substrates into the soil through heterotrophic bacterial activity (Cruz-Martínez *et al.*, 2012). Deng et al. (2017) showed that *PPT* increases the phospholipid fatty acid biomass of soil. Thus it can be concluded that there exists a close relationship among *PPT*, *Fh*, and *Hb* in the agricultural fields.

28.4.3 Sensitive factors of cowpea growth at vegetative, reproductive, and mature phases

The *ANN* model shows that the *ANf* is the most sensitive group at the vegetative phase among all the bacterial groups considered in the study. In the model, 55 training sets for bacteria and 62 training sets for Pcf were used to get the final results with the minimum error. Most of the biological nitrogen fixation is done by the diazotrophs in association with legumes. However, the *ANf* group has a remarkable role in fixing nitrogen under certain soil conditions (Bürgmann *et al.*, 2004). This is true specifically in organically managed soils, where the available nitrogen content is low (Rosen & Allan, 2007). In the present study, the average baseline nitrogen was far below the critical limit (240 kg/ha). In addition to nitrogen fixation, the *ANf* group promotes the secretion of hormones such as auxins, gibberellins, and cytokinin (Akhetar et al., 2012). Thus the *ANf* group of bacteria was relatively important during the vegetative growth of cowpeas.

Soil polysaccharide is composed of starch and cellulose (Murakami, 1988). Starch is frequently used as a model substrate to study the hydrolysis of polysaccharides (Karahan *et al.*, 2006b; Karahan, Van Loosdrecht, & Orhon, 2006a; Ubukata, 1999). The starch hydrolyzing bacteria release amylase, an exoenzyme that degrades starch into a smaller disaccharide, maltose (Karahan *et al.*, 2006b; San Pedro, Mino, & Matsuo, 1994; Ubukata, 1999). Further, the maltase enzyme acts on maltose and converts it into glucose, which is utilized by soil microbes (Thomsen, Kong, & Nielsen, 2007). The *ANN* model results revealed that the bacterial group *Sh* was abundant and relatively important during the reproductive phase of cowpea growth. Besides the role of photoperiod, temperature and phytohormones that regulate floral evocation and morphogenesis, the *Sh* group of bacteria was assumed to play some role during the reproductive phase of cowpea.

Phosphate solubilization is one of the important plant growth-promoting traits of soil bacteria (Souza, Ambrosini, & Passaglia, 2015). Salim, Joseph, and Jisha (2008) studied the role of *Ps* on growth parameters, nodulation, and nutrient uptake of cowpea and found that there was a significant increase in the nutrient uptake by seeds, nodule number, and number of seeds per pod. They concluded that this increase was attributed to *Ps* addition to soils. Singh, Singh, and Khanna (2016) studied the influence of *Ps* on the growth of lentil plants and found that *Ps* was responsible for the increase in seed weight, pod maturity, and number of pods per plant. The present study was in agreement with the above statements. The model showed that the *Ps* group was relatively more important than the other bacterial groups at the mature phase of cowpea growth.

Crops require basic *AT* to complete a specific phenophase or the whole life cycle. Among the Pcf, *AT* appeared to be the most sensitive factor during the vegetative and mature phases. Covell et al. (1986) experimented on four varieties of legumes, including cowpea, and suggested that the optimal temperature for the seed germination of legumes varies in all varieties. The rate of production of nodes and leaves and leaf appearance rate during the vegetative phase of cowpeas showed a linear relationship with average *AT*, higher than the base *AT*, with no effect of photoperiod (Craufurd, Subedi, &

Summerfield, 1997). Butler et al. (2014) studied the effect of *AT* on legume germination and found that the greatest germination of seeds occurred at around 25°C for warm-season legumes. In the present study, the *AT* during the vegetative phase was considered to be ideal for cowpea seed germination in the tropical region. Luo (2011) identified temperature thresholds and the impact of supra-optimal temperature thresholds on crop yield and growth components for various phenological stages of legume crops; the study reviewed different *ATs* for the maturity of legumes. For peanuts or groundnut, the optimum temperature for the fruit set was found to be <33°C (Prasad *et al.*, 2003). In the present study, the average *AT* for all the treatment plots was found to be 31.9 ± 1.67°C, which was ideal for pod maturity (physiological as well as harvest maturity) in cowpea.

Stewart and Summerfield (1978) showed the existence of a direct relationship between *RT* and reproductive growth in cowpea. They showed that the variation in diurnal *RT* affects the floral morphology. In addition, they found that floral abnormalities are more prominent when the *RT* difference between day and night is greater. Normal flowers appear when this difference is less. In the present study, a narrow difference was found between day and night root zone temperatures during the reproductive phase, as the season approaches summer in the tropics. Greer et al. (2006) studied the effect of root zone temperature on bud break and flowering of *Malus domestica* and apples and found that the relationship between the formation of floral clusters and bud break was directly proportional to root zone temperatures. The *ANN* model showed that the *RT* was the most important environmental factor during the reproductive growth of cowpeas. The values of *RT* observed during the field survey were in a close range for control, chemical, and organically treated plots throughout the reproductive phase.

Overall, the soil bacterial groups were found to be more effective than the physico-chemical groups in regulating cowpea growth. In the present study, the values of *N*, *P*, *K*, and *Org C* were not used in the *ANN* model because there was no significant difference between the values of the aforementioned environmental variables on two consecutive sampling dates during the period of study. Similarly, among the growth parameters, root biomass, and root length, nodules per plant were not included in the model construction due to the inadequacy of the data sets. An attempt was made to understand the effect of physical, chemical, and biological components of the environment on the cowpea growth phases. In the near future, more elaborate studies are required to understand the intricate processes involved in the growth of legumes. The present study was based on the local climate situation in the tropics. The *ANN* model may behave differently if the same model is applied to other latitudes of the globe.

28.5 Conclusions

Apart from the endosymbionts and rhizobia, the Pcf and the association of various nonsymbiotic rhizospheric soil bacterial groups play a pivotal role in the growth regulation of cowpeas. Among the bacterial groups, the *ANf*, *Sh*, and *Ps* were relatively important in terms of activity at the vegetative, reproductive, and mature phases, respectively. Among the Pcf, *AT* and *RT* were the most sensitive factors during cowpea growth in control, chemical, and organically treated plots. The *PCA* study revealed a close association among the bacterial groups *ANf* and *Sh* with the Pcf, *SWC*, and *H*. A similar relationship was observed between the *IR* and *Ps* groups of bacteria. During the run of *ANN* models, the average sum square errors of the nonsymbiotic soil bacterial groups were far less when compared to the average sum square errors of Pcf considered in the model. This indicates that the phenophases of cowpea growth were largely regulated by soil bacterial groups over the Pcf. Moreover, the study helps to understand the underlying role of soil bacteria that shapes the legume growth phases.

Acknowledgments

The author (Priya Chatterjee) is thankful to the Department of Zoology at the University of Burdwan for giving all sorts of laboratory facilities to conduct this research. One of the authors (Sudipto Mandal) greatly acknowledges the support from SERB, DST, Govt. of India, New Delhi (Project No. EMR/2016/002618) to carry out the research work. Thanks are due to the agricultural scientists of the Department of Agriculture, Govt. of West Bengal, and Dr. Sabhyasachi Patra, CRSMF, the University of Burdwan for extending their help to carry out the research. All the authors thank Mr. Debomay Chanda for editing the figures during manuscript preparation.

References

Abbasi, M. K., & Khizar, A. (2012). Microbial biomass carbon and nitrogen transformations in a loam soil amended with organic-inorganic N sources and their effect on growth and N-uptake in maize. *Ecological Engineering*, *39*, 123–132. Available from https://doi.org/10.1016/j.ecoleng.2011.12.027.

Agbulu, N. O., & Elaigwu, O. A. (2013). *Environmental education and management for students in tertiary institutions in Nigeria*. Makurdi: SAP Publishing House.

Akhetar, M. S., Hossain, S. J., Hossain, S. A., & Datta, R. K. (2012). Isolation and characterization of salinity tolerant Azotobacter sp. *Greener Journal of Biological Sciences*, *2*, 43–51.

Arif, M., Jalal, F., Jan, M. T., & Muhammad, D. (2014). Integration of biochar and legumes in summer gap for enhancing productivity of cereal based cropping system. *Sarhad Journal of Agriculture*, *30*(4), 393–403.

Azmi, S. A., & Chatterjee, S. (2016). Population dynamics of soil bacteria in some areas of Midnapore coastal belt, West Bengal, India. *3 Biotech*, *6*(1), 1–7. Available from https://doi.org/10.1007/s13205-015-0361-y, http://www.springerlink.com/content/2190-572x/.

Azraf-ul-Haq, A., Ahmad, R., & Naeem, M. (2007). Production potential and quality of mixed sorghum forage under different intercropping systems and planting patterns. *Pakistan Journal of Agricultural Sciences.*, *44*, 203–207.

Azuaga, C. I. J., Wombo, A. B., & Agbulu, O. N. (2017). Influence of organic and inorganic fertilizers on the production of cowpea (Vigna unguiculata) in Taraba State. Nigeria. *Journal of Agricultural and Food Sciences*, *5*(4), 53–59.

Barroso, L. M. A., Teodoro, P. E., Nascimento, M., Torres, F. E., Nascimento, A. C. C., Azevedo, C. F., & Teixeira, F. R. F. (2016). Using artificial neural networks to select upright cowpea (Vigna unguiculata) genotypes with high productivity and phenotypic stability, Fundacao de Pesquisas Cientificas de Ribeirao Preto, Brazil*Genetics and Molecular Research*, *15*(4). Available from https://doi.org/10.4238/gmr15049049, http://www.geneticsmr.com/sites/default/files/articles/year2016/vol15-4/pdf/gmr-15-gmr15049049.pdf.

Beijerinck, M. W. (1890). Künstliche infection von Vicia Faba mit *Bacillus radicicola*. Ernährungsbedingungen dieser Bacterie. *Bot. Ztg*, *52*, 837–843.

Borowik, A., & Wyszkowska, J. (2016). Soil moisture as a factor affecting the microbiological and biochemical activity of soil. *Czech Academy of Agricultural Sciences, Poland Plant, Soil and Environment*, *62*(6), 250–255. Available from https://doi.org/10.17221/158/2016-PSE, http://www.agriculturejournals.cz/publicFiles/187620.pdf.

Brady, D. J., Edwards, D. G., & Asher, C. J. (1994). Effects of aluminium on the peanut (*Arachis hypogaea* L.)/ Bradyrhizobium symbiosis. *Plant and Soil*, *159*(2), 265–276. Available from https://doi.org/10.1007/BF00009289.

Bray, R. H., & Kurtz, L. T. (1945). Determination of total, organic, and available forms of phosphorus in soils. *Soil Science*, *59*(1), 39–45. Available from https://doi.org/10.1097/00010694-194501000-00006.

Butler, T. J., Celen, A. E., Webb, S. L., Krstic, D., & Interrante, S. M. (2014). Temperature affects the germination of forage legume seeds. *Crop Science Society of America, United States Crop Science*, *54*(6), 2846–2853. Available from https://doi.org/10.2135/cropsci2014.01.0063, https://www.crops.org/publications/cs/pdfs/54/6/2846.

Bürgmann, H., Widmer, F., Von Sigler, W., & Zeyer, J. (2004). New molecular screening tools for analysis of free-living diazotrophs in soil. *Applied and Environmental Microbiology*, *70* (1), 240–247. Available from https://doi.org/10.1128/AEM.70.1.240-247.2004.

Chatterjee, P., Mandal, S., Mandal, S., & Chatterjee, S. (2019). Vermicompost and cow dung admixture increases rhizosphere bacterial population and promotes rapid physiological maturity in maize (*Zea mays* L.). *NASS Journal of Agricultural Sciences*, *1*(2), 27–36.

Chatterjee, S. N., Syed, A. A., & Mukhopadhyay, B. (2014). Diversity of soil bacteria in some village areas adjoining to Joypur forest of Bankura District of West Bengal, India. *International Journal of Environmental Biology*, *4*(1), 67–70.

Covell, S., Ellis, R. H., Roberts, E. H., & Summerfield, R. J. (1986). The influence of temperature on seed germination rate in grain legumes: I. A comparison of chickpea, lentil, soyabean and cowpea at constant temperatures. *Journal of Experimental Botany*, *37*(5), 705–715. Available from https://doi.org/10.1093/jxb/37.5.705.

Craufurd, P. Q., Subedi, M., & Summerfield, R. J. (1997). Leaf appearance in cowpea: Effects of temperature and photoperiod. *Crop Science Society of America, Nigeria Crop Science. 0011183X.*, *37*(1), 167–171. Available from https://doi.org/10.2135/cropsci1997.0011183X003700010029x, https://www.crops.org/publications/cs.

Cruz-Martínez, K., Rosling, A., Zhang, Y., Song, M., Andersen, G. L., & Banfield, J. F. (2012). Effect of rainfall-induced soil geochemistry dynamics on grassland soil microbial communities. *Applied and Environmental Microbiology*, *78*(21), 7587–7595. Available from https://doi.org/10.1128/AEM.00203-12, http://aem.asm.org/content/78/21/7587.full.pdf.

Deepa, C. K., Dastager, S. G., & Pandey, A. (2010). Isolation and characterization of plant growth promoting bacteria from non-rhizospheric soil and their effect on cowpea (*Vigna unguiculata* (L.) Walp.) seedling growth. *World Journal of Microbiology and Biotechnology*, *26*(7), 1233–1240. Available from https://doi.org/10.1007/s11274-009-0293-y.

Deng, Q., Hui, D., Chu, G., Han, X., & Zhang, Q. (2017). Rain-induced changes in soil CO2 flux and microbial community composition in a tropical forest of China. *Nature Publishing Group, China Scientific Reports*, *7*(1). Available from https://doi.org/10.1038/s41598-017-06345-2, http://www.nature.com/srep/index.html.

Diamantopoulou, M. J. (2005). Artificial neural networks as an alternative tool in pine bark volume estimation. *Computers and Electronics in Agriculture*, *48*(3), 235–244. Available from https://doi.org/10.1016/j.compag.2005.04.002.

Dimkpa, C., Svatoš, A., Merten, D., Büchel, G., & Kothe, E. (2008). Hydroxamate siderophores produced by Streptomyces acidiscabies E13 bind nickel and promote growth in cowpea (*Vigna unguiculata* L.) under nickel stress. *Canadian Journal of Microbiology*, *54*(3), 163–172. Available from https://doi.org/10.1139/W07-130Germany, http://article.pubs.nrc-cnrc.gc.ca/RPAS/RPViewDoc?_handler_ = HandleInitialGet&calyLang = eng&journal = cjm&volume = 54&articleFile = w07-130.pdf.

Fereidooni, M., Raiesi, F., & Fallah, S. (2013). Ecological restoration of soil respiration, microbial biomass and enzyme activities through broiler litter application in a calcareous soil cropped with silage maize. *Ecological Engineering*, *58*, 266–277. Available from https://doi.org/10.1016/j.ecoleng.2013.06.032.

Ferreira, M. S., & Galo, Md. L. B. T. (2013). Chlorophyll a spatial inference using artificial neural network from multispectral images and in situ measurements. *Anais da Academia Brasileira de Ciencias*, *85*(2), 519–532. Available from https://doi.org/10.1590/S0001-37652013005000037Brazil, http://www.scielo.br/pdf/aabc/v85n2/0001-3765-aabc-00-00-3913.pdf.

Frank, B. (1889). Über die Pilzsymbiose der Leguminosen. *Berichte der Deutschen Botanischen Gesellschaft*, *7*, 332–346.

Gago, J., Martínez-Núñez, L., Landín, M., & Gallego, P. P. (2010). Artificial neural networks as an alternative to the traditional statistical methodology in plant research. *Journal of Plant Physiology*, *167*(1), 23–27. Available from https://doi.org/10.1016/j.jplph.2009.07.007.

Gilliam, F. S., Galloway, J. E., & Sarmiento, J. S. (2015). Variation with slope aspect in effects of temperature on nitrogen mineralization and nitrification in mineral soil of mixed hardwood forests. *Canadian Science Publishing, United States Canadian Journal of Forest Research*, *45*(7), 958–962. Available from https://doi.org/10.1139/cjfr-2015-0087, http://pubs.nrc-cnrc.gc.ca/rp-ps/inDetail.jsp?jcode = cjfr&lang = eng&vol = 39&is = 1.

Girvan, M. S., Bullimore, J., Pretty, J. N., Osborn, A. M., & Ball, A. S. (2003). Soil type is the primary determinant of the composition of the total and active bacterial communities in arable soils. *Applied and Environmental Microbiology*, *69*(3), 1800–1809. Available from https://doi.org/10.1128/AEM.69.3.1800-1809.2003.

Goltsev, V., Zaharieva, I., Chernev, P., Kouzmanova, M., Kalaji, H. M., Yordanov, I., … Strasser, R. J. (2012). Bulgaria drought-induced modifications of photosynthetic electron transport in intact leaves: Analysis and use of neural networks as a tool for a rapid non-invasive estimation. *Biochimica et Biophysica Acta – Bioenergetics*, *1817*(8), 1490–1498. Available from https://doi.org/10.1016/j.bbabio.2012.04.018.

Greer, D. H., Wunsche, J. N., Norling, C. L., & Wiggins, H. N. (2006). Root-zone temperatures affect phenology of bud break, flower cluster development, shoot extension growth and gas exchange of 'Braeburn' (Malus domestica) apple trees. *Tree Physiology*, *26*(1), 105–111. Available from https://doi.org/10.1093/treephys/26.1.105.

Jiang, M., & Zhang, J. (2002). Water stress-induced abscisic acid accumulation triggers the increased generation of reactive oxygen species and up-regulates the activities of antioxidant enzymes in maize leaves. *Journal of Experimental Botany*, *53*(379), 2401–2410. Available from https://doi.org/10.1093/jxb/erf090, http://jxb.oxfordjournals.org/.

Karahan, O., Van Loosdrecht, M. C. M., & Orhon, D. (2006a). Modeling the utilization of starch by activated sludge for simultaneous substrate storage and microbial growth. *Biotechnology and Bioengineering*, *94*(1), 43–53. Available from https://doi.org/10.1002/bit.20793.

Karahan, O., Martins, A., Orhon, D., & Loosdrecht, M. C. M. V. (2006b). Experimental evaluation of starch utilization mechanism by activated sludge. *Biotechnology and Bioengineering*, *93*(5), 964–970. Available from https://doi.org/10.1002/bit.20795.

Kunert, K. J., Vorster, B. J., Fenta, B. A., Kibido, T., Dionisio, G., & Foyer, C. H. (2016). Drought stress responses in soybean roots and nodules. *Frontiers in Plant Science*, *7*(2016). Available from https://doi.org/10.3389/fpls.2016.01015, http://journal.frontiersin.org/article/10.3389/fpls.2016.01015/full.

Lacey, L. A. (1997). *Manual of techniques in insect pathology*. Academic Press.

P. Legendre, L. Legendre, Numerical ecology. (1998),.

Leite, J., Fischer, D., Rouws, L. F. M., Fernandes-Júnior, P. I., Hofmann, A., Kublik, S., Schloter, M., Xavier, G. R., & Radl, V. (2017). Cowpea nodules harbor non-rhizobial bacterial communities that are shaped by soil type rather than plant genotype. *Frontiers in Plant Science*, *7*. Available from https://doi.org/10.3389/fpls.2016.02064, http://journal.frontiersin.org/article/10.3389/fpls.2016.02064/full.

Li, X., & He, Y. (2008). Discriminating varieties of tea plant based on Vis/NIR spectral characteristics and using artificial neural networks. *Biosystems Engineering*, *99*(3), 313–321. Available from https://doi.org/10.1016/j.biosystemseng.2007.11.007.

Liu, Z., Rong, Q., Zhou, W., Liang, G., & Paz-Ferreiro, J. (2017). Effects of inorganic and organic amendment on soil chemical properties, enzyme activities, microbial community and soil quality in yellow clayey soil. *PLoS One*, *12*(3), e0172767. Available from https://doi.org/10.1371/journal.pone.0172767.

Luo, Q. (2011). Temperature thresholds and crop production: A review. *Climatic Change*, *109*(3-4), 583–598. Available from https://doi.org/10.1007/s10584-011-0028-6.

Maia, F. M. M., Oliveira, J. T. A., Matos, M. R. T., Moreira, R. A., & Vasconcelos, I. M. (2000). Proximate composition, amino acid content and haemagglutinating and trypsin-inhibiting activities of some Brazilian *Vigna unguiculata* (L) Walp cultivars. *Journal of the Science of Food and Agriculture*, *80*(4), 453–458, 10.1002/(SICI)1097-0010(200003)80:4 < 453::AID-JSFA548 > 3.0.CO;2-X.

Marschner, P., Yang, C. H., Lieberei, R., & Crowley, D. E. (2001). Soil and plant specific effects on bacterial community composition in the rhizosphere. *Soil Biology and Biochemistry*, *33*(11), 1437–1445. Available from https://doi.org/10.1016/S0038-0717(01)00052-9.

De Meyer, S. E., De Beuf, K., Vekeman, B., & Willems, A. (2015). A large diversity of non-rhizobial endophytes found in legume root nodules in Flanders (Belgium). *Soil Biology and Biochemistry*, *83*, 1–11. Available from https://doi.org/10.1016/j.soilbio.2015.01.002, http://www.elsevier.com/inca/publications/store/3/3/2.

Murakami K. 1988 Treatment, disposal and utilization of sewage sludge in Japan: Current practice and future direction, Proceedings of 3rd WPCFIJSWA Joint Technical Seminar on Sewage Treatment Technology, 371-429.

Nielsen S.S. Ohler T.A. Mitchell C.A. 1997 Cowpea leaves for human consumption: Production, utilization and nutrient composition, Advances in Cowpea Research. Co-publication of International Institute of Tropical Agriculture (IITA) and Japan International Research Centre for Agricultural Science, 326-332.

Ogbodo, E. N. (2013). Impact of the use of inorganic fertilizers to the soils of the Ebonyi State Agro-Ecology, South Eastern Nigeria. *Journal of Environmental & Earth Sciences.*, *3*(7), 34–38.

D.A. Olal, Determining quantity of cowpea (*Vigna unguiculata*) leaf yield under different manure application regimes and cropping systems in Western Kenya. (2015).

Patil, G. P., & Taillie, C. (1977). Diversity as a concept and its implications for random communities. *Bulletin of the International Statistical Institute*, *47*, 497–515.

Peix, A., Ramírez-Bahena, M. H., Velázquez, E., & Bedmar, E. J. (2015). Bacterial associations with legumes. *Critical Reviews in Plant Sciences*, *34*, 17–42. Available from https://doi.org/10.1080/07352689.2014.897899, http://www.tandf.co.uk/journals/titles/07352689.asp.

Prasad, P. V. V., Boote, K. J., Allen, L. H., & Thomas, J. M. G. (2003). Super-optimal temperatures are detrimental to peanut (*Arachis hypogaea* L.) reproductive processes and yield at both ambient and elevated carbon dioxide. *Global Change Biology*, *9*(12), 1775–1787. Available from https://doi.org/10.1046/j.1365-2486.2003.00708.x.

Rich, C. I. (1965). *Methods of soil analysis. Part 2. Chemical and microbiological properties* (pp. 849–864). American Society of Agronomy.

Rosen, C. J., & Allan, D. L. (2007). American Society for Horticultural Science United States exploring the benefits of organic nutrient sources for crop production and soil quality. *Horttechnology*, *17*(4), 422–430. Available from https://doi.org/10.21273/horttech.17.4.422, https://journals.ashs.org/horttech/view/journals/horttech/horttech-overview.xml?rskey = 6oxk1u.

Sahrawat, K. L. (2008). Factors affecting nitrification in soils. *Communications in soil science and plant analysis*, *39*(9-10), 1436–1446. Available from https://doi.org/10.1080/00103620802004235.

Salim, L. M., Joseph, S., & Jisha, M. S. (2008). Response of cowpea [*Vigna unguiculata* (L.) Walp.] to phosphate solubilizing bacteria isolated from rhizosphere. *Asian Journal of Biological Sciences*, *3*, 95–98.

San Pedro, D. C., Mino, T., & Matsuo, T. (1994). Evaluation of the rate of hydrolysis of slowly biodegradable COD (SBCOD) using starch as substrate under anaerobic, anoxic and aerobic conditions. *Water Science and Technology*, *30*(11), 191–199. Available from https://doi.org/10.2166/wst.1994.0559, https://iwaponline.com/wst.

Savci, S. (2012). An agricultural pollutant: Chemical fertilizer. *International Journal of Environmental Science and Development*, 73–80. Available from https://doi.org/10.7763/IJESD.2012.V3.191.

Shrestha, R. K., Lal, R., & Rimal, B. (2013). Soil carbon fluxes and balances and soil properties of organically amended no-till corn production systems. *Geoderma. 00167061.* ***197-198*** *(*, 177–185. Available from https://doi.org/10.1016/j.geoderma.2013.01.005.

Singh, N., Singh, G., & Khanna, V. (2016). Growth of lentil (Lens culinaris medikus) as influenced by phosphorus, rhizobium and plant growth promoting rhizobacteria. *Indian Journal of Agricultural Research*, *50*(6), 567–572. Available from https://doi.org/10.18805/ijare.

v0iOF.4573, http://www.indianjournals.com/ijor.aspx?target = ijor:ijar2&volume = 50&issue = 6&article = 010&type = pdf.

Souza, R. de, Ambrosini, A., & Passaglia, L. M. P. (2015). Plant growth-promoting bacteria as inoculants in agricultural soils. *Genetics and Molecular Biology*, *38*(4), 401–419. Available from https://doi.org/10.1590/s1415-475738420150053.

Stewart, K. A., & Summerfield, R. J. (1978). Effect of root temperature on floral morphology in cowpea (*Vigna unguiculata* (L.) Walp.) cv K 2809. *Plant and Soil*, *49*(2), 443–448. Available from https://doi.org/10.1007/BF02149754.

Subbiah, B. V., & Asija, G. L. (1956). A rapid procedure for determination of available nitrogen in soils. *Current Science*, *25*, 259–260.

Tarawali S.A. Peters M. Blade S.F. 1997 Cowpea haulms as fodder Advances in Cowpea Research. Co-publication of International Institute of Tropical Agricultural Sciences and the JIRCAS, 313-325.

Thomsen, T. R., Kong, Y., & Nielsen, P. H. (2007). Ecophysiology of abundant denitrifying bacteria in activated sludge. *FEMS Microbiology Ecology*, *60*(3), 370–382. Available from https://doi.org/10.1111/j.1574-6941.2007.00309.x.

Tiwari, K. R., Nyborg, I. L. P., Sitaula, B. K., & Paudel, G. S. (2008). Analysis of the sustainability of upland farming systems in the Middle Mountains region of Nepal. *International Journal of Agricultural Sustainability*, *6*(4), 289–306. Available from https://doi.org/10.3763/ijas.2008.0390Norway, http://www.earthscanjournals.com/ijas/006/0289/0060289.pdf.

Tóthmérész, B. (1995). Comparison of different methods for diversity ordering. *Journal of Vegetation Science*, *6*(2), 283–290. Available from https://doi.org/10.2307/3236223.

Ubukata, Y. (1999). Kinetics and fundamental mechanisms of starch removal by activated sludge: Hydrolysis of starch to maltose and maltotriose is the rate-determining step. *Water Science and Technology*, *40*(1), 61–68. Available from https://doi.org/10.1016/S0273-1223(99)00364-9.

Walker, T. S., Bais, H. P., Grotewold, E., & Vivanco, J. M. (2003). Root exudation and rhizosphere biology. *Plant Physiology*, *132*(1), 44–51. Available from https://doi.org/10.1104/pp.102.019661, http://www.plantphysiol.org/.

Walkley, A., & Black, I. A. (1934). An examination of the degtjareff method for determining soil organic matter, and a proposed modification of the chromic acid titration method. *Soil Science*, *37*(1), 29–38. Available from https://doi.org/10.1097/00010694-193401000-00003.

Yadav, A. N., Verma, P., Kumar, M., Pal, K. K., Dey, R., Gupta, A., Padaria, J. C., Gujar, G. T., Kumar, S., Suman, A., Prasanna, R., & Saxena, A. K. (2015). Diversity and phylogenetic profiling of niche-specific Bacilli from extreme environments of India. *Annals of Microbiology*, *65*(2), 611–629. Available from https://doi.org/10.1007/s13213-014-0897-9, http://springerlink.com/content/121707/.

Yang, J., Kloepper, J. W., & Ryu, C. M. (2009). Rhizosphere bacteria help plants tolerate abiotic stress. *Trends in Plant Science*, *14*(1), 1–4. Available from https://doi.org/10.1016/j.tplants.2008.10.004.

Zerihun, A., & Haile, D. (2017). The effect of organic and inorganic fertilizers on the yield of two contrasting soybean varieties and residual nutrient effects on a subsequent finger millet crop. *Agronomy*, *7*(2). Available from https://doi.org/10.3390/agronomy7020042, http://www.mdpi.com/2073-4395/7/2/42/pdf.

Index

Note: Page numbers followed by "*f*" and "*t*" refer to figures and tables, respectively.

B

D

E

F

G

H

I

J

K

N

O

S

V

W

X

Y

9780443236655